Web site www.wiley.com/college/cutnell

The *Physics,* 6th edition, Web site offers a range of study aids that support your study of physics. Most of these are cross-referenced in the text so that you always know the right time to go to the Web site. The following pages will give you an idea of the resources available to you:

- **Solutions to selected end-of-chapter problems**

- **Interactive Solutions**

- **Interactive LearningWare examples**

- **Self-Assessment Tests**

- **Concept Simulations**

- **MCAT Self-Quizzes**

- **Web Links to other physics tutorials**

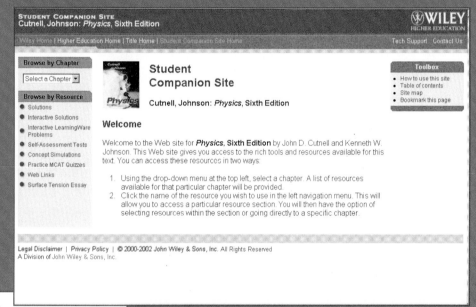

CHAPTER 2 | KINEMATICS IN ONE DIMENSION

9.   **REASONING** Since the woman runs for a known distance at a known constant speed, we can find the time it takes for her to reach the water from Equation 2.1. We can then use Equation 2.1 to determine the total distance traveled by the dog in this time.

SOLUTION The time required for the woman to reach the water is

$$\text{Elapsed time} = \frac{d_{\text{woman}}}{v_{\text{woman}}} = \left(\frac{4.0 \text{ km}}{2.5 \text{ m/s}}\right)\left(\frac{1000 \text{ m}}{1.0 \text{ km}}\right) = 1600 \text{ s}$$

In 1600 s, the dog travels a total distance of

$$d_{\text{dog}} = v_{\text{dog}}t = (4.5 \text{ m/s})(1600 \text{ s}) = \boxed{7.2 \times 10^3 \text{ m}}$$

● **Solutions** These solutions to selected end-of-chapter problems, taken from the Student Solutions Manual, are identified with a **www** icon in the text. They have been written by the authors of the text and follow the problem-solving methodology presented in the text.

● **Interactive Solutions** These solutions are designed to serve as models for an additional 111 end-of-chapter problems, all labeled in the text. The solutions are structured in an interactive tutorial format and follow the problem-solving methodology presented in the text. You can find them easily on the Web site because the numbers of the Interactive Solutions match the numbers of the corresponding problems.

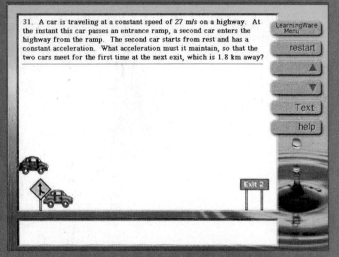

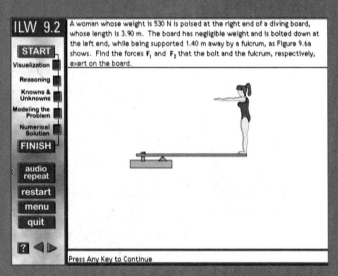

● **Interactive LearningWare** These interactive calculational examples will help you develop your problem-solving skills. References to these examples are contained in nearly every chapter of the text. Their user-friendly clarity allows you to see how conceptual understanding and quantitative analysis work together in a problem-solving environment.

● **Self-Assessment Tests** The Self-Assessment Tests are linked directly to specific sections of the book. After you study these text sections, the tests allow you to assess your understanding of the material immediately. Each test contains thought-provoking conceptual questions as well as calculational problems. You can take the tests as often as you wish. They provide extensive feedback in the form of helpful hints and suggestions when incorrect answers are selected.

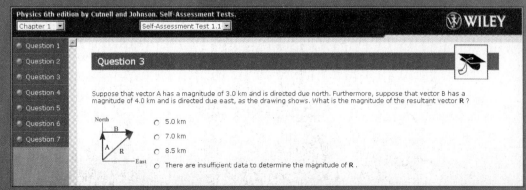

Conversion Factors

Length

1 in. = 2.54 cm

1 ft = 0.3048 m

1 mi = 5280 ft = 1.609 km

1 m = 3.281 ft

1 km = 0.6214 mi

1 angstrom (Å) = 10^{-10} m

Mass

1 slug = 14.59 kg

1 kg = 1000 grams = 6.852×10^{-2} slug

1 atomic mass unit (u) = 1.6605×10^{-27} kg

(1 kg has a weight of 2.205 lb where the
   acceleration due to gravity is 32.174 ft/s^2)

Time

1 day = 24 h = 1.44×10^3 min = 8.64×10^4 s

1 yr = 365.24 days = 3.156×10^7 s

Speed

1 mi/h = 1.609 km/h = 1.467 ft/s = 0.4470 m/s

1 km/h = 0.6214 mi/h = 0.2778 m/s = 0.9113 ft/s

Force

1 lb = 4.448 N

1 N = 10^5 dynes = 0.2248 lb

Work and Energy

1 J = 0.7376 ft·lb = 10^7 ergs

1 kcal = 4186 J

1 Btu = 1055 J

1 kWh = 3.600×10^6 J

1 eV = 1.602×10^{-19} J

Power

1 hp = 550 ft·lb/s = 745.7 W

1 W = 0.7376 ft·lb/s

Pressure

1 Pa = 1 N/m^2 = 1.450×10^{-4} $lb/in.^2$

1 $lb/in.^2$ = 6.895×10^3 Pa

1 atm = 1.013×10^5 Pa = 1.013 bar =
   14.70 $lb/in.^2$ = 760 torr

Volume

1 liter = 10^{-3} m^3 = 1000 cm^3 = 0.03531 ft^3

1 ft^3 = 0.02832 m^3 = 7.481 U.S. gallons

1 U.S. gallon = 3.785×10^{-3} m^3 = 0.1337 ft^3

Angle

1 radian = 57.30°

1° = 0.01745 radian

Standard Prefixes Used to Denote Multiples of Ten

Prefix	Symbol	Factor
Tera	T	10^{12}
Giga	G	10^{9}
Mega	M	10^{6}
Kilo	k	10^{3}
Hecto	h	10^{2}
Deka	da	10^{1}
Deci	d	10^{-1}
Centi	c	10^{-2}
Milli	m	10^{-3}
Micro	μ	10^{-6}
Nano	n	10^{-9}
Pico	p	10^{-12}
Femto	f	10^{-15}

Basic Mathematical Formulae

Area of a circle = πr^2

Circumference of a circle = $2\pi r$

Surface area of a sphere = $4\pi r^2$

Volume of a sphere = $\frac{4}{3}\pi r^3$

Pythagorean theorem: $h^2 = h_o^2 + h_a^2$

Sine of an angle: $\sin \theta = h_o/h$

Cosine of an angle: $\cos \theta = h_a/h$

Tangent of an angle: $\tan \theta = h_o/h_a$

Law of cosines: $c^2 = a^2 + b^2 - 2ab \cos \gamma$

Law of sines: $a/\sin \alpha = b/\sin \beta = c/\sin \gamma$

Quadratic formula:

If $ax^2 + bx + c = 0$, then, $x = (-b \pm \sqrt{b^2 - 4ac})/(2a)$

● **Concept Simulations** This section consists of 61 interactive simulations. Many concepts in physics—such as relative velocities, collisions, and ray tracing—can be better understood when they are simulated. With each simulation, you can experiment, since one or more parameters are under user control.

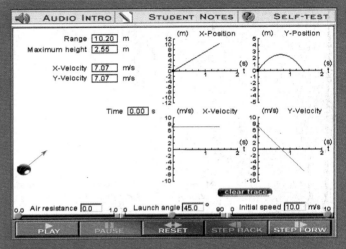

Air pressure decreases with altitude. In fact, by assuming that air has the same temperature at all altitudes, it is possible to derive the *law of atmospheres*:

$$p = p_0 e^{-h/a}.$$

where p_0 denotes the air pressure at sea level, h denotes the altitude above sea level, a denotes a constant, and p denotes the pressure at height h. For instance, many people have difficulty breathing on Mt. Everest, due in part to the lower air pressure.

1. According to the law of atmospheres, which graph best expresses the relationship between air pressure and altitude above sea level?

A.

B.

C.

D.

○ A. A
○ B. B
○ C. C
○ D. D

● **MCAT Self-Quizzes** These questions are similar to those you will find on the MCAT exam. They are presented in a multiple-choice, self-test format.

● **Web Links** These links, organized by chapter, provide a rich resource of on-line study aids.

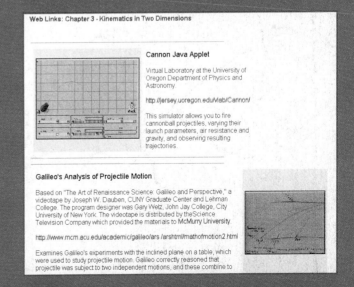

Web Links: Chapter 3 - Kinematics in Two Dimensions

Cannon Java Applet

Virtual Laboratory at the University of Oregon Department of Physics and Astronomy.

http://jersey.uoregon.edu/vlab/Cannon/

This simulator allows you to fire cannonball projectiles, varying their launch parameters, air resistance and gravity, and observing resulting trajectories.

Galileo's Analysis of Projectile Motion

Based on "The Art of Renaissance Science: Galileo and Perspective," a videotape by Joseph W. Dauben, CUNY Graduate Center and Lehman College. The program designer was Gary Welz, John Jay College, City University of New York. The videotape is distributed by theScience Television Company which provided the materials to McMurry University.

http://www.mcm.acu.edu/academic/galileo/ars /arshtml/mathofmotion2.html

Examines Galileo's experiments with the inclined plane on a table, which were used to study projectile motion. Galileo correctly reasoned that projectile was subject to two independent motions, and these combine to

Volume Two

Physics

Sixth Edition

WILEY

John Wiley & Sons, Inc.

John D. Cutnell
Kenneth W. Johnson

Southern Illinois University at Carbondale

Volume Two

Physics

Sixth Edition

ACQUISITION EDITOR	*Stuart Johnson*
DIRECTOR OF DEVELOPMENT	*Barbara Heaney*
MARKETING MANAGER	*Robert Smith*
PRODUCTION EDITOR	*Barbara Russiello*
COVER AND TEXT DESIGNER	*Madelyn Lesure*
SENIOR ILLUSTRATION EDITOR	*Sigmund Malinowski*
ELECTRONIC ILLUSTRATIONS	*Precision Graphics*
PHOTO EDITOR	*Hilary Newman*
COVER PHOTO	*© John Kelly/The Image Bank/Getty Images*

This book was set in 10/12 Times Roman by Progressive Information
Technologies and printed and bound by Von Hoffmann Press, Inc. The cover was
printed by Von Hoffmann Press.

This book is printed on acid free paper.

Main text (hardcover)	ISBN 0471-15183-1
Volume 1 (paperback)	ISBN 0-471-20940-6
Volume 2 (paperback)	ISBN 0-471-20939-2
Wiley International Edition	ISBN 0471-44895-8

Printed in the United States of America

10 9 8 7 6 5 4 3 2

To my wife Joan Cutnell,
a patient friend and my support
throughout this project.

To Anne Johnson,
my wonderful wife, a caring person,
and a great friend.

Brief Contents

Contents

Preface

We have written this text for students and teachers who are partners in a one-year algebra-based physics course. The single most important pedagogical issue in this course is the synergistic relationship between conceptual understanding and problem solving; we have substantially strengthened this relationship in the sixth edition. Our efforts were helped immeasurably by the many insights and suggestions provided by users of the fifth edition, as well as by the work of physics-education researchers. In taking a fresh look at the text, we have added new features and material, refined areas in need of improvement, and worked toward an even greater clarity in the design of the book. In particular, the high level of student activity on our Web site indicates that students are quite interested in on-line help, so we have greatly increased our on-line learning aids and facilitated their use with carefully placed text references to them.

Goals

Conceptual Understanding Helping students develop a conceptual understanding of physics principles is a primary goal of this text. This is a challenging task, because physics is often regarded as a collection of equations that can be used blindly to solve problems. However, a good problem-solving technique does not begin with equations. It starts with a firm grasp of the concepts and how they fit together to provide a coherent description of the physical world. The features in the text that work toward this goal are *Check Your Understanding* questions (a new feature), *Self-Assessment Tests* (a new on-line feature), *Concept Simulations* (a new online feature), *Concepts & Calculations* sections, *Conceptual Examples*, and *Concepts-at-a-Glance* charts.

Reasoning The ability to reason in an organized manner is essential to solving problems, and helping students to improve their reasoning ability is also one of our primary goals. To this end, we have included *Interactive LearningWare* (an on-line tutorial that steps students through 60 different examples), *Interactive Solutions* (111 on-line solutions that explore the reasoning pertinent to certain homework problems), *explicit reasoning steps* in all examples, and *Reasoning Strategies* that summarize the logical steps involved in solving certain classes of problems. A strong reasoning ability, combined with firm conceptual understanding, is the legacy that we wish to leave to our students.

Relevance Finally, we want to show students that physics principles come into play over and over again in their lives. It is always easier to learn something new if it is directly related to day-to-day living. Direct applications of physics principles are identified in the margin by the label *The Physics of....* Many of these applications are biomedical in nature and deal with human physiology. Others, such as CDs and DVDs and digital photography, deal with modern technology. Still others deal with things that we take for granted in our lives, such as household plumbing. We have also incorporated real-world situations into many of the worked-out examples and the homework material at the end of each chapter.

Organization and Coverage

The text consists of 32 chapters and is organized in a fairly standard fashion according to the following sequence: *Mechanics, Thermal Physics, Wave Motion, Electricity and Magnetism, Light and Optics,* and *Modern Physics.* The text is available in two formats: in a complete volume, consisting of all 32 chapters, and in two volumes (Chapters 1 – 17 and Chapters 18 – 32). **Chapter sections marked with an asterisk can be omitted with little impact on the overall development of the material.**

Based on feedback from users and reviewers, we have added new material, made judicial deletions, and revised other material. In particular, a number of the examples have been modified to improve clarity and present topics of current interest. The new material includes completely rewritten chapter-ending summaries for each chapter, a discussion of the relation between relativistic total energy and momentum (Section 28.6), a discussion of light emitting diodes (Section 23.5), and a table of the isotopes (Appendix F). We have also updated our applications of physics principles and added new ones. Because many students are interested in health-related careers, a number of the new applications are biomedical in nature, such as wireless capsule endoscopy, transcranial magnetic stimulation, a fetal oxygen monitor, and exercise thallium heart scans. We have also presented a variety of new applications relating to modern technology, including, for example, ion propulsion drives for space probes, special relativity and the global positioning system, 3-D films, and light emitting diodes.

For instructors who wish to cover surface tension, we have included a module on our Web site (**www.wiley.com/college/cutnell**). This module discusses the nature of surface tension, capillary action, and the pressure inside a soap bubble and a liquid drop. Homework problems are also included for assignment.

Multimedia Version of the Sixth Edition

The Multimedia Book This hypertext version of the text is linked to the Study Guide, the Student Solutions Manual, the Self-Assessment Tests, the Interactive LearningWare, the Interactive Solutions, the Concept Simulations, and other multimedia. The Multimedia Book can be used by students in many ways, but it is preeminently a *very powerful study aid.* The algebra-based introductory physics course is a problem-solving course, and the Multimedia Book provides more problem-solving help than any other source available.

 eGrade Plus By combining two powerful Web-based programs—eGrade and the Multimedia Book—we can offer physics teachers and physics students a unique teaching and learning environment.

Helping Teachers Teach: eGrade is a homework management system that allows the teacher to assign, collect, and grade homework automatically. eGrade keeps the students "on task" without the teacher having to invest hours every week grading homework.

Helping Students Learn: All problems assigned in eGrade have a link to an on-line version of the complete text and the associated student supplements. The links give the students quick access to focused problem-solving help. Context-sensitive help—just a click away.

Features of the Sixth Edition

 Check Your Understanding This feature consists of 130 questions found at key locations throughout the text. Each is a carefully selected question that is designed to test a student's understanding of the concepts. The questions are in either a multiple-choice or free-response format, and the answers are provided at the end of the book.

> ✔ **Check Your Understanding 1**
>
> A suitcase is hanging straight down from your hand as you ride an escalator. Your hand exerts a force on the suitcase, and this force does work. Which one of the following statements is correct? (a) The work is negative when you ride up the escalator and positive when you ride down the escalator. (b) The work is positive when you ride up the escalator and negative when you ride down the escalator. (c) The work is positive irrespective of whether you ride up or down the escalator. (d) The work is negative irrespective of whether you ride up or down the escalator. *(The answer is given at the end of the book.)*
>
> ***Background:*** This question deals with work and its relationship to force and displacement. Pay close attention to the directions of the force and displacement vectors.
>
> ***For similar questions (including calculational counterparts), consult Self-Assessment Test 6.1. This test is described at the end of Section 6.2.***

Self-Assessment Tests The self-assessment tests are on-line tests (**www.wiley.com/college/cutnell**) that are linked directly to specific sections of the book. There are about two self-assessment tests per chapter, each containing 8–10 questions and problems. The tests are designed so that students, after studying the designated sections, can immediately assess their understanding of the material. Each contains thought-provoking conceptual questions, like the Check-Your-Understanding questions, as well as calculational problems. The Self-Assessment tests are user-friendly, for they can be taken as often as the student wishes, and they provide extensive feedback in the form of helpful hints and suggestions when incorrect answers are selected.

>  **Self-Assessment Test 2.3**
>
> Test your understanding of the material in Sections 2.6 and 2.7:
>
> • Freely Falling Bodies   • Graphical Analysis of Velocity and Acceleration
>
> Go to **www.wiley.com/college/cutnell**

Concept Simulation 4.1

Three horizontally directed forces are applied to an object, and the resulting motion of the object is displayed, along with synchronized graphs of position, velocity, and acceleration versus time. The mass of the object and the magnitudes and directions of the forces are under your control, so that you can explore how these quantities affect the motion.

Related Homework: Problems 2, 64

Go to
www.wiley.com/college/cutnell

NEW! *Concept Simulations* Often, many concepts in physics—such as relative velocities, collisions, and ray tracing—can be better understood when they are simulated. At appropriate locations throughout the text, we have placed references to 61 on-line simulations (**www.wiley.com/college/cutnell**). With each simulation, the student can experiment, since one or more parameters are under user control. Many of the simulations provide references to related homework material, which, in turn, is cross-referenced to the simulation.

NEW! *Concept Summaries* All chapter-ending summaries have been rewritten so that they present an abridged, but complete, version of the material on a section-by-section basis, including important equations. To help students review the material, the discussion of each section is flanked by two lists. On one side, a list of topics provides a quick-reference option for locating a given topic. On the other side, a color-coded list of the text examples and the on-line learning aids (Concept Simulations, Interactive LearningWare, and Interactive Solutions) shows students the related learning aids. The summary also identifies the on-line Self-Assessment Tests that are keyed to the sections, so that students can evaluate their understanding of the chapter topics.

Concept Summary

This summary presents an abridged version of the chapter, including the important equations and all available learning aids. For convenient reference, the learning aids (including the text's examples) are placed next to or immediately after the relevant equation or discussion. The following Learning Aids may be found on-line at **www.wiley.com/college/cutnell**:

Topic	Discussion	Learning Aids
	5.1 Uniform Circular Motion	
	Uniform circular motion is the motion of an object traveling at a constant (uniform) speed on a circular path.	
Period of the motion	The period T is the time required for the object to travel once around the circle. The speed v of the object is related to the period and the radius r of the circle by $$v = \frac{2\pi r}{T} \qquad (5.1)$$	**Example 1**
	5.2 Centripetal Acceleration	
	An object in uniform circular motion experiences an acceleration, known as centripetal acceleration. The magnitude a_c of the centripetal acceleration is	**Examples 2, 3, 4, 15**
Centripetal acceleration (magnitude)	$$a_c = \frac{v^2}{r} \qquad (5.2)$$	**Interactive LearningWare 5.1** **Interactive Solution 5.3**
Centripetal acceleration (direction)	where v is the speed of the object and r is the radius of the circle. The direction of the centripetal acceleration vector always points toward the center of the circle and continually changes as the object moves.	
	5.3 Centripetal Force	
	To produce a centripetal acceleration, a net force pointing toward the center of the circle is required. This net force is called the centripetal force, and its magnitude F_c is	**Examples 5, 6, 7, 16** **Interactive LearningWare 5.2**
Centripetal force (magnitude)	$$F_c = \frac{mv^2}{r} \qquad (5.3)$$	**Concept Simulations 5.1, 5.2**
Centripetal force (direction)	where m and v are the mass and speed of the object, and r is the radius of the circle. The direction of the centripetal force vector, like that of the centripetal acceleration vector, always points toward the center of the circle.	**Interactive Solution 5.19**

   Use Self-Assessment Test 5.1 to evaluate your understanding of Sections 5.1–5.3.

Interactive Solutions Often, students experience difficulty in getting started with the more difficult homework problems. The Interactive Solutions feature consists of 111 on-line problems (**www.wiley.com/college/cutnell**), each of which is allied with a particular homework problem in the text. Each Interactive Solution is worked out by the student in an interactive manner, and the solution is designed to serve as a model for the associated homework problem. Those homework problems that are associated with Interactive Solutions are identified in the text.

Concepts & Calculations

Conceptual understanding is an essential part of a good problem-solving technique. To emphasize the role of conceptual under-standing, every chapter includes a Concepts & Calculations sec-tion. These sections are organ-ized around a special type of ex-ample, each of which begins with several conceptual ques-tions that are answered before the quantitative problem is worked out. The purpose of the questions is to put into sharp fo-cus the concepts with which the problem deals. We hope that these examples also provide mini-reviews of material studied earlier in the chapter and in pre-vious chapters.

Concepts & Calculations Example 20
Velocity, Acceleration, and Newton's Second Law of Motion

Figure 4.37 shows two forces, $F_1 = +3000$ N and $F_2 = +5000$ N acting on a spacecraft, where the plus signs indicate that the forces are directed along the $+x$ axis. A third force F_3 also acts on the spacecraft but is not shown in the drawing. The craft is moving with a constant velocity of $+850$ m/s. Find the magnitude and direction of F_3.

Concept Questions and Answers Suppose the spacecraft were stationary. What would be the direction of F_3?

  Answer If the spacecraft is stationary, its acceleration is zero. According to Newton's second law, the acceleration of an object is proportional to the net force acting on it. Thus, the net force must also be zero. But the net force is the vector sum of the three forces in this case. Therefore, the force F_3 must have a direction such that it balances to zero the forces F_1 and F_2. Since F_1 and F_2 point along the $+x$ axis in Figure 4.37, F_3 must then point along the $-x$ axis.

When the spacecraft is moving at a constant velocity of $+850$ m/s, what is the direction of F_3?

  Answer Since the velocity is constant, the acceleration is still zero. As a result, every-thing we said in the stationary case applies again here. The net force is zero, and the force F_3 must point along the $-x$ axis in Figure 4.37.

Solution Since the velocity is constant, the acceleration is zero. The net force must also be zero, so that

$$\Sigma F_x = F_1 + F_2 + F_3 = 0$$

Solving for F_3 yields

$$F_3 = -(F_1 + F_2) = -(3000 \text{ N} + 5000 \text{ N}) = \boxed{-8000 \text{ N}}$$

Conceptual Examples A good problem-solving technique begins with a foundation of conceptual understanding. Therefore, every chapter includes examples that are entirely conceptual in nature. These examples are worked out in a rigorous but qualitative fashion, with no (or very few) equations. We have added art to many of these as a way to help stu-dents visualize the concepts. Our intent is to provide students with explicit models of how to "think through" a problem be-fore attempting to solve it nu-merically. The *Conceptual Ex-amples* deal with a wide range of topics and a large number of is-sues that often confuse students. Wherever possible, we have fo-cused on real-world situations and have structured the exam-ples so that they lead naturally to homework material found at the ends of the chapters. The related homework material is explicitly identified, and that material con-tains a cross reference that en-courages students to review the pertinent *Conceptual Example*.

Conceptual Example 7   Mass Versus Weight

A vehicle is being designed for use in exploring the moon's surface and is being tested on earth, where it weighs roughly six times more than it will on the moon. In one test, the accel-eration of the vehicle along the ground is measured. To achieve the same acceleration on the moon, will the net force acting on the vehicle be greater than, less than, or the same as that re-quired on earth?

Reasoning and Solution The net force ΣF required to accelerate the vehicle is specified by Newton's second law as $\Sigma F = ma$, where m is the vehicle's mass and a is the acceleration along the ground. For a given acceleration, the net force depends only on the mass. But the mass is an intrinsic property of the vehicle and is the same on the moon as it is on the earth. Therefore, the same net force would be required for a given acceleration on the moon as on the earth. Do not be misled by the fact that the vehicle weighs more on earth. The greater weight occurs only because the earth's mass and radius are different than the moon's. In any event, *in Newton's second law, the net force is proportional to the vehicle's mass, not its weight.*

Related Homework: *Problems 21, 22*

Need more practice?

🖱 **Interactive LearningWare 6.1**
A cyclist coasts down a hill, losing 20.0 m in height, and then coasts up an identical hill to the same initial vertical height. The bicycle and rider have a mass of 73.5 kg. Two forces act on the bicycle/rider system: the weight of the bicycle and rider and the normal force exerted on the tires by the ground. Determine the work done by each of these forces during (a) the downward part of the motion and (b) the upward part of the motion.

Related Homework: *Problem 6*

Go to
www.wiley.com/college/cutnell
for an interactive solution.

Interactive LearningWare This feature consists of interactive calculational examples presented on our Web site (**www.wiley.com/college/cutnell**). References to the Learning-Ware are contained in nearly every chapter. Each example is designed in an interactive format that consists of five steps. We believe that these steps are essential to the development of problem-solving skills and that students will benefit from seeing them explicitly identified in each example.

1. The *Visualization* step uses animations and drawings to help students understand the general nature of the situation being analyzed.

2. The *Reasoning* step establishes the conceptual basis for the solution with the aid of multiple-choice questions and answers.

3. The *Knowns-and-Unknowns* step deals with the important process of identifying all the variables and the symbols that represent them. This information is not simply given to the students but, rather, is interactively developed in the form of tables that evolve as the example proceeds.

4. In the *Modeling* step, the student, using the results of the previous steps, models the problem with equations developed in the text.

5. Finally, in the *Solution* step, the solution to the problem is obtained.

In summary, the hallmark of these Interactive LearningWare examples is a user-friendly clarity that allows the student to see how conceptual understanding and quantitative analysis work together in a problem-solving environment. Related homework material is identified and contains a cross-reference to the LearningWare.

Explicit Reasoning Steps Since reasoning is the cornerstone of problem solving, we believe that students will benefit from seeing the reasoning stated explicitly. Therefore, the format in which examples are worked out includes an explicit reasoning step. In this step we explain what motivates our procedure for solving the problem before any algebraic or numerical work is done. In the *Concepts & Calculations* examples, the reasoning is presented in a question-and-answer format.

Reasoning Strategy

Applying the Equations of Kinematics

1. Make a drawing to represent the situation being studied. A drawing helps us to see what's happening.

2. Decide which directions are to be called positive $(+)$ and negative $(-)$ relative to a conveniently chosen coordinate origin. Do not change your decision during the course of a calculation.

3. In an organized way, write down the values (with appropriate plus and minus signs) that are given for any of the five kinematic variables (x, a, v, v_0, and t). Be on the alert for "implied data," such as the phrase "starts from rest," which means that the value of the initial velocity is $v_0 = 0$ m/s. The data summary boxes used in the examples in the text are a good way of keeping track of this information. In addition, identify the variables that you are being asked to determine.

4. Before attempting to solve a problem, verify that the given information contains values for at least three of the five kinematic variables. Once the three known variables are identified along with the desired unknown variable, the appropriate relation from Table 2.1 can be selected. Remember that the motion of two objects may be interrelated, so they may share a common variable. The fact that the motions are interrelated is an important piece of information. In such cases, data for only two variables need be specified for each object.

5. When the motion of an object is divided into segments, as in Example 9, remember that the final velocity of one segment is the initial velocity for the next segment.

6. Keep in mind that there may be two possible answers to a kinematics problem as, for instance, in Example 8. Try to visualize the different physical situations to which the answers correspond.

Reasoning Strategies A number of the examples in the text deal with well-defined strategies for solving certain types of problems. In such cases, we have included summaries of the steps involved. These summaries, which are titled *Reasoning Strategies,* encourage frequent review of the techniques used and help students focus on the related concepts.

Concepts at a Glance To provide a coherent picture of how new concepts are built upon previous ones and to reinforce fundamental unifying ideas, the text contains 66 Concepts-at-a-Glance flowcharts. These charts show in a visual way the conceptual development of physics principles. Within a chart, new concepts are placed in gold panels, and previously introduced and related concepts are placed in light blue panels. Each chart is discussed in a separate and highlighted paragraph that helps students connect the new concept with the real world.

CONCEPTS AT A GLANCE

Equilibrium Nonequilibrium

External Forces

1. **Gravitational Force** (Section 4.7)
2. **Normal Force** (Section 4.8)
3. **Frictional Forces** (Section 4.9)
4. **Tension Force** (Section 4.10)

Equilibrium (Section 4.11)

$a = 0 \text{ m/s}^2$

Newton's Second Law

$\Sigma F = ma$

$a \neq 0 \text{ m/s}^2$

Nonequilibrium (Section 4.12)

The Physics of . . . This edition contains 239 real-world applications that reflect our commitment to show students how prevalent physics is in their lives. Each application is identified in the margin with the label *The Physics of*, and those that deal with biomedical material are further marked with an icon in the shape of a caduceus. A complete list of the applications can be found after the Acknowledgments section.

The physics of catapulting a jet from an aircraft carrier.

Problem Solving Insights To reinforce the problem-solving techniques illustrated in the worked-out examples, we have included short statements in the margins, identified by the label *Problem Solving Insight*. These statements help students to develop good problem-solving skills by providing the kind of advice that a teacher gives when explaining a calculation in detail.

Problem solving insight "Implied data" are important. For instance, in Example 6 the phrase "starting from rest" means that the initial velocity is zero ($v_0 = 0$ m/s).

Homework Problems and Conceptual Questions The sixth edition contains an even greater amount of homework material than does the fifth edition; about 17% of the problems are new or modified. In providing this large number of problems and questions, we have used a wide variety of real-world situations with realistic data. The problems are ranked according to difficulty, with the most difficult problems marked with a double asterisk (**) and those of intermediate difficulty marked with a single asterisk (*). The easiest problems are unmarked. Those whose solutions appear in the *Student Solutions Manual* are identified with the label **ssm**. Those whose solutions are available on the Web (**www.wiley.com/college/cutnell**) are marked with the label **www**. Problems and questions that are biomedical in nature are identified with an icon in the shape of a caduceus. Since teachers sometimes want to assign homework without identifying a particular section, we have provided a group of problems without references to sections under the heading *Additional Problems. Concepts & Calculations—Group Learning Problems* were introduced in the fifth edition and have been very well received. Therefore, we have expanded this grouping to include about 30% more problems. These problems are modeled on the examples in the *Concepts & Calculations* section found at the end of each chapter and consist of a series of concept questions followed by a related quantitative problem. The questions are designed to help students identify the concepts that are used in solving the problem and provide a focus for group discussions.

In spite of our best efforts to produce an error-free book, errors no doubt remain. They are solely our responsibility, and we would appreciate hearing of any that you find. We hope that this text makes learning and teaching physics easier and more enjoyable and look forward to hearing about your experiences with it. Please feel free to write us care of Physics Editor, Higher Education Division, John Wiley & Sons, Inc., 111 River Street, Hoboken, NJ 07030 or contact us at **www.wiley.com/college/cutnell.**

xxvi   *Supplements*

Supplements

An extensive package of supplements to accompany *Physics,* sixth edition, is available to assist both the teacher and the student.

Instructor's Supplements
Helping Teachers Teach

Instructor's Resource Guide by David T. Marx, Southern Illinois University of Carbondale. This guide contains an extensive listing of Web-based physics education resources. It also includes teaching ideas, lecture notes, demonstration suggestions, alternative syllabi for courses of different lengths and emphasis, as well as conversion notes that allow instructors to use their class notes from other texts. A Problem Locator Guide provides an easy way to correlate fifth-edition problem numbers with the corresponding sixth-edition numbers.

Instructor's Solutions Manual by John D. Cutnell and Kenneth W. Johnson. This manual provides worked-out solutions for all end-of-chapter Conceptual Questions and Problems. Volume 1 contains the solutions for Chapter 1-17. Volume 2 contains the solutions for Chapters 18–32.

Test Bank by David T. Marx, Southern Illinois University at Carbondale. This manual includes more than 2200 multiple-choice questions. These items are also available in the *Computerized Test Bank* (see below).

Instructor's Resource CD-ROM This CD contains:

- The entire *Instructor's Solutions Manual* in both Microsoft Word © (Windows and Macintosh) and PDF files.

- A *Computerized Test Bank,* for use with both Windows and Macintosh PC's with full editing features to help you customize tests.

- Multiple-choice format for 1200 of the end-of-chapter problems in the text. This material allows instructors to assign homework in a machine-gradable format.

- All text illustrations, suitable for classroom projection, printing, and web posting.

Wiley Physics Simulations CD-ROM This CD contains 50 interactive simulations (Java applets) that can be used for classroom demonstrations.

Transparencies More than 200 four-color illustrations from the text are provided in a form suitable for projection in the classroom.

On-line Homework and Quizzing *Physics,* sixth edition, supports WebAssign, CAPA, and eGrade, which are programs that give instructors the ability to deliver and grade homework and quizzes over the Internet.

WebCT and Blackboard Many of the instructor and student supplements have been coded for easy integration into WebCT and Blackboard, which are course management programs that allow instructors to set up a complete on-line course with chat rooms, bulletin boards, quizzing, built-in student tracking tools, etc.

Student's Supplements

Helping Students Learn

Student Web Site This Web site (**www.wiley.com/college/cutnell**) was developed specifically for *Physics,* sixth edition, and is designed to assist students further in the study of physics. At this site, students can access the following resources; the first few pages of the text show some examples of these resources.

- Self-Assessment Tests

- Solutions to selected end-of-chapter problems that are identified with a **www** icon in the text.

- Interactive Solutions

- Concept Simulations

- Interactive LearningWare examples

- Review quizzes for the MCAT exam

- Links to other Web sites that offer physics tutorial help

Student Study Guide by John D. Cutnell and Kenneth W. Johnson. This student study guide consists of traditional print materials; with the Student Web site, it provides a rich, interactive environment for review and study.

Student Solutions Manual by John D. Cutnell and Kenneth W. Johnson. This manual provides students with complete worked-out solutions for approximately 600 of the odd-numbered end-of-chapter problems. These problems are indicated in the text with an **ssm** icon.

Acknowledgments

We have had the help of many individuals in putting together this text. Working with them and witnessing their talents, dedication, and professionalism has been an exhilarating experience. It is a genuine pleasure to acknowledge their contributions.

Special thanks go to our editor, Stuart Johnson, for his expert and on-target advice at all stages of the project. We are most fortunate to have his help and encouragement. His keen understanding of college publishing and his vision for its future are assets to be treasured.

Having Barbara Heaney, Director of Product and Market Development, as our developmental editor is like winning the lottery. Her impact on the project has been enormous, including an infusion of numerous ideas, an unwavering commitment to excellence, and a careful attention to detail.

We are especially indebted to Barbara Russiello, Production Editor, for coordinating the all-electronic production of the book. She was always there for us with answers to our questions, help, and encouragement. Her tireless efforts made it possible to produce a book that meets the highest production standards.

In modifying and adding to the illustrations for this edition, we worked with Sigmund Malinowski of the illustrations department. He was always ready and willing to help with advice and suggestions. We are grateful for his efforts and enjoy working with him immensely.

The design of the text and its cover is the work of Maddy Lesure, Design Director. The clear and pleasing presentation of the text's many features and elements is due to her immense talent in coordinating color, design, and fonts. It has been a pleasure, Maddy.

We are grateful to Hilary Newman, Photo Department Manager, for the photos in the text. She was able to find great photos that express just the right physics. We think that you will like the photos as much as we do.

Our admiration and appreciation are extended to Martin Batey and Tom Kulesa, for their work on the many media components; they have made this complex process much easier for us.

Outstanding proofreaders are essential in producing an error-free book. Our thanks go to Gloria Hamilton for an extraordinary job in catching errors of all types, from awkwardly worded sentences to wrong fonts.

Our sincere gratitude goes to Geraldine Osnato for coordinating the extensive supplements package, to Alec Borenstein, Editorial Assistant, and to Helen Walden for copyediting the manuscript.

The sales representatives of John Wiley & Sons, Inc. are a very special group. We deeply appreciate their efforts on our behalf as they are in constant contact with physics departments throughout the country.

Many physicists have generously shared their ideas with us about good pedagogy and helped us by pointing out errors in our presentation. For all of their suggestions we are grateful. They have helped us to write more clearly and accurately and have influenced markedly the evolution of this text. To the reviewers of this and previous editions, we especially owe a large debt of gratitude. Specifically, we thank:

Joseph Alward,
University of the Pacific

Zaven Altounian,
McGill University

Chi Kwan Au,
University of South Carolina

Donald Ballegeer,
University of Wisconsin

David Bannon,
Oregon State University

William A. Barker, *formerly at*
Santa Clara University

Paul D. Beale,
University of Colorado at Boulder

Edward E. Beasley,
Gallaudet University

Roger Bland,
San Francisco State University

Edward R. Borchardt,
Mankato State University

Treasure Brasher,
West Texas State University

Robert Brehme,
Wake Forest University

Michael Bretz,
University of Michigan at Ann Arbor

Carl Bromberg,
Michigan State University

Michael E. Browne,
University of Idaho

Ronald W. Canterna,
University of Wyoming

Neal Cason,
University of Wyoming

Marvin Chester,
University of California at Los Angeles

William S. Chow, *emeritus*
University of Cincinnati

Lowell Christensen, *formerly at*
American River College

Albert C. Claus,
Loyola University of Chicago

Thomas Berry Cobb,
Bowling Green State University

Lawrence Coleman,
University of California at Davis

Lattie F. Collins,
East Tennessee State University

Mark Comella, *formerly at*
Duquesne University

Biman Das,
SUNY-Potsdam

Doyle Davis,
New Hampshire Community
Technical College

Steven Davis,
University of Arkansas at Little Rock

James E. Dixon,
Iowa State University

Duane Doty,
California State University at Northridge

Miles J. Dressler, *emeritus*
Washington State University

Dewey Dykstra,
Boise State University

Robert J. Endorf,
University of Cincinnati

Lewis Ford,
Texas A&M University

Greg Francis,
Montana State University

Roger Freedman,
University of California of Santa Barbara

Roger J. Friauf,
University of Kansas

C. Sherman Frye, Jr.,
Northern Virginia Community College

John Gagliardi,
Rutgers University

Simon George,
California State University at Long Beach

James B. Gerhart,
University of Washington

John Gieniec, *formerly at*
Central Missouri State University

Barry Gilbert,
Rhode Island College

Roy Goodrich,
Louisiana State University

D. Wayne Green,
Knox College

William Gregg,
Louisiana State University

David Griffiths,
Oregon State University

Grant Hart,
Brigham Young University

Thomas Herrmann,
East Oregon University

Lawrence A. Hitchingham, *formerly at*
Jackson Community College

Paul R. Holody,
Henry Ford Community College

Darrell O. Huwe, *formerly at*
Ohio University

Mario Iona, *emeritus*
University of Denver

David A. Jerde,
St. Cloud State University

Larry Josbeno,
Corning Community College

R. Lee Kernell, *emeritus*
Old Dominion University

Gary Kessler, *formerly at*
Illinois Wesleyan University

Randy Kobes,
University of Winnipeg

K. Kothari,
Tuskegee University

Robert A. Kromhout,
Florida State University

Theodore Kruse,
Rutgers University

Pradeep Kumar,
University of Florida at Gainesville

Rubin H. Landau,
Oregon State University

Christopher P. Landee,
Clark University

Alfredo Louro,
University of Calgary

Daines Lund, *formerly at*
Utah State University

R. Wayne Major, *formerly at*
University of Richmond

A. John Mallinckrodt,
California State Polytechnic University,
Pomona

Thomas P. Marvin,
Southern Oregon State College

John McClain, *formerly at*
Southern Illinois University at Carbondale

John McCullen,
University of Arizona

Laurence McIntyre,
University of Arizona

B. Wieb Van Der Meer,
Western Kentucky University

Donald D. Miller,
Central Missouri State University

James Miller, *formerly at*
East Tennessee State University

Paul Morris,
Abilene Christian University

Robert Morris,
Coppin State College

Richard A. Morrow,
University of Maine

Kenneth Mucker,
Bowling Green State University

Rod Nave,
Georgia State University

David Newton,
DeAnza College

R. Chris Olsen,
Naval Postgraduate School, Monterey

Robert F. Petry, *formerly at*
University of Oklahoma

Peter John Polito,
Springfield College

Jon Pumplin,
Michigan State University

Talat Rahman,
Kansas State University

Michael Ram,
State University of New York at Buffalo

Vallabhaneni Rao, *formerly at*
Memorial University of Newfoundland

Jacobo Rapaport,
Ohio University

Wayne W. Repko,
Michigan State University

Barry C. Robertson,
Queens University (Ontario)

Harold Romero, *formerly at*
University of Southern Mississippi

Larry Rowan,
University of North Carolina at Chapel Hill

Roy S. Rubins,
University of Texas at Arlington

O.M.P. Rustgi, *formerly at*
State University of New York at Buffalo

Al Sanders,
University of Tennessee at Knoxville

Charles Scherr, *formerly at*
University of Texas at Austin

Wesley Shanholtzer,
Marshall University

Marc Sher,
College of William & Mary

James Simmons, *formerly at*
Concordia College

John J. Sinai,
University of Louisville

Virgil Stubblefield,
John A. Logan College

Ronald G. Tabak,
Youngstown State University

Patrick Tam,
Humboldt State University

Robert Tyson,
University of North Carolina at Charlotte

Timothy Usher,
California State University at San Bernardino

Rolf Vatne,
Portland Community College

Howard G. Voss,
Arizona State University

James M. Wallace, *formerly at*
Jackson Community College

Henry White,
University of Missouri

Jerry H. Willson, *formerly at*
Metropolitan State College

Linn Van Woerkom,
Ohio State University

The physics of

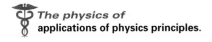

The physics of
applications of physics principles.

To show students that physics has a widespread impact on their lives, we have included a large number of applications of physics principles. Many of these applications are not found in other texts. The most important ones are listed below along with the page number locating the corresponding discussion. They are identified in the margin of the page on which they occur with the label "*The physics of*" Biomedical applications are marked with an icon in the shape of a caduceus ⚕. The discussions are integrated into the text, so that they occur as a natural part of the physics being presented. It should be noted that the list is not a complete list of all the applications of physics principles to be found in the text. There are many additional applications that are discussed only briefly or that occur in the homework questions and problems.

Chapter 18

Electric Forces and Electric Fields

Lightning, which is the flow of electric charge in the atmosphere, is nature's most spectacular display of electricity. These multiple lightning strikes occurred in Cape Town, South Africa. (© Jan Frederick/ Taxi/ Getty Images)

18.1 The Origin of Electricity

The electrical nature of matter is inherent in atomic structure. An atom consists of a small, relatively massive nucleus that contains particles called protons and neutrons. A proton has a mass of 1.673×10^{-27} kg, and a neutron has a slightly greater mass of 1.675×10^{-27} kg. Surrounding the nucleus is a diffuse cloud of orbiting particles called electrons, as Figure 18.1 suggests. An electron has a mass of 9.11×10^{-31} kg. Like mass, *electric charge* is an intrinsic property of protons and electrons, and only two types of charge have been discovered, positive and negative. A proton has a positive charge, and an electron has a negative charge. A neutron has no net electric charge.

Experiment reveals that the magnitude of the charge on the proton *exactly equals* the magnitude of the charge on the electron; the proton carries a charge $+e$, and the electron carries a charge $-e$. The SI unit for measuring the magnitude of an electric charge is the *coulomb** (C), and e has been determined experimentally to have the value

$$e = 1.60 \times 10^{-19} \text{ C}$$

The symbol e represents only the magnitude of the charge on a proton or an electron and does not include the algebraic sign that indicates whether the charge is positive or negative. In nature, atoms are normally found with equal numbers of protons and electrons. Usually, then, an atom carries no net charge because the algebraic sum of the positive charge of the nucleus and the negative charge of the electrons is zero. When an atom, or any object, carries no net charge, the object is said to be *electrically neutral*. The neutrons in the nucleus are electrically neutral particles.

The charge on an electron or a proton is the *smallest* amount of free charge that has been discovered. Charges of larger magnitude are built up on an object by adding or removing electrons. Thus, any charge of magnitude q is an integer multiple of e; that is, $q = Ne$, where N is an integer. Because any electric charge q occurs in integer multiples of elementary, indivisible charges of magnitude e, electric charge is said to be *quantized*. Example 1 emphasizes the quantized nature of electric charge.

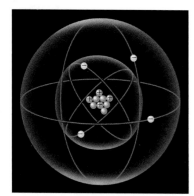

○ electron
⊕ proton
○ neutron

Figure 18.1 An atom contains a small, positively charged nucleus, about which the negatively charged electrons move. The closed-loop paths shown here are symbolic only. In reality, the electrons do not follow discrete paths, as Section 30.5 discusses.

Example 1 A Lot of Electrons

How many electrons are there in one coulomb of negative charge?

Reasoning The negative charge is due to the presence of excess electrons, since they carry negative charge. Because an electron has a charge whose magnitude is $e = 1.60 \times 10^{-19}$ C, the number of electrons is equal to the charge q divided by the charge e on each electron.

Solution The number N of electrons is

$$N = \frac{q}{e} = \frac{1.00 \text{ C}}{1.60 \times 10^{-19} \text{ C}} = \boxed{6.25 \times 10^{18}}$$

18.2 Charged Objects and the Electric Force

Electricity has many useful applications, and they are related to the fact that it is possible to transfer electric charge from one object to another. Usually electrons are transferred, and the body that gains electrons acquires an excess of negative charge. The body that loses electrons has an excess of positive charge. Such separation of charge occurs often when two unlike materials are rubbed together. For example, when an ebonite (hard, black rubber) rod is rubbed against animal fur, some of the electrons from atoms of the fur are transferred to the rod. The ebonite becomes negatively charged, and the fur becomes positively charged, as Figure 18.2 indicates. Similarly, if a glass rod is rubbed with a silk cloth, some of the electrons are removed from the atoms of the glass and deposited on the silk, leaving the silk negatively charged and the glass positively charged. There are many familiar examples of charge separation, as when you walk across a nylon rug or run a comb through dry hair. In each case, objects become "electrified" as surfaces rub against one another.

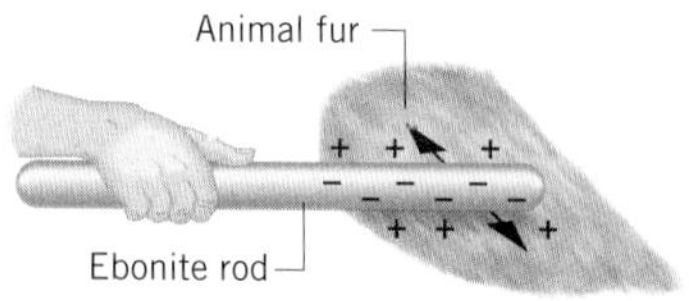

Animal fur

Ebonite rod

Figure 18.2 When an ebonite rod is rubbed against animal fur, electrons from atoms of the fur are transferred to the rod. This transfer gives the rod a negative charge (−) and leaves a positive charge (+) on the fur.

*The definition of the coulomb depends on electric currents and magnetic fields, concepts that will be discussed later. Therefore, we postpone its definition until Section 21.7.

When an ebonite rod is rubbed with animal fur, the rubbing process serves only to separate electrons and protons already present in the materials. No electrons or protons are created or destroyed. Whenever an electron is transferred to the rod, a proton is left behind on the fur. Since the charges on the electron and proton have identical magnitudes but opposite signs, the algebraic sum of the two charges is zero, and the transfer does not change the net charge of the fur/rod system. If each material contains an equal number of protons and electrons to begin with, the net charge of the system is zero initially and remains zero at all times during the rubbing process.

Electric charges play a role in many situations other than rubbing two surfaces together. They are involved, for instance, in chemical reactions, electric circuits, and radioactive decay. A great number of experiments have verified that in any situation, the *law of conservation of electric charge* is obeyed.

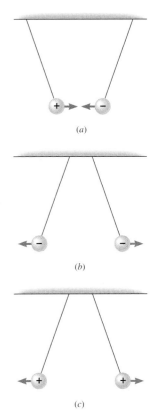

■ LAW OF CONSERVATION OF ELECTRIC CHARGE

During any process, the net electric charge of an isolated system remains constant (is conserved).

It is easy to demonstrate that two electrically charged objects exert a force on one another. Consider Figure 18.3*a*, which shows two small balls that have been *oppositely charged* and are light and free to move. The balls attract each other. On the other hand, balls with the *same* type of charge, either both positive or both negative, repel each other, as parts *b* and *c* of the drawing indicate. The behavior depicted in Figure 18.3 illustrates the following fundamental characteristic of electric charges:

Like charges repel and unlike charges attract each other.

▶ CONCEPTS AT A GLANCE Like other forces that we have encountered, the electric force (also called the electrostatic force) can alter the motion of an object. It can do so by contributing to the net external force $\Sigma\mathbf{F}$ that acts on the object. Newton's second law, $\Sigma\mathbf{F} = m\mathbf{a}$, specifies the acceleration $\mathbf{a}$ that arises because of the net external force. The Concepts-at-a-Glance chart in Figure 18.4 is an expanded version of the charts in Figures 4.9, 10.4, and 11.5, and emphasizes that any external electric force that acts on an object must be included when determining the net external force to be used in the second law. ◀

Figure 18.3 (*a*) A positive charge (+) and a negative charge (−) attract each other. (*b*) Two negative charges repel each other. (*c*) Two positive charges repel each other.

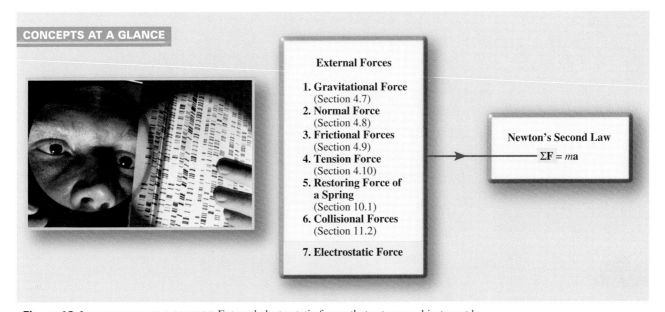

CONCEPTS AT A GLANCE

External Forces

1. **Gravitational Force** (Section 4.7)
2. **Normal Force** (Section 4.8)
3. **Frictional Forces** (Section 4.9)
4. **Tension Force** (Section 4.10)
5. **Restoring Force of a Spring** (Section 10.1)
6. **Collisional Forces** (Section 11.2)

7. **Electrostatic Force**

Newton's Second Law
$$\Sigma\mathbf{F} = m\mathbf{a}$$

Figure 18.4 CONCEPTS AT A GLANCE External electrostatic forces that act on an object must be included in the net external force when using Newton's second law to determine acceleration. In obtaining a DNA "fingerprint" (see photograph) using electrophoresis, an electrostatic force accelerates electrically charged DNA fragments to different terminal speeds in an agarose gel. (© Geoff Tompkinson/Photo Researchers)

The physics of...
electronic ink.

A new technology based on the electric force may revolutionize the way books and other printed matter are made. This technology, called electronic ink, allows letters and graphics on a page to be changed instantly, much like the symbols displayed on a computer monitor. Figure 18.5*a* illustrates the essential features of electronic ink. It consists of millions of clear microcapsules, each having the diameter of a human hair and filled with a dark inky liquid. Inside each microcapsule are several dozen extremely tiny white beads that carry a slightly negative charge. The microcapsules are sandwiched between two sheets, an opaque base layer and a transparent top layer, at which the reader looks. When a positive charge is applied to a small region of the base layer, as shown in part *b* of the drawing, the negatively charged white beads are drawn to it, leaving dark ink at the top layer. Thus, a viewer sees only the dark liquid. When a negative charge is applied to a region of the base layer, the negatively charged white beads are repelled from it and are forced to the top of the microcapsules; now a viewer sees a white area due to the beads. Thus, electronic ink is based on the principle that like charges repel and unlike charges attract each other; a positive charge causes one color to appear, and a negative charge causes another color to appear. Each small region, whether dark or light, is known as a pixel (short for "picture element"). Computer chips provide the instructions to produce the negative and positive charges on the base layer of each pixel. Letters and graphics are produced by the patterns generated with the two colors.

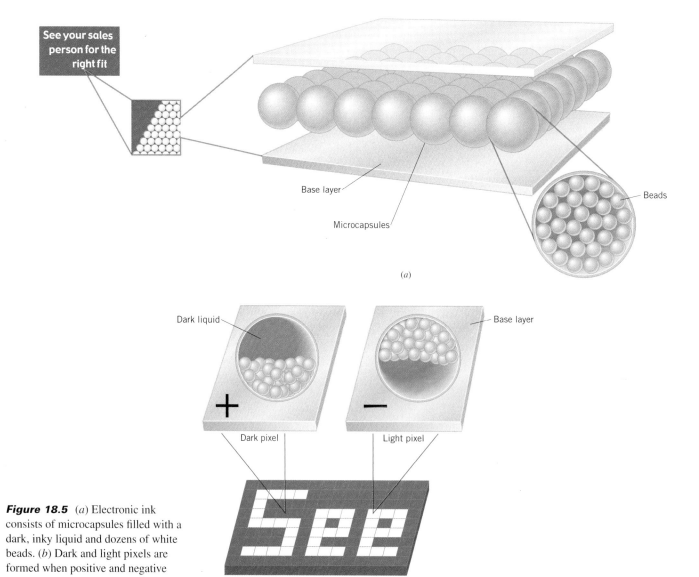

Figure 18.5 (*a*) Electronic ink consists of microcapsules filled with a dark, inky liquid and dozens of white beads. (*b*) Dark and light pixels are formed when positive and negative charges are placed in the base layer by electronic circuitry.

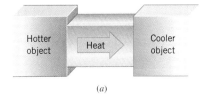 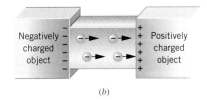

Figure 18.6 (*a*) Heat is conducted from the hotter end of the metal bar to the cooler end. (*b*) Electrons are conducted from the negatively charged end of the metal bar to the positively charged end.

18.3 *Conductors and Insulators*

Not only can electric charge exist *on an object,* but it can also move *through an object.* However, materials differ vastly in their abilities to allow electric charge to move or be conducted through them. To help illustrate such differences in conductivity, Figure 18.6*a* recalls the conduction of heat through a bar of material whose ends are maintained at different temperatures. As Section 13.2 discusses, metals conduct heat readily and, therefore, are known as thermal conductors. On the other hand, substances that conduct heat poorly are referred to as thermal insulators.

A situation analogous to the conduction of heat arises when a metal bar is placed between two charged objects, as in Figure 18.6*b*. Electrons are conducted through the bar from the negatively charged object toward the positively charged object. Substances that readily conduct electric charge are called *electrical conductors.* Although there are exceptions, good thermal conductors are generally good electrical conductors. Metals such as copper, aluminum, silver, and gold are excellent electrical conductors and, therefore, are used in electrical wiring. Materials that conduct electric charge *poorly* are known as *electrical insulators.* In many cases, thermal insulators are also electrical insulators. Common electrical insulators are rubber, many plastics, and wood. Insulators, such as the rubber or plastic that coats electrical wiring, prevent electric charge from going where it is not wanted.

The difference between electrical conductors and insulators is related to atomic structure. As electrons orbit the nucleus, those in the outer orbits experience a weaker force of attraction to the nucleus than do those in the inner orbits. Consequently, the outermost electrons (also called the valence electrons) can be dislodged more easily than the inner ones. In a good conductor, some valence electrons become detached from a parent atom and wander more or less freely throughout the material, belonging to no one atom in particular. The exact number of electrons detached from each atom depends on the nature of the material, but is usually between one and three. When one end of a conducting bar is placed in contact with a negatively charged object and the other end in contact with a positively charged object, as in Figure 18.6*b*, the "free" electrons are able to move readily away from the negative end and toward the positive end. The ready movement of electrons is the hallmark of a good conductor. In an insulator the situation is different, for there are very few electrons free to move throughout the material. Virtually every electron remains bound to its parent atom. Without the "free" electrons, there is very little flow of charge when the material is placed between two oppositely charged bodies, so the material is an electrical insulator.

18.4 *Charging by Contact and by Induction*

When a negatively charged ebonite rod is rubbed on a metal object, such as the sphere in Figure 18.7*a*, some of the excess electrons from the rod are transferred to the object. Once the electrons are on the metal sphere, where they can move readily, they repel one another and spread out over the sphere's surface. The insulated stand prevents them from flowing to the earth, where they could spread out even more. When the rod is removed, as in part *b* of the picture, the sphere is left with a negative charge distributed over its surface. In a similar manner, the sphere would be left with a positive charge after being rubbed with a positively charged rod. In this case, electrons from the sphere would be transferred to the rod. The process of giving one object a net electric charge by placing it in contact with another object that is already charged is known as *charging by contact.*

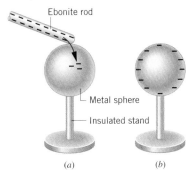

Figure 18.7 (*a*) Electrons are transferred by rubbing the negatively charged rod on the metal sphere. (*b*) When the rod is removed, the electrons distribute themselves over the surface of the sphere.

It is also possible to charge a conductor in a way that does not involve contact. In Figure 18.8, a negatively charged rod is brought close to, *but does not touch,* a metal sphere. In the sphere, the free electrons closest to the rod move to the other side, as part *a* of the drawing indicates. As a result, the part of the sphere nearest the rod becomes positively charged and the part farthest away becomes negatively charged. These positively and negatively charged regions have been "induced"or "persuaded" to form because of the repulsive force between the negative rod and the free electrons in the sphere. If the rod were removed, the free electrons would return to their original places, and the charged regions would disappear.

Under most conditions the earth is a good electrical conductor. So when a metal wire is attached between the sphere and the ground, as in Figure 18.8*b*, some of the free electrons leave the sphere and distribute themselves over the much larger earth. If the grounding wire is then removed, followed by the ebonite rod, the sphere is left with a positive net charge, as part *c* of the picture shows. The process of giving one object a net electric charge *without touching* the object to a second charged object is called ***charging by induction.*** The process could also be used to give the sphere a negative net charge, if a positively charged rod were used. Then, electrons would be drawn up from the ground through the grounding wire and onto the sphere.

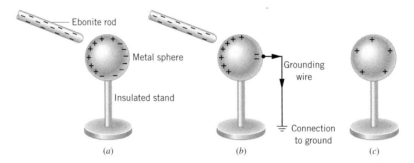

Figure 18.8 (*a*) When a charged rod is brought near the metal sphere without touching it, some of the positive and negative charges in the sphere are separated. (*b*) Some of the electrons leave the sphere through the grounding wire, with the result (*c*) that the sphere acquires a positive net charge.

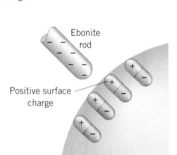

Figure 18.9 The negatively charged rod induces a slight positive surface charge on the plastic.

If the sphere in Figure 18.8 were made from an insulating material like plastic, instead of metal, the method of producing a net charge by induction would not work, because very little charge would flow through the insulating material and down the grounding wire. However, the electric force of the charged rod would have some effect, as Figure 18.9 illustrates. The electric force would cause the positive and negative charges in the molecules of the insulating material to separate slightly, with the negative charges being "pushed" away from the negative rod. Although no net charge is created, the surface of the plastic does acquire a slight induced positive charge and is attracted to the negative rod. For a similar reason, one piece of cloth can stick to another in the phenomenon known as "static cling," which occurs when an article of clothing has acquired an electric charge while being tumbled about in a clothes dryer.

✔ **Check Your Understanding 1**

Two metal spheres are identical. They are electrically neutral and are touching. An electrically charged ebonite rod is then brought near the spheres without touching them, as the drawing shows. After a while, with the rod held in place, the spheres are separated, and the rod is then removed. The following statements refer to the masses m_A and m_B of the spheres after they are separated and the rod is removed. Which one or more of the statements is true? (a) $m_A = m_B$, (b) $m_A > m_B$ if the rod is positive, (c) $m_A < m_B$ if the rod is positive, (d) $m_A > m_B$ if the rod is negative, (e) $m_A < m_B$ if the rod is negative. *[The answer(s) is (are) given at the end of the book.]*

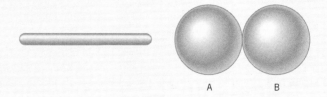

(Continues)

Background: Mass and electric charge are both properties of electrons and protons. In a metal, which move, electrons or protons? An object can be given an electric charge by contact with another charged object or by induction.

For similar questions (including calculational counterparts), consult Self-Assessment Test 18.1, which is described at the end of Section 18.5.

18.5 Coulomb's Law

THE FORCE THAT POINT CHARGES EXERT ON EACH OTHER

The electrostatic force that stationary charged objects exert on each other depends on the amount of charge on the objects and the distance between them. Experiments reveal that the greater the charge and the closer together they are, the greater is the force. To set the stage for explaining these features in more detail, Figure 18.10 shows two charged bodies. These objects are so small, compared to the distance r between them, that they can be regarded as mathematical points. The "point charges" have magnitudes* $|q_1|$ and $|q_2|$. If the charges have *unlike* signs, as in part a of the picture, each object is *attracted* to the other by a force that is directed along the line between them; $+\mathbf{F}$ is the electric force exerted on object 1 by object 2 and $-\mathbf{F}$ is the electric force exerted on object 2 by object 1. If, as in part b, the charges have the *same* sign (both positive or both negative), each object is repelled from the other. The repulsive forces, like the attractive forces, act along the line between the charges. Whether attractive or repulsive, the two forces are equal in magnitude but opposite in direction. These forces always exist as a pair, each one acting on a different object, in accord with Newton's action–reaction law.

The French physicist Charles-Augustin Coulomb (1736–1806) carried out a number of experiments to determine how the electric force that one point charge applies to another depends on the amount of each charge and the separation between them. His result, now known as ***Coulomb's law,*** is stated below.

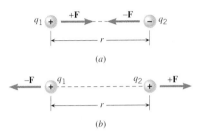

Figure 18.10 Each point charge exerts a force on the other. Regardless of whether the forces are (a) attractive or (b) repulsive, they are directed along the line between the charges and have equal magnitudes.

■ **COULOMB'S LAW**

The magnitude F of the electrostatic force exerted by one point charge q_1 on another point charge q_2 is directly proportional to the magnitudes $|q_1|$ and $|q_2|$ of the charges and inversely proportional to the square of the distance r between them:

$$F = k\frac{|q_1||q_2|}{r^2} \tag{18.1}$$

where k is a proportionality constant whose value in SI units is $k = 8.99 \times 10^9 \ \text{N} \cdot \text{m}^2/\text{C}^2$. The electrostatic force is directed along the line joining the charges, and it is attractive if the charges have unlike signs and repulsive if the charges have like signs.

It is common practice to express k in terms of another constant ϵ_0, by writing $k = 1/(4\pi\epsilon_0)$; ϵ_0 is called the ***permittivity of free space*** and has a value of $\epsilon_0 = 1/(4\pi k) = 8.85 \times 10^{-12} \ \text{C}^2/(\text{N} \cdot \text{m}^2)$. Equation 18.1 gives the magnitude of the electrostatic force that each point charge exerts on the other. When using this equation, then, it is important to remember to substitute only the charge magnitudes (without algebraic signs) for $|q_1|$ and $|q_2|$, as Example 2 illustrates.

*The magnitude of a variable is sometimes called the absolute value and is symbolized by a vertical bar to the left and to the right of the variable. Thus, $|q|$ denotes the magnitude or absolute value of the variable q, which is the value of q without its algebraic plus or minus sign. For example, if $q = -2.0$ C, then $|q| = 2.0$ C.

Need more practice?

Interactive LearningWare 18.1
An object of mass of 215 kg is located just above the surface of the earth (radius = 6.38×10^6 m). Suppose that the object and the earth have an identical charge q. Therefore, two forces—the gravitational force and the electrostatic force—act on the object. Assuming that the earth's charge is located at the center of the earth, determine q (magnitude and sign) when the object is in equilibrium.

Related Homework: Problem 11

Go to
www.wiley.com/college/cutnell
for an interactive solution.

Example 2 A Large Attractive Force

Two objects, whose charges are $+1.0$ and -1.0 C, are separated by 1.0 km. Compared to 1.0 km, the sizes of the objects are small. Find the magnitude of the attractive force that either charge exerts on the other.

Reasoning Considering that the sizes of the objects are small compared to the separation distance, we can treat the charges as point charges. Coulomb's law may then be used to find the magnitude of the attractive force, provided that only the *magnitudes of the charges* are used for the symbols $|q_1|$ and $|q_2|$ that appear in the law.

Solution The magnitude of the force is

$$F = k\frac{|q_1||q_2|}{r^2} = \frac{(8.99 \times 10^9 \text{ N} \cdot \text{m}^2/\text{C}^2)(1.0 \text{ C})(1.0 \text{ C})}{(1.0 \times 10^3 \text{ m})^2} = \boxed{9.0 \times 10^3 \text{ N}} \quad (18.1)$$

The force calculated in Example 2 corresponds to about 2000 pounds and is so large because charges of ± 1.0 C are enormous. Such large charges are encountered only in the most severe conditions, as in a lightning bolt, where as much as 25 coulombs can be transferred between the cloud and the ground. The typical charges produced in the laboratory are much smaller and are measured conveniently in microcoulombs (1 microcoulomb = 1 μC = 10^{-6} C).

Coulomb's law has a form remarkably similar to Newton's law of gravitation ($F = Gm_1m_2/r^2$). The force in both laws depends on the inverse square ($1/r^2$) of the distance between the two objects and is directed along the line between them. In addition, the force is proportional to the product of an intrinsic property of each of the objects, the magnitudes of the charges $|q_1|$ and $|q_2|$ in Coulomb's law and the masses m_1 and m_2 in the gravitation law. But there is a major difference between the two laws. The electrostatic force can be either repulsive or attractive, depending on whether or not the charges have the same sign; in contrast, the gravitational force is always an attractive force.

Section 5.5 discusses how the gravitational attraction between the earth and a satellite provides the centripetal force that keeps a satellite in orbit. Example 3 illustrates that the electrostatic force of attraction plays a similar role in a famous model of the atom created by the Danish physicist Niels Bohr (1885–1962).

Example 3 A Model of the Hydrogen Atom

In the Bohr model of the hydrogen atom, the electron ($-e$) is in orbit about the nuclear proton ($+e$) at a radius of $r = 5.29 \times 10^{-11}$ m, as Figure 18.11 shows. Determine the speed of the electron, assuming the orbit to be circular.

Reasoning Recall from Section 5.2 that any object moving with speed v on a circular path of radius r has a centripetal acceleration of $a_c = v^2/r$. This acceleration is directed toward the center of the circle. Newton's second law specifies that the net force ΣF needed to create this acceleration is $\Sigma F = ma_c = mv^2/r$, where m is the mass of the object. This equation can be solved for the speed: $v = \sqrt{(\Sigma F)r/m}$. Since the mass of the electron is $m = 9.11 \times 10^{-31}$ kg and the radius is given, we can calculate the speed, once a value for the net force is available. For the electron in the hydrogen atom, the net force is provided almost exclusively by the electrostatic force, as given by Coulomb's law. This force points toward the center of the circle, since the electron and the proton have opposite signs. The electron is also pulled toward the proton by the gravitational force. However, the gravitational force is negligible in comparison to the electrostatic force.

Figure 18.11 In the Bohr model of the hydrogen atom, the electron ($-e$) orbits the proton ($+e$) at a distance of $r = 5.29 \times 10^{-11}$ m. The velocity of the electron is **v**.

Problem solving insight
When using Coulomb's law ($F = k|q_1||q_2|/r^2$), remember that the symbols $|q_1|$ and $|q_2|$ stand for the charge magnitudes. Do not substitute negative numbers for these symbols.

Solution The electron experiences an electrostatic force of attraction because of the proton, and the magnitude of this force is

$$F = k\frac{|q_1||q_2|}{r^2} = \frac{(8.99 \times 10^9 \text{ N} \cdot \text{m}^2/\text{C}^2)(1.60 \times 10^{-19} \text{ C})(1.60 \times 10^{-19} \text{ C})}{(5.29 \times 10^{-11} \text{ m})^2}$$

$$= 8.22 \times 10^{-8} \text{ N}$$

Using this value for the net force, we find

$$v = \sqrt{\frac{(\Sigma F)r}{m}} = \sqrt{\frac{(8.22 \times 10^{-8} \text{ N})(5.29 \times 10^{-11} \text{ m})}{9.11 \times 10^{-31} \text{ kg}}} = \boxed{2.18 \times 10^6 \text{ m/s}}$$

This orbital speed is almost five million miles per hour.

Since the electrostatic force depends on the inverse square of the distance between the charges, it becomes larger for smaller distances, such as those involved when a strip of adhesive tape is stuck to a smooth surface. Electrons shift over the small distances between the tape and the surface. As a result, the materials become oppositely charged. Since the distance between the charges is relatively small, the electrostatic force of attraction is large enough to contribute to the adhesive bond. Figure 18.12 shows an image of the sticky surface of a piece of tape after it has been pulled off a metal surface. The image was obtained using an atomic-force microscope and reveals the tiny pits left behind when microscopic portions of the adhesive remain stuck to the metal because of the strong adhesive bonding forces.

The physics of...
adhesion.

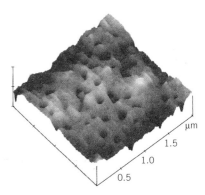

Figure 18.12 After a strip of adhesive tape has been pulled off a metal surface, there are tiny pits (approximately one ten-millionth of a meter in diameter) in the sticky surface of the tape, as this image shows. It was obtained using an atomic-force microscope. (Courtesy Louis Scudiero and J. Thomas Dickinson, Washington State University.)

THE FORCE ON A POINT CHARGE DUE TO TWO OR MORE OTHER POINT CHARGES

Up to now, we have been discussing the electrostatic force on a point charge (magnitude $|q_1|$) due to another point charge (magnitude $|q_2|$). Suppose that a third point charge (magnitude $|q_3|$) is also present. What would be the net force on q_1 due to both q_2 and q_3? It is convenient to deal with such a problem in parts. First, find the magnitude and direction of the force exerted on q_1 by q_2 (ignoring q_3). Then, determine the force exerted on q_1 by q_3 (ignoring q_2). The *net force* on q_1 is the *vector sum* of these forces. Examples 4 and 5 illustrate this approach when the charges lie along a straight line and on a plane, respectively.

Example 4 Three Charges on a Line

Figure 18.13a shows three point charges that lie along the x axis in a vacuum. Determine the magnitude and direction of the net electrostatic force on q_1.

Reasoning Part b of the drawing shows a free-body diagram of the forces that act on q_1. Since q_1 and q_2 have opposite signs, they attract one another. Thus, the force exerted on q_1 by q_2 is $\mathbf{F}_{12}$, and it points to the left. Similarly, the force exerted on q_1 by q_3 is $\mathbf{F}_{13}$ and is also an attractive force. It points to the right in Figure 18.13b. The magnitudes of these forces can be obtained from Coulomb's law. The net force is the vector sum of $\mathbf{F}_{12}$ and $\mathbf{F}_{13}$.

Solution The magnitudes of the forces are

$$F_{12} = k\frac{|q_1||q_2|}{r_{12}^2} = \frac{(8.99 \times 10^9 \text{ N} \cdot \text{m}^2/\text{C}^2)(3.0 \times 10^{-6} \text{ C})(4.0 \times 10^{-6} \text{ C})}{(0.20 \text{ m})^2} = 2.7 \text{ N}$$

$$F_{13} = k\frac{|q_1||q_3|}{r_{13}^2} = \frac{(8.99 \times 10^9 \text{ N} \cdot \text{m}^2/\text{C}^2)(3.0 \times 10^{-6} \text{ C})(7.0 \times 10^{-6} \text{ C})}{(0.15 \text{ m})^2} = 8.4 \text{ N}$$

Since $\mathbf{F}_{12}$ points in the negative x direction, and $\mathbf{F}_{13}$ points in the positive x direction, the net force $\mathbf{F}$ is

$$\mathbf{F} = \mathbf{F}_{12} + \mathbf{F}_{13} = (-2.7 \text{ N}) + (8.4 \text{ N}) = \boxed{+5.7 \text{ N}}$$

The plus sign in the answer indicates that the net force points to the right in the drawing.

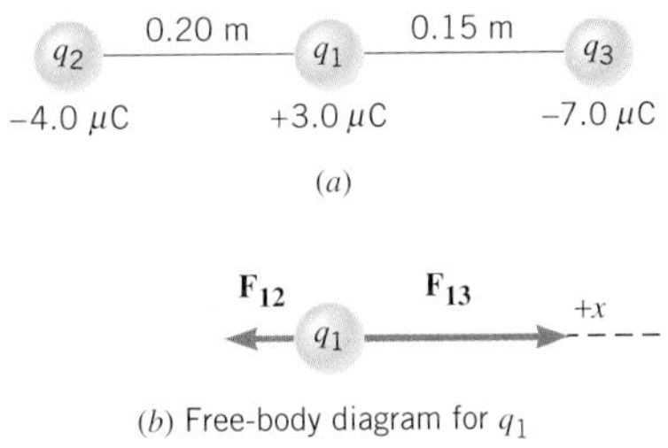

q_2 0.20 m q_1 0.15 m q_3
$-4.0 \ \mu\text{C}$ $+3.0 \ \mu\text{C}$ $-7.0 \ \mu\text{C}$

(a)

$\mathbf{F}_{12}$ $\mathbf{F}_{13}$ $+x$
q_1

(b) Free-body diagram for q_1

Figure 18.13 (a) Three charges lying along the x axis. (b) The force exerted on q_1 by q_2 is $\mathbf{F}_{12}$, while the force exerted on q_1 by q_3 is $\mathbf{F}_{13}$.

Example 5 Three Charges in a Plane

Figure 18.14a shows three point charges that lie in the x, y plane in a vacuum. Find the magnitude and direction of the net electrostatic force on q_1.

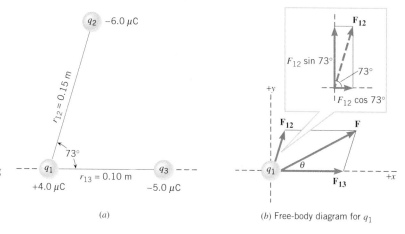

Figure 18.14 (*a*) Three charges lying in a plane. (*b*) The net force acting on q_1 is $\mathbf{F} = \mathbf{F}_{12} + \mathbf{F}_{13}$. The angle that $\mathbf{F}$ makes with the $+x$ axis is θ.

(*a*)

(*b*) Free-body diagram for q_1

Need more practice?

Interactive LearningWare 18.2
A positive charge is placed on the x axis ($q = +7.00\ \mu C$, $x = 0.600$ m), and another positive charge is placed on the y axis ($q = +9.00\ \mu C$, $y = 0.400$ m). A third charge ($q = -6.00\ \mu C$, $m = 5.00 \times 10^{-8}$ kg) is placed at the coordinate origin. If the charge at the origin were free to move, what would be the magnitude of its acceleration?

Related Homework: *Problem 18*

Go to
www.wiley.com/college/cutnell
for an interactive solution.

Reasoning The force exerted on q_1 by q_2 is $\mathbf{F}_{12}$ and is an attractive force because the two charges have opposite signs. It points along the line between the charges. The force exerted on q_1 by q_3 is $\mathbf{F}_{13}$ and is also an attractive force. It points along the line between q_1 and q_3. Coulomb's law specifies the magnitudes of these forces. Since the forces point in different directions (see Figure 18.14*b*), we will use vector components to find the net force.

Solution The magnitudes of the forces are

$$F_{12} = k\frac{|q_1||q_2|}{r_{12}^2} = \frac{(8.99 \times 10^9\ \text{N} \cdot \text{m}^2/\text{C}^2)(4.0 \times 10^{-6}\ \text{C})(6.0 \times 10^{-6}\ \text{C})}{(0.15\ \text{m})^2} = 9.6\ \text{N}$$

$$F_{13} = k\frac{|q_1||q_3|}{r_{13}^2} = \frac{(8.99 \times 10^9\ \text{N} \cdot \text{m}^2/\text{C}^2)(4.0 \times 10^{-6}\ \text{C})(5.0 \times 10^{-6}\ \text{C})}{(0.10\ \text{m})^2} = 18\ \text{N}$$

The net force $\mathbf{F}$ is the vector sum of $\mathbf{F}_{12}$ and $\mathbf{F}_{13}$, as part *b* of the drawing shows. The components of $\mathbf{F}$ that lie in the x and y directions are $\mathbf{F}_x$ and $\mathbf{F}_y$, respectively. Our approach to finding $\mathbf{F}$ is the same as that used in Chapters 1 and 4. The forces $\mathbf{F}_{12}$ and $\mathbf{F}_{13}$ are resolved into x and y components. Then, the x components are combined to give $\mathbf{F}_x$, and the y components are combined to give $\mathbf{F}_y$. Once $\mathbf{F}_x$ and $\mathbf{F}_y$ are known, the magnitude and direction of $\mathbf{F}$ can be determined using trigonometry.

Force	x component	y component
$\mathbf{F}_{12}$	$+(9.6\ \text{N})\cos 73° = +2.8\ \text{N}$	$+(9.6\ \text{N})\sin 73° = +9.2\ \text{N}$
$\mathbf{F}_{13}$	$+18\ \text{N}$	$0\ \text{N}$
$\mathbf{F}$	$F_x = +21\ \text{N}$	$F_y = +9.2\ \text{N}$

The magnitude F and the angle θ of the net force are

$$F = \sqrt{F_x^2 + F_y^2} = \sqrt{(21\ \text{N})^2 + (9.2\ \text{N})^2} = \boxed{23\ \text{N}}$$

$$\theta = \tan^{-1}\left(\frac{F_y}{F_x}\right) = \tan^{-1}\left(\frac{9.2\ \text{N}}{21\ \text{N}}\right) = \boxed{24°}$$

Problem solving insight
The electrostatic force is a vector and has a direction as well as a magnitude. When adding electrostatic forces, take into account the directions of all forces, using vector components as needed.

Concept Simulation 18.1

This simulation allows you to change the magnitude and algebraic sign of each charge in Figure 18.14. The forces acting on charge q_1 and the resultant of these forces are displayed to scale as changes are made, so that you can see the effects of your choices.

Related Homework: *Problem 61*

Go to **www.wiley.com/college/cutnell**

✓ Check Your Understanding 2

The drawing shows three point charges arranged in different ways. The charges are $+q$, $-q$, and $-q$; each has the same magnitude, one being positive and the other two negative. In each part of the drawing the distance d is the same. Rank the arrangements in descending order (largest first) according to the magnitude of the net electrostatic force that acts on the positive charge. (*The answer is given at the end of the book.*)   (*Continues*)

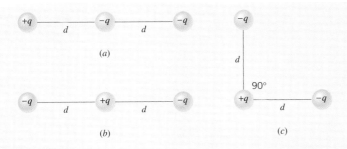

Background: The electrostatic force that one point charge exerts on another is described by Coulomb's law. Like any force, this force is a vector quantity, and the rules of vector addition apply.

For similar questions (including calculational counterparts), consult Self-Assessment Test 18.1, which is described next.

Self-Assessment Test 18.1

Test your understanding of the material in Sections 18.1–18.5:

• The Origin of Electricity   • Conductors and Insulators
• Charging by Contact and by Induction   • Coulomb's Law

Go to **www.wiley.com/college/cutnell**

18.6 *The Electric Field*

DEFINITION

As we know, a charge can experience an electrostatic force due to the presence of other charges. For instance, the positive charge q_0 in Figure 18.15 experiences a force **F**, which is the vector sum of the forces exerted by the charges on the rod and the two spheres. It is useful to think of q_0 as a ***test charge*** for determining the extent to which the surrounding charges generate a force. However, in using a test charge, we must be careful to select one with a very small magnitude, so that it does not alter the locations of the other charges. The next example illustrates how the concept of a test charge is applied.

Example 6   A Test Charge

The positive test charge in Figure 18.15 is $q_0 = +3.0 \times 10^{-8}$ C and experiences a force $F = 6.0 \times 10^{-8}$ N in the direction shown in the drawing. (a) Find the *force per coulomb* that the test charge experiences. (b) Using the result of part (a), predict the force that a charge of $+12 \times 10^{-8}$ C would experience if it replaced q_0.

Reasoning The charges in the environment apply a force **F** to the test charge q_0. The force per coulomb experienced by the test charge is $\mathbf{F}/q_0$. If q_0 is replaced by a new charge q, then the force on this new charge is the force per coulomb times q.

Solution

(a) The force per coulomb of charge is

$$\frac{F}{q_0} = \frac{6.0 \times 10^{-8}\text{ N}}{3.0 \times 10^{-8}\text{ C}} = \boxed{2.0 \text{ N/C}}$$

(b) The result from part (a) indicates that the surrounding charges can exert 2.0 newtons of force per coulomb of charge. Thus, a charge of $+12 \times 10^{-8}$ C would experience a force whose magnitude is

$$F = (2.0 \text{ N/C})(12 \times 10^{-8}\text{ C}) = \boxed{24 \times 10^{-8}\text{ N}}$$

The direction of this force would be the same as that experienced by the test charge, since both have the same positive sign.

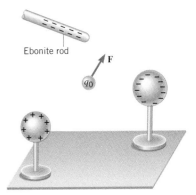

Figure 18.15 A positive charge q_0 experiences an electrostatic force **F** due to the surrounding charges on the ebonite rod and the two spheres.

CONCEPTS AT A GLANCE

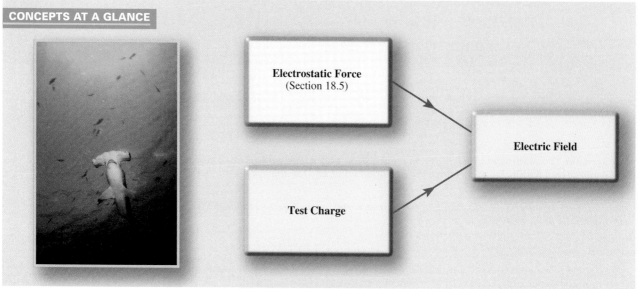

Figure 18.16 CONCEPTS AT A GLANCE A test charge and the electrostatic force that acts on it are brought together to define the concept of an electric field. Sharks have specialized organs that can detect electric fields. In the hammerhead shark shown here, the organs are located in the wide head. (© Henry C. Wolcott III/National Geographic/Getty Images)

▶ CONCEPTS AT A GLANCE The electric force per coulomb, $\mathbf{F}/q_0$, calculated in Example 6(a) is one illustration of an idea that is very important in the study of electricity. The idea is called the *electric field.* As the Concepts-at-a-Glance chart in Figure 18.16 illustrates, the notion of an electric field emerges when we combine the electrostatic force with the idea of a test charge. Equation 18.2 presents the definition of the electric field. ◀

■ **DEFINITION OF THE ELECTRIC FIELD**

The electric field $\mathbf{E}$ that exists at a point is the electrostatic force $\mathbf{F}$ experienced by a small test charge* q_0 placed at that point divided by the charge itself:

$$\mathbf{E} = \frac{\mathbf{F}}{q_0} \tag{18.2}$$

The electric field is a vector, and its direction is the same as the direction of the force $\mathbf{F}$ on a positive test charge.

SI Unit of Electric Field: newton per coulomb (N/C)

Equation 18.2 indicates that the unit for the electric field is that of force divided by charge, which is a newton/coulomb (N/C) in SI units.

It is the surrounding charges that create an electric field at a given point. Any positive or negative charge placed at the point interacts with the field and, as a result, experiences a force, as the next example indicates.

Example 7   An Electric Field Leads to a Force

In Figure 18.17a the charges on the two metal spheres and the ebonite rod create an electric field $\mathbf{E}$ at the spot indicated. This field has a magnitude of 2.0 N/C and is directed as in the drawing. Determine the force on a charge placed at that spot, if the charge has a value of (a) $q_0 = +18 \times 10^{-8}$ C and (b) $q_0 = -24 \times 10^{-8}$ C.

---

*As long as the test charge is small enough that it does not disturb the surrounding charges, it may be either positive or negative. Compared to a positive test charge, a negative test charge of the same magnitude experiences a force of the same magnitude that points in the opposite direction. However, the same electric field is given by Equation 18.2, in which $\mathbf{F}$ is replaced by $-\mathbf{F}$ and q_0 is replaced by $-q_0$.

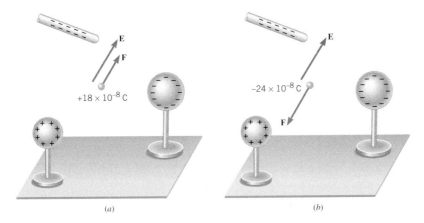

Figure 18.17 The electric field **E** that exists at a given spot can exert a variety of forces. The force exerted depends on the magnitude and sign of the charge placed at that spot. (*a*) The force on a positive charge points in the same direction as **E**, while (*b*) the force on a negative charge points opposite to **E**.

Reasoning The electric field at a given spot can exert a variety of forces, depending on the magnitude and sign of the charge placed there. The charge is assumed to be small enough that it does not alter the locations of the surrounding charges that create the field.

Solution

(a) The magnitude of the force is the product of the magnitudes of q_0 and **E**:

$$F = |q_0| E = (18 \times 10^{-8}\ \text{C})(2.0\ \text{N/C}) = \boxed{36 \times 10^{-8}\ \text{N}} \qquad (18.2)$$

Since q_0 is positive, the force points in the same direction as the electric field, as part *a* of the drawing indicates.

(b) In this case, the magnitude of the force is

$$F = |q_0| E = (24 \times 10^{-8}\ \text{C})(2.0\ \text{N/C}) = \boxed{48 \times 10^{-8}\ \text{N}} \qquad (18.2)$$

The force on the negative charge points in the direction *opposite* to the force on the positive charge—that is, opposite to the electric field (see part *b* of the drawing).

At a particular point in space, each of the surrounding charges contributes to the net electric field that exists there. To determine the net field, it is necessary to obtain the various contributions separately and then find the vector sum of them all. Such an approach is an illustration of the principle of linear superposition, as applied to electric fields. (This principle is introduced in Section 17.1, in connection with waves.) Example 8 emphasizes the vector nature of the electric field.

Example 8 Electric Fields Add as Vectors Do

Figure 18.18 shows two charged objects, *A* and *B*. Each contributes as follows to the net electric field at point *P*: E_A = 3.00 N/C directed to the right, and E_B = 2.00 N/C directed downward. Thus, E_A and E_B are perpendicular. What is the net electric field at *P*?

Reasoning The net electric field **E** is the vector sum of E_A and E_B: $\mathbf{E} = \mathbf{E_A} + \mathbf{E_B}$. As illustrated in Figure 18.18, E_A and E_B are perpendicular, so **E** is the diagonal of the rectangle shown in the drawing. Thus, we can use the Pythagorean theorem to find the magnitude of **E** and trigonometry to find the directional angle θ.

Solution The magnitude of the net electric field is

$$E = \sqrt{E_A{}^2 + E_B{}^2} = \sqrt{(3.00\ \text{N/C})^2 + (2.00\ \text{N/C})^2} = \boxed{3.61\ \text{N/C}}$$

The direction of **E** is given by the angle θ in the drawing:

$$\theta = \tan^{-1}\left(\frac{E_B}{E_A}\right) = \tan^{-1}\left(\frac{2.00\ \text{N/C}}{3.00\ \text{N/C}}\right) = \boxed{33.7°}$$

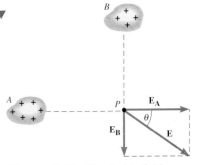

Figure 18.18 The electric field contributions E_A and E_B, which come from the two charge distributions, are added vectorially to obtain the net field **E** at point *P*.

POINT CHARGES

A more complete understanding of the electric field concept can be gained by considering the field created by a point charge, as in the following example.

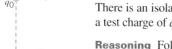

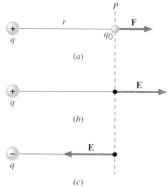

Figure 18.19 (*a*) At location P, a positive test charge q_0 experiences a repulsive force **F** due to the positive point charge q. (*b*) At P the electric field **E** is directed to the right. (*c*) If the charge q were negative rather than positive, the electric field would have the same magnitude as in (*b*) but point to the left.

Example 9   The Electric Field of a Point Charge

There is an isolated point charge of $q = +15\ \mu\text{C}$ in a vacuum at the left in Figure 18.19*a*. Using a test charge of $q_0 = +0.80\ \mu\text{C}$, determine the electric field at point P, which is 0.20 m away.

Reasoning Following the definition of the electric field, we place the test charge q_0 at point P, determine the force acting on the test charge, and then divide the force by the test charge.

Solution Coulomb's law (Equation 18.1), gives the magnitude of the force:

$$F = k\frac{|q_0||q|}{r^2} = \frac{(8.99 \times 10^9\ \text{N} \cdot \text{m}^2/\text{C}^2)(0.80 \times 10^{-6}\ \text{C})(15 \times 10^{-6}\ \text{C})}{(0.20\ \text{m})^2} = 2.7\ \text{N}$$

Equation 18.2 gives the magnitude of the electric field:

$$E = \frac{F}{|q_0|} = \frac{2.7\ \text{N}}{0.80 \times 10^{-6}\ \text{C}} = \boxed{3.4 \times 10^6\ \text{N/C}}$$

The electric field **E** points in the *same direction* as the force **F** on the positive test charge. Since the test charge experiences a force of repulsion directed to the right, the electric field vector also points to the right, as Figure 18.19*b* shows.

The electric field produced by a point charge q can be obtained in general terms from Coulomb's law. First, note that the magnitude of the force exerted by the charge q on a test charge q_0 is $F = k|q||q_0|/r^2$. Then, divide this value by $|q_0|$ to obtain the magnitude of the field. Since $|q_0|$ is eliminated algebraically from the result, *the electric field does not depend on the test charge:*

Point charge q
$$E = \frac{k|q|}{r^2} \qquad (18.3)$$

As in Coulomb's law, the symbol $|q|$ denotes the magnitude of q in Equation 18.3, without regard to whether q is positive or negative. If q is positive, then **E** is directed away from q, as in Figure 18.19*b*. On the other hand, if q is negative, then **E** is directed toward q, since a negative charge attracts a positive test charge. For instance, Figure 18.19*c* shows the electric field that would exist at P if there were a charge of $-q$ instead of $+q$ at the left of the drawing. Example 10 reemphasizes the fact that all the surrounding charges make a contribution to the electric field that exists at a given place.

Example 10   The Electric Fields from Separate Charges May Cancel

Two positive point charges, $q_1 = +16\ \mu\text{C}$ and $q_2 = +4.0\ \mu\text{C}$, are separated in a vacuum by a distance of 3.0 m, as Figure 18.20 illustrates. Find the spot on the line between the charges where the net electric field is zero.

Reasoning Between the charges the two field contributions have opposite directions, and the net electric field is zero at the place where the magnitude of $\mathbf{E}_1$ equals that of $\mathbf{E}_2$. However, since q_2 is smaller than q_1, this location must be *closer* to q_2, in order that the field of the smaller charge can balance the field of the larger charge. In the drawing, the cancellation spot is labeled P, and its distance from q_1 is d.

Solution At P, $E_1 = E_2$, and using the expression $E = k|q|/r^2$, we have

$$\frac{k(16 \times 10^{-6}\ \text{C})}{d^2} = \frac{k(4.0 \times 10^{-6}\ \text{C})}{(3.0\ \text{m} - d)^2}$$

Rearranging this expression shows that $4.0(3.0\ \text{m} - d)^2 = d^2$. Taking the square root of each side of this equation reveals that

$$2.0(3.0\ \text{m} - d) = \pm d$$

The plus and minus signs on the right occur because either the positive or negative root can be taken. Therefore, there are two possible values for d: $+2.0$ m and $+6.0$ m. The value $+6.0$ m corresponds to a location off to the right of both charges, where the magnitudes of $\mathbf{E}_1$ and $\mathbf{E}_2$ are equal, but where the directions are the same. Thus, $\mathbf{E}_1$ and $\mathbf{E}_2$ do not cancel at this spot. The other value for d corresponds to the location shown in the drawing and is the zero-field location: $\boxed{d = +2.0\ \text{m}}$.

Figure 18.20 The two point charges q_1 and q_2 create electric fields $\mathbf{E}_1$ and $\mathbf{E}_2$ that cancel at a location P on the line between the charges.

Problem solving insight
Equation 18.3 gives only the magnitude of the electric field produced by a point charge. Therefore, do not use negative numbers for the symbol $|q|$ in this equation.

When point charges are arranged in a symmetrical fashion, it is often possible to deduce useful information about the magnitude and direction of the electric field by taking advantage of the symmetry. Conceptual Example 11 illustrates the use of this technique.

Conceptual Example 11 Symmetry and the Electric Field

Figure 18.21 shows point charges fixed to the corners of a rectangle in two different ways. The charges all have the same magnitudes, but they have different signs. Consider the net electric field at the center *C* of the rectangle in each case. Which field is larger?

Reasoning and Solution In Figure 18.21*a*, the charges at corners 1 and 3 are both $+q$. The positive charge at corner 1 produces an electric field at *C* that points toward corner 3. In contrast, the positive charge at corner 3 produces an electric field at *C* that points toward corner 1. Thus, the two fields have opposite directions. The magnitudes of the fields are identical because the charges have the same magnitude and are equally far from the center. Therefore, the fields from the two positive charges cancel.

Now, let's look at the electric field produced by the charges on corners 2 and 4 in Figure 18.21*a*. The electric field due to the negative charge at corner 2 points toward corner 2, and the field due to the positive charge at corner 4 points the same way. Furthermore, the magnitudes of these fields are equal because each charge has the same magnitude and is located at the same distance from the center of the rectangle. As a result, the two fields combine to give the net electric field $\mathbf{E}_{24}$ shown in the drawing.

In Figure 18.21*b*, the charges on corners 2 and 4 are identical to those in part *a* of the drawing, so this pair gives rise to the electric field labeled $\mathbf{E}_{24}$. The charges on corners 1 and 3 are identical to those on corners 2 and 4, so they give rise to an electric field labeled $\mathbf{E}_{13}$, which has the same magnitude as $\mathbf{E}_{24}$. The net electric field $\mathbf{E}$ is the vector sum of $\mathbf{E}_{24}$ and $\mathbf{E}_{13}$ and is also shown in the drawing. Clearly, this sum is greater than $\mathbf{E}_{24}$ alone. Therefore, *the net field in part b is larger than that in part a.*

Related Homework: *Conceptual Question 12, Problem 34*

Need more practice?

Interactive LearningWare 18.3
A constant electric field exists in a region of space. The field has a magnitude of 1600 N/C and points due north. A point charge of $+4.0 \times 10^{-9}$ C is then placed in this electric field. (*a*) What is the magnitude and direction of the net electric field at a spot 13 cm due east of the charge? Specify the direction as an angle relative to due east. (*b*) If a $-7.00\ \mu$C charge were placed at this spot, what would be the electrostatic force (magnitude and direction) exerted on it?

Related Homework: *Problem 31*

Go to **www.wiley.com/college/cutnell** for an interactive solution.

THE PARALLEL PLATE CAPACITOR

Equation 18.3, which gives the electric field of a point charge, is a very useful result. With the aid of integral calculus, this equation can be applied in a variety of situations where point charges are distributed over one or more surfaces. One such example that has considerable practical importance is the **parallel plate capacitor**. As Figure 18.22 shows, this device consists of two parallel metal plates, each with area *A*. A charge $+q$ is spread uniformly over one plate, while a charge $-q$ is spread uniformly over the other plate. In the region between the plates and away from the edges, the electric field points from the positive plate toward the negative plate and is perpendicular to both. Using Gauss' law (see Section 18.9), it can be shown that the electric field has a magnitude of

Parallel plate capacitor
$$E = \frac{q}{\epsilon_0 A} = \frac{\sigma}{\epsilon_0} \qquad (18.4)$$

where ϵ_0 is the permittivity of free space. In this expression the Greek symbol sigma (σ) denotes the charge per unit area ($\sigma = q/A$) and is sometimes called the charge density. Except in the region near the edges, the field has the same value at all places between the plates. The field does *not* depend on the distance from the charges, in distinct contrast to the field created by an isolated point charge.

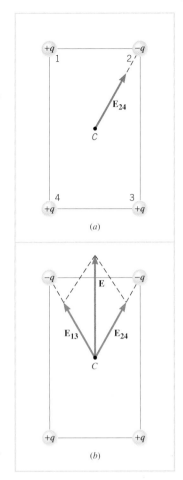

Figure 18.21 Charges of identical magnitude, but different signs, are placed at the corners of a rectangle. The charges give rise to different electric fields at the center *C* of the rectangle, depending on the signs of the charges.

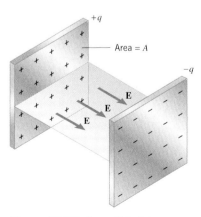

Figure 18.22 A parallel plate capacitor.

✔ Check Your Understanding 3

A positive point charge $+q$ is fixed in position at the center of a square, as the drawing shows. A second point charge is fixed to either corner B, corner C, or corner D. The net electric field at corner A is zero. (a) At which corner is the second charge located? (b) Is the second charge positive or negative? (c) Does the second charge have a greater, a smaller, or the same magnitude as the charge at the center? *(The answers are given at the end of the book.)*

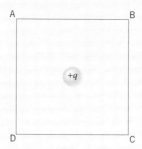

Background: The electric field is a vector quantity. Its magnitude and direction depend on the magnitudes and algebraic signs of the charges that create it.

For similar questions (including calculational counterparts), consult Self-Assessment Test 18.2, which is described at the end of Section 18.9.

18.7 Electric Field Lines

As we have seen, electric charges create an electric field in the space surrounding them. It is useful to have a kind of "map" that gives the direction and indicates the strength of the field at various places. The great English physicist Michael Faraday (1791–1867) proposed an idea that provides such a "map," the idea of *electric field lines.* Since the electric field is the electric force per unit charge, the electric field lines are sometimes called *lines of force.*

To introduce the electric field line concept, Figure 18.23a shows a positive point charge $+q$. At the locations numbered 1–8, a positive test charge would experience a repulsive force, as the arrows in the drawing indicate. Therefore, the electric field created by the charge $+q$ is directed radially outward. The electric field lines are lines drawn to show this direction, as part b of the drawing illustrates. They begin on the charge $+q$ and point radially *outward.* Figure 18.24 shows the field lines in the vicinity of a negative charge $-q$. In this case they are directed radially *inward* because the force on a positive test charge is one of attraction, indicating that the electric field points inward. In general, *electric field lines are always directed away from positive charges and toward negative charges.*

Figure 18.23 (*a*) At any of the eight marked spots around a positive point charge $+q$, a positive test charge would experience a repulsive force directed radially outward. (*b*) The electric field lines are directed radially outward from a positive point charge $+q$.

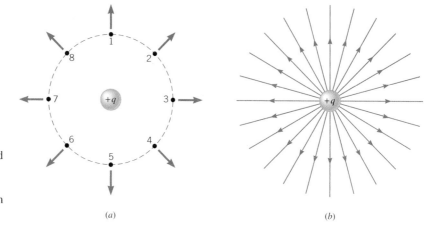

(a)   (b)

The electric field lines in Figures 18.23 and 18.24 are drawn in only two dimensions, as a matter of convenience. Field lines radiate from the charges in three dimensions, and an infinite number of lines could be drawn. However, for clarity only a small number is ever included in pictures. The number is chosen to be proportional to the magnitude of the charge; thus, five times as many lines would emerge from a $+5q$ charge as from a $+q$ charge.

The pattern of electric field lines also provides information about the magnitude or strength of the field. Notice in Figures 18.23 and 18.24 that near the charges, where the electric field is stronger, the lines are closer together. At distances far from the charges, where the electric field is weaker, the lines are more spread out. It is true in general that the electric field is stronger in regions where the field lines are closer together. In fact, no matter how many charges are present, the number of lines per unit area passing perpendicularly through a surface is proportional to the magnitude of the electric field.

In regions where the electric field lines are equally spaced, there is the same number of lines per unit area everywhere, and the electric field has the same strength at all points. For example, Figure 18.25 shows that the field lines between the plates of a parallel plate capacitor are parallel and equally spaced, except near the edges where they bulge outward. The equally spaced, parallel lines indicate that the electric field has the same magnitude and direction at all points in the central region of the capacitor.

Often, electric field lines are curved, as in the case of an *electric dipole.* An electric dipole consists of two separated point charges that have the same magnitude but opposite signs. The electric field of a dipole is proportional to the product of the magnitude of one of the charges and the distance between the charges. This product is called the *dipole moment.* Many molecules, such as H_2O and HCl, have dipole moments. Figure 18.26 depicts the field lines in the vicinity of a dipole. For a curved field line, the electric field vector at a point is *tangent* to the line at that point (see points 1, 2, and 3 in the drawing). The pattern of the lines for the dipole indicates that the electric field is greatest in the region between and immediately surrounding the two charges, since the lines are closest together there.

Notice in Figure 18.26 that any given field line starts on the positive charge and ends on the negative charge. In general, *electric field lines always begin on a positive charge and end on a negative charge and do not start or stop in midspace. Furthermore, the number of lines leaving a positive charge or entering a negative charge is proportional to the magnitude of the charge.* This means, for example, that if 100 lines are drawn leaving a $+4$ μC charge, then 75 lines would have to end on a -3 μC charge and 25 lines on a -1 μC charge. Thus, 100 lines leave the charge of $+4$ μC and end on a total charge of -4 μC, so the lines begin and end on equal amounts of total charge.

The electric field lines are also curved in the vicinity of two identical charges. Figure 18.27 shows the pattern associated with two positive point charges and reveals that there

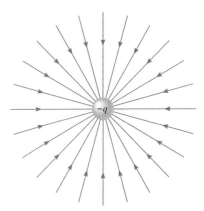

Figure 18.24 The electric field lines are directed radially inward toward a negative point charge $-q$.

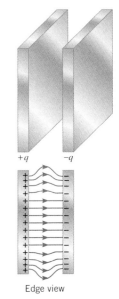

Figure 18.25 In the central region of a parallel plate capacitor, the electric field lines are parallel and evenly spaced, indicating that the electric field there has the same magnitude and direction at all points.

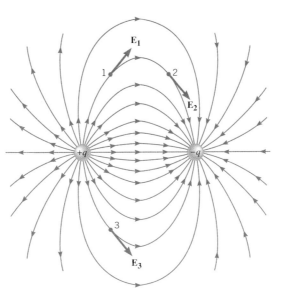

Figure 18.26 The electric field lines of an electric dipole are curved and extend from the positive to the negative charge. At any point, such as 1, 2, or 3, the field created by the dipole is tangent to the line through the point.

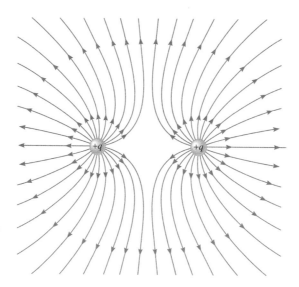

Figure 18.27 The electric field lines for two identical positive point charges. If the charges were both negative, the directions of the lines would be reversed.

is an absence of lines in the region between the charges. The absence of lines indicates that the electric field is relatively weak between the charges.

Some of the important properties of electric field lines are reexamined in Conceptual Example 12.

Conceptual Example 12   Drawing Electric Field Lines

Figure 18.28*a* shows three negative point charges ($-q$, $-q$, and $-2q$) and one positive point charge ($+4q$), along with some electric field lines drawn between the charges. There are three things wrong with this drawing. What are they?

Reasoning and Solution One aspect of Figure 18.28*a* that is incorrect is that electric field lines cross at point P. Field lines can never cross, and here's why. An electric charge placed at P experiences a single net force due to the presence of the other charges in its environment. Therefore, there is only one value for the electric field (which is the force per unit charge) at that point. If two field lines intersected, there would be two electric fields at the point of intersection, one associated with each line. Since there can be only one value of the electric field at any point, there can be only one electric field line passing through that point.

Another mistake in Figure 18.28*a* is the number of electric field lines that end on the negative charges. Remember that the number of field lines leaving a positive charge or entering a negative charge is proportional to the magnitude of the charge. The $-2q$ charge has half the magnitude of the $+4q$ charge. Therefore, since 8 lines leave the $+4q$ charge, 4 of them (one-half of them) must enter the $-2q$ charge. Of the remaining 4 lines that leave the positive charge, 2 enter each of the $-q$ charges, according to a similar line of reasoning.

The third error in Figure 18.28*a* is the way in which the electric field lines are drawn between the $+4q$ charge and the $-q$ charge at the left of the drawing. As drawn, the lines are parallel and evenly spaced. This would indicate that the electric field everywhere in this region has a constant magnitude and direction, as is the case in the central region of a parallel plate capacitor. But the electric field between the $+4q$ and $-q$ charges is not constant everywhere. It certainly is stronger in places close to the $+4q$ or $-q$ charge than it is midway between them. The field lines, therefore, should be drawn with a curved nature, similar (but not identical) to those that surround a dipole. Figure 18.28*b* shows more nearly correct representations of the field lines for the four charges.

Related Homework: *Problems 26, 55*

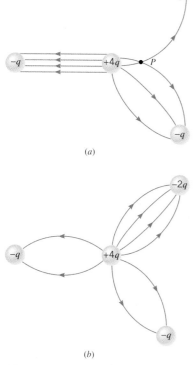

Figure 18.28 (*a*) Incorrectly and (*b*) correctly drawn electric field lines.

Concept Simulation 18.2

You can explore the electric field lines created by one or two point charges in this simulation. The magnitudes and algebraic signs of the charges are under your control, and either the field lines of the individual charges or the net field lines of both charges can be displayed.

Related Homework: *Problem 55*

Go to **www.wiley.com/college/cutnell**

18.8 The Electric Field Inside a Conductor: Shielding

In conducting materials such as copper, electric charges move readily in response to the forces that electric fields exert. This property of conducting materials has a major effect on the electric field that can exist within and around them. Suppose that a piece of copper carries a number of excess electrons somewhere within it, as in Figure 18.29a. Each electron would experience a force of repulsion because of the electric field of its neighbors. And, since copper is a conductor, the excess electrons move readily in response to that force. In fact, as a consequence of the $1/r^2$ dependence on distance in Coulomb's law, they rush to the surface of the copper. Once static equilibrium is established with all of the excess charge on the surface, no further movement of charge occurs, as part b of the drawing indicates. Similarly, excess positive charge also moves to the surface of a conductor. In general, ***at equilibrium under electrostatic conditions, any excess charge resides on the surface of a conductor.***

Now consider the interior of the copper in Figure 18.29b. The interior is electrically neutral, although there are still free electrons that can move under the influence of an electric field. The absence of a net movement of these free electrons indicates that there is no net electric field present within the conductor. In fact, the excess charges arrange themselves on the conductor surface precisely in the manner needed to make the electric field zero within the material. Thus, ***at equilibrium under electrostatic conditions, the electric field is zero at any point within a conducting material.*** This fact has some fascinating implications.

Figure 18.30a shows an uncharged, solid, cylindrical conductor at equilibrium in the central region of a parallel plate capacitor. Induced charges on the surface of the cylinder alter the electric field lines of the capacitor. Since an electric field cannot exist within the conductor under these conditions, the electric field lines do not penetrate the cylinder. Instead, they end or begin on the induced charges. Consequently, a test charge placed *inside* the conductor would feel no force due to the presence of the charges on the capacitor. In other words, ***the conductor shields any charge within it from electric fields created outside the conductor.*** The shielding results from the induced charges on the conductor surface.

Since the electric field is zero inside the conductor, nothing is disturbed if a cavity is cut from the interior of the material, as in part b of the drawing. Thus, the interior of the cavity is also shielded from external electric fields, a fact that has important applications, particularly for shielding electronic circuits. "Stray" electric fields are produced by various electrical appliances (e.g., hair dryers, blenders, and vacuum cleaners), and these fields can interfere with the operation of sensitive electronic circuits, such as those in stereo amplifiers, televisions, and computers. To eliminate such interference, circuits are often enclosed within metal boxes that provide shielding from external fields.

The blowup in Figure 18.30a shows another aspect of how conductors alter the electric field lines created by external charges. The lines are altered because ***the electric field just outside the surface of a conductor is perpendicular to the surface at equilibrium under electrostatic conditions.*** If the field were not perpendicular, there would be a com-

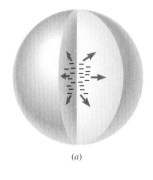

(a)

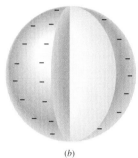

(b)

Figure 18.29 (a) Excess charge within a conductor (copper) moves quickly (b) to the surface.

The physics of...
shielding electronic circuits.

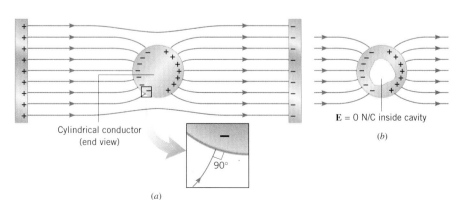

Cylindrical conductor
(end view)

90°

(a)

E = 0 N/C inside cavity

(b)

Figure 18.30 (a) A cylindrical conductor (shown as an end view) is placed between the oppositely charged plates of a capacitor. The electric field lines do not penetrate the conductor. The blowup shows that, just outside the conductor, the electric field lines are perpendicular to its surface. (b) The electric field is zero in a cavity within the conductor.

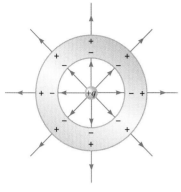

Figure 18.31 A positive charge $+q$ is suspended at the center of a hollow spherical conductor that is electrically neutral. Induced charges appear on the inner and outer surfaces of the conductor. The electric field within the conductor itself is zero.

ponent of the field parallel to the surface. Since the free electrons on the surface of the conductor can move, they would do so under the force exerted by that parallel component. In reality, however, no electron flow occurs at equilibrium. Therefore, there can be no parallel component, and the electric field is perpendicular to the surface.

The preceding discussion deals with features of the electric field within and around a conductor at equilibrium under electrostatic conditions. These features are related to the fact that conductors contain mobile free electrons and *do not apply to insulators,* which contain very few free electrons. Example 13 further explores the behavior of a conducting material in the presence of an electric field.

Conceptual Example 13 A Conductor in an Electric Field

A charge $+q$ is suspended at the center of a hollow, electrically neutral, spherical conductor, as Figure 18.31 illustrates. Show that this charge induces (a) a charge of $-q$ on the interior surface and (b) a charge of $+q$ on the exterior surface of the conductor.

Reasoning and Solution

(a) Electric field lines emanate from the positive charge $+q$. Since the electric field inside the metal conductor must be zero at equilibrium under electrostatic conditions, each field line ends when it reaches the conductor, as the picture shows. Since field lines terminate only on negative charges, there must be an induced *negative* charge on the interior surface of the conductor. Furthermore, the lines begin and end on equal amounts of charge, so the magnitude of the total induced charge is the same as the magnitude of the charge at the center. Thus, the total induced charge on the interior surface is $-q$.

(b) Before the charge $+q$ is introduced, the conductor is electrically neutral. Therefore, it carries no net charge. We have also seen that there can be no excess charge within the metal. Thus, since an induced charge of $-q$ appears on the interior surface, a charge of $+q$ must be induced on the outer surface. The positive charge on the outer surface generates field lines that radiate outward (see the drawing) as if they originated from the central charge and the conductor were absent. The conductor does not shield the outside from the field produced by the charge on the inside.

Related Homework: *Problem 57*

Figure 18.32 CONCEPTS AT A GLANCE The electric field and the surface through which it passes are brought together to define the concept of electric flux. It is in terms of electric flux that Gauss' law is formulated. When the electric field is produced by a distribution of charges, such as that on the silvery metal sphere in the photograph, Gauss' law relates the electric flux to the charge. (© Charles D. Winters/Photo Researchers)

18.9 Gauss' Law

Section 18.6 discusses how a point charge creates an electric field in the space around the charge. There are also many situations in which an electric field is produced by charges that are spread out over a region, rather than by a single point charge. Such an extended collection of charges is called a charge distribution. For example, the electric field within the parallel plate capacitor in Figure 18.22 is produced by positive charges spread uni-

CONCEPTS AT A GLANCE

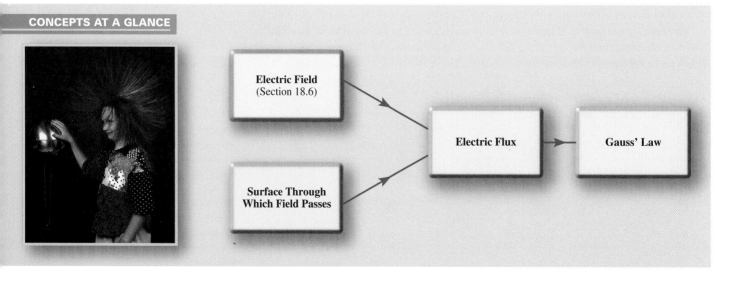

formly over one plate and an equal number of negative charges spread over the other plate. As we will see, Gauss' law describes the relationship between a charge distribution and the electric field it produces. This law was formulated by the German mathematician and physicist Carl Friedrich Gauss (1777–1855).

▶ **CONCEPTS AT A GLANCE** In presenting Gauss' law, it will be necessary to introduce a new idea called electric flux, and the Concepts-at-a-Glance chart in Figure 18.32 outlines the approach we will follow. This figure is an extended version of the concept chart in Figure 18.16 because the idea of flux involves both the electric field and the surface through which it passes. By bringing together the electric field and the surface through which it passes, we will be able to define electric flux and then present Gauss' law ◀

We begin by developing a version of Gauss' law that applies only to a point charge, which we assume to be positive. The electric field lines for a positive point charge radiate outward in all directions from the charge, as Figure 18.23b indicates. The magnitude E of the electric field at a distance r from the charge is $E = kq/r^2$, according to Equation 18.3, in which we have replaced the symbol $|q|$ with the symbol q since we are assuming that the charge is positive. As mentioned in Section 18.5, the constant k can be expressed as $k = 1/(4\pi\epsilon_0)$, where ϵ_0 is the permittivity of free space. With this substitution, the magnitude of the electric field becomes $E = q/(4\pi\epsilon_0 r^2)$. We now place this point charge at the center of an imaginary spherical surface of radius r, as Figure 18.33 shows. Such a hypothetical closed surface is called a *Gaussian surface,* although in general it need not be spherical. The surface area A of a sphere is $A = 4\pi r^2$, and the magnitude of the electric field can be written in terms of this area as $E = q/(A\epsilon_0)$, or

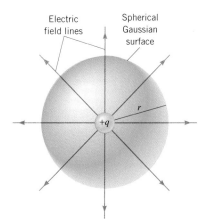

Figure 18.33 A positive point charge is located at the center of an imaginary spherical surface of radius r. Such a surface is one example of a Gaussian surface. Here the electric field is perpendicular to the surface and has the same magnitude everywhere on it.

Gauss' law for a point charge

$$\underbrace{EA}_{\substack{\text{Electric} \\ \text{flux, } \Phi_E}} = \frac{q}{\epsilon_0} \qquad (18.5)$$

The left side of Equation 18.5 is the product of the magnitude E of the electric field at any point on the Gaussian surface and the area A of the surface. In Gauss' law this product is especially important and is called the *electric flux,* Φ_E: $\Phi_E = EA$. (It will be necessary to modify this definition of flux when we consider the general case of a Gaussian surface with an arbitrary shape.)

Equation 18.5 is the result we have been seeking, for it is the form of Gauss' law that applies to a point charge. This result indicates that, aside from the constant ϵ_0, the electric flux Φ_E depends only on the charge q within the Gaussian surface and is independent of the radius r of the surface. We will now see how to generalize Equation 18.5 to account for distributions of charges and Gaussian surfaces with arbitrary shapes.

Figure 18.34 shows a charge distribution whose net charge is labeled Q. The charge distribution is surrounded by a Gaussian surface—that is, an imaginary closed surface. The surface can have *any arbitrary shape* (it need not be spherical), but *it must be closed* (an open surface would be like that of half an eggshell). The direction of the electric field is not necessarily perpendicular to the Gaussian surface. Furthermore, the magnitude of the electric field need not be constant on the surface, but can vary from point to point.

To determine the electric flux through such a surface, we divide the surface into many tiny sections with areas ΔA_1, ΔA_2, and so on. Each section is so small that it is essentially flat and the electric field **E** is a constant (both in magnitude and direction) over it. For reference, a dashed line called the "normal" is drawn perpendicular to each section on the outside of the surface. To determine the electric flux for each of the sections, we use only the component of **E** that is perpendicular to the surface—that is, the component of the electric field that passes through the surface. From the drawing it can be seen that this component has a magnitude of $E \cos \phi$, where ϕ is the angle between the electric field and the normal. The electric flux through any one section is then $(E \cos \phi)\Delta A$. The electric flux Φ_E that passes through the entire Gaussian surface is the sum of all of these individual fluxes: $\Phi_E = (E_1 \cos \phi_1)\Delta A_1 + (E_2 \cos \phi_2)\Delta A_2 + \cdots$, or

$$\Phi_E = \Sigma(E \cos \phi)\Delta A \qquad (18.6)$$

where, as usual, the symbol Σ means "the sum of." Gauss' law relates the electric flux Φ_E to the net charge Q enclosed by the arbitrarily shaped Gaussian surface.

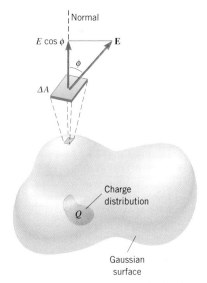

Figure 18.34 The charge distribution Q is surrounded by an arbitrarily shaped Gaussian surface. The electric flux Φ through any tiny segment of the surface is the product of $E \cos \phi$ and the area ΔA of the segment: $\Phi = (E \cos \phi) \Delta A$. The angle ϕ is the angle between the electric field and the normal to the surface.

■ **GAUSS' LAW**

The electric flux Φ_E through a Gaussian surface is equal to the net charge Q enclosed by the surface divided by ϵ_0, the permittivity of free space:

$$\underbrace{\sum (E \cos \phi)\Delta A}_{\text{Electric flux, } \Phi_E} = \frac{Q}{\epsilon_0} \qquad (18.7)$$

SI Unit of Electric Flux: $\text{N} \cdot \text{m}^2/\text{C}$

Gauss' law is often used to find the magnitude of the electric field produced by a distribution of charges. The law is most useful when the distribution is uniform and symmetrical. In the next two examples we will see how to apply Gauss' law in such situations.

Example 14 The Electric Field of a Charged Thin Spherical Shell

Figure 18.35 shows a thin spherical shell of radius R. A positive charge q is spread uniformly over the shell. Find the magnitude of the electric field at any point (a) outside the shell and (b) inside the shell.

Reasoning Because the charge is distributed uniformly over the spherical shell, the electric field is symmetrical. This means that the electric field is directed radially outward in all directions, and its magnitude is the same at all points that are equidistant from the shell. All such points lie on a sphere, so the symmetry is called *spherical symmetry.* With this symmetry in mind, we will use a spherical Gaussian surface to evaluate the electric flux Φ_E. We will then use Gauss' law to determine the magnitude of the electric field.

Solution

(a) To find the magnitude of the electric field outside the charged shell, we evaluate the electric flux $\Phi_E = \Sigma(E \cos \phi)\Delta A$ by using a spherical Gaussian surface of radius r ($r > R$) that is concentric with the shell. See the surface labeled S in Figure 18.35. Since the electric field **E** is everywhere perpendicular to the Gaussian surface, $\phi = 0°$ and $\cos \phi = 1$. In addition, E has the same value at all points on the surface, since they are equidistant from the charged shell. Being constant over the surface, E can be factored outside the summation, with the result that

$$\Phi_E = \Sigma(E \cos 0°)\Delta A = E \underbrace{(\Sigma \Delta A)}_{\substack{\text{Area of} \\ \text{Gaussian} \\ \text{surface}}} = E \underbrace{(4\pi r^2)}_{\substack{\text{Surface area} \\ \text{of sphere}}}$$

The term $\Sigma \Delta A$ is just the sum of the tiny areas that make up the Gaussian surface. Since the area of a spherical surface is $4\pi r^2$, we have $\Sigma \Delta A = 4\pi r^2$. Setting the electric flux equal to Q/ϵ_0, as specified by Gauss' law, yields $E(4\pi r^2) = Q/\epsilon_0$. Since the only charge within the Gaussian surface is the charge q on the shell, it follows that the net charge within the Gaussian surface is $Q = q$. Thus, we can solve for E and find that

$$\boxed{E = \frac{q}{4\pi\epsilon_0 r^2} \qquad \text{(for } r > R\text{)}}$$

This is a surprising result, for it is the same as that for a point charge (see Equation 18.3 with $|q| = q$). Thus, the electric field outside a uniformly charged spherical shell is the same as if all the charge q were concentrated as a point charge at the center of the shell.

(b) To find the magnitude of the electric field inside the charged shell, we select a spherical Gaussian surface that lies inside the shell and is concentric with it. See the surface labeled S_1 in Figure 18.35. Inside the charged shell, the electric field (if it exists) must also have spherical symmetry. Therefore, using reasoning like that in part (a), the electric flux through the Gaussian surface is $\Phi_E = \Sigma(E \cos \phi)\Delta A = E(4\pi r_1^2)$. In accord with Gauss' law, the electric flux must be equal to Q/ϵ_0, where Q is the net charge *inside* the Gaussian surface. But now $Q = 0$ C, since all the charge lies on the shell that is *outside* the surface S_1. Consequently, we have $E(4\pi r_1^2) = Q/\epsilon_0 = 0$, or

$$\boxed{E = 0 \text{ N/C} \qquad \text{(for } r < R\text{)}}$$

Gauss' law allows us to deduce that there is no electric field inside a uniform spherical shell of charge. An electric field exists only on the outside.

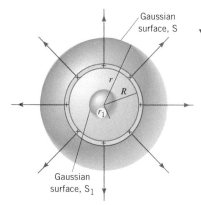

Figure 18.35 A uniform distribution of positive charge resides on a thin spherical shell of radius R. The spherical Gaussian surfaces S and S_1 are used in Example 14 to evaluate the electric flux outside and inside the shell, respectively.

Concept Simulation 18.3

This simulation, in effect, is an extension of Example 14. In addition to the spherical shell treated in that example, the simulation includes the options of adding a sphere within the shell and a point charge at the center of the sphere. The magnitudes and algebraic signs of the charges on the three objects are under your control. Therefore, you can explore the effect that these charges have on the electric field lines, make predictions based on Gauss' law, and see whether your predictions are consistent with the pattern of the field lines.

Related Homework: *Problem 52*

Go to
www.wiley.com/college/cutnell

Example 15 The Electric Field Inside a Parallel Plate Capacitor

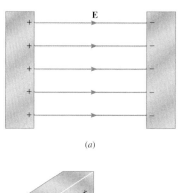

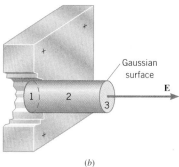

(a)

According to Equation 18.4, the electric field inside a parallel plate capacitor, and away from the edges, is constant and has a magnitude of $E = \sigma/\epsilon_0$, where σ is the charge density (the charge per unit area) on a plate. Use Gauss' law to prove this result.

Reasoning Figure 18.36a shows the electric field inside a parallel plate capacitor. Because the positive and negative charges are distributed uniformly over the surfaces of the plates, symmetry requires that the electric field be perpendicular to the plates. We will take advantage of this symmetry by choosing our Gaussian surface to be a small cylinder whose axis is perpendicular to the plates (see part b of the figure). With this choice, we will be able to evaluate the electric flux and then, with the aid of Gauss' law, determine E.

Solution Figure 18.36b shows that we have placed our Gaussian cylinder so that its left end is inside the positive metal plate, and the right end is in the space between the plates. To determine the electric flux through this Gaussian surface, we evaluate the flux through each of the three parts—labeled 1, 2, and 3 in the drawing—that make up the total surface of the cylinder and then add up the fluxes.

Surface 1—the flat left end of the cylinder—is embedded inside the positive metal plate. As discussed in Section 18.8, the electric field is zero everywhere inside a conductor that is in equilibrium under electrostatic conditions. Since $E = 0$ N/C, the electric flux through this surface is also zero:

$$\Phi_1 = \Sigma(E \cos \phi)\Delta A = \Sigma[(0 \text{ N/C}) \cos \phi]\Delta A = 0$$

Surface 2—the curved wall of the cylinder—is everywhere parallel to the electric field between the plates, so that $\cos \phi = \cos 90° = 0$. Therefore, the electric flux through this surface is also zero:

$$\Phi_2 = \Sigma(E \cos \phi)\Delta A = \Sigma(E \cos 90°)\Delta A = 0$$

Surface 3—the flat right end of the cylinder—is perpendicular to the electric field between the plates, so $\cos \phi = \cos 0° = 1$. The electric field is constant over this surface, so E can be taken outside the summation in Equation 18.6. Noting that $\Sigma\Delta A = A$ is the area of surface 3, we find that the electric flux through this surface is

$$\Phi_3 = \Sigma(E \cos \phi)\Delta A = \Sigma(E \cos 0°)\Delta A = E(\Sigma\Delta A) = EA$$

The electric flux through the entire Gaussian cylinder is the sum of the three fluxes determined above:

$$\Phi_E = \Phi_1 + \Phi_2 + \Phi_3 = 0 + 0 + EA = EA$$

According to Gauss' law, we set the electric flux equal to Q/ϵ_0, where Q is the net charge *inside* the Gaussian cylinder: $EA = Q/\epsilon_0$. But Q/A is the charge per unit area, σ, on the plate. Therefore, we arrive at the value of the electric field inside a parallel plate capacitor: $\boxed{E = \sigma/\epsilon_0}$. Notice that the distance of the right end of the Gaussian cylinder from the positive plate does not appear in this result, indicating that the electric field has the same value everywhere between the plates.

Figure 18.36 (a) A side view of a parallel plate capacitor, showing some of the electric field lines. (b) The Gaussian surface is a cylinder oriented so its axis is perpendicular to the positive plate and its left end is inside the plate.

✔ **Check Your Understanding 4**

The drawing shows an arrangement of three charges. In parts (a) and (b) different Gaussian surfaces are shown. Through which surface, if either, does the greater electric flux pass? *(The answer is given at the end of the book.)*

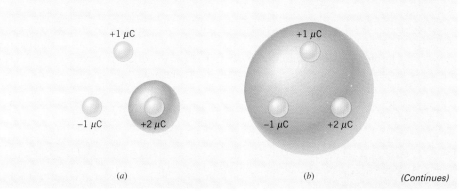

(a)   (b)   *(Continues)*

Background: Electric flux is defined in terms of the electric field and the surface area through which the field passes. Gauss' law is formulated in terms of electric flux and relates the flux to the net electric charge.

For similar questions (including calculational counterparts), consult Self-Assessment Test 18.2, which is described next.

Self-Assessment Test 18.2

Test your understanding of the material in Sections 18.6–18.9:

• The Electric Field   • Electric Field Lines   • Gauss' Law

Go to **www.wiley.com/college/cutnell**

18.10 Copiers and Computer Printers

The physics of...
xerography.

The electrostatic force that charged particles exert on one another plays the central role in an office copier. The copying process is called *xerography,* from the Greek *xeros* and *graphos,* meaning "dry writing." The heart of a copier is the xerographic drum, an aluminum cylinder coated with a layer of selenium (see Figure 18.37a). Aluminum is an excellent electrical conductor. Selenium, on the other hand, is a photoconductor; it is an insulator in the dark but becomes a conductor when exposed to light. Consequently, a positive charge deposited on the selenium surface will remain there, provided the selenium is kept in the dark. When the drum is exposed to light, however, electrons from the aluminum pass through the conducting selenium and neutralize the positive charge.

Figure 18.37 (a) This cutaway view shows the essential elements of a copying machine. (b) The five steps in the xerographic process.

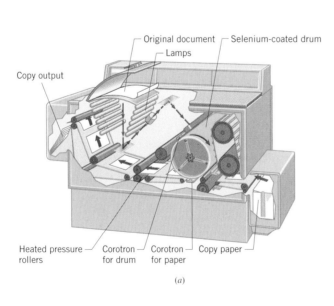

(a)

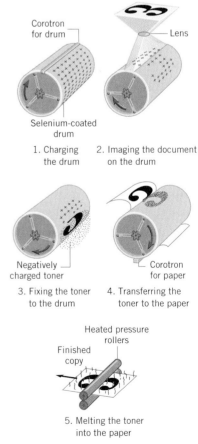

(b)

The photoconductive property of selenium is critical to the xerographic process, as Figure 18.37*b* illustrates. First, an electrode called a *corotron* gives the entire selenium surface a positive charge in the dark. Second, a series of lenses and mirrors focuses an image of a document onto the revolving drum. The dark and light areas of the document produce corresponding areas on the drum. The dark areas retain their positive charge, but the light areas become conducting and lose their positive charge, ending up neutralized. Thus, a positive-charge image of the document remains on the selenium surface. In the third step, a special dry black powder, called the *toner,* is given a negative charge and then spread onto the drum, where it adheres selectively to the positively charged areas. The fourth step involves transferring the toner onto a blank piece of paper. However, the attraction of the positive-charge image holds the toner to the drum. To transfer the toner, the paper is given a *greater positive charge* than that of the image, with the aid of another corotron. Last, the paper and adhering toner pass through heated pressure rollers, which melt the toner into the fibers of the paper and produce the finished copy.

A laser printer is used with computers to provide high-quality copies of text and graphics. It is similar in operation to the xerographic machine, except that the information to be reproduced is not on paper. Instead, the information is transferred from the computer's memory to the printer, and laser light is used to copy it onto the selenium–aluminum drum. A laser beam, focused to a fine point, is scanned rapidly from side to side across the rotating drum, as Figure 18.38 indicates. While the light remains on, the positive charge on the drum is neutralized. As the laser beam moves, the computer turns the beam off at the right moments during each scan to produce the desired positive-charge image, which is the letter "A" in the picture.

An inkjet printer is another type of printer that uses electric charges in its operation. While shuttling back and forth across the paper, the inkjet printhead ejects a thin stream of ink. Figure 18.39 illustrates the elements of one type of printhead. The ink is forced out of a small nozzle and breaks up into extremely small droplets, with diameters that can be as small as 9×10^{-6} m. About 150 000 droplets leave the nozzle each second and travel with a speed of approximately 18 m/s toward the paper. During their flight, the droplets pass through two electrical components, an *electrode* and the *deflection plates* (a parallel plate capacitor). When the printhead moves over regions of the paper that are not to be inked, the charging control is turned on and an electric field is established between the printhead and the electrode. As the drops pass through the electric field, they acquire a net charge by the process of induction. The deflection plates divert the charged droplets into a gutter and thus prevent them from reaching the paper. Whenever ink is to be placed on the paper, the charging control, responding to instructions from the computer, turns off the electric field. The uncharged droplets fly straight through the deflection plates and strike the paper.

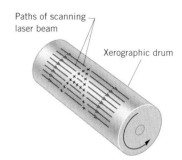

Figure 18.38 As the laser beam scans back and forth across the surface of the xerographic drum, a positive-charge image of the letter "A" is created.

The physics of...
a laser printer.

The physics of...
an inkjet printer.

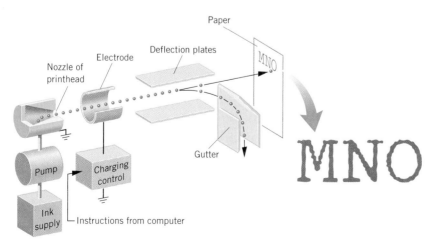

Figure 18.39 An inkjet printhead ejects a steady flow of ink droplets. Charged droplets are deflected into a gutter by the deflection plates, while uncharged droplets fly straight onto the paper. Letters formed by an inkjet printer look normal, except when greatly enlarged and the patterns from the drops become apparent.

18.11 Concepts & Calculations

In this chapter we have studied electric forces and electric fields. We conclude now by presenting some examples that review important features of these concepts. The three-part format of the examples stresses the role of conceptual understanding in problem solving. First, the problem statement is given. Then, there is a concept question-and-answer section, followed by the solution section. The purpose of the concept question-and-answer section is to provide help in understanding the solution and to illustrate how a review of the concepts can help in anticipating some of the characteristics of the numerical answers.

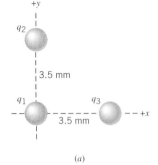

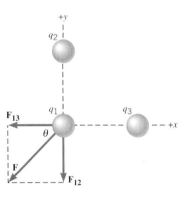

Figure 18.40 (*a*) Three equal charges lie on the *x* and *y* axes. (*b*) The net force exerted on q_1 by the other two charges is **F**.

Concepts & Calculations Example 16
The Vector Nature of Electric Forces

The charges on three identical metal spheres are -12 μC, $+4.0$ μC, and $+2.0$ μC. The spheres are brought together so they simultaneously touch each other. They are then separated and placed on the *x* and *y* axes, as in Figure 18.40*a*. What is the net force (magnitude and direction) exerted on the sphere at the origin? Treat the spheres as if they were particles.

Concept Questions and Answers Is the net charge on the system comprised of the three spheres the same before and after touching?

> *Answer* Yes. The conservation of electric charge states that, during any process, the net electric charge of an isolated system remains constant (is conserved). Therefore, the net charge on the three spheres before they touch (-12.0 μC $+ 4.0$ μC $+ 2.0$ μC $= -6.0$ μC) is the same as the net charge after they touch.

After the spheres touch and are separated, do they have identical charges?

> *Answer* Yes. Since the spheres are identical, the net charge (-6.0 μC) distributes itself equally over the three spheres. After they are separated, each has one-third of the net charge: $q_1 = q_2 = q_3 = \frac{1}{3}(-6.0 \text{ μC}) = -2.0 \text{ μC}$.

Do q_2 and q_3 exert forces of equal magnitude on q_1?

> *Answer* Yes. The charges q_2 and q_3 have equal magnitudes and are the same distance from q_1. According to Coulomb's law, then, they exert forces of equal magnitude on q_1.

Is the magnitude of the net force exerted on q_1 equal to $2F$, where F is the magnitude of the force that either q_2 or q_3 exerts on q_1?

> *Answer* No. Although the two forces that act on q_1 have equal magnitudes, they have different directions. The forces are repulsive forces, since all of the charges in part *a* of the drawing are identical. Figure 18.40*b* shows the force $\mathbf{F}_{12}$ exerted on q_1 by q_2 and the force $\mathbf{F}_{13}$ exerted on q_1 by q_3. To obtain the net force $\mathbf{F}$, we must take these directions into account by using vector addition.

Solution
The magnitude F_{12} of the force exerted on q_1 by q_2 is given by Coulomb's law, Equation 18.1, as

$$F_{12} = k\frac{|q_1||q_2|}{r_{12}^2} = \frac{(8.99 \times 10^9 \text{ N} \cdot \text{m}^2/\text{C}^2)(2.0 \times 10^{-6} \text{ C})(2.0 \times 10^{-6} \text{ C})}{(3.5 \times 10^{-3} \text{ m})^2}$$

$$= 2.9 \times 10^3 \text{ N}$$

Note that we have used the magnitudes of q_1 and q_2 in Coulomb's law. As mentioned previously, the magnitude of the force $\mathbf{F}_{13}$ exerted on q_1 by q_3 has the same value as F_{12}, so $F_{13} = 2.9 \times 10^3 \text{ N}$. Since the forces $\mathbf{F}_{12}$ and $\mathbf{F}_{13}$ are perpendicular to each other, we may use the Pythagorean theorem to find the magnitude F of the net force:

$$F = \sqrt{F_{12}^2 + F_{13}^2} = \sqrt{(2.9 \times 10^3 \text{ N})^2 + (2.9 \times 10^3 \text{ N})^2} = \boxed{4.1 \times 10^3 \text{ N}}$$

The angle θ that the net force makes with the $-x$ axis (see part *b* of the drawing) is

$$\theta = \tan^{-1}\left(\frac{F_{12}}{F_{13}}\right) = \tan^{-1}\left(\frac{2.9 \times 10^3 \text{ N}}{2.9 \times 10^3 \text{ N}}\right) = \boxed{45°}$$

Problem solving insight
Often charge magnitudes are specified in microcoulombs (μC). When using Coulomb's law, be sure to convert microcoulombs into coulombs (1 μ $= 10^{-6}$ C) before substituting for the charge magnitudes $|q_1|$ and $|q_2|$.

Concepts & Calculations Example 17
Becoming Familiar with Electric Fields

Two point charges are lying on the *y* axis in Figure 18.41*a*: $q_1 = -4.00 \ \mu\text{C}$ and $q_2 = +4.00 \ \mu\text{C}$. They are equidistant from the point *P*, which lies on the *x* axis. (a) What is the net electric field at *P*? (b) A small object of charge $q_0 = +8.00 \ \mu\text{C}$ and mass $m = 1.20$ g is placed at *P*. When it is released, what is its acceleration?

Concept Questions and Answers There is no charge at *P* in part (a). Is there an electric field at *P*?

Answer Yes. An electric field is produced by the charges q_1 and q_2, and it exists throughout the entire region that surrounds them. If a test charge were placed at this point, it would experience a force due to the electric field. The force would be the product of the charge and the electric field.

The charge q_1 produces an electric field at the point *P*. What is the direction of this field?

Answer The electric field created by a charge always points away from a positive charge and toward a negative charge. Since q_1 is negative, the electric field $\mathbf{E_1}$ points toward it (see Figure 18.41*b*).

What is the direction of the electric field produced by q_2 at *P*?

Answer Since q_2 is positive, the electric field $\mathbf{E_2}$ that it produces points away from q_2, as shown in the drawing.

Is the magnitude of the net electric field equal to $E_1 + E_2$, where E_1 and E_2 are the magnitudes of the electric fields produced by q_1 and q_2?

Answer No, because the electric fields have different directions. We must add the individual fields as vectors to obtain the net electric field. Only then can we determine its magnitude.

Solution

(a) The magnitude of the electric fields that q_1 and q_2 produce at *P* are given by Equation 18.3, where the distances are specified in the drawing:

$$E_1 = \frac{k|q_1|}{r_1^2} = \frac{(8.99 \times 10^9 \ \text{N} \cdot \text{m}^2/\text{C}^2)(4.00 \times 10^{-6} \ \text{C})}{(0.700 \ \text{m})^2} = 7.34 \times 10^4 \ \text{N/C}$$

$$E_2 = \frac{k|q_2|}{r_2^2} = \frac{(8.99 \times 10^9 \ \text{N} \cdot \text{m}^2/\text{C}^2)(4.00 \times 10^{-6} \ \text{C})}{(0.700 \ \text{m})^2} = 7.34 \times 10^4 \ \text{N/C}$$

The *x* and *y* components of these fields and the total field **E** are given in the following table:

Electric field	*x* component	*y* component
$\mathbf{E_1}$	$-E_1 \cos 31.0° = -6.29 \times 10^4$ N/C	$+E_1 \sin 31.0° = +3.78 \times 10^4$ N/C
$\mathbf{E_2}$	$+E_2 \cos 31.0° = +6.29 \times 10^4$ N/C	$+E_2 \sin 31.0° = +3.78 \times 10^4$ N/C
E	$E_x = 0$ N/C	$\mathbf{E}_y = +7.56 \times 10^4$ N/C

The net electric field **E** has only a component along the +*y* axis. So,

$$\boxed{\mathbf{E} = 7.56 \times 10^4 \ \text{N/C, directed along the } +y \text{ axis}}$$

(b) According to Newton's second law, Equation 4.2, the acceleration **a** of an object placed at this point is equal to the net force acting on it divided by its mass. The net force **F** is the product of the charge and the net electric field, $\mathbf{F} = q_0 \mathbf{E}$, as indicated by Equation 18.2. Thus, the acceleration is

$$\mathbf{a} = \frac{\mathbf{F}}{m} = \frac{q_0 \mathbf{E}}{m}$$

$$= \frac{(8.00 \times 10^{-6} \ \text{C})(7.56 \times 10^4 \ \text{N/C})}{1.20 \times 10^{-3} \ \text{kg}} = \boxed{5.04 \times 10^2 \ \text{m/s}^2 \text{, along the } +y \text{ axis}}$$

At the end of the problem set for this chapter, you will find homework problems that contain both conceptual and quantitative parts. These problems are grouped under the heading *Concepts & Calculations, Group Learning Problems*. They are designed for use by students working alone or in small learning groups. The conceptual part of each problem provides a convenient focus for group discussions.

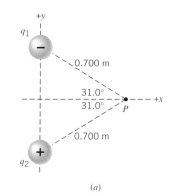

(a)

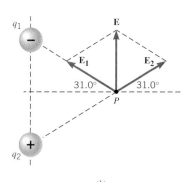

(b)

Figure 18.41 (*a*) Two charges q_1 and q_2 produce an electric field at the point *P*. (*b*) The electric fields $\mathbf{E_1}$ and $\mathbf{E_2}$ add to give the net electric field **E**.

Concept Summary

This summary presents an abridged version of the chapter, including the important equations and all available learning aids. For convenient reference, the learning aids (including the text's examples) are placed next to or immediately after the relevant equation or discussion. The following learning aids may be found on-line at **www.wiley.com/college/cutnell**:

Interactive LearningWare examples are solved according to a five-step interactive format that is designed to help you develop problem-solving skills.	Concept Simulations are animated versions of text figures or animations that illustrate important concepts. You can control parameters that affect the display, and we encourage you to experiment.
Interactive Solutions offer specific models for certain types of problems in the chapter homework. The calculations are carried out interactively.	Self-Assessment Tests include both qualitative and quantitative questions. Extensive feedback is provided for both incorrect and correct answers, to help you evaluate your understanding of the material.

Topic	Discussion	Learning Aids
	18.1 The Origin of Electricity	
The coulomb (C)	There are two kinds of electric charge: positive and negative. The SI unit of electric charge is the coulomb (C). The magnitude of the charge on an electron or a proton is	
Magnitude of charge on electron or proton	$$e = 1.60 \times 10^{-19}\,\text{C}$$	
	Since the symbol e denotes a magnitude, it has no algebraic sign. Thus, the electron carries a charge of $-e$, and the proton carries a charge of $+e$.	
Charge is quantized	The charge on any object, whether positive or negative, is quantized, in the sense that the charge consists of an integer number of protons or electrons.	Example 1
	18.2 Charged Objects and the Electric Force	
Law of conservation of electric charge	The law of conservation of electric charge states that the net electric charge of an isolated system remains constant during any process.	
Electrical repulsion and attraction	Like charges repel and unlike charges attract each other.	
	18.3 Conductors and Insulators	
Conductor	An electrical conductor is a material, such as copper, that conducts electric charge readily.	
Insulator	An electrical insulator is a material, such as rubber, that conducts electric charge poorly.	
	18.4 Charging by Contact and by Induction	
Charging by contact	Charging by contact is the process of giving one object a net electric charge by placing it in contact with an object that is already charged.	
Charging by induction	Charging by induction is the process of giving an object a net electric charge without touching it to a charged object.	
	18.5 Coulomb's Law	
Point charge	A point charge is a charge that occupies so little space that it can be regarded as a mathematical point.	
	Coulomb's law gives the magnitude F of the electric force that two point charges q_1 and q_2 exert on each other:	Examples 2–5 Interactive LearningWare 18.1 Concept Simulation 18.1
Coulomb's law	$$F = k\frac{|q_1||q_2|}{r^2} \qquad (18.1)$$	Interactive LearningWare 18.2 Example 16 Interactive Solution 18.15
	where $|q_1|$ and $|q_2|$ are the magnitudes of the charges and have no algebraic sign. The term k is a constant and has the value $k = 8.99 \times 10^9\,\text{N} \cdot \text{m}^2/\text{C}^2$. The force specified by Equation 18.1 acts along the line between the two charges.	
Permittivity of free space	The permittivity of free space ϵ_0 is defined by the relation	
	$$k = \frac{1}{4\pi\epsilon_0}$$	

Topic	Discussion	Learning Aids

 Use *Self-Assessment Test 18.1* **to evaluate your understanding of Sections 18.1–18.5.**

18.6 The Electric Field

The electric field **E** at a given spot is a vector and is the electrostatic force **F** experienced by a very small test charge q_0 placed at that spot divided by the charge itself:

Electric field

$$\mathbf{E} = \frac{\mathbf{F}}{q_0} \qquad (18.2)$$

Examples 6, 7, 8
Interactive Solution 18.65

The direction of the electric field is the same as the direction of the force on a positive test charge. The SI unit for the electric field is the Newton per coulomb (N/C). The source of the electric field at any spot is the charged objects surrounding that spot.

The magnitude of the electric field created by a point charge q is

Electric field of a point charge

$$E = \frac{k|q|}{r^2} \qquad (18.3)$$

Examples 9, 10, 11, 17
Interactive LearningWare 18.3
Interactive Solution 18.41

where $|q|$ is the magnitude of the charge and has no algebraic sign and r is the distance from the charge.

For a parallel plate capacitor that has a charge per unit area of σ on each plate, the magnitude of the electric field between the plates is

Electric field of a parallel plate capacitor

$$E = \frac{\sigma}{\epsilon_0} \qquad (18.4)$$

18.7 Electric Field Lines

Electric field lines

Electric field lines are lines that can be thought of as a "map," insofar as the lines provide information about the direction and strength of the electric field. The lines are directed away from positive charges and toward negative charges.

Example 12

Direction of electric field

The direction of the lines gives the direction of the electric field, since the electric field vector at a point is tangent to the line at that point. The electric field is strongest in regions where the number of lines per unit area passing perpendicularly through a surface is the greatest—that is, where the lines are packed together most tightly.

Concept Simulation 18.2

Strength of electric field

18.8 The Electric Field Inside a Conductor: Shielding

Excess charge carried by a conductor at equilibrium

Excess negative or positive charge resides on the surface of a conductor at equilibrium under electrostatic conditions. In such a situation, the electric field at any point within the conducting material is zero, and the electric field just outside the surface of the conductor is perpendicular to the surface.

Example 13

18.9 Gauss' Law

The electric flux Φ_E through a surface is related to the magnitude E of the electric field, the area A of the surface, and the angle ϕ that specifies the direction of the field relative to the normal to the surface:

Electric flux

$$\Phi_E = \Sigma (E \cos \phi)\Delta A \qquad (18.6)$$

Gauss' law states that the electric flux through a closed surface (a Gaussian surface) is equal to the net charge Q enclosed by the surface divided by ϵ_0, the permittivity of free space:

Examples 14, 15
Concept Simulation 18.3

Gauss' law

$$\Phi_E = \Sigma (E \cos \phi)\Delta A = \frac{Q}{\epsilon_0} \qquad (18.7)$$

Interactive Solution 18.53

 Use *Self-Assessment Test 18.2* **to evaluate your understanding of Sections 18.6–18.9.**

Conceptual Questions

1. In Figure 18.8 the grounding wire is removed first, followed by the rod, and the sphere is left with a positive charge. If the rod were removed first, followed by the grounding wire, would the sphere be left with a charge? Account for your answer.

2. A metallic object is given a positive charge by the process of induction, as illustrated in Figure 18.8. (a) Does the mass of the object increase, decrease, or remain the same? Why? (b) What happens to the mass of the object if it is given a negative charge by induction? Explain.

3. A rod made from insulating material carries a net charge, while a copper sphere is neutral. The rod and the sphere do not touch. Is it possible for the rod and the sphere to (a) attract one another and (b) repel one another? Explain.

4. On a dry day, just after washing your hair to remove natural oils and drying it thoroughly, run a plastic comb through it. Small bits of paper will be attracted to the comb. Explain why.

5. Blow up a balloon and rub it against your shirt a number of times. In so doing you give the balloon a net electric charge. Now touch the balloon to the ceiling. On being released, the balloon will remain stuck to the ceiling. Why?

6. A proton and an electron are held in place on the x axis. The proton is at $x = -d$, while the electron is at $x = +d$. They are released simultaneously, and the only force that affects their motions is the electrostatic force of attraction that each applies to the other. Which particle reaches the origin first? Give your reasoning.

7. A particle is attached to a spring and is pushed so that the spring is compressed more and more. As a result, the spring exerts a greater and greater force on the particle. Similarly, a charged particle experiences a greater and greater force when pushed closer and closer to another particle that is fixed in position and has a charge of the same polarity. In spite of the similarity, the charged particle will *not* exhibit simple harmonic motion on being released, as will the particle on the spring. Explain why not.

8. Identical point charges are fixed to opposite corners of a square. Where does a third point charge experience the greater force, at one of the empty corners or at the center of the square? Account for your answer.

9. On a thin, nonconducting rod, positive charges are spread evenly, so that there is the same amount of charge per unit length at every point. On another identical rod, positive charges are spread evenly over only the left half, and the same amount of negative charges are spread evenly over the right half. For each rod, deduce the *direction* of the electric field at a point that is located directly above the midpoint of the rod. Give your reasoning.

10. There is an electric field at point P. A very small charge is placed at this point and experiences a force. Another very small charge is then placed at this point and experiences a force that differs in both magnitude and direction from that experienced by the first charge. How can these two different forces result from the single electric field that exists at point P?

11. Three point charges are fixed to the corners of a square, one to a corner, in such a way that the net electric field at the empty corner is zero. Do these charges all have (a) the same sign and (b) the same magnitude (but, possibly, different signs)? Justify your answers.

12. Review Conceptual Example 11 as an aid in answering this question. Suppose in Figure 18.21 that charges $+q$ are placed on corners 1 and 3 of the rectangle, and charges $-q$ are placed on corners 2 and 4. What is the net electric field at the center C of the rectangle?

13. In Figure 18.26 there is no place on the line through the charges where the electric field is zero, neither to the left of the positive charge, nor between the charges, nor to the right of the negative charge. Now, suppose the magnitude of the negative charge were greater than the magnitude of the positive charge. Is there any place on the line through the charges where the electric field is zero? Justify your answer.

14. Drawings I and II show two examples of electric field lines. Decide which of the following statements are true and which are false, defending your choice in each case. (a) In both I and II the electric field is the same everywhere. (b) As you move from left to right in each case, the electric field becomes stronger. (c) The electric field in I is the same everywhere but becomes stronger in II as you move from left to right. (d) The electric fields in both I and II could be created by negative charges located somewhere on the left and positive charges somewhere on the right. (e) Both I and II arise from a single positive point charge located somewhere on the left.

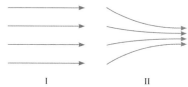

15. A positively charged particle is moving horizontally when it enters the region between the plates of a capacitor, as the drawing illustrates. (a) Draw the trajectory that the particle follows in moving through the capacitor. (b) When the particle is within the capacitor, which of the following four vectors, if any, are *parallel* to the electric field inside the capacitor: the particle's displacement, its velocity, its linear momentum, its acceleration? For each vector explain why the vector is, or is not, parallel to the electric field of the capacitor.

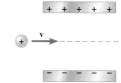

16. Refer to Figure 18.27. Imagine a plane that is perpendicular to the line between the charges, midway between them, and is half into and half out of the paper. The electric flux through this plane is zero. Explain why.

17. Two charges, $+q$ and $-q$, are inside a Gaussian surface. Since the net charge inside the Gaussian surface is zero, Gauss' law states that the electric flux through the surface is also zero; that is, $\Phi_E = 0$. Does the fact that $\Phi_E = 0$ imply that the electric field **E** at any point on the Gaussian surface is also zero? Justify your answer. *(Hint: Imagine a Gaussian surface that encloses the two charges in Figure 18.26.)*

18. The drawing shows three charges, labeled q_1, q_2, and q_3. A Gaussian surface is drawn around q_1 and q_2. (a) Which charges determine the electric flux through the Gaussian surface? (b) Which charges produce the electric field at the point P? Justify your answers.

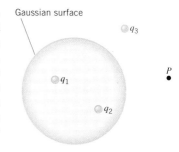

Gaussian surface

19. A charge $+q$ is placed inside a spherical Gaussian surface. The charge is *not* located at the center of the sphere. (a) Can Gauss' law tell us exactly where the charge is located inside the sphere? Justify your answer. (b) Can Gauss' law tell us about the magnitude of the electric flux through the Gaussian surface? Why?

Problems

Problems that are not marked with a star are considered the easiest to solve. Problems that are marked with a single star (*) are more difficult, while those marked with a double star (**) are the most difficult. Note: All charges are point charges, unless specified otherwise.

ssm **Solution is in the Student Solutions Manual.**   www **Solution is available on the World Wide Web at www.wiley.com/college/cutnell**
♆ **This icon represents a biomedical application.**

Section 18.1 The Origin of Electricity,
Section 18.2 Charged Objects and the Electric Force,
Section 18.3 Conductors and Insulators,
Section 18.4 Charging by Contact and by Induction

1. ssm How many electrons must be removed from an electrically neutral silver dollar to give it a charge of $+2.4 \ \mu C$?

2. A metal sphere has a charge of $+8.0 \ \mu C$. What is the net charge after 6.0×10^{13} electrons have been placed on it?

3. A plate carries a charge of $-3.0 \ \mu C$, while a rod carries a charge of $+2.0 \ \mu C$. How many electrons must be transferred from the plate to the rod, so that both objects have the same charge?

4. Object A is metallic and electrically neutral. It is charged by induction so that it acquires a charge of -3.0×10^{-6} C. Object B is identical to object A and is also electrically neutral. It is charged by induction so that it acquires a charge of $+3.0 \times 10^{-6}$ C. Find the *difference* in mass between the charged objects and state which has the greater mass.

5. ssm Consider three identical metal spheres, A, B, and C. Sphere A carries a charge of $+5q$. Sphere B carries a charge of $-q$. Sphere C carries no net charge. Spheres A and B are touched together and then separated. Sphere C is then touched to sphere A and separated from it. Last, sphere C is touched to sphere B and separated from it. (a) How much charge ends up on sphere C? What is the total charge on the three spheres (b) before they are allowed to touch each other and (c) after they have touched?

*6. Water has a mass per mole of 18.0 g/mol, and each water molecule (H_2O) has 10 electrons. (a) How many electrons are there in one liter ($1.00 \times 10^{-3} \, \text{m}^3$) of water? (b) What is the net charge of all these electrons?

Section 18.5 Coulomb's Law

7. ssm Two charges attract each other with a force of 1.5 N. What will be the force if the distance between them is reduced to one-ninth of its original value?

8. The nucleus of the helium atom contains two protons that are separated by about 3.0×10^{-15} m. Find the magnitude of the electrostatic force that each proton exerts on the other. (The protons remain together in the nucleus because the repulsive electrostatic force is balanced by an attractive force called the strong nuclear force.)

9. The force of repulsion that two like charges exert on each other is 3.5 N. What will the force be if the distance between the charges is increased to five times its original value?

10. In a vacuum, two particles have charges of q_1 and q_2, where $q_1 = +3.5 \ \mu C$. They are separated by a distance of 0.26 m, and particle 1 experiences an attractive force of 3.4 N. What is q_2 (magnitude and sign)?

11. ssm Interactive LearningWare 18.1 at **www.wiley.com/college/cutnell** offers some perspective on this problem. Two tiny spheres have the same mass and carry charges of the same magnitude. The mass of each sphere is 2.0×10^{-6} kg. The gravitational force that each sphere exerts on the other is balanced by the electric force. (a) What algebraic signs can the charges have? (b) Determine the charge magnitude.

12. Two tiny conducting spheres are identical and carry charges of $-20.0 \ \mu C$ and $+50.0 \ \mu C$. They are separated by a distance of 2.50 cm. (a) What is the magnitude of the force that each sphere experiences, and is the force attractive or repulsive? (b) The spheres are brought into contact and then separated to a distance of 2.50 cm. Determine the magnitude of the force that each sphere now experiences, and state whether the force is attractive or repulsive.

13. ssm www Two particles, with identical positive charges and a separation of 2.60×10^{-2} m, are released from rest. Immediately after the release, particle 1 has an acceleration $\mathbf{a_1}$ whose magnitude is 4.60×10^3 m/s², while particle 2 has an acceleration $\mathbf{a_2}$ whose magnitude is 8.50×10^3 m/s². Particle 1 has a mass of 6.00×10^{-6} kg. Find (a) the charge on each particle and (b) the mass of particle 2.

14. A charge of $-3.00 \ \mu C$ is fixed at the center of a compass. Two additional charges are fixed on the circle of the compass (radius $= 0.100$ m). The charges on the circle are $-4.00 \ \mu C$ at the position due north and $+5.00 \ \mu C$ at the position due east. What is the magnitude and direction of the net electrostatic force acting on the charge at the center? Specify the direction relative to due east.

15. Interactive Solution 18.15 at **www.wiley.com/college/cutnell** provides a model for solving this type of problem. Two small objects, A and B, are fixed in place and separated by 3.00 cm in a vacuum. Object A has a charge of $+2.00 \ \mu C$, and object B has a charge of $-2.00 \ \mu C$. How many electrons must be removed from A and put onto B to make the electrostatic force that acts on each object an attractive force whose magnitude is 68.0 N?

*16. The drawing shows an equilateral triangle, each side of which has a length of 2.00 cm. Point charges are fixed to each corner, as shown. The 4.00 μC charge experiences a net force due to the charges q_A and q_B. This net force points vertically downward in the drawing and has a magnitude of 405 N. Determine the magnitudes and algebraic signs of the charges q_A and q_B.

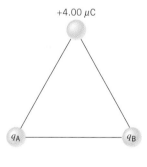

*17. A point charge of $-0.70 \ \mu C$ is fixed to one corner of a square. An identical charge is fixed to the diagonally opposite corner. A point charge q is fixed to each of the remaining corners. The net force acting on either of the charges q is zero. Find the magnitude and algebraic sign of q.

*18. Interactive LearningWare 18.2 at **www.wiley.com/college/cutnell** provides one approach to solving problems such as this one. The drawing shows three point charges fixed in place. The charge at the coordinate origin has a value of $q_1 = +8.00 \ \mu C$; the other two have identical magnitudes, but opposite signs: $q_2 = -5.00 \ \mu C$ and $q_3 = +5.00 \ \mu C$. (a) Determine the net force (magnitude and direction) exerted on q_1 by the other two charges. (b) If q_1 had a mass of 1.50 g and it were free to move, what would be its acceleration?

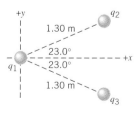

* **19. ssm** In the rectangle
in the drawing, a charge is
to be placed at the empty
corner to make the net
force on the charge at corner *A* point along the ver-

tical direction. What charge (magnitude and algebraic sign) must be
placed at the empty corner?

* **20.** Four point charges have equal magnitudes. Three are positive,
and one is negative, as the drawing shows. They are fixed in place on
the same straight line, and adjacent charges are equally separated by
a distance *d*. Consider the net electrostatic force acting on each
charge. Calculate the ratio of the largest to the smallest net force.

* **21.** An electrically neutral model airplane is flying in a horizontal
circle on a 3.0-m guideline, which is nearly parallel to the ground.
The line breaks when the kinetic energy of the plane is 50.0 J. Re-
consider the same situation, except that now there is a point charge
of $+q$ on the plane and a point charge of $-q$ at the other end of the
guideline. In this case, the line breaks when the kinetic energy of the
plane is 51.8 J. Find the magnitude of the charges.

** **22.** Two objects are identical and small enough that their sizes can
be ignored relative to the distance between them, which is 0.200 m.
In a vacuum, each object carries a different charge, and they attract
each other with a force of 1.20 N. The objects are brought into contact, so the net charge is shared equally, and then they are returned to
their initial positions. Now it is found that the objects repel one another with a force whose magnitude is equal to that of the initial attractive force. What is the initial charge on each object? Note that
there are two answers.

** **23. ssm** A small spherical insulator of mass 8.00×10^{-2} kg and
charge $+0.600$ μC is hung by a
thin wire of negligible mass. A
charge of -0.900 μC is held
0.150 m away from the sphere
and directly to the right of it, so
the wire makes an angle θ with the

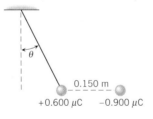

vertical (see the drawing). Find (a) the angle θ and (b) the tension in
the wire.

** **24.** There are four charges, each with a magnitude of 2.0 μC. Two
are positive and two are negative. The charges are fixed to the corners of a 0.30-m square, one to a corner, in such a way that the net
force on any charge is directed toward the center of the square. Find
the magnitude of the net electrostatic force experienced by any
charge.

Section 18.6 The Electric Field,
Section 18.7 Electric Field Lines,
Section 18.8 The Electric Field Inside a Conductor: Shielding

25. An electric field of 260 000 N/C points due west at a certain
spot. What are the magnitude and direction of the force that acts on
a charge of -7.0 μC at this spot?

26. Review Conceptual Example 12 as an aid in working this problem. Charges of $-4q$ are fixed to diagonally opposite corners of a
square. A charge of $+5q$ is fixed to one of the remaining corners,
and a charge of $+3q$ is fixed to the last corner. Assuming that ten
electric field lines emerge from the $+5q$ charge, sketch the field
lines in the vicinity of the four charges.

27. ssm www A tiny ball (mass = 0.012 kg) carries a charge of
-18 μC. What electric field (magnitude and direction) is needed to
cause the ball to float above the ground?

28. At a distance r_1 from a point charge, the magnitude of the
electric field created by the charge is 248 N/C. At a distance r_2
from the charge, the field has a magnitude of 132 N/C. Find the ratio r_2/r_1.

29. Two charges are placed on the *x* axis. One charge ($q_1 =
+8.5$ μC) is at $x_1 = +3.0$ cm and the other ($q_2 = -21$ μC) is at
$x_1 = +9.0$ cm. Find the net electric field (magnitude and direction)
at (a) $x = 0$ cm and (b) $x = +6.0$ cm.

30. Two charges, -16 and $+4.0$ μC, are fixed in place and separated by 3.0 m. (a) At what spot along a line through the charges is
the net electric field zero? Locate this spot relative to the positive
charge. (*Hint: The spot does not necessarily lie between the two
charges.*) (b) What would be the force on a charge of $+14$ μC
placed at this spot?

31. ssm Background pertinent to this problem is available in
Interactive LearningWare 18.3 at **www.wiley.com/college/cutnell**.
A 3.0-μC point charge is placed in an external uniform electric field
of 1.6×10^4 N/C. At what distance from the charge is the net electric field zero?

32. A charge of $q = +7.50$ μC is located in an electric field. The *x*
and *y* components of the electric field are $E_x = 6.00 \times 10^3$ N/C and
$E_y = 8.00 \times 10^3$ N/C, respectively. (a) What is the magnitude of the
force on the charge? (b) Determine the angle that the force makes
with the $+x$ axis.

33. ssm A small drop of water is suspended motionless in air by a
uniform electric field that is directed upward and has a magnitude
of 8480 N/C. The mass of the water drop is 3.50×10^{-9} kg. (a) Is
the excess charge on the water drop positive or negative? Why?
(b) How many excess electrons or protons reside on the drop?

34. Review Conceptual Example 11 before attempting this problem.
The magnitude of each of the charges in Figure 18.21 is $8.60 \times
10^{-12}$ C. The lengths of the sides of the rectangles are 3.00 cm and
5.00 cm. Find the magnitude of the electric field at the center of the
rectangle in Figures 18.21*a* and *b*.

* **35.** Two charges are located on the *x* axis: $q_1 = +6.0$ μC at $x_1 =
+4.0$ cm, and $q_2 = +6.0$ μC at $x_2 = -4.0$ cm. Two other charges
are located on the *y* axis: $q_3 = +3.0$ μC at $y_3 = +5.0$ cm, and $q_4 =
-8.0$ μC at $y_4 = +7.0$ cm. Find the net electric field (magnitude
and direction) at the origin.

* **36.** Two parallel plate capacitors have circular plates. The magnitude of the charge on these plates is the same. However, the electric
field between the plates of the first capacitor is 2.2×10^5 N/C, while
the field within the second capacitor is 3.8×10^5 N/C. Determine
the ratio r_2/r_1 of the plate radius for the second capacitor to the plate
radius for the first capacitor.

* **37.** A proton is moving parallel to a uniform electric field. The electric field accelerates the proton and increases its linear momentum to
5.0×10^{-23} kg·m/s from 1.5×10^{-23} kg·m/s in a time of $6.3 \times
10^{-6}$ s. What is the magnitude of the electric field?

* **38.** An electron is released from rest at the negative plate of a parallel plate capacitor. The charge per unit area on each plate is $\sigma =
1.8 \times 10^{-7}$ C/m², and the plates are separated by a distance of
1.5×10^{-2} m. How fast is the electron moving just before it reaches
the positive plate?

* **39. ssm www** A rectangle has a length of $2d$ and a height of d.
Each of the following three charges is located at a corner of the rec-

tangle: $+q_1$ (upper left corner), $+q_2$ (lower right corner), and $-q$ (lower left corner). The net electric field at the (empty) upper right corner is zero. Find the magnitudes of q_1 and q_2. Express your answers in terms of q.

* **40.** A uniform electric field has a magnitude of 2.3×10^3 N/C. In a vacuum, a proton begins with a speed of 2.5×10^4 m/s and moves in the direction of this field. Find the speed of the proton after it has moved a distance of 2.0 mm.

* **41.** Review **Interactive Solution 18.41** at **www.wiley.com/college/cutnell** for help with this problem. The drawing shows two positive charges q_1 and q_2 fixed to a circle. At the center of the circle they produce a net electric field that is directed upward along the vertical axis. Determine the ratio $|q_2|/|q_1|$ of the charge magnitudes.

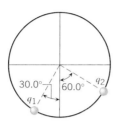

** **42.** The drawing shows an electron entering the lower left side of a parallel plate capacitor and exiting at the upper right side. The initial speed of the electron is 7.00×10^6 m/s. The capacitor is 2.00 cm long, and its plates are separated by 0.150 cm. Assume that the electric field between the plates is uniform everywhere and find its magnitude.

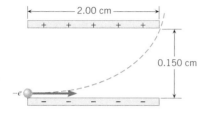

** **43.** A small plastic ball of mass 6.50×10^{-3} kg and charge $+0.150$ μC is suspended from an insulating thread and hangs between the plates of a capacitor (see the drawing). The ball is in equilibrium, with the thread making an angle of $30.0°$ with respect to the vertical. The area of each plate is 0.0150 m^2. What is the magnitude of the charge on each plate?

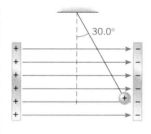

** **44.** The magnitude of the electric field between the plates of a parallel plate capacitor is 480 N/C. A silver dollar is placed between the plates and oriented parallel to the plates. (a) Ignoring the edges of the coin, find the induced charge density σ on each face of the coin. (b) Assuming the coin has a radius of 1.9 cm, find the magnitude of the total charge on each face of the coin.

** **45.** **ssm** Two point charges of the same magnitude but opposite signs are fixed to either end of the base of an isosceles triangle, as the drawing shows. The electric field at the midpoint M between the charges has a magnitude E_M. The field directly above the midpoint at point P has a magnitude E_P. The ratio of these two field magnitudes is $E_M/E_P = 9.0$. Find the angle α in the drawing.

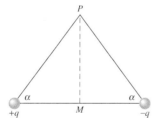

Section 18.9 Gauss' Law

46. A rectangular surface (0.16 m $\times$ 0.38 m) is oriented in a uniform electric field of 580 N/C. What is the maximum possible electric flux through the surface?

47. **ssm** The drawing shows an edge-on view of two planar surfaces that intersect and are mutually perpendicular. Surface 1 has an area of 1.7 m^2, while surface 2 has an area of 3.2 m^2. The electric field **E** in the drawing is uniform and has a magnitude of 250 N/C. Find the electric flux through (a) surface 1 and (b) surface 2.

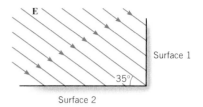

48. A spherical surface completely surrounds a collection of charges. Find the electric flux through the surface if the collection consists of (a) a single $+3.5 \times 10^{-6}$ C charge, (b) a single -2.3×10^{-6} C charge, and (c) both of the charges in (a) and (b).

49. A surface completely surrounds a $+2.0 \times 10^{-6}$ C charge. Find the electric flux through this surface when the surface is (a) a sphere with a radius of 0.50 m, (b) a sphere with a radius of 0.25 m, and (c) a cube with edges that are 0.25 m long.

50. A vertical wall (5.9 m $\times$ 2.5 m) in a house faces due east. A uniform electric field has a magnitude of 150 N/C. This field is parallel to the ground and points $35°$ north of east. What is the electric flux through the wall?

* **51.** **ssm** A cube is located with one corner at the origin of an x, y, z coordinate system. One of the cube's faces lies in the x, y plane, another in the y, z plane, and another in the x, z plane. In other words, the cube is in the first octant of the coordinate system. The edges of the cube are 0.20 m long. A uniform electric field is parallel to the x, y plane and points in the direction of the $+y$ axis. The magnitude of the field is 1500 N/C. (a) Find the electric flux through each of the six faces of the cube. (b) Add the six values obtained in part (a) to show that the electric flux through the cubical surface is zero, as Gauss' law predicts, since there is no net charge within the cube.

* **52.** Refer to **Concept Simulation 18.3** at **www.wiley.com/college/cutnell** for a perspective that is useful in solving this problem. Two spherical shells have a common center. A -1.6×10^{-6} C charge is spread uniformly over the inner shell, which has a radius of 0.050 m. A $+5.1 \times 10^{-6}$ C charge is spread uniformly over the outer shell, which has a radius of 0.15 m. Find the magnitude and direction of the electric field at a distance (measured from the common center) of (a) 0.20 m, (b) 0.10 m, and (c) 0.025 m.

* **53.** **Interactive Solution 18.53** at **www.wiley.com/college/cutnell** offers help with this problem in an interactive environment. A solid nonconducting sphere has a positive charge q spread uniformly throughout its volume. The charge density or charge per unit volume, therefore, is $\dfrac{q}{\frac{4}{3}\pi R^3}$. Use Gauss' law to show that the electric field at a point within the sphere at a radius r has a magnitude of $\dfrac{qr}{4\pi\epsilon_0 R^3}$. *(Hint: For a Gaussian surface, use a sphere of radius r centered within the solid sphere. Note that the net charge within any volume is the charge density times the volume.)*

** **54.** A long, thin, straight wire of length L has a positive charge Q distributed uniformly along it. Use Gauss' law to show that the electric field created by this wire at a radial distance r has a magnitude of $E = \lambda/(2\pi\epsilon_0 r)$, where $\lambda = Q/L$. *(Hint: For a Gaussian surface, use a cylinder aligned with its axis along the wire and note that the cylinder has a flat surface at either end, as well as a curved surface.)*

Additional Problems

55. Concept Simulation 18.2 at **www.wiley.com/college/cutnell** provides background concerning the electric field lines that are the focus of this problem. Review the important features of electric field lines discussed in Conceptual Example 12. Three point charges ($+q$, $+2q$, and $-3q$) are at the corners of an equilateral triangle. Sketch in six electric field lines between the three charges.

56. ssm www Two very small spheres are initially neutral and separated by a distance of 0.50 m. Suppose that 3.0×10^{13} electrons are removed from one sphere and placed on the other. (a) What is the magnitude of the electrostatic force that acts on each sphere? (b) Is the force attractive or repulsive? Why?

57. Conceptual Example 13 deals with the hollow spherical conductor in Figure 18.31. The conductor is initially electrically neutral, and then a charge $+q$ is placed at the center of the hollow space. Suppose the conductor initially has a net charge of $+2q$ instead of being neutral. What is the total charge on the interior and on the exterior surface when the $+q$ charge is placed at the center?

58. Four point charges have the same magnitude of 2.4×10^{-12} C and are fixed to the corners of a square that is 4.0 cm on a side. Three of the charges are positive and one is negative. Determine the magnitude of the net electric field that exists at the center of the square.

59. ssm A charge of $+3.0 \times 10^{-5}$ C is located at a place where there is an electric field that points due east and has a magnitude of 15 000 N/C. What are the magnitude and direction of the force acting on the charge?

60. A charge $+q$ is located at the origin, while an identical charge is located on the x axis at $x = +0.50$ m. A third charge of $+2q$ is located on the x axis at such a place that the net electrostatic force on the charge at the origin doubles, its direction remaining unchanged. Where should the third charge be located?

61. ssm Consult Concept Simulation 18.1 at **www.wiley.com/college/cutnell** for insight into this problem. Three charges are fixed to an x, y coordinate system. A charge of $+18$ μC is on the y axis at $y = +3.0$ m. A charge of -12 μC is at the origin. Last, a charge of $+45$ μC is on the x axis at $x = +3.0$ m. Determine the magnitude and direction of the net electrostatic force on the charge at $x = +3.0$ m. Specify the direction relative to the $-x$ axis.

62. A long, thin rod (length = 4.0 m) lies along the x axis, with its midpoint at the origin. In a vacuum, a $+8.0$ μC point charge is fixed to one end of the rod, and a -8.0 μC point charge is fixed to the other end. Everywhere in the x, y plane there is a constant external electric field (magnitude = 5.0×10^3 N/C) that is perpendicular to the rod. With respect to the z axis, find the magnitude of the net torque applied to the rod.

* **63.** Two charges are placed between the plates of a parallel plate capacitor. One charge is $+q_1$ and the other is $q_2 = +5.00$ μC. The charge per unit area on each plate has a magnitude of $\sigma = 1.30 \times 10^{-4}$ C/m^2. The force on q_1 due to q_2 equals the force on q_1 due to the electric field of the parallel plate capacitor. What is the distance r between the two charges?

* **64.** Two small charged objects are attached to a horizontal spring, one at each end. The magnitudes of the charges are equal, and the spring constant is 220 N/m. The spring is observed to be stretched by 0.020 m relative to its unstrained length of 0.32 m. Determine (a) the possible algebraic signs and (b) the magnitude of the charges.

* **65.** Interactive Solution 18.65 at **www.wiley.com/college/cutnell** provides a model for problems of this kind. A small object has a mass of 3.0×10^{-3} kg and a charge of -34 μC. It is placed at a certain spot where there is an electric field. When released, the object experiences an acceleration of 2.5×10^3 m/s^2 in the direction of the $+x$ axis. Determine the magnitude and direction of the electric field.

* **66.** A small object, which has a charge $q = 7.5$ μC and mass $m = 9.0 \times 10^{-5}$ kg, is placed in a constant electric field. Starting from rest, the object accelerates to a speed of 2.0×10^3 m/s in a time of 0.96 s. Determine the magnitude of the electric field.

* **67.** ssm Two spheres are mounted on identical horizontal springs and rest on a frictionless table, as in the drawing. When the spheres are uncharged, the spacing between them is 0.0500 m, and the springs are unstrained. When each sphere has a charge of $+1.60$ μC, the spacing doubles. Assuming that the spheres have a negligible diameter, determine the spring constant of the springs.

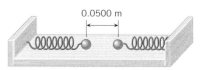

** **68.** Two identical small insulating balls are suspended by separate 0.25-m threads that are attached to a common point on the ceiling. Each ball has a mass of 8.0×10^{-4} kg. Initially the balls are uncharged and hang straight down. They are then given identical positive charges and, as a result, spread apart with an angle of 36° between the threads. Determine (a) the charge on each ball and (b) the tension in the threads.

Concepts & Calculations Group Learning Problems

Note: Each of these problems consists of Concept Questions followed by a related quantitative Problem. They are designed for use by students working alone or in small learning groups. The Concept Questions involve little or no mathematics and are intended to stimulate group discussions. They focus on the concepts with which the problems deal. Recognizing the concepts is the essential initial step in any problem-solving technique.

69. Concept Questions Two identical metal spheres have charges of q_1 and q_2. They are brought together so they touch, and then they are separated. (a) How is the net charge on the two spheres before they touch related to the net charge after they touch? (b) After they touch and are separated, is the charge on each sphere the same? Why?

Problem Four identical metal spheres have charges of $q_A = -8.0$ μC, $q_B = -2.0$ μC, $q_C = +5.0$ μC, and $q_D = +12.0$ μC. (a) Two of the spheres are brought together so they touch and then they are separated. Which spheres are they, if the final charge on each of the two is $+5.0$ μC? (b) In a similar manner, which three spheres are brought together and then separated, if the final charge on each of the three is $+3.0$ μC? (c) How many electrons would have to be added to one of the spheres in part (b) to make it electrically neutral?

70. Concept Questions The drawings show three charges that have the same magnitude, but different signs. In all cases the distance between charges 1 and 2 and between 2 and 3 is the same. (a) Draw the electrical force that each charge exerts on charge 2. Each force should be drawn in the correct direction, and its magnitude should be correct relative to that of the other force. (b) Rank the magnitudes of the net electrical force on charge 2, largest first. Explain.

Problem The magnitude of the charges is $q = 8.6 \ \mu C$, and the distance between them is 3.8 mm. Determine the magnitude of the net force on charge 2 for each of the three drawings. Verify that your answers are consistent with your answers to the Concept Questions.

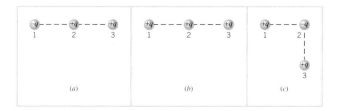

(a)   (b)   (c)

71. Concept Questions Suppose you want to neutralize the gravitational attraction between the earth and the moon by placing equal amounts of charge on each. (a) Should the charges be both positive, both negative, or one positive and the other negative? Why? (b) Do you need to know the distance between the earth and the moon to find the magnitude of the charge? Why or why not?

Problem The masses of the earth and moon are 5.98×10^{24} and 7.35×10^{22} kg, respectively. Identical amounts of charge are placed on each body, such that the net force (gravitational plus electrical) on each is zero. What is the magnitude of the charge?

72. Concept Questions Suppose you want to determine the electric field in a certain region of space. You have a small object of known charge and an instrument that measures the magnitude and direction of the force exerted on the object by the electric field. How would you determine the magnitude and direction of the electric field if the object were (a) positively charged and (b) negatively charged?

Problem (a) The object has a charge of $+20.0 \ \mu C$ and the instrument indicates that the electric force exerted on it is 40.0 μN, due east. What is the magnitude and direction of the electric field? (b) What is the magnitude and direction of the electric field if the object has a charge of $-10.0 \ \mu C$ and the instrument indicates that the force is 20.0 μN due west?

73. Concept Question The drawing shows two situations in which charges are placed on the *x* and *y* axes. They are all located at the same distance from the origin. Without doing any calculations, does

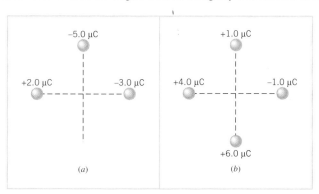

(a)   (b)

the electric field at the origin in part (a) have a magnitude that is greater than, less than, or equal to the magnitude of the field at the origin in part (b)? Justify your answer.

Problem The distance between each of the charges and the origin is 6.1 cm. For each of the situations shown in the drawing, determine the magnitude of the electric field at the origin. Check to see that your results are consistent with your answer to the Concept Question.

74. Concept Questions A proton and an electron are moving due east in a constant electric field that also points due east. (a) Does each experience an electric force of the same magnitude and direction? (b) What is the direction of the proton's acceleration and the direction of the electron's acceleration? (c) Is the magnitude of the proton's acceleration greater than, less than, or the same as that of the electron's acceleration? Explain your answers.

Problem The electric field points due east and has a magnitude of 8.0×10^4 N/C. Determine the magnitude of the acceleration of the proton and the electron. Check that your answers are consistent with part (c) of the Concept Questions.

* **75. Concept Questions** Three point charges have equal magnitudes, two being positive and one negative. These charges are fixed to the corners of an equilateral triangle, as the drawing shows. (a) The charge at any one corner experiences forces from the charges at the other corners. Do the individual forces exerted by the charges have the same or different magnitudes? (b) At which one or more corners does (do) the charge(s) experience a net force that has the greatest magnitude? (c) At which one or more corners does (do) the charge(s) experience a net force that has the smallest magnitude?

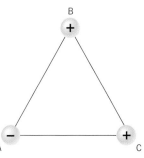

Problem The magnitude of each of the charges is 5.0 μC, and the lengths of the sides of the triangle are 3.0 cm. Calculate the magnitude of the net force that each charge experiences. Be sure that your answers are consistent with your answers to the Concept Questions.

* **76. Concept Questions** The drawing shows a positive point charge $+q_1$, a second point charge q_2 that may be positive or negative, and a spot labeled *P*, all on the same straight line. The distance *d* between the two charges is the same as the distance between q_1 and the point *P*. With q_2 present, the magnitude of the net electric field at *P* is twice what it is when q_1 is present alone. (a) When the second charge is positive, is its magnitude smaller than, equal to, or greater than the magnitude of q_1? Explain your reasoning. (b) When the second charge is negative, is its magnitude smaller than, equal to, or greater than that in question (a)? Account for your answer.

Problem Given that $q_1 = +0.50 \ \mu C$, determine q_2 when it is (a) positive and (b) negative. Verify that your answers are consistent with your answers to the Concept Questions.

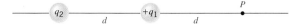

Chapter 19

Electric Potential Energy and the Electric Potential

The wires on these transmission towers deliver electric energy throughout the country in a controlled manner; lightning, on the other hand, is an uncontrolled manifestation of electric energy. Electric potential energy, and the related concept of electric potential, are the subjects of this chapter. (© Harald Sund/The Image Bank/Getty Images)

19.1 Potential Energy

In Chapter 18 we discussed the electrostatic force that two point charges exert on each other, the magnitude of which is $F = k|q_1||q_2|/r^2$. The form of this equation is similar to that for the gravitational force that two particles exert on each other, which is $F = Gm_1m_2/r^2$, according to Newton's law of universal gravitation (see Section 4.7). Both of these forces are conservative and, as Section 6.4 explains, a potential energy can be associated with a conservative force. Thus, an electric potential energy exists that is analogous to the gravitational potential energy. To set the stage for a discussion of the electric potential energy, let's review some of the important aspects of the gravitational counterpart.

Figure 19.1, which is essentially Figure 6.10, shows a basketball of mass m falling from point A to point B. The gravitational force, $m\mathbf{g}$, is the only force acting on the ball, where g is the magnitude of the acceleration due to gravity. As Section 6.3 discusses, the work W_{AB} done by the gravitational force when the ball falls from a height of h_A to a height of h_B is

$$W_{AB} = \underbrace{mgh_A}_{\substack{\text{Initial} \\ \text{gravitational} \\ \text{potential energy,} \\ \text{GPE}_A}} - \underbrace{mgh_B}_{\substack{\text{Final} \\ \text{gravitational} \\ \text{potential energy,} \\ \text{GPE}_B}} = \text{GPE}_A - \text{GPE}_B \qquad (6.4)$$

Recall that the quantity mgh is the gravitational potential energy* of the ball, $\text{GPE} = mgh$ (Equation 6.5), and represents the energy that the ball has by virtue of its position relative to the surface of the earth. Thus, the work done by the gravitational force equals the initial gravitational potential energy minus the final gravitational potential energy.

Figure 19.2 clarifies the analogy between electric and gravitational potential energies. In this drawing a positive test charge $+q_0$ is situated at point A between two oppositely charged plates. Because of the charges on the plates, an electric field $\mathbf{E}$ exists in the region between them. Consequently, the test charge experiences an electric force, $\mathbf{F} = q_0\mathbf{E}$, that is directed downward, toward the lower plate. (The gravitational force is being neglected here.) As the charge moves from A to B, work is done by this force, in a fashion analogous to the work done by the gravitational force in Figure 19.1. The work W_{AB} done by the electric force equals the difference between the electric potential energy EPE at A and that at B:

$$W_{AB} = \text{EPE}_A - \text{EPE}_B \qquad (19.1)$$

This expression is similar to Equation 6.4. The path along which the test charge moves from A to B is of no consequence because the electric force is a conservative force, and so the work W_{AB} is the same for all paths.

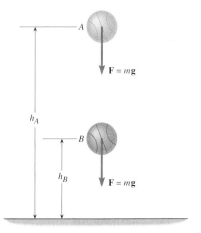

Figure 19.1 Gravity exerts a force, $\mathbf{F} = m\mathbf{g}$, on the basketball of mass m. Work is done by the gravitational force as the ball falls from A to B.

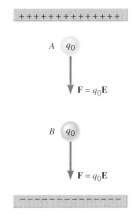

Figure 19.2 Because of the electric field $\mathbf{E}$, an electric force, $\mathbf{F} = q_0\mathbf{E}$, is exerted on a positive test charge $+q_0$. Work is done by the force as the charge moves from A to B.

*The gravitational potential energy is now being denoted by GPE to distinguish it from the electric potential energy EPE.

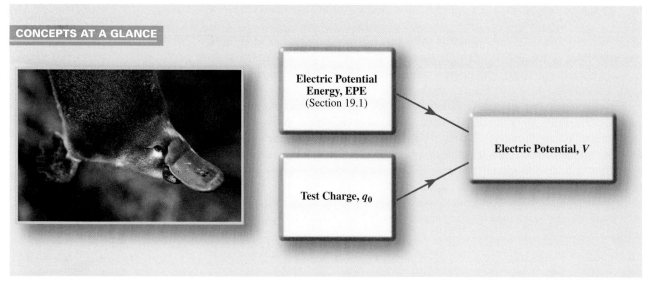

Figure 19.3 CONCEPTS AT A GLANCE The electric potential energy and the notion of a test charge are combined to produce the concept of electric potential. The duck-billed platypus is an aquatic freshwater animal that hunts by using electro-receptors in its bill to detect minute electric potentials produced by the muscles of its prey (shrimp, worms, and insect larvae). (© Nicole Duplaix/ National Geographic/Getty Images)

19.2 The Electric Potential Difference

▶ CONCEPTS AT A GLANCE   Since the electric force is $\mathbf{F} = q_0\mathbf{E}$, the work that it does as the charge moves from A to B in Figure 19.2 depends on the charge q_0. It is useful, therefore, to express this work on a per-unit-charge basis, by dividing both sides of Equation 19.1 by the charge:

$$\frac{W_{AB}}{q_0} = \frac{\text{EPE}_A}{q_0} - \frac{\text{EPE}_B}{q_0} \tag{19.2}$$

Notice that the right-hand side of this equation is the difference between two terms, each of which is an electric potential energy divided by the test charge, EPE/q_0. The quantity EPE/q_0 is the electric potential energy per unit charge and is an important concept in electricity. It is called the ***electric potential*** or, simply, the ***potential*** and is referred to with the symbol V, as in Equation 19.3. The Concepts-at-a-Glance chart in Figure 19.3 illustrates the fact that the electric potential is a concept that incorporates both the electric potential energy and the idea of a test charge. ◀

■ **DEFINITION OF ELECTRIC POTENTIAL**

The electric potential V at a given point is the electric potential energy EPE of a small test charge q_0 situated at that point divided by the charge itself:

$$V = \frac{\text{EPE}}{q_0} \tag{19.3}$$

SI Unit of Electric Potential: joule/coulomb = volt (V)

The SI unit of electric potential is a joule per coulomb, a quantity known as a *volt*. The name honors Alessandro Volta (1745–1827), who invented the voltaic pile, the forerunner of the battery. In spite of the similarity in names, the electric potential energy EPE and the electric potential V are *not* the same. The electric potential energy, as its name implies, is an *energy* and, therefore, is measured in joules. In contrast, the electric potential is an *energy per unit charge* and is measured in joules per coulomb, or volts.

We can now relate the work W_{AB} done by the electric force when a charge q_0 moves from A to B to the potential difference $V_B - V_A$ between the points. Combining Equations 19.2 and 19.3, we have:

$$V_B - V_A = \frac{\text{EPE}_B}{q_0} - \frac{\text{EPE}_A}{q_0} = \frac{-W_{AB}}{q_0} \tag{19.4}$$

Often the "delta" notation is used to express the difference (final value minus initial value) in potentials and in potential energies: $\Delta V = V_B - V_A$ and $\Delta(\text{EPE}) = \text{EPE}_B - \text{EPE}_A$. In terms of this notation, Equation 19.4 takes the following more compact form:

$$\Delta V = \frac{\Delta(\text{EPE})}{q_0} = \frac{-W_{AB}}{q_0} \qquad (19.4)$$

Neither the potential V nor the potential energy EPE can be determined in an absolute sense, because only the *differences* ΔV and $\Delta(\text{EPE})$ are measurable in terms of the work W_{AB}. The gravitational potential energy has this same characteristic, since only the value at one height relative to that at some reference height has any significance. Example 1 emphasizes the relative nature of the electric potential.

The energy used to operate this 3-wheel, single-seat electric car comes from the charging station and is stored in the vehicle's battery pack. The factors that determine how much electric energy the car receives are the voltage of the battery pack and the magnitude of the charge passing through it. (© Mehau Kulyk/Science Photo Library/Photo Researchers)

Example 1 Work, Electric Potential Energy, and Electric Potential

In Figure 19.2, the work done by the electric force as the test charge ($q_0 = +2.0 \times 10^{-6}$ C) moves from A to B is $W_{AB} = +5.0 \times 10^{-5}$ J. (a) Find the difference, $\Delta(\text{EPE}) = \text{EPE}_B - \text{EPE}_A$, in the electric potential energies of the charge between these points. (b) Determine the potential difference, $\Delta V = V_B - V_A$, between the points.

Reasoning The work done by the electric force when the charge moves from A to B is $W_{AB} = \text{EPE}_A - \text{EPE}_B$, according to Equation 19.1. Therefore, the difference in the electric potential energies (final value minus initial value) is $\Delta(\text{EPE}) = \text{EPE}_B - \text{EPE}_A = -W_{AB}$. The potential difference, $\Delta V = V_B - V_A$, is the difference in the electric potential energies divided by the charge q_0, according to Equation 19.4.

Solution

(a) The difference in the electric potential energies of the charge at points A and B is

$$\underbrace{\text{EPE}_B - \text{EPE}_A}_{= \Delta(\text{EPE})} = -W_{AB} = \boxed{-5.0 \times 10^{-5}\ \text{J}} \qquad (19.1)$$

Thus the charge has a higher electric potential energy at A than at B.

(b) The potential difference ΔV between A and B is

$$\underbrace{V_B - V_A}_{= \Delta V} = \frac{\text{EPE}_B - \text{EPE}_A}{q_0} = \frac{-5.0 \times 10^{-5}\ \text{J}}{2.0 \times 10^{-6}\ \text{C}} = \boxed{-25\ \text{V}} \qquad (19.4)$$

We can see, then, that the electric potential is higher at A than at B.

In Figure 19.1 the speed of the basketball increases as it falls from A to B. Since point A has a greater gravitational potential energy than point B, we see that an object of mass m accelerates when it moves from a region of higher potential energy toward a region of lower potential energy. Likewise, the positive charge in Figure 19.2 accelerates as it moves from A to B because of the electric repulsion from the upper plate and the attraction to the lower plate. Since point A has a higher electric potential than point B, we conclude that ***a positive charge accelerates from a region of higher electric potential toward a region of lower electric potential.*** On the other hand, a negative charge placed between the plates in Figure 19.2 behaves in the opposite fashion, since the electric force acting on the negative charge is directed opposite to that on the positive charge. ***A negative charge accelerates from a region of lower potential toward a region of higher potential.*** The next example illustrates the way positive and negative charges behave.

Problem solving insight

Problem solving insight

Conceptual Example 2
The Accelerations of Positive and Negative Charges

Three points, A, B, and C, are located along a horizontal line, as Figure 19.4 illustrates. A positive test charge is released from rest at A and accelerates toward B. Upon reaching B, the test charge continues to accelerate toward C. Assuming that only motion along the line is possible, what will a negative test charge do when it is released from rest at B?

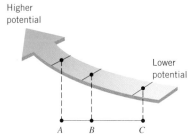

Figure 19.4 The electric potentials at points A, B, and C are different. Under the influence of these potentials, positive and negative charges accelerate in opposite directions.

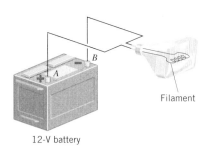

Figure 19.5 A headlight connected to a 12-V battery.

Reasoning and Solution A negative charge will accelerate from a region of lower potential toward a region of higher potential. Therefore, before we can decide what the negative test charge will do, it is necessary to know how the electric potentials compare at points A, B, and C. This information can be deduced from the behavior of the positive test charge, which accelerates from a region of higher toward a region of lower potential. Since the positive test charge accelerates from A to B, the potential at A must exceed that at B. And since the positive test charge accelerates from B to C, the potential at B must exceed that at C. The potential at point B, then, must lie between that at points A and C, as Figure 19.4 illustrates. When the negative test charge is released from rest at B, it will accelerate toward the region of higher potential. In other words, *it will begin moving toward A.*

As a familiar application of electric potential energy and electric potential, Figure 19.5 shows a 12-V automobile battery with a headlight connected between its terminals. The positive terminal, point A, has a potential that is 12 V higher than the potential at the negative terminal, point B; in other words, $V_A - V_B = 12$ V. Positive charges are repelled from the positive terminal and travel through the wires and headlight toward the negative terminal.* As the charges pass through the headlight, virtually all their potential energy is converted into heat, which causes the filament to glow "white hot" and emit light. When the charges reach the negative terminal, they no longer have any potential energy. The battery then gives the charges an additional "shot" of potential energy by moving them to the higher-potential positive terminal, and the cycle is repeated. In raising the potential energy of the charges, the battery does work on them and draws from its reserve of chemical energy to do so. Example 3 illustrates the concepts of electric potential energy and electric potential as applied to a battery.

Example 3   Operating a Headlight

Determine the number of particles, each carrying a charge of 1.60×10^{-19} C (the magnitude of the charge on an electron), that pass between the terminals of a 12-V car battery when a 60.0-W headlight burns for one hour.

Reasoning To obtain the number of charged particles, we determine the total charge needed to provide the energy consumed by the headlight in one hour. Dividing the total charge by the charge on each particle gives the number of particles.

Solution Using energy at a rate of 60.0 joules per second (60.0 watts) for one hour, the headlight uses a total energy of

$$\text{Energy} = \text{Power} \times \text{Time} = (60.0 \text{ W})(3600 \text{ s}) = 2.2 \times 10^5 \text{ J} \qquad (6.10\text{b})$$

Since this energy comes from the battery, it represents the difference between the electrical potential energy of the charges at point A and that at point B: 2.2×10^5 J $= \text{EPE}_A - \text{EPE}_B$. According to Equation 19.4, we have

$$\frac{\text{EPE}_A - \text{EPE}_B}{q_0} = V_A - V_B$$

$$q_0 = \frac{\text{EPE}_A - \text{EPE}_B}{V_A - V_B} = \frac{2.2 \times 10^5 \text{ J}}{12 \text{ V}} = 1.8 \times 10^4 \text{ C}$$

The number of particles whose individual charges combine to provide this total charge is $(1.8 \times 10^4 \text{ C})/(1.60 \times 10^{-19} \text{ C}) = \boxed{1.1 \times 10^{23}}$.

As used in connection with batteries, the volt is a familiar unit for measuring electric potential difference. The word "volt" also appears in another context, as part of a unit that

Need more practice?

Interactive LearningWare 19.1
The energy in a lightning bolt is enormous. Consider, for example, a lightning bolt in which 29 C of charge moves through a potential difference of 1.4×10^8 V. With the amount of energy in this bolt, how many days could a household function that uses 75 kWh of energy per day?

Go to
www.wiley.com/college/cutnell
for an interactive solution.

*Historically, it was believed that positive charges flow in the wires of an electric circuit. Today, it is known that negative charges flow in wires from the negative toward the positive terminal. Here, however, we follow the customary practice of describing the flow of negative charges by specifying the opposite but equivalent flow of positive charges. This hypothetical flow of positive charges is called the "conventional electric current," as we will see in Section 20.1.

is used to measure energy, particularly the energy of an atomic particle, such as an electron or a proton. This energy unit is called the *electron volt* (eV). **One electron volt is the magnitude of the amount by which the potential energy of an electron changes when the electron moves through a potential difference of one volt.** Since the magnitude of the change in potential energy is $|q_0 \Delta V| = |(-1.60 \times 10^{-19} \text{ C}) \times (1.00 \text{ V})| = 1.60 \times 10^{-19}$ J, it follows that

$$1 \text{ eV} = 1.60 \times 10^{-19} \text{ J}$$

Problem solving insight

One million (10^{+6}) electron volts of energy is referred to as one MeV, and one billion (10^{+9}) electron volts of energy is one GeV, where the "G" stands for the prefix "giga" (pronounced "jig'a").

▶ CONCEPTS AT A GLANCE   The Concepts-at-a-Glance chart in Figure 19.6 emphasizes that the total energy of an object, which is the sum of its kinetic and potential energies, is an important concept. Its significance lies in the fact that the total energy remains the same (is conserved) during the object's motion, provided that nonconservative forces, such as friction, are either absent or do no net work. While the sum of the energies at each instant remains constant, energy may be converted from one form to another; for example, gravitational potential energy is converted into kinetic energy as a ball falls. As Figure 19.6 illustrates, we now include the electric potential energy EPE as part of the total energy that an object can have:

$$E \;=\; \tfrac{1}{2}mv^2 \;+\; \tfrac{1}{2}I\omega^2 \;+\; mgh \;+\; \tfrac{1}{2}kx^2 \;+\; \text{EPE}$$

Total energy	Translational kinetic energy	Rotational kinetic energy	Gravitational potential energy	Elastic potential energy	Electric potential energy

If the total energy is conserved as the object moves, then its final energy E_f is equal to its initial energy E_0, or $E_f = E_0$. Example 4 illustrates how the conservation of energy is applied to a charge moving in an electric field. ◀

Figure 19.6 CONCEPTS AT A GLANCE The electric potential energy is another type of energy. The total energy is conserved when nonconservative forces do no net work ($W_{nc} = 0$ J). The patient in this photograph is undergoing an X-ray angiography examination of the heart. To produce X-rays, electrons are accelerated from rest and allowed to collide with a target material within the X-ray machine. As each electron is accelerated, electric potential energy is converted into kinetic energy; the total energy is conserved during the process. (© Rosenfeld Images, Ltd/Science Photo Library/Photo Researchers)

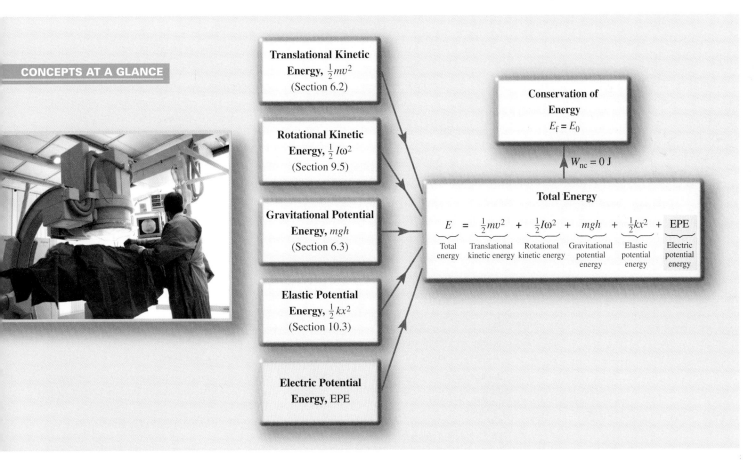

CONCEPTS AT A GLANCE

Translational Kinetic Energy, $\tfrac{1}{2}mv^2$ (Section 6.2)

Rotational Kinetic Energy, $\tfrac{1}{2}I\omega^2$ (Section 9.5)

Gravitational Potential Energy, mgh (Section 6.3)

Elastic Potential Energy, $\tfrac{1}{2}kx^2$ (Section 10.3)

Electric Potential Energy, EPE

Conservation of Energy $E_f = E_0$

$W_{nc} = 0$ J

Total Energy

$$E \;=\; \tfrac{1}{2}mv^2 \;+\; \tfrac{1}{2}I\omega^2 \;+\; mgh \;+\; \tfrac{1}{2}kx^2 \;+\; \text{EPE}$$

Total energy · Translational kinetic energy · Rotational kinetic energy · Gravitational potential energy · Elastic potential energy · Electric potential energy

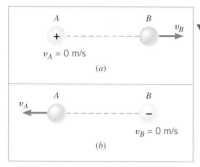

Figure 19.7 (a) A positive charge starts from rest at point A and accelerates toward point B. (b) A negative charge starts from rest at B and accelerates toward A.

Problem solving insight
A positive charge accelerates from a region of higher potential toward a region of lower potential. In contrast, a negative charge accelerates from a region of lower potential toward a region of higher potential.

Example 4 The Conservation of Energy

(a) A particle has a mass of $m = 1.8 \times 10^{-5}$ kg and a positive charge of $q_0 = +3.0 \times 10^{-5}$ C. It is released from rest at point A and accelerates horizontally until it reaches point B, as Figure 19.7a shows. During transit, the particle does not rotate. The only force acting on the particle is an electric force (not shown in the drawing), and the electric potential at A is 25 V greater than that at B. In other words, $V_A - V_B = 25$ V. What is the speed v_B of the particle when it reaches point B? (b) If the same particle had a *negative* charge and were released from rest at B (see Figure 19.7b), what would be its speed v_A at A?

Reasoning Since the only force acting on the particle is the conservative electric force, the total energy of the particle is conserved. This means that the total energy is the same at points B and A:

$$\tfrac{1}{2}mv_B^2 + \underbrace{\tfrac{1}{2}I\omega_B^2}_{\substack{=0\text{ J, since}\\ \omega_B = 0\text{ rad/s}}} + mgh_B + \underbrace{\tfrac{1}{2}kx_B^2}_{\substack{=0\text{ J, since no}\\ \text{elastic forces}\\ \text{are present}}} + \text{EPE}_B$$

$$= \tfrac{1}{2}mv_A^2 + \underbrace{\tfrac{1}{2}I\omega_A^2}_{\substack{=0\text{ J, since}\\ \omega_A = 0\text{ rad/s}}} + mgh_A + \underbrace{\tfrac{1}{2}kx_A^2}_{\substack{=0\text{ J, since no}\\ \text{elastic forces}\\ \text{are present}}} + \text{EPE}_A$$

In this expression the angular speeds ω_A and ω_B are zero because the particle does not rotate. Moreover, we set $h_A = h_B$ (the particle moves horizontally) and note that $\text{EPE}_A - \text{EPE}_B = q_0(V_A - V_B)$; the conservation of energy equation then reduces to

$$\tfrac{1}{2}mv_B^2 = \tfrac{1}{2}mv_A^2 + q_0(V_A - V_B)$$

We will use this equation to determine the final speeds.

Solution

(a) The positively charged particle starts from rest at A ($v_A = 0$ m/s). Its speed v_B at B can be found from the equation above:

$$v_B = \sqrt{\frac{2q_0(V_A - V_B)}{m}}$$

$$= \sqrt{\frac{2(+3.0 \times 10^{-5}\text{ C})(25\text{ V})}{1.8 \times 10^{-5}\text{ kg}}} = \boxed{9.1\text{ m/s}}$$

(b) The negatively charged particle starts from rest at B ($v_B = 0$ m/s). Its speed v_A at A can also be found from the last equation in the reasoning section:

$$v_A = \sqrt{\frac{-2q_0(V_A - V_B)}{m}}$$

$$= \sqrt{\frac{-2(-3.0 \times 10^{-5}\text{ C})(25\text{ V})}{1.8 \times 10^{-5}\text{ kg}}} = \boxed{9.1\text{ m/s}}$$

✓ **Check Your Understanding 1**

An ion, starting from rest in a constant electric field, accelerates from point A to point B. Does the electric potential energy of the ion at point B depend on (a) the magnitude of its charge and (b) its mass? Does the speed of the ion at B depend on (c) the magnitude of its charge and (d) its mass? *(The answers are given at the end of the book.)*

Background: This question deals with how a charged particle moves between two locations that have different electric potential energies. The answers also require an understanding of how the kinetic energy of a particle depends on its mass and speed.

For similar questions (including calculational counterparts), consult Self-Assessment Test 19.1, which is described at the end of Section 19.3.

19.3 The Electric Potential Difference Created by Point Charges

A positive point charge $+q$ creates an electric potential in a way that Figure 19.8 helps explain. This picture shows two locations A and B, at distances r_A and r_B from the charge. At any position between A and B an electrostatic force of repulsion $\mathbf{F}$ acts on a positive test charge $+q_0$. The magnitude of the force is given by Coulomb's law as $F = kq_0q/r^2$, where we assume for convenience that q_0 and q are positive, so that $|q_0| = q_0$ and $|q| = q$. When the test charge moves from A to B, work is done by this force. Since r varies between r_A and r_B, the force F also varies, and the work is not the product of the force and the distance between the points. (Recall from Section 6.1 that work is force times distance only if the force is constant.) However, the work W_{AB} can be found with the methods of integral calculus. The result is

$$W_{AB} = \frac{kqq_0}{r_A} - \frac{kqq_0}{r_B}$$

This result is valid whether q is positive or negative, and whether q_0 is positive or negative. The potential difference, $V_B - V_A$, between A and B can now be determined by substituting this expression for W_{AB} into Equation 19.4:

$$V_B - V_A = \frac{-W_{AB}}{q_0} = \frac{kq}{r_B} - \frac{kq}{r_A} \qquad (19.5)$$

As point B is located farther and farther from the charge q, r_B becomes larger and larger. In the limit that r_B is infinitely large, the term kq/r_B becomes zero, and it is customary to set V_B equal to zero also. In this limit, Equation 19.5 becomes $V_A = kq/r_A$, and it is standard convention to omit the subscripts and write the potential in the following form:

Potential of a point charge $$V = \frac{kq}{r} \qquad (19.6)$$

The symbol V in this equation does not refer to the potential in any absolute sense. Rather, $V = kq/r$ stands for the amount by which the potential at a distance r from a point charge differs from the potential at an infinite distance away. In other words, V refers to a potential difference with the arbitrary assumption that the potential at infinity is zero.

With the aid of Equation 19.6, we can describe the effect that a point charge q has on the surrounding space. When q is positive, the value of $V = kq/r$ is also positive, indicating that the positive charge has everywhere raised the potential above the zero reference value. Conversely, when q is negative, the potential V is also negative, indicating that the negative charge has everywhere decreased the potential below the zero reference value. The next example deals with these effects quantitatively.

Example 5 The Potential of a Point Charge

Using a zero reference potential at infinity, determine the amount by which a point charge of 4.0×10^{-8} C alters the electric potential at a spot 1.2 m away when the charge is (a) positive and (b) negative.

Reasoning A point charge q alters the potential at every location in the surrounding space. In the expression $V = kq/r$, the effect of the charge in increasing or decreasing the potential is conveyed by the algebraic sign for the value of q.

Solution

(a) Figure 19.9a shows the potential when the charge is positive:

$$V = \frac{kq}{r} = \frac{(8.99 \times 10^9 \text{ N} \cdot \text{m}^2/\text{C}^2)(+4.0 \times 10^{-8} \text{ C})}{1.2 \text{ m}} = \boxed{+300 \text{ V}} \qquad (19.6)$$

(b) Part b of the drawing illustrates the results when the charge is negative. A calculation similar to the one in part (a) shows that the potential is now negative: $\boxed{-300 \text{ V}}$.

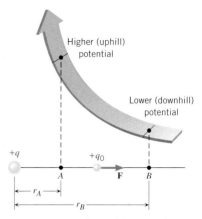

Figure 19.8 The positive test charge $+q_0$ experiences a repulsive force $\mathbf{F}$ due to the positive point charge $+q$. As a result, work is done by this force when the test charge moves from A to B. Consequently, the electric potential is higher (uphill) at A and lower (downhill) at B.

🖱️ Concept Simulation 19.1

This simulation displays the electric force that a positive charge q exerts on a positive test charge q_0 as a function of the distance between them. Also shown is a graph of the electric potential created by q as a function of distance. The user can vary the magnitude of each charge and see the effect on the force and the electric potential.

Go to
www.wiley.com/college/cutnell

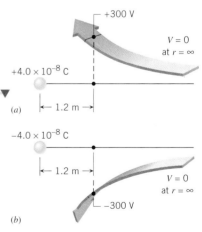

Figure 19.9 A point charge of 4.0×10^{-8} C alters the potential at a spot 1.2 m away. The potential is (a) increased by 300 V when the charge is positive and (b) decreased by 300 V when the charge is negative, relative to a zero reference potential at infinity.

A single point charge raises or lowers the potential at a given location, depending on whether the charge is positive or negative. ***When two or more charges are present, the potential due to all the charges is obtained by adding together the individual potentials,*** as the next two examples show.

Example 6 The Total Electric Potential

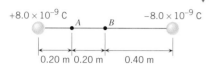

+8.0 × 10⁻⁹ C   −8.0 × 10⁻⁹ C

0.20 m 0.20 m   0.40 m

Figure 19.10 Both the positive and negative charges affect the electric potential at locations *A* and *B*.

At locations *A* and *B* in Figure 19.10, find the total electric potential due to the two point charges.

Reasoning At each location, each charge contributes to the total electric potential. We obtain the individual contributions by using $V = kq/r$ and find the total potential by adding the individual contributions algebraically. The two charges have the same magnitude, but different signs. Thus, at *A* the total potential is positive because this spot is closer to the positive charge, whose effect dominates over that of the more distant negative charge. At *B*, midway between the charges, the total potential is zero, since the potential of one charge exactly offsets that of the other.

Solution

Location	Contribution from + Charge		Contribution from − Charge	Total Potential
A	$\dfrac{(8.99 \times 10^9 \text{ N} \cdot \text{m}^2/\text{C}^2)(+8.0 \times 10^{-9} \text{ C})}{0.20 \text{ m}}$	+	$\dfrac{(8.99 \times 10^9 \text{ N} \cdot \text{m}^2/\text{C}^2)(-8.0 \times 10^{-9} \text{ C})}{0.60 \text{ m}}$ =	+240 V
B	$\dfrac{(8.99 \times 10^9 \text{ N} \cdot \text{m}^2/\text{C}^2)(+8.0 \times 10^{-9} \text{ C})}{0.40 \text{ m}}$	+	$\dfrac{(8.99 \times 10^9 \text{ N} \cdot \text{m}^2/\text{C}^2)(-8.0 \times 10^{-9} \text{ C})}{0.40 \text{ m}}$ =	0 V

Conceptual Example 7 Where Is the Potential Zero?

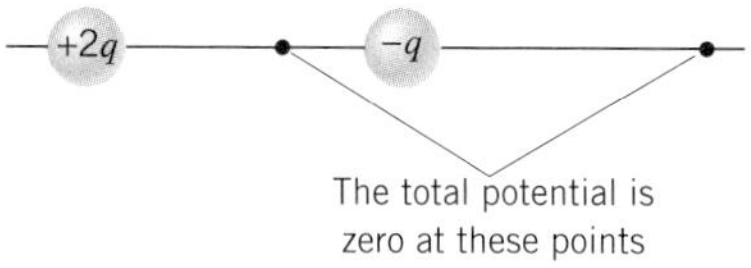

+2*q*   −*q*

The total potential is
zero at these points

Figure 19.11 Two point charges, one positive and one negative. The positive charge, +2*q*, has twice the magnitude of the negative charge, −*q*.

At any location, the total electric potential is the algebraic sum of the individual potentials created by each point charge that is present.

Two point charges are fixed in place, as in Figure 19.11. The positive charge is +2*q* and has twice the magnitude of the negative charge, which is −*q*. On the line that passes through the charges, how many places are there at which the total potential is zero?

Reasoning and Solution The total potential is the algebraic sum of the individual potentials created by each charge. It will be zero if the potential due to the positive charge is exactly offset by the potential due to the negative charge. The potential of a point charge is directly proportional to the charge and inversely proportional to the distance from the charge. The positive charge has the larger magnitude and, in the region to its left, is closer to any spot than the negative charge is. As a result, the potential of the positive charge here always dominates over that of the negative charge. Therefore, the total potential cannot be zero anywhere to the left of the positive charge.

Between the charges there is a location where the individual potentials do balance. We saw a similar situation in Example 6, where the balance occurred at the midpoint between two charges that had equal magnitudes. Now the charges have unequal magnitudes, so the balance does not occur at the midpoint. Instead, it occurs at a location that is closer to the charge with the smaller magnitude—namely, the negative charge. Then, since the potential of a point charge is inversely proportional to the distance from the charge, the effect of the smaller charge will be able to offset that of the more distant larger charge.

To the right of the negative charge, there is another location at which the individual potentials exactly offset one another. All places on this section of the line are closer to the negative than to the positive charge. In this region, therefore, a location exists at which the potential of the smaller negative charge again exactly offsets that of the more distant and larger positive charge.

Related Homework: *Problem 18*

In Example 6 we determined the total potential at a spot due to several point charges. In Example 8 we now extend this technique to find the total potential energy of three charges.

Example 8 The Potential Energy of a Group of Charges

Figure 19.12 shows three point charges; initially, they are infinitely far apart. They are then brought together and placed at the corners of an equilateral triangle. Each side of the triangle has a length of 0.50 m. Determine the electric potential energy of the triangular group. In other words, determine the amount by which the electric potential energy of the group differs from that of the three charges in their initial, infinitely separated locations.

Reasoning We will proceed in steps by adding charges to the triangle, one at a time, and then determining the electric potential energy at each step. According to Equation 19.3, EPE = q_0V, the electric potential energy is the product of the charge and the electric potential at the spot where the charge is placed. The total electric potential energy of the triangular group is the sum of the energies of each step in assembling the group.

Solution The order in which the charges are put on the triangle does not matter; we begin with the charge of +5.0 μC. When this charge is placed at a corner of the triangle, it has no electric potential energy, according to EPE = q_0V. This is because the total potential V produced by the other two charges is zero at this corner, since they are infinitely far away. Once the charge is in place, the potential it creates at either empty corner ($r = 0.50$ m) is

$$V = \frac{kq}{r} = \frac{(8.99 \times 10^9 \text{ N} \cdot \text{m}^2/\text{C}^2)(+5.0 \times 10^{-6} \text{ C})}{0.50 \text{ m}} = +9.0 \times 10^4 \text{ V} \quad (19.6)$$

Therefore, when the +6.0-μC charge is placed at the second corner of the triangle, its electric potential energy is

$$\text{EPE} = qV = (+6.0 \times 10^{-6} \text{ C})(+9.0 \times 10^4 \text{ V}) = +0.54 \text{ J} \quad (19.3)$$

The electric potential at the remaining empty corner is the sum of the potentials due to the two charges that are already in place:

$$V = \frac{(8.99 \times 10^9 \text{ N} \cdot \text{m}^2/\text{C}^2)(+5.0 \times 10^{-6} \text{ C})}{0.50 \text{ m}}$$
$$+ \frac{(8.99 \times 10^9 \text{ N} \cdot \text{m}^2/\text{C}^2)(+6.0 \times 10^{-6} \text{ C})}{0.50 \text{ m}} = +2.0 \times 10^5 \text{ V}$$

When the third charge, −2.0 μC, is placed at the remaining empty corner, its electric potential energy is

$$\text{EPE} = qV = (-2.0 \times 10^{-6} \text{ C})(+2.0 \times 10^5 \text{ V}) = -0.40 \text{ J} \quad (19.3)$$

The total potential energy of the triangular group differs from that of the widely separated charges by an amount that is the sum of the potential energies calculated above:

$$\text{Total potential energy} = 0 \text{ J} + 0.54 \text{ J} - 0.40 \text{ J} = \boxed{+0.14 \text{ J}}$$

This energy originates in the work done to bring the charges together.

Figure 19.12 Three point charges are placed on the corners of an equilateral triangle. Example 8 illustrates how to determine the total electric potential energy of this group of charges.

Problem solving insight
Be careful to distinguish between the concepts of potential V and electric potential energy EPE. Potential is electric potential energy per unit charge: $V = \text{EPE}/q$.

 Check Your Understanding 2

The figure shows four arrangements of two point charges. In each arrangement consider the total electric potential that the charges produce at location P. Rank the values of the total potential in descending order (largest first). *(The answer is given at the end of the book.)*

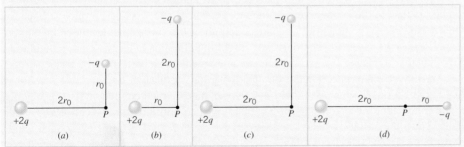

(a) (b) (c) (d)

Background: The electric potential produced by a point charge is the key concept here. The potential may be positive or negative, and its magnitude depends on the magnitude of the charge as well as the distance between it and the location P.

For similar questions (including calculational counterparts), consult Self-Assessment Test 19.1, which is described next.

Self-Assessment Test 19.1

Test your understanding of the material in Sections 19.1–19.3:

• Electric Potential Energy • Electric Potential Difference • Electric Potential Difference Created by Point Charges • Electric Potential Energy of a Group of Point Charges

Go to **www.wiley.com/college/cutnell**

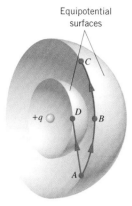

Figure 19.13 The equipotential surfaces that surround the point charge $+q$ are spherical. The electric force does no work as a charge moves on a path that lies on an equipotential surface, such as the path *ABC*. However, work is done by the electric force when a charge moves between two equipotential surfaces, as along the path *AD*.

Problem solving insight

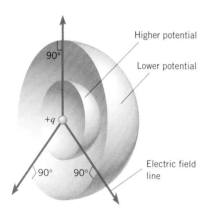

Figure 19.14 The radially directed electric field of a point charge is perpendicular to the spherical equipotential surfaces that surround the charge. The electric field points in the direction of *decreasing* potential.

19.4 *Equipotential Surfaces and Their Relation to the Electric Field*

An *equipotential surface* is a surface on which the electric potential is the same everywhere. The easiest equipotential surfaces to visualize are those that surround an isolated point charge. According to Equation 19.6, the potential at a distance r from a point charge q is $V = kq/r$. Thus, wherever r is the same, the potential is the same, and the equipotential surfaces are spherical surfaces centered on the charge. There are an infinite number of such surfaces, one for every value of r, and Figure 19.13 illustrates two of them. The larger the distance r, the smaller is the potential of the equipotential surface.

The net electric force does no work as a charge moves on an equipotential surface. This important characteristic arises because when an electric force does work W_{AB} as a charge moves from A to B, the potential changes according to Equation 19.4, $V_B - V_A = -W_{AB}/q_0$. Since the potential remains the same on an equipotential surface, $V_A = V_B$, and we see that $W_{AB} = 0$ J. In Figure 19.13, for instance, the electric force does no work as a test charge moves along the circular arc *ABC*, which lies on an equipotential surface. In contrast, the electric force does work when a charge moves *between* equipotential surfaces, as from A to D in the picture.

The spherical equipotential surfaces that surround an isolated point charge illustrate another characteristic of such surfaces. Figure 19.14 shows two of the surfaces around a positive point charge, along with some electric field lines. The electric field lines give the direction of the electric field, and for a positive point charge, the electric field is directed radially outward. Therefore, at each location on an equipotential sphere the electric field is perpendicular to the surface and points outward in the direction of decreasing potential, as the drawing emphasizes. This perpendicular relation is valid whether or not the equipotential surfaces result from a positive charge or have a spherical shape; *the electric field created by any charge or group of charges is everywhere perpendicular to the associated equipotential surfaces and points in the direction of decreasing potential.* For example, Figure 19.15 shows the electric field lines around an electric dipole, along with some equipotential surfaces (shown in cross section). Since the field lines are not simply radial, the equipotential surfaces are no longer spherical but, instead, have the shape necessary to be everywhere perpendicular to the field lines.

To see why an equipotential surface must be perpendicular to the electric field, consider Figure 19.16, which shows a hypothetical situation in which the perpendicular relation does *not* hold. If **E** were not perpendicular to the equipotential surface, there would be a component of **E** parallel to the surface. This field component would exert an electric force on a test charge placed on the surface. As the charge moved along the surface, work would be done by this component of the electric force. The work, according to Equation 19.4, would cause the potential to change, and, thus, the surface could not be an equipotential surface as assumed. The only way out of the dilemma is for the electric field to be perpendicular to the surface, so there is no component of the field parallel to the surface.

We have already encountered one equipotential surface. In Section 18.8, we found that the direction of the electric field just outside an electrical conductor is perpendicular to the conductor's surface, when the conductor is at equilibrium under electrostatic conditions. Thus, the surface of any conductor is an equipotential surface under such conditions. In fact, since the electric field is zero everywhere inside a conductor whose charges are in equilibrium, the entire conductor can be regarded as an equipotential volume.

There is a quantitative relation between the electric field and the equipotential surfaces. One example that illustrates this relation is the parallel plate capacitor in Figure 19.17. As Section 18.6 discusses, the electric field **E** between the metal plates is perpendicular to them and is the same everywhere, ignoring fringe fields at the edges. To be perpendicular to the electric field, the equipotential surfaces must be planes that are parallel to the plates, which themselves are equipotential surfaces. The potential difference between the plates is given by Equation 19.4 as $\Delta V = V_B - V_A = -W_{AB}/q_0$, where A is a point on the positive plate and B is a point on the negative plate. The work done by the electric force as a positive test charge q_0 moves from A to B is $W_{AB} = F\Delta s$, where F refers to the electric force and Δs to the displacement along a line perpendicular to the plates. The force equals the product of the charge and the electric field E ($F = q_0 E$), so the work becomes $W_{AB} = F\Delta s = q_0 E \Delta s$.

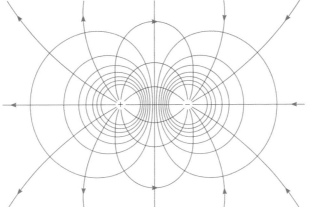

Figure 19.15 A cross-sectional view of the equipotential surfaces (in blue) of an electric dipole. The surfaces are drawn so that at every point they are perpendicular to the electric field lines (in red) of the dipole.

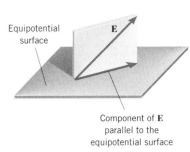

Figure 19.16 In this hypothetical situation, the electric field **E** is not perpendicular to the equipotential surface. As a result, there is a component of **E** parallel to the surface.

Therefore, the potential difference between the capacitor plates can be written in terms of the electric field as $\Delta V = - W_{AB}/q_0 = - q_0 E \Delta s/q_0$, or

$$E = -\frac{\Delta V}{\Delta s} \tag{19.7}$$

The quantity $\Delta V/\Delta s$ is referred to as the *potential gradient* and has units of volts per meter. In general, the relation $E = -\Delta V/\Delta s$ gives only the component of the electric field along the displacement Δs; it does not give the perpendicular component. The next example deals further with the equipotential surfaces between the plates of a capacitor.

Example 9 The Electric Field and Potential Are Related

The plates of the capacitor in Figure 19.17 are separated by a distance of 0.032 m, and the potential difference between them is $\Delta V = V_B - V_A = -64$ V. Between the two equipotential surfaces shown in color there is a potential difference of -3.0 V. Find the spacing between the two colored surfaces.

Reasoning The electric field is $E = -\Delta V/\Delta s$. To find the spacing between the two colored equipotential surfaces, we solve this equation for Δs, with $\Delta V = -3.0$ V and E equal to the electric field between the plates of the capacitor. A value for E can be obtained by using the values given for the distance and potential difference between the plates.

Solution The electric field between the capacitor plates is

$$E = -\frac{\Delta V}{\Delta s} = -\frac{-64\ \text{V}}{0.032\ \text{m}} = 2.0 \times 10^3\ \text{V/m} \tag{19.7}$$

The spacing between the colored equipotential surfaces can now be determined:

$$\Delta s = -\frac{\Delta V}{E} = -\frac{-3.0\ \text{V}}{2.0 \times 10^3\ \text{V/m}} = \boxed{1.5 \times 10^{-3}\ \text{m}}$$

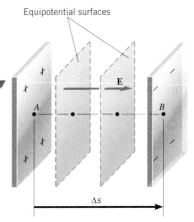

Figure 19.17 The metal plates of a parallel plate capacitor are equipotential surfaces. Two additional equipotential surfaces are shown between the plates. These two equipotential surfaces are parallel to the plates and are perpendicular to the electric field **E** between the plates.

✔ Check Your Understanding 3

The drawing shows a cross-sectional view of two spherical equipotential surfaces and two electric field lines that are perpendicular to these surfaces. When an electron moves from point *A* to point *B* (against the electric field), the electric force does $+3.2 \times 10^{-19}$ J of work. What are the electric potential differences (a) $V_B - V_A$, (b) $V_C - V_B$, and (c) $V_C - V_A$? *(The answers are given at the end of the book.)*

Background: Two ideas are important here: (1) the relationship between the work done by the electric force and the electric potential difference, and (2) equipotential surfaces.

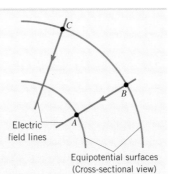

For similar questions (including conceptual counterparts), consult Self-Assessment Test 19.2, which is described at the end of Section 19.5.

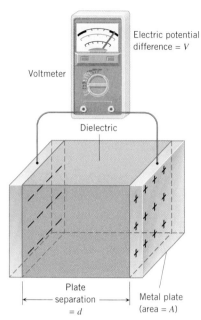

Voltmeter

Electric potential difference = V

Dielectric

Plate separation = d

Metal plate (area = A)

Figure 19.18 A parallel plate capacitor consists of two metal plates, one carrying a charge $+q$ and the other a charge $-q$. The potential of the positive plate exceeds that of the negative plate by an amount V. The region between the plates is filled with a dielectric.

19.5 Capacitors and Dielectrics

THE CAPACITANCE OF A CAPACITOR

In Section 18.6 we saw that a parallel plate capacitor consists of two parallel metal plates placed near one another but not touching. This type of capacitor is only one among many. In general, a **capacitor** consists of two conductors of any shape placed near one another without touching. For a reason that will become clear later on, it is common practice to fill the region between the conductors or plates with an electrically insulating material called a **dielectric,** as Figure 19.18 illustrates.

A capacitor stores electric charge. Each capacitor plate carries a charge of the *same magnitude,* one positive and the other negative. Because of the charges, the electric potential of the positive plate exceeds that of the negative plate by an amount V, as Figure 19.18 indicates. Experiment shows that when the magnitude q of the charge on each plate is doubled, the electric potential difference V is also doubled, so q is proportional to V: $q \propto V$. Equation 19.8 expresses this proportionality with the aid of a proportionality constant C, which is the **capacitance** of the capacitor.

■ **THE RELATION BETWEEN CHARGE AND POTENTIAL DIFFERENCE FOR A CAPACITOR**

The magnitude q of the charge on each plate of a capacitor is directly proportional to the magnitude V of the potential difference between the plates:

$$q = CV \qquad (19.8)$$

where C is the capacitance.

SI Unit of Capacitance: coulomb/volt = farad (F)

Equation 19.8 shows that the SI unit of capacitance is the coulomb per volt (C/V). This unit is called the *farad* (F), named after the English scientist Michael Faraday (1791–1867). One farad is an enormous capacitance. Usually smaller amounts, such as a microfarad (1 μF = 10^{-6} F) or a picofarad (1 pF = 10^{-12} F), are used in electric circuits. The capacitance reflects the ability of the capacitor to store charge, in the sense that a larger capacitance C allows more charge q to be put onto the plates for a given value of the potential difference V.

The physics of random-access memory (RAM) chips.

The ability of a capacitor to store charge lies at the heart of the random-access memory (RAM) chips used in computers, where information is stored in the form of the "ones" and "zeros" that comprise binary numbers. Figure 19.19 illustrates the role of a capacitor in a RAM chip. The capacitor is connected to a transistor switch, to which two lines are connected, an address line and a data line. A single RAM chip often contains millions of such transistor–capacitor units. The address line is used by the computer to locate a particular transistor–capacitor combination, and the data line carries the data to be stored. A pulse on the address line turns on the transistor switch. With the switch turned on, a pulse coming in on the data line can cause the capacitor to charge. A charged capacitor means that a "one" has been stored, whereas an uncharged capacitor means that a "zero" has been stored.

THE DIELECTRIC CONSTANT

If a dielectric is inserted between the plates of a capacitor, the capacitance can increase markedly because of the way in which the dielectric alters the electric field between the plates. Figure 19.20 shows how this effect comes about. In part a, the region between the charged plates is empty. The field lines point from the positive toward the negative plate. In part b, a dielectric has been inserted between the plates. Since the capacitor is not connected to anything, the charge on the plates remains constant as the dielectric is inserted. In many materials (e.g., water) the molecules possess permanent dipole moments, even though the molecules are electrically neutral. The dipole moment exists be-

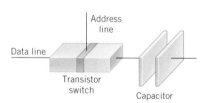

Data line

Address line

Transistor switch

Capacitor

Figure 19.19 A transistor–capacitor combination is part of a RAM chip used in computer memories.

cause one end of a molecule has a slight excess of negative charge while the other end has a slight excess of positive charge. When such molecules are placed between the charged plates of the capacitor, the negative ends are attracted to the positive plate and the positive ends are attracted to the negative plate. As a result, the dipolar molecules tend to orient themselves end to end, as in part *b*. Whether or not a molecule has a permanent dipole moment, the electric field can cause the electrons to shift position within a molecule, making one end slightly negative and the opposite end slightly positive. Because of the end-to-end orientation, the left surface of the dielectric becomes positively charged, and the right surface becomes negatively charged. The surface charges are shown in red in the picture.

Because of the surface charges on the dielectric, not all the electric field lines generated by the charges on the plates pass through the dielectric. As Figure 19.20*c* shows, some of the field lines end on the negative surface charges and begin again on the positive surface charges. Thus, the electric field inside the dielectric is less strong than the electric field inside the empty capacitor, assuming the charge on the plates remains constant. This reduction in the electric field is described by the ***dielectric constant*** κ, which is the ratio of the field magnitude E_0 without the dielectric to the field magnitude E inside the dielectric:

$$\kappa = \frac{E_0}{E} \tag{19.9}$$

Being a ratio of two field strengths, the dielectric constant is a number without units. Moreover, since the field $\mathbf{E}_0$ without the dielectric is greater than the field $\mathbf{E}$ inside the dielectric, the dielectric constant is greater than unity. The value of κ depends on the nature of the dielectric material, as Table 19.1 indicates.

THE CAPACITANCE OF A PARALLEL PLATE CAPACITOR

The capacitance of a capacitor is affected by the geometry of the plates and the dielectric constant of the material between them. For example, Figure 19.18 shows a parallel plate capacitor in which the area of each plate is A and the separation between the plates is d. The magnitude of the electric field inside the dielectric is given by Equation 19.7 (without the minus sign) as $E = V/d$, where V is the magnitude of the potential difference between the plates. If the charge on each plate is kept fixed, the electric field inside the dielectric is related to the electric field in the absence of the dielectric via Equation 19.9. Therefore,

$$E = \frac{E_0}{\kappa} = \frac{V}{d}$$

Since the electric field within an empty capacitor is $E_0 = q/(\epsilon_0 A)$ (see Equation 18.4), it follows that $q/(\kappa \epsilon_0 A) = V/d$, which can be solved for q to give

$$q = \left(\frac{\kappa \epsilon_0 A}{d}\right) V$$

A comparison of this expression with $q = CV$ (Equation 19.8) reveals that the capacitance C is

Parallel plate capacitor filled with a dielectric
$$C = \frac{\kappa \epsilon_0 A}{d} \tag{19.10}$$

Notice that only the geometry of the plates (A and d) and the dielectric constant κ affect the capacitance. With C_0 representing the capacitance of the empty capacitor ($\kappa = 1$), Equation 19.10 shows that $C = \kappa C_0$. In other words, the capacitance with the dielectric present is increased by a factor of κ over the capacitance without the dielectric. It can be shown that the relation $C = \kappa C_0$ applies to any capacitor, not just to a parallel plate capacitor. One reason, then, that capacitors are filled with dielectric materials is to increase the capacitance. Example 10 illustrates the effect that increasing the capacitance has on the charge stored by a capacitor.

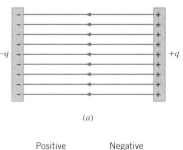

(a)

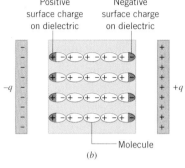

(b)

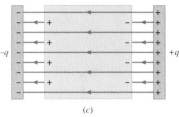

(c)

Figure 19.20 (*a*) The electric field lines inside an empty capacitor. (*b*) The electric field produced by the charges on the plates aligns the molecular dipoles within the dielectric end to end. (*c*) The surface charges on the dielectric reduce the electric field inside the dielectric. The space between the dielectric and the plates is added for clarity. In reality, the dielectric fills the region between the plates.

Table 19.1 *Dielectric Constants of Some Common Substances*[a]

Substance	Dielectric Constant, κ
Vacuum	1
Air	1.000 54
Teflon	2.1
Benzene	2.28
Paper (royal gray)	3.3
Ruby mica	5.4
Neoprene rubber	6.7
Methyl alcohol	33.6
Water	80.4

[a]Near room temperature.

Example 10 Storing Electric Charge

The capacitance of an empty capacitor is 1.2 μF. The capacitor is connected to a 12-V battery and charged up. With the capacitor connected to the battery, a slab of dielectric material is inserted between the plates. As a result, 2.6×10^{-5} C of *additional* charge flows from one plate, through the battery, and on to the other plate. What is the dielectric constant κ of the material?

Reasoning The charge stored by a capacitor is $q = CV$, according to Equation 19.8. The battery maintains a constant potential difference of $V = 12$ volts between the plates of the capacitor while the dielectric is inserted. Inserting the dielectric causes the capacitance C to increase, so that with V held constant, the charge q must increase. Thus, additional charge flows onto the plates. To find the dielectric constant, we apply Equation 19.8 to the empty capacitor and then to the capacitor filled with the dielectric material.

Solution The empty capacitor has a capacitance of $C_0 = 1.2$ μF and, according to Equation 19.8, stores an amount of charge $q_0 = C_0V$. With the dielectric material in place, the capacitor has a capacitance $C = \kappa C_0$ and stores an amount of charge $q = (\kappa C_0)V$. Thus, the additional charge that the battery supplies is

$$q - q_0 = (\kappa C_0)V - C_0V$$

Solving for the dielectric constant, we find that

$$\kappa = \frac{q - q_0}{C_0V} + 1 = \frac{2.6 \times 10^{-5} \text{ C}}{(1.2 \times 10^{-6} \text{ F})(12 \text{ V})} + 1 = \boxed{2.8}$$

In Example 10 the capacitor remains connected to the battery while the dielectric is inserted between the plates. The next example discusses what happens if the capacitor is disconnected from the battery before the dielectric is inserted.

Conceptual Example 11
The Effect of a Dielectric When a Capacitor Has a Constant Charge

An empty capacitor is connected to a battery and charged up. The capacitor is then disconnected from the battery, and a slab of dielectric material is inserted between the plates. Does the voltage across the plates increase, remain the same, or decrease?

Reasoning and Solution Our reasoning is guided by the following fact: once the capacitor is disconnected from the battery, the charge on its plates remains constant, for there is no longer any way for charge to be added or removed. According to Equation 19.8, the charge q stored by the capacitor is $q = CV$. Inserting the dielectric causes the capacitance C to increase. Therefore, *the voltage V across the plates must decrease in order for q to remain unchanged.* The amount by which the voltage decreases from the value initially established by the battery depends on the dielectric constant of the slab.

Related Homework: *Problem 46*

Capacitors are used often in electronic devices, and Example 12 deals with one familiar application.

Example 12 A Computer Keyboard

The physics of a computer keyboard.

One common kind of computer keyboard is based on the idea of capacitance. Each key is mounted on one end of a plunger, the other end being attached to a movable metal plate (see Figure 19.21). The movable plate is separated from a fixed plate, the two plates forming a capacitor. When the key is pressed, the movable plate is pushed closer to the fixed plate, and the capacitance increases. Electronic circuitry enables the computer to detect the *change* in capacitance, thereby recognizing which key has been pressed. The separation of the plates is normally 5.00×10^{-3} m but decreases to 0.150×10^{-3} m when a key is pressed. The plate area is 9.50×10^{-5} m^2, and the capacitor is filled with a material whose dielectric constant is 3.50. Determine the change in capacitance that is detected by the computer.

Reasoning We can use Equation 19.10 directly to find the capacitance of the key, since the dielectric constant κ, the plate area A, and the plate separation d are known. We will use this relation twice, once to find the capacitance when the key is pressed and once when it is not pressed. The change in capacitance will be the difference between these two values.

Solution When the key is pressed, the capacitance is

$$C = \frac{\kappa\epsilon_0 A}{d} = \frac{(3.50)[8.85 \times 10^{-12}\ \text{C}^2/(\text{N}\cdot\text{m}^2)](9.50 \times 10^{-5}\ \text{m}^2)}{0.150 \times 10^{-3}\ \text{m}}$$

$$= 19.6 \times 10^{-12}\ \text{F} \quad (19.6\ \text{pF}) \qquad (19.10)$$

A similar calculation reveals that when the key is *not* pressed, the capacitance has a value of 0.589×10^{-12} F (0.589 pF). The *change* in capacitance is an increase of $\boxed{19.0 \times 10^{-12}\ \text{F}\ (19.0\ \text{pF})}$. The *change* in the capacitance is greater with the dielectric present, which makes it easier for the circuitry within the computer to detect it. ▲

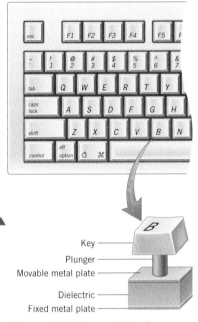

ENERGY STORAGE IN A CAPACITOR

When a capacitor stores charge, it also stores energy. In charging up a capacitor, for example, a battery does work in transferring an increment of charge from one plate of the capacitor to the other plate. The work done is equal to the product of the charge increment and the potential difference between the plates. However, as each increment of charge is moved, the potential difference increases slightly, and a larger amount of work is needed to move the next increment. The total work W done in completely charging the capacitor is the product of the total charge q transferred and the average potential difference $\bar{V}$; $W = q\bar{V}$. Since the average potential difference is one-half the final potential V, or $\bar{V} = \frac{1}{2}V$, the total work done by the battery is $W = \frac{1}{2}qV$. This work does not disappear but is stored as electric potential energy in the capacitor, so that EPE $= \frac{1}{2}qV$. Since $q = CV$, the energy stored becomes

$$\text{Energy} = \tfrac{1}{2}(CV)V = \tfrac{1}{2}CV^2 \qquad (19.11)$$

Figure 19.21 In one kind of computer keyboard, each key, when pressed, changes the separation between the plates of a capacitor.

Need more practice?

🖱 *Interactive LearningWare 19.2*
The bulb in a flashlight uses a power of 1.2 W. A 550-μF capacitor stores the energy that this bulb uses in ten minutes of operation. (a) What is the voltage across the plates of this capacitor and (b) what is the charge on one of those plates?

Related Homework: Problem 43

Go to **www.wiley.com/college/cutnell** for an interactive solution.

It is also possible to regard the energy as being stored in the electric field between the plates. The relation between energy and field strength can be obtained for a parallel plate capacitor by substituting $V = Ed$ (Equation 19.7 without the minus sign) and $C = \kappa\epsilon_0 A/d$ (Equation 19.10) into Equation 19.11:

$$\text{Energy} = \frac{1}{2}\left(\frac{\kappa\epsilon_0 A}{d}\right)(Ed)^2$$

Since the area A times the separation d is the volume between the plates, the energy per unit volume or **energy density** is

$$\text{Energy density} = \frac{\text{Energy}}{\text{Volume}} = \tfrac{1}{2}\kappa\epsilon_0 E^2 \qquad (19.12)$$

It can be shown that this expression is valid for any electric field strength, not just that between the plates of a capacitor.

The energy-storing capability of a capacitor is often put to good use in electronic circuits. For example, in an electronic flash attachment for a camera, energy from the battery pack is stored in a capacitor. The capacitor is then discharged between the electrodes of the flash tube, which converts the energy into light. Flash duration times range from

The physics of
**an electronic flash attachment
for a camera.**

1/200 to 1/1 000 000 second or less, with the shortest flashes being used in high-speed photography (see Figure 19.22). Some flash attachments automatically control the flash duration by monitoring the light reflected from the photographic subject and quickly stopping or quenching the capacitor discharge when the reflected light reaches a predetermined level.

During a heart attack, the heart produces a rapid, unregulated pattern of beats, a condition known as cardiac fibrillation. Cardiac fibrillation can often be stopped by sending a very fast discharge of electrical energy through the heart. Emergency medical personnel use defibrillators, such as the one shown in Figure 19.23. A paddle is connected to each plate of a large capacitor, and the paddles are placed on the chest near the heart. The capacitor is charged to a potential difference of about a thousand volts. The capacitor is then discharged in a few thousandths of a second; the discharge current passes through a paddle, the heart, and the other paddle. Within a few seconds, the heart often returns to its normal beating pattern.

The physics of a defibrillator.

Check Your Understanding 4

An empty parallel plate capacitor is connected to a battery that maintains a constant potential difference between the plates. With the battery connected, a dielectric is then inserted between the plates. Do the following quantities decrease, remain the same, or increase when the dielectric is inserted? (a) The electric field between the plates; (b) the capacitance; (c) the charge on the plates; (d) the energy stored in the capacitor. *(The answers are given at the end of the book.)*

Background: This question emphasizes the importance of the dielectric material in capacitors.

For similar questions (including calculational counterparts), consult Self-Assessment Test 19.2, which is described next.

Self-Assessment Test 19.2

Test your understanding of the material in Sections 19.4 and 19.5:

• Equipotential Surfaces and Their Relation to the Electric Field
• Capacitors and Dielectrics

Go to **www.wiley.com/college/cutnell**

Figure 19.22 This time-lapse photo of a bike rider was obtained using a camera with an electronic flash attachment. The energy for each "flash" comes from the electrical energy stored in a capacitor. (© Charlie Samuels/Corbis Images)

Figure 19.23 A defibrillator uses the electrical energy stored in a capacitor to deliver a controlled electric current that can restore normal heart rhythm in a heart attack victim. Airplane personnel are now being trained to use portable units such as the one shown here to save lives in the air. (Courtesy Heartstream Operation/Philips Medical Systems)

*19.6 Biomedical Applications of Electric Potential Differences

CONDUCTION OF ELECTRICAL SIGNALS IN NEURONS

The human nervous system is remarkable for its ability to transmit information in the form of electrical signals. These signals are carried by the nerves, and the concept of electric potential difference plays an important role in the process. For example, sensory information from our eyes and ears is carried to the brain by the optic nerve and auditory nerves, respectively. Other nerves transmit signals from the brain or spinal column to muscles, causing them to contract. Still other nerves carry signals within the brain.

A nerve consists of a bundle of *axons,* and each axon is one part of a nerve cell, or *neuron.* As Figure 19.24 illustrates, a typical neuron consists of a cell body with numerous extensions, called *dendrites,* and a single axon. The dendrites convert stimuli, such as pressure or heat, into electrical signals that travel through the neuron. The axon sends the signal to the nerve endings, which transmit the signal across a gap (called a *synapse*) to the next neuron or to a muscle.

The fluid inside a cell, the intracellular fluid, is quite different from that outside the cell, the extracellular fluid. Both fluids contain concentrations of positive and negative ions. However, the extracellular fluid is rich in sodium (Na^+) and chlorine (Cl^-) ions, and the intracellular fluid is rich in potassium (K$^+$) ions and negatively charged proteins. These concentration differences between the fluids are extremely important to the life of the cell. If the cell membrane were freely permeable, the ions would diffuse across it until the concentrations on both sides were equal. (See Section 14.4 for a review of diffusion.) This does not happen, because a living cell has a selectively permeable membrane. Ions can enter or leave the cell only through membrane channels, and the permeability of the channels varies markedly from one ion to another. For example, it is much easier for K$^+$ ions to diffuse out of the cell than it is for Na$^+$ to enter the cell. As a result of selective membrane permeability, there is a small buildup of negative charges just inside the membrane and an equal amount of positive charges on the outside (see Figure 19.25). The buildup of charge occurs very close to the membrane, so the membrane acts like a capacitor (see problems 38 and 51). Elsewhere in the intracellular and extracellular fluids, there are equal numbers of positive and negative ions, so the fluids are overall electrically neutral. Such a separation of positive and negative charges gives rise to an electric potential difference across the membrane, called the *resting membrane potential*. In neurons, the resting membrane potential ranges from −40 to −90 mV, with a typical value being −70 mV. The minus sign indicates that the inside of the membrane is negative relative to the outside.

A "resting" neuron is one that is not conducting an electrical signal. The *change* in the resting membrane potential is the key factor in the initiation and conduction of a signal. When a sufficiently strong stimulus is applied to a given point on the neuron, "gates" in the membrane open and sodium ions flood into the cell, as Figure 19.26 illustrates. The sodium ions are driven into the cell by attraction to the negative ions on the inside of the membrane as well as by the relatively high concentration of sodium ions outside the cell. The large influx of Na$^+$ ions first neutralizes the negative ions on the interior of the membrane and then causes it to become positively charged. As a result, the membrane potential in this localized region goes from −70 mV, the resting potential, to about +30 mV in a very short time (see Figure 19.27). The sodium gates then close, and the cell membrane quickly returns to its normal resting potential. This change in potential, from −70 mV to +30 mV and back to −70 mV, is known as the *action potential*. The action potential lasts for a few milliseconds, and it is the electrical signal that propagates down the axon, typically at a speed of about 50 m/s, to the next neuron or to a muscle cell.

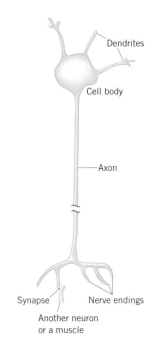

Figure 19.24 The anatomy of a typical neuron.

The physics of an action potential.

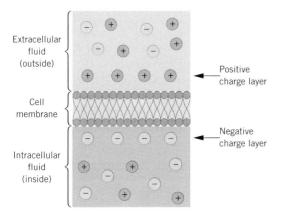

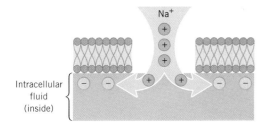

Figure 19.25 Positive and negative charge layers form on the outside and inside surfaces of a membrane during its resting state.

Figure 19.26 When a stimulus is applied to the cell, positive sodium ions (Na$^+$) rush into the cell, causing the interior surface of the membrane to become momentarily positive.

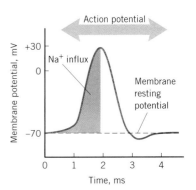

Figure 19.27 The action potential is caused by the rush of positive sodium ions into the cell and a subsequent return of the cell to its resting potential.

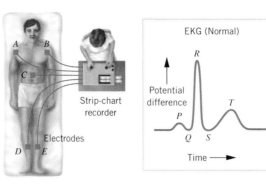

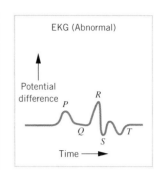

Figure 19.28 The potential differences generated by heart muscle activity provide the basis for electrocardiography. The normal and abnormal EKG patterns correspond to one heartbeat.

MEDICAL DIAGNOSTIC TECHNIQUES

Several important medical diagnostic techniques depend on the fact that the surface of the human body is *not* an equipotential surface. Between various points on the body there are small potential differences (approximately 30–500 μV), which provide the basis for electrocardiography, electroencephalography, and electroretinography. The potential differences can be traced to the electrical characteristics of muscle cells and nerve cells. In carrying out their biological functions, these cells utilize positively charged sodium and potassium ions and negatively charged chlorine ions that exist within the cells and in the intercellular fluid. As a result of such charged particles, electric fields are generated that extend to the surface of the body and lead to small potential differences.

The physics of **electrocardiography.**

Figure 19.28 shows some locations on the body where electrodes are placed to measure potential differences in electrocardiography. The potential difference between two locations changes as the heart beats and forms a repetitive pattern. The recorded pattern of potential difference versus time is called an electrocardiogram (ECG or EKG), and its shape depends on which pair of points in the picture (*A* and *B*, *B* and *C*, etc.) is used to locate the electrodes. The figure also shows some EKGs and indicates the regions (*P, Q, R, S,* and *T*) that can be associated with specific parts of the heart's beating cycle. The distinct differences between the EKGs of healthy (normal) and damaged (abnormal) hearts provide physicians with a valuable diagnostic tool.

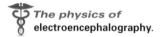
The physics of **electroencephalography.**

In electroencephalography the electrodes are placed at specific locations on the head, as Figure 19.29 indicates, and they record the potential differences that characterize brain behavior. The graph of potential difference versus time is known as an electroencephalogram (EEG). The various parts of the patterns in an EEG are often referred to as "waves" or "rhythms." The drawing shows an example of the main resting rhythm of the brain, the so-called alpha rhythm, and also illustrates the distinct differences that are found between the EEGs generated by healthy (normal) and diseased (abnormal) tissue.

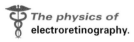
The physics of **electroretinography.**

The electrical characteristics of the retina of the eye lead to the potential differences measured in electroretinography. Figure 19.30 shows a typical electrode placement used to record the pattern of potential difference versus time that occurs when the eye is stimulated by a flash of light. One electrode is mounted on a contact lens, while the other is often placed on the forehead. The recorded pattern is called an electroretinogram (ERG), parts of the pattern being referred to as the "*A* wave" and the "*B* wave." As the graphs show, the ERGs of normal and diseased (abnormal) eyes can differ markedly.

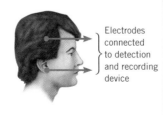

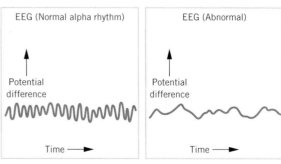

Figure 19.29 In electroencephalography the potential differences created by the electrical activity of the brain are used for diagnosing abnormal behavior.

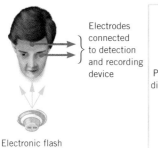
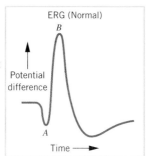
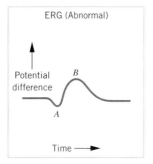

Electrodes connected to detection and recording device

Electronic flash

Potential difference

ERG (Normal)

Time →

Potential difference

ERG (Abnormal)

Time →

Figure 19.30 The electrical activity of the retina of the eye generates the potential differences used in electroretinography.

19.7 *Concepts & Calculations*

The conservation of energy (Chapter 6) and the conservation of linear momentum (Chapter 7) are two of the most broadly applicable principles in all of science. In this chapter, we have seen that electrically charged particles obey the conservation-of-energy principle, provided that the electric potential energy is taken into account. The behavior of electrically charged particles, however, must also be consistent with the conservation-of-momentum principle, as the first example in this section emphasizes.

Concepts & Calculations Example 13 Conservation Principles

Particle 1 has a mass of $m_1 = 3.6 \times 10^{-6}$ kg, while particle 2 has a mass of $m_2 = 6.2 \times 10^{-6}$ kg. Each has the same electric charge. These particles are initially held at rest, and the two-particle system has an initial electric potential energy of 0.150 J. Suddenly, the particles are released and fly apart because of the repulsive electric force that acts on each one (see Figure 19.31). The effects of the gravitational force are negligible, and no other forces act on the particles. At one instant following the release, the speed of particle 1 is measured to be $v_1 = 170$ m/s. What is the electric potential energy of the two-particle system at this instant?

Concept Questions and Answers What type of energy does the two-particle system have initially?

> *Answer* Initially, the particles are at rest, so they have no kinetic energy. However, they do have electric potential energy. They also have gravitational potential energy, but it is negligible.

What type of energy does the two-particle system have at the instant illustrated in part *b* of the drawing?

> *Answer* Since each particle is moving, each has kinetic energy. The two-particle system also has electric potential energy at this instant. Gravitational potential energy remains negligible.

Does the principle of conservation of energy apply?

> *Answer* Yes. Since the gravitational force is negligible, the only force acting here is the conservative electric force. Nonconservative forces are absent. Thus, the principle of conservation of energy applies.

Does the principle of conservation of linear momentum apply to the two particles as they fly apart?

> *Answer* Yes. This principle states that the total linear momentum of an isolated system remains constant. An isolated system is one for which the vector sum of the external forces acting on the system is zero. There are no external forces acting here. The only appreciable force acting on each of the two particles is the force of electric repulsion, which is an internal force.

Solution The conservation-of-energy principle indicates that the total energy of the two-particle system is the same at the later instant as it was initially. Considering that only kinetic energy and electric potential energy are present, the principle can be stated as follows:

$$\underbrace{\text{KE}_f + \text{EPE}_f}_{\text{Final total energy}} = \underbrace{\text{KE}_0 + \text{EPE}_0}_{\text{Initial total energy}} \quad \text{or} \quad \text{EPE}_f = \text{EPE}_0 - \text{KE}_f$$

q q

m_1 m_2

(*a*) Initial (at rest)

v_{1f}

q

m_1

v_{2f}

q

m_2

(*b*) Final

Figure 19.31 (*a*) Two particles have different masses, but the same electric charge q. They are initially held at rest. (*b*) At an instant following the release of the particles, they are flying apart due to the mutual force of electric repulsion.

We have used the fact that KE_0 is zero, since the particles are initially at rest. The final kinetic energy of the system is $KE_f = \frac{1}{2}m_1v_{f1}^2 + \frac{1}{2}m_2v_{f2}^2$. In this expression, we have values for both masses and the speed v_{f1}. We can find the speed v_{f2} by using the principle of conservation of momentum:

$$\underbrace{m_1v_{f1} + m_2v_{f2}}_{\substack{\text{Final total}\\\text{momentum}}} = \underbrace{m_1v_{01} + m_2v_{02}}_{\substack{\text{Initial total}\\\text{momentum}}} \quad \text{or} \quad v_{f2} = -\frac{m_1}{m_2}v_{f1}$$

We have again used the fact that the particles are initially at rest, so that $v_{01} = v_{02} = 0$ m/s. Using this result for v_{f2}, we can now obtain the final electric potential energy from the conservation-of-energy equation:

$$EPE_f = EPE_0 - \left[\frac{1}{2}m_1v_{f1}^2 + \frac{1}{2}m_2\left(-\frac{m_1}{m_2}v_{f1}\right)^2\right]$$

$$= EPE_0 - \frac{1}{2}m_1\left(1 + \frac{m_1}{m_2}\right)v_{f1}^2$$

$$= 0.150\ \text{J} - \frac{1}{2}(3.6 \times 10^{-6}\ \text{kg})\left(1 + \frac{3.6 \times 10^{-6}\ \text{kg}}{6.2 \times 10^{-6}\ \text{kg}}\right)(170\ \text{m/s})^2$$

$$= \boxed{0.068\ \text{J}}$$

Chapter 18 introduces the electric field, and the present chapter introduces the electric potential. These two concepts are central to the study of electricity, and it is important to distinguish between them, as the next example emphasizes.

Concepts & Calculations Example 14
Electric Field and Electric Potential

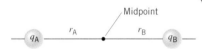

Figure 19.32 Example 14 determines the electric field and the electric potential at the midpoint ($r_A = r_B$) between the identical charges ($q_A = q_B$).

Two identical point charges ($+2.4 \times 10^{-9}$ C) are fixed in place, separated by 0.50 m. (see Figure 19.32). Find the electric field and the electric potential at the midpoint of the line between the charges q_A and q_B.

Concept Questions and Answers The electric field is a vector and has a direction. At the midpoint, what are the directions of the individual electric-field contributions from q_A and q_B?

Answer   Since both charges are positive, the individual electric-field contributions from q_A and q_B point away from each charge. Thus, at the midpoint they point in opposite directions.

Is the magnitude of the net electric field at the midpoint greater than, less than, or equal to zero?

Answer   The magnitude of the net electric field is zero. This is because the electric field from q_A cancels the electric field from q_B. These charges have the same magnitude q, and the midpoint is the same distance r from each of them. According to $E = k|q|/r^2$ (Equation 18.3), then, each charge produces a field of the same strength. Since the individual electric-field contributions from q_A and q_B point in opposite directions at the midpoint, their vector sum is zero.

Is the total electric potential at the midpoint positive, negative, or zero?

Answer   The total electric potential at the midpoint is the algebraic sum of the individual contributions from each charge. According to $V = kq/r$ (Equation 19.6), each contribution is positive, since each charge is positive. Thus, the total electric potential is also positive.

Does the total electric potential have a direction associated with it?

Answer   No. The electric potential is a scalar quantity, not a vector quantity. Therefore, it has no direction associated with it.

Solution As discussed in the second concept question, the total electric field at the midpoint is $\boxed{E = 0\ \text{N/C}}$. Using $V = kq/r$ (Equation 19.6), we find that the total potential at the midpoint is

$$V_{\text{Total}} = \frac{kq_A}{r_A} + \frac{kq_B}{r_B}$$

Since $q_A = q_B = +2.4 \times 10^{-9}$ C and $r_A = r_B = 0.25$ m, we have

$$V = \frac{2kq_A}{r_A} = \frac{2(8.99 \times 10^9 \text{ N} \cdot \text{m}^2/\text{C}^2)(+2.4 \times 10^{-9} \text{ C})}{0.25 \text{ m}} = \boxed{170 \text{ V}}$$

▲

At the end of the problem set for this chapter, you will find homework problems that contain both conceptual and quantitative parts. These problems are grouped under the heading *Concepts & Calculations, Group Learning Problems.* They are designed for use by students working alone or in small learning groups. The conceptual part of each problem provides a convenient focus for group discussions.

Concept Summary

This summary presents an abridged version of the chapter, including the important equations and all available learning aids. For convenient reference, the learning aids (including the text's examples) are placed next to or immediately after the relevant equation or discussion. The following learning aids may be found on-line at **www.wiley.com/college/cutnell**:

Interactive LearningWare examples are solved according to a five-step interactive format that is designed to help you develop problem-solving skills.	**Concept Simulations** are animated versions of text figures or animations that illustrate important concepts. You can control parameters that affect the display, and we encourage you to experiment.
Interactive Solutions offer specific models for certain types of problems in the chapter homework. The calculations are carried out interactively.	**Self-Assessment Tests** include both qualitative and quantitative questions. Extensive feedback is provided for both incorrect and correct answers, to help you evaluate your understanding of the material.

Topic	*Discussion*	*Learning Aids*

19.1 Potential Energy

Work and electric potential energy

When a positive test charge $+q_0$ moves from point A to point B in an electric field, work W_{AB} is done by the electric force. The work equals the electric potential energy (EPE) at A minus that at B:

$$W_{AB} = \text{EPE}_A - \text{EPE}_B \qquad (19.1)$$

Path independence

The electric force is a conservative force, so the path along which the test charge moves from A to B is of no consequence, for the work W_{AB} is the same for all paths.

19.2 The Electric Potential Difference

The electric potential V at a given point is the electric potential energy of a small test charge q_0 situated at that point divided by the charge itself:

Electric potential

$$V = \frac{\text{EPE}}{q_0} \qquad (19.3)$$

The SI unit of electric potential is the joule per coulomb (J/C) or volt (V).

The electric potential difference between two points A and B is

Electric potential difference

$$V_B - V_A = \frac{\text{EPE}_B}{q_0} - \frac{\text{EPE}_A}{q_0} = \frac{-W_{AB}}{q_0} \qquad (19.4)$$

Examples 1, 3
Interactive LearningWare 19.1
Interactive Solution 19.7

Acceleration of positive and negative charges

A positive charge accelerates from a region of higher potential toward a region of lower potential. Conversely, a negative charge accelerates from a region of lower potential toward a region of higher potential.

Example 2

Electron volt

An electron volt (eV) is a unit of energy. The relation between electron volts and joules is 1 eV $= 1.60 \times 10^{-19}$ J.

The total energy E of a system is the sum of its translational ($\frac{1}{2}mv^2$) and rotational ($\frac{1}{2}I\omega^2$) kinetic energies, gravitational potential energy (mgh), elastic potential energy ($\frac{1}{2}kx^2$), and electric potential energy (EPE):

Total energy

$$E = \tfrac{1}{2}mv^2 + \tfrac{1}{2}I\omega^2 + mgh + \tfrac{1}{2}kx^2 + \text{EPE}$$

Topic	Discussion	Learning Aids
Conservation of energy	If external nonconservative forces like friction do no net work, the total energy of the system is conserved. That is, the final total energy E_f is equal to the initial total energy E_0; $E_f = E_0$.	Examples 4, 13

19.3 The Electric Potential Difference Created by Point Charges

The electric potential V at a distance r from a point charge q is

Electric potential of a point charge	$$V = \frac{kq}{r} \qquad (19.6)$$	Example 5 Concept Simulation 19.1 Interactive Solution 19.21

where $k = 8.99 \times 10^9$ N·m²/C². This expression for V assumes that the electric potential is zero at an infinite distance away from the charge.

Total electric potential	The total electric potential at a given location due to two or more charges is the algebraic sum of the potentials due to each charge.	Examples 6, 7, 14
Total potential energy of a group of charges	The total potential energy of a group of charges is the amount by which the electric potential energy of the group differs from its initial value when the charges are infinitely far apart. It is also equal to the work required to assemble the group, one charge at a time, starting with the charges infinitely far apart.	Example 8

 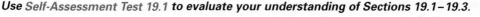 **Use *Self-Assessment Test 19.1* to evaluate your understanding of Sections 19.1–19.3.**

19.4 Equipotential Surfaces and Their Relation to the Electric Field

Equipotential surface	An equipotential surface is a surface on which the electric potential is the same everywhere. The electric force does no work as a charge moves on an equipotential surface, because the force is always perpendicular to the displacement of the charge.	
	The electric field created by any group of charges is everywhere perpendicular to the associated equipotential surfaces and points in the direction of decreasing potential.	

The electric field is related to two equipotential surfaces by

Relation between the electric field and the potential gradient	$$E = -\frac{\Delta V}{\Delta s} \qquad (19.7)$$	Example 9 Interactive Solution 19.33

where ΔV is the potential difference between the surfaces and Δs is the displacement. The term $\Delta V/\Delta s$ is called the potential gradient.

19.5 Capacitors and Dielectrics

A capacitor	A capacitor is a device that stores charge and energy. It consists of two conductors or plates that are near one another, but not touching. The magnitude q of the charge on each plate is given by	
Relation between charge and potential difference	$$q = CV \qquad (19.8)$$	

where V is the magnitude of the potential difference between the plates and C is the capacitance. The SI unit for capacitance is the coulomb per volt (C/V) or farad (F).

The insulating material included between the plates of a capacitor is called a dielectric. The dielectric constant κ of the material is defined as

Dielectric constant	$$\kappa = \frac{E_0}{E} \qquad (19.9)$$	

where E_0 and E are, respectively, the magnitudes of the electric fields between the plates without and with a dielectric, assuming the charge on the plates is kept fixed.

The capacitance of a parallel plate capacitor filled with a dielectric is

Capacitance of a parallel plate capacitor	$$C = \frac{\kappa \epsilon_0 A}{d} \qquad (19.10)$$	Examples 10,11,12 Interactive Solution 19.57

where $\epsilon_0 = 8.85 \times 10^{-12}$ C²/(N·m²) is the permittivity of free space, A is the area of each plate, and d is the distance between the plates.

Topic	Discussion	Learning Aids
Energy stored in a capacitor	The electric potential energy stored in a capacitor is $$\text{Energy} = \tfrac{1}{2}CV^2 \qquad (19.11)$$	Interactive LearningWare 19.2 Interactive Solution 19.41
Energy density	The energy density is the energy stored per unit volume and is related to the magnitude E of the electric field as follows: $$\text{Energy density} = \tfrac{1}{2}\kappa\epsilon_0 E^2 \qquad (19.12)$$	

 Use *Self-Assessment Test 19.2* to evaluate your understanding of Sections 19.4 and 19.5.

Conceptual Questions

1. The drawing shows three possibilities for the potentials at two points, A and B. In each case, the same positive charge is moved from A to B. In which case, if any, is the most work done on the positive charge by the electric force? Account for your answer.

A $\quad$ B	A $\quad$ B	A $\quad$ B
150 V $\quad$ 100 V	25 V $\quad$ −25 V	−10 V $\quad$ −60 V
Case 1	Case 2	Case 3

2. A positive point charge and a negative point charge have equal magnitudes. One charge is fixed to one corner of a square, and the other is fixed to another corner. On which corners should the charges be placed, so that the same potential exists at the empty corners? Give your reasoning.

3. Three point charges have identical magnitudes, but two of the charges are positive and one is negative. These charges are fixed to the corners of a square, one to a corner. No matter how the charges are arranged, the potential at the empty corner is positive. Explain why.

4. What point charges, all having the same magnitude, would you place at the corners of a square (one charge per corner), so that both the electric field and the electric potential (assuming a zero reference value at infinity) are zero at the center of the square? Account for the fact that the charge distribution gives rise to *both* a zero field and a zero potential.

5. Positive charge is spread uniformly around a circular ring, as the drawing illustrates. Equation 19.6 gives the correct potential at points along the line perpendicular to the plane of the ring at its center. However, the equation does not give the correct potential at points that do not lie on this line. In the equation, q represents the total charge on the ring. Why does Equation 19.6 apply for points on the line, but not for points off the line?

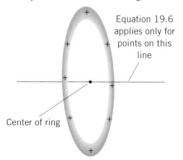

Equation 19.6 applies only for points on this line

Center of ring

6. The electric field at a single location is zero. Does this fact necessarily mean that the electric potential at the same place is zero? Use a spot on the line between two identical point charges as an example to support your reasoning.

7. An electric potential energy exists when two protons are separated by a certain distance. Does the electric potential energy increase, decrease, or remain the same when (a) both protons are replaced by electrons, and (b) only one of the protons is replaced by an electron? Justify your answers.

8. A proton is fixed in place. An electron is released from rest and allowed to collide with the proton. Then the roles of the proton and electron are interchanged, and the same experiment is repeated. Which is traveling faster when the collision occurs, the proton or the electron? Justify your answer.

9. The potential is constant throughout a given region of space. Is the electric field zero or nonzero in this region? Explain.

10. In a region of space where the electric field is constant everywhere, as it is inside a parallel plate capacitor, is the potential constant everywhere? Account for your answer.

11. A positive test charge is placed in an electric field. In what direction should the charge be moved relative to the field, such that the charge experiences a constant electric potential? Explain.

12. The location marked P in the drawing lies midway between the point charges $+q$ and $-q$. The blue lines labeled A, B, and C are edge-on views of three planes. Which one of these planes is an equipotential surface? Why?

Question 12

13. Imagine that you are moving a positive test charge along the line between two identical point charges. With regard to the electric potential, is the midpoint on the line analogous to the top of a mountain or the bottom of a valley when the two point charges are (a) positive and (b) negative? In each case, explain your answer.

14. Repeat question 13, assuming that you are moving a negative instead of a positive test charge.

15. The potential at a point in space has a certain value, which is not zero. Is the electric potential energy the same for every charge that is placed at that point? Give your reasoning.

16. A proton and an electron are released from rest at the midpoint between the plates of a charged parallel plate capacitor. Except for these particles, nothing else is between the plates. Ignore the attraction between the proton and the electron, and decide which particle strikes a capacitor plate first. Why?

17. A parallel plate capacitor is charged up by a battery. The battery is then disconnected, but the charge remains on the plates. The plates are then pulled apart. Explain whether each of the following quantities increases, decreases, or remains the same as the distance between the plates increases: (a) the capacitance of the capacitor, (b) the potential difference between the plates, (c) the electric field between the plates, and (d) the electric potential energy stored by the capacitor. Give reasons for your answers.

Problems

Note: All charges are assumed to be point charges unless specified otherwise.

ssm　Solution is in the Student Solutions Manual.　　**www**　Solution is available on the World Wide Web at **www.wiley.com/college/cutnell**

　This icon represents a biomedical application.

Section 19.1 Potential Energy,
Section 19.2 The Electric Potential Difference

1.　**ssm** Suppose that the electric potential outside a living cell is higher than that inside the cell by 0.070 V. How much work is done by the electric force when a sodium ion (charge = $+e$) moves from the outside to the inside?

2. An electric force moves a charge of $+1.80 \times 10^{-4}$ C from point A to point B and performs 5.80×10^{-3} J of work on the charge. (a) What is the difference ($EPE_A - EPE_B$) between the electric potential energies of the charge at the two points? (b) Determine the potential difference ($V_A - V_B$) between the two points. (c) State which point is at the higher potential.

3. The anode (positive terminal) of an X-ray tube is at a potential of $+125\ 000$ V with respect to the cathode (negative terminal). (a) How much work (in joules) is done by the electric force when an electron is accelerated from the cathode to the anode? (b) If the electron is initially at rest, what kinetic energy does the electron have when it arrives at the anode?

4. During a particular thunderstorm, the electric potential difference between a cloud and the ground is $V_{cloud} - V_{ground} = 1.3 \times 10^8$ V, with the cloud being at the higher potential. What is the change in an electron's electric potential energy when the electron moves from the ground to the cloud?

5.　**ssm** In a television picture tube, electrons strike the screen after being accelerated from rest through a potential difference of $25\ 000$ V. The speeds of the electrons are quite large, and for accurate calculations of the speeds, the effects of special relativity must be taken into account. Ignoring such effects, find the electron speed just before the electron strikes the screen.

6. Point A is at a potential of $+250$ V, and point B is at a potential of -150 V. An α-particle is a helium nucleus that contains two protons and two neutrons; the neutrons are electrically neutral. An α-particle starts from rest at A and accelerates toward B. When the α-particle arrives at B, what kinetic energy (in electron volts) does it have?

7. Consult **Interactive Solution 19.7** at **www.wiley.com/college/cutnell** to review one method for modeling this problem. An electric car accelerates for 7.0 s by drawing energy from its 290-V battery pack. During this time, 1200 C of charge pass through the battery pack. Find the minimum horsepower rating of the car.

* 8. A typical 12-V car battery can deliver about 7.5×10^5 C of charge before dying. This is not very much. To get a feel for this, calculate the maximum number of kilograms of water (100 °C) that could be boiled into steam (100 °C) using energy from this battery.

* 9.　**ssm www** The potential at location A is 452 V. A positively charged particle is released there from rest and arrives at location B with a speed v_B. The potential at location C is 791 V, and when released from rest from this spot, the particle arrives at B with twice the speed it previously had, or $2v_B$. Find the potential at B.

** 10. A particle is uncharged and is thrown vertically upward from ground level with a speed of 25.0 m/s. As a result, it attains a max-

imum height h. The particle is then given a positive charge $+q$ and reaches the same maximum height h when thrown vertically upward with a speed of 30.0 m/s. The electric potential at the height h exceeds the electric potential at ground level. Finally, the particle is given a negative charge $-q$. Ignoring air resistance, determine the speed with which the negatively charged particle must be thrown vertically upward, so that it attains exactly the maximum height h. In all three situations, be sure to include the effect of gravity.

Section 19.3 The Electric Potential Difference Created by Point Charges

11.　**ssm** There is an electric potential of $+130$ V at a spot that is 0.25 m away from a charge. Find the magnitude and sign of the charge.

12. Two point charges, $+3.40$ μC and -6.10 μC, are separated by 1.20 m. What is the electric potential midway between them?

13. An electron and a proton are initially very far apart (effectively an infinite distance apart). They are then brought together to form a hydrogen atom, in which the electron orbits the proton at an average distance of 5.29×10^{-11} m. What is $EPE_{final} - EPE_{initial}$, which is the change in the electric potential energy?

14. Location A is 3.00 m to the right of a point charge q. Location B lies on the same line and is 4.00 m to the right of the charge. The potential difference between the two locations is $V_B - V_A = 45.0$ V. What is the magnitude and sign of the charge?

15.　**ssm www** Two identical point charges are fixed to diagonally opposite corners of a square that is 0.500 m on a side. Each charge is $+3.0 \times 10^{-6}$ C. How much work is done by the electric force as one of the charges moves to an empty corner?

16. The drawing shows four point charges. The value of q is 2.0 μC, and the distance d is 0.96 m. Find the total potential at the location P. Assume that the potential of a point charge is zero at infinity.

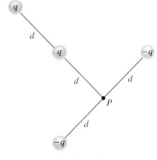

Problem 16

17. A charge of $+9q$ is fixed to one corner of a square, while a charge of $-8q$ is fixed to the diagonally opposite corner. Expressed in terms of q, what charge should be fixed to the center of the square, so the potential is zero at each of the two empty corners?

18. Review Conceptual Example 7 as background for this problem. Two charges are fixed in place with a separation d. One charge is positive and has twice the magnitude of the other charge, which is negative. The positive charge lies to the left of the negative charge, as in Figure 19.11. Relative to the negative charge, locate the two spots on the line through the charges where the total potential is zero.

* **19. ssm** Determine the electric potential energy for the array of three charges shown in the drawing, relative to its value when the charges are infinitely far away.

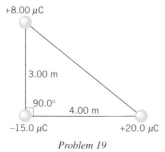

Problem 19

* **20.** Four identical charges (+2.0 μC each) are brought from infinity and fixed to a straight line. The charges are located 0.40 m apart. Determine the electric potential energy of this group.

* **21.** Refer to **Interactive Solution 19.21** at **www.wiley.com/college/ cutnell** to review one way in which this problem can be solved. Two protons are moving directly toward one another. When they are very far apart, their initial speeds are 3.00×10^6 m/s. What is the distance of closest approach?

* **22.** Identical point charges of +1.7 μC are fixed to diagonally opposite corners of a square. A third charge is then fixed at the center of the square, such that it causes the potentials at the empty corners to change signs without changing magnitudes. Find the sign and magnitude of the third charge.

* **23. ssm** A charge of −3.00 μC is fixed in place. From a horizontal distance of 0.0450 m, a particle of mass 7.20×10^{-3} kg and charge −8.00 μC is fired with an initial speed of 65.0 m/s directly toward the fixed charge. How far does the particle travel before its speed is zero?

** **24.** A positive charge of $+q_1$ is located 3.00 m to the left of a negative charge $-q_2$. The charges have different magnitudes. On the line through the charges, the net *electric field* is zero at a spot 1.00 m to the right of the negative charge. On this line there are also two spots where the potential is zero. Locate these two spots relative to the negative charge.

** **25.** Charges q_1 and q_2 are fixed in place, q_2 being located at a distance d to the right of q_1. A third charge q_3 is then fixed to the line joining q_1 and q_2 at a distance d to the right of q_2. The third charge is chosen so the potential energy of the group is zero; that is, the potential energy has the same value as that of the three charges when they are widely separated. Determine q_3, assuming that (a) $q_1 = q_2 = q$ and (b) $q_1 = q$ and $q_2 = -q$. Express your answers in terms of q.

** **26.** One particle has a mass of 3.00×10^{-3} kg and a charge of +8.00 μC. A second particle has a mass of 6.00×10^{-3} kg and the same charge. The two particles are initially held in place and then released. The particles fly apart, and when the separation between them is 0.100 m, the speed of the 3.00×10^{-3}-kg particle is 125 m/s. Find the initial separation between the particles.

Section 19.4 Equipotential Surfaces and Their Relation to the Electric Field

27. ssm An equipotential surface that surrounds a $+3.0 \times 10^{-7}$-C point charge has a radius of 0.15 m. What is the potential of this surface?

28. Two equipotential surfaces surround a $+1.50 \times 10^{-8}$-C point charge. How far is the 190-V surface from the 75.0-V surface?

29. A spark plug in an automobile engine consists of two metal conductors that are separated by a distance of 0.75 mm. When an electric spark jumps between them, the magnitude of the electric field is 4.7×10^7 V/m. What is the magnitude of the potential difference ΔV between the conductors?

30. The inner and outer surfaces of a cell membrane carry a negative and positive charge, respectively. Because of these charges, a potential difference of about 0.070 V exists across the membrane. The thickness of the membrane is 8.0×10^{-9} m. What is the magnitude of the electric field in the membrane?

31. ssm www When you walk across a rug on a dry day, your body can become electrified, and its electric potential can change. When the potential becomes large enough, a spark of negative charges can jump between your hand and a metal surface. A spark occurs when the electric field strength created by the charges on your body reaches the dielectric strength of the air. The dielectric strength of the air is 3.0×10^6 N/C and is the electric field strength at which the air suffers electrical breakdown. Suppose a spark 3.0 mm long jumps between your hand and a metal doorknob. Assuming that the electric field is uniform, find the potential difference ($V_{knob} - V_{hand}$) between your hand and the doorknob.

* **32.** The drawing shows a uniform electric field that points in the negative y direction; the magnitude of the field is 3600 N/C. Determine the electric potential difference (a) $V_B - V_A$ between points A and B, (b) $V_C - V_B$ between points B and C, and (c) $V_A - V_C$ between points C and A.

* **33.** Refer to **Interactive Solution 19.33** at **www.wiley.com/ college/cutnell** to review a method by which this problem can be solved. The electric field has a constant value of 4.0×10^3 V/m and is directed downward. The field is the same everywhere. The potential at a point P within this region is 155 V. Find the potential at the following points: (a) 6.0×10^{-3} m directly above P, (b) 3.0×10^{-3} m directly below P, (c) 8.0×10^{-3} m directly to the right of P.

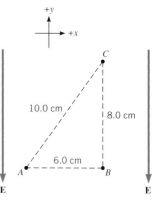

Problem 32

* **34.** The drawing shows the electric potential as a function of distance along the x axis. Determine the magnitude of the electric field in the region (a) A to B, (b) B to C, and (c) C to D.

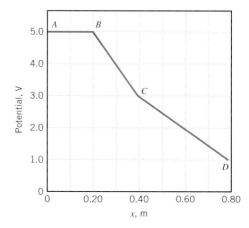

* **35. ssm** Equipotential surface A has a potential of 5650 V, while equipotential surface B has a potential of 7850 V. A particle has a mass of 5.00×10^{-2} kg and a charge of $+4.00 \times 10^{-5}$ C. The particle has a speed of 2.00 m/s on surface A. An outside force is applied to the particle, and it moves to surface B, arriving there with a speed of 3.00 m/s. How much work is done by the outside force in moving the particle from A to B?

Section 19.5 Capacitors and Dielectrics

36. What voltage is required to store 7.2×10^{-5} C of charge on the plates of a 6.0-μF capacitor?

37. ssm The electric potential energy stored in the capacitor of a defibrillator is 73 J, and the capacitance is 120 μF. What is the potential difference across the capacitor plates?

38. An axon is the relatively long tail-like part of a neuron, or nerve cell. The outer surface of the axon membrane (dielectric constant = 5, thickness = 1×10^{-8} m) is charged positively, and the inner portion is charged negatively. Thus, the membrane is a kind of capacitor. Assuming that an axon can be treated like a parallel plate capacitor with a plate area of 5×10^{-6} m², what is its capacitance?

39. ssm A parallel plate capacitor has a capacitance of 7.0 μF when filled with a dielectric. The area of each plate is 1.5 m² and the separation between the plates is 1.0×10^{-5} m. What is the dielectric constant of the dielectric?

40. A capacitor has a capacitance of 2.5×10^{-8} F. In the charging process, electrons are removed from one plate and placed on the other plate. When the potential difference between the plates is 450 V, how many electrons have been transferred?

41. Refer to **Interactive Solution 19.41** at **www.wiley.com/college/cutnell** for one approach to this problem. The electronic flash attachment for a camera contains a capacitor for storing the energy used to produce the flash. In one such unit, the potential difference between the plates of an 850-μF capacitor is 280 V. (a) Determine the energy that is used to produce the flash in this unit. (b) Assuming that the flash lasts for 3.9×10^{-3} s, find the effective power or "wattage" of the flash.

42. Two capacitors are identical, except that one is empty and the other is filled with a dielectric ($\kappa = 4.50$). The empty capacitor is connected to a 12.0-V battery. What must be the potential difference across the plates of the capacitor filled with a dielectric such that it stores the same amount of electrical energy as the empty capacitor?

* **43.** ssm **Interactive LearningWare 19.2** at **www.wiley.com/college/cutnell** reviews the concepts pertinent to this problem.

What is the potential difference between the plates of a 3.3-F capacitor that stores sufficient energy to operate a 75-W light bulb for one minute?

* **44.** The dielectric strength of an insulating material is the maximum electric field strength to which the material can be subjected without electrical breakdown occurring. Suppose a parallel plate capacitor is filled with a material whose dielectric constant is 3.5 and whose dielectric strength is 1.4×10^7 N/C. If this capacitor is to store 1.7×10^{-7} C of charge on each plate without suffering breakdown, what must be the radius of its circular plates?

* **45.** Two hollow metal spheres are concentric with each other. The inner sphere has a radius of 0.1500 m and a potential of 85.0 V. The radius of the outer sphere is 0.1520 m and its potential is 82.0 V. If the region between the spheres is filled with Teflon, find the electric energy contained in this space.

* **46.** Review Conceptual Example 11 before attempting this problem. An empty capacitor is connected to a 12.0-V battery and charged up. The capacitor is then disconnected from the battery, and a slab of dielectric material ($\kappa = 2.8$) is inserted between the plates. Find the amount by which the potential difference across the plates changes. Specify whether the change is an increase or a decrease.

** **47.** ssm The drawing shows a parallel plate capacitor. One-half of the region between the plates is filled with a material that has a dielectric constant κ_1. The other half is filled with a material that has a dielectric constant κ_2. The area of each plate is A, and the plate separation is d. The potential difference across the plates is V. Note especially that the charge stored by the capacitor is $q_1 + q_2 = CV$, where q_1 and q_2 are the charges on the area of the plates in contact with materials 1 and 2, respectively. Show that $C = \epsilon_0 A(\kappa_1 + \kappa_2)/(2d)$.

** **48.** The plate separation of a charged capacitor is 0.0800 m. A proton and an electron are released from rest at the midpoint between the plates. Ignore the attraction between the two particles, and determine how far the proton has traveled by the time the electron strikes the positive plate.

Additional Problems

49. Two charges A and B are fixed in place, at different distances from a certain spot. At this spot the potentials due to the two charges are equal. Charge A is 0.18 m from the spot, while charge B is 0.43 m from it. Find the ratio q_B/q_A of the charges.

50. A particle has a charge of $+1.5$ μC and moves from point A to point B, a distance of 0.20 m. The particle experiences a constant electric force, and its motion is along the line of action of the force. The difference between the particle's electric potential energy at A and B is $EPE_A - EPE_B = +9.0 \times 10^{-4}$ J. (a) Find the magnitude and direction of the electric force that acts on the particle. (b) Find the magnitude and direction of the electric field that the particle experiences.

51. ssm The membrane that surrounds a certain type of living cell has a surface area of 5.0×10^{-9} m² and a thickness of 1.0×10^{-8} m. Assume that the membrane behaves like a parallel plate capacitor and has a dielectric constant of 5.0. (a) The potential on the outer surface of the membrane is $+60.0$ mV greater than that on the inside surface. How much charge resides on the outer surface? (b) If the charge in part (a) is due to K$^+$ ions (charge $+e$), how many such ions are present on the outer surface?

52. A particle with a charge of -1.5 μC and a mass of 2.5×10^{-6} kg is released from rest at point A and accelerates toward point

B, arriving there with a speed of 42 m/s. (a) What is the potential difference $V_B - V_A$ between A and B? (b) Which point is at the higher potential? Give your reasoning.

53. A capacitor stores 5.3×10^{-5} C of charge when connected to a 6.0-V battery. How much charge does the capacitor store when connected to a 9.0-V battery?

54. The drawing shows six point charges arranged in a rectangle. The value of q is 9.0 μC, and the distance d is 0.13 m. Find the total electric potential at location P, which is at the center of the rectangle.

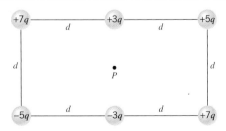

55. ssm Two points, A and B, are separated by 0.016 m. The potential at A is $+95$ V, and that at B is $+28$ V. Find the magnitude and direction of the constant electric field between the points.

* **56.** The drawing shows the potential at five points on a set of axes. Each of the four outer points is 6.0×10^{-3} m from the point at the origin. From the data shown, find the magnitude and direction of the electric field in the vicinity of the origin.

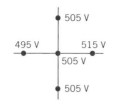

* **57.** Refer to **Interactive Solution 19.57** at **www.wiley.com/college/cutnell** for help in solving this problem. An empty capacitor has a capacitance of 3.2 μF and is connected to a 12-V battery. A dielectric material ($\kappa = 4.5$) is inserted between the plates of this capacitor. What is the magnitude of the surface charge on the dielectric that is adjacent to either plate of the capacitor? *(Hint: The surface charge is equal to the difference in the charge on the plates with and without the dielectric.)*

** **58.** Two particles each have a mass of 6.0×10^{-3} kg. One has a charge of $+5.0 \times 10^{-6}$ C, and the other has a charge of -5.0×10^{-6} C. They are initially held at rest at a distance of 0.80 m apart. Both are then released and accelerate toward each other. How fast is each particle moving when the separation between them is one-half its initial value?

** **59. ssm www** The potential difference between the plates of a capacitor is 175 V. Midway between the plates, a proton and an electron are released. The electron is released from rest. The proton is projected perpendicularly toward the negative plate with an initial speed. The proton strikes the negative plate at the same instant that the electron strikes the positive plate. Ignore the attraction between the two particles, and find the initial speed of the proton.

** **60.** A positive charge $+q_1$ is located to the left of a negative charge $-q_2$. On a line passing through the two charges, there are two places where the total potential is zero. The first place is between the charges and is 4.00 cm to the left of the negative charge. The second place is 7.00 cm to the right of the negative charge. (a) What is the distance between the charges? (b) Find q_1/q_2, the ratio of the magnitudes of the charges.

Concepts & Calculations   Group Learning Problems

Note: Each of these problems consists of Concept Questions followed by a related quantitative Problem. They are designed for use by students working alone or in small learning groups. The Concept Questions involve little or no mathematics and are intended to stimulate group discussions. They focus on the concepts with which the problems deal. Recognizing the concepts is the essential initial step in any problem-solving technique.

61. Concept Questions The drawing shows a square, on two corners of which are fixed different positive charges. A third charge that is negative is brought to one of the empty corners. (a) At which of the empty corners, A or B, is the potential greater? (b) Is the electric potential energy of the third charge positive or negative? (c) For which location of the third charge, corner A or B, is the magnitude of the electric potential energy greater? Explain your answers.

Problem The length of a side of the square is $L = 0.25$ m. Find the electric potential energy of a charge $q_3 = -6.0 \times 10^{-9}$ C placed at corner A and then at corner B. Compare your answers for consistency with your answers to the Concept Questions.

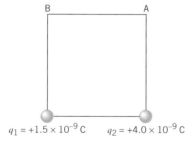

$q_1 = +1.5 \times 10^{-9}$ C $q_2 = +4.0 \times 10^{-9}$ C

62. Concept Questions Charges of $-q$ and $+2q$ are fixed in place, with a distance d between them. A dashed line is drawn through the negative charge, perpendicular to the line between the charges. On the dashed line, at a distance L from the negative charge, there is at least one spot where the total potential is zero. (a) At this spot, is the magnitude of the potential from the positive charge greater than, less than, or equal to the magnitude of the potential from the negative charge? (b) Is the distance from the positive charge to the zero-potential spot greater than, less than, or equal to L? (c) How many spots on the dashed line are there where the total potential is zero? Account for your answers.

Problem The distance between the charges is $d = 2.00$ m. Find L.

63. Concept Questions An electron and a proton, starting from rest, are accelerated through an electric potential difference of the same magnitude. In the process, the electron acquires a speed v_e, while the proton acquires a speed v_p. (a) As each particle accelerates from rest, it gains kinetic energy. Does it gain or lose electric potential energy? (b) Does the electron gain more, less, or the same amount of kinetic energy as the proton does? (c) Is v_e greater than, less than, or equal to v_p? Justify your answers.

Problem Find the ratio v_e/v_p. Verify that your answer is consistent with your answers to the Concept Questions.

64. Concept Questions A positive point charge is surrounded by an equipotential surface A, which has a radius of r_A. A positive test charge moves from surface A to another equipotential surface B, which has a radius of r_B. In the process, the electric force does negative work. (a) Does the electric force acting on the test charge have the same or opposite direction as the displacement of the test charge? (b) Is r_B greater than or less than r_A? Explain your answers.

Problem The positive point charge is $q = +7.2 \times 10^{-8}$ C, and the test charge is $q_0 = +4.5 \times 10^{-11}$ C. The work done as the test charge moves from surface A to surface B is $W_{AB} = -8.1 \times 10^{-9}$ J. The radius of surface A is $r_A = 1.8$ m. Find r_B. Check to see that your answer is consistent with your answers to the Concept Questions.

65. Concept Questions Two capacitors have the same plate separation. However, one has square plates, while the other has circular plates. The square plates are a length L on each side, and the diameter of the circular plates is L. (a) If the same dielectric material were between the plates in each capacitor, which one would have the greater capacitance? (b) By putting different dielectric materials between the capacitor plates, we can make the two capacitors have the same capacitance. Which capacitor should contain the dielectric material with the greater dielectric constant? Give your reasoning in each case.

Problem The capacitors have the same capacitance because they contain different dielectric materials. The dielectric constant of the material between the square plates has a value of $\kappa_{\text{square}} = 3.00$. What is the dielectric constant κ_{circle} of the material between the circular plates? Be sure that your answer is consistent with your answers to the Concept Questions.

66. Concept Questions Capacitor A and capacitor B each have the same voltage across their plates. However, the energy of capacitor A can melt m kilograms of ice at 0 °C, while the energy of capacitor B can boil away the same amount of water at 100 °C. (a) Which requires more energy, melting the ice or boiling the water? (b) Which capacitor has the greater capacitance? Explain your answers.

Problem The capacitance of capacitor A is 9.3 μF. What is the capacitance of capacitor B? Be sure that your answer is consistent with your answers to the Concept Questions.

* **67. Concept Questions** During a lightning flash, a potential difference $V_{cloud} - V_{ground}$ exists between a cloud and the ground. As a result of this potential difference, electric charge is transferred from the ground to the cloud (the charge is negative). (a) As the charge moves from the ground to the cloud, work is done on the charge by the electric force. How is this work related to the potential difference and the charge q? (b) If this work could be used to accelerate an automobile of mass m from rest, what would be the automobile's final speed? Express your answer in terms of the potential difference, the charge, and the mass of the automobile. (c) If this work could be converted entirely into heat, what mass m of water at 0 °C could be heated to 100 °C? Write your answer in terms of the potential difference, the charge, the mass of the water, and the specific heat capacity c of water.

Problem Suppose a potential difference of $V_{cloud} - V_{ground} = 1.2 \times 10^9$ V exists between the cloud and the ground, and $q = -25$ C of charge is transferred from the ground to the cloud. (a) How much work $W_{ground\text{-}cloud}$ is done on the charge by the electric force? (b) If the work done by the electric force were used to accelerate a 1100-kg

automobile from rest, what would be its final speed? (c) If the work done by the electric force were converted into heat, how many kilograms of water at 0 °C could be heated to 100 °C?

* **68. Concept Questions** An electron is released at the negative plate of a parallel plate capacitor and accelerates to the positive plate (see the drawing). (a) As the electron gains kinetic energy, does its electric potential energy increase or decrease? Why? (b) The difference in the electron's electric potential energy between the positive and negative plates is $EPE_{positive} - EPE_{negative}$. How is this difference related to the charge on the electron ($-e$) and to the difference $V_{positive} - V_{negative}$ in the electric potential between the plates? (c) How is the potential difference $V_{positive} - V_{negative}$ related to the electric field within the capacitor and the displacement of the positive plate relative to the negative plate?

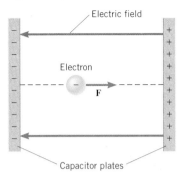
Electric field
Electron
F
Capacitor plates

Problem The plates of a parallel plate capacitor are separated by a distance of 1.2 cm, and the electric field within the capacitor has a magnitude of 2.1×10^6 V/m. An electron starts from rest at the negative plate and accelerates to the positive plate. What is the kinetic energy of the electron just as the electron reaches the positive plate?

Chapter 20

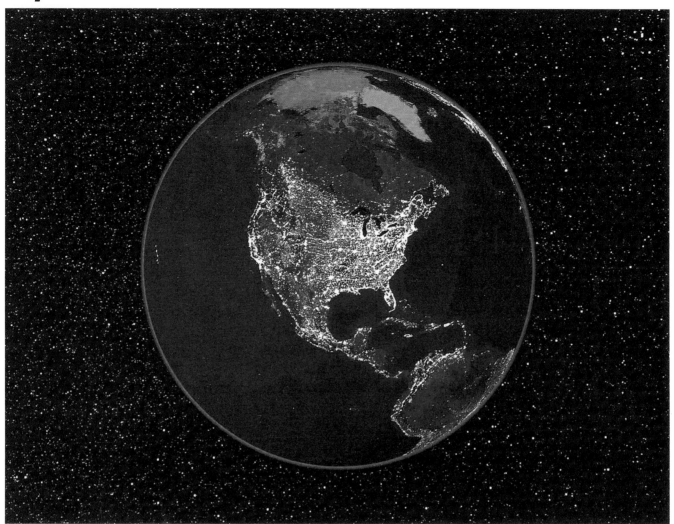

Electric Circuits

This image was computer-generated from various data collected by satellites orbiting the earth. It shows the distribution of light emanating at night from North America. The brightest light comes from the population centers in the eastern part of the United States. (© AP/Wide World Photos/NASA Goddard Space Flight Center)

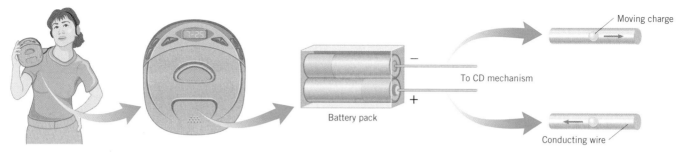

Figure 20.1 In an electric circuit, energy is transferred from a source (the battery pack) to a device (the CD player) by charges that move through a conducting wire.

20.1 *Electromotive Force and Current*

Look around you. Chances are that there is an electrical device nearby—a radio, a hair dryer, a computer—something that uses electrical energy to operate. The energy needed to run a portable CD player, for instance, comes from batteries, as Figure 20.1 illustrates. The transfer of energy takes place via an electric circuit, in which the energy source (the battery pack) and the energy-consuming device (the CD player) are connected by conducting wires, through which electric charges move.

Within a battery, a chemical reaction occurs that transfers electrons from one terminal (leaving it positively charged) to another terminal (leaving it negatively charged). Figure 20.2 shows the two terminals of a car battery and a flashlight battery. The drawing also illustrates the symbol $\left(\dfrac{+}{}\middle|\dfrac{-}{}\right)$ used to represent a battery in circuit drawings. Because of the positive and negative charges on the battery terminals, an electric potential difference exists between them. The maximum potential difference is called the ***electromotive force* (emf)*** of the battery, for which the symbol $\mathscr{E}$ is used. In a typical car battery, the chemical reaction maintains the potential of the positive terminal at a maximum of 12 volts (12 joules/coulomb) higher than the potential of the negative terminal, so the emf is $\mathscr{E} = 12$ V. Thus, one coulomb of charge emerging from the battery and entering a circuit has at most 12 joules of energy. In a typical flashlight battery the emf is 1.5 V. In reality, the potential difference between the terminals of a battery is somewhat less than the maximum value indicated by the emf, for reasons that Section 20.9 discusses.

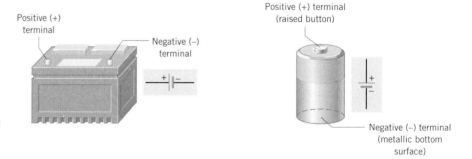

Figure 20.2 Typical batteries and the symbol $\left(\dfrac{+}{}\middle|\dfrac{-}{}\right)$ used to represent them in electric circuits.

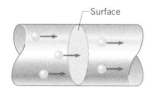

Figure 20.3 The electric current is the amount of charge per unit time that passes through a surface that is perpendicular to the motion of the charges.

▶ **CONCEPTS AT A GLANCE** In a circuit such as that in Figure 20.1, the battery creates an electric field within and parallel to the wire, directed from the positive toward the negative terminal. The electric field exerts a force on the free electrons in the wire, and they respond by moving. Figure 20.3 shows charges moving inside a wire and crossing an imaginary surface that is perpendicular to their motion. This flow of charge is known as an ***electric current.*** The Concepts-at-a-Glance chart in Figure 20.4 illustrates that we will formulate this new concept of electric current by bringing together the amount of charge Δq that passes through the surface and the time interval Δt during which it does so. ◀

---

* The word "force" appears in this context for historical reasons, even though it is incorrect. As we have seen in Section 19.2, electric potential is energy per unit charge, which is not force.

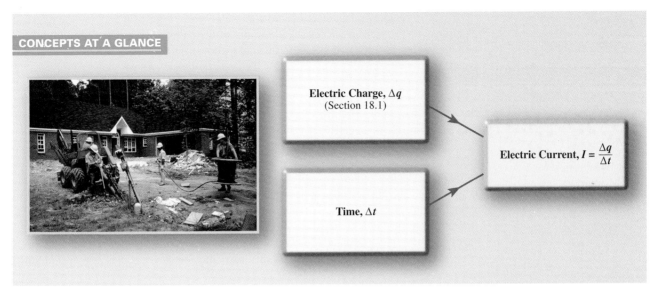

CONCEPTS AT A GLANCE

Electric Charge, Δq
(Section 18.1)

Time, Δt

Electric Current, $I = \dfrac{\Delta q}{\Delta t}$

Figure 20.4 CONCEPTS AT A GLANCE The concept of electric current is formulated from the concept of electric charge and the time during which it flows in a circuit. One of the essential steps in constructing a new house is installing the wiring that will carry the electric current used by the lights, heating and air conditioning system, and various household appliances. The photograph shows workmen installing the outside part of the wiring that connects the house to the supply from the electric utility company. (© Chris Hamilton/Corbis Images)

The electric current I is defined as the amount of charge per unit time that crosses the imaginary surface in Figure 20.3, in much the same sense that a river current is the amount of water flowing per unit of time. If the rate is constant, the current is

$$I = \frac{\Delta q}{\Delta t} \qquad (20.1)$$

If the rate of flow is not constant, then Equation 20.1 gives the average current. Since the units for charge and time are the coulomb (C) and the second (s), the SI unit for current is a coulomb per second (C/s). One coulomb per second is referred to as an ***ampere*** (A), after the French mathematician André-Marie Ampère (1775–1836).

If the charges move around a circuit in the same direction at all times, the current is said to be ***direct current (dc),*** which is the kind produced by batteries. In contrast, the current is said to be ***alternating current (ac)*** when the charges move first one way and then the opposite way, changing direction from moment to moment. Many energy sources produce alternating current—for example, generators at power companies and microphones. Example 1 deals with direct current.

Example 1 A Pocket Calculator

The current from the 3.0-V battery pack of a pocket calculator is 0.17 mA. In one hour of operation, (a) how much charge flows in the circuit and (b) how much energy does the battery deliver to the calculator circuit?

Reasoning Since current is defined as charge per unit time, the charge that flows in one hour is the product of the current and the time (3600 s). The charge that leaves the 3.0-V battery pack has 3.0 joules of energy per coulomb of charge. Thus, the total energy delivered to the calculator circuit is the charge (in coulombs) times the energy per unit charge (in volts or joules/coulomb).

Solution

(a) The charge that flows in one hour can be determined from Equation 20.1:

$$\Delta q = I(\Delta t) = (0.17 \times 10^{-3}\,\text{A})(3600\,\text{s}) = \boxed{0.61\,\text{C}}$$

(b) The energy delivered to the calculator circuit is

$$\text{Energy} = \text{Charge} \times \underbrace{\frac{\text{Energy}}{\text{Charge}}}_{\text{Battery emf}} = (0.61\,\text{C})(3.0\,\text{V}) = \boxed{1.8\,\text{J}}$$

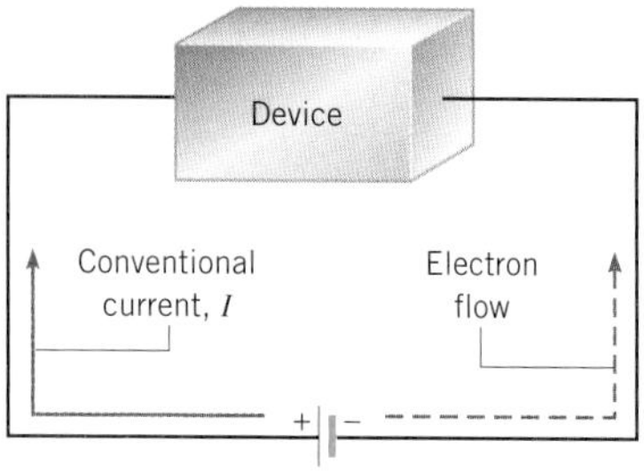

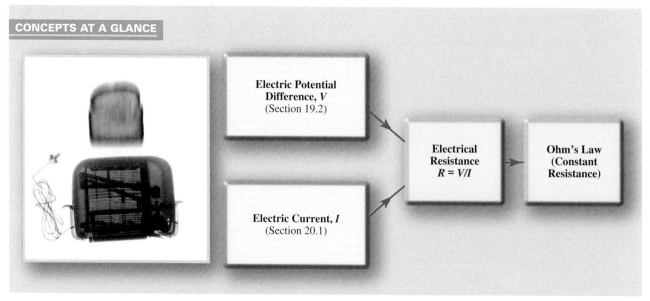

Figure 20.5 In a circuit, electrons actually flow through the metal wires. However, it is customary to use a conventional current *I* to describe the flow of charges.

Today, it is known that electrons flow in metal wires. Figure 20.5 shows the negative electrons emerging from the negative terminal of the battery and moving around the circuit toward the positive terminal. It is customary, however, *not* to use the flow of electrons when discussing circuits. Instead, a so-called ***conventional current*** is used, for reasons that date to the time when it was believed that positive charges moved through metal wires. Conventional current is the hypothetical flow of positive charges that would have the same effect in the circuit as the movement of negative charges that actually does occur. In Figure 20.5, negative electrons leave the negative terminal of the battery, pass through the device, and arrive at the positive terminal. The same effect would have been achieved if an equivalent amount of positive charge had left the positive terminal, passed through the device, and arrived at the negative terminal. Therefore, the drawing shows the conventional current originating from the positive terminal. A conventional current of hypothetical positive charges is consistent with our earlier use of a positive test charge for defining electric fields and potentials. The direction of conventional current is always from a point of higher potential toward a point of lower potential—that is, from the positive toward the negative terminal. In this text, the symbol *I* stands for conventional current.

20.2 *Ohm's Law*

Figure 20.6 CONCEPTS AT A GLANCE The concepts of electric potential difference or voltage *V* and the electric current *I* are brought together to define the electrical resistance *R*. Ohm's law specifies that the resistance remains constant as the voltage and current change. The current in a toaster, for example, causes the internal heating element to glow red-hot because the wire from which the element is made has resistance. The photograph shows a computer-processed X-ray of a toaster, which has been color-enhanced so that the heater element (pink) is obvious. (© David Arky/Corbis Images)

The current that a battery can push through a wire is analogous to the water flow that a pump can push through a pipe. Greater pump pressures lead to larger water flow rates, and, similarly, greater battery voltages* lead to larger electric currents. In the simplest case, the current *I* is directly proportional to the voltage *V*; that is, $I \propto V$. Thus, a 12-V battery leads to twice as much current as a 6-V battery, when each is connected to the same circuit.

▶ CONCEPTS AT A GLANCE In a water pipe, the flow rate is not only determined by the pump pressure but is also affected by the length and diameter of the pipe. Longer and narrower pipes offer higher resistance to the moving water and lead to smaller flow rates for a given pump pressure. A similar situation exists in electric circuits, and to deal with it we introduce the concept of electrical resistance. As the Concepts-at-a-Glance chart in Figure 20.6 indicates, electrical resistance is defined in terms of two ideas that have already been discussed—the electric potential difference, or voltage (see Section 19.2), and the electric current (see Section 20.1). ◀

CONCEPTS AT A GLANCE

Electric Potential Difference, *V* (Section 19.2)

Electric Current, *I* (Section 20.1)

Electrical Resistance $R = V/I$

Ohm's Law (Constant Resistance)

---

* The potential difference between two points, such as the terminals of a battery, is commonly called the voltage between the points.

The *resistance (R)* is defined as the ratio of the voltage V applied across a piece of material to the current I through the material, or $R = V/I$. When only a small current results from a large voltage, there is a high resistance to the moving charge. For many materials (e.g., metals), the ratio V/I is the same for a given piece of material over a wide range of voltages and currents. In such a case, the resistance is a constant. Then, as Figure 20.6 indicates, the relation $R = V/I$ is referred to as **Ohm's law,** after the German physicist Georg Simon Ohm (1789–1854), who discovered it.

■ **OHM'S LAW**

The ratio V/I is a constant, where V is the voltage applied across a piece of material (such as a wire) and I is the current through the material:

$$\frac{V}{I} = R = \text{constant} \quad \text{or} \quad V = IR \qquad (20.2)$$

R is the resistance of the piece of material.

SI Unit of Resistance: volt/ampere (V/A) = ohm (Ω)

Concept Simulation 20.1

In this simulation the battery voltage and the filament resistance of the bulb in Figure 20.7 can be adjusted, so that the application of Ohm's law to the circuit can be studied.

Go to
www.wiley.com/college/cutnell

The SI unit for resistance is a volt per ampere, which is called an *ohm* and is represented by the Greek capital letter omega (Ω). Ohm's law is not a fundamental law of nature like Newton's laws of motion. It is only a statement of the way certain materials behave in electric circuits.

To the extent that a wire or an electrical device offers resistance to the flow of charges, it is called a *resistor.* The resistance can have a wide range of values. The copper wires in a television set, for instance, have a very small resistance. On the other hand, commercial resistors can have resistances up to many kilohms (1 kΩ = 10^3 Ω) or megohms (1 MΩ = 10^6 Ω). Such resistors play an important role in electric circuits, where they are used to limit the amount of current and establish proper voltage levels.

In drawing electric circuits we follow the usual conventions: (1) a zigzag line ($-\wedge\wedge\wedge-$) represents a resistor and (2) a straight line ($\longrightarrow$) represents an ideal conducting wire, or one with a negligible resistance. Example 2 illustrates an application of Ohm's law to the circuit in a flashlight.

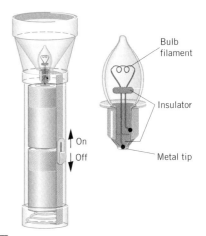

Example 2 A Flashlight

The filament in a light bulb is a resistor in the form of a thin piece of wire. The wire becomes hot enough to emit light because of the current in it. Figure 20.7 shows a flashlight that uses two 1.5-V batteries (effectively a single 3.0-V battery) to provide a current of 0.40 A in the filament. Determine the resistance of the glowing filament.

Reasoning The filament resistance is assumed to be the only resistance in the circuit. The potential difference applied across the filament is that of the 3.0-V battery. The resistance, given by Equation 20.2, is equal to this potential difference divided by the current.

Solution The resistance of the filament is

$$R = \frac{V}{I} = \frac{3.0 \text{ V}}{0.40 \text{ A}} = \boxed{7.5 \ \Omega} \qquad (20.2)$$

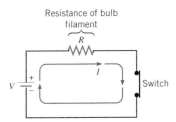

Figure 20.7 The circuit in this flashlight consists of a resistor (the filament of the light bulb) connected to a 3.0-V battery (two 1.5-V batteries).

20.3 *Resistance and Resistivity*

In a water pipe, the length and cross-sectional area of the pipe determine the resistance that the pipe offers to the flow of water. Longer pipes with smaller cross-sectional areas offer greater resistance. Analogous effects are found in the electrical case. For a wide range of materials, the resistance of a piece of material of length L and cross-sectional area A is

$$R = \rho \frac{L}{A} \qquad (20.3)$$

Table 20.1 *Resistivities^a of Various Materials*

Material	Resistivity ρ ($\Omega \cdot m$)	Material	Resistivity ρ ($\Omega \cdot m$)
Conductors		**Semiconductors**	
Aluminum	2.82×10^{-8}	Carbon	3.5×10^{-5}
Copper	1.72×10^{-8}	Germanium	0.5^b
Gold	2.44×10^{-8}	Silicon	$20-2300^b$
Iron	9.7×10^{-8}	**Insulators**	
Mercury	95.8×10^{-8}	Mica	$10^{11}-10^{15}$
Nichrome (alloy)	100×10^{-8}	Rubber (hard)	$10^{13}-10^{16}$
Silver	1.59×10^{-8}	Teflon	10^{16}
Tungsten	5.6×10^{-8}	Wood (maple)	3×10^{10}

^a The values pertain to temperatures near 20 °C.
^b Depending on purity.

where ρ is a proportionality constant known as the **resistivity** of the material. It can be seen from Equation 20.3 that the unit for resistivity is the ohm · meter ($\Omega \cdot m$), and Table 20.1 lists values for various materials. All the conductors in the table are metals and have small resistivities. Insulators such as rubber have large resistivities. Materials like germanium and silicon have intermediate resistivity values and are, accordingly, called *semiconductors*.

Resistivity is an inherent property of a material, inherent in the same sense that density is an inherent property. Resistance, on the other hand, depends on both the resistivity and the geometry of the material. Thus, two wires can be made from copper, which has a resistivity of 1.72×10^{-8} $\Omega \cdot m$, but Equation 20.3 indicates that a short wire with a large cross-sectional area has a smaller resistance than does a long, thin wire. Wires that carry large currents, such as main power cables, are thick rather than thin so that the resistance of the wires is kept as small as possible. Similarly, electric tools that are to be used far away from wall sockets require thicker extension cords, as Example 3 illustrates.

Example 3   Longer Extension Cords

The physics of electrical extension cords.

The instructions for an electric lawn mower suggest that a 20-gauge extension cord can be used for distances up to 35 m, but a thicker 16-gauge cord should be used for longer distances, to keep the resistance of the wire as small as possible. The cross-sectional area of 20-gauge wire is 5.2×10^{-7} m², while that of 16-gauge wire is 13×10^{-7} m². Determine the resistance of (a) 35 m of 20-gauge copper wire and (b) 75 m of 16-gauge copper wire.

Reasoning According to Equation 20.3, the resistance of a copper wire depends on the resistivity of copper and the length and cross-sectional area of the wire. The resistivity can be obtained from Table 20.1, while the length and cross-sectional area are given in the problem statement.

Solution According to Table 20.1 the resistivity of copper is 1.72×10^{-8} $\Omega \cdot m$. The resistance of the wires can be found using Equation 20.3:

20-gauge wire

$$R = \frac{\rho L}{A} = \frac{(1.72 \times 10^{-8} \, \Omega \cdot m)(35 \, m)}{5.2 \times 10^{-7} \, m^2} = \boxed{1.2 \, \Omega}$$

16-gauge wire

$$R = \frac{\rho L}{A} = \frac{(1.72 \times 10^{-8} \, \Omega \cdot m)(75 \, m)}{13 \times 10^{-7} \, m^2} = \boxed{0.99 \, \Omega}$$

Even though it is more than twice as long, the thicker 16-gauge wire has less resistance than the thinner 20-gauge wire. It is necessary to keep the resistance as low as possible to minimize heating of the wire, thereby reducing the possibility of a fire, as Conceptual Example 7 in Section 20.5 emphasizes.

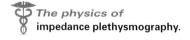
The physics of
impedance plethysmography.

Equation 20.3 provides the basis for an important medical diagnostic technique known as impedance (or resistance) plethysmography. Figure 20.8 shows how the technique is applied to diagnose blood clotting in the veins (deep venous thrombosis) near the knee. A pressure cuff, like that used in blood pressure measurements, is placed around the midthigh, while electrodes are attached around the calf. The two outer electrodes are connected to a source of a small amount of ac current. The two inner electrodes are separated by a distance L, and the voltage between them is measured. The voltage divided by the current gives the resistance. The key to this technique is the fact that resistance can be related to the volume V_{calf} of the calf between the inner electrodes. The volume is the product of the length L and the cross-sectional area A of the calf, or $V_{calf} = LA$. Solving for A and substituting in Equation 20.3 shows that

$$R = \rho \frac{L}{A} = \rho \frac{L}{V_{calf}/L} = \rho \frac{L^2}{V_{calf}}$$

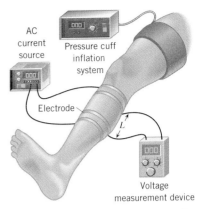

Thus, resistance is inversely proportional to volume, a fact that is exploited in diagnosing deep venous thrombosis. Blood flows from the heart into the calf through arteries in the leg and returns through the system of veins. The pressure cuff in Figure 20.8 is inflated to the point where it cuts off the venous flow, but not the arterial flow. As a result, more blood enters than leaves the calf, the volume of the calf increases, and the electrical resistance decreases. When the cuff pressure is removed suddenly, the volume returns to a normal value, and so does the electrical resistance. With healthy (unclotted) veins, there is a rapid return to normal values. A slow return, however, reveals the presence of clotting.

Figure 20.8 Using the technique of impedance plethysmography, the electrical resistance of the calf can be measured to diagnose deep venous thrombosis (blood clotting in the veins).

The resistivity of a material depends on temperature. In metals, the resistivity increases with increasing temperature, whereas in semiconductors the reverse is true. For many materials and limited temperature ranges, it is possible to express the temperature dependence of the resistivity as follows:

$$\rho = \rho_0[1 + \alpha(T - T_0)] \tag{20.4}$$

In this expression ρ and ρ_0 are the resistivities at temperatures T and T_0, respectively. The term α has the unit of reciprocal temperature and is the ***temperature coefficient of resistivity.*** When the resistivity increases with increasing temperature, α is positive, as it is for metals. When the resistivity decreases with increasing temperature, α is negative, as it is for the semiconductors carbon, germanium, and silicon. Since resistance is given by $R = \rho L/A$, both sides of Equation 20.4 can be multiplied by L/A to show that resistance depends on temperature according to

$$R = R_0[1 + \alpha(T - T_0)] \tag{20.5}$$

The next example illustrates the role of the resistivity and its temperature coefficient in determining the electrical resistance of a piece of material.

Example 4 The Heating Element of an Electric Stove

Figure 20.9*a* shows a cherry-red heating element on an electric stove. The element contains a wire (length = 1.1 m, cross-sectional area = 3.1×10^{-6} m^2) through which electric charge flows. As Figure 20.9*b* shows, this wire is embedded within an electrically insulating material that is contained within a metal casing. The wire becomes hot in response to the flowing charge and heats the casing. The material of the wire has a resistivity of $\rho_0 = 6.8 \times 10^{-5}$ $\Omega \cdot$m at $T_0 = 320$ °C and a temperature coefficient of resistivity of $\alpha = 2.0 \times 10^{-3}$ (C°)$^{-1}$. Determine the resistance of the heater wire at an operating temperature of 420 °C.

Reasoning Equation 20.3 ($R = \rho L/A$) can be used to find the resistance of the wire at 420 °C, once the resistivity ρ is determined at this temperature. Since the resistivity at 320 °C is given, Equation 20.4 can be employed to find the resistivity at 420 °C.

Solution At the operating temperature of 420 °C, the material of the wire has a resistivity of

$$\rho = \rho_0[1 + \alpha(T - T_0)]$$
$$\rho = (6.8 \times 10^{-5}\ \Omega \cdot m)\{1 + [2.0 \times 10^{-3}\ (\text{C}°)^{-1}](420\ °C - 320\ °C)\}$$
$$= 8.2 \times 10^{-5}\ \Omega \cdot m$$

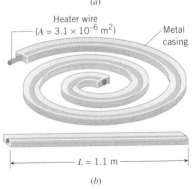

(a)

Heater wire ($A = 3.1 \times 10^{-6}$ m^2) — Metal casing

$L = 1.1$ m

(b)

Figure 20.9 A heating element from an electric stove. (© David Chasey/PhotoDisc, Inc./Getty Images)

The physics of
a heating element on an electric
stove.

This value of the resistivity can be used along with the given length and cross-sectional area to find the resistance of the heater wire:

$$R = \frac{\rho L}{A} = \frac{(8.2 \times 10^{-5} \, \Omega \cdot m)(1.1 \, m)}{3.1 \times 10^{-6} \, m^2} = \boxed{29 \, \Omega} \tag{20.3}$$

There is an important class of materials whose resistivity suddenly goes to zero below a certain temperature T_c, which is called the *critical temperature* and is commonly a few degrees above absolute zero. Below this temperature, such materials are called *superconductors*. The name derives from the fact that with zero resistivity, these materials offer no resistance to electric current and are, therefore, perfect conductors. One of the remarkable properties of zero resistivity is that once a current is established in a superconducting ring, it continues indefinitely without the need for an emf. Currents have persisted in superconductors for many years without measurable decay. In contrast, the current in a nonsuperconducting material drops to zero almost immediately after the emf is removed.

Many metals become superconductors only at very low temperatures, such as aluminum ($T_c = 1.18$ K), tin ($T_c = 3.72$ K), lead ($T_c = 7.20$ K), and niobium ($T_c = 9.25$ K). Materials involving copper oxide complexes have been made that undergo the transition to the superconducting state at 175 K. Superconductors have many technological applications, including magnetic resonance imaging (Section 21.7), magnetic levitation of trains (Section 21.9), cheaper transmission of electric power, powerful (yet small) electric motors, and faster computer chips.

✔ **Check Your Understanding 1**

A resistor is connected between the terminals of a battery. This resistor is a wire, and the following table gives various possibilities for its length and radius. For which one or more of the possibilities is the current in the resistor a minimum? *[The answer(s) is (are) given at the end of the book.]*

	Length	Radius
A	L_0	r_0
B	$\frac{1}{2}L_0$	$\frac{1}{2}r_0$
C	$2L_0$	$2r_0$
D	$2L_0$	r_0
E	$8L_0$	$2r_0$

Background: Ohm's law specifies the relationship between current, voltage, and resistance. The resistance depends on the resistivity of the material from which the wire is made and on the geometry of the wire.

For similar questions (including calculational counterparts), consult Self-Assessment Test 20.1, which is described at the end of Section 20.7.

20.4 Electric Power

Suppose that an amount of charge Δq emerges from a battery in a time Δt and that the potential difference between the battery terminals is V. According to the definition of electric potential given in Equation 19.3, the energy of this charge is the product of the charge and the potential difference, or $(\Delta q)V$. Since the change in energy per unit time is the power P (see Equation 6.10b), the electric power provided to the circuit is

$$P = \frac{(\Delta q)V}{\Delta t} = \underbrace{\frac{\Delta q}{\Delta t}}_{\text{Current } I} V$$

Electric utility companies are planning for the replacement of power lines with high-temperature superconducting cables. These workmen are feeding a test cable through an existing passage at a power substation. (© Andrew Sacks/ saxpix.com)

The term $\Delta q/\Delta t$ is the charge per unit time, or the current I in the circuit, according to Equation 20.1. Thus, the electric power is the product of current and voltage.

■ **ELECTRIC POWER**

When there is a current I in a circuit as a result of a voltage V, the electric power P delivered to the circuit is

$$P = IV \qquad\qquad (20.6)$$

SI Unit of Power: watt (W)

Power is measured in watts, and Equation 20.6 indicates that the product of an ampere and a volt is equal to a watt.

Many electrical devices are essentially resistors that become hot when provided with sufficient electric power: toasters, irons, space heaters, heating elements on electric stoves, and incandescent light bulbs, to name a few. In such cases, it is possible to obtain two additional, but equivalent, expressions for the power. These two expressions follow directly upon substituting $V = IR$, or equivalently $I = V/R$, into the relation $P = IV$:

$$P = IV \qquad\qquad (20.6a)$$
$$P = I(IR) = I^2R \qquad\qquad (20.6b)$$
$$P = \left(\frac{V}{R}\right)V = \frac{V^2}{R} \qquad\qquad (20.6c)$$

Example 5 deals with the electric power delivered to the bulb of a flashlight.

Example 5 The Power and Energy Used in a Flashlight

In the flashlight in Figure 20.7, the current is 0.40 A, and the voltage is 3.0 V. Find (a) the power delivered to the bulb and (b) the energy dissipated in the bulb in 5.5 minutes of operation.

Reasoning The electric power delivered to the bulb is the product of the current and voltage. Since power is energy per unit time, the energy delivered to the bulb is the product of the power and time.

Solution

(a) The power is

$$P = IV = (0.40\text{ A})(3.0\text{ V}) = \boxed{1.2\text{ W}} \qquad\qquad (20.6a)$$

The "wattage" rating of this bulb would, therefore, be 1.2 W.

(b) The energy consumed in 5.5 minutes (330 s) follows from the definition of power as energy per unit time:

$$\text{Energy} = P\,\Delta t = (1.2\text{ W})(330\text{ s}) = \boxed{4.0 \times 10^2\text{ J}}$$

Monthly electric bills specify the cost for the energy consumed during the month. Energy is the product of power and time, and electric companies compute your energy consumption by expressing power in kilowatts and time in hours. Therefore, a commonly used unit for energy is the *kilowatt · hour* (kWh). For instance, if you used an average power of 1440 watts (1.44 kW) for 30 days (720 h), your energy consumption would be (1.44 kW)(720 h) = 1040 kWh. At a cost of $0.10 per kWh, your monthly bill would be $104. As shown in Example 10 in Chapter 12, 1 kWh = 3.60×10^6 J of energy.

20.5 Alternating Current

Many electric circuits use batteries and involve direct current (dc). However, there are considerably more circuits that operate with alternating current (ac), in which the charge

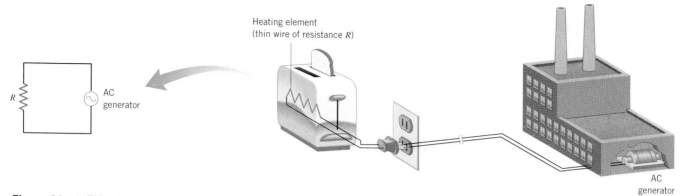

Heating element
(thin wire of resistance R)

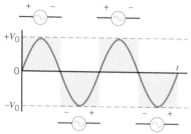

Figure 20.10 This circuit consists of a toaster (resistance $= R$) and an ac generator at the electric power company.

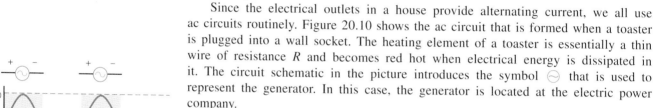

flow reverses direction periodically. The common generators that create ac electricity depend on magnetic forces for their operation and are discussed in Chapter 22. In an ac circuit, these generators serve the same purpose as a battery serves in a dc circuit; that is, they give energy to the moving electric charges. This section deals with ac circuits that contain only resistance.

Since the electrical outlets in a house provide alternating current, we all use ac circuits routinely. Figure 20.10 shows the ac circuit that is formed when a toaster is plugged into a wall socket. The heating element of a toaster is essentially a thin wire of resistance R and becomes red hot when electrical energy is dissipated in it. The circuit schematic in the picture introduces the symbol $\ominus$ that is used to represent the generator. In this case, the generator is located at the electric power company.

Figure 20.11 shows a graph that records the voltage V produced between the terminals of the ac generator in Figure 20.10 at each moment of time t. This is the most common type of ac voltage. It fluctuates sinusoidally between positive and negative values as a function of time:

$$V = V_0 \sin 2\pi ft \tag{20.7}$$

Figure 20.11 In the most common case, the ac voltage is a sinusoidal function of time. The relative polarity of the generator terminals during the positive and negative parts of the sine wave is indicated.

where V_0 is the maximum or peak value of the voltage, and f is the frequency (in cycles/s or Hz) at which the voltage oscillates. The angle $2\pi ft$ in Equation 20.7 is expressed in radians, so a calculator must be set to its radian mode before the sine of this angle is evaluated. In the United States, the voltage at most home wall outlets has a *peak value* of approximately $V_0 = 170$ volts and oscillates with a frequency of $f = 60$ Hz. Thus, the period of each cycle is $\frac{1}{60}$ s, and the polarity of the generator terminals reverses twice during each cycle, as Figure 20.11 indicates.

The current in an ac circuit also oscillates. In circuits that contain only resistance, the current reverses direction each time the polarity of the generator terminals reverses. Thus, the current in a circuit like that in Figure 20.10 would have a frequency of 60 Hz and would change direction twice during each cycle. Substituting $V = V_0 \sin 2\pi ft$ into $V = IR$ shows that the current can be represented as

$$I = \frac{V_0}{R} \sin 2\pi ft = I_0 \sin 2\pi ft \tag{20.8}$$

The peak current is given by $I_0 = V_0/R$, so it can be determined if the peak voltage and the resistance are known.

The power delivered to an ac circuit by the generator is given by $P = IV$, just as it is in a dc circuit. However, since both I and V depend on time, the power fluctuates as time passes. Substituting Equations 20.7 and 20.8 for V and I into $P = IV$ gives

$$P = I_0 V_0 \sin^2 2\pi ft \tag{20.9}$$

This expression is plotted in Figure 20.12.

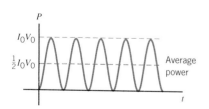

Figure 20.12 In an ac circuit, the power P delivered to a resistor oscillates between zero and a peak value of $I_0 V_0$, where I_0 and V_0 are the peak current and voltage, respectively.

Since the power fluctuates in an ac circuit, it is customary to consider the average power $\overline{P}$, which is one-half the peak power, as Figure 20.12 indicates:

$$\overline{P} = \tfrac{1}{2}I_0V_0 \tag{20.10}$$

On the basis of this expression, a kind of average current and average voltage can be introduced that are very useful when discussing ac circuits. A slight rearrangement of Equation 20.10 reveals that

$$\overline{P} = \left(\frac{I_0}{\sqrt{2}}\right)\left(\frac{V_0}{\sqrt{2}}\right) = I_{rms}V_{rms} \tag{20.11}$$

I_{rms} and V_{rms} are called the ***root mean square (rms)*** current and voltage, respectively, and may be calculated from the peak values by dividing them by $\sqrt{2}$*:

$$I_{rms} = \frac{I_0}{\sqrt{2}} \tag{20.12}$$

$$V_{rms} = \frac{V_0}{\sqrt{2}} \tag{20.13}$$

Since the maximum ac voltage at a home wall socket in the United States is typically $V_0 = 170$ volts, the corresponding rms voltage is $V_{rms} = (170 \text{ volts})/\sqrt{2} = 120$ volts. Thus, when the instructions for an electrical appliance specify 120 V, it is an rms voltage that is meant. Similarly, when we specify an ac voltage or current in this text, it is an rms value, unless indicated otherwise. Likewise, when we specify ac power, it is an average power, unless stated otherwise.

Except for dealing with average quantities, the relation $\overline{P} = I_{rms}V_{rms}$ has the same form as Equation 20.6a ($P = IV$). Moreover, Ohm's law can be written conveniently in terms of rms quantities:

$$V_{rms} = I_{rms}R \tag{20.14}$$

Substituting Equation 20.14 into $\overline{P} = I_{rms}V_{rms}$ shows that the average power can be expressed in the following ways:

$$\overline{P} = I_{rms}V_{rms} \tag{20.15a}$$
$$\overline{P} = I_{rms}^2R \tag{20.15b}$$
$$\overline{P} = \frac{V_{rms}^2}{R} \tag{20.15c}$$

These expressions are completely analogous to $P = IV = I^2R = V^2/R$ for dc circuits. Example 6 deals with the average power in one familiar ac circuit.

Example 6 Electrical Power Sent to a Loudspeaker

A stereo receiver applies a peak ac voltage of 34 V to a speaker. The speaker behaves approximately† as if it had a resistance of 8.0 Ω, as the circuit in Figure 20.13 indicates. Determine (a) the rms voltage, (b) the rms current, and (c) the average power for this circuit.

Reasoning The rms voltage is, by definition, equal to the peak voltage divided by $\sqrt{2}$. Furthermore, we are assuming that the circuit contains only a resistor. Therefore, we can use Ohm's law (Equation 20.14) to calculate the rms current as the rms voltage divided by the resistance, and we can then determine the average power as the rms current times the rms voltage (Equation 20.15a).

Solution

(a) The peak value of the voltage is $V_0 = 34$ V, so the corresponding rms value is

$$V_{rms} = \frac{V_0}{\sqrt{2}} = \frac{34 \text{ V}}{\sqrt{2}} = \boxed{24 \text{ V}} \tag{20.13}$$

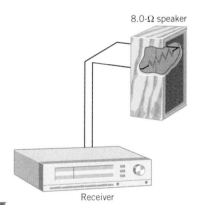

8.0-Ω speaker

Receiver

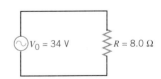

$V_0 = 34$ V $R = 8.0$ Ω

Figure 20.13 A receiver applies an ac voltage (peak value = 34 V) to an 8.0-Ω speaker.

Problem solving insight
The rms values of the ac voltage and the ac current, V_{rms} and I_{rms}, respectively, are not the same as the peak values V_0 and I_0. The rms values are always smaller than the peak values by a factor of $\sqrt{2}$.

---

* Equations 20.12 and 20.13 apply only for sinusoidal current and voltage.
† Other factors besides resistance can affect the current and voltage in ac circuits; they are discussed in Chapter 23.

(b) The rms current can be obtained from Ohm's law:

$$I_{rms} = \frac{V_{rms}}{R} = \frac{24\ V}{8.0\ \Omega} = \boxed{3.0\ A} \qquad (20.14)$$

(c) The average power is

$$\overline{P} = I_{rms}V_{rms} = (3.0\ A)(24\ V) = \boxed{72\ W} \qquad (20.15a)$$

The electric power dissipated in a resistor causes it to heat up. Excessive power can lead to a potential fire hazard, as Conceptual Example 7 discusses.

Conceptual Example 7   Extension Cords and a Potential Fire Hazard

Figure 20.14 When an extension cord is used with a space heater, the cord must have a resistance that is sufficiently small to prevent overheating of the cord.

During the winter, many people use portable electric space heaters to keep warm. Sometimes, however, the heater must be located far from a 120-V wall receptacle, so an extension cord must be used (see Figure 20.14). However, manufacturers often warn against using an extension cord. If one must be used, they recommend a certain wire gauge, or smaller. Why the warning, and why are smaller-gauge wires better than larger-gauge wires?

Reasoning and Solution An electric space heater contains a heater element that is a piece of wire of resistance R, which is heated to a high temperature. The heating occurs because of the power $I_{rms}^2 R$ dissipated in the heater element. A typical heater uses a relatively large current I_{rms} of about 12 A. On its way to the heater, this current passes through the wires of the extension cord. Since these additional wires offer resistance to the current, the extension cord can also heat up, just as the heater element does. The extent of the heating depends on the resistance of the extension cord. If the heating becomes excessive, the insulation around the wire can melt, and a fire can result. Thus, ***manufacturers of space heaters warn about extension cords in order to prevent a fire hazard.*** To keep the heating of the extension cord to a safe level, the resistance of the wire must be kept small. Recall from Section 20.3 that the resistance of a wire depends inversely on its cross-sectional area, so a large cross-sectional area gives rise to a small resistance. A wire with a large cross-sectional area is one that has a small gauge number, as discussed in Example 3. ***Smaller-gauge wires are better because they offer less resistance than larger-gauge wires.***

Related Homework: *Problem 34*

✔ **Check Your Understanding 2**

An ac circuit contains only a generator and a resistor. Which one of the following changes leads to the greatest average power being delivered to the circuit? (a) Double the peak voltage of the generator and double the resistance. (b) Double the resistance. (c) Double the peak voltage of the generator and reduce the resistance by a factor of two. (d) Double the peak voltage of the generator. *(The answer is given at the end of the book.)*

Background: The average power delivered to an ac circuit is related to the rms voltage applied to the circuit and the rms current in the circuit. When only a resistor is present in an ac circuit, Ohm's law applies and can be expressed in terms of rms quantities.

For similar questions (including calculational counterparts), consult Self-Assessment Test 20.1, which is described at the end of Section 20.7.

20.6 Series Wiring

Thus far, we have dealt with circuits that include only a single device, such as a light bulb or a loudspeaker. There are, however, many circuits in which more than one device is connected to a voltage source. This section introduces one method by which such connections may be made—namely, series wiring. **Series wiring means that the devices** *are connected in such a way that there is the same electric current through each device.* Figure 20.15 shows a circuit in which two different devices, represented by resistors R_1 and R_2, are connected in series with a battery. Note that if the current in one resistor is interrupted, the current in the other is too. This could occur, for example, if two light bulbs were connected in series and the filament of one bulb broke. Because of the series wiring, the voltage V supplied by the battery is divided between the two resistors. The drawing indicates that the portion of the voltage across R_1 is V_1, while the portion across R_2 is V_2, so $V = V_1 + V_2$. For each resistance, the definition of resistance is $R = V/I$, which may be solved for the corresponding voltage $V = IR$. Therefore, we have

$$V = V_1 + V_2 = IR_1 + IR_2 = I(R_1 + R_2) = IR_S$$

where R_S is called the *equivalent resistance* of the series circuit. Thus, two resistors in series are equivalent to a single resistor whose resistance is $R_S = R_1 + R_2$, in the sense that there is the same current through R_S as there is through the series combination of R_1 and R_2. This line of reasoning can be extended to any number of resistors in series, with the result that

Series resistors $$R_S = R_1 + R_2 + R_3 + \cdots \qquad (20.16)$$

Example 8 illustrates the concept of equivalent resistance in a series circuit.

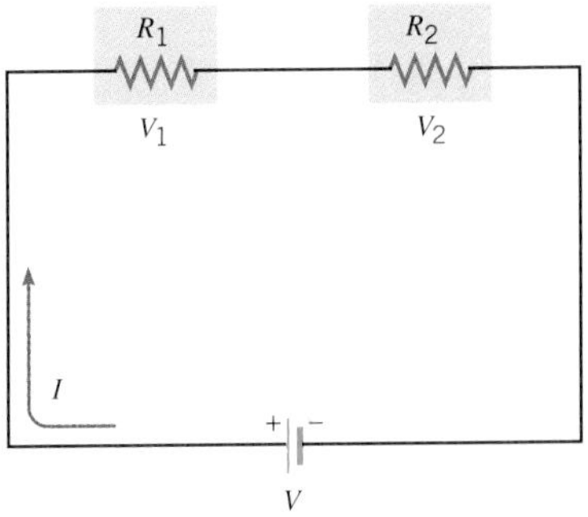

Figure 20.15 When two resistors are connected in series, the same current I is in both of them.

Example 8   Resistors in a Series Circuit

A 6.00-Ω resistor and a 3.00-Ω resistor are connected in series with a 12.0-V battery, as Figure 20.16 indicates. Assuming that the battery contributes no resistance to the circuit, find (a) the current, (b) the power dissipated in each resistor, and (c) the total power delivered to the resistors by the battery.

Reasoning The current I can be obtained from Ohm's law as $I = V/R_S$, where $R_S = R_1 + R_2$ is the equivalent resistance of the two resistors. The power delivered to each resistor is given by Equation 20.6b as $P = I^2R$, where R is the resistance of the resistor being considered and I is the current through it. The total power delivered by the battery is the power delivered to the 6.00-Ω resistor plus the power delivered to the 3.00-Ω resistor.

Solution

(a) The equivalent resistance is

$$R_S = 6.00~\Omega + 3.00~\Omega = 9.00~\Omega \qquad (20.16)$$

Ohm's law indicates that the current is

$$I = \frac{V}{R_S} = \frac{12.0~\text{V}}{9.00~\Omega} = \boxed{1.33~\text{A}} \qquad (20.2)$$

(b) Now that the current is known, the power dissipated in each resistor can be obtained from $P = I^2R$:

6.00-Ω resistor $$P = I^2R = (1.33~\text{A})^2(6.00~\Omega) = \boxed{10.6~\text{W}} \qquad (20.6\text{b})$$

3.00-Ω resistor $$P = I^2R = (1.33~\text{A})^2(3.00~\Omega) = \boxed{5.31~\text{W}}$$

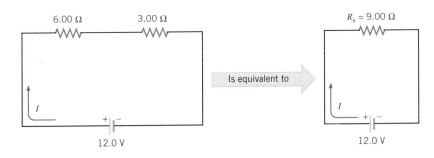

Figure 20.16 A 6.00-Ω and a 3.00-Ω resistor connected in series are equivalent to a single 9.00-Ω resistor.

(**c**) The total power delivered by the battery is the sum of the contributions in part (b): $P = 10.6 \text{ W} + 5.31 \text{ W} = 15.9 \text{ W}$. Alternatively, the total power can be obtained directly by using the equivalent resistance $R_S = 9.00 \text{ }\Omega$ and the current from part (a):

$$P = I^2 R_S = (1.33 \text{ A})^2 (9.00 \text{ }\Omega) = \boxed{15.9 \text{ W}}$$

Problem solving insight

In general, *the total power delivered to any number of resistors in series is equal to the power delivered to the equivalent resistor.*

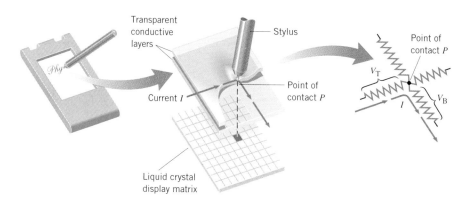

Figure 20.17 The pressure pad on which the user writes in a personal electronic assistant is based on the use of resistances that are connected in series.

The physics of
personal electronic assistants.

Pressure-sensitive pads form the heart of computer input devices that function as personal electronic assistants or personal digital assistants, and they offer an interesting application of series resistors. These devices are simple to use. You write directly on the pad with a plastic stylus that itself contains no electronics (see Figure 20.17). The writing appears as the stylus is moved, and recognition software interprets it as input for the built-in computer. The pad utilizes two transparent conductive layers that are separated by a small distance, except where pressure from the stylus brings them into contact (see point P in the drawing). Current I enters the positive side of the top layer, flows into the bottom layer through point P, and exits that layer through its negative side. Each layer provides resistance to the current, the amounts depending on where the point P is located. As the right side of the drawing indicates, the resistances from the layers are in series, since the same current flows through both of them. The voltage across the top-layer resistance is V_T, and the voltage across the bottom-layer resistance is V_B. In each case, the voltage is the current times the resistance. These two voltages are used to locate the point P and to activate (darken) one of the elements or pixels in a liquid crystal display matrix that lies beneath the transparent layers (see Section 24.6 for a discussion of liquid crystal displays). As the stylus is moved, the writing becomes visible as one element after another in the display matrix is activated.

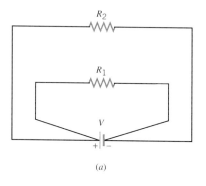

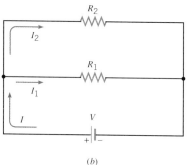

Figure 20.18 (*a*) When two resistors are connected in parallel, the same voltage V is applied across each resistor. (*b*) This circuit drawing is equivalent to that in part *a*. I_1 and I_2 are, respectively, the currents in R_1 and R_2.

20.7 *Parallel Wiring*

Parallel wiring is another method of connecting electrical devices. *Parallel wiring means that the devices are connected in such a way that the same voltage is applied across each device.* Figure 20.18 shows two resistors connected in parallel between the terminals of a battery. Part *a* of the picture is drawn so as to emphasize that the entire voltage of the battery is applied across each resistor. Actually, parallel connections are rarely drawn in this manner; instead they are drawn as in part *b*, where the dots indicate the points where the wires for the two branches are joined together. Parts *a* and *b* are equivalent representations of the same circuit.

Parallel wiring is very common. For example, when an electrical appliance is plugged into a wall socket, the appliance is connected in parallel with other appliances, as in Figure 20.19, where the entire voltage of 120 V is applied across the television, the stereo, and the light bulb. The presence of the unused socket or other devices that are turned off does not affect the operation of those devices that are turned on. Moreover, if the current in one device is interrupted (perhaps by an opened switch or a broken wire), the current in the other devices is not interrupted. In contrast, if household appliances were connected in series, there would be no current through any appliance if the current in the circuit were halted at any point.

When two resistors R_1 and R_2 are connected as in Figure 20.18, each receives current from the battery as if the other were not present. Therefore, R_1 and R_2 together draw more current from the battery than does either resistor alone. According to the definition of resistance, $R = V/I$, a larger current implies a smaller resistance. Thus, the two parallel resistors behave as a single equivalent resistance that is *smaller* than either R_1 or R_2. Figure 20.20 returns to the water-flow analogy to provide additional insight into this important feature of parallel wiring. In part *a*, two sections of pipe that have the same length are connected in parallel with a pump. In part *b* these two sections have been replaced with a single pipe of the same length, whose cross-sectional area equals the combined cross-sectional areas of section 1 and section 2. The pump can push more water per second through the wider pipe in part *b* than it can through *either* of the narrower pipes in part *a*. In effect, the wider pipe offers less resistance to the flow than either of the narrower pipes offers individually.

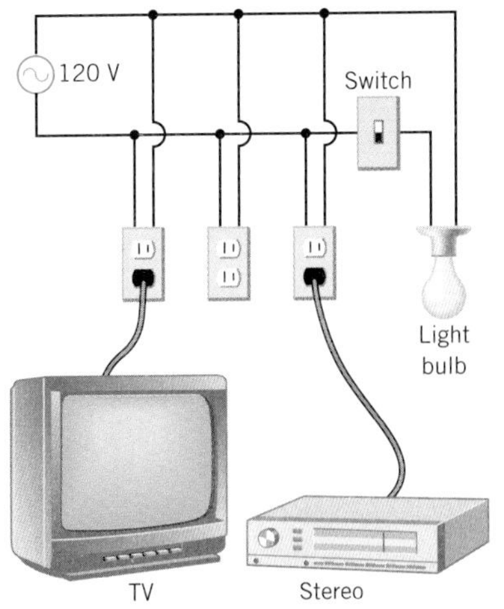

As in a series circuit, it is possible to replace a parallel combination of resistors with an equivalent resistor that results in the same total current and power for a given voltage as the original combination. To determine the equivalent resistance for the two resistors in Figure 20.18*b*, note that the total current I from the battery is the sum of I_1 and I_2, where I_1 is the current in resistor R_1 and I_2 is the current in resistor R_2: $I = I_1 + I_2$. Since the same voltage V is applied across each resistor, the definition of resistance indicates that $I_1 = V/R_1$ and $I_2 = V/R_2$. Therefore,

$$I = I_1 + I_2 = \frac{V}{R_1} + \frac{V}{R_2} = V\left(\frac{1}{R_1} + \frac{1}{R_2}\right) = V\left(\frac{1}{R_P}\right)$$

where R_P is the equivalent resistance. Hence, when two resistors are connected in parallel, they are equivalent to a single resistor whose resistance R_P can be obtained from $1/R_P = 1/R_1 + 1/R_2$. For any number of resistors in parallel, a similar line of reasoning reveals that

Parallel resistors
$$\frac{1}{R_P} = \frac{1}{R_1} + \frac{1}{R_2} + \frac{1}{R_3} + \cdots \qquad (20.17)$$

The next example deals with a parallel combination of resistors that occurs in a stereo system.

Figure 20.19 This drawing shows some of the parallel connections found in a typical home. Each wall socket provides 120 V to the appliance connected to it. In addition, 120 V is applied to the light bulb when the switch is turned on.

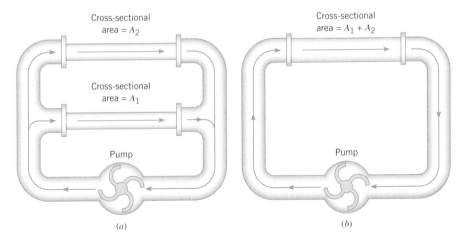

Cross-sectional area = A_2

Cross-sectional area = A_1

Pump

(a)

Cross-sectional area = $A_1 + A_2$

Pump

(b)

Figure 20.20 (*a*) Two equally long pipe sections, with cross-sectional areas A_1 and A_2, are connected in parallel to a water pump. (*b*) The two parallel pipe sections in part *a* are equivalent to a single pipe of the same length whose cross-sectional area is $A_1 + A_2$.

Example 9 Main and Remote Stereo Speakers

Most receivers allow the user to connect "remote" speakers (to play music in another room, for instance) in addition to the main speakers. Figure 20.21 shows that the remote speaker and the main speaker for the right stereo channel are connected to the receiver in parallel

The physics of
main and remote stereo speakers.

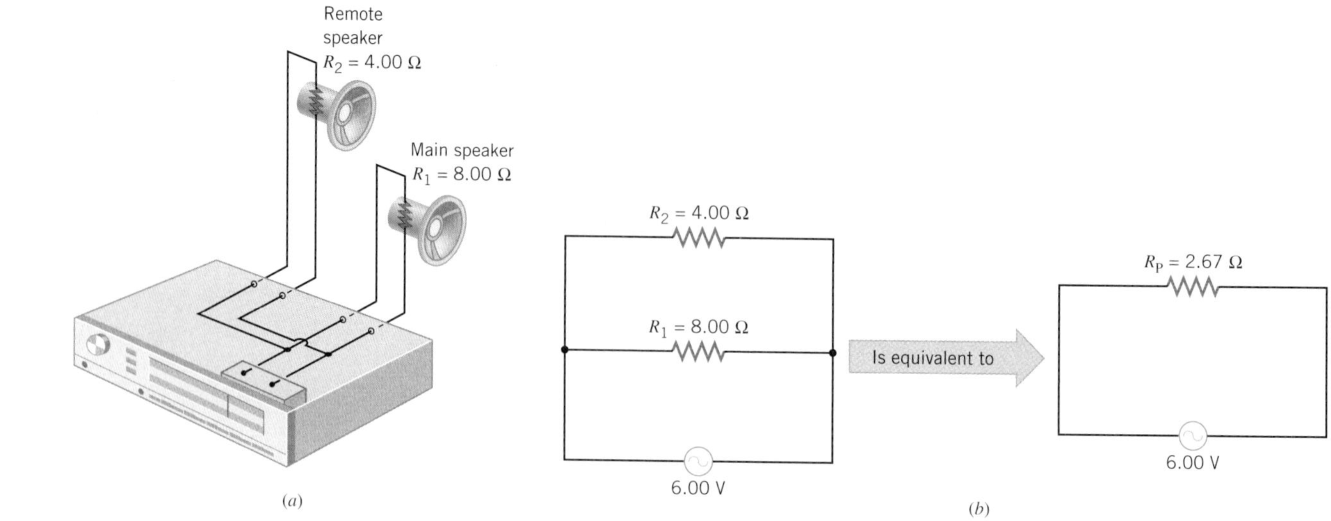

Figure 20.21 (*a*) The main and remote speakers in a stereo system are connected in parallel to the receiver. (*b*) The circuit schematic shows the situation when the ac voltage across the speakers is 6.00 V.

(for clarity, the speakers for the left channel are not shown). At the instant represented in the picture, the ac voltage across the speakers is 6.00 V. The main-speaker resistance is 8.00 Ω, and the remote-speaker resistance is 4.00 Ω.* Determine (a) the equivalent resistance of the two speakers, (b) the total current supplied by the receiver, (c) the current in each speaker, (d) the power dissipated in each speaker, and (e) the total power delivered by the receiver.

Reasoning The total current supplied to the two speakers by the receiver can be calculated as $I_{\text{rms}} = V_{\text{rms}}/R_{\text{P}}$, where R_{P} is the equivalent resistance of the two speakers in parallel and can be obtained from $1/R_{\text{P}} = 1/R_1 + 1/R_2$. The current in each speaker is different, however, since the speakers have different resistances. The average power delivered to a given speaker is the product of its current and voltage. In the parallel connection the same voltage is applied to each speaker.

Solution

(a) According to Equation 20.17, the equivalent resistance of the two speakers is given by

$$\frac{1}{R_{\text{P}}} = \frac{1}{8.00\ \Omega} + \frac{1}{4.00\ \Omega} = \frac{3}{8.00\ \Omega} \quad \text{or} \quad R_{\text{P}} = \frac{8.00\ \Omega}{3} = \boxed{2.67\ \Omega}$$

This result is illustrated in part *b* of the drawing.

(b) Using the equivalent resistance in Ohm's law shows that the total current is

$$I_{\text{rms}} = \frac{V_{\text{rms}}}{R_{\text{P}}} = \frac{6.00\ \text{V}}{2.67\ \Omega} = \boxed{2.25\ \text{A}} \tag{20.14}$$

(c) Applying Ohm's law to each speaker gives the individual speaker currents:

8.00-Ω speaker $I_{\text{rms}} = \dfrac{V_{\text{rms}}}{R} = \dfrac{6.00\ \text{V}}{8.00\ \Omega} = \boxed{0.750\ \text{A}}$

4.00-Ω speaker $I_{\text{rms}} = \dfrac{V_{\text{rms}}}{R} = \dfrac{6.00\ \text{V}}{4.00\ \Omega} = \boxed{1.50\ \text{A}}$

The sum of these currents is equal to the total current obtained in part (b).

(d) The average power dissipated in each speaker can be calculated using $\overline{P} = I_{\text{rms}} V_{\text{rms}}$ with the individual currents obtained in part (c):

8.00-Ω speaker $\overline{P} = (0.750\ \text{A})(6.00\ \text{V}) = \boxed{4.50\ \text{W}}$ (20.15a)

4.00-Ω speaker $\overline{P} = (1.50\ \text{A})(6.00\ \text{V}) = \boxed{9.00\ \text{W}}$ (20.15a)

Problem solving insight
The equivalent resistance R_{P} of a number of resistors in parallel has a reciprocal given by $R_{\text{P}}^{-1} = R_1^{-1} + R_2^{-1} + R_3^{-1} + \cdots$, where R_1, R_2, and R_3 are the individual resistances. After adding together the reciprocals R_1^{-1}, R_2^{-1}, and R_3^{-1}, do not forget to take the reciprocal of the result to find R_{P}.

* In reality, frequency-dependent characteristics of the speaker (see Chapter 23) play a role in the operation of a loudspeaker. We assume here, however, that the frequency of the sound is low enough that the speakers behave as pure resistances.

(e) The total power delivered by the receiver is the sum of the individual values that were found in part (d), $\overline{P}$ = 4.50 W + 9.00 W = 13.5 W. Alternatively, the total power can be obtained from the equivalent resistance R_P = 2.67 Ω and the total current in part (b):

$$\overline{P} = I_{rms}^2 R_P = (2.25 \text{ A})^2(2.67 \text{ } \Omega) = \boxed{13.5 \text{ W}} \qquad (20.15b)$$

In general, *the total power delivered to any number of resistors in parallel is equal to the power delivered to the equivalent resistor.*

Problem solving insight

▲

In a parallel combination of resistances, it is the *smallest* resistance that has the largest impact in determining the equivalent resistance. In fact, if one resistance approaches zero, then according to Equation 20.17, the equivalent resistance also approaches zero. In such a case, the near-zero resistance is said to *short out* the other resistances by providing a near-zero resistance path for the current to follow as a shortcut around the other resistances.

An interesting application of parallel wiring occurs in a three-way light bulb, as Conceptual Example 10 discusses.

Conceptual Example 10   A Three-Way Light Bulb and Parallel Wiring

Three-way light bulbs are popular because they can provide three levels of illumination (e.g., 50 W, 100 W, and 150 W) using a 120-V socket. The socket, however, must be equipped with a special three-way switch that enables one to select the illumination level. This switch does not select different voltages, because a three-way bulb uses a single voltage of 120 V. Within the bulb are two separate filaments. When one of these burns out (i.e., vaporizes), the bulb can produce only one level of illumination (either its lowest or its intermediate level) but not the highest level. How are the two filaments connected, in series or in parallel? And how can two filaments be used to produce three different illumination levels?

The physics of a three-way light bulb.

Reasoning and Solution If the filaments were wired in series and one of them burned out, no current would pass through the bulb and none of the illumination levels would be available, contrary to what is observed. Therefore, *the filaments must be connected in parallel,* as in Figure 20.22.

As to how two filaments can be used to produce three illumination levels, we note that the power dissipated in a resistance R is $\overline{P} = V_{rms}^2/R$, according to Equation 20.15c. With a single value of 120 V for the voltage V_{rms}, three different wattage ratings for the bulb can be obtained only if three different values for the resistance R can occur. In a 50-W/100-W/150-W three-way bulb, for example, one resistance R_{50} is provided by the 50-W filament, and the second resistance R_{100} comes from the 100-W filament. The third resistance R_{150} is the parallel combination of the other two and can be obtained from $1/R_{150} = 1/R_{50} + 1/R_{100}$. Figure 20.22 illustrates a simplified version of how the three-way switch operates to provide these three alternatives. In the first position of the switch, contact A is closed and contact B is open, so current passes only through the 50-W filament. In the second position, contact A is open and contact B is closed, so current passes only through the 100-W filament. In the third position, both contacts A and B are closed, and both filaments light up, giving 150 W of illumination.

Related Homework: *Problem 50*

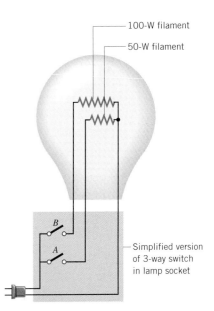

100-W filament

50-W filament

B

A

Simplified version of 3-way switch in lamp socket

Figure 20.22 A three-way light bulb uses two filaments connected in parallel. The filaments can be turned on one at a time or both together.

Self-Assessment Test 20.1

Test your understanding of the material in Sections 20.1–20.7:

- Electromotive Force and Current   • Ohm's Law   • Resistance and Resistivity
- Electric Power   • Alternating Current   • Series Wiring   • Parallel Wiring

Go to **www.wiley.com/college/cutnell**

20.8   *Circuits Wired Partially in Series and Partially in Parallel*

Often an electric circuit is wired partially in series and partially in parallel. The key to determining the current, voltage, and power in such a case is to deal with the circuit in parts, with the resistances in each part being either in series or in parallel with each other. Example 11 shows how such an analysis is carried out.

Example 11   A Four-Resistor Circuit

Figure 20.23 shows a circuit composed of a 24-V battery and four resistors, whose resistances are 110, 180, 220, and 250 Ω. Find (a) the total current supplied by the battery and (b) the voltage between points A and B in the circuit.

Reasoning The total current supplied by the battery can be obtained from Ohm's law, $I = V/R$, where R is the equivalent resistance of the four resistors. The equivalent resistance can be calculated by dealing with the circuit in parts. The voltage V_{AB} between the points A and B is also given by Ohm's law, $V_{AB} = IR_{AB}$, where I is the current and R_{AB} is the equivalent resistance between the two points.

Solution

(a) The 220-Ω resistor and the 250-Ω resistor are in series, so they are equivalent to a single resistor whose resistance is 220 Ω + 250 Ω = 470 Ω (see Figure 20.23*a*). The 470-Ω resistor is in parallel with the 180-Ω resistor. Their equivalent resistance can be obtained from Equation 20.17:

$$\frac{1}{R_{AB}} = \frac{1}{470\ \Omega} + \frac{1}{180\ \Omega} = 0.0077\ \Omega^{-1} \quad \text{or} \quad R_{AB} = \frac{1}{0.0077\ \Omega^{-1}} = 130\ \Omega$$

The circuit is now equivalent to a circuit containing a 110-Ω resistor in series with a 130-Ω resistor (see Figure 20.23*b*). This combination acts like a single resistor whose resistance is $R = 110\ \Omega + 130\ \Omega = 240\ \Omega$ (see Figure 20.23*c*). The total current from the battery is, then,

$$I = \frac{V}{R} = \frac{24\ \text{V}}{240\ \Omega} = \boxed{0.10\ \text{A}}$$

(b) The current $I = 0.10$ A passes through the resistance between points A and B. Therefore, Ohm's law indicates that the voltage across the 130-Ω resistor between points A and B is

$$V_{AB} = IR_{AB} = (0.10\ \text{A})(130\ \Omega) = \boxed{13\ \text{V}}$$

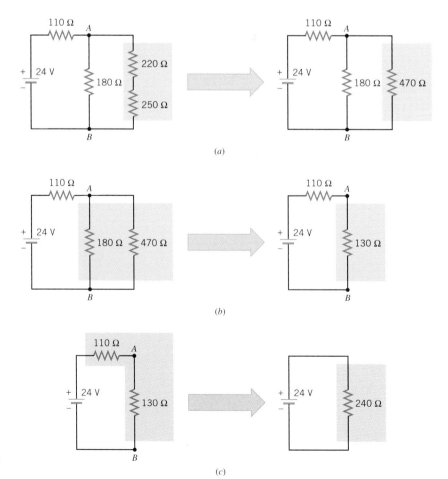

Figure 20.23 The circuits shown in this picture are equivalent.

Concept Simulation 20.2

In this simulation you can connect up to three resistors to a battery in a variety of ways and set the values of the resistances and the voltage of the battery. Graphs of the current versus the voltage for each resistor are displayed. In addition, the voltage across and the current in each resistor are given, along with the total current and the equivalent resistance.

Go to **www.wiley.com/college/cutnell**

Check Your Understanding 3

You have three identical resistors. By connecting all three together in various ways, how many different values of the total resistance can you obtain, and what are those values? Express your answers in terms of R, the resistance of one of the resistors. Note that some of the values are greater than R and some are less than R. *(The answers are given at the end of the book.)*

Background: The concepts of series and parallel wiring hold the key to this question. A given combination of resistors may be part series and part parallel.

For similar questions (including calculational counterparts), consult Self-Assessment Test 20.2, which is described at the end of Section 20.13.

20.9 *Internal Resistance*

So far, the circuits we have considered include batteries or generators that contribute only their emfs to a circuit. In reality, however, such devices also add some resistance. This resistance is called the ***internal resistance*** of the battery or generator because it is located inside the device. In a battery, the internal resistance is due to the chemicals within the battery. In a generator, the internal resistance is the resistance of wires and other components within the generator.

Figure 20.24 shows a schematic representation of the internal resistance r of a battery. The drawing emphasizes that when an external resistance R is connected to the battery, the resistance is connected *in series* with the internal resistance. The internal resistance of a functioning battery is typically small (several thousandths of an ohm for a new car battery). Nevertheless, the effect of the internal resistance may not be negligible. Example 12 illustrates that when current is drawn from a battery, the internal resistance causes the voltage between the terminals to drop below the maximum value specified by the battery's emf. The actual voltage between the terminals of a battery is known as the ***terminal voltage.***

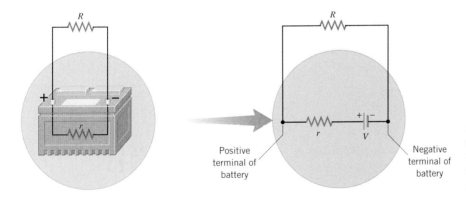

Positive terminal of battery

Negative terminal of battery

Figure 20.24 When an external resistance R is connected between the terminals of a battery, the resistance is connected in series with the internal resistance r of the battery.

Example 12 The Terminal Voltage of a Battery

Figure 20.25 shows a car battery whose emf is 12.0 V and whose internal resistance is 0.010 Ω. This resistance is relatively large because the battery is old and the terminals are corroded. What is the terminal voltage when the current I drawn from the battery is (a) 10.0 A and (b) 100.0 A?

The physics of automobile batteries.

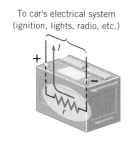

To car's electrical system
(ignition, lights, radio, etc.)

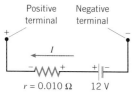

Positive terminal   Negative terminal

$r = 0.010\ \Omega$ 12 V

Figure 20.25 A car battery whose emf is 12 V and whose internal resistance is r.

Reasoning The voltage between the terminals is not the entire 12.0-V emf, because part of the emf is needed to make the current go through the internal resistance. The amount of voltage needed can be determined from Ohm's law as the current I through the battery times the internal resistance r. For larger currents, a larger amount of voltage is needed, leaving less of the emf between the terminals.

Solution

(a) The voltage needed to make a current of $I = 10.0$ A go through an internal resistance of $r = 0.010\ \Omega$ is $V = Ir = (10.0\ \text{A})(0.010\ \Omega) = 0.10$ V. To find the terminal voltage, remember that the direction of conventional current is always from a higher toward a lower potential. To emphasize this fact in the drawing, plus and minus signs have been included at the right and left ends, respectively, of the resistance r. The terminal voltage can be calculated by starting at the negative terminal and keeping track of how the voltage increases and decreases as we move toward the positive terminal. The voltage rises by 12.0 V due to the battery's emf. However, the voltage drops by 0.10 V because of the potential difference across the internal resistance. Therefore, the terminal voltage is 12.0 V − 0.10 V = $\boxed{11.9\ \text{V}}$.

(b) When the current through the battery is 100.0 A, the voltage needed to make the current go through the internal resistance is $V = (100.0\ \text{A})(0.010\ \Omega) = 1.0$ V. The terminal voltage now decreases to 12.0 V − 1.0 V = $\boxed{11.0\ \text{V}}$.

Example 12 indicates that the terminal voltage of a battery is smaller when the current drawn from the battery is larger, an effect that any car owner can demonstrate. Turn the headlights on before starting your car, so the current through the battery is about 10 A, as in part (a) of Example 12. Then start the car. The starter motor draws a large amount of additional current from the battery, momentarily increasing the total current by an appreciable amount. Consequently, the terminal voltage of the battery decreases, causing the headlights to dim.

20.10 Kirchhoff's Rules

Electric circuits that contain a number of resistors can often be analyzed by combining individual groups of resistors in series and parallel, as Section 20.8 discusses. However, there are many circuits in which no two resistors are in series or in parallel. To deal with

Figure 20.26 CONCEPTS AT A GLANCE Kirchhoff's rules are applications to electric circuits of the law of conservation of electric charge and the principle of conservation of energy. They can be used to analyze complex circuits. Examples of complex circuits can be found in a portable radio, as the color-enhanced X-ray photograph illustrates. The circular green object is the loudspeaker. (© D. Roberts/Science Photo Library/Photo Researchers)

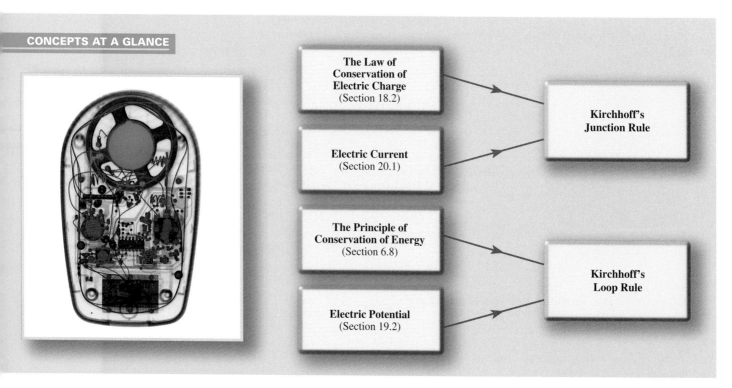

CONCEPTS AT A GLANCE

The Law of Conservation of Electric Charge (Section 18.2)

Electric Current (Section 20.1)

Kirchhoff's Junction Rule

The Principle of Conservation of Energy (Section 6.8)

Electric Potential (Section 19.2)

Kirchhoff's Loop Rule

such circuits it is necessary to employ methods other than the series–parallel method. One alternative is to take advantage of Kirchhoff's rules, named after their developer Gustav Kirchhoff (1824–1887).

▶ CONCEPTS AT A GLANCE There are two rules, the *junction rule* and the *loop rule.* The Concepts-at-a-Glance chart in Figure 20.26 emphasizes that these rules arise from principles and ideas that we have encountered earlier. The junction rule is an application of the law of conservation of electric charge (see Section 18.2) to the electric current in a circuit. The loop rule is an application of the principle of conservation of energy (see Section 6.8) to the electric potential (see Section 19.2) that exists at various places in a circuit. ◀

Figure 20.27 illustrates in greater detail the basic idea behind Kirchhoff's junction rule. The picture shows a junction where several wires are connected together. As Section 18.2 discusses, electric charge is conserved. Therefore, since there is no accumulation of charges at the junction itself, the total charge per second flowing into the junction must equal the total charge per second flowing out of it. In other words, *the junction rule states that the total current directed into a junction must equal the total current directed out of the junction,* or 7 A = 5 A + 2 A for the specific case shown in the picture.

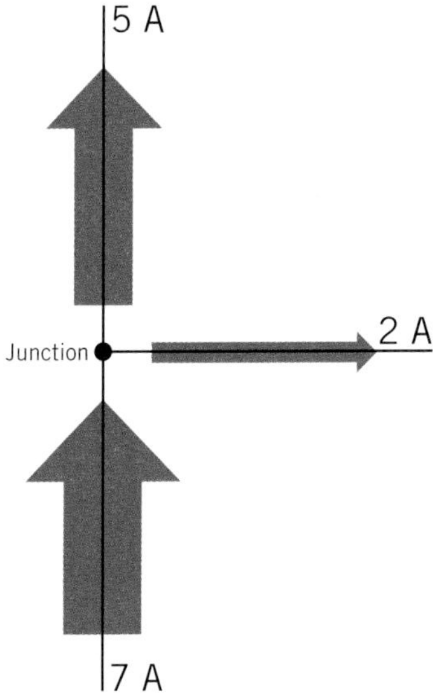

To help explain Kirchhoff's loop rule, Figure 20.28 shows a circuit in which a 12-V battery is connected to a series combination of a 5-Ω and a 1-Ω resistor. The plus and minus signs associated with each resistor remind us that, outside a battery, conventional current is directed from a higher toward a lower potential. From left to right, there is a potential drop of 10 V across the first resistor and another drop of 2 V across the second resistor. Keeping in mind that potential is the electric potential energy per unit charge, let us follow a positive test charge clockwise* around the circuit. Starting at the negative terminal of the battery, we see that the test charge gains potential energy because of the 12-V rise in potential due to the battery. The test charge then loses potential energy because of the 10-V and 2-V drops in potential across the resistors, ultimately arriving back at the negative terminal. In traversing the closed-circuit loop, the test charge is like a skier gaining gravitational potential energy in going up a hill on a chair lift and then losing it to friction in coming down and stopping. When the skier returns to the starting point, the gain equals the loss, so there is no net change in potential energy. Similarly, when the test charge arrives back at its starting point, there is no net change in electric potential energy, the gains matching the losses. *The loop rule expresses this example of energy conservation in terms of the electric potential and states that for a closed-circuit loop, the total of all the potential rises is the same as the total of all the potential drops,* or 12 V = 10 V + 2 V for the specific case in Figure 20.28.

Kirchhoff's rules can be applied to any circuit, even when the resistors are not in series or in parallel. The two rules are summarized below, and Examples 13 and 14 illustrate how to use them.

KIRCHHOFF'S RULES

Junction rule. The sum of the magnitudes of the currents directed into a junction equals the sum of the magnitudes of the currents directed out of the junction.

Loop rule. Around any closed-circuit loop, the sum of the potential drops equals the sum of the potential rises.

Figure 20.27 A junction is a point in a circuit where a number of wires are connected together. If 7 A of current is directed into the junction, then a total of 7 A (5 A + 2 A) of current must be directed out of the junction.

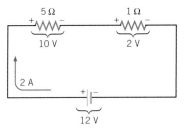

Figure 20.28 Following a positive test charge clockwise around the circuit, we see that the total voltage drop of 10 V + 2 V across the two resistors equals the voltage rise of 12 V due to the battery. The plus and minus signs on the resistors emphasize that, outside a battery, conventional current is directed from a higher potential (+) toward a lower potential (−).

Example 13 Using Kirchhoff's Loop Rule

Figure 20.29 shows a circuit that contains two batteries and two resistors. Determine the current *I* in the circuit.

Reasoning The first step is to draw the current, which we have chosen to be clockwise around the circuit. The choice of direction is *arbitrary,* and if it is incorrect, *I* will turn out to be negative.

---

* The choice of the clockwise direction is arbitrary.

The second step is to mark the resistors with plus and minus signs, which serve as an aid in identifying the potential drops and rises for Kirchhoff's loop rule. *Remember that, outside a battery, conventional current is always directed from a higher potential (+) toward a lower potential (−).* Thus, we *must* mark the resistors as indicated in Figure 20.29, to be consistent with the clockwise direction chosen for the current.

We may now apply Kirchhoff's loop rule to the circuit, starting at corner *A*, proceeding clockwise around the loop, and identifying the potential drops and rises as we go. The potential across each resistor is given by Ohm's law as *V* = *IR*. The clockwise direction is arbitrary, and the same result is obtained with a counterclockwise path.

Solution Starting at corner *A* and moving clockwise around the loop, there is

1. A potential drop (+ to −) of *IR* = *I*(12 Ω) across the 12-Ω resistor

2. A potential drop (+ to −) of 6.0 V across the 6.0-V battery

3. A potential drop (+ to −) of *IR* = *I*(8.0 Ω) across the 8.0-Ω resistor

4. A potential rise (− to +) of 24 V across the 24-V battery

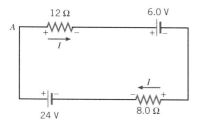

Figure 20.29 A single-loop circuit that contains two batteries and two resistors.

Setting the sum of the potential drops equal to the sum of the potential rises, as required by Kirchhoff's loop rule, gives

$$\underbrace{I(12\ \Omega) + 6.0\ \text{V} + I(8.0\ \Omega)}_{\text{Potential drops}} = \underbrace{24\ \text{V}}_{\text{Potential rises}}$$

Solving this equation for the current yields $\boxed{I = 0.90\ \text{A}}$. The current is a positive number, indicating that our initial choice for the direction (clockwise) of the current was correct.

Need more practice?

Interactive LearningWare 20.1
The circuit in the drawing contains three resistors and two batteries. Find the currents in the 3.0-Ω resistor and the 4.0-Ω resistor.

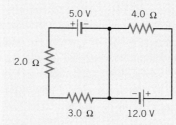

Related Homework: *Problem 76*

Go to **www.wiley.com/college/cutnell** for an interactive solution.

Example 14   The Electrical System of a Car

In a car, the headlights are connected to the battery and would discharge the battery if it were not for the alternator, which is run by the engine. Figure 20.30 indicates how the car battery, headlights, and alternator are connected. The circuit includes an internal resistance of 0.0100 Ω for the car battery and its leads and a resistance of 1.20 Ω for the headlights. For the sake of simplicity, the alternator is approximated as an additional 14.00-V battery with an internal resistance of 0.100 Ω. Determine the currents through the car battery (I_B), the headlights (I_H), and the alternator (I_A).

Reasoning The drawing shows the directions chosen arbitrarily for the currents I_B, I_H, and I_A. If any direction is incorrect, the analysis will reveal a negative value for the corresponding current.

Next, we mark the resistors with the plus and minus signs that serve as an aid in identifying the potential drops and rises for the loop rule, recalling that, outside a battery, conventional current is always directed from a higher potential (+) toward a lower potential (−). Thus, given the directions selected for I_B, I_H, and I_A, the plus and minus signs *must* be those indicated in Figure 20.30.

Kirchhoff's junction and loop rules can now be used.

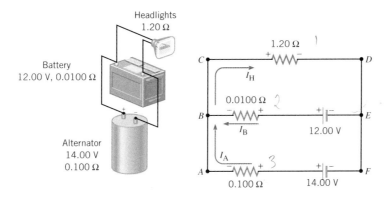

Figure 20.30 A circuit showing the headlight(s), battery, and alternator of a car.

Solution The junction rule can be applied to junction *B* or junction *E*. In either case, the same equation results:

*Junction rule
applied at B*

$$\underbrace{I_A + I_B}_{\substack{\text{Into} \\ \text{junction}}} = \underbrace{I_H}_{\substack{\text{Out of} \\ \text{junction}}}$$

In applying the loop rule to the lower loop *BEFA,* we start at point *B*, move clockwise around the loop, and identify the potential drops and rises. There is a potential rise (− to +) of I_B (0.0100 Ω) across the 0.0100-Ω resistor and then a drop (+ to −) of 12.00 V due to the car battery. Continuing around the loop, we find a 14.00-V rise (− to +) across the alternator, followed by a potential drop (+ to −) of I_A (0.100 Ω) across the 0.100-Ω resistor. Setting the sum of the potential drops equal to the sum of the potential rises gives the following result:

Loop rule: BEFA $\underbrace{I_A(0.100\ \Omega) + 12.00\ \text{V}}_{\text{Potential drops}} = \underbrace{I_B(0.0100\ \Omega) + 14.00\ \text{V}}_{\text{Potential rises}}$

Since there are three unknown variables in this problem, I_B, I_H, and I_A, a third equation is needed for a solution. To obtain the third equation, we apply the loop rule to the upper loop *CDEB*, choosing a clockwise path for convenience. The result is

Loop rule: CDEB $\underbrace{I_B(0.0100\ \Omega) + I_H(1.20\ \Omega)}_{\substack{\text{Potential} \\ \text{drops}}} = \underbrace{12.00\ \text{V}}_{\text{Potential rises}}$

These three equations can be solved simultaneously to show that

$$\boxed{I_B = -9.0\ \text{A}, I_H = 10.1\ \text{A}, I_A = 19.1\ \text{A}}$$

The negative answer for I_B indicates that the current through the battery is not directed from right to left, as drawn in Figure 20.30. Instead, the 9.0-A current is directed from left to right, opposite to the way current would be directed if the alternator were not connected. It is the left-to-right current created by the alternator that keeps the battery charged.

Concept Simulation 20.3

Kirchhoff's rules are applied to a circuit containing three resistors and three batteries. The values for each of the resistances and battery voltages are under your control, as are the directions chosen for the currents in the resistors. The equations that result from applying Kirchhoff's rules are displayed, along with the solutions for the unknown currents.

Related Homework: Problem 78

Go to
www.wiley.com/college/cutnell

Reasoning Strategy

Applying Kirchhoff's Rules

1. Draw the current in each branch of the circuit. Choose any direction. If your choice is incorrect, the value obtained for the current will turn out to be a negative number.

2. Mark each resistor with a plus sign at one end and a minus sign at the other end, in a way that is consistent with your choice for the current direction in Step 1. Outside a battery, conventional current is always directed from a higher potential (the end marked +) toward a lower potential (the end marked −).

3. Apply the junction rule and the loop rule to the circuit, obtaining in the process as many independent equations as there are unknown variables.

4. Solve the equations obtained in Step 3 simultaneously for the unknown variables. (See Appendix C.3.)

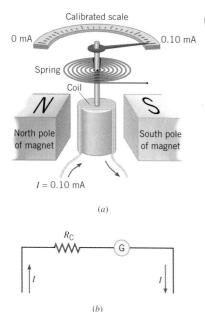

Figure 20.31 (a) A dc galvanometer. The coil of wire and pointer rotate when there is a current in the wire. (b) A galvanometer with a coil resistance of R_C is represented in a circuit diagram as shown here.

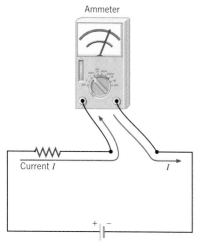

Figure 20.32 An ammeter must be inserted into a circuit so that the current passes directly through it.

The physics of an ammeter.

✓ **Check Your Understanding 4**

The drawing shows a circuit containing three resistors and three batteries. In preparation for applying Kirchhoff's rules, the currents in each resistor have been drawn. For these currents, write down the equations that result from applying Kirchhoff's junction rule and loop rule. Apply the loop rule to loops *ABCD* and *BEFC*. (*The answers are given at the end of the book.*)

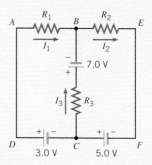

Background: In the circuit in the drawing no two resistors are in series or in parallel. Kirchhoff's rules do apply, however.

For similar questions (including calculational counterparts), consult Self-Assessment Test 20.2, which is described at the end of Section 20.13.

20.11 The Measurement of Current and Voltage

Current and voltage can be measured with devices known, respectively, as ammeters and voltmeters. There are two types of such devices: those that use digital electronics and those that do not. The essential feature of nondigital devices is the dc *galvanometer.* As Figure 20.31a illustrates, a galvanometer consists of a magnet, a coil of wire, a spring, a pointer, and a calibrated scale. The coil is mounted in such a way that it can rotate, which causes the pointer to move in relation to the scale. As Section 21.6 will discuss, the coil rotates in response to the torque applied by the magnet when there is a current in the coil. The coil stops rotating when this torque is balanced by the torque of the spring.

A galvanometer has two important characteristics that must be considered when it is used as part of a measurement device. First, the amount of dc current that causes a full-scale deflection of the pointer indicates the sensitivity of the galvanometer. For instance, Figure 20.31a shows an instrument that deflects full scale when the current in the coil is 0.10 mA. The second important characteristic is the resistance R_C of the wire in the coil. Figure 20.31b shows how a galvanometer with a coil resistance of R_C is represented in a circuit diagram.

Since an *ammeter* is an instrument that measures current, it must be inserted in the circuit so the current passes directly through it, as Figure 20.32 shows. An ammeter includes a galvanometer and one or more *shunt resistors,* which are connected in parallel with the galvanometer and provide a bypass for current in excess of the galvanometer's full-scale limit. The bypass allows the ammeter to be used to measure a current exceeding the full-scale limit. In Figure 20.33, for instance, a current of 60.0 mA enters terminal *A* of an ammeter, which uses a galvanometer with a full-scale current of 0.100 mA. The shunt resistor R can be selected so that 0.100 mA of current enters the galvanometer, while 59.9 mA bypasses it. In such a case, the measurement scale on the ammeter would be labeled 0 to 60.0 mA. To determine the shunt resistance, it is necessary to know the coil resistance R_C (see problem 80).

When an ammeter is inserted into a circuit, the equivalent resistance of the coil and the shunt resistor adds to the circuit resistance. Any increase in circuit resistance causes a reduction in current, and this is a problem, because an ammeter should only measure the current, not change it. Therefore, an *ideal* ammeter would have zero resistance. In practice, a good ammeter is designed with an equivalent resistance small enough so there is only a negligible reduction of the current in the circuit when the ammeter is inserted.

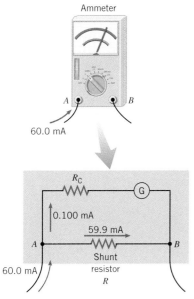

Ammeter

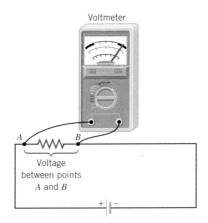

Voltmeter

Figure 20.33 If a galvanometer with a full-scale limit of 0.100 mA is to be used to measure a current of 60.0 mA, a shunt resistance R must be used, so the excess current of 59.9 mA can detour around the galvanometer coil.

Figure 20.34 To measure the voltage between two points A and B in a circuit, a voltmeter is connected between the points.

A ***voltmeter*** is an instrument that measures the voltage between two points, A and B, in a circuit. Figure 20.34 shows that the voltmeter must be connected between the points and is *not* inserted into the circuit as an ammeter is. A voltmeter includes a galvanometer whose scale is calibrated in volts. Suppose, for instance, that the galvanometer in Figure 20.35 has a full-scale current of 0.1 mA and a coil resistance of 50 Ω. Under full-scale conditions, the voltage across the coil would be $V = IR_C = (0.1 \times 10^{-3} \text{ A})(50 \; \Omega) = 0.005$ V. Thus, this galvanometer could be used to register voltages in the range $0 - 0.005$ V. A voltmeter, then, is a galvanometer used in this fashion, along with some provision for adjusting the range of voltages to be measured. To adjust the range, an additional resistance R is connected in series with the coil resistance R_C. The value of R needed to extend the measurement range to a full-scale voltage V can be determined from the full-scale current I by applying Ohm's law in the form $V = I(R + R_C)$ (see problem 81).

Ideally, the voltage registered by a voltmeter should be the same as the voltage that exists when the voltmeter is not connected. However, a voltmeter takes some current from a circuit and, thus, alters the circuit voltage to some extent. An *ideal* voltmeter would have infinite resistance and draw away only an infinitesimal amount of current. In reality, a good voltmeter is designed with a resistance that is large enough so the unit does not appreciably alter the voltage in the circuit to which it is connected.

The physics of a voltmeter.

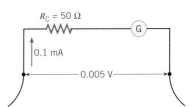

Figure 20.35 The galvanometer shown has a full-scale deflection of 0.1 mA and a coil resistance of 50 Ω.

20.12 *Capacitors in Series and Parallel*

Figure 20.36 shows two different capacitors connected in parallel to a battery. Since the capacitors are in parallel, they have the same voltage V across their plates. However, the capacitors *contain different amounts of charge.* The charge stored by a capacitor is $q = CV$ (Equation 19.8), so $q_1 = C_1V$ and $q_2 = C_2V$.

As with resistors, it is always possible to replace a parallel combination of capacitors with an *equivalent capacitor* that stores the same charge and energy for a given voltage as the combination does. To determine the equivalent capacitance C_P, note that the total charge q stored by the two capacitors is

$$q = q_1 + q_2 = C_1V + C_2V = (C_1 + C_2)V = C_PV$$

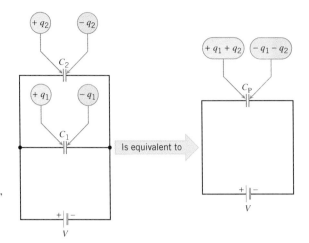

Figure 20.36 In a parallel combination of capacitances C_1 and C_2, the voltage V across each capacitor is the same, but the charges q_1 and q_2 on each capacitor are different.

This result indicates that two capacitors in parallel can be replaced by an equivalent capacitor whose capacitance is $C_P = C_1 + C_2$. For any number of capacitors in parallel, the equivalent capacitance is

Parallel capacitors $\qquad\qquad C_P = C_1 + C_2 + C_3 + \cdots$ $\qquad\qquad$ (20.18)

Capacitances in parallel simply add together to give an equivalent capacitance. This behavior contrasts with that of resistors in parallel, which combine as reciprocals, according to Equation 20.17.

The equivalent capacitor not only stores the same amount of charge as the parallel combination of capacitors, but also stores the same amount of energy. For instance, the energy stored in a single capacitor is $\frac{1}{2}CV^2$ (Equation 19.11), so the total energy stored by two capacitors in parallel is

$$\text{Total energy} = \tfrac{1}{2}C_1V^2 + \tfrac{1}{2}C_2V^2 = \tfrac{1}{2}(C_1 + C_2)V^2 = \tfrac{1}{2}C_PV^2$$

which is equal to the energy stored in the equivalent capacitor C_P.

Problem solving insight

When capacitors are connected in series, the equivalent capacitance is different than when they are in parallel. As an example, Figure 20.37 shows two capacitors in series and reveals the following important fact: ***All capacitors in series, regardless of their capacitances, contain charges of the same magnitude, $+q$ and $-q$, on their plates.*** The battery places a charge of $+q$ on plate a of capacitor C_1, and this charge induces a charge of $+q$ to depart from the opposite plate a', leaving behind a charge $-q$. The $+q$ charge that leaves plate a' is deposited on plate b of capacitor C_2 (since these two plates are connected by a wire), where it induces a $+q$ charge to move away from the opposite plate b', leaving behind a charge of $-q$. Thus, all capacitors in series contain charges of the same magnitude on their plates. Note the difference between charging capacitors in parallel and in series. When charging parallel capacitors, the battery moves a charge q that is the sum of the charges moved for each of the capacitors: $q = q_1 + q_2 + q_3 + \cdots$. In contrast, when charging a series combination of n capacitors, the battery only moves a charge q, not nq, because the charge q passes by induction from one capacitor directly to the next one in line.

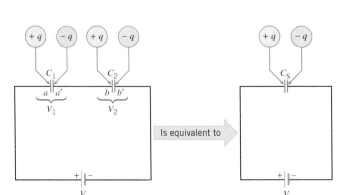

Figure 20.37 In a series combination of capacitances C_1 and C_2, the same amount of charge q is on the plates of each capacitor, but the voltages V_1 and V_2 across each capacitor are different.

The equivalent capacitance C_S for the series connection in Figure 20.37 can be determined by observing that the battery voltage V is shared by the two capacitors. The drawing indicates that the voltages across C_1 and C_2 are V_1 and V_2, so that $V = V_1 + V_2$. Since the voltages across the capacitors are $V_1 = q/C_1$ and $V_2 = q/C_2$, we find that

$$V = V_1 + V_2 = \frac{q}{C_1} + \frac{q}{C_2} = q\left(\frac{1}{C_1} + \frac{1}{C_2}\right) = q\left(\frac{1}{C_S}\right)$$

Thus, two capacitors in series can be replaced by an equivalent capacitor whose capacitance C_S can be obtained from $1/C_S = 1/C_1 + 1/C_2$. For any number of capacitors connected in series the equivalent capacitance is given by

Series capacitors $\qquad\qquad \dfrac{1}{C_S} = \dfrac{1}{C_1} + \dfrac{1}{C_2} + \dfrac{1}{C_3} + \cdots$ (20.19)

Equation 20.19 indicates that capacitances in series combine as reciprocals and do not simply add together as resistors in series do. It is left as an exercise (problem 91) to show that the equivalent series capacitance stores the same electrostatic energy as the sum of the energies of the individual capacitors.

It is possible to simplify circuits containing a number of capacitors in the same general fashion as that outlined for resistors in Example 11 and Figure 20.23. The capacitors in a parallel grouping can be combined according to Equation 20.18, and those in a series grouping can be combined according to Equation 20.19.

20.13 RC Circuits

Many electric circuits contain both resistors and capacitors. Figure 20.38 illustrates an example of a resistor–capacitor or *RC* circuit. Part *a* of the drawing shows the circuit at a time *t* after the switch has been closed and the battery has begun to charge up the capacitor plates. The charge on the plates builds up gradually to its equilibrium value of $q_0 = CV_0$, where V_0 is the voltage of the battery. Assuming that the capacitor is uncharged at time $t = 0$ s when the switch is closed, it can be shown that the magnitude q of the charge on the plates at time *t* is

Capacitor charging $\qquad\qquad q = q_0[1 - e^{-t/(RC)}]$ (20.20)

where the exponential *e* has the value of 2.718. . . . Part *b* of the drawing shows a graph of this expression, which indicates that the charge is $q = 0$ C when $t = 0$ s and increases gradually toward the equilibrium value of $q_0 = CV_0$. The voltage *V* across the capacitor at any time can be obtained from Equation 20.20 by dividing the charges q and q_0 by the capacitance *C*, since $V = q/C$ and $V_0 = q_0/C$.

The term *RC* in the exponent in Equation 20.20 is called the **time constant** τ of the circuit:

$$\tau = RC \qquad (20.21)$$

The time constant is measured in seconds; verification of the fact that an ohm times a farad is equivalent to a second is left as an exercise (see conceptual question 16). The time constant is the amount of time required for the capacitor to accumulate 63.2% of its equilibrium charge, as can be seen by substituting $t = \tau = RC$ in Equation 20.20; $q_0(1 - e^{-1}) = q_0(0.632)$. The charge approaches its equilibrium value rapidly when the time constant is small and slowly when the time constant is large.

Figure 20.39 shows a circuit at a time *t* after a switch is closed to allow a charged capacitor to begin discharging. There is no battery in this circuit, so the charge $+q$ on the left plate of the capacitor can flow counterclockwise through the resistor and neutralize the charge $-q$ on the right plate. Assuming that the capacitor has a charge q_0 at time $t = 0$ s when the switch is closed, it can be shown that

Capacitor discharging $\qquad\qquad q = q_0 e^{-t/(RC)}$ (20.22)

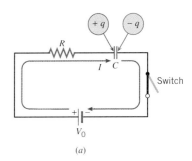
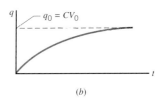

Figure 20.38 Charging a capacitor.

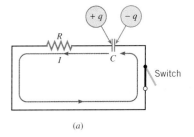
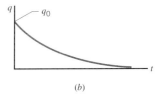

Figure 20.39 Discharging a capacitor.

where q is the amount of charge remaining on either plate at time t. The graph of this expression in part b of the drawing shows that the charge begins at q_0 when $t = 0$ s and decreases gradually toward zero. Smaller values of the time constant RC lead to a more rapid discharge. Equation 20.22 indicates that when $t = \tau = RC$, the magnitude of the charge remaining on each plate is $q_0e^{-1} = q_0(0.368)$. Therefore, the time constant is also the amount of time required for a charged capacitor to *lose* 63.2% of its charge.

The charging/discharging of a capacitor has many applications. Heart pacemakers, for instance, incorporate RC circuits to control the timing of voltage pulses that are delivered to a malfunctioning heart to regulate its beating cycle. The pulses are delivered by electrodes attached externally to the chest or located internally near the heart when the pacemaker is implanted surgically (see Figure 20.40). A voltage pulse is delivered when the capacitor discharges to a preset level, following which the capacitor is recharged rapidly and the cycle repeats. The value of the time constant RC controls the pulsing rate, which is about once per second.

The charging/discharging of a capacitor is also used in automobiles that have windshield wipers equipped for intermittent operation during a light drizzle. In this mode of operation, the wipers remain off for a while and then turn on briefly. The timing of the on–off cycle is determined by the time constant of a resistor–capacitor combination.

☩ *The physics of* **heart pacemakers.**

The physics of **windshield wipers.**

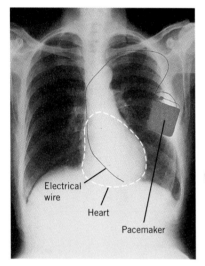

Figure 20.40 This X-ray photograph shows a heart pacemaker that has been implanted surgically. (© RNHRD NHS Trust/Stone/Getty Images)

Concept Simulation 20.4

The current in an *RC* circuit and the voltage across the capacitor are plotted as a function of time in this simulation. You can change the values of resistance and capacitance and see how the changes affect the charging of the capacitor.

Related Homework: Problem 98

Go to **www.wiley.com/college/cutnell**

✔ **Check Your Understanding 5**

The drawings show two different resistor–capacitor arrangements. The time constant for arrangement A is 0.20 s. What is the time constant for arrangement B? *(The answer is given at the end of the book.)*

Background: Capacitors, like resistors, can be connected in series and in parallel. The time constant for a given connection depends on the resistance and the equivalent capacitance of the capacitors.

For similar questions (including conceptual counterparts), consult Self-Assessment Test 20.2, which is described next.

Self-Assessment Test 20.2

Test your understanding of the material in Sections 20.8–20.13:

• Circuits Wired Partially in Series and Partially in Parallel   • Internal Resistance
• Kirchhoff's Rules   • The Measurement of Current and Voltage
• Capacitors in Series and Parallel   • *RC* Circuits

Go to **www.wiley.com/college/cutnell**

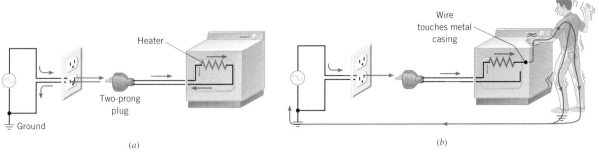

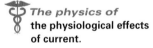

Figure 20.41 (*a*) A normally operating clothes dryer that is connected to a wall socket via a two-prong plug. (*b*) An internal wire accidentally touches the metal casing, and a person who touches the casing receives an electrical shock.

20.14 *Safety and the Physiological Effects of Current*

Electric circuits, although very useful, can also be hazardous. To reduce the danger inherent in using circuits, proper *electrical grounding* is necessary. The next two figures help to illustrate what electrical grounding means and how it is achieved.

Figure 20.41*a* shows a clothes dryer connected to a wall socket via a two-prong plug. The dryer is operating normally; that is, the wires inside are insulated from the metal casing of the dryer, so no charge flows through the casing itself. Notice that one terminal of the ac generator is customarily connected to ground ( ) by the electric power company. Part *b* of the drawing shows the hazardous result that occurs if a wire comes loose and contacts the metal casing. A person touching it receives a shock, since electric charge flows through the casing, the person's body, and the ground on the way back to the generator.

Figure 20.42 shows the same appliance connected to a wall socket via a three-prong plug that provides safe electrical grounding. The third prong connects the metal casing directly to a copper rod driven into the ground or to a copper water pipe that is in the ground. This arrangement protects against electrical shock in the event that a broken wire touches the metal casing. In this event, charge would flow through the casing, through the third prong, and into the ground, returning eventually to the generator. No charge would flow through the person's body, because the copper rod provides much less electrical resistance than does the body.

The physics of **safe electrical grounding.**

Serious and sometimes fatal injuries can result from electrical shock. The severity of the injury depends on the magnitude of the current and the parts of the body through which the moving charges pass. The amount of current that causes a mild tingling sensation is about 0.001 A. Currents on the order of 0.01–0.02 A can lead to muscle spasms, in which a person "can't let go" of the object causing the shock. Currents of approximately 0.2 A are potentially fatal because they can make the heart fibrillate, or beat in an uncontrolled manner. Substantially larger currents stop the heart completely. However, since the heart often begins beating normally again after the current ceases, the larger currents can be less dangerous than the smaller currents that cause fibrillation.

The physics of **the physiological effects of current.**

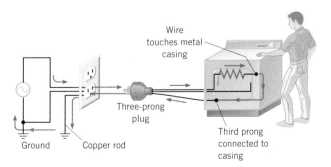

Figure 20.42 When the dryer malfunctions, a person touching it receives no shock, since electric charge flows through the third prong and into the ground via a copper rod, rather than through the person's body.

20.15 *Concepts & Calculations*

Series and parallel wiring are two common ways in which devices, such as light bulbs, can be connected to a circuit. The next example reviews the concepts of voltage, current, resistance, and power in the context of these two types of circuits.

Concepts & Calculations Example 15
The Brightness of Light Bulbs Wired in Series and in Parallel

A circuit contains a 48-V battery and a single light bulb whose resistance is 240 Ω. A second, identical, light bulb can be wired either in series or parallel with the first one (see Figure 20.43). Determine the power in a single bulb when the circuit contains (a) only one bulb, (b) two bulbs in series, and (c) two bulbs in parallel. Assume that the battery has no internal resistance.

Concept Questions and Answers How is the power P consumed by a light bulb related to the bulb's resistance R and the voltage V across it?

Answer Equation 20.6c gives the answer as $P = V^2/R$.

When there is only one light bulb in the circuit, what is the voltage V?

Answer Since there is only one bulb in the circuit, V is the voltage of the battery, $V = 48$ V.

The more power dissipated in a light bulb, the brighter it is. When two bulbs are wired in series, does the brightness of each bulb increase, decrease, or remain the same relative to the brightness when only a single bulb is in the circuit?

Answer Since the bulbs are identical and are wired in series, each receives one-half the battery voltage V. The power consumed by each bulb, then, is

$$P = \frac{(\frac{1}{2}V)^2}{R} = \frac{1}{4}\left(\frac{V^2}{R}\right)$$

This result shows that the power dissipated in each bulb in the series circuit is only one-fourth the power dissipated in the single-bulb circuit. Thus, the brightness of each bulb decreases.

When two bulbs are wired in parallel, does the brightness of each bulb increase, decrease, or remain the same relative to the brightness when only a single bulb is in the circuit?

Answer Since the bulbs are wired in parallel, each receives the full battery voltage V. Thus, the power consumed by each bulb remains the same as when only one bulb is in the circuit, so the brightness of each bulb does not change.

Solution

(a) When only one bulb is in the circuit, the power it consumes is

$$P = \frac{V^2}{R} = \frac{(48 \text{ V})^2}{240 \text{ }\Omega} = \boxed{9.6 \text{ W}} \tag{20.6c}$$

(b) When the two bulbs are wired in series, the current in the circuit is given by Equation 20.2 as $I = V/R_S$, where $R_S = R + R$ is the equivalent resistance of the series circuit. The power consumed by one of the light bulbs can be expressed using Equation 20.6b:

$$P = I^2R = \left(\frac{V}{R + R}\right)^2 R = \frac{V^2}{4R} = \frac{(48 \text{ V})^2}{4(240 \text{ }\Omega)} = \boxed{2.4 \text{ W}}$$

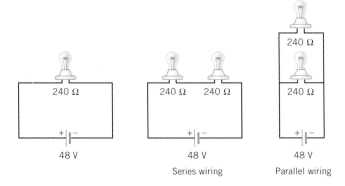

Figure 20.43 When a second identical light bulb is connected to the circuit on the left, it can be connected either in series or in parallel with the original bulb.

As expected, the power dissipated in each bulb is only one-fourth the power dissipated when there is only one bulb in the circuit.

(c) When the two bulbs are wired in parallel, the voltage across each is the same as the voltage of the battery. Therefore, the power dissipated in each bulb is given by

$$P = \frac{V^2}{R} = \frac{(48 \text{ V})^2}{240 \text{ }\Omega} = \boxed{9.6 \text{ W}} \qquad (20.6c)$$

As expected, the power dissipated, and hence the brightness, does not change relative to that in the single-bulb circuit.

Concept Simulation 20.5

This simulation explores the brightness of identical light bulbs in a 2-bulb series circuit with those in a 3-bulb circuit. For each of the circuits, the current and power in each bulb are displayed.

Related Homework: *Problem 64*

Go to
www.wiley.com/college/cutnell

Kirchhoff's junction rule and loop rule are important tools for analyzing the currents and voltages in complex circuits. The rules are easy to use, once some of the subtleties are understood. The next example explores these subtleties in a two-loop circuit.

Concepts & Calculations Example 16 Using Kirchhoff's Rules

For the circuit in Figure 20.44, use Kirchhoff's junction rule and loop rule to find the currents through the three resistors.

Concept Questions and Answers Notice that there are two loops, labeled 1 and 2, in this circuit. Does it matter that there is no battery in loop 1, but only two resistors?

Answer No, it doesn't matter. A loop can have any number of batteries, including none at all.

The currents through the three resistors are labeled as I_1, I_2 and I_3. Does it matter which direction, left-to-right or right-to-left, has been chosen for each current?

Answer No. If we initially select the wrong direction for a current, there is no problem. The value obtained for that particular current will turn out to be a negative number, indicating that the actual current is in the opposite direction.

When we place the + and − signs at the ends of each resistor, does it matter which end is + and which is −?

Answer Yes, it does. Once the direction of the current has been selected, the + and − signs must be chosen so that the current goes from the + end toward the − end of the resistor. Notice that this is the case for the three resistors in Figure 20.44.

When we evaluate the potential drops and rises around a closed loop, does it matter which direction, clockwise or counterclockwise, is chosen for the evaluation?

Answer No, the direction is arbitrary. If we choose a clockwise direction, for example, we will have a certain number of potential drops and rises. If we choose a counterclockwise direction, all the drops become rises, and vice versa. Since we always set the potential drops equal to the potential rises, it does not matter which direction is picked for evaluating them. In the Solution that follows, we will use a counterclockwise direction for loop 1 and a clockwise direction for loop 2.

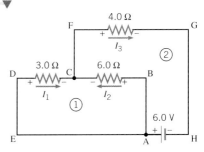

Figure 20.44 The three currents in this circuit are evaluated using Kirchhoff's loop and junction rules.

Solution Let's begin by expressing the current I_3 in terms of I_1 and I_2. Notice that I_1 and I_2 flow into junction C and I_3 flows out of it. Kirchhoff's junction rule states that

$$\underbrace{I_1 + I_2}_{\substack{\text{Currents} \\ \text{into junction}}} = \underbrace{I_3}_{\substack{\text{Currents} \\ \text{out of junction}}} \qquad (20.23)$$

We now apply the loop rule to loop 1, starting at point A and proceeding counterclockwise around the closed circuit (ABCDEA):

$$\underbrace{I_2(6.0 \text{ }\Omega)}_{\text{Potential drops}} = \underbrace{I_1(3.0 \text{ }\Omega)}_{\text{Potential rises}} \qquad (20.24)$$

The two equations above contain three unknown variables, so we now apply the loop rule to loop 2 to obtain another equation that contains these variables. Starting at point A, we will move clockwise around loop 2 (ABCFGHA):

$$\underbrace{I_2(6.0 \text{ }\Omega) + I_3(4.0 \text{ }\Omega)}_{\text{Potential drops}} = \underbrace{6.0 \text{ V}}_{\text{Potential rises}} \qquad (20.25)$$

Substituting Equation 20.23 for I_3 into Equation 20.25 gives

$$I_2(6.0 \ \Omega) + (I_1 + I_2)(4.0 \ \Omega) = 6.0 \ \text{V} \qquad (20.26)$$

Solving Equation 20.24 for I_2 in terms of I_1 and substituting the result into Equation 20.26 show that

$$\tfrac{1}{2}I_1 \ (6.0 \ \Omega) + (I_1 + \tfrac{1}{2}I_1)(4.0 \ \Omega) = 6.0 \ \text{V}$$

Solving this result for I_1 yields $I_1 = \boxed{\tfrac{2}{3} \ \text{A}}$. Substituting this result back into Equation 20.24 and solving for I_2 gives $I_2 = \boxed{\tfrac{1}{3} \ \text{A}}$. Equation 20.23 indicates, then, that $I_3 = I_1 + I_2 = \boxed{1.0 \ \text{A}}$.

▲

At the end of the problem set for this chapter, you will find homework problems that contain both conceptual and quantitative parts. These problems are grouped under the heading *Concepts &Calculations, Group Learning Problems*. They are designed for use by students working alone or in small learning groups. The conceptual part of each problem provides a convenient focus for group discussions.

Concept Summary

This summary presents an abridged version of the chapter, including the important equations and all available learning aids. For convenient reference, the learning aids (including the text's examples) are placed next to or immediately after the relevant equation or discussion. The following learning aids may be found on-line at **www.wiley.com/college/cutnell:**

Interactive LearningWare examples are solved according to a five-step interactive format that is designed to help you develop problem-solving skills.	**Concept Simulations** are animated versions of text figures or animations that illustrate important concepts. You can control parameters that affect the display, and we encourage you to experiment.
Interactive Solutions offer specific models for certain types of problems in the chapter homework. The calculations are carried out interactively.	**Self-Assessment Tests** include both qualitative and quantitative questions. Extensive feedback is provided for both incorrect and correct answers, to help you evaluate your understanding of the material.

Topic	*Discussion*	*Learning Aids*

20.1 Electromotive Force and Current

Electromotive force

There must be at least one source or generator of electrical energy in an electric circuit. The electromotive force (emf) of a generator, such as a battery, is the maximum potential difference (in volts) that exists between the terminals of the generator.

Current

The rate of flow of charge is called the electric current. If the rate is constant, the current I is given by

$$I = \frac{\Delta q}{\Delta t} \qquad (20.1) \quad \textbf{Example 1}$$

The ampere

where Δq is the magnitude of the charge crossing a surface in a time Δt, the surface being perpendicular to the motion of the charge. The SI unit for current is the coulomb per second (C/s), which is referred to as an ampere (A).

Direct current
Alternating current

When the charges flow only in one direction around a circuit, the current is called direct current (dc). When the direction of charge flow changes from moment to moment, the current is known as alternating current (ac).

Conventional current

Conventional current is the hypothetical flow of positive charges that would have the same effect in a circuit as the movement of negative charges that actually does occur.

20.2 Ohm's Law

Resistance

The definition of electrical resistance R is $R = V/I$, where V (in volts) is the voltage applied across a piece of material and I (in amperes) is the current through the material. Resistance is measured in volts per ampere, a unit called

The ohm

an ohm (Ω).

Topic	Discussion	Learning Aids
Ohm's law	If the ratio of the voltage to the current is constant for all values of voltage and current, the resistance is constant. In this event, the definition of resistance becomes Ohm's law as follows: $$\frac{V}{I} = R = \text{constant} \quad \text{or} \quad V = IR \qquad (20.2)$$	Example 2 Concept Simulation 20.1

20.3 Resistance and Resistivity

The resistance of a piece of material of length L and cross-sectional area A is

Resistance in terms of length and area	$$R = \rho \frac{L}{A} \qquad (20.3)$$	Example 3
Resistivity	where ρ is the resistivity of the material.	

Temperature dependence of resistivity	The resistivity of a material depends on the temperature. For many materials and limited temperature ranges, the temperature dependence is given by $$\rho = \rho_0[1 + \alpha(T - T_0)] \qquad (20.4)$$	Example 4
Temperature coefficient of resistivity	where ρ and ρ_0 are the resistivities at temperatures T and T_0, respectively, and α is the temperature coefficient of resistivity.	
Temperature dependence of resistance	The temperature dependence of the resistance R is given by $$R = R_0[1 + \alpha(T - T_0)] \qquad (20.5)$$ where R and R_0 are the resistances at temperatures T and T_0, respectively.	

20.4 Electric Power

In a circuit in which a current I results from a voltage V, the electric power delivered to the circuit is

Electric power	$$P = IV \qquad (20.6a)$$	Example 5
	For a resistor, Ohm's law applies, and it follows that the power dissipated in the resistance is also given by either of the following two equations:	
Power dissipated in a resistor	$$P = I^2R \qquad (20.6b)$$ $$P = \frac{V^2}{R} \qquad (20.6c)$$	

20.5 Alternating Current

The alternating voltage between the terminals of an ac generator can be represented by

Alternating voltage	$$V = V_0 \sin 2\pi f t \qquad (20.7)$$	
	where V_0 is the peak value of the voltage, t is the time, and f is the frequency (in Hertz) at which the voltage oscillates.	
	Correspondingly, in a circuit containing only resistance, the ac current is	
Alternating current	$$I = I_0 \sin 2\pi f t \qquad (20.8)$$	
	where I_0 is the peak value of the current and is related to the peak voltage via $I_0 = V_0/R$.	
	The root mean square (rms) current and voltage are related to the peak values according to the following equations:	
Root mean square current	$$I_{\text{rms}} = \frac{I_0}{\sqrt{2}} \qquad (20.12)$$	Examples 6, 7
Root mean square voltage	$$V_{\text{rms}} = \frac{V_0}{\sqrt{2}} \qquad (20.13)$$	
	The power in an ac circuit is the product of the current and the voltage and oscillates in time. The average power is	
Average ac power	$$\overline{P} = I_{\text{rms}}V_{\text{rms}} \qquad (20.15a)$$	

Topic	Discussion	Learning Aids
	For a resistor, Ohm's law applies, so that $V_{rms} = I_{rms}R$ and the average power dissipated in the resistor is also given by the following two equations:	
Average ac power dissipated in a resistor	$$\overline{P} = I_{rms}^2 R \qquad (20.15b)$$ $$\overline{P} = \frac{V_{rms}^2}{R} \qquad (20.15c)$$	

20.6 Series Wiring

When devices are connected in series, there is the same current through each device. The equivalent resistance R_S of a series combination of resistances (R_1, R_2, R_3, etc.) is

Equivalent series resistance

$$R_S = R_1 + R_2 + R_3 + \cdots \qquad (20.16)$$

The equivalent resistance dissipates the same total power as the series combination.

Interactive Solution 20.45

Example 8

20.7 Parallel Wiring

When devices are connected in parallel, the same voltage is applied across each device. In general, devices wired in parallel carry different currents. The reciprocal of the equivalent resistance R_P of a parallel combination of resistances (R_1, R_2, R_3, etc.) is

Equivalent parallel resistance

$$\frac{1}{R_P} = \frac{1}{R_1} + \frac{1}{R_2} + \frac{1}{R_3} + \cdots \qquad (20.17)$$

The equivalent resistance dissipates the same total power as the parallel combination.

Interactive Solution 20.49

Examples 9, 10

Use *Self-Assessment Test 20.1* to evaluate your understanding of Sections 20.1–20.7.

20.8 Circuits Wired Partially in Series and Partially in Parallel

Sometimes, one section of a circuit is wired in series, while another is wired in parallel. In such cases the circuit can be analyzed in parts, according to the respective series and parallel equivalent resistances of the various sections.

Example 11

Concept Simulations 20.2, 20.5

20.9 Internal Resistance

Internal resistance
Terminal voltage

The internal resistance of a battery or generator is the resistance within the battery or generator. The terminal voltage is the voltage between the terminals of a battery or generator and is equal to the emf only when there is no current through the device. When there is a current I, the internal resistance r causes the terminal voltage to be less than the emf by an amount Ir.

Example 12

20.10 Kirchhoff's Rules

The junction rule

Kirchhoff's junction rule states that the sum of the magnitudes of the currents directed into a junction equals the sum of the magnitudes of the currents directed out of the junction.

Example 13

Interactive LearningWare 20.1

The loop rule

Kirchhoff's loop rule states that, around any closed-circuit loop, the sum of the potential drops equals the sum of the potential rises. The Reasoning Strategy given at the end of Section 20.10 explains how these two rules can be applied to analyze any circuit.

Example 14

Concept Simulation 20.3

20.11 The Measurement of Current and Voltage

Galvanometer
Ammeter

Voltmeter

A galvanometer is a device that responds to electric current and is used in nondigital ammeters and voltmeters. An ammeter is an instrument that measures current and must be inserted into a circuit in such a way that the current passes directly through the ammeter. A voltmeter is an instrument for measuring the voltage between two points in a circuit. A voltmeter must be connected between the two points and is not inserted into a circuit as an ammeter is.

Topic	Discussion	Learning Aids
	20.12 *Capacitors in Series and Parallel*	
Equivalent parallel capacitance	The equivalent capacitance C_P for a parallel combination of capacitances (C_1, C_2, C_3, etc.) is $$C_P = C_1 + C_2 + C_3 + \cdots \qquad (20.18)$$ In general, each capacitor in a parallel combination carries a different amount of charge. The equivalent capacitor carries the same total charge and stores the same total energy as the parallel combination.	
Equivalent series capacitance	The reciprocal of the equivalent capacitance C_S for a series combination (C_1, C_2, C_3, etc.) of capacitances is $$\frac{1}{C_S} = \frac{1}{C_1} + \frac{1}{C_2} + \frac{1}{C_3} + \cdots \qquad (20.19)$$ The equivalent capacitor carries the same amount of charge as *any one* of the capacitors in the combination and stores the same total energy as the series combination.	
	20.13 *RC Circuits*	
Charging a capacitor	The charging or discharging of a capacitor in a dc series circuit (resistance R, capacitance C) does not occur instantaneously. The charge on a capacitor builds up gradually, as described by the following equation: $$q = q_0[1 - e^{-t/(RC)}] \qquad (20.20)$$ where q is the charge on the capacitor at time t and q_0 is the equilibrium value of the charge. The time constant τ of the circuit is	**Interactive Solution 20.99**
Time constant	$$\tau = RC \qquad (20.21)$$	**Concept Simulation 20.4**
Discharging a capacitor	The discharging of a capacitor through a resistor is described as follows: $$q = q_0 e^{-t/(RC)} \qquad (20.22)$$ where q_0 is the charge on the capacitor at time $t = 0$ s.	

 Use *Self-Assessment Test 20.2* to evaluate your understanding of Sections 20.8–20.13.

Conceptual Questions

1. The drawing shows a circuit in which a light bulb is connected to the household ac voltage via two switches S_1 and S_2. This is the kind of wiring, for example, that allows you to turn a carport light on and off from either inside the house or out in the carport. Explain which position A or B of S_2 turns the light on when S_1 is set to (a) position A and (b) position B.

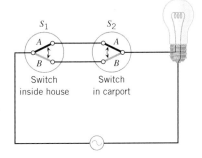

2. When an incandescent light bulb is turned on, the tungsten filament becomes white hot. The temperature coefficient of resistivity for tungsten is a positive number. What happens to the power delivered to the bulb as the filament heats up? Does the power increase, remain the same, or decrease? Justify your answer.

3. Two materials have different resistivities. Two wires of the same length are made, one from each of the materials. Is it possible for each wire to have the same resistance? Explain.

4. Does the resistance of a copper wire increase or decrease when both the length and the diameter of the wire are doubled? Justify your answer.

5. One electrical appliance operates with a voltage of 120 V, while another operates with 240 V. Based on this information alone, is it correct to say that the second appliance uses more power than the first? Give your reasoning.

6. Two light bulbs are designed for use at 120 V and are rated at 75 W and 150 W. Which light bulb has the greater filament resistance? Why?

7. Often, the instructions for an electrical appliance do not state how many watts of power the appliance uses. Instead, a statement such as "10 A, 120 V" is given. Explain why this statement is equivalent to telling you the power consumption.

8. The drawing shows a circuit that includes a bimetallic strip (made from brass and steel, see Section 12.4) with a resistance heater wire wrapped around it. When the switch is initially closed, a current appears in the circuit, because charges flow through the heater wire (which becomes hot), the strip itself, the contact point, and the light bulb. The bulb glows in response. As long as the switch remains closed, does the bulb continue to glow, eventually turn off permanently, or flash on and off? Account for your answer.

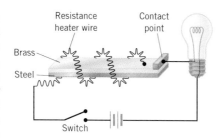

9. The power rating of a 1000-W heater specifies the power consumed when the heater is connected to an ac voltage of 120 V. Explain why the power consumed by two of these heaters connected in series with a voltage of 120 V is not 2000 W.

10. A number of light bulbs are to be connected to a single electrical outlet. Will the bulbs provide more brightness if they are connected in series or in parallel? Why?

11. A car has two headlights. The filament of one burns out. However, the other headlight stays on. Draw a circuit diagram that shows how the lights are connected to the battery. Give your reasoning.

12. In one of the circuits in the drawing, none of the resistors is in series or in parallel. Which is it? Explain.

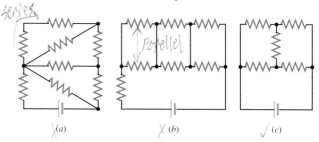

13. You have four identical resistors, each with a resistance of *R*. You are asked to connect these four together so that the equivalent resistance of the resulting combination is *R*. How many ways can you do it? There is more than one way. Justify your answers.

14. Compare the resistance of an ideal ammeter with the resistance of an ideal voltmeter and explain why the resistances are so different.

15. Describe what would happen to the current in a circuit if a voltmeter, inadvertently mistaken for an ammeter, were inserted into the circuit.

16. The time constant of a series *RC* circuit is $\tau = RC$. Verify that an ohm times a farad is equivalent to a second.

Problems

Note: For problems that involve ac conditions, the current and voltage are rms values and the power is an average value, unless indicated otherwise.

ssm Solution is in the Student Solutions Manual.   **www** Solution is available on the World Wide Web at www.wiley.com/college/cutnell
This icon represents a biomedical application.

Section 20.1 Electromotive Force and Current,
Section 20.2 Ohm's Law

1. ssm A fax machine uses 0.110 A of current in its normal mode of operation, but only 0.067 A in the standby mode. The machine uses a potential difference of 120 V. (a) In one minute, how much more charge passes through the machine in the normal mode versus the standby mode, and (b) how much more energy is used?

2. A defibrillator is used during a heart attack to restore the heart to its normal beating pattern (see Section 19.5). A defibrillator passes 18 A of current through the torso of a person in 2.0 ms. (a) How much charge moves during this time? (b) How many electrons pass through the wires connected to the patient?

3. The filament of a light bulb has a resistance of 580 Ω. A voltage of 120 V is connected across the filament. How much current is in the filament?

4. A battery charger is connected to a dead battery and delivers a current of 6.0 A for 5.0 hours, keeping the voltage across the battery terminals at 12 V in the process. How much energy is delivered to the battery?

5. ssm In the Arctic, electric socks are useful. A pair of socks uses a 9.0-V battery pack for each sock. A current of 0.11 A is drawn from each battery pack by wire woven into the socks. Find the resistance of the wire in one sock.

* **6.** A car battery has a rating of 220 ampere · hours (A · h). This rating is one indication of the *total charge* that the battery can provide to a circuit before failing. (a) What is the total charge (in coulombs) that this battery can provide? (b) Determine the maximum current that the battery can provide for 38 minutes.

* **7.** The resistance of a bagel toaster is 14 Ω. To prepare a bagel, the toaster is operated for one minute from a 120-V outlet. How much energy is delivered to the toaster?

* **8.** A resistor is connected across the terminals of a 9.0-V battery, which delivers 1.1×10^5 J of energy to the resistor in six hours. What is the resistance of the resistor?

** **9. ssm** A beam of protons is moving toward a target in a particle accelerator. This beam constitutes a current whose value is 0.50 μA. (a) How many protons strike the target in 15 seconds? (b) Each proton has a kinetic energy of 4.9×10^{-12} J. Suppose the target is a 15-gram block of aluminum, and all the kinetic energy of the protons goes into heating it up. What is the change in temperature of the block at the end of 15 s?

Section 20.3 Resistance and Resistivity

10. High-voltage power lines are a familiar sight throughout the country. The aluminum wire used for some of these lines has a cross-sectional area of 4.9×10^{-4} m². What is the resistance of ten kilometers of this wire?

11. ssm www Two wires have the same length and the same resistance. One is made from aluminum and the other from copper. Obtain the ratio of the cross-sectional area of the aluminum wire to that of the copper wire.

12. Two wires are identical, except that one is aluminum and one is copper. The aluminum wire has a resistance of 0.20 Ω. What is the resistance of the copper wire?

13. A cylindrical copper cable carries a current of 1200 A. There is a potential difference of 1.6×10^{-2} V between two points on the cable that are 0.24 m apart. What is the radius of the cable?

14. A coil of wire has a resistance of 38.0 Ω at 25 °C and 43.7 Ω at 55 °C. What is the temperature coefficient of resistivity?

15. In Section 12.3 it was mentioned that temperatures are often measured with electrical resistance thermometers made of platinum wire. Suppose that the resistance of a platinum resistance thermometer is 125 Ω when its temperature is 20.0 °C. The wire is then immersed in boiling chlorine, and the resistance drops to 99.6 Ω. The temperature coefficient of resistivity of platinum is $\alpha = 3.72 \times 10^{-3}$ (C°)$^{-1}$. What is the temperature of the boiling chlorine?

* **16.** The temperature coefficient of resistivity for gold is 0.0034 (C°)$^{-1}$, and for tungsten it is 0.0045 (C°)$^{-1}$. The resistance of a gold wire increases by 7.0% due to an increase in temperature. For the same increase in temperature, what is the percentage increase in the resistance of a tungsten wire?

* **17. ssm www** A wire has a resistance of 21.0 Ω. It is melted down, and from the same volume of metal a new wire is made that is three times longer than the original wire. What is the resistance of the new wire?

* **18.** An iron wire has a resistance of 5.90 Ω at 20.0 °C, and a gold wire has a resistance of 6.70 Ω at the same temperature. The temperature coefficient of resistivity for iron is 0.0050 (C°)$^{-1}$, and for gold it is 0.0034 (C°)$^{-1}$. At what temperature do the wires have the same resistance?

* **19. ssm www** Two wires have the same cross-sectional area and are joined end to end to form a single wire. One is tungsten, which has a temperature coefficient of resistivity of $\alpha = 0.0045$ (C°)$^{-1}$. The other is carbon, for which $\alpha = -0.0005$ (C°)$^{-1}$. The total resistance of the composite wire is the sum of the resistances of the pieces. The total resistance of the composite *does not change with temperature*. What is the ratio of the lengths of the tungsten and carbon sections? Ignore any changes in length due to thermal expansion.

** **20.** An aluminum wire is hung between two towers and has a length of 175 m. A current of 125 A exists in the wire, and the potential difference between the ends of the wire is 0.300 V. The density of aluminum is 2700 kg/m^3. Find the mass of the wire.

Section 20.4 Electric Power

21. The heating element in an iron has a resistance of 24 Ω. The iron is plugged into a 120-V outlet. What is the power dissipated by the iron?

22. A cigarette lighter in a car is a resistor that, when activated, is connected across the 12-V battery. Suppose a lighter dissipates 33 W of power. Find (a) the resistance of the lighter and (b) the current that the battery delivers to the lighter.

23. ssm In doing a load of clothes, a clothes dryer uses 16 A of current at 240 V for 45 min. A personal computer, in contrast, uses 2.7 A of current at 120 V. With the energy used by the clothes dryer, how long (in hours) could you use this computer to "surf" the Internet?

24. There are approximately 110 million TVs in the United States. Each uses, on average, 75 W of power and is turned on for 6.0 hours a day. If electrical energy costs $0.10 per KWh, how much money is spent every day in keeping the TVs turned on?

25. A blow-dryer and a vacuum cleaner each operate with a voltage of 120 V. The current rating of the blow-dryer is 11 A, and that of the vacuum cleaner is 4.0 A. Determine the power consumed by (a) the blow-dryer and (b) the vacuum cleaner. (c) Determine the ratio of the energy used by the blow-dryer in 15 minutes to the energy used by the vacuum cleaner in one-half hour.

* **26.** An electric heater is used to boil small amounts of water and consists of a 15-Ω coil that is immersed directly in the water. It operates from a 120-V socket. How much time is required for this heater to raise the temperature of 0.50 kg of water from 13 °C to the normal boiling point?

* **27. ssm** Tungsten has a temperature coefficient of resistivity of 0.0045 (C°)$^{-1}$. A tungsten wire is connected to a source of constant voltage via a switch. At the instant the switch is closed, the temperature of the wire is 28 °C, and the initial power dissipated in the wire is P_0. At what wire temperature has the power dissipated in the wire decreased to $\frac{1}{2}P_0$?

* **28.** A piece of Nichrome wire has a radius of 6.5×10^{-4} m. It is used in a laboratory to make a heater that dissipates 4.00×10^2 W of power when connected to a voltage source of 120 V. Ignoring the effect of temperature on resistance, estimate the necessary length of wire.

** **29.** An iron wire has a resistance of 12 Ω at 20.0 °C and a mass of 1.3×10^{-3} kg. A current of 0.10 A is sent through the wire for one minute and causes the wire to become hot. Assuming that all the electrical energy is dissipated in the wire and remains there, find the final temperature of the wire. *[Hint: Use the average resistance of the wire during the heating process, and see Table 12.2 for the specific heat capacity of iron. Note that $\alpha = 0.0050$ (C°)$^{-1}$.]*

Section 20.5 Alternating Current

30. According to Equation 20.7, an ac voltage V is given as a function of time t by $V = V_0 \sin 2\pi f t$, where V_0 is the peak voltage and f is the frequency (in hertz). For a frequency of 60.0 Hz, what is the smallest value of the time at which the voltage equals one-half of the peak value?

31. The current in a circuit is ac and has a peak value of 2.50 A. Determine the rms current.

32. The average power dissipated in a stereo speaker is 55 W. Assuming that the speaker can be treated as a 4.0-Ω resistance, find the peak value of the ac voltage applied to the speaker.

33. ssm The heating element in an iron has a resistance of 16 Ω and is connected to a 120-V wall socket. (a) What is the average power consumed by the iron, and (b) the peak power?

34. Review Conceptual Example 7 as an aid in solving this problem. A portable electric heater uses 18 A of current. The manufacturer recommends that an extension cord attached to the heater dissipate no more than 2.0 W of power per meter of length. What is the smallest radius of copper wire that can be used in the extension cord? *(Note: An extension cord contains two wires.)*

35. An electric furnace runs nine hours a day to heat a house during January (31 days). The heating element has a resistance of 5.3 Ω and carries a current of 25 A. The cost of electricity is $0.10/kWh. Find the cost of running the furnace for the month of January.

* **36.** A light bulb is connected to a 120.0-V wall socket. The current in the bulb depends on the time t according to the relation $I = (0.707 \text{ A})\sin[(314 \text{ Hz})t]$. (a) What is the frequency of the alternating current? (b) Determine the resistance of the bulb's filament. (c) What is the average power consumed by the light bulb?

* **37. ssm** The *recovery time* of a hot water heater is the time required to heat all the water in the unit to the desired temperature. Suppose that a 52-gal (1.00 gal = 3.79×10^{-3} m^3) unit starts with cold water at 11 °C and delivers hot water at 53 °C. The unit is electric and utilizes a resistance heater (120 V ac, 3.0 Ω) to heat the water. Assuming that no heat is lost to the environment, determine the recovery time (in hours) of the unit.

** **38.** To save on heating costs, the owner of a greenhouse keeps 660 kg of water around in barrels. During a winter day, the water is heated by the sun to 10.0 °C. During the night the water freezes into ice at 0.0 °C in nine hours. What is the minimum ampere rating of an electric heating system (240 V) that would provide the same heating effect as the water does?

Section 20.6 Series Wiring

39. ssm The current in a series circuit is 15.0 A. When an additional 8.00-Ω resistor is inserted in series, the current drops to 12.0 A. What is the resistance in the original circuit?

40. The current in a 47-Ω resistor is 0.12 A. This resistor is in series with a 28-Ω resistor, and the series combination is connected across a battery. What is the battery voltage?

41. A 36.0-Ω resistor and an 18.0-Ω resistor are connected in series across a 15.0-V battery. What is the voltage across (a) the 36.0-Ω resistor and (b) the 18.0-Ω resistor?

42. A 60.0-W lamp is placed in series with a resistor and a 120.0-V source. If the voltage across the lamp is 25 V, what is the resistance R of the resistor?

43. ssm Three resistors, 25, 45, and 75 Ω, are connected in series, and a 0.51-A current passes through them. What is (a) the equivalent resistance and (b) the potential difference across the three resistors?

* **44.** A 47-Ω resistor can dissipate up to 0.25 W of power without burning up. What is the smallest number of such resistors that can be connected in series across a 9.0-V battery without any one of them burning up?

* **45. Interactive Solution 20.45** at **www.wiley.com/college/cutnell** provides one approach to problems like this one. Three resistors are connected in series across a battery. The value of each resistance and its maximum power rating are as follows: 2.0 Ω and 4.0 W, 12.0 Ω and 10.0 W, and 3.0 Ω and 5.0 W. (a) What is the greatest voltage that the battery can have without one of the resistors burning up? (b) How much power does the battery deliver to the circuit in (a)?

* **46.** One heater uses 340 W of power when connected by itself to a battery. Another heater uses 240 W of power when connected by itself to the same battery. How much total power do the heaters use when they are both connected in series across the battery?

** **47.** Two resistances, R_1 and R_2, are connected in series across a 12-V battery. The current increases by 0.20 A when R_2 is removed, leaving R_1 connected across the battery. However, the current increases by just 0.10 A when R_1 is removed, leaving R_2 connected across the battery. Find (a) R_1 and (b) R_2.

Section 20.7 Parallel Wiring

48. What resistance must be placed in parallel with a 155-Ω resistor to make the equivalent resistance 115 Ω?

49. Help with problems of this kind is available in **Interactive Solution 20.49** at **www.wiley.com/college/cutnell.** A coffee cup heater and a lamp are connected in parallel to the same 120-V outlet. Together, they use a total of 111 W of power. The resistance of the heater is 4.0×10^2 Ω. Find the resistance of the lamp.

50. For the three-way bulb (50 W, 100 W, 150 W) discussed in Conceptual Example 10, find the resistance of each of the two filaments. Assume that the wattage ratings are not limited by significant figures and ignore any heating effects on the resistances.

51. ssm A 16-Ω loudspeaker and an 8.0-Ω loudspeaker are connected in parallel across the terminals of an amplifier. Assuming the speakers behave as resistors, determine the equivalent resistance of the two speakers.

52. A wire whose resistance is R is cut into three equally long pieces, which are then connected in parallel. In terms of R, what is the resistance of the parallel combination?

53. ssm Two resistors, 42.0 and 64.0 Ω, are connected in parallel. The current through the 64.0-Ω resistor is 3.00 A. (a) Determine the current in the other resistor. (b) What is the total power consumed by the two resistors?

* **54.** Two resistors have resistances R_1 and R_2. When the resistors are connected in series to a 12.0-V battery, the current from the battery is 2.00 A. When the resistors are connected in parallel to the battery, the total current from the battery is 9.00 A. Determine R_1 and R_2.

* **55. ssm** The total current delivered to a number of devices connected in parallel is the sum of the individual currents in each device. Circuit breakers are resettable automatic switches that protect against a dangerously large total current by "opening" to stop the current at a specified safe value. A 1650-W toaster, a 1090-W iron, and a 1250-W microwave oven are turned on in a kitchen. As the drawing shows, they are all connected through a 20-A circuit breaker to an ac voltage of 120 V. (a) Find the equivalent resistance of the three devices. (b) Obtain the total current delivered by the source and determine whether the breaker will "open" to prevent an accident.

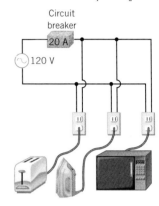

* **56.** A resistor (resistance = R) is connected first in parallel and then in series with a 2.00-Ω resistor. A battery delivers five times as much current to the parallel combination as it does to the series combination. Determine the two possible values for R.

** **57.** The rear window defogger of a car consists of thirteen thin wires (resistivity = 88.0×10^{-8} $\Omega \cdot$m) embedded in the glass. The wires are connected in parallel to the 12.0-V battery, and each has a length of 1.30 m. The defogger can melt 2.10×10^{-2} kg of ice at 0 °C into water at 0 °C in two minutes. Assume that all the power dissipated in the wires is used immediately to melt the ice. Find the cross-sectional area of each wire.

Section 20.8 Circuits Wired Partially in Series and Partially in Parallel

58. Find the equivalent resistance between points A and B in the drawing.

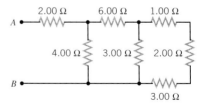

59. ssm For the combination of resistors shown in the drawing, determine the equivalent resistance between points A and B.

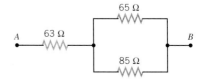

60. A 14-Ω coffee maker and a 16-Ω frying pan are connected in series across a 120-V source of voltage. A 23-Ω bread maker is also connected across the 120-V source and is in parallel with the series combination. Find the total current supplied by the source of voltage.

61. ssm Determine the equivalent resistance between the points A and B for the group of resistors in the drawing.

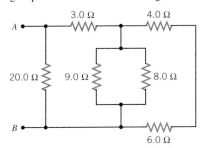

62. A 60.0-Ω resistor is connected in parallel with a 120.0-Ω resistor. This parallel group is connected in series with a 20.0-Ω resistor. The total combination is connected across a 15.0-V battery. Find (a) the current and (b) the power dissipated in the 120.0-Ω resistor.

* **63.** The circuit in the drawing contains five identical resistors. The 45-V battery delivers 58 W of power to the circuit. What is the resistance R of each resistor?

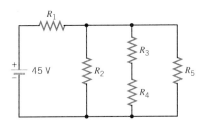

* **64. Concept Simulation 20.5** at **www.wiley.com/college/cutnell** provides some background pertinent to this problem. Determine the power dissipated in each of the resistors in the drawing.

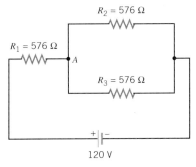

* **65. ssm** Eight different values of resistance can be obtained by connecting together three resistors (1.00, 2.00, and 3.00 Ω) in all possible ways. What are they?

** **66.** Three identical resistors are connected in parallel. The equivalent resistance increases by 700 Ω when one resistor is removed and connected in series with the remaining two, which are still in parallel. Find the resistance of each resistor.

Section 20.9 Internal Resistance

67. A new "D" battery has an emf of 1.5 V. When a wire of negligible resistance is connected between the terminals of the battery, a current of 28 A is produced. Find the internal resistance of the battery.

68. A battery has an internal resistance of 0.012 Ω and an emf of 9.00 V. What is the maximum current that can be drawn from the battery without the terminal voltage dropping below 8.90 V?

69. ssm A battery has an internal resistance of 0.50 Ω. A number of identical light bulbs, each with a resistance of 15 Ω, are connected in parallel across the battery terminals. The terminal voltage of the battery is observed to be one-half the emf of the battery. How many bulbs are connected?

70. A battery has an emf of 12.0 V and an internal resistance of 0.15 Ω. What is the terminal voltage when the battery is connected to a 1.50-Ω resistor?

* **71.** A battery delivering a current of 55.0 A to a circuit has a terminal voltage of 23.4 V. The electric power being dissipated by the internal resistance of the battery is 34.0 W. Find the emf of the battery.

** **72.** A resistor has a resistance R, and a battery has an internal resistance r. When the resistor is connected across the battery, ten percent less power is dissipated in R than would be dissipated if the battery had no internal resistance. Find the ratio r/R.

Section 20.10 Kirchhoff's Rules

73. ssm Consider the circuit in the drawing. Determine (a) the magnitude of the current in the circuit and (b) the magnitude of the voltage between the points labeled A and B. (c) State which point, A or B, is at the higher potential.

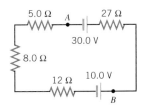

74. A current of 2.0 A exists in the partial circuit shown in the drawing. What is the magnitude of the potential difference between the points (a) A and B, and (b) A and C?

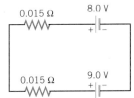

75. Two batteries, each with an internal resistance of 0.015 Ω, are connected as in the drawing. In effect, the 9.0-V battery is being used to charge the 8.0-V battery. What is the current in the circuit?

76. Interactive LearningWare 20.1 at **www.wiley.com/college/cutnell** provides background for this problem. Find the magnitude and direction of the current in the 2.0-Ω resistor in the drawing.

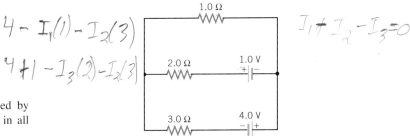

$$4 - I_1(1) - I_2(3)$$
$$4 + 1 - I_3(2) - I_2(3)$$
$$I_1 + I_2 - I_3 = 0$$

* **77. ssm** Determine the voltage across the 5.0-Ω resistor in the drawing. Which end of the resistor is at the higher potential?

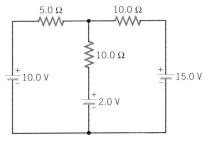

* 78. Concept Simulation 20.3 at www.wiley.com/college/cutnell allows you to verify your answer for this problem. Find the current in the 4.00-Ω resistor in the drawing. Specify the direction of the current.

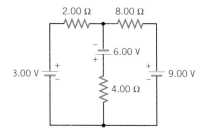

** 79. The circuit in the drawing is known as a Wheatstone bridge circuit. Find the voltage between points B and D, and state which point is at the higher potential.

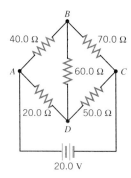

Section 20.11 The Measurement of Current and Voltage

80. A galvanometer has a full-scale current of 0.100 mA and a coil resistance of 50.0 Ω. This instrument is used with a shunt resistor to form an ammeter that will register full scale for a current of 60.0 mA. Determine the resistance of the shunt resistor.

81. ssm A voltmeter utilizes a galvanometer that has a 180-Ω coil resistance and a full-scale current of 8.30 mA. The voltmeter measures voltages up to 30.0 V. Determine the resistance that is connected in series with the galvanometer.

82. Voltmeter A has an equivalent resistance of 2.40×10^5 Ω and a full-scale voltage of 50.0 V. Voltmeter B, using the same galvanometer as voltmeter A, has an equivalent resistance of 1.44×10^5 Ω. What is its full-scale voltage?

83. A galvanometer with a coil resistance of 12.0 Ω and a full-scale current of 0.150 mA is used with a shunt resistor to make an ammeter. The ammeter registers a maximum current of 4.00 mA. Find the equivalent resistance of the ammeter.

* 84. Two scales on a voltmeter measure voltages up to 20.0 and 30.0 V, respectively. The resistance connected in series with the galvanometer is 1680 Ω for the 20.0-V scale and 2930 Ω for the 30.0-V scale. Determine the coil resistance and the full-scale current of the galvanometer that is used in the voltmeter.

** 85. ssm In measuring a voltage, a voltmeter uses some current from the circuit. Consequently, the voltage measured is only an approximation to the voltage present when the voltmeter is not connected. Consider a circuit consisting of two 1550-Ω resistors connected in series across a 60.0-V battery. (a) Find the voltage across one of the resistors. (b) A voltmeter has a full-scale voltage of 60.0 V and uses a galvanometer with a full-scale deflection of 5.00 mA. Determine the voltage that this voltmeter registers when it is connected across the resistor used in part (a).

Section 20.12 Capacitors in Series and Parallel

86. Two capacitors are connected in parallel across the terminals of a battery. One has a capacitance of 2.0 μF and the other a capacitance of 4.0 μF. These two capacitors together store 5.4×10^{-5} C of charge. What is the voltage of the battery?

87. Determine the equivalent capacitance between A and B for the group of capacitors in the drawing.

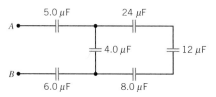

88. Three capacitors (4.0, 6.0, and 12.0 μF) are connected in series across a 50.0-V battery. Find the voltage across the 4.0-μF capacitor.

89. ssm Three capacitors (3.0, 7.0, and 9.0 μF) are connected in series. What is their equivalent capacitance?

90. Three capacitors have identical geometries. One is filled with a material whose dielectric constant is 2.50. Another is filled with a material whose dielectric constant is 4.00. The third capacitor is filled with a material whose dielectric constant κ is such that this single capacitor has the same capacitance as the series combination of the other two. Determine κ.

91. Suppose two capacitors (C_1 and C_2) are connected in series. Show that the sum of the energies stored in these capacitors is equal to the energy stored in the equivalent capacitor. [*Hint: The energy stored in a capacitor can be expressed as $q^2/(2C)$.*]

* 92. A 3.00-μF and a 5.00-μF capacitor are connected in series across a 30.0-V battery. A 7.00-μF capacitor is then connected in parallel across the 3.00-μF capacitor. Determine the voltage across the 7.00-μF capacitor.

* 93. ssm A sheet of gold foil (negligible thickness) is placed between the plates of a capacitor and has the same area as each of the plates. The foil is parallel to the plates, at a position one-third of the way from one to the other. Before the foil is inserted, the capacitance is C_0. What is the capacitance after the foil is in place? Express your answer in terms of C_0.

** 94. The drawing shows two fully charged capacitors ($C_1 = 2.00$ μF, $q_1 = 6.00$ μC; $C_2 = 8.00$ μF, $q_2 = 12.0$ μC). The switch is closed, and charge flows until equilibrium is reestablished (i.e., until both capacitors have the same voltage across their plates). Find the resulting voltage across either capacitor.

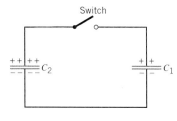

Section 20.13 *RC* Circuits

95. The circuit in the drawing contains two resistors and two capacitors that are connected to a battery via a switch. When the switch is closed, the capacitors begin to charge up. What is the time constant for the charging process?

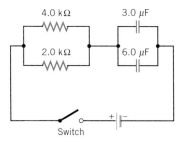

96. An electronic flash attachment for a camera produces a flash by using the energy stored in a 750-μF capacitor. Between flashes, the capacitor recharges through a resistor whose resistance is chosen so the capacitor recharges with a time constant of 3.0 s. Determine the value of the resistance.

97. ssm In a heart pacemaker, a pulse is delivered to the heart 81 times per minute. The capacitor that controls this pulsing rate discharges through a resistance of $1.8 \times 10^6\ \Omega$. One pulse is delivered every time the fully charged capacitor loses 63.2% of its original charge. What is the capacitance of the capacitor?

* **98.** Concept Simulation 20.4 at **www.wiley.com/college/cutnell** provides background for this problem and gives you the opportunity to verify your answer graphically. How many time constants must elapse before a capacitor in a series RC circuit is charged to 80.0% of its equilibrium charge?

* **99.** For one approach to problems like this one, consult Interactive Solution 20.99 at **www.wiley.com/college/cutnell**. Four identical capacitors are connected with a resistor in two different ways. When they are connected as in part a of the drawing, the time constant to charge up this circuit is 0.72 s. What is the time constant when they are connected with the same resistor as in part b?

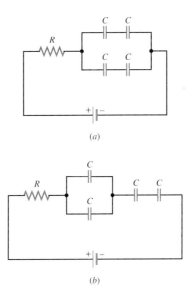
(a)

(b)

Additional Problems

100. An automobile battery is being charged at a voltage of 12.0 V and a current of 19.0 A. How much power is being produced by the charger?

101. ssm A 3.0-μF capacitor and a 4.0-μF capacitor are connected in series across a 40.0-V battery. A 10.0-μF capacitor is also connected directly across the battery terminals. Find the total charge that the battery delivers to the capacitors.

102. A lightning bolt delivers a charge of 35 C to the ground in a time of 1.0×10^{-3} s. What is the current?

103. ssm The heating element of a clothes dryer has a resistance of 11 Ω and is connected across a 240-V electrical outlet. What is the current in the heating element?

104. For the circuit shown in the drawing, find the current I through the 2.00-Ω resistor and the voltage V of the battery to the left of this resistor.

105. A wire of unknown composition has a resistance of $R_0 = 35.00\ \Omega$ when immersed in water at 20.0 °C. When the wire is placed in boiling water, its resistance rises to 47.60 Ω. What is the temperature on a hot summer day when the wire has a resistance of 37.80 Ω?

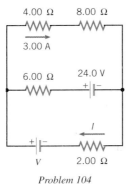

Problem 104

106. The circuit in the drawing shows two resistors, a capacitor, and a battery. When the capacitor is fully charged, what is the magnitude q of the charge on one of its plates?

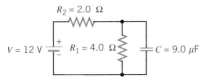

107. The coil of a galvanometer has a resistance of 20.0 Ω, and its meter deflects full scale when a current of 6.20 mA passes through it. To make the galvanometer into an ammeter, a 24.8-mΩ shunt resistor is added to it. What is the maximum current that this ammeter can read?

108. Three resistors, 9.0, 5.0, and 1.0 Ω, are connected in series across a 24-V battery. Find (a) the current in, (b) the voltage across, and (c) the power dissipated in each resistor.

* **109.** ssm www The current in the 8.00-Ω resistor in the drawing is 0.500 A. Find the current in (a) the 20.0-Ω resistor and in (b) the 9.00-Ω resistor.

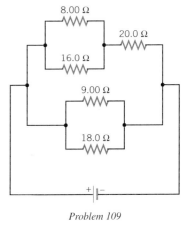

Problem 109

* **110.** A cylindrical aluminum pipe of length 1.50 m has an inner radius of 2.00×10^{-3} m and an outer radius of 3.00×10^{-3} m. The interior of the pipe is completely filled with copper. What is the resistance of this unit? *(Hint: Imagine that the pipe is connected between the terminals of a battery and decide whether the aluminum and copper parts of the pipe are in series or in parallel.)*

* **111.** ssm An extension cord is used with an electric weed trimmer that has a resistance of 15.0 Ω. The extension cord is made of copper wire that has a cross-sectional area of 1.3×10^{-6} m². The combined length of the two wires in the extension cord is 92 m. (a) Determine the resistance of the extension cord. (b) The extension cord is plugged into a 120-V socket. What voltage is applied to the trimmer itself?

* **112.** A 75.0-Ω and a 45.0-Ω resistor are connected in parallel. When this combination is connected across a battery, the current delivered by the battery is 0.294 A. When the 45.0-Ω resistor is disconnected, the current from the battery drops to 0.116 A. Determine (a) the emf and (b) the internal resistance of the battery.

* **113.** A toaster uses a Nichrome heating wire. When the toaster is turned on at 20 °C, the initial current is 1.50 A. A few seconds later, the toaster warms up and the current has a value of 1.30 A. The average temperature coefficient of resistivity for Nichrome wire is $4.5 \times 10^{-4}\ (\text{C}°)^{-1}$. What is the temperature of the heating wire?

* **114.** Two cylindrical rods, one copper and the other iron, are identical in lengths and cross-sectional areas. They are joined end to end to form one long rod. A 12-V battery is connected across the free ends of the copper–iron rod. What is the voltage between the ends of the copper rod?

* **115. ssm** A 7.0-μF and a 3.0-μF capacitor are connected in series across a 24-V battery. What voltage is required to charge a parallel combination of the two capacitors to the same total energy?

** **116.** A digital thermometer uses a thermistor as the temperature-sensing element. A thermistor is a kind of semiconductor and has a large negative temperature coefficient of resistivity α. Suppose $\alpha = -0.060$ $(\text{C}°)^{-1}$ for the thermistor in a digital thermometer used to measure the temperature of a patient. The resistance of the thermistor decreases to 85% of its value at the normal body temperature of 37.0 °C. What is the patient's temperature?

Concepts & Calculations Group Learning Problems

Note: Each of these problems consists of Concept Questions followed by a related quantitative Problem. They are designed for use by students working alone or in small learning groups. The Concept Questions involve little or no mathematics and are intended to stimulate group discussions. They focus on the concepts with which the problems deal. Recognizing the concepts is the essential initial step in any problem-solving technique.

117. Concept Questions The resistance and the magnitude of the current depend on the path that the current takes. The drawing shows three situations in which the current takes different paths through a piece of material. Rank them according to (a) resistance and (b) current, largest first. Give your reasoning.

Problem Each of the rectangular pieces is made from a material whose resistivity is $\rho = 1.50 \times 10^{-2}$ $\Omega \cdot \text{m}$, and the unit of length in the drawing is $L_0 = 5.00$ cm. If the material is connected to a 3.00-V battery, find (a) the resistance and (b) the current in each case. Verify that your answers are consistent with your answers to the Concept Questions.

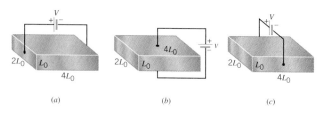

(a)   (b)   (c)

118. Concept Questions Each of the four circuits in the drawing consists of a single resistor whose resistance is either R or $2R$, and a single battery whose voltage is either V or $2V$. Rank the circuits according to (a) the power and (b) the current delivered to the resistor, largest to smallest. Explain your answers.

Problem The unit of voltage in each circuit is $V = 12.0$ V and the unit of resistance is $R = 6.00$ Ω. Determine (a) the power dissipated in each resistor and (b) the current delivered to each resistor. Check to see that your answers are consistent with your answers to the Concept Questions.

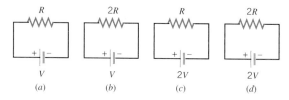

(a)   (b)   (c)   (d)

119. Concept Questions The drawing shows three different resistors in two different circuits. The resistances are such that $R_1 > R_2 > R_3$. (a) For the circuit on the left, rank the current through each resistor and the voltage across each one, largest first. (b) Repeat part (a) for the circuit on the right. Justify your answers.

Problem The battery has a voltage of $V = 24.0$ V, and the resistors have values of $R_1 = 50.0$ Ω, $R_2 = 25.0$ Ω, and $R_3 = 10.0$ Ω. (a) For the circuit on the left, determine the current through and the voltage across each resistor. (b) Repeat part (a) for the circuit on the right. Be sure your answers are consistent with your answers to the Concept Questions.

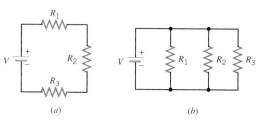

(a)   (b)

120. Concept Question The circuit in the drawing contains three identical resistors. Consider the equivalent resistance between the two points a and b, b and c, and a and c. Rank the equivalent resistances in decreasing order. Explain your reasoning.

Problem Each resistor has a value of 10.0 Ω. Determine the equivalent resistance between the points a and b, b and c, and a and c. Check to see that your answers are consistent with your answer to the Concept Question.

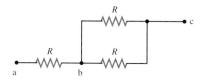

121. Concept Question Parts a and b of the drawing contain a resistor and a battery. Using the direction of the current as a guide, label the ends of each resistor with + and − signs. In both cases, proceed from point A to point B and determine the potential drops and potential rises. Express the drops and rises in terms of the current I, the resistance R, and the battery voltage V.

	Potential drops	Potential rises
Part a		
Part b		

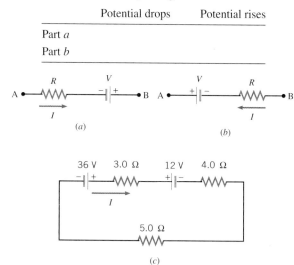

(a)   (b)

(c)

Problem Using Kirchhoff's loop rule, find the value of the current I in part c of the drawing.

122. Concept Question Two capacitors, C_1 and C_2 are connected to a battery whose voltage is V. Recall from Section 19.5 that the electrical energy stored by each capacitor is $\frac{1}{2}C_1V_1^2$ and $\frac{1}{2}C_2V_2^2$, where V_1 and V_2 are, respectively, the voltages across C_1 and C_2. If the capacitors are connected in series, is the total energy stored by them greater than, less than, or equal to the total energy stored when they are connected in parallel? Justify your answer.

Problem The battery voltage is $V = 60.0$ V, and the capacitances are $C_1 = 2.00\ \mu$F and $C_2 = 4.00\ \mu$F. Determine the total energy stored by the two capacitors when they are wired (a) in parallel and (b) in series. Check to make sure that your answer is consistent with your answer to the Concept Question.

123. Concept Questions The drawing shows two circuits, and the same battery is used in each. The two resistances R_A in circuit A are the same, and the two resistances R_B in circuit B are the same. (a) How is the total power delivered by the battery related to the equivalent resistance connected between the battery terminals and to the battery voltage? (b) When two resistors are connected in series, is the equivalent resistance of the combination greater than, smaller than, or equal to the resistance of either resistor alone? (c) When two resistors are connected in parallel, is the equivalent resistance of the combination greater than, smaller than, or equal to the resistance of either resistor alone? (d) The same total power is delivered by the battery in circuits A and B. Is R_B greater than, smaller than, or equal to R_A?

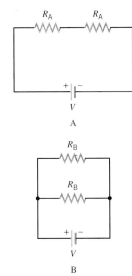

Problem Knowing that the same total power is delivered in each case, find the ratio R_B/R_A for the circuits in the drawing. Verify that your answer is consistent with your answer to Concept Question (d).

124. Concept Questions Each resistor in the three circuits in the drawing has the same resistance R, and the batteries have the same voltage V. (a) How is the total power delivered by the battery related to the equivalent resistance connected between the battery terminals and to the battery voltage? (b) Rank the equivalent resistances of the circuits in descending order (largest first). (c) Rank the three values of the total power delivered by the batteries in descending order (largest first).

Problem The values for R and V in the drawing are 9.0 Ω and 6.0 V, respectively. Determine the total power delivered by the battery in each of the three circuits. Be sure that your answer is consistent with your answer to Concept Question (c).

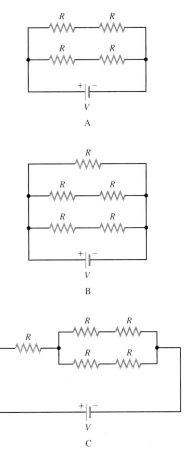

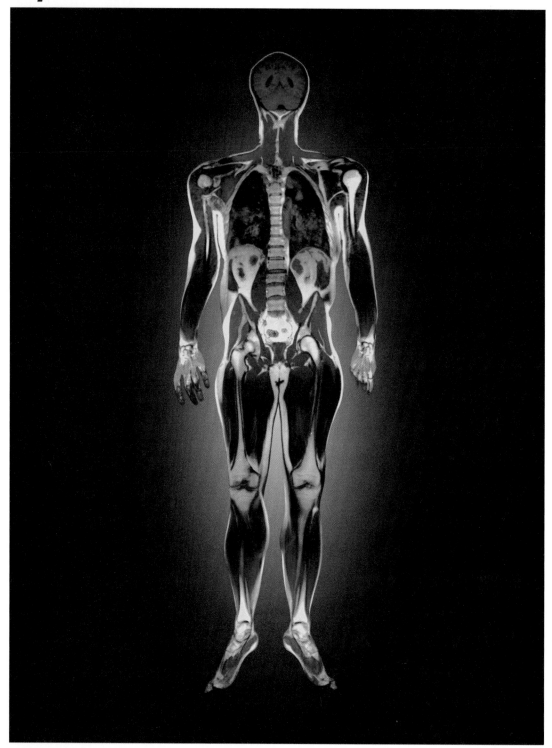

Chapter 21

Magnetic Forces and Magnetic Fields

Magnetic resonance imaging (MRI) uses a magnetic field in a noninvasive way to provide physicians with a powerful diagnostic tool. MRI can provide detailed three-dimensional whole-body images, such as the one shown here. Magnetic forces and magnetic fields are the subjects of this chapter. (© David Job/Stone/Getty Images)

21.1 Magnetic Fields

Permanent magnets have long been used in navigational compasses. As Figure 21.1 illustrates, the compass needle is a permanent magnet supported so it can rotate freely in a plane. When the compass is placed on a horizontal surface, the needle rotates until one end points approximately to the north. The end of the needle that points north is labeled the *north magnetic pole;* the opposite end is the *south magnetic pole.*

Magnets can exert forces on each other. Figure 21.2 shows that the magnetic forces between north and south poles have the property that

> *like poles repel each other, and unlike poles attract.*

This behavior is similar to that of like and unlike electric charges. However, there is a significant difference between magnetic poles and electric charges. It is possible to separate positive from negative electric charges and produce isolated charges of either kind. In contrast, no one has found a magnetic monopole (an isolated north or south pole). Any attempt to separate north and south poles by cutting a bar magnet in half fails, because each piece becomes a smaller magnet with its own north and south poles.

Surrounding a magnet, there is a *magnetic field.* The magnetic field is analogous to the electric field that exists in the space around electric charges. Like the electric field, the magnetic field has both a magnitude and a direction. We postpone a discussion of the magnitude until Section 21.2, concentrating our attention here on the direction. *The direction of the magnetic field at any point in space is the direction indicated by the north pole of a small compass needle placed at that point.* In Figure 21.3 the compass needle is symbolized by an arrow, the head of the arrow being the north pole. The drawing shows how compasses can be used to map out the magnetic field in the space around a bar magnet. Since like poles repel and unlike poles attract, the needle of each compass becomes aligned relative to the magnet in the manner shown in the picture. The compass needles provide a visual picture of the magnetic field that the bar magnet creates.

To help visualize the electric field, we introduced electric field lines in Section 18.7. In a similar fashion, it is possible to draw magnetic field lines, and Figure 21.4a illustrates some of the lines around a bar magnet. The lines appear to originate from the north pole and end on the south pole; they do not start or stop in midspace. A visual image of the magnetic field lines can be created by sprinkling finely ground iron filings on a piece of paper that covers the magnet. Iron filings in a magnetic field behave like tiny compasses and align themselves along the field lines, as the photo in Figure 21.4b shows.

As is the case with electric field lines, the magnetic field at any point is tangent to the magnetic field line at that point. Furthermore, the strength of the field is proportional to the number of lines per unit area that passes through a surface oriented perpendicular to the lines. Thus, the magnetic field is stronger in regions where the field lines are relatively close together and weaker where they are relatively far apart. For instance, in Figure 21.4a the lines are closest together near the north and south poles, reflecting the fact that the strength of the field is greatest in these regions. Away from the poles, the magnetic field becomes weaker. Notice in part c of the drawing that the field lines in the gap between the poles of the horseshoe magnet are nearly parallel and equally spaced, indicating that the magnetic field there is approximately constant.

Figure 21.1 The needle of a compass is a permanent magnet that has a north magnetic pole (N) at one end and a south magnetic pole (S) at the other.

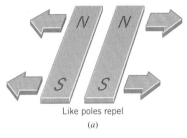

Like poles repel

(a)

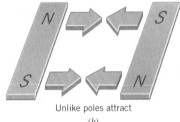

Unlike poles attract

(b)

Figure 21.2 Bar magnets have a north magnetic pole at one end and a south magnetic pole at the other end. (*a*) Like poles repel each other, and (*b*) unlike poles attract.

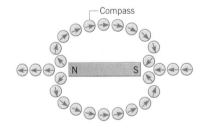

Figure 21.3 At any location in the vicinity of a magnet, the north pole (the arrowhead in this drawing) of a small compass needle points in the direction of the magnetic field at that location.

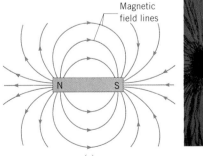

Magnetic field lines

(a)

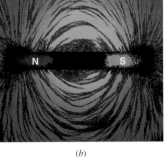

(b)

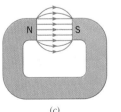

(c)

Figure 21.4 (*a*) The magnetic field lines and (*b*) the pattern of iron filings (dark, curved regions) in the vicinity of a bar magnet. (*c*) The magnetic field lines in the gap of a horseshoe magnet. (© Yoav Levy/Phototake)

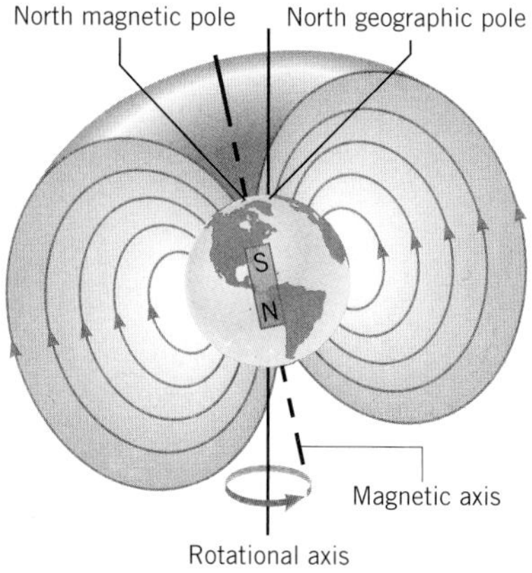

North magnetic pole   North geographic pole

Magnetic axis

Rotational axis

Figure 21.5 The earth behaves magnetically almost as if a bar magnet were located near its center. The axis of this fictitious bar magnet does not coincide with the earth's rotational axis; the two axes are currently about 11.5° apart.

Although the north pole of a compass needle points northward, it does not point exactly at the north geographic pole. The north geographic pole is that point where the earth's axis of rotation crosses the surface in the northern hemisphere (see Figure 21.5). Measurements of the magnetic field surrounding the earth show that the earth behaves magnetically almost as if it were a bar magnet.* As the drawing illustrates, the orientation of this fictitious bar magnet defines a magnetic axis for the earth. The location where the magnetic axis crosses the surface in the northern hemisphere is known as the north magnetic pole. The north magnetic pole is so named because it is the location toward which the north end of a compass needle points. Since unlike poles attract, the south pole of the earth's fictitious bar magnet lies beneath the north magnetic pole, as Figure 21.5 indicates.

The north magnetic pole does not coincide with the north geographic pole but, instead, lies at a latitude of nearly 80°, just northwest of Ellef Ringnes Island in extreme northern Canada. It is interesting to note that the position of the north magnetic pole is not fixed, but moves over the years. Pointing as it does at the north magnetic pole, a compass needle deviates from the north geographic pole. The angle that a compass needle deviates is called the *angle of declination*. For New York City, the present angle of declination is about 12° west, meaning that a compass needle points 12° west of geographic north.

Figure 21.5 shows that the earth's magnetic field lines are not parallel to the surface at all points. For instance, near the north magnetic pole the field lines are almost perpendicular to the surface of the earth. The angle that the magnetic field makes with respect to the surface at any point is known as the *angle of dip*.

21.2 The Force That a Magnetic Field Exerts on a Moving Charge

▶ CONCEPTS AT A GLANCE When a charge is placed in an electric field, it experiences an electric force, as Section 18.6 discusses. When a charge is placed in a magnetic field, it also experiences a force, provided that certain conditions are met, as we will see. The **magnetic force,** like all the forces we have studied (e.g., the gravitational, elastic, and electric forces), may contribute to the net force that causes an object to accelerate. Thus, when present, the magnetic force must be included in Newton's second law, as the Concepts-at-a-Glance chart in Figure 21.6 illustrates. This chart is an extension of the charts in Figures 4.9, 10.4, 11.5, and 18.4 and emphasizes that the second law is remarkable for taking into account the forces that arise in such a wide variety of situations. ◀

The following two conditions must be met for a charge to experience a magnetic force when placed in a magnetic field:

1. The charge must be moving, for no magnetic force acts on a stationary charge.

2. The velocity of the moving charge must have a component that is perpendicular to the direction of the magnetic field.

To examine the second condition more closely, consider Figure 21.7, which shows a positive test charge $+q_0$ moving with a velocity **v** through a magnetic field. In the drawing, the magnetic field vector is labeled by the symbol **B**. The field is produced by an arrangement of magnets not shown in the drawing and is assumed to be constant in both magnitude and direction. If the charge moves *parallel or antiparallel* to the field, as in part *a* of the drawing, the charge experiences *no magnetic force*. If, on the other hand, the charge moves *perpendicular* to the field, as in part *b*, the charge experiences the *maximum possi-*

---

* At present it is not known with certainty what causes the earth's magnetic field. The magnetic field seems to arise from electric currents that circulate within the liquid outer region of the earth's core. Section 21.7 discusses how a current produces a magnetic field.

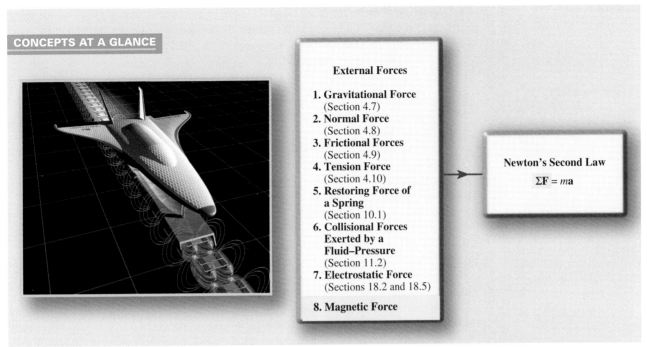

CONCEPTS AT A GLANCE

External Forces

1. **Gravitational Force**
   (Section 4.7)
2. **Normal Force**
   (Section 4.8)
3. **Frictional Forces**
   (Section 4.9)
4. **Tension Force**
   (Section 4.10)
5. **Restoring Force of a Spring**
   (Section 10.1)
6. **Collisional Forces Exerted by a Fluid–Pressure**
   (Section 11.2)
7. **Electrostatic Force**
   (Sections 18.2 and 18.5)
8. **Magnetic Force**

Newton's Second Law
$\Sigma \mathbf{F} = m\mathbf{a}$

Figure 21.6 CONCEPTS AT A GLANCE The magnetic force, like other forces we have encountered, may contribute to the net force $\Sigma \mathbf{F}$ that acts on an object. According to Newton's second law, the acceleration **a** of an object is directly proportional to the net force. This computer-aided drawing illustrates a future spacecraft being launched by a system of magnets (the magnetic fields are shown as red loops). (© NASA/Science Photo Library/Photo Researchers)

ble force $\mathbf{F}_{max}$. In general, if a charge moves at an angle $\theta*$ with respect to the field (see part *c* of the drawing), only the velocity component $v \sin \theta$, which is perpendicular to the field, gives rise to a magnetic force. This force **F** is smaller than the maximum possible force. The component of the velocity that is parallel to the magnetic field yields no force.

Figure 21.7 shows that the direction of the magnetic force **F** is perpendicular to both the velocity **v** and the magnetic field **B**; in other words, **F** is perpendicular to the plane defined by **v** and **B**. As an aid in remembering the direction of the force, it is convenient to use *Right-Hand Rule No. 1 (RHR-1),* as Figure 21.8 illustrates:

Figure 21.7 (*a*) No magnetic force acts on a charge moving with a velocity **v** that is parallel or antiparallel to a magnetic field **B**. (*b*) The charge experiences a maximum force $\mathbf{F}_{max}$ when the charge moves perpendicular to the field. (*c*) If the charge travels at an angle θ with respect to **B**, only the velocity component perpendicular to **B** gives rise to a magnetic force **F**, which is smaller than $\mathbf{F}_{max}$. This component is $v \sin \theta$.

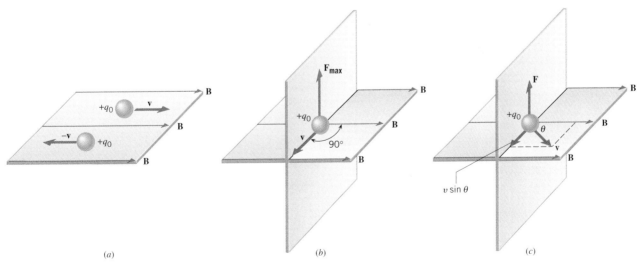

(a) (b) (c)

* The angle θ between the velocity of the charge and the magnetic field is chosen so that it lies in the range $0 \le \theta \le 180°$.

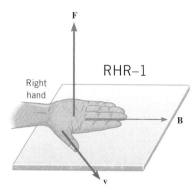

Figure 21.8 Right-Hand Rule No. 1 is illustrated. When the right hand is oriented so the fingers point along the magnetic field **B** and the thumb points along the velocity **v** of a positively charged particle, the palm faces in the direction of the magnetic force **F** applied to the particle.

Right-Hand Rule No. 1. Extend the right hand so the fingers point along the direction of the magnetic field **B** and the thumb points along the velocity **v** of the charge. The palm of the hand then faces in the direction of the magnetic force **F** that acts on a positive charge.

It is as if the open palm of the right hand pushes on the positive charge in the direction of the magnetic force. ***If the moving charge is negative instead of positive, the direction of the magnetic force is opposite to that predicted by RHR-1.*** Thus, there is an easy method for finding the force on a moving negative charge. First, assume that the charge is positive and use RHR-1 to find the direction of the force. Then, reverse this direction to find the direction of the force acting on the negative charge.

▶ CONCEPTS AT A GLANCE We will now use what we know about the magnetic force to define the magnetic field, in a procedure that is analogous to that used in Section 18.6 to define the electric field. The Concepts-at-a-Glance chart in Figure 18.16 illustrates that a test charge and the electrostatic force acting on it are brought together to define the electric field. The Concepts-at-a-Glance chart in Figure 21.9 illustrates that a similar procedure is used in the magnetic case. Recall that the electric field at any point in space is the force per unit charge that acts on a test charge q_0 placed at that point. In other words, to determine the electric field **E**, we divide the electrostatic force **F** by the charge q_0: $\mathbf{E} = \mathbf{F}/q_0$. In the magnetic case, however, the test charge is moving, and the force depends not only on the charge q_0, but also on the velocity component $v \sin \theta$ that is perpendicular to the magnetic field. Therefore, to determine the magnitude of the magnetic field, we divide the magnitude of the magnetic force not only by q_0, but also by $v \sin \theta$, according to the following definition. ◀

Figure 21.9 CONCEPTS AT A GLANCE The magnetic field is defined in terms of the magnetic force and a moving test charge. This chart is analogous to the chart in Figure 18.16, which illustrates how the electric field is defined. This photograph shows an aurora borealis ("northern lights") display over silhouetted trees near Fairbanks, Alaska. Charged particles from the sun are captured by the earth's magnetic field. When the particles collide with the gas molecules in the upper atmosphere, curtains of colorful light are often formed. (© Chris Madeley/ Science Photo Library/Photo Researchers)

■ **DEFINITION OF THE MAGNETIC FIELD**

The magnitude B of the magnetic field at any point in space is defined as

$$B = \frac{F}{q_0(v \sin \theta)} \tag{21.1}$$

where F is the magnitude of the magnetic force on a positive test charge q_0 and **v** is the velocity of the charge and makes an angle θ ($0 \leq \theta \leq 180°$) with the direction of the magnetic field. The magnetic field **B** is a vector, and its direction can be determined by using a small compass needle.

SI Unit of Magnetic Field: $\dfrac{\text{newton} \cdot \text{second}}{\text{coulomb} \cdot \text{meter}} = 1$ tesla (T)

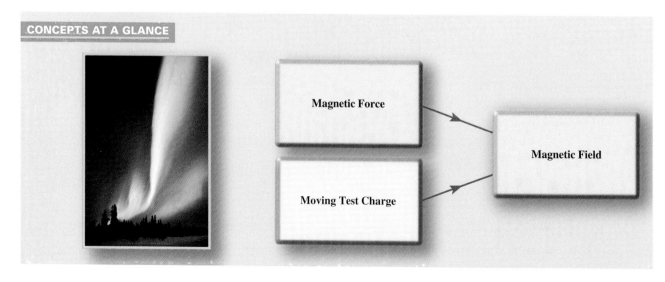

CONCEPTS AT A GLANCE

Magnetic Force

Moving Test Charge

Magnetic Field

The unit of magnetic field strength that follows from Equation 21.1 is the $N \cdot s/(C \cdot m)$. This unit is called the *tesla* (T), a tribute to the Croatian-born American engineer Nikola Tesla (1856–1943). Thus, one tesla is the strength of the magnetic field in which a unit test charge, traveling perpendicular to the magnetic field with a speed of one meter per second, experiences a force of one newton. Because a coulomb per second is an ampere (1 C/s = 1 A, see Section 20.1), the tesla is often written as $1 \text{ T} = 1 \text{ N}/(A \cdot m)$.

In many situations the magnetic field has a value that is considerably less than one tesla. For example, the strength of the magnetic field near the earth's surface is approximately 10^{-4} T. In such circumstances, a magnetic field unit called the *gauss* (G) is sometimes used. Although not an SI unit, the gauss is a convenient size for many applications involving magnetic fields. The relation between the gauss and the tesla is

$$1 \text{ gauss} = 10^{-4} \text{ tesla}$$

Example 1 deals with the magnetic force exerted on a moving proton and on a moving electron.

Example 1 Magnetic Forces on Charged Particles

A proton in a particle accelerator has a speed of 5.0×10^{6} m/s. The proton encounters a magnetic field whose magnitude is 0.40 T and whose direction makes an angle of $\theta = 30.0°$ with respect to the proton's velocity (see Figure 21.7c). Find (a) the magnitude and direction of the magnetic force on the proton and (b) the acceleration of the proton. (c) What would be the force and acceleration if the particle were an electron instead of a proton?

Reasoning For both the proton and the electron, the magnitude of the magnetic force is given by Equation 21.1. The magnetic forces that act on these particles have opposite directions, however, because the charges have opposite signs. In either case, the acceleration is given by Newton's second law, which applies to the magnetic force just as it does to any force. In using the second law, we must take into account the fact that the masses of the proton and the electron are different.

Solution

(a) The positive charge on a proton is 1.60×10^{-19} C, and according to Equation 21.1, the magnitude of the magnetic force is $F = q_0 v B \sin \theta$. Therefore,

$$F = (1.60 \times 10^{-19} \text{ C})(5.0 \times 10^{6} \text{ m/s})(0.40 \text{ T})(\sin 30.0°) = \boxed{1.6 \times 10^{-13} \text{ N}}$$

The direction of the magnetic force is given by RHR-1 and is directed ***upward*** in Figure 21.7c, with the magnetic field pointing to the right.

(b) The magnitude a of the proton's acceleration follows directly from Newton's second law as the magnitude of the net force divided by the mass m_p of the proton. Since the only force acting on the proton is the magnetic force F, it is the net force. Thus,

$$a = \frac{F}{m_p} = \frac{1.6 \times 10^{-13} \text{ N}}{1.67 \times 10^{-27} \text{ kg}} = \boxed{9.6 \times 10^{13} \text{ m/s}^2} \tag{4.1}$$

The direction of the acceleration is the same as the direction of the net force (the magnetic force).

(c) The magnitude of the magnetic force on the electron is the same as that on the proton, since both have the same speed and charge magnitude. However, the direction of the force on the electron is opposite to that on the proton, or ***downward*** in Figure 21.7c, since the electron charge is negative. Furthermore, the electron has a smaller mass m_e and, therefore, experiences a significantly greater acceleration:

$$a = \frac{F}{m_e} = \frac{1.6 \times 10^{-13} \text{ N}}{9.11 \times 10^{-31} \text{ kg}} = \boxed{1.8 \times 10^{17} \text{ m/s}^2}$$

The direction of this acceleration is downward in Figure 21.7c.

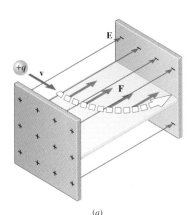

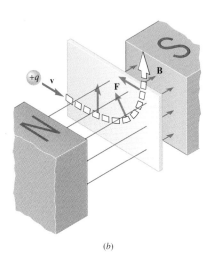

Figure 21.10 (*a*) The electric force **F** that acts on a positive charge is parallel to the electric field **E** and causes the particle's trajectory to bend in a horizontal plane. (*b*) The magnetic force **F** is perpendicular to both the magnetic field **B** and the velocity **v** and causes the particle's trajectory to bend in a vertical plane.

The physics of a velocity selector.

✔ **Check Your Understanding 1**

Two particles, having the same charge but different velocities, are moving in a constant magnetic field (see the drawing, where the velocity vectors are drawn to scale). Which particle, if either, experiences the greater magnetic force? (a) Particle 1 experiences the greater force, because it is moving perpendicular to the magnetic field. (b) Particle 2 experiences the greater force, because it has the greater speed. (c) Particle 2 experiences the greater force, because a component of its velocity is parallel to the magnetic field. (d) Both particles experience the same magnetic force, because the component of each velocity that is perpendicular to the magnetic field is the same. *(The answer is given at the end of the book.)*

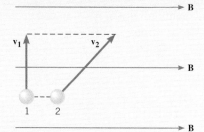

Background: The magnetic force depends on the electric charge on the object, the velocity of the object, and the magnetic field. In particular, the force depends on how the velocity is oriented relative to the field.

For similar questions (including calculational counterparts), consult Self-Assessment Test 21.1, which is described at the end of Section 21.4.

21.3 *The Motion of a Charged Particle in a Magnetic Field*

COMPARING PARTICLE MOTION IN ELECTRIC AND MAGNETIC FIELDS

The motion of a charged particle in an electric field is noticeably different from the motion in a magnetic field. For example, Figure 21.10*a* shows a positive charge moving between the plates of a parallel plate capacitor. Initially, the charge is moving perpendicular to the direction of the electric field. Since the direction of the electric force on a positive charge is in the same direction as the electric field, the particle is deflected sideways. Part *b* of the drawing shows the same particle traveling initially at right angles to a magnetic field. An application of RHR-1 shows that when the charge enters the field, the charge is deflected upward (not sideways) by the magnetic force. As the charge moves upward, the direction of the magnetic force changes, always remaining perpendicular to both the magnetic field and the velocity. Conceptual Example 2 focuses on the difference in how electric and magnetic fields apply forces to a moving charge.

Conceptual Example 2   **A Velocity Selector**

▼

A velocity selector is a device for measuring the velocity of a charged particle. The device operates by applying electric and magnetic forces to the particle in such a way that these forces balance. Figure 21.11*a* shows a particle with a positive charge $+q$ and a velocity **v**, which is perpendicular to a constant magnetic field* **B**. How should an electric field **E**

_____

* In many instances it is convenient to orient the magnetic field **B** so its direction is perpendicular to the page. In these cases it is customary to use a dot to symbolize the magnetic field pointing out of the page (toward the reader); this dot symbolizes the tip of the arrow representing the **B** vector. A region where a magnetic field is directed *into the page* is drawn as a series of crosses that indicate the tail feathers of the arrows representing the **B** vectors. Therefore, regions where a magnetic field is directed out of the page or into the page are drawn as shown below:

Out of page   Into page

be directed so that the force it applies to the particle can balance the magnetic force produced by **B**?

Reasoning and Solution If the electric and magnetic forces are to balance, they must have opposite directions. By applying RHR-1, we find that the magnetic force acting on the positively charged particle in Figure 21.11*a* is directed upward, toward the top of the page. Therefore, the electric force must be directed downward. But the force applied to a positive charge by an electric field has the same direction as the field itself. Therefore, ***the electric field must point downward*** if the electric force is to balance the magnetic force. In other words, the electric and magnetic fields are perpendicular. Figure 21.11*b* shows a velocity selector that uses perpendicular electric and magnetic fields. The device is a cylindrical tube located within the magnetic field **B**. Inside the tube is a parallel plate capacitor that produces the electric field **E**. The charged particle enters the left end of the tube perpendicular to both fields. If the strengths of the fields **E** and **B** are adjusted properly, the electric and magnetic forces acting on the particle will cancel each other. With no net force acting on the particle, the velocity remains unchanged, according to Newton's first law. As a result, the particle moves in a straight line at a constant speed and exits at the right end of the tube. The magnitude of the velocity "selected" can be determined from a knowledge of the strengths of the electric and magnetic fields. Particles with velocities different from the one "selected" are deflected and do not exit at the right end of the tube.

Related Homework: *Problems 16, 20, 73*

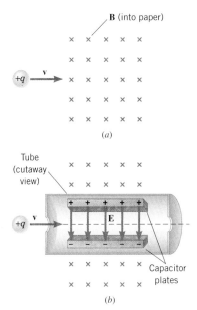

Figure 21.11 (*a*) A particle with a positive charge *q* and velocity **v** moves perpendicularly into a magnetic field **B**. (*b*) A velocity selector is a tube in which an electric field **E** is perpendicular to a magnetic field, and the field magnitudes are adjusted so that the electric and magnetic forces acting on the particle balance.

We have seen that a charged particle traveling in a magnetic field experiences a magnetic force that is always perpendicular to the field. In contrast, the force applied by an electric field is always parallel (or antiparallel) to the field direction. Because of the difference in the way that electric and magnetic fields exert forces, the work done on a charged particle by each field is different, as we now discuss.

THE WORK DONE ON A CHARGED PARTICLE MOVING THROUGH ELECTRIC AND MAGNETIC FIELDS

In Figure 21.10*a* an electric field applies a force to a positively charged particle, and, consequently, the path of the particle bends in the direction of the force. Because there is a component of the particle's displacement in the direction of the electric force, the force does work on the particle, according to Equation 6.1. This work increases the kinetic energy and, hence, the speed of the particle, in accord with the work–energy theorem (see Section 6.2). In contrast, the magnetic force in Figure 21.10*b* always acts in a direction that is perpendicular to the motion of the charge. Consequently, the displacement of the moving charge never has a component in the direction of the magnetic force. As a result, *the magnetic force cannot do work and change the kinetic energy of the charged particle* in Figure 21.10*b*. Thus, the speed of the particle *does not* change, although the force does alter the direction of the motion.

THE CIRCULAR TRAJECTORY

To describe the motion of a charged particle in a constant magnetic field more completely, let's discuss the special case in which the velocity of the particle is perpendicular to a uniform magnetic field. As Figure 21.12 illustrates, the magnetic force serves to move the particle in a circular path. To understand why the path is circular, consider two points on the circumference labeled 1 and 2. When the positively charged particle is at point 1, the magnetic force **F** is perpendicular to the velocity **v** and points directly upward in the drawing. This force causes the trajectory to bend upward. When the particle reaches point 2, the magnetic force still remains perpendicular to the velocity but is now directed to the left in the drawing. ***The magnetic force always remains perpendicular to the velocity and is directed toward the center of the circular path.***

To find the radius of the path in Figure 21.12, we use the concept of centripetal force from Section 5.3. The centripetal force is the net force, directed toward the center of the circle, that is needed to keep a particle moving along a circular path. The magnitude F_c of

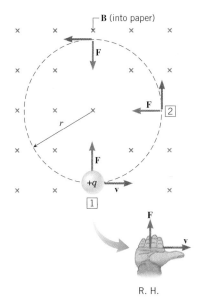

Figure 21.12 A positively charged particle is moving perpendicular to a constant magnetic field. The magnetic force **F** causes the particle to move on a circular path (R.H. = right hand).

Problem solving insight

Concept Simulation 21.1

The motion of a charged particle in a magnetic field is illustrated in this simulation. The user can vary the charge (both magnitude and sign), velocity, and mass of the particle, as well as the magnitude of the magnetic field. In addition, an electric field can be included, so the motion due to the combined magnetic and electric forces can be observed.

Related Homework: Conceptual Question 5

Go to
www.wiley.com/college/cutnell

this force depends on the speed v and mass m of the particle, as well as the radius r of the circle:

$$F_c = \frac{mv^2}{r} \qquad (5.3)$$

In the present situation, the magnetic force furnishes the centripetal force. Being perpendicular to the velocity, the magnetic force does no work in keeping the charge $+q$ on the circular path. According to Equation 21.1, the magnitude of the magnetic force is $qvB \sin 90°$, so $qvB = mv^2/r$ or

$$r = \frac{mv}{qB} \qquad (21.2)$$

This result shows that the radius of the circle is inversely proportional to the magnitude of the magnetic field, with stronger fields producing "tighter" circular paths. Example 3 illustrates an application of Equation 21.2.

Example 3   The Motion of a Proton

A proton is released from rest at point A, which is located next to the positive plate of a parallel plate capacitor (see Figure 21.13). The proton then accelerates toward the negative plate, leaving the capacitor at point B through a small hole in the plate. The electric potential of the positive plate is 2100 V greater than that of the negative plate, so $V_A - V_B = 2100$ V. Once outside the capacitor, the proton travels at a constant velocity until it enters a region of constant magnetic field of magnitude 0.10 T. The velocity is perpendicular to the magnetic field, which is directed out of the page in Figure 21.13. Find (a) the speed v_B of the proton when it leaves the negative plate of the capacitor, and (b) the radius r of the circular path on which the proton moves in the magnetic field.

Reasoning The only force that acts on the proton (charge $= +e$) while it is between the capacitor plates is the conservative electric force. Thus, we can use the conservation of energy to find the speed of the proton when it leaves the negative plate. The total energy of the proton is the sum of its kinetic energy, $\frac{1}{2}mv^2$, and electric potential energy, EPE. Following Example 4 in Chapter 19, we set the total energy at point B equal to the total energy at point A:

$$\underbrace{\tfrac{1}{2}mv_B{}^2 + \text{EPE}_B}_{\text{Total energy at }B} = \underbrace{\tfrac{1}{2}mv_A{}^2 + \text{EPE}_A}_{\text{Total energy at }A}$$

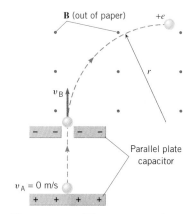

Figure 21.13 The proton, starting from rest at the positive plate of the capacitor, accelerates toward the negative plate. After leaving the capacitor, the proton enters a magnetic field, where it moves on a circular path of radius r.

We note that $v_A = 0$ m/s, since the proton starts from rest, and use Equation 19.3 to set $\text{EPE}_A - \text{EPE}_B = e(V_A - V_B)$. Then the conservation of energy reduces to $\frac{1}{2}mv_B{}^2 = e(V_A - V_B)$. Solving for v_B gives $v_B = \sqrt{2e(V_A - V_B)/m}$. The proton enters the magnetic field with this speed and moves on a circular path with a radius that is given by Equation 21.2.

Solution

(a) The speed of the proton is

$$v_B = \sqrt{\frac{2e(V_A - V_B)}{m}} = \sqrt{\frac{2(1.60 \times 10^{-19}\ \text{C})(2100\ \text{V})}{1.67 \times 10^{-27}\ \text{kg}}} = \boxed{6.3 \times 10^5\ \text{m/s}}$$

(b) When the proton moves in the magnetic field, the radius of the circular path is

$$r = \frac{mv_B}{eB} = \frac{(1.67 \times 10^{-27}\ \text{kg})(6.3 \times 10^5\ \text{m/s})}{(1.60 \times 10^{-19}\ \text{C})(0.10\ \text{T})} = \boxed{6.6 \times 10^{-2}\ \text{m}} \qquad (21.2)$$

Need more practice?

Interactive LearningWare 21.1
A particle has a charge of 3.3×10^{-7} C. In a 0.20-T magnetic field, it travels on a circular path that has a radius of 0.18 m. The centripetal acceleration of the particle is 2.0×10^{10} m/s². What is the magnitude of the centripetal force that acts on the particle?

Go to
www.wiley.com/college/cutnell
for an interactive solution.

One of the exciting areas in physics today is the study of elementary particles, which are the basic building blocks from which all matter is constructed. Important information about an elementary particle can be obtained from its motion in a magnetic field, with the aid of a device known as a bubble chamber. A bubble chamber contains a superheated liquid such as hydrogen, which will boil and form bubbles readily. When an electrically charged particle passes through the chamber, a thin track of bubbles is left in its wake. This track can be photographed to show how a magnetic field affects the particle motion.

Conceptual Example 4 illustrates how physicists deduce information from such photographs.

Conceptual Example 4   Particle Tracks in a Bubble Chamber

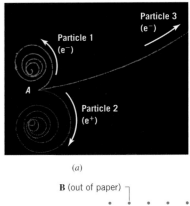

Figure 21.14*a* shows the bubble-chamber tracks resulting from an event that begins at point *A*. At this point a gamma ray (emitted by certain radioactive substances) travels in from the left, spontaneously transforms into two charged particles. There is no track from the gamma ray itself. These particles move away from point *A*, producing the two spiral tracks. A third charged particle is knocked out of a hydrogen atom and moves forward, producing the long track with the slight upward curvature. Each of the three particles has the same mass and carries a charge of the same magnitude. A uniform magnetic field is directed out of the paper toward you. Guided by RHR-1 and Equation 21.2, deduce the sign of the charge carried by each particle, identify which particle is moving most rapidly, and account for the two spiral paths.

Reasoning and Solution To help in our reasoning, Figure 21.14*b* shows a positively charged particle traveling with a velocity **v** that is perpendicular to a magnetic field. The field is directed out of the paper. RHR-1 indicates that the magnetic force points downward. Therefore, downward-curving tracks in the photograph indicate a positive charge, while upward-curving tracks indicate a negative charge. *Particles 1 and 3, then, must carry a negative charge.* They are, in fact, electrons (e⁻). In contrast, *particle 2 must have a positive charge.* It is called a positron (e⁺), an elementary particle that has the same mass as an electron but an opposite charge.

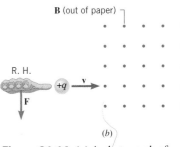

In Equation 21.2 the mass *m*, the charge magnitude *q*, and the magnetic field strength *B* are the same for each particle. Therefore, the radius *r* is proportional to the speed *v*, and a greater radius means a greater speed. The track for particle 3 has the greatest radius, so *particle 3 has the greatest speed.*

Each spiral indicates that the radius is decreasing as the particle moves. Since the radius is proportional to the speed, the speeds of particles 1 and 2 must be decreasing as these particles move. Correspondingly, the kinetic energies of these particles must be decreasing. *Particles 1 and 2 are losing energy each time they collide with a hydrogen atom in the bubble chamber.*

Related Homework: *Conceptual Question 5, Problem 22*

Figure 21.14 (*a*) A photograph of tracks in a bubble chamber. A magnetic field is directed out of the paper. At point *A* a gamma ray (not visible) spontaneously transforms into an electron (e⁻) and a positron (e⁺), which produce the spirals. In addition, an electron is knocked forward out of a hydrogen atom in the chamber. (© Lawrence Berkeley Laboratory/Photo Researchers). (*b*) In accord with RHR-1, the magnetic field applies a downward force to a positively charged particle that moves to the right.

✔ **Check Your Understanding 2**

Three particles have identical charges and masses. They enter a constant magnetic field and follow the paths shown in the drawing. Rank the speeds of the particles, largest to smallest. *(The answer is given at the end of the book.)*

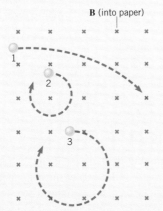

Background: When a charged particle enters a magnetic field, the particle experiences a magnetic force that causes it to move in a circular path, the radius of which depends on the particle's speed, mass, and charge, and also on the magnitude of the field.

For similar questions (including calculational counterparts), consult Self-Assessment Test 21.1, which is described at the end of Section 21.4.

21.4 The Mass Spectrometer

The physics of a mass spectrometer.

Physicists use mass spectrometers for determining the relative masses and abundances of isotopes.* Chemists use these instruments to help identify unknown molecules produced in chemical reactions. Mass spectrometers are also used during surgery, where they give the anesthesiologist information on the gases, including the anesthetic, in the patient's lungs.

In the type of mass spectrometer illustrated in Figure 21.15, the atoms or molecules are first vaporized and then ionized by the ion source. The ionization process removes one electron from the particle, leaving it with a net positive charge of $+e$. The positive ions are then accelerated through the potential difference V, which is applied between the ion source and the metal plate. With a speed v, the ions pass through a hole in the plate and enter a region of constant magnetic field $\mathbf{B}$, where they are deflected in semicircular paths. Only those ions following a path with the proper radius r strike the detector, which records the number of ions arriving per second.

The mass m of the detected ions can be expressed in terms of r, B, and v by recalling that the radius of the path followed by a particle of charge $+e$ is $r = mv/(eB)$ (Equation 21.2). In addition, the Reasoning section in Example 3 shows that the ion speed v can be expressed in terms of the potential difference V as $v = \sqrt{2eV/m}$. This expression for v is the same at that used in Example 3, except that, for convenience, we have replaced the potential difference, $V_A - V_B$, by the symbol V. Eliminating v from these two equations algebraically and solving for the mass gives

$$m = \left(\frac{er^2}{2V}\right)B^2$$

This result shows that the mass of each ion reaching the detector is proportional to B^2. Experimentally changing the value of B and keeping the term in the parentheses constant will allow ions of different masses to enter the detector. A plot of the detector output as a function of B^2 then gives an indication of what masses are present and the abundance of each mass.

Figure 21.16 shows a record obtained by a mass spectrometer for naturally occurring neon gas. The results show that the element neon has three isotopes whose atomic mass numbers are 20, 21, and 22. These isotopes occur because neon atoms exist with different numbers of neutrons in the nucleus. Notice that the isotopes have different abundances, with neon-20 being the most abundant.

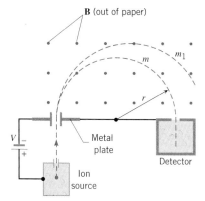

Figure 21.15 In this mass spectrometer the dashed lines are the paths traveled by ions of different masses. Ions with mass m follow the path of radius r and enter the detector. Ions with the larger mass m_1 follow the outer path and miss the detector.

Self-Assessment Test 21.1

Test your understanding of the material in Sections 21.1–21.4:

• Magnetic Fields   • The Force That a Magnetic Field Exerts on a Moving Charge
• The Motion of a Charged Particle in a Magnetic Field   • The Mass Spectrometer

Go to **www.wiley.com/college/cutnell**

21.5 The Force on a Current in a Magnetic Field

As we have seen, a charge moving through a magnetic field can experience a magnetic force. Since an electric current is a collection of moving charges, a current in the presence of a magnetic field can also experience a magnetic force. In Figure 21.17, for instance, a current-carrying wire is placed between the poles of a magnet. When the direction of the current I is as shown, the moving charges experience a magnetic force that pushes the wire to the right in the drawing. The direction of the force is determined in the usual manner by using RHR-1, with the minor modification that the direction of the velocity of a positive charge is replaced by the direction of the conventional current I. If the current in

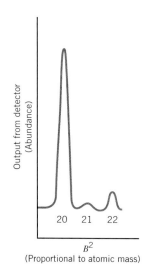

Figure 21.16 The mass spectrum (not to scale) of naturally occurring neon, showing three isotopes whose atomic mass numbers are 20, 21, and 22. The larger the peak, the more abundant the isotope.

---

* Isotopes are atoms that have the same atomic number but different atomic masses due to the presence of different numbers of neutrons in the nucleus. They are discussed in Section 31.1.

the drawing were reversed by switching the leads to the battery, the direction of the force would be reversed, and the wire would be pushed to the left.

When a charge moves through a magnetic field, the magnitude of the force that acts on the charge is $F = qvB \sin \theta$ (Equation 21.1). With the aid of Figure 21.18, this expression can be put into a form that is more suitable for use with an electric current. The drawing shows a wire of length L that carries a current I. The wire is oriented at an angle θ with respect to a magnetic field **B**. This picture is similar to Figure 21.7c, except that now the charges move in a wire. The magnetic force exerted on this length of wire is the net force acting on the total amount of charge moving in the wire. Suppose that an amount of charge Δq travels the length of the wire in a time interval Δt. The magnitude of the magnetic force on this amount of charge is given by Equation 21.1 as $F = (\Delta q)vB \sin \theta$. Multiplying and dividing the right side of this equation by Δt, we find that

$$F = \underbrace{\left(\frac{\Delta q}{\Delta t}\right)}_{I} \underbrace{(v \, \Delta t)}_{L} B \sin \theta$$

According to Equation 20.1, the term $\Delta q/\Delta t$ is the current I in the wire, and the term $v\Delta t$ is the length L of the wire. With these two substitutions, the expression for the magnetic force exerted on a current-carrying wire becomes

Magnetic force on a current-carrying wire of length L
$$F = ILB \sin \theta \qquad (21.3)$$

As in the case of a single charge traveling in a magnetic field, the magnetic force on a current-carrying wire is a maximum when the wire is oriented perpendicular to the field ($\theta = 90°$) and vanishes when the current is parallel or antiparallel to the field ($\theta = 0°$ or $180°$). The direction of the magnetic force is given by RHR-1, as Figure 21.18 indicates.

✓ **Check Your Understanding 3**

The same current-carrying wire is placed in the same magnetic field **B** in four different orientations (see the drawing). Rank the orientations according to the magnitude of the magnetic force exerted on the wire, largest to smallest. *(The answer is given at the end of the book.)*

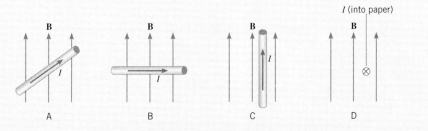

Background: The magnetic force exerted on a current-carrying wire depends on the current, the length of the wire, the magnetic field, and the orientation of the wire relative to the field.

For similar questions (including calculational counterparts), consult Self-Assessment Test 21.2, which is described at the end of Section 21.8.

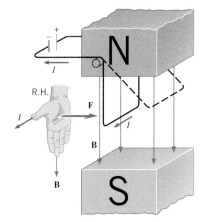

Figure 21.17 The wire carries a current I, and the bottom segment of the wire is oriented perpendicular to a magnetic field **B**. A magnetic force deflects the wire to the right.

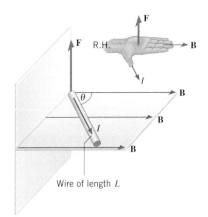

Figure 21.18 The current I in the wire, oriented at an angle θ with respect to a magnetic field **B**, is acted upon by a magnetic force **F**.

Most loudspeakers operate on the principle that a magnetic field exerts a force on a current-carrying wire. Figure 21.19a shows a speaker design that consists of three basic parts: a cone, a voice coil, and a permanent magnet. The cone is mounted so it can vibrate back and forth. When vibrating, it pushes and pulls on the air in front of it, thereby creating sound waves. Attached to the apex of the cone is the voice coil, which is a hollow cylinder around which coils of wire are wound. The voice coil is slipped over one of the poles of the stationary permanent magnet (the north pole in the drawing) and can move freely. The two ends of the voice-coil wire are connected to the speaker terminals on the back panel of a receiver.

The physics of a loudspeaker.

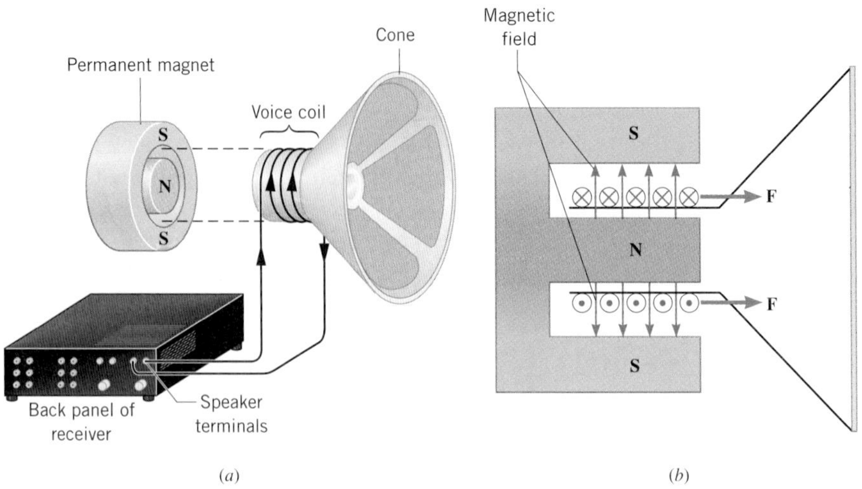

Figure 21.19 (a) An "exploded" view of one type of speaker design, which shows a cone, a voice coil, and a permanent magnet. (b) Because of the current in the voice coil (shown as ⊗ and ⊙), the magnetic field causes a force **F** to be exerted on the voice coil and cone.

The receiver acts as an ac generator, sending an alternating current to the voice coil. The alternating current interacts with the magnetic field to generate an alternating force that pushes and pulls on the voice coil and the attached cone. To see how the magnetic force arises, consider Figure 21.19b, which is a cross-sectional view of the voice coil and the magnet. In the cross-sectional view, the current is directed into the page in the upper half of the voice coil (⊗⊗⊗) and out of the page in the lower half (⊙⊙⊙). In both cases the magnetic field is perpendicular to the current, so the maximum possible force is exerted on the wire. An application of RHR-1 to both the upper and lower halves of the voice coil shows that the magnetic force **F** in the drawing is directed to the right, causing the cone to accelerate in that direction. One-half of a cycle later when the current is reversed, the direction of the magnetic force is also reversed, and the cone accelerates to the left. If, for example, the alternating current from the receiver has a frequency of 1000 Hz, the alternating magnetic force causes the cone to vibrate back and forth at the same frequency, and a 1000-Hz sound wave is produced. Thus, it is the magnetic force on a current-carrying wire that is responsible for converting an electrical signal into a sound wave. In Example 5 a typical force and acceleration in a loudspeaker are determined.

Problem solving insight
Whenever the current in a wire reverses direction, the force exerted on the wire by a given magnetic field also reverses direction.

Example 5   The Force and Acceleration in a Loudspeaker

The voice coil of a speaker has a diameter of $d = 0.025$ m, contains 55 turns of wire, and is placed in a 0.10-T magnetic field. The current in the voice coil is 2.0 A. (a) Determine the magnetic force that acts on the coil and cone. (b) The voice coil and cone have a combined mass of 0.020 kg. Find their acceleration.

Reasoning The magnetic force that acts on the current-carrying voice coil is given by Equation 21.3 as $F = ILB \sin \theta$. The effective length L of the wire in the voice coil is very nearly the number of turns N times the circumference (πd) of one turn: $L = N\pi d$. The acceleration of the voice coil and cone is given by Newton's second law as the magnetic force divided by the combined mass.

Solution

(a) Since the magnetic field acts perpendicular to all parts of the wire, $\theta = 90°$ and the force on the voice coil is

$$F = ILB \sin \theta = (2.0 \text{ A})[55\pi(0.025 \text{ m})](0.10 \text{ T})\sin 90° = \boxed{0.86 \text{ N}} \qquad (21.3)$$

(b) According to Newton's second law, the acceleration of the voice coil and cone is

$$a = \frac{F}{m} = \frac{0.86 \text{ N}}{0.020 \text{ kg}} = \boxed{43 \text{ m/s}^2} \qquad (4.1)$$

This acceleration is more than four times the acceleration due to gravity.

Need more practice?

Interactive LearningWare 21.2
Two conducting rails are 1.6 m apart and are parallel to the ground at the same height. As the drawing shows, a 0.20-kg aluminum rod is lying on top of the rails, and a 0.050-T magnetic field points upward, perpendicular to the ground. There is a current I in the rod, directed as in the drawing. The coefficient of static friction between the rod and each rail is $\mu_S = 0.45$. How much current is needed to make the rod begin moving, and in which direction will it move?

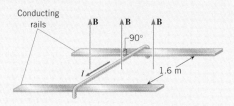

Go to **www.wiley.com/college/cutnell** for an interactive solution.

Magnetohydrodynamic (MHD) propulsion is a revolutionary type of propulsion system that does not use propellers to power ships and submarines, but, instead, uses a magnetic force on a current. The method eliminates motors, drive shafts, and gears, as well as propellers, so it promises to be a low-noise system with great reliability at relatively low cost. Figure 21.20*a* shows the first ship to use MHD technology, the *Yamato 1*. In part *b* the schematic side view shows one of the two MHD propulsion units that are mounted underneath the vessel. Seawater enters the front of the unit and is expelled from the rear. A jet engine uses air in a similar fashion, taking it in in the front and pushing it out at the back to propel the plane forward.

Figure 21.20*c* presents an enlarged view of a propulsion unit. An electromagnet (see Section 21.7) within the vessel uses superconducting wire to produce a strong magnetic field. Electrodes (metal plates) mounted on either side of the unit are attached to a dc electrical generator. The generator sends an electric current through the seawater, from one electrode to the other, perpendicular to the magnetic field. Consistent with RHR-1, the field applies a force **F** to the current, as the drawing shows. The magnetic force pushes the seawater, similar to the way in which it pushes the wire in Figure 21.17. As a result, a water jet is expelled from the tube. Since the MHD unit exerts a magnetic force on the water, the water exerts a force on the propulsion unit and the vessel attached to it. According to Newton's third law, this "reaction" force is equal in magnitude, but opposite in direction, to the magnetic force, and it is the reaction force that provides the thrust to drive the vessel.

The physics of **magnetohydrodynamic propulsion.**

Figure 21.20 (*a*) The *Yamato 1* is the first ship to use magnetohydrodynamic (MHD) propulsion (© Dennis Budd Gray). (*b*) A schematic side view of the *Yamato 1*. (*c*) In the MHD unit, the magnetic force exerted on the current forces water out the back.

(*a*)

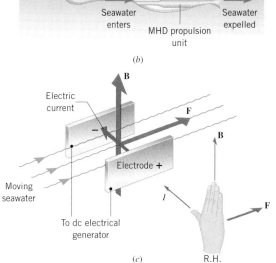

(*b*)

(*c*)

Figure 21.21 (a) A current-carrying loop of wire, which can rotate about a vertical shaft, is situated in a magnetic field. (b) A top view of the loop. The current in side 1 is directed out of the page (⊙), while that in side 2 is directed into the page (⊗). The current in side 1 experiences a force **F** that is opposite to the force exerted on side 2. The two forces produce a clockwise torque about the shaft.

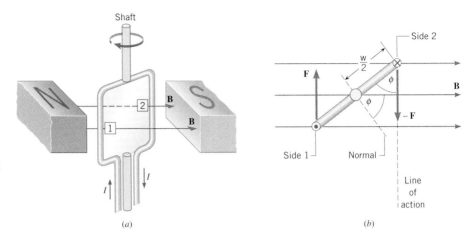

(a)

(b)

21.6 The Torque on a Current-Carrying Coil

We have seen that a current-carrying wire can experience a force when placed in a magnetic field. If a loop of wire is suspended properly in a magnetic field, the magnetic force produces a torque that tends to rotate the loop. This torque is responsible for the operation of a widely used type of electric motor.

Figure 21.21a shows a rectangular loop of wire attached to a vertical shaft. The shaft is mounted such that it is free to rotate in a uniform magnetic field. When there is a current in the loop, the loop rotates because magnetic forces act on the two vertical sides, labeled 1 and 2 in the drawing. Part b shows a top view of the loop and the magnetic forces **F** and −**F** that act on the two sides. These two forces have the same magnitude, but an application of RHR-1 shows that they point in opposite directions, so the loop experiences no net force. The loop does, however, experience a net torque that tends to rotate it in a clockwise fashion about the vertical shaft. Figure 21.22a shows that the torque is maximum when the normal to the plane of the loop is perpendicular to the field. In contrast, part b shows that the torque is zero when the normal is parallel to the field. **When a current-carrying loop is placed in a magnetic field, the loop tends to rotate such that its normal becomes aligned with the magnetic field.** In this respect, a current loop behaves like a magnet (e.g., a compass needle) suspended in a magnetic field, since a magnet also rotates to align itself with the magnetic field.

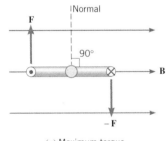

(a) Maximum torque

It is possible to determine the magnitude of the torque on the loop. From Equation 21.3 the magnetic force on each vertical side has a magnitude of $F = ILB \sin 90°$, where L is the length of side 1 or side 2, and $\theta = 90°$ because the current I always remains perpendicular to the magnetic field as the loop rotates. As Section 9.1 discusses, the torque produced by a force is the product of the magnitude of the force and the lever arm. In Figure 21.21b the lever arm is the perpendicular distance from the line of action of the force to the shaft. This distance is given by $(w/2) \sin \phi$, where w is the width of the loop, and ϕ is the angle between the normal to the plane of the loop and the direction of the magnetic field. The net torque is the sum of the torques on the two sides, so

$$\text{Net torque} = \tau = ILB \left(\tfrac{1}{2}w \sin \phi\right) + ILB \left(\tfrac{1}{2}w \sin \phi\right) = IAB \sin \phi$$

In this result the product Lw has been replaced by the area A of the loop. If the wire is wrapped so as to form a coil containing N loops, each of area A, the force on each side is N times larger, and the torque becomes proportionally greater:

$$\tau = NIAB \sin \phi \qquad (21.4)$$

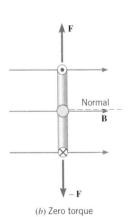

(b) Zero torque

Figure 21.22 (a) Maximum torque occurs when the normal to the plane of the loop is perpendicular to the magnetic field. (b) The torque is zero when the normal is parallel to the field.

Equation 21.4 has been derived for a rectangular coil, but it is valid for any shape of flat coil, such as a circular coil. The torque depends on the geometric properties of the coil and the current in it through the quantity NIA. This quantity is known as the **magnetic moment** of the coil, and its unit is ampere · meter^2. The greater the magnetic moment of a current-carrying coil, the greater the torque that the coil experiences when placed in a magnetic field. Example 6 discusses the torque that a magnetic field applies to such a coil.

Example 6 The Torque Exerted on a Current-Carrying Coil

A coil of wire has an area of 2.0×10^{-4} m^2, consists of 100 loops or turns, and contains a current of 0.045 A. The coil is placed in a uniform magnetic field of magnitude 0.15 T. (a) Determine the magnetic moment of the coil. (b) Find the maximum torque that the magnetic field can exert on the coil.

Reasoning and Solution

(a) The magnetic moment of the coil is

$$\text{Magnetic moment} = NIA = (100)(0.045 \text{ A})(2.0 \times 10^{-4} \text{ m}^2) = \boxed{9.0 \times 10^{-4} \text{ A} \cdot \text{m}^2}$$

(b) According to Equation 21.4, the torque is the product of the magnetic moment NIA and $B \sin \phi$. However, the maximum torque occurs when $\phi = 90°$, so

$$\tau = \underbrace{(NIA)}_{\substack{\text{Magnetic} \\ \text{moment}}}(B \sin 90°) = (9.0 \times 10^{-4} \text{ A} \cdot \text{m}^2)(0.15 \text{ T}) = \boxed{1.4 \times 10^{-4} \text{ N} \cdot \text{m}}$$

The electric motor is found in many devices, such as CD players, tape decks, automobiles, washing machines, and air conditioners. Figure 21.23 shows that a direct-current (dc) motor consists of a coil of wire placed in a magnetic field and free to rotate about a vertical shaft. The coil of wire contains many turns and is wrapped around an iron cylinder that rotates with the coil, although these features have been omitted to simplify the drawing. The coil and iron cylinder assembly is known as the armature. Each end of the wire coil is attached to a metallic half-ring. Rubbing against each of the half-rings is a graphite contact called a brush. While the half-rings rotate with the coil, the graphite brushes remain stationary. The two half-rings and the associated brushes are referred to as a split-ring commutator, the purpose of which will be explained shortly.

The operation of a motor can be understood by considering Figure 21.24. In part *a* the current from the battery enters the coil through the left brush and half-ring, goes around the coil, and then leaves through the right half-ring and brush. Consistent with RHR-1, the directions of the magnetic forces **F** and $-$**F** on the two sides of the coil are as shown in the drawing. These forces produce the torque that turns the coil. Eventually the coil reaches the position shown in part *b* of the drawing. In this position the half-rings momentarily lose electrical contact with the brushes, so that there is no current in the coil and no applied torque. However, like any moving object, the rotating coil does not stop immediately, for its inertia carries it onward. When the half-rings reestablish contact with the brushes, there again is a current in the coil, and a magnetic torque again rotates the coil in the same direction. The split-ring commutator ensures that the current is always in the proper direction to yield a torque that produces a continuous rotation of the coil.

The physics of
a direct-current electric motor.

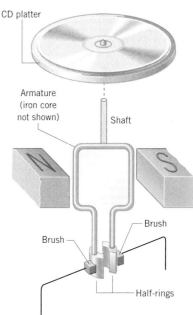

Figure 21.23 The basic components of a dc motor. A CD platter is shown as it might be attached to the motor.

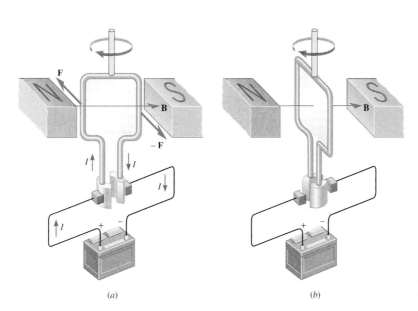

Figure 21.24 (*a*) When a current exists in the coil, the coil experiences a torque. (*b*) Because of its inertia, the coil continues to rotate when there is no current.

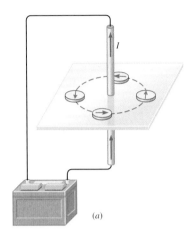

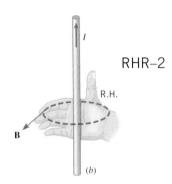

RHR-2

Figure 21.25 (a) A very long, straight current-carrying wire produces magnetic field lines that are circular about the wire. One such circular line is indicated by the compass needles. (b) If the thumb of the right hand (R.H.) is pointed in the direction of the current I, the curled fingers point in the direction of the magnetic field, according to RHR-2.

21.7 Magnetic Fields Produced by Currents

We have seen that a current-carrying wire can experience a magnetic force when placed in a magnetic field that is produced by an external source, such as a permanent magnet. *A current-carrying wire also produces a magnetic field of its own,* as we will see in this section. Hans Christian Oersted (1777–1851) first discovered this effect in 1820 when he observed that a current-carrying wire influences the orientation of a nearby compass needle. The compass needle aligns itself with the net magnetic field produced by the current and the magnetic field of the earth. Oersted's discovery, which linked the motion of electric charges with the creation of a magnetic field, marked the beginning of an important discipline called *electromagnetism.*

A LONG, STRAIGHT WIRE

Figure 21.25a illustrates Oersted's discovery with a very long, straight wire. When a current is present, the compass needles point in a circular pattern about the wire. The pattern indicates that the magnetic field lines produced by the current are circles centered on the wire. If the direction of the current is reversed, the needles also reverse their directions, indicating that the direction of the magnetic field has reversed. The direction of the field can be obtained by using *Right-Hand Rule No. 2 (RHR-2),* as part b of the drawing indicates:

> *Right-Hand Rule No. 2.* Curl the fingers of the right hand into the shape of a half-circle. Point the thumb in the direction of the conventional current I, and the tips of the fingers will point in the direction of the magnetic field **B**.

Experimentally, it is found that the magnitude B of the magnetic field produced by an infinitely long, straight wire is directly proportional to the current I and inversely proportional to the radial distance r from the wire: $B \propto I/r$. As usual, this proportionality is converted into an equation by introducing a proportionality constant, which, in this instance, is written as $\mu_0/(2\pi)$. Thus, the magnitude of the magnetic field is

Infinitely long, straight wire
$$B = \frac{\mu_0 I}{2\pi r} \qquad (21.5)$$

The constant μ_0 is known as the *permeability of free space,* and its value is $\mu_0 = 4\pi \times 10^{-7}$ T·m/A. The magnetic field becomes stronger nearer the wire, where r is smaller. Therefore, the field lines near the wire are closer together than those located farther away, where the field is weaker. Figure 21.26 shows the pattern of field lines.

The magnetic field that surrounds a current-carrying wire can exert a force on a moving charge, as the next example illustrates.

Example 7 A Current Exerts a Magnetic Force on a Moving Charge

Figure 21.27 shows a very long, straight wire carrying a current of $I = 3.0$ A. A particle of charge $q_0 = +6.5 \times 10^{-6}$ C is moving parallel to the wire at a distance of $r = 0.050$ m; the speed of the particle is $v = 280$ m/s. Determine the magnitude and direction of the magnetic force exerted on the moving charge by the current in the wire.

Reasoning The current generates a magnetic field in the space around the wire. A charge moving through this field experiences a magnetic force **F** whose magnitude is given by Equation 21.1 as $F = q_0 v B \sin\theta$, where θ is the angle between the magnetic field and the velocity of the charge. The magnitude of the magnetic field follows from Equation 21.5 as $B = \mu_0 I/(2\pi r)$. Thus, the magnitude of the magnetic force can be expressed as

$$F = q_0 v B \sin\theta = q_0 v \left(\frac{\mu_0 I}{2\pi r} \right) \sin\theta$$

The direction of the magnetic force is predicted by RHR-1 (see Section 21.2).

Figure 21.26 The magnetic field becomes stronger as the radial distance r decreases, so the field lines are closer together near the wire.

Solution Figure 21.27 shows that the magnetic field **B** lies in the plane that is perpendicular to both the wire and velocity **v** of the particle. Thus, the angle between **B** and **v** is $\theta = 90°$, and the magnitude of the magnetic force is

$$F = q_0 v \left(\frac{\mu_0 I}{2\pi r} \right) \sin 90°$$

$$F = (6.5 \times 10^{-6} \text{ C})(280 \text{ m/s}) \left[\frac{(4\pi \times 10^{-7} \text{ T} \cdot \text{m/A})(3.0 \text{ A})}{2\pi(0.050 \text{ m})} \right]$$

$$= \boxed{2.2 \times 10^{-8} \text{ N}}$$

The direction of the magnetic force is predicted by RHR-1 and, as the drawing shows, is radially inward toward the wire.

Need more practice?

Interactive LearningWare 21.3
A long, straight wire is oriented in the north–south direction, and the current in the wire is directed to the north. The horizontal component of the earth's magnetic field is 4.5×10^{-5} T and points due north. A small horizontal compass needle is located directly below the wire, 1.9 cm from it. The compass needle points 35° north of west. What is the current in the wire?

Go to **www.wiley.com/college/cutnell** for an interactive solution.

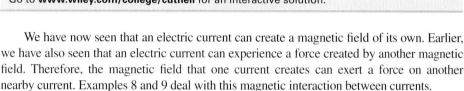

Figure 21.27 The moving positive charge q_0 experiences a magnetic force **F** because of the magnetic field **B** produced by the current in the wire.

We have now seen that an electric current can create a magnetic field of its own. Earlier, we have also seen that an electric current can experience a force created by another magnetic field. Therefore, the magnetic field that one current creates can exert a force on another nearby current. Examples 8 and 9 deal with this magnetic interaction between currents.

Example 8
Two Current-Carrying Wires Exert Magnetic Forces on One Another

Figure 21.28 shows two parallel straight wires that are very long. The wires are separated by a distance of $r = 0.065$ m and carry currents of $I_1 = 15$ A and $I_2 = 7.0$ A. Find the magnitude and direction of the force that the magnetic field of wire 1 applies to a 1.5-m length of wire 2 when the currents are (a) in opposite directions and (b) in the same direction.

Reasoning The current I_2 in wire 2 is situated in the magnetic field produced by the current in wire 1. The magnitude F of the magnetic force experienced by a length L of wire 2 is given by Equation 21.3 as $F = I_2 LB \sin \theta$. Here B is the magnitude of the magnetic field produced by wire 1 and is given by Equation 21.5 as $B = \mu_0 I_1/(2\pi r)$. The direction of the magnetic force can be determined by using RHR-1.

Solution

(a) At wire 2, the magnitude of the magnetic field created by wire 1 is

$$B = \frac{\mu_0 I_1}{2\pi r} = \frac{(4\pi \times 10^{-7} \text{ T} \cdot \text{m/A})(15 \text{ A})}{2\pi(0.065 \text{ m})} = 4.6 \times 10^{-5} \text{ T} \qquad (21.5)$$

The direction of this field is upward at the location of wire 2, as part *a* of the figure shows. The direction can be obtained using RHR-2 (thumb of right hand along I_1, curled fingers point upward at wire 2 and indicate the direction of **B**). The magnetic field is perpendicular to wire 2 ($\theta = 90°$), so the magnitude of the force on a 1.5-m section of its length is

$$F = I_2 LB \sin \theta = (7.0 \text{ A})(1.5 \text{ m})(4.6 \times 10^{-5} \text{ T}) \sin 90° = \boxed{4.8 \times 10^{-4} \text{ N}} \quad (21.3)$$

The direction of the magnetic force on wire 2 is away from wire 1, as part *a* of the drawing indicates; the force direction is found by using RHR-1 (fingers of the right hand extended upward along **B**, thumb points along I_2, palm pushes in the direction of the force **F**).

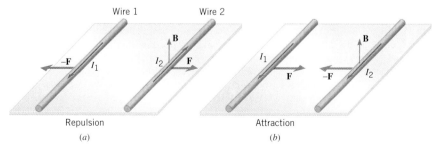

Figure 21.28 (*a*) Two long, parallel wires carrying currents I_1 and I_2 in opposite directions repel each other. (*b*) The wires attract each other when the currents are in the same direction.

In a like manner, the current in wire 2 also creates a magnetic field that produces a force on wire 1. Reasoning similar to that above shows that a 1.5-m length of wire 1 is repelled from wire 2 with a force that also has a magnitude of 4.8×10^{-4} N. Thus, each wire generates a force on the other, and, if the currents are in *opposite* directions, the wires *repel* each other. The fact that the two wires exert equal but oppositely directed forces on each other is consistent with Newton's third law, the action–reaction law.*

(**b**) If the current in wire 2 is reversed, as part *b* of the drawing indicates, wire 2 is attracted to wire 1 because the direction of the magnetic force is reversed. However, the magnitude of the force is the same as that calculated in part *a* above. Likewise, wire 1 is attracted to wire 2. Two parallel wires carrying currents in the *same* direction *attract* each other.

Conceptual Example 9
The Net Force That a Current-Carrying Wire Exerts on a Current-Carrying Coil

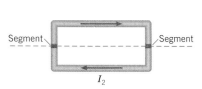

Figure 21.29 A very long, straight wire carries a current I_1, and a rectangular coil carries a current I_2. The dashed line is parallel to the wire and locates a small segment on each short side of the coil.

Figure 21.29 shows a very long, straight wire carrying a current I_1 and a rectangular coil carrying a current I_2. The wire and the coil lie in the same plane, with the wire parallel to the long sides of the rectangle. Is the coil attracted to or repelled from the wire?

Reasoning and Solution The current in the straight wire exerts a force on each of the four sides of the coil. The net force acting on the coil is the vector sum of these four forces. To decide whether the net force is attractive or repulsive, we need to consider the directions and magnitudes of the individual forces. In the long side of the coil near the wire, the current I_2 has the same direction as the current I_1, and we have just seen in Example 8 that two such currents attract each other. In the long side of the coil farthest from the wire, I_2 has a direction opposite to that of I_1 and, according to Example 8, they repel one another. However, the attractive force is stronger than the repulsive force because the magnetic field produced by the current I_1 is stronger at shorter distances than at greater distances. Consequently, we reach the preliminary conclusion that the coil is attracted to the wire.

But what about the forces that act on the two short sides? Consider a small segment of each of the short sides, located at the same distance from the straight wire, as indicated by the dashed line in Figure 21.29. Each of these segments experiences the same magnetic field from the current I_1. RHR-2 shows that this field is directed downward into the plane of the paper, so that it is perpendicular to the current I_2 in each segment. But the directions of I_2 in the segments are opposite. As a result, RHR-1 reveals that the magnetic field from the straight wire applies a force to one segment that is opposite to that applied to the other segment. Thus, the forces acting on the two short sides of the coil cancel, and our preliminary conclusion is valid; *the coil is attracted to the wire.*

Related Homework: *Problem 54*

Concept Simulation 21.2

The magnetic field produced at any point by a very long, straight wire depends on the current in the wire and the distance from the point to the wire. With this simulation you can alter the strength and direction of the current, and the program calculates the magnitude and direction of the magnetic field at any point in the region surrounding the wire. In addition, a second wire, oriented parallel to the first, can be included, and the strength and direction of its current can be changed. The simulation shows either the magnetic field lines produced by the individual wires or the net magnetic field that surrounds the two wires. The two current-carrying wires exert forces on one another, and the program displays the force per unit length that each wire exerts on the other.

Go to
www.wiley.com/college/cutnell

A LOOP OF WIRE

If a current-carrying wire is bent into a circular loop, the magnetic field lines around the loop have the pattern shown in Figure 21.30a. At the *center* of a loop of radius R, the magnetic field is perpendicular to the plane of the loop and has the value $B = \mu_0 I/(2R)$, where I is the current in the loop. Often, the loop consists of N turns of wire that are wound sufficiently close together that they form a flat coil with a single radius. In this case, the magnetic fields of the individual turns add together to give a net field that is N times greater than that of a single loop. For such a coil the magnetic field at the center is

Center of a circular loop
$$B = N\frac{\mu_0 I}{2R}$$
(21.6)

The direction of the magnetic field at the center of the loop can be determined with the

* In this example the currents in the two wires and the distance between them are known; therefore, the magnetic force that one wire exerts on the other can be calculated. If, instead, the force and the distance were known and the wires carried the same current, that current could be calculated. This is, in fact, the procedure used to define the ampere, which is the unit for electric current. This procedure is chosen because force and distance are quantities that can be measured with a high degree of precision. One ampere is the amount of electric current in each of two long, parallel wires that gives rise to a magnetic force per unit length of 2×10^{-7} N/m on each wire when the wires are separated by one meter. With the ampere defined in terms of force and distance, the coulomb is defined as the quantity of electrical charge that passes a given point in one second when the current is one ampere or one coulomb per second.

help of RHR-2. If the thumb of the right hand is pointed in the direction of the current and the curled fingers are placed at the center of the loop, as in Figure 21.30*b*, the fingers indicate that the magnetic field points from right to left.

Concept Simulation 21.3

In this simulation you can explore the magnetic field created by a current in a loop of wire. How does the magnetic field depend on the magnitude of the current? What happens to the magnetic field if the direction of the current is reversed? Where is the magnetic field the strongest? All these questions can be answered by the simulation, because it displays the magnitude and direction of the magnetic field at any point in the vicinity of the loop. The magnetic field also depends on the radius of the loop, and you can examine this effect by selecting two loops of different sizes.

Go to **www.wiley.com/college/cutnell**

Example 10 shows how the magnetic fields produced by the current in a loop of wire and the current in a long, straight wire combine to form a net magnetic field.

Example 10 Finding the Net Magnetic Field

A long, straight wire carries a current of $I_1 = 8.0$ A. As Figure 21.31*a* illustrates, a circular loop of wire lies immediately to the right of the straight wire. The loop has a radius of $R = 0.030$ m and carries a current of $I_2 = 2.0$ A. Assuming that the thickness of the wires is negligible, find the magnitude and direction of the net magnetic field at the center *C* of the loop.

Reasoning The net magnetic field at the point *C* is the sum of two contributions: (1) the field **B₁** produced by the long, straight wire, and (2) the field **B₂** produced by the circular loop. An application of RHR-2 shows that at point *C* the field **B₁** is directed upward, perpendicular to the plane containing the straight wire and the loop (see part *b* of the drawing). Similarly, RHR-2 shows that the magnetic field **B₂** is directed downward, opposite to the direction of **B₁**.

Solution If we choose the upward direction in Figure 21.31*b* as positive, the net magnetic field at the point *C* is

$$B = \underbrace{\frac{\mu_0 I_1}{2\pi r}}_{\substack{\text{Long, straight}\\\text{wire}}} - \underbrace{\frac{\mu_0 I_2}{2R}}_{\substack{\text{Center of a}\\\text{circular loop}}} = \frac{\mu_0}{2}\left(\frac{I_1}{\pi r} - \frac{I_2}{R}\right)$$

$$B = \frac{(4\pi \times 10^{-7}\ \text{T}\cdot\text{m/A})}{2}\left[\frac{8.0\ \text{A}}{\pi(0.030\ \text{m})} - \frac{2.0\ \text{A}}{0.030\ \text{m}}\right] = \boxed{1.1 \times 10^{-5}\ \text{T}}$$

The net field is positive, so it is directed upward, perpendicular to the plane.

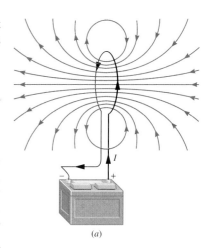

(a)

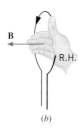

(b)

Figure 21.30 (*a*) The magnetic field lines in the vicinity of a current-carrying circular loop. (*b*) The direction of the magnetic field at the center of the loop is given by RHR-2.

Problem solving insight
Do not confuse the formula for the magnetic field produced at the center of a circular loop with that of a very long, straight wire. The formulas are similar, differing only by a factor of π in the denominator.

✔ Check Your Understanding 4

Each of the four drawings shows the same three concentric loops of wire. The currents in the loops have the same magnitude *I* and have the directions shown. Rank the magnitude of the net magnetic field produced at the center of each of the four drawings, largest to smallest. *(The answer is given at the end of the book.)*

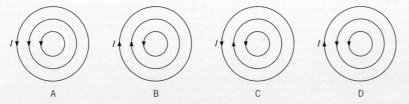

Background: A circular loop produces a magnetic field at its center, the magnitude of which depends on the current and the radius of the loop. The direction of the magnetic field can be found with the aid of Right-Hand Rule No. 2.

For similar questions (including calculational counterparts), consult Self-Assessment Test 21.2, which is described at the end of Section 21.8.

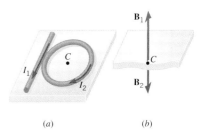

(a)   *(b)*

Figure 21.31 (*a*) A long, straight wire carrying a current I_1 lies next to a circular loop that carries a current I_2. (*b*) The magnetic fields at the center *C* of the loop produced by the straight wire (**B₁**) and the loop (**B₂**).

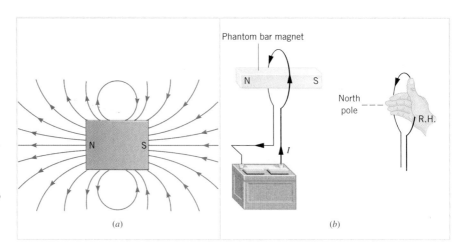

Figure 21.32 (*a*) The field lines around the bar magnet resemble those around the loop in Figure 21.30*a*. (*b*) The current loop can be imagined to be a phantom bar magnet with a north pole and a south pole.

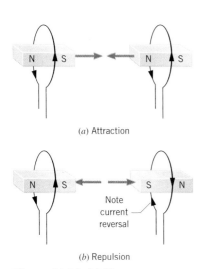

(*a*) Attraction

Note current reversal

(*b*) Repulsion

Figure 21.33 (*a*) The two current loops attract each other if the directions of the currents are the same and (*b*) repel each other if the directions are opposite. The "phantom" magnets help explain the attraction and repulsion.

A comparison of the magnetic field lines around the current loop in Figure 21.30*a* with those in the vicinity of the short bar magnet in Figure 21.32*a* shows that the two patterns are quite similar. Not only are the patterns similar, but the loop itself behaves as a bar magnet with a "north pole" on one side and a "south pole" on the other side. To emphasize that the loop may be imagined to be a bar magnet, Figure 21.32*b* includes a "phantom" bar magnet at the center of the loop. The side of the loop that acts like a north pole can be determined with the aid of RHR-2; curl the fingers of the right hand into the shape of a half-circle, point the thumb along the current *I*, and place the curled fingers at the center of the loop. The fingers not only point in the direction of **B**, but they also point toward the north pole.

Because a current-carrying loop acts like a bar magnet, two adjacent loops can be either attracted to or repelled from each other, depending on the relative directions of the currents. Figure 21.33 includes a "phantom" magnet for each loop and shows that the loops are attracted to each other when the currents are in the same direction and repelled from each other when the currents are in opposite directions. This behavior is analogous to that of the two long, straight wires discussed in Example 8 (see Figure 21.28).

A SOLENOID

A solenoid is a long coil of wire in the shape of a helix (see Figure 21.34). If the wire is wound so the turns are packed close to each other and the solenoid is long compared to its diameter, the magnetic field lines have the appearance shown in the drawing. Notice that the field inside the solenoid and away from its ends is nearly constant in magnitude and directed parallel to the axis. The direction of the field inside the solenoid is given by

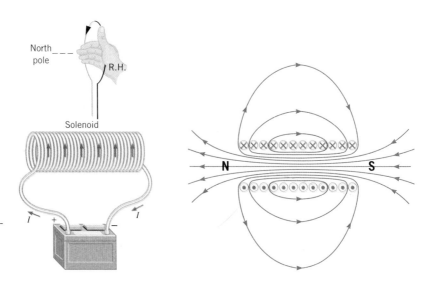

Figure 21.34 A solenoid and a cross-sectional view of it, showing the magnetic field lines and the north and south poles.

RHR-2, just as it is for a circular current loop. The magnitude of the magnetic field in the interior of a long solenoid is

***Interior of a
long solenoid***

$$B = \mu_0 n I \qquad (21.7)$$

where n is the number of turns per unit length of the solenoid and I is the current. If, for example, the solenoid contains 100 turns and has a length of 0.05 m, the number of turns per unit length is $n = (100 \text{ turns})/(0.05 \text{ m}) = 2000$ turns/m. The magnetic field outside the solenoid is not constant and is much weaker than the interior field. In fact, the magnetic field outside is nearly zero if the length of the solenoid is much greater than its diameter.

Concept Simulation 21.4

This simulation explores the magnetic field created by the current in a solenoid. You can see how the magnetic field depends on the magnitude and direction of the current. Notice, in particular, how relatively uniform the magnetic field is inside the solenoid when compared to the field near the center of a single loop (see Concept Simulation 21.3). Different places inside the solenoid can be sampled to test the uniformity of the magnetic field. Also, one can explore how the magnetic field depends on the radius of the solenoid.

Go to **www.wiley.com/college/cutnell**

As with a single loop of wire, a solenoid can also be imagined to be a bar magnet, for the solenoid is just an array of connected current loops. And, as with a circular current loop, the location of the north pole can be determined with RHR-2. Figure 21.34 shows that the left end of the solenoid acts as a north pole, and the right end behaves as a south pole. Solenoids are often referred to as *electromagnets,* and they have several advantages over permanent magnets. For one thing, the strength of the magnetic field can be altered by changing the current and/or the number of turns per unit length. Furthermore, the north and south poles of an electromagnet can be readily switched by reversing the current.

Applications of the magnetic field produced by a current-carrying solenoid are widespread. An exciting medical application is in the technique of magnetic resonance imaging (MRI). With this technique, detailed pictures of the internal parts of the body can be obtained in a noninvasive way that involves none of the risks inherent in the use of X-rays. Figure 21.35 shows a patient who has just been removed from a magnetic resonance imaging machine. The opening behind the patient provides access to the interior of a solenoid, which is typically made from superconducting wire. The superconducting wire facilitates the use of large currents to produce a strong magnetic field. In the presence of this field, the nuclei of certain atoms can be made to behave as tiny radio transmitters and emit radio waves similar to those used by FM stations. The hydrogen atom, which is so prevalent in the human body, can be made to behave in this fashion. The strength of the magnetic field determines where a given collection of hydrogen atoms will "broadcast" on an imaginary FM dial. With a magnetic field that has a slightly different strength at different places, it is possible to associate the location on this imaginary FM dial with a physical location within the body. Computer processing of these locations produces the magnetic resonance image. When hydrogen atoms are used in this way, the image is essentially a map showing their distribution within the body. Remarkably detailed magnetic resonance images can now be obtained, such as those shown in Figure 21.36. They provide doctors with a powerful diagnostic tool that complements those available from X-ray and other techniques. Surgeons can now perform operations more accurately by stepping inside specially designed MRI scanners along with the patient and seeing live MRI images of the area into which they are cutting.

Television sets and computer display monitors use electromagnets (solenoids) to produce images by exerting magnetic forces on moving electrons. An evacuated glass tube, called a cathode-ray tube (CRT), contains an electron gun that sends a narrow beam of high-speed electrons toward the screen of the tube, as Figure 21.37a illustrates. The inner surface of the screen is covered with a phosphor coating, and when the electrons strike it, they generate a spot of visible light. This spot is called a pixel (a contraction of "picture element").

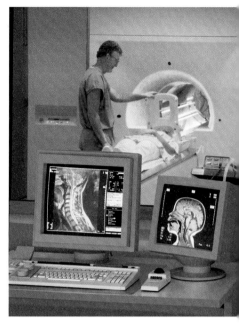

Figure 21.35 A magnetic resonance imaging (MRI) machine. The patient has just been removed from the large semicircular opening in the background, which is the interior of a solenoid. MRI scans of the spinal cord and head can be seen on the monitors. (© Lester Lefkowitz/Taxi/Getty Images)

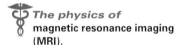

**The physics of
magnetic resonance imaging
(MRI).**

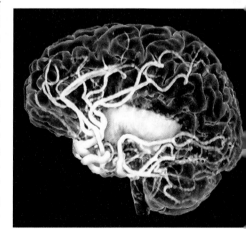

Figure 21.36 Magnetic resonance imaging provides one way to diagnose brain disorders. This three-dimensional magnetic resonance angiogram scan shows a human brain after a stroke. Major arteries are white. The central region (yellow) is an area of bleeding. (© Zephyr/Photo Researchers)

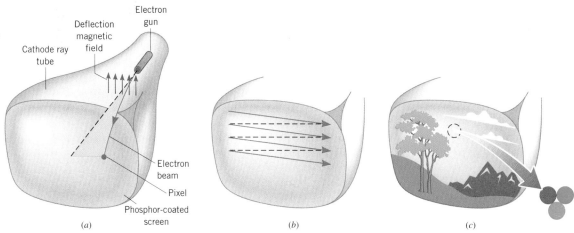

(a)   *(b)*   *(c)*

Figure 21.37 *(a)* A cathode-ray tube contains an electron gun, a magnetic field for deflecting the electron beam, and a phosphor-coated screen. A color TV actually uses three guns, although only one is shown here for clarity. *(b)* The image is formed by scanning the electron beam across the screen. *(c)* The red, green, and blue phosphors of a color TV.

The physics of
television screens and computer display monitors.

To create a black-and-white picture, the electron beam is scanned rapidly from left to right across the screen. As the beam makes each horizontal scan, the number of electrons per second striking the screen is changed by electronics controlling the electron gun, making the scan line brighter in some places and darker in others. When the beam reaches the right side of the screen, it is turned off and returned to the left side slightly below where it started (see part *b* of the figure). The beam is then scanned across the next line, and so on. In current TV sets, a complete picture consists of 525 scan lines (or 625 in Europe) and is formed in $\frac{1}{30}$ of a second. High-definition TV sets have about 1100 scan lines, giving a much sharper, more detailed picture.

The electron beam is deflected by a pair of electromagnets placed around the neck of the tube, between the electron gun and the screen. One electromagnet is responsible for producing the horizontal deflection of the beam and the other for the vertical deflection. For clarity, Figure 21.37*a* shows the net magnetic field generated by the electromagnets at one instant, and not the electromagnets themselves. The electric current in the electromagnets produces a net magnetic field that exerts a force on the moving electrons, causing their trajectories to bend and reach different points on the screen. Changing the current changes the field, so the electrons can be deflected to any point on the screen.

A color TV operates with three electron guns instead of one. And the single phosphor of a black-and-white TV is replaced by a large number of three-dot clusters of phosphors that glow red, green, and blue when struck by an electron beam, as indicated in Figure 21.37*c*. Each red, green, and blue color in a cluster is produced when electrons from one of the three guns strike the corresponding phosphor dot. The three dots are so close together that, from a normal viewing distance, they cannot be separately distinguished. Red, green, and blue are primary colors, so virtually all other colors can be created by varying the intensities of the three beams focused on a cluster.

21.8 Ampere's Law

▶ **CONCEPTS AT A GLANCE** We have seen that an electric current creates a magnetic field. However, the magnitude and direction of the field at any point in space depends on the specific geometry of the wire carrying the current. For instance, distinctly different magnetic fields surround a long, straight wire, a circular loop of wire, and a solenoid. Although different, each of these fields can be obtained from a general law known as *Ampere's law,* which is valid for a wire of any geometrical shape. The Concepts-at-a-Glance chart in Figure 21.38, which is an extension of the chart in Figure 21.9, illustrates that Ampere's law specifies the relationship between a current and its associated magnetic field. ◀

To see how Ampere's law is stated, consider Figure 21.39, which shows two wires carrying currents I_1 and I_2. In general, there may be any number of currents. Around the wires we construct an arbitrarily shaped but closed path. This path encloses a surface and is constructed from a large number of short segments, each of length $\Delta \ell$. Ampere's law

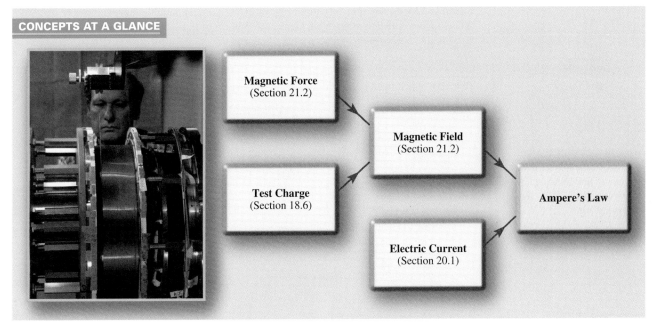

CONCEPTS AT A GLANCE

Magnetic Force (Section 21.2)

Test Charge (Section 18.6)

Magnetic Field (Section 21.2)

Electric Current (Section 20.1)

Ampere's Law

Figure 21.38 CONCEPTS AT A GLANCE As the concept chart in Figure 21.9 illustrates, the magnetic field is defined in terms of the magnetic force that acts on a moving test charge; this earlier chart is reproduced here in the three upper-left panels. Electric currents produce magnetic fields, and the relation between the two is expressed by Ampere's law. The photograph shows a technician winding a superconducting wire onto a coil for use as an electromagnet. (© Maximilian Stock, Ltd./Science Photo Library/Photo Researchers)

deals with the product of $\Delta\ell$ and $B_{\parallel}$ for each segment, where $B_{\parallel}$ is the component of the magnetic field that is *parallel* to $\Delta\ell$ (see the blow-up view in the drawing). For magnetic fields that do not change as time passes, the law states that the sum of all the $B_{\parallel}\Delta\ell$ terms is proportional to the net current I passing through the surface enclosed by the path. For the specific example in Figure 21.39, we see that $I = I_1 + I_2$. Ampere's law is stated in equation form as follows:

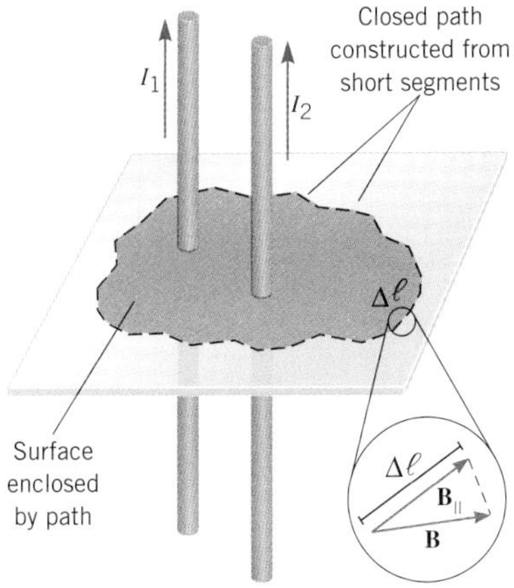

■ **AMPERE'S LAW FOR STATIC MAGNETIC FIELDS**

For any current geometry that produces a magnetic field that does not change in time,

$$\Sigma B_{\parallel}\Delta\ell = \mu_0 I \qquad (21.8)$$

where $\Delta\ell$ is a small segment of length along a closed path of arbitrary shape around the current, $B_{\parallel}$ is the component of the magnetic field parallel to $\Delta\ell$, I is the net current passing through the surface bounded by the path, and μ_0 is the permeability of free space. The symbol Σ indicates that the sum of all $B_{\parallel}\Delta\ell$ terms must be taken around the closed path.

To illustrate the use of Ampere's law, we apply it in Example 11 to the special case of the current in a long, straight wire and show that it leads to the proper expression for the magnetic field.

Figure 21.39 This setup is used in the text to explain Ampere's law.

Closed path constructed from short segments

I_1 I_2

$\Delta\ell$

Surface enclosed by path

$\Delta\ell$ $B_{\parallel}$ B

Example 11 An Infinitely Long, Straight, Current-Carrying Wire

Use Ampere's law to obtain the magnetic field produced by the current in an infinitely long, straight wire.

Reasoning Figure 21.25*a* shows that compass needles point in a circular pattern around the wire, so we know that the magnetic field lines are circular. Therefore, it is convenient to use a circular path of radius *r* when applying Ampere's law, as Figure 21.40 indicates.

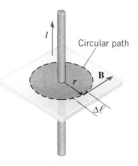

Figure 21.40 Example 11 uses Ampere's law to find the magnetic field in the vicinity of this long, straight, current-carrying wire.

Solution Along the circular path in Figure 21.40, the magnetic field is everywhere parallel to $\Delta\ell$ and has a constant magnitude, since each point is at the same distance from the wire. Thus, $B_\parallel = B$ and, according to Ampere's law, we have

$$\Sigma\, B_\parallel\, \Delta\ell = B(\Sigma\Delta\ell) = \mu_0 I$$

But $\Sigma\Delta\ell$ is just the circumference $2\pi r$ of the circle, so Ampere's law reduces to

$$B\,(\Sigma\Delta\ell) = B\,(2\pi r) = \mu_0 I$$

Dividing both sides by $2\pi r$ shows that $\boxed{B = \mu_0 I/(2\pi r)}$, as given earlier in Equation 21.5.

Self-Assessment Test 21.2

Test your understanding of the material in Sections 21.5–21.8:

- The Force on a Current in a Magnetic Field
- The Torque on a Current-Carrying Coil   • Magnetic Fields Produced by Currents
- Ampere's Law

Go to **www.wiley.com/college/cutnell**

21.9 *Magnetic Materials*

FERROMAGNETISM

The similarity between the magnetic field lines in the neighborhood of a bar magnet and those around a current loop suggests that the magnetism in each case arises from a common cause. The field that surrounds the loop is created by the charges moving in the wire. The magnetic field around a bar magnet is also due to the motion of charges, but the motion is not that of a bulk current through the magnetic material. Instead, the motion responsible for the magnetism is that of the electrons within the atoms of the material.

The magnetism produced by electrons within an atom can arise from two motions. First, each electron orbiting the nucleus behaves like an atomic-sized loop of current that generates a small magnetic field, similar to the field created by the current loop in Figure 21.30a. Second, each electron possesses a spin that also gives rise to a magnetic field. The net magnetic field created by the electrons within an atom is due to the combined fields created by their orbital and spin motions.

In most substances the magnetism produced at the atomic level tends to cancel out, with the result that the substance is nonmagnetic overall. However, there are some materials, known as *ferromagnetic materials,* in which the cancellation does not occur for groups of approximately $10^{16} - 10^{19}$ neighboring atoms, because they have electron spins that are naturally aligned parallel to each other. This alignment results from a special type of quantum mechanical* interaction between the spins. The result of the interaction is a small but highly magnetized region of about 0.01 to 0.1 mm in size, depending on the nature of the material; this region is called a *magnetic domain.* Each domain behaves as a small magnet with its own north and south poles. Common ferromagnetic materials are iron, nickel, cobalt, chromium dioxide, and alnico (an *al*uminum–*ni*ckel–*co*balt alloy).

INDUCED MAGNETISM

Often the magnetic domains in a ferromagnetic material are arranged randomly, as Figure 21.41a illustrates for a piece of iron. In such a situation, the magnetic fields of the domains cancel each other, so the iron displays little, if any, overall magnetism. However, an unmagnetized piece of iron can be magnetized by placing it in an external magnetic field provided by a permanent magnet or an electromagnet. The external magnetic field penetrates the unmagnetized iron and *induces* (or brings about) a state of magnetism in the iron by causing two effects. Those domains whose magnetism is parallel or nearly paral-

---

* The branch of physics called quantum mechanics is mentioned in Section 29.5, although a detailed discussion of it is beyond the scope of this book.

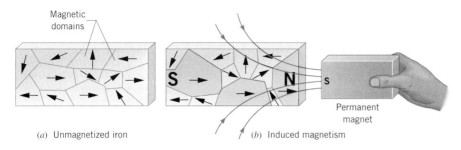

(a) Unmagnetized iron   (b) Induced magnetism

Permanent magnet

Magnetic domains

Figure 21.41 (*a*) Each magnetic domain is a highly magnetized region that behaves like a small magnet (represented by an arrow whose head indicates a north pole). An unmagnetized piece of iron consists of many domains that are randomly aligned. The size of each domain is exaggerated for clarity. (*b*) The external magnetic field of the permanent magnet causes those domains that are parallel or nearly parallel to the field to grow in size (shown in gold).

lel to the external magnetic field grow in size at the expense of other domains that are not so oriented. Part *b* of the drawing shows the growing domains in gold. In addition, the magnetic alignment of some domains may rotate and become more oriented in the direction of the external field. The resulting preferred alignment of the domains gives the iron an overall magnetism, so the iron behaves like a magnet with associated north and south poles. In some types of ferromagnetic materials, such as the chromium dioxide used in cassette tapes, the domains remain aligned for the most part when the external magnetic field is removed, and the material thus becomes permanently magnetized.

The magnetism induced in a ferromagnetic material can be surprisingly large, even in the presence of a weak external field. For instance, it is not unusual for the induced magnetic field to be a hundred to a thousand times stronger than the external field that causes the alignment. For this reason, high-field electromagnets are constructed by wrapping the current-carrying wire around a solid core made from iron or other ferromagnetic material.

Induced magnetism explains why a permanent magnet sticks to a refrigerator door and why an electromagnet can pick up scrap iron at a junkyard. The photo in Figure 21.42 illustrates yet another example of induced magnetism. Notice in Figure 21.41*b* that there is a north pole at the end of the iron that is closest to the south pole of the permanent magnet. The net result is that the two opposite poles give rise to an attraction between the iron and the permanent magnet. Conversely, the north pole of the permanent magnet would also attract the piece of iron by inducing a south pole in the nearest side of the iron. In nonferromagnetic materials, such as aluminum and copper, the formation of magnetic domains does not occur, so magnetism cannot be induced into these substances. Consequently, magnets do not stick to aluminum cans or to copper pennies.

Magnetic grippers

Figure 21.42 Because of induced magnetism, powerful magnetic grippers enable this worker to climb the metal bulkhead of a ship. (© Trident Technologies. Reproduced with permission.)

The physics of
magnetic tape recording.

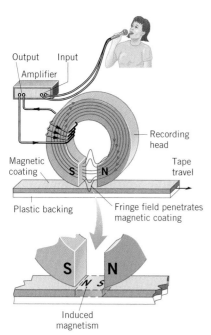

Figure 21.43 The magnetic fringe field of the recording head penetrates the magnetic coating on the tape and magnetizes it.

The physics of
a magnetically levitated train.

MAGNETIC TAPE RECORDING

The process of magnetic tape recording uses induced magnetism, as Figure 21.43 illustrates. The weak electrical signal from a microphone is routed to an amplifier where it is amplified. The current from the output of the amplifier is then sent to the recording head, which is a coil of wire wrapped around an iron core. The iron core has the approximate shape of a horseshoe with a small gap between the two ends. The ferromagnetic iron substantially enhances the magnetic field produced by the current in the wire.

When there is a current in the coil, the recording head becomes an electromagnet with a north pole at one end and a south pole at the other end. The magnetic field lines pass through the iron core and cross the gap. Within the gap, the lines are directed from the north pole to the south pole. Some of the field lines in the gap "bow outward," as Figure 21.43 indicates, the bowed region of magnetic field being called the *fringe field.* The fringe field penetrates the magnetic coating on the tape and induces magnetism in the coating. This induced magnetism is retained when the tape leaves the vicinity of the recording head and, thus, provides a means for storing audio information. Audio information is stored, because at any instant in time the way in which the tape is magnetized depends on the amount and direction of current in the recording head. The current, in turn, depends on the sound picked up by the microphone, so that changes in the sound that occur from moment to moment are preserved as changes in the tape's induced magnetism.

MAGLEV TRAINS

A magnetically levitated train—maglev, for short—uses forces that arise from induced magnetism to levitate or float above a guideway. Since it rides a few centimeters above the guideway, a maglev does not need wheels. Freed from friction with the guideway, the train can achieve significantly greater speeds than do conventional trains. For example, the Transrapid maglev in Figure 21.44 has achieved speeds of 110 m/s (250 mph).

Figure 21.44*a* shows that the Transrapid maglev achieves levitation with electromagnets mounted on arms that extend around and under the guideway. When a current is sent to an electromagnet, the resulting magnetic field creates induced magnetism in a rail mounted in the guideway. The upward attractive force from the induced magnetism is balanced by the weight of the train, so the train moves without touching the rail or the guideway.

Magnetic levitation only lifts the train and does not move it forward. Figure 21.44*b* illustrates how magnetic propulsion is achieved. In addition to the levitation electromag-

Figure 21.44 (*a*) The Transrapid maglev (a German train) has achieved speeds of 110 m/s (250 mph). The levitation electromagnets are drawn up toward the rail in the guideway, levitating the train. (*b*) The magnetic propulsion system. (© Courtesy Deutsche Bundesbahn/Transrapid)

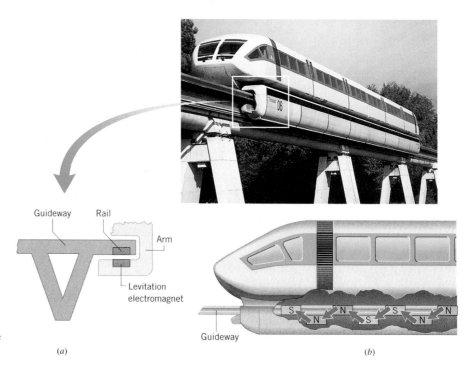

(*a*)   (*b*)

nets, propulsion electromagnets are also placed along the guideway. By controlling the direction of the currents in the train and guideway electromagnets, it is possible to create an unlike pole in the guideway just ahead of each electromagnet on the train and a like pole just behind. Each electromagnet on the train is thus both pulled and pushed forward by electromagnets in the guideway. By adjusting the timing of the like and unlike poles in the guideway, the speed of the train can be adjusted. Reversing the poles in the guideway electromagnets serves to brake the train.

21.10 Concepts & Calculations

Both magnetic fields and electric fields can apply forces to an electric charge. However, there are distinct differences in the way the two types of fields apply their forces. This chapter discusses the magnetic field, and Chapter 18 deals with the electric field. The following example serves to review how the two types of fields behave.

Concepts & Calculations Example 12 Magnetic and Electric Fields

Figure 21.45 shows a negatively charged particle whose charge magnitude is $q = 2.80 \times 10^{-6}$ C. It is moving along the $+y$ axis at a speed of $v = 4.80 \times 10^6$ m/s. A magnetic field of magnitude 3.35×10^{-5} T is directed along the $+z$ axis, and an electric field of magnitude 123 N/C points along the $-x$ axis. Determine the magnitude and direction of the net force that acts on the particle.

Concept Questions and Answers What is the net force?

Answer   The net force is the vector sum of all the forces acting on the particle, which in this case are the magnetic force and the electric force.

How do you determine the direction of the magnetic force acting on the negative charge?

Answer   First, remember that the direction for either a positive or negative charge is perpendicular to the plane formed by the magnetic field vector **B** and the velocity vector **v**. Then, apply Right-Hand Rule No. 1 (RHR-1) to find the direction of the force as if the charge were positive. Finally, reverse the direction indicated by RHR-1 to determine the direction of the force on the negative charge.

How do you determine the direction of the electric force acting on the negative charge?

Answer   The direction for either a positive or negative charge is along the line of the electric field. The electric force on a positive charge is in the same direction as the electric field, while the force on a negative charge is in the direction opposite to the electric field.

Does the fact that the charge is moving affect the values of the magnetic and electric forces?

Answer   The fact that the charge is moving affects the value of the magnetic force, for without motion, there is no magnetic force. The electric force, in contrast, has the same value whether or not the charge is moving.

Solution An application of RHR-1 (see Figure 21.8) reveals that if the charge were positive, the magnetic force would point along the $+x$ axis. But the charge is negative, so the magnetic force points along the $-x$ axis. According to Equation 21.1, the magnitude of the magnetic force is $F_{\text{magnetic}} = qvB \sin \theta$, where q is the magnitude of the charge. Since the velocity and the magnetic field are perpendicular, $\theta = 90°$, and we have $F_{\text{magnetic}} = qvB$.

The electric field points along the $-x$ axis and would apply a force in that same direction if the charge were positive. But the charge is negative, so the electric force points in the opposite direction, or along the $+x$ axis. According to Equation 18.2, the magnitude of the electric force is $F_{\text{electric}} = qE$.

Using plus and minus signs to take into account the different directions of the magnetic and electric forces, we find that the net force is

$$\Sigma F = +F_{\text{electric}} - F_{\text{magnetic}} = +qE - qvB$$
$$= +(2.80 \times 10^{-6} \text{ C})(123 \text{ N/C}) - (2.80 \times 10^{-6} \text{ C})(4.80 \times 10^6 \text{ m/s})$$
$$\times (3.35 \times 10^{-5} \text{ T})$$
$$= \boxed{-1.06 \times 10^{-4} \text{ N}}$$

The fact that the answer is negative means that the net force points along the $-x$ axis.

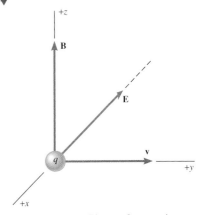

Figure 21.45 The net force acting on the moving charge in this drawing is calculated in Example 12.

Problem solving insight
The direction of the magnetic force exerted on a negative charge is opposite to that exerted on a positive charge, assuming both charges are moving in the same direction in the same magnetic field.

A number of forces can act simultaneously on a rigid object, some of them producing torques, as Chapter 9 discusses. If, as a result of the forces and torques, the object has no acceleration of any kind, it is in equilibrium. Section 9.2 discusses the equilibrium of rigid objects. The next example illustrates that the magnetic force can be one of the forces keeping an object in equilibrium.

Concepts & Calculations Example 13   Equilibrium

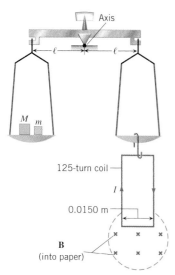

Figure 21.46 This setup is the focus of Example 13.

A 125-turn rectangular coil of wire is hung from one arm of a balance, as Figure 21.46 shows. With the magnetic field turned off, a mass M is added to the pan on the other arm to balance the weight of the coil. When a constant 0.200-T magnetic field is turned on and there is a current of 8.50 A in the coil, how much additional mass m must be added to regain the balance?

Concept Questions and Answers In a balanced or equilibrium condition the device has no angular acceleration. What does this imply about the net torque acting on the device?

Answer   According to Newton's second law as applied to rotational motion (Equation 9.7), when the angular acceleration is zero, the net torque is zero. Counterclockwise torques must balance clockwise torques for an object that is in equilibrium.

What is a torque?

Answer   According to Equation 9.1, the torque produced by a force is the magnitude of the force times the lever arm of the force. The lever arm is the perpendicular distance between the line of action of the force and the axis of rotation. Counterclockwise torques are positive, and clockwise torques are negative.

In calculating the torques acting on an object in equilibrium, where do you locate the axis of rotation?

Answer   The location of the axis is arbitrary because, if there is no angular acceleration, the torques must add up to zero with respect to any axis whatsoever. Choose an axis that simplifies your calculations.

Solution In a condition of equilibrium the sum of the torques acting on the device is zero. For the purpose of calculating torques, we use an axis that is perpendicular to the page at the knife-edge supporting the balance arms (see the drawing). The forces that can produce torques are the weights of the pans, the masses in the left pan, the coil hanging from the right pan, and the magnetic force on the coil. The weight of each pan is W_{pan}, and the weight of the coil is W_{coil}. The weights of the masses M and m are Mg and mg, where g is the magnitude of the acceleration due to gravity. The lever arm for each of these forces is ℓ (see the drawing). When the magnetic field is turned off and the mass m is not present, the sum of the torques is

Field off $\Sigma \tau = \underbrace{(+W_{pan}\ell + Mg\ell)}_{\text{Left pan}} + \underbrace{(-W_{pan}\ell - W_{coil}\ell)}_{\text{Right pan}} = Mg\ell - W_{coil}\ell = 0$

When the field is turned on, a magnetic force is applied to the 0.0150-m segment of the coil that is in the field. Since the magnetic field points into the page and the current is from right to left, an application of RHR-1 reveals that the magnetic force points downward. According to Equation 21.3, the magnitude of the magnetic force is $F = ILB \sin \theta$, where I is the current and L is the length of the wire in the field. Since the current and the field are perpendicular, $\theta = 90°$, and we have $F = ILB$. The sum of the torques with the field turned on and the mass m added to the left pan is

Field on $\Sigma \tau = \underbrace{(+W_{pan}\ell + Mg\ell + mg\ell)}_{\text{Left pan}} + \underbrace{(-W_{pan}\ell - W_{coil}\ell - ILB\ell)}_{\text{Right pan}}$

$= Mg\ell + mg\ell - W_{coil}\ell - ILB\ell = 0$

From the field-off equation, it follows that $Mg\ell - W_{coil}\ell = 0$. With this substitution, the field-on equation simplifies to the following:

$$mg\ell - ILB\ell = 0$$

Solving this expression for the additional mass m gives

$$m = \frac{ILB}{g} = \frac{(8.50 \text{ A})[125(0.0150 \text{ m})](0.200 \text{ T})}{9.80 \text{ m/s}^2} = \boxed{0.325 \text{ kg}}$$

In this calculation we have used the fact that the coil has 125 turns, so the length of the wire in the field is 125 (0.0150 m).

▲

At the end of the problem set for this chapter, you will find homework problems that contain both conceptual and quantitative parts. These problems are grouped under the heading *Concepts &Calculations, Group Learning Problems*. They are designed for use by students working alone or in small learning groups. The conceptual part of each problem provides a convenient focus for group discussions.

Concept Summary

This summary presents an abridged version of the chapter, including the important equations and all available learning aids. For convenient reference, the learning aids (including the text's examples) are placed next to or immediately after the relevant equation or discussion. The following learning aids may be found on-line at **www.wiley.com/college/cutnell**:

Interactive LearningWare examples are solved according to a five-step interactive format that is designed to help you develop problem-solving skills.	**Concept Simulations** are animated versions of text figures or animations that illustrate important concepts. You can control parameters that affect the display, and we encourage you to experiment.
Interactive Solutions offer specific models for certain types of problems in the chapter homework. The calculations are carried out interactively.	**Self-Assessment Tests** include both qualitative and quantitative questions. Extensive feedback is provided for both incorrect and correct answers, to help you evaluate your understanding of the material.

Topic	*Discussion*	*Learning Aids*

21.1 *Magnetic Fields*

North and south poles

A magnet has a north pole and a south pole. The north pole is the end that points toward the north magnetic pole of the earth when the magnet is freely suspended. Like magnetic poles repel each other, and unlike poles attract each other.

Direction of the magnetic field

A magnetic field exists in the space around a magnet. The magnetic field is a vector whose direction at any point is the direction indicated by the north pole of a small compass needle placed at that point.

Magnetic field lines

As an aid in visualizing the magnetic field, magnetic field lines are drawn in the vicinity of a magnet. The lines appear to originate from the north pole and end on the south pole. The magnetic field at any point in space is tangent to the magnetic field line at the point. Furthermore, the strength of the magnetic field is proportional to the number of lines per unit area that passes through a surface oriented perpendicular to the lines.

21.2 *The Force That a Magnetic Field Exerts on a Moving Charge*

Magnetic force (direction)

The direction of the magnetic force acting on a charge moving with a velocity **v** in a magnetic field **B** is perpendicular to both **v** and **B**. For a positive charge the direction can be determined with the aid of Right-Hand Rule No. 1 (see below). The magnetic force on a moving negative charge is opposite to the force on a moving positive charge.

Right-Hand Rule No. 1

Extend the right hand so the fingers point along the direction of the magnetic field **B** and the thumb points along the velocity **v** of the charge. The palm of the hand then faces in the direction of the magnetic force **F** that acts on a positive charge.

The magnitude B of the magnetic field at any point in space is defined as

Magnetic field (magnitude)

$$B = \frac{F}{q_0 v \sin \theta}$$

(21.1) **Example 1**

The tesla and the gauss

where F is the magnitude of the magnetic force that acts on a positive charge q_0 whose velocity **v** makes an angle θ with respect to the magnetic field. The SI unit for the magnetic field is the tesla (T). Another, smaller unit for the magnetic field is the gauss; 1 gauss $= 10^{-4}$ tesla. The gauss is not an SI unit.

Topic	Discussion	Learning Aids

21.3 The Motion of a Charged Particle in a Magnetic Field

When a charged particle moves in a region that contains both magnetic and electric fields, the net force on the particle is the vector sum of the magnetic and electric forces.

Examples 2, 12

A constant magnetic force does no work

A magnetic force does no work on a particle, because the direction of the force is always perpendicular to the motion of the particle. Being unable to do work, the magnetic force cannot change the kinetic energy, and hence the speed, of the particle; however, the magnetic force does change the direction in which the particle moves.

When a particle of charge q and mass m moves with speed v perpendicular to a uniform magnetic field of magnitude B, the magnetic force causes the particle to move on a circular path of radius

Radius of circular path

$$r = \frac{mv}{qB} \qquad (21.2)$$

Concept Simulation 21.1
Examples 3, 4
Interactive LearningWare 21.1
Interactive Solution 21.21

21.4 The Mass Spectrometer

The mass spectrometer is an instrument for measuring the abundance of ionized atoms or molecules that have different masses. The atoms or molecules are ionized ($+e$), accelerated to a speed v by a potential difference V, and sent into a uniform magnetic field of magnitude B. The magnetic field causes the particles (each with a mass m) to move on a circular path of radius r. The relation between m and B is

Relation between mass and magnetic field

$$m = \left(\frac{er^2}{2V}\right)B^2$$

Use *Self-Assessment Test 21.1* **to evaluate your understanding of Sections 21.1–21.4.**

21.5 The Force on a Current in a Magnetic Field

An electric current, being composed of moving charges, can experience a magnetic force when placed in a magnetic field of magnitude B. For a straight wire that has a length L and carries a current I, the magnetic force has a magnitude of

Force on a current (magnitude)

$$F = ILB \sin \theta \qquad (21.3)$$

Examples 5, 13
Interactive LearningWare 21.2
Interactive Solution 21.34

Force on a current (direction)

where θ is the angle between the directions of the current and the magnetic field. The direction of the force is perpendicular to both the current and the magnetic field and is given by Right-Hand Rule No. 1.

21.6 The Torque on a Current-Carrying Coil

Magnetic forces can exert a torque on a current-carrying loop of wire and thus cause the loop to rotate. When a current I exists in a coil of wire with N turns, each of area A, in the presence of a magnetic field of magnitude B, the coil experiences a net torque of magnitude

Torque on a current-carrying coil

$$\tau = NIAB \sin \phi \qquad (21.4)$$

Example 6
Interactive Solution 21.43

where ϕ is the angle between the direction of the magnetic field and the normal to the plane of the coil.

Magnetic moment

The quantity NIA is known as the magnetic moment of the coil.

21.7 Magnetic Fields Produced by Currents

An electric current produces a magnetic field, with different current geometries giving rise to different field patterns. For an infinitely long, straight wire, the magnetic field lines are circles centered on the wire, and their direction is given by Right-Hand Rule No. 2 (see page 651). The magnitude of the magnetic field at a radial distance r from the wire is

Topic	Discussion	Learning Aids
Long, straight wire	$$B = \frac{\mu_0 I}{2\pi r} \quad (21.5)$$	Examples 7, 8, 9 Interactive LearningWare 21.3 Concept Simulation 21.2
Permeability of free space	where I is the current in the wire and μ_0 is a constant known as the permeability of free space ($\mu_0 = 4\pi \times 10^{-7}\,\mathrm{T\cdot m/A}$).	
Right-Hand Rule No. 2	Curl the fingers of the right hand into the shape of a half-circle. Point the thumb in the direction of the conventional current I, and the tips of the fingers will point in the direction of the magnetic field **B**.	
Center of a circular loop	The magnitude of the magnetic field at the center of a flat circular loop consisting of N turns, each of radius R, is $$B = \frac{N\mu_0 I}{2R} \quad (21.6)$$ The loop has associated with it a north pole on one side and a south pole on the other side. The side of the loop that behaves like a north pole can be predicted by using Right-Hand Rule No. 2.	Concept Simulation 21.3 Example 10 Interactive Solution 21.55
Interior of a solenoid	A solenoid is a coil of wire wound in the shape of a helix. Inside a long solenoid the magnetic field is nearly constant and has a magnitude of $$B = \mu_0 n I \quad (21.7)$$ where n is the number of turns per unit length of the solenoid. One end of the solenoid behaves like a north pole, and the other end like a south pole. The end that is the north pole can be predicted by using Right-Hand Rule No. 2.	Concept Simulation 21.4
Ampere's law	**21.8 Ampere's Law** Ampere's law specifies the relationship between a current and its associated magnetic field. For any current geometry that produces a magnetic field that does not change in time, Ampere's law states that $$\Sigma B_\parallel \Delta\ell = \mu_0 I \quad (21.8)$$ where $\Delta\ell$ is a small segment of length along a closed path of arbitrary shape around the current, $B_\parallel$ is the component of the magnetic field parallel to $\Delta\ell$, I is the net current passing through the surface bounded by the path, and μ_0 is the permeability of free space. The symbol Σ indicates that the sum of all $B_\parallel\Delta\ell$ terms must be taken around the closed path.	Example 11 Interactive Solution 21.62

 Use Self-Assessment Test 21.2 to evaluate your understanding of Sections 21.5–21.8.

Ferromagnetic materials	**21.9 Magnetic Materials** Ferromagnetic materials, such as iron, are made up of tiny regions called magnetic domains, each of which behaves as a small magnet. In an unmagnetized ferromagnetic material, the domains are randomly aligned. In a permanent magnet, many of the domains are aligned, and a high degree of magnetism results. An unmagnetized ferromagnetic material can be induced into becoming magnetized by placing it in an external magnetic field.	

Conceptual Questions

1. Magnetic field lines, like electric field lines, never intersect. Suppose it were possible for two magnetic field lines to intersect at a point in space. Discuss what this would imply about the force(s) that act on a charge moving through such a point, thereby ruling out the possibility of field lines crossing.

2. Suppose you accidentally use your left hand, instead of your right hand, to determine the direction of the magnetic force on a positive charge moving in a magnetic field. Do you get the correct answer? If not, what direction do you get?

3. A charged particle, passing through a certain region of space, has a velocity whose magnitude and direction remain constant. (a) If it is known that the external magnetic field is zero everywhere in this region, can you conclude that the external electric field is also zero? Explain. (b) If it is known that the external electric field is zero everywhere, can you conclude that the external magnetic field is also zero? Explain.

4. Suppose that the positive charge in Figure 21.10*a* were launched from the negative plate toward the positive plate, in a direction opposite to the electric field. A sufficiently strong electric field would prevent the charge from striking the positive plate. Suppose the positive charge in Figure 21.10*b* were launched from the south pole toward the north pole, in a direction opposite to the magnetic field. Would a sufficiently strong magnetic field prevent the charge from reaching the north pole? Account for your answer.

5. Review Conceptual Example 4 and Concept Simulation 21.1 at **www.wiley.com/college/cutnell** as background for this question. Three particles move through a constant magnetic field and follow the paths shown in the drawing. Determine whether each particle is positively charged, negatively charged, or neutral. Give a reason for each answer.

Question 5

6. Refer to Figure 21.12. Assume that the particle in the picture is a proton. If an electron is projected at point 1 with the same velocity **v**, it will not follow the same path as the proton, unless the magnetic field is adjusted. Explain how the magnitude and direction of the field must be changed.

7. The drawing shows a top view of four interconnected chambers. A negative charge is fired into chamber 1. By turning on separate magnetic fields in each chamber, the charge can be made to exit from chamber 4, as shown. (a) Describe how the magnetic field in each chamber should be directed. (b) If the speed of the charge is v when it enters chamber 1, what is the speed of the charge when it exits chamber 4? Why?

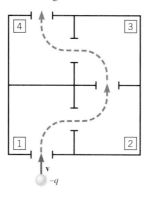

8. A positive charge moves along a circular path under the influence of a magnetic field. The magnetic field is perpendicular to the plane of the circle, as in Figure 21.12. If the velocity of the particle is reversed at some point along the path, will the particle retrace its path? If not, draw the new path. Explain.

9. When one end of a bar magnet is placed near a TV screen, the picture becomes distorted. Why?

10. The drawing shows a particle carrying a positive charge $+q$ at the coordinate origin, as well as a target located in the third quadrant. A uniform magnetic field is directed perpendicularly into the plane of the paper. The charge can be projected in the plane of the paper only, along the positive or negative x or y axis. Thus, there are four possible directions for the initial velocity of the particle. The particle can be made to hit the target for only two of the

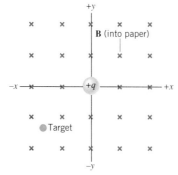

four directions. Which two are they? Give your reasoning, and draw the two paths that the particle can follow on its way to the target.

11. Refer to Figure 21.17. (a) What happens to the direction of the magnetic force if the current is reversed? (b) What happens to the direction of the force if *both* the current *and* the magnetic poles are reversed? Explain your answers.

12. The drawing shows a conducting wire wound into a helical shape. The helix acts like a spring and expands back toward its original shape after its coils are squeezed together and released. The bottom end of the wire just barely touches the mercury (a good electrical conductor) in the cup. After the switch is closed, current in the circuit causes the light bulb to glow. Does the bulb glow continually, glow briefly and then go out, or repeatedly turn on and off like a turn signal on a car? Explain.

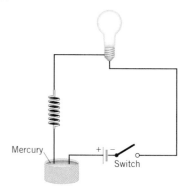

13. In Figure 21.28, assume that the current I_1 is larger than the current I_2. In parts *a* and *b*, decide whether there are places where the total magnetic field is zero. State whether they are located to the left of both wires, between the wires, or to the right of both wires. Give your reasoning.

14. The drawing shows an end-on view of three parallel wires that are perpendicular to the plane of the paper. In two of the wires the current is directed into the paper, while in the remaining wire the current is directed out of the paper. The two outermost wires are held rigidly in place. Which way will the middle wire move? Explain.

15. For each electromagnet at the left of the drawing, explain whether it will be attracted to or repelled from the permanent magnet at the right.

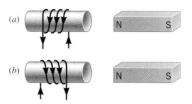

16. For each electromagnet at the left of the drawing, explain whether it will be attracted to or repelled from the adjacent electromagnet at the right.

17. Refer to Figure 21.5. If the earth's magnetism is assumed to originate from a large circular loop of current within the earth, how is the plane of this current loop oriented relative to the magnetic axis, and what is the direction of the current around the loop?

18. There are four wires viewed end-on in the drawing. They are long, straight, and perpendicular to the plane of the paper. Their cross sections lie at the corners of a square. Currents of the same magnitude are in each of these wires. Choose the direction of the current for each wire, so that when any single current is turned off, the total magnetic field at point *P* (the center of the square) is directed toward a corner of the square. Account for your answer.

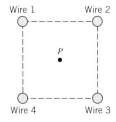

19. Suppose you have two bars, one of which is a permanent magnet and the other of which is not a magnet, but is made from a fer- romagnetic material like iron. The two bars look exactly alike. (a) Using a third bar, which is known to be a magnet, how can you determine which of the look-alike bars is the permanent magnet and which is not? (b) Can you determine the identities of the look-alike bars with the aid of a third bar that is not a magnet, but is made from a ferromagnetic material? Give reasons for your answers.

20. In a TV commercial that advertises a soda pop, a strong electromagnet picks up a delivery truck carrying cans of the soft drink. The picture switches to the interior of the truck, where cans are seen to fly upward and stick to the roof just beneath the electromagnet. Are these cans made entirely of aluminum? Justify your answer.

Problems

ssm Solution is in the Student Solutions Manual.   www Solution is available on the World Wide Web at www.wiley.com/college/cutnell

⚕ This icon represents a biomedical application.

Section 21.1 Magnetic Fields,
Section 21.2 The Force That a Magnetic Field Exerts on a Moving Charge

1. ssm Due to friction with the air, an airplane has acquired a net charge of 1.70×10^{-5} C. The plane moves with a speed of 2.80×10^2 m/s at an angle θ with respect to the earth's magnetic field, the magnitude of which is 5.00×10^{-5} T. The magnetic force on the airplane has a magnitude of 2.30×10^{-7} N. Find the angle θ. (There are two possible angles.)

2. A particle with a charge of $+8.4$ μC and a speed of 45 m/s enters a uniform magnetic field whose magnitude is 0.30 T. For each of the cases in the drawing, find the magnitude and direction of the magnetic force on the particle.

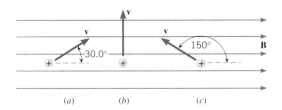

(a)   (b)   (c)

3. In a television set, electrons are accelerated from rest through a potential difference of 19 kV. The electrons then pass through a 0.28-T magnetic field that deflects them to the appropriate spot on the screen. Find the magnitude of the maximum magnetic force that an electron can experience.

4. Two charged particles move in the same direction with respect to the same magnetic field. Particle 1 travels three times faster than particle 2. However, each particle experiences a magnetic force of the same magnitude. Find the ratio q_1/q_2 of the magnitudes of the charges.

5. ssm At a certain location, the horizontal component of the earth's magnetic field is 2.5×10^{-5} T, due north. A proton moves eastward with just the right speed, so the magnetic force on it balances its weight. Find the speed of the proton.

6. In New England, the horizontal component of the earth's magnetic field has a magnitude of 1.6×10^{-5} T. An electron is shot vertically straight up from the ground with a speed of 2.1×10^6 m/s. What is the magnitude of the acceleration caused by the magnetic force? Ignore the gravitational force acting on the electron.

7. An electron is moving through a magnetic field whose magnitude is 8.70×10^{-4} T. The electron experiences only a magnetic force and has an acceleration of magnitude 3.50×10^{14} m/s^2. At a certain instant, it has a speed of 6.80×10^6 m/s. Determine the angle θ (less than 90°) between the electron's velocity and the magnetic field.

*8. The drawing shows a parallel plate capacitor that is moving with a speed of 32 m/s through a 3.6-T magnetic field. The velocity **v** is perpendicular to the magnetic field. The electric field within the capacitor has a value of 170 N/C, and each plate has an area of 7.5×10^{-4} m^2. What is the magnetic force (magnitude and direction) exerted on the positive plate of the capacitor?

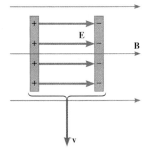

*9. ssm www The electrons in the beam of a television tube have a kinetic energy of 2.40×10^{-15} J. Initially, the electrons move horizontally from west to east. The vertical component of the earth's magnetic field points down, toward the surface of the earth, and has a magnitude of 2.00×10^{-5} T. (a) In what direction are the electrons deflected by this field component? (b) What is the acceleration of an electron in part (a)?

Section 21.3 The Motion of a Charged Particle in a Magnetic Field,
Section 21.4 The Mass Spectrometer

10. A magnetic field has a magnitude of 1.2×10^{-3} T, and an electric field has a magnitude of 4.6×10^3 N/C. Both fields point in the same direction. A positive 1.8-μC charge moves at a speed of 3.1×10^6 m/s in a direction that is perpendicular to both fields. Determine the magnitude of the net force that acts on the charge.

11. An electron moves at a speed of 6.0×10^6 m/s perpendicular to a constant magnetic field. The path is a circle of radius 1.3×10^{-3} m. (a) Draw a sketch showing the magnetic field and the electron's path. (b) What is the magnitude of the field? (c) Find the magnitude of the electron's acceleration.

12. A charged particle enters a uniform magnetic field and follows the circular path shown in the drawing. (a) Is the particle positively or negatively charged? Why? (b) The particle's speed is

140 m/s, the magnitude of the magnetic field is 0.48 T, and the radius of the path is 960 m. Determine the mass of the particle, given that its charge has a magnitude of 8.2×10^{-4} C.

Problem 12

13. ssm www A charged particle with a charge-to-mass ratio of $q/m = 5.7 \times 10^8$ C/kg travels on a circular path that is perpendicular to a magnetic field whose magnitude is 0.72 T. How much time does it take for the particle to complete one revolution?

14. The solar wind is a thin, hot gas given off by the sun. Charged particles in this gas enter the magnetic field of the earth and can experience a magnetic force. Suppose a charged particle traveling with a speed of 9.0×10^6 m/s encounters the earth's magnetic field at an altitude where the field has a magnitude of 1.2×10^{-7} T. Assuming that the particle's velocity is perpendicular to the magnetic field, find the radius of the circular path on which the particle would move if it were (a) an electron and (b) a proton.

15. ssm A beam of protons moves in a circle of radius 0.25 m. The protons move perpendicular to a 0.30-T magnetic field. (a) What is the speed of each proton? (b) Determine the magnitude of the centripetal force that acts on each proton.

16. Review Conceptual Example 2 before attempting this problem. Derive an expression for the magnitude v of the velocity "selected" by the velocity selector. This expression should give v in terms of the strengths E and B of the electric and magnetic fields, respectively.

17. Two isotopes of carbon, carbon-12 and carbon-13, have masses of 19.93×10^{-27} kg and 21.59×10^{-27} kg, respectively. These two isotopes are singly ionized ($+e$) and each is given a speed of 6.667×10^5 m/s. The ions then enter the bending region of a mass spectrometer where the magnetic field is 0.8500 T. Determine the spatial separation between the two isotopes after they have traveled through a half-circle.

* **18.** The ion source in a mass spectrometer produces both singly and doubly ionized species, X^+ and X^{2+}. The difference in mass between these species is too small to be detected. Both species are accelerated through the same electric potential difference, and both experience the same magnetic field, which causes them to move on circular paths. The radius of the path for the species X^+ is r_1, while the radius for species X^{2+} is r_2. Find the ratio r_1/r_2 of the radii.

* **19. ssm** A particle of charge $+7.3$ μC and mass 3.8×10^{-8} kg is traveling perpendicular to a 1.6-T magnetic field, as the drawing shows. The speed of the particle is 44 m/s. (a) What is the value of the angle θ, such that the particle's subsequent path will intersect the y axis at the greatest possible value of y? (b) Determine this value of y.

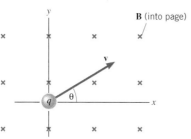

* **20.** Review Conceptual Example 2 as background for this problem. A charged particle moves through a velocity selector at a constant speed in a straight line. The electric field of the velocity selector is

3.80×10^3 N/C, while the magnetic field is 0.360 T. When the electric field is turned off, the charged particle travels on a circular path whose radius is 4.30 cm. Find the charge-to-mass ratio of the particle.

* **21.** Consult **Interactive Solution 21.21** at **www.wiley.com/college/cutnell** to review a model for solving this problem. A proton with a speed of 3.5×10^6 m/s is shot into a region between two plates that are separated by a distance of 0.23 m. As the drawing shows, a magnetic field exists between the plates, and it is perpendicular to the velocity of the proton. What must be the magnitude of the magnetic field, so the proton just misses colliding with the opposite plate?

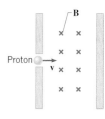

* **22.** Conceptual Example 4 provides background pertinent to this problem. An electron has a kinetic energy of 2.0×10^{-17} J. It moves on a circular path that is perpendicular to a uniform magnetic field of magnitude 5.3×10^{-5} T. Determine the radius of the path.

* **23. ssm www** A particle of mass 6.0×10^{-8} kg and charge $+7.2$ μC is traveling due east. It enters perpendicularly a magnetic field whose magnitude is 3.0 T. After entering the field, the particle completes one-half of a circle and exits the field traveling due west. How much time does the particle spend in the magnetic field?

* **24.** A positively charged particle of mass 7.2×10^{-8} kg is traveling due east with a speed of 85 m/s and enters a 0.31-T uniform magnetic field. The particle moves through one-quarter of a circle in a time of 2.2×10^{-3} s, at which time it leaves the field heading due south. All during the motion the particle moves perpendicular to the magnetic field. (a) What is the magnitude of the magnetic force acting on the particle? (b) Determine the magnitude of its charge.

** **25.** Refer to Conceptual Question 10 (not Problem 10) before starting this problem. Suppose that the target discussed there is located at the coordinates $x = -0.10$ m and $y = -0.10$ m. In addition, suppose that the particle is a proton and the magnetic field has a magnitude of 0.010 T. The speed at which the particle is projected is the same for either of the two paths leading to the target. Find the speed.

Section 21.5 The Force on a Current in a Magnetic Field

26. A 45-m length of wire is stretched horizontally between two vertical posts. The wire carries a current of 75 A and experiences a magnetic force of 0.15 N. Find the magnitude of the earth's magnetic field at the location of the wire, assuming the field makes an angle of $60.0°$ with respect to the wire.

27. ssm An electric power line carries a current of 1400 A in a location where the earth's magnetic field is 5.0×10^{-5} T. The line makes an angle of $75°$ with respect to the field. Determine the magnitude of the magnetic force on a 120-m length of line.

28. A wire carries a current of 0.66 A. This wire makes an angle of $58°$ with respect to a magnetic field of magnitude 4.7×10^{-5} T. The wire experiences a magnetic force of magnitude 7.1×10^{-5} N. What is the length of the wire?

29. A square coil of wire containing a single turn is placed in a uniform 0.25-T magnetic field, as the drawing shows. Each side has a length of 0.32 m, and the current in the coil is 12 A. Determine the magnitude of the magnetic force on each of the four sides.

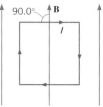

30. The triangular loop of wire shown in the drawing carries a current of I = 4.70 A. A uniform magnetic field is directed parallel to side AB of the triangle and has a magnitude of 1.80 T. (a) Find the magnitude and direction of the magnetic force exerted on each side of the triangle. (b) Determine the magnitude of the net force exerted on the triangle.

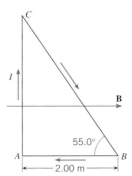

31. ssm A wire of length 0.655 m carries a current of 21.0 A. In the presence of a 0.470-T magnetic field, the wire experiences a force of 5.46 N. What is the angle (less than 90°) between the wire and the magnetic field?

32. Two insulated wires, each 2.40 m long, are taped together to form a two-wire unit that is 2.40 m long. One wire carries a current of 7.00 A; the other carries a smaller current I in the opposite direction. The two-wire unit is placed at an angle of 65.0° relative to a magnetic field whose magnitude is 0.360 T. The magnitude of the net magnetic force experienced by the two-wire unit is 3.13 N. What is the current I?

* 33. ssm A copper rod of length 0.85 m is lying on a frictionless table (see the drawing). Each end of the rod is attached to a fixed wire by an unstretched spring whose spring constant is k = 75 N/m. A magnetic field with a strength of 0.16 T is oriented perpendicular to the surface of the table. (a) What must be the direction of the current in the copper rod that causes the springs to stretch? (b) If the current is 12 A, by how much does each spring stretch?

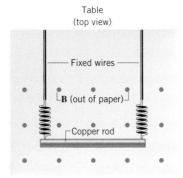

* 34. Consult **Interactive Solution 21.34** at **www.wiley.com/college/cutnell** to explore a model for solving this problem. The drawing shows a thin, uniform rod, which has a length of 0.45 m and a mass of 0.094 kg. This rod lies in the plane of the paper and is attached to the floor by a hinge at point P. A uniform magnetic field of 0.36 T is directed perpendicularly into the plane of the paper. There is a current I = 4.1 A in the rod, which does not rotate clockwise or counterclockwise. Find the angle θ. (*Hint: The magnetic force may be taken to act at the center of gravity.*)

** 35. The two conducting rails in the drawing are tilted upward so they each make an angle of 30.0° with respect to the ground. The vertical magnetic field has a magnitude of 0.050 T. The 0.20-kg aluminum rod (length = 1.6 m) slides *without friction* down the rails at a constant velocity. How much current flows through the bar?

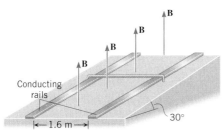

Section 21.6 The Torque on a Current-Carrying Coil

36. The 1200-turn coil in a dc motor has an area per turn of 1.1×10^{-2} m². The design for the motor specifies that the magnitude of the maximum torque is 5.8 N·m when the coil is placed in a 0.20-T magnetic field. What is the current in the coil?

37. ssm A circular coil of wire has a radius of 0.10 m. The coil has 50 turns and a current of 15 A, and is placed in a magnetic field whose magnitude is 0.20 T. (a) Determine the magnetic moment of the coil. (b) What is the maximum torque the coil can experience in this field?

38. Two coils have the same number of circular turns and carry the same current. Each rotates in a magnetic field as in Figure 21.21. Coil 1 has a radius of 5.0 cm and rotates in a 0.18-T field. Coil 2 rotates in a 0.42-T field. Each coil experiences the same maximum torque. What is the radius (in cm) of coil 2?

39. The maximum torque experienced by a coil in a 0.75-T magnetic field is 8.4×10^{-4} N·m. The coil is circular and consists of only one turn. The current in the coil is 3.7 A. What is the length of the wire from which the coil is made?

40. Two pieces of the same wire have the same length. From one piece, a square coil containing a single loop is made. From the other, a circular coil containing a single loop is made. The coils carry different currents. When placed in the same magnetic field with the same orientation, they experience the same torque. What is the ratio I_{square}/I_{circle} of the current in the square coil to that in the circular coil?

41. ssm www The rectangular loop in the drawing consists of 75 turns and carries a current of I = 4.4 A. A 1.8-T magnetic field is directed along the $+y$ axis. The loop is free to rotate about the z axis. (a) Determine the magnitude of the net torque exerted on the loop and (b) state whether the 35° angle will increase or decrease.

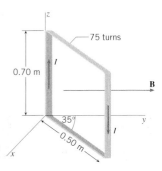

42. A coil carries a current and experiences a torque due to a magnetic field. The value of the torque is 80.0% of the maximum possible torque. (a) What is the smallest angle between the magnetic field and the normal to the plane of the coil? (b) Make a drawing, showing how this coil would be oriented relative to the magnetic field. Be sure to include the angle in the drawing.

* 43. Consult **Interactive Solution 21.43** at **www.wiley.com/college/cutnell** to see how this problem can be solved. The coil in Figure 21.22a contains 410 turns and has an area per turn of 3.1×10^{-3} m². The magnetic field is 0.23 T, and the current in the coil is 0.26 A. A brake shoe is pressed perpendicularly against the shaft to keep the coil from turning. The coefficient of static friction between the shaft and the brake shoe is 0.76. The radius of the shaft is 0.012 m. What is the magnitude of the minimum normal force that the brake shoe exerts on the shaft?

* 44. A square coil and a rectangular coil are each made from the same length of wire. Each contains a single turn. The long sides of the rectangle are twice as long as the short sides. Find the ratio $\tau_{square}/\tau_{rectangle}$ of the maximum torques that these coils experience in the same magnetic field when they contain the same current.

** 45. ssm A charge of 4.0×10^{-6} C is placed on a small conducting sphere that is located at the end of a thin insulating rod whose length is 0.20 m. The rod rotates with an angular speed of ω = 150 rad/s about an axis that passes perpendicularly through its other end. Find the magnetic moment of the rotating charge. (*Hint: The charge travels around a circle in a time equal to the period of the motion.*)

Section 21.7 Magnetic Fields Produced by Currents

46. A long, straight wire carries a current of 48 A. The magnetic field produced by this current at a certain point is 8.0×10^{-5} T. How far is the point from the wire?

47. In a lightning bolt, 15 C of charge flows in a time of 1.5×10^{-3} s. Assuming that the lightning bolt can be represented as a long, straight line of current, what is the magnitude of the magnetic field at a distance of 25 m from the bolt?

48. What must be the radius of a circular loop of wire so the magnetic field at its center is 1.8×10^{-4} T when the loop carries a current of 12 A?

49. ssm A long solenoid has 1400 turns per meter of length, and it carries a current of 3.5 A. A small circular coil of wire is placed inside the solenoid with the normal to the coil oriented at an angle of 90.0° with respect to the axis of the solenoid. The coil consists of 50 turns, has an area of 1.2×10^{-3} m^2, and carries a current of 0.50 A. Find the torque exerted on the coil.

50. A $+6.00 \ \mu$C charge is moving with a speed of 7.50×10^4 m/s parallel to a very long, straight wire. The wire is 5.00 cm from the charge and carries a current of 67.0 A in a direction opposite to that of the moving charge. Find the magnitude and direction of the force on the charge.

51. ssm Two rigid rods are oriented parallel to each other and to the ground. The rods carry the same current in the same direction. The length of each rod is 0.85 m, while the mass of each is 0.073 kg. One rod is held in place above the ground, and the other floats beneath it at a distance of 8.2×10^{-3} m. Determine the current in the rods.

52. Two circular loops of wire, each containing a single turn, have the same radius of 4.0 cm and a common center. The planes of the loops are perpendicular. Each carries a current of 1.7 A. What is the magnitude of the net magnetic field at the common center?

53. A circular loop of wire and a long, straight wire carry currents of I_1 and I_2 (see the drawing), where $I_2 = 6.6I_1$. The loop and the straight wire lie in the same plane. The net magnetic field at the center of the loop is zero. Find the distance H, expressing your answer in terms of R, the radius of the loop.

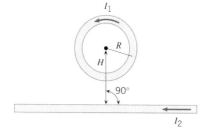

54. As background for this problem, review Conceptual Example 9. A rectangular current loop is located near a long, straight wire that carries a current of 12 A (see the drawing). The current in the loop is 25 A. Determine the magnitude of the net magnetic force that acts on the loop.

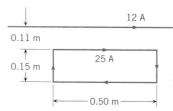

55. Review **Interactive Solution 21.55** at **www.wiley.com/college/cutnell** for one approach to this problem. Two circular coils are concentric and lie in the same plane. The inner coil contains 140 turns of wire, has a radius of 0.015 m, and carries a current of 7.2 A. The outer coil contains 180 turns and has a radius of 0.023 m. What must be the magnitude and direction (relative to the current in the inner coil) of the current in the outer coil, such that the net magnetic field at the common center of the two coils is zero?

56. Two long, straight, parallel wires *A* and *B* are separated by a distance of one meter. They carry currents in opposite directions, and the current in wire *A* is one-third of that in wire *B*. On a line drawn perpendicular to the wires, find the point where the net magnetic field is zero. Determine this point relative to wire *A*.

57. ssm A piece of copper wire has a resistance per unit length of 5.90×10^{-3} Ω/m. The wire is wound into a thin, flat coil of many turns that has a radius of 0.140 m. The ends of the wire are connected to a 12.0-V battery. Find the magnetic field strength at the center of the coil.

58. The drawing shows two wires that carry the same current of $I = 85.0$ A and are oriented perpendicular to the plane of the paper. The current in one wire is directed out of the paper, while the current in the other is directed into the paper. Find the magnitude and direction of the net magnetic field at point *P*.

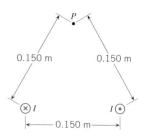

59. The drawing shows an end-on view of three wires. They are long, straight, and perpendicular to the plane of the paper. Their cross sections lie at the corners of a square. The currents in wires 1 and 2 are $I_1 = I_2 = I$ and are directed into the paper. What is the direction of the current in wire 3, and what is the ratio I_3/I, such that the net magnetic field at the empty corner is zero?

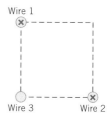

Section 21.8 Ampere's Law

60. Suppose a uniform magnetic field is everywhere perpendicular to this page. The field points directly upward toward you. A circular path is drawn on the page. Use Ampere's law to show that there can be no net current passing through the circular surface.

61. ssm The wire in Figure 21.40 carries a current of 12 A. Suppose that a second long, straight wire is placed right next to this wire. The current in the second wire is 28 A. Use Ampere's law to find the magnitude of the magnetic field at a distance of $r = 0.72$ m from the wires when the currents are (a) in the same direction and (b) in opposite directions.

62. Refer to **Interactive Solution 21.62** at **www.wiley.com/college/cutnell** for help with problems like this one. A very long, hollow cylinder is formed by rolling up a thin sheet of copper. Electric charges flow along the copper sheet parallel to the axis of the cylinder. The arrangement is, in effect, a hollow tube of current *I*. Use Ampere's law to show that the magnetic field (a) is $\mu_0 I/(2\pi r)$ outside the cylinder at a distance *r* from the axis and (b) is zero at any point within the hollow interior of the cylinder. (*Hint: For closed paths, use circles perpendicular to and centered on the axis of the cylinder.*)

63. A long, cylindrical conductor is solid throughout and has a radius *R*. Electric charges flow parallel to the axis of the cylinder and pass uniformly through the entire cross section. The arrangement is, in effect, a solid tube of current I_0. The current per unit cross-sectional area (i.e., the current density) is $I_0/(\pi R^2)$. Use Ampere's law to show that the magnetic field inside the conductor at a distance *r* from the axis is $\mu_0 I_0 r/(2\pi R^2)$. (*Hint: For a closed path, use a circle of radius r perpendicular to and centered on the axis. Note that the current through any surface is the area of the surface times the current density.*)

Additional Problems

64. A charge $q_1 = 25.0$ μC moves with a speed of 4.50×10^3 m/s perpendicular to a uniform magnetic field. The charge experiences a magnetic force of 7.31×10^{-3} N. A second charge $q_2 = 5.00$ μC travels at an angle of $40.0°$ with respect to the same magnetic field and experiences a 1.90×10^{-3}-N force. Determine (a) the magnitude of the magnetic field and (b) the speed of q_2.

65. ssm A long solenoid consists of 1400 turns of wire and has a length of 0.65 m. There is a current of 4.7 A in the wire. What is the magnitude of the magnetic field within the solenoid?

66. When a charged particle moves at an angle of $25°$ with respect to a magnetic field, it experiences a magnetic force of magnitude F. At what angle (less than $90°$) with respect to this field will this particle, moving at the same speed, experience a magnetic force of magnitude $2F$?

67. A 0.50-m length of wire is formed into a single-turn, square loop in which there is a current of 12 A. The loop is placed in a magnetic field of 0.12 T, as in Figure 21.22a. What is the maximum torque that the loop can experience?

68. Suppose in Figure 21.28a that $I_1 = I_2 = 25$ A and that the separation between the wires is 0.016 m. By applying an external magnetic field (created by a source other than the wires) it is possible to cancel the mutual repulsion of the wires. This external field must point along the vertical direction. (a) Does the external field point up or down? Explain. (b) What is the magnitude of the external field?

69. ssm A charge of 12 μC, traveling with a speed of 9.0×10^6 m/s in a direction perpendicular to a magnetic field, experiences a magnetic force of 8.7×10^{-3} N. What is the magnitude of the field?

70. The x, y, and z components of a magnetic field are $B_x = 0.10$ T, $B_y = 0.15$ T, and $B_z = 0.17$ T. A 25-cm wire is oriented along the z axis and carries a current of 4.3 A. What is the magnitude of the magnetic force that acts on this wire?

71. Suppose that an ion source in a mass spectrometer produces *doubly* ionized gold ions (Au^{2+}), each with a mass of 3.27×10^{-25} kg. The ions are accelerated from rest through a potential difference of 1.00 kV. Then, a 0.500-T magnetic field causes the ions to follow a circular path. Determine the radius of the path.

72. Two long, straight wires are separated by 0.120 m. The wires carry currents of 8.0 A in opposite directions, as the drawing indicates. Find the magnitude of the net magnetic field at the points labeled (a) A and (b) B.

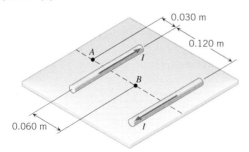

73. ssm Review Conceptual Example 2 as an aid in understanding this problem. The drawing shows a positively charged particle entering a 0.52-T magnetic field. The particle has a speed of 270 m/s and moves perpendicular to the magnetic field. Just as the particle enters the magnetic field, an electric field is turned on. What must be the magnitude and direction of the electric field such that the *net* force on the particle is twice the magnetic force?

74. Two parallel rods are each 0.50 m in length. They are attached at their centers to either end of a spring (spring constant = 150 N/m) that is initially neither stretched nor compressed. When 950 A of current is in each rod in the same direction, the spring is observed to be compressed by 2.0 cm. Treat the rods as long, straight wires and find the separation between them when the current is present.

75. One component of a magnetic field has a magnitude of 0.048 T and points along the $+x$ axis, while the other component has a magnitude of 0.065 T and points along the $-y$ axis. A particle carrying a charge of $+2.0 \times 10^{-5}$ C is moving along the $+z$ axis at a speed of 4.2×10^3 m/s. (a) Find the magnitude of the net magnetic force that acts on the particle. (b) Determine the angle that the net force makes with respect to the $+x$ axis.

76. The drawing shows two long, straight wires that are suspended from a ceiling. The mass per unit length of each wire is 0.050 kg/m. Each of the four strings suspending the wires has a length of 1.2 m. When the wires carry identical currents in opposite directions, the angle between the strings holding the two wires is $15°$. What is the current in each wire?

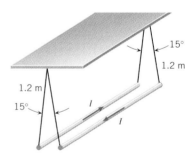

77. ssm In the model of the hydrogen atom due to Niels Bohr, the electron moves around the proton at a speed of 2.2×10^6 m/s in a circle of radius 5.3×10^{-11} m. Considering the orbiting electron to be a small current loop, determine the magnetic moment associated with this motion. (*Hint: The electron travels around the circle in a time equal to the period of the motion.*)

Concepts & Calculations Group Learning Problems

Note: Each of these problems consists of Concept Questions followed by a related quantitative Problem. They are designed for use by students working alone or in small learning groups. The Concept Questions involve little or no mathematics and are intended to stimulate group discussions. They focus on the concepts with *which the problems deal. Recognizing the concepts is the essential initial step in any problem-solving technique.*

78. Concept Questions (a) A charge moves along the $+x$ axis and experiences no magnetic force, although there is a magnetic field.

What can you conclude about the direction of the magnetic field? (b) A moving charge experiences the maximum possible magnetic force when moving in a magnetic field. What can you conclude about the angle θ that the charge's velocity makes with respect to the magnetic field? Explain your answers.

Problem A particle that has an 8.2-μC charge moves with a velocity of magnitude 5.0×10^5 m/s. When the velocity points along the +x axis, the particle experiences no magnetic force, although there is a magnetic field present. The maximum possible magnetic force that the charge could experience has a magnitude of 0.48 N. Find the magnitude and direction of the magnetic field. Note that there are two possible answers for the direction of the field. Make sure that your answers are consistent with your answers to the Concept Questions.

79. Concept Questions A proton is projected perpendicularly into a magnetic field with a certain velocity and follows a circular path. Then an electron is projected perpendicularly into the same magnetic field with the same velocity. (a) Does the electron follow the exact same circular path that the proton followed? (b) To make the electron follow the exact same circular path, should the field direction be kept the same or reversed, and (c) should the field magnitude be increased, reduced, or kept the same? Account for your answers.

Problem A proton is projected perpendicularly into a magnetic field that has a magnitude of 0.50 T. The field is then adjusted so that an electron will follow the exact same circular path when it is projected perpendicularly into the field with the same velocity that the proton had. What is the magnitude of the field used for the electron? Verify that your answer is consistent with your answers to the Concept Questions.

80. Concept Questions Particle 1 and particle 2 carry the same charge q, but particle 1 has a smaller mass than particle 2. These two particles accelerate from rest through the same electric potential difference V and enter the same magnetic field, which has a magnitude B. The particles travel perpendicular to the field on circular paths. Upon entering the field region, which particle, if either, has the greater (a) kinetic energy and (b) speed? Give your reasoning.

Problem The masses of the particles are $m_1 = 2.3 \times 10^{-8}$ kg and $m_2 = 5.9 \times 10^{-8}$ kg. The radius of the circular path for particle 1 is $r_1 = 12$ cm. What is the radius of the circular path for particle 2?

81. Concept Question The drawing shows a wire comprised of three segments, AB, BC, and CD. There is a current I in the wire. There is also a magnetic field **B** that is the same everywhere and points in the direction of the +z axis. Rank the wire segments according to the magnetic force (largest first) that they experience. Justify your ranking.

Problem The lengths of the wire segments are $L_{AB} = 1.1$ m, $L_{BC} = 0.55$ m, and $L_{CD} = 0.55$ m. The current is $I = 2.8$ A, and the magnitude of the magnetic field is $B = 0.26$ T. Find the magnitude of the force that acts on each segment. Be sure your answers are consistent with your answer to the Concept Question.

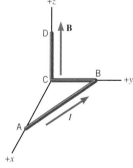

82. Concept Question You have a wire of length L from which to make the square coil of a dc motor. In a given magnetic field a coil of N turns, each with area A, produces more torque when its total effective area of NA is greater rather than smaller. This follows directly from Equation 21.4. Is more torque obtained by using the length of wire to make a single-turn coil or a two-turn coil? Explain.

Problem The length of the wire is $L = 1.00$ m. The current in the coil is $I = 1.7$ A, and the magnetic field of the motor is 0.34 T. Find the maximum torque when the wire is used to make a single-turn

square coil and a two-turn square coil. Verify that your answers are consistent with your answer to the Concept Question.

83. Concept Questions A particle has a charge q and is located at the coordinate origin. An electric field E_x points along the +x axis. A magnetic field also exists, and its components B_x and B_y point along the +x and +y axis, respectively (see the drawing). Which, if any, of the fields (E_x, B_x, B_y) exert a force on the particle if it is (a) stationary and (b) moving along the +x axis? Justify your answers. (c) None of the three fields points along the +z axis. If the particle is moving along this axis, does it experience a force due to any of the fields? Provide a reason for your answer.

Problem The electric and magnetic fields in the drawing are $E_x = +245$ N/C, $B_x = +1.80$ T, and $B_y = +1.40$ T, and the charge is $q = +5.60$ μC. Calculate the force (magnitude and direction) exerted on the particle by each of the three fields when it is (a) stationary, (b) moving along the +x axis at a speed of 375 m/s, and (c) moving along the +z axis at a speed of 375 m/s. Be sure your answers are consistent with those in the Concept Questions.

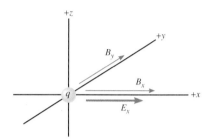

* **84. Concept Questions** The drawing shows two perpendicular, long, straight wires, both of which lie in the plane of the paper. Each wire carries the same current I. What is the direction of the net magnetic field at (a) point A and (b) point B? (c) Is the magnitude of the net field at point A greater than, less than, or equal to the magnitude of the net field at point B?

Problem The current in each of the wires is $I = 5.6$ A. Find the magnitudes of the net fields at points A and B. Verify that your answers are consistent with your answers to the Concept Questions.

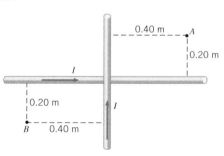

* **85. Concept Questions** A horizontal wire is hung from the ceiling of a room by two massless strings. A uniform magnetic field is directed from the ceiling to the floor. When a current exists in the wire, the wire swings upward and makes an angle ϕ with respect to the vertical, as the drawing shows. (a) How is the magnitude of the magnetic force related to the magnetic field, the length of the wire, and the current in the wire? (b) What is the direction of the magnetic force? (c) The wire is stationary, so it is in equilibrium. What are the forces that keep it in equilibrium? (d) What must be true about the sum of the forces in the horizontal direction and the sum of the forces in the vertical direction?

Problem The wire has a length of 0.20 m and a mass of 0.080 kg and carries a current of 42 A. The magnitude of the magnetic field is 0.070 T. Find (a) the angle ϕ and (b) the tension in each of the two strings.

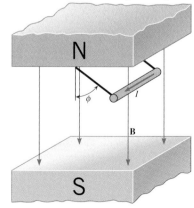

Chapter 22

Electromagnetic Induction

Electric guitars, such as the one being played by Carlos Santana in this photograph, use electromagnetic induction in producing their sound. This phenomenon is the central topic of this chapter. The specific example of an electric guitar pickup will be discussed in detail. (© All Over Press/ Getty Images News and Sport Services)

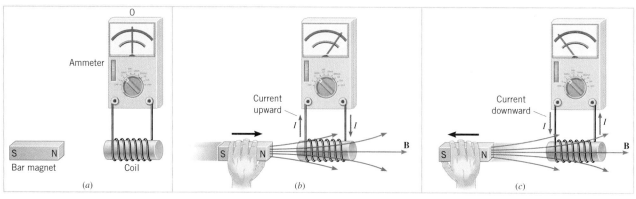

Figure 22.1 (*a*) When there is no relative motion between the coil of wire and the bar magnet, there is no current in the coil. (*b*) A current is created in the coil when the magnet moves toward the coil. (*c*) A current also exists when the magnet moves away from the coil, but the direction of the current is opposite to that in (*b*).

The physics of an automobile cruise control device.

22.1 *Induced Emf and Induced Current*

There are a number of ways a magnetic field can be used to generate an electric current, and Figure 22.1 illustrates one of them. This drawing shows a bar magnet and a helical coil of wire to which an ammeter is connected. When there is no *relative* motion between the magnet and the coil, as in part *a* of the drawing, the ammeter reads zero, indicating that no current exists. However, when the magnet moves toward the coil, as in part *b*, a current *I* appears. As the magnet approaches, the magnetic field that it creates at the location of the coil becomes stronger and stronger, and it is this *changing* field that produces the current. When the magnet moves away from the coil, as in part *c*, a current is also produced, but with a reversed direction. Now the magnetic field at the coil becomes weaker as the magnet moves away, and once again it is the *changing* field that generates the current.

A current would also be created in Figure 22.1 if the magnet were held stationary and the coil were moved, because the magnetic field at the coil would be changing as the coil approached or receded from the magnet. Only relative motion between the magnet and the coil is needed to generate a current; it does not matter which one moves.

The current in the coil is called an ***induced current*** because it is brought about (or "induced") by a changing magnetic field. Since a source of emf is always needed to produce a current, the coil itself behaves as if it were a source of emf. This emf is known as an ***induced emf.*** Thus, a changing magnetic field induces an emf in the coil, and the emf leads to an induced current.

Induced emf and induced current are frequently used in the cruise control devices found in many cars. Figure 22.2 illustrates how a cruise control device operates. Usually two magnets are mounted on opposite sides of the vehicle's drive shaft, with a stationary sensing coil positioned nearby. As the shaft turns, the magnets pass by the coil

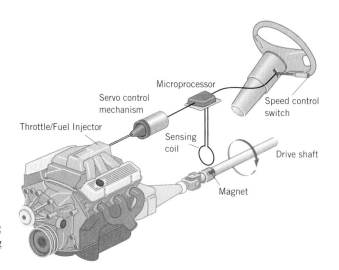

Figure 22.2 Induced emf lies at the heart of an automobile cruise control device. The emf is induced in a sensing coil by magnets attached to the rotating drive shaft.

and cause an induced emf and current to appear in it. A microprocessor (the "brain" of a computer) counts the pulses of current and, with the aid of its internal clock and a knowledge of the shaft's radius, determines the rotational speed of the drive shaft. The rotational speed, in turn, is related to the car's speed. Thus, once the driver sets the desired cruising speed with the speed control switch (mounted near the steering wheel), the microprocessor can compare it with the measured speed. To the extent that the selected cruising speed and the measured speed differ, a signal is sent to a servo or control mechanism, which causes the throttle/fuel injector to send more or less fuel to the engine. The car speeds up or slows down accordingly, until the desired cruising speed is reached.

Figure 22.3 shows another way to induce an emf and a current in a coil. An emf can be induced by *changing the area* of a coil in a constant magnetic field. Here the shape of the coil is being distorted so as to reduce the area. As long as the area is changing, an induced emf and current exist; they vanish when the area is no longer changing. If the distorted coil is returned to its original shape, thereby increasing the area, an oppositely directed current is generated while the area is changing.

In each of the previous examples, both an emf and a current are induced in the coil because the coil is part of a complete, or closed, circuit. If the circuit were open—perhaps because of an open switch—there would be no induced current. However, an emf would still be induced in the coil, whether the current exists or not.

Changing a magnetic field and changing the area of a coil are methods that can be used to create an induced emf. The phenomenon of producing an induced emf with the aid of a magnetic field is called *electromagnetic induction.* The next section discusses yet another method by which an induced emf can be created.

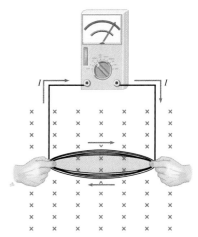

Figure 22.3 While the area of the coil is changing, an induced emf and current are generated.

22.2 *Motional Emf*

THE EMF INDUCED IN A MOVING CONDUCTOR

When a conducting rod moves through a constant magnetic field, an emf is induced in the rod. This special case of electromagnetic induction arises as a result of the magnetic force (see Section 21.2) that acts on a moving charge. Consider the metal rod of length L moving to the right in Figure 22.4a. The velocity $\mathbf{v}$ of the rod is constant and is perpendicular to a uniform magnetic field $\mathbf{B}$. Each charge q within the rod also moves with a velocity $\mathbf{v}$ and experiences a magnetic force of magnitude $F = qvB$, according to Equation 21.1. By using RHR-1, it can be seen that the mobile, free electrons are driven to the bottom of the rod, leaving behind an equal amount of positive charge at the top. (Remember to reverse the direction of the force that RHR-1 predicts, since the electrons have a negative charge. See Section 21.2.) The positive and negative charges accumulate until the attractive electric force that they exert on each other becomes equal in magnitude to the magnetic force. When the two forces balance, equilibrium is reached and no further charge separation occurs.

The separated charges on the ends of the moving conductor give rise to an induced emf, called a ***motional emf*** because it originates from the motion of charges through a magnetic field. The emf exists as long as the rod moves. If the rod is brought to a halt, the magnetic force vanishes, with the result that the attractive electric force reunites the positive and negative charges and the emf disappears. The emf of the moving rod is analogous to that between the terminals of a battery. However, the emf of a battery is produced by chemical reactions, whereas the motional emf is created by the agent that moves the rod through the magnetic field (like the hand in Figure 22.4b.)

The fact that the electric and magnetic forces balance at equilibrium in Figure 22.4a can be used to determine the magnitude of the motional emf $\mathscr{E}$. Acccording to Equation 18.2, the magnitude of the electric force acting on the positive charge q at the top of the rod is Eq, where E is the magnitude of the electric field due to the separated charges. And according to Equation 19.7 (without the minus sign), the electric field magnitude is given by the voltage between the ends of the rod (the emf $\mathscr{E}$) divided by the length L of the rod. Thus, the electric force is $Eq = (\mathscr{E}/L)q$. The magnetic force is qvB, according to Equation

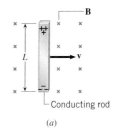

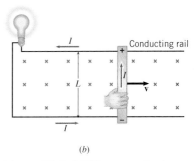

Figure 22.4 (*a*) When a conducting rod moves at right angles to a constant magnetic field, the magnetic force causes opposite charges to appear at the ends of the rod, giving rise to an induced emf. (*b*) The induced emf causes an induced current I to appear in the circuit.

21.1, because the charge q moves perpendicular to the magnetic field. Since these two forces balance, it follows that $(\mathcal{E}/L)q = qvB$. The emf, then, is

Motional emf when **v, B,**
and L are mutually $\mathcal{E} = vBL$ (22.1)
perpendicular

As expected, $\mathcal{E} = 0$ V when $v = 0$ m/s, because no motional emf is developed in a stationary rod. Greater speeds and stronger magnetic fields lead to greater emfs for a given length L. As with batteries, $\mathcal{E}$ is expressed in volts. In Figure 22.4b the rod is sliding on conducting rails that form part of a closed circuit, and L is the length of the rod between the rails. Due to the emf, electrons flow in a clockwise direction around the circuit. Positive charge would flow in the direction opposite to the electron flow, so the conventional current I is drawn counterclockwise in the picture. Example 1 illustrates how to determine the electrical energy that the motional emf delivers to a device such as the light bulb in the drawing.

Example 1 Operating a Light Bulb with Motional Emf

Suppose the rod in Figure 22.4b is moving at a speed of 5.0 m/s in a direction perpendicular to a 0.80-T magnetic field. The rod has a length of 1.6 m and a negligible electrical resistance. The rails also have negligible resistance. The light bulb, however, has a resistance of 96 Ω. Find (a) the emf produced by the rod, (b) the current induced in the circuit, (c) the electric power delivered to the bulb, and (d) the energy used by the bulb in 60.0 s.

Reasoning The moving rod acts like an imaginary battery and supplies a motional emf of vBL to the circuit. The induced current can be determined from Ohm's law as the motional emf divided by the resistance of the bulb. The electric power delivered to the bulb is the product of the induced current and the potential difference across the bulb (which, in this case, is the motional emf). The energy used is the product of the power and the time.

Solution

(a) The motional emf is given by Equation 22.1 as

$$\mathcal{E} = vBL = (5.0 \text{ m/s})(0.80 \text{ T})(1.6 \text{ m}) = \boxed{6.4 \text{ V}}$$

(b) According to Ohm's law, the induced current is equal to the motional emf divided by the resistance of the circuit:

$$I = \frac{\mathcal{E}}{R} = \frac{6.4 \text{ V}}{96 \text{ }\Omega} = \boxed{0.067 \text{ A}}$$ (20.2)

(c) The electric power P delivered to the light bulb is the product of the current I and the potential difference across the bulb:

$$P = I\mathcal{E} = (0.067 \text{ A})(6.4 \text{ V}) = \boxed{0.43 \text{ W}}$$ (20.6a)

(d) Since power is energy per unit time, the energy E used in 60.0 s is the product of the power and the time:

$$E = Pt = (0.43 \text{ W})(60.0 \text{ s}) = \boxed{26 \text{ J}}$$ (6.10b)

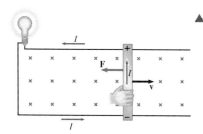

Figure 22.5 A magnetic force **F** is exerted on the current I in the moving rod and is opposite to the velocity **v**. The rod will slow down unless a counterbalancing force is applied by the hand.

MOTIONAL EMF AND ELECTRICAL ENERGY

Motional emf arises because a magnetic force acts on the charges in a conductor that is moving through a magnetic field. Whenever this emf causes a current, a second magnetic force enters the picture. In Figure 22.4b, for instance, the second force arises because the current I in the rod is perpendicular to the magnetic field. The current, and hence the rod, experiences a magnetic force **F** whose magnitude is given by Equation 21.3 as $F = ILB \sin 90°$. Using the values of I, L, and B from Example 1, we see that $F = (0.067 \text{ A})(1.6 \text{ m})(0.80 \text{ T}) = 0.086 \text{ N}$. The direction of **F** is specified by RHR-1 and is *opposite* to the velocity **v** of the rod, as Figure 22.5 shows. By itself, **F** would *slow down* the rod, and here lies the crux of the matter. To keep the rod moving to the right with a constant velocity, a counterbalancing force must be applied to the rod by an external agent, such as the hand in the picture. The counterbalancing force must have a magnitude of 0.086 N and must be directed opposite to the magnetic force **F**. If the counterbalancing force were removed, the rod would decelerate under the influence of **F** and eventually come to rest. During the deceleration, the motional emf would decrease and the light bulb would eventually go out.

We can now answer an important question—Who or what provides the 26 J of electrical energy that the light bulb in Example 1 uses in sixty seconds? The provider is the external agent that applies the 0.086-N counterbalancing force needed to keep the rod moving. This agent does work, and Example 2 shows that the work done is equal to the electrical energy used by the bulb.

Example 2 The Work Needed to Keep the Light Bulb Burning

In Example 1, an external agent (the hand in Figures 22.4*b* and 22.5) supplies a 0.086-N force that keeps the rod moving at a constant speed of 5.0 m/s. Determine the work done in 60.0 s by the external agent.

Reasoning The work W done by the hand in Figure 22.5 is given by Equation 6.1 as $W = (F \cos \theta)x$. In this equation F is the magnitude of the external force, θ is the angle between the external force and the displacement of the rod ($\theta = 0°$), and x is the distance that the rod moves in a time t. The distance is the product of the speed of the rod and the time, $x = vt$, so the work can be expressed as $W = (F \cos 0°)vt = Fvt$.

Solution The work done by the external agent is

$$W = F\,vt = (0.086 \text{ N})(5.0 \text{ m/s})(60.0 \text{ s}) = \boxed{26 \text{ J}}$$

The 26 J of work done on the rod by the external agent is the same as the 26 J of energy used by the bulb. Hence, the moving rod and the magnetic force convert mechanical work into electrical energy, much as a battery converts chemical energy into electrical energy.

Need more practice?

Interactive LearningWare 22.1
A person is twirling a 1.2-m metal chain in a horizontal circle about his head at an angular speed of 6.3 rad/s. The vertical component of the earth's magnetic field points straight down toward the earth and has a magnitude of 1.8×10^{-5} T. (a) What is the magnitude of the motional emf induced in the rotating chain? (b) When viewed from above, the chain is twirled in a counterclockwise manner. Which end of the chain is positive?

Go to
www.wiley.com/college/cutnell
for an interactive solution.

It is important to realize that the direction of the current in Figure 22.5 is consistent with the principle of conservation of energy. Consider what would happen if the direction of the current were reversed, as in Figure 22.6. With the direction of the current reversed, the direction of the magnetic force **F** would also be reversed and point in the direction of the velocity **v** of the rod. As a result, the force would cause the rod to accelerate rather than decelerate. The rod would accelerate without the need for an external force (like that provided by the hand in Figure 22.5) and create a motional emf that supplies energy to the light bulb. Thus, this hypothetical generator would produce energy out of nothing, since there is no external agent. Such a device cannot exist because it violates the principle of conservation of energy, which states that energy cannot be created or destroyed, but can only be converted from one form to another. Therefore, the current cannot be directed clockwise around the circuit, as in Figure 22.6. In situations such as that in Examples 1 and 2, when a motional emf leads to an induced current, a magnetic force always appears that opposes the motion, in accord with the principle of conservation of energy. Conceptual Example 3 deals further with the important issue of energy conservation.

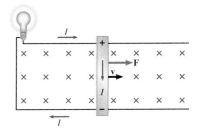

Figure 22.6 The current cannot be directed clockwise in this circuit, because the magnetic force **F** exerted on the rod would then be in the same direction as the velocity **v**. The rod would accelerate to the right and create energy on its own, violating the principle of conservation of energy.

Conceptual Example 3 Conservation of Energy

Figure 22.7*a* illustrates a conducting rod that is free to slide down between two vertical copper tracks. There is no kinetic friction between the rod and the tracks, although the rod maintains electrical contact with the tracks during its fall. A constant magnetic field **B** is directed perpendicular to the motion of the rod, as the drawing shows. Because there is no friction, the only force that acts on the rod is its weight **W**, so the rod falls with an acceleration equal to the acceleration due to gravity, $a = 9.8 \text{ m/s}^2$. Suppose that a resistance R is connected between the tops of the tracks, as in part *b* of the drawing. (a) Does the rod now fall with the acceleration due to gravity? (b) How does the principle of conservation of energy apply to what happens in Figure 22.7*b*?

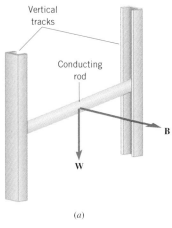

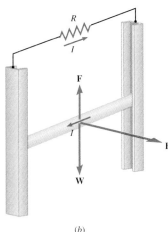

Figure 22.7 (*a*) Because there is no kinetic friction between the falling rod and the tracks, the only force acting on the rod is its weight **W**. (*b*) When an induced current *I* exists in the circuit, a magnetic force **F** also acts on the rod.

Reasoning and Solution

(**a**) As the rod falls perpendicular to the magnetic field, a motional emf is induced between its ends. This emf is induced whether or not the resistance *R* is attached between the tracks. However, when *R* is present, a complete circuit is formed, and the emf produces an induced current *I* that is perpendicular to the field. The direction of this current is such that the rod experiences an upward magnetic force **F**, opposite to the falling motion and the weight of the rod (see part *b* of the drawing and use RHR-1). The net force acting downward on the rod is **W** + **F**, which is *less* than the weight, since **F** points upward and the weight **W** points downward. In accord with Newton's second law of motion, the downward acceleration is proportional to the net force, so ***the rod falls with an acceleration that is less than the acceleration due to gravity.*** In fact, as the speed of the rod in Figure 22.7*b* increases during the descent, the magnetic force becomes larger and larger. There will be a time when the magnitude of the magnetic force becomes equal to the magnitude of the rod's weight. When this occurs, the net force on the rod will be zero, and the rod's acceleration will also be zero. From then on, the rod will fall at a constant velocity.

(**b**) For the freely falling rod in Figure 22.7*a*, gravitational potential energy (GPE) is converted into kinetic energy (KE) as the rod picks up speed on the way down. In Figure 22.7*b*, however, the rod always has a smaller speed than that of a freely falling rod at the same place. This is because the acceleration is less than the acceleration due to gravity. In terms of energy conservation, the smaller speed arises because only part of the GPE is converted into KE, with part also being dissipated as heat in the resistance *R*. In fact, when the rod reaches the point when it falls with a constant velocity, none of the GPE is converted into KE because all of it is dissipated as heat in the resistance.

Related Homework: *Problem 9*

✔ **Check Your Understanding 1**

Consider the induced emf being generated in Figure 22.4. Suppose the length of the rod is reduced by a factor of four. For the induced emf to be the same, what should be done? (a) Without changing the speed of the rod, increase the strength of the magnetic field by a factor of four. (b) Without changing the magnetic field, increase the speed of the rod by a factor of four. (c) Increase both the speed of the rod and the strength of the magnetic field by a factor of two. (d) All of the previous three methods may be used. *(The answer is given at the end of the book.)*

Background: Motional emf is one type of induced emf. It depends on the speed and length of the moving conductor, as well as the strength of the magnetic field, according to Equation 22.1.

For similar questions (including calculational counterparts), consult Self-Assessment Test 22.1, which is described at the end of Section 22.5.

22.3 *Magnetic Flux*

MOTIONAL EMF AND MAGNETIC FLUX

▶ **CONCEPTS AT A GLANCE** Motional emf, as well as any other type of induced emf, can be described in terms of a concept called *magnetic flux*. Magnetic flux is analogous to electric flux, which deals with the electric field and the surface through which it passes (see Section 18.9 and Figure 18.32). Magnetic flux is defined in a similar way. The Concepts-at-a-Glance chart in Figure 22.8 outlines the ideas involved. The upper three boxes in the middle of the chart are the same as those in Figure 21.9, and they indicate that a magnetic field is defined in terms of the magnetic force that acts on a moving test charge. By bringing together the magnetic field and the surface through which it passes, we will see how the concept of magnetic flux arises. ◀

Now let us see how the motional emf given in Equation 22.1 ($\mathscr{E} = vBL$) can be written in terms of the magnetic flux. Figure 22.9*a* shows the rod used to derive Equation 22.1; it is moving through a magnetic field between a time $t = 0$ s and a later time t_0. During this interval, it has moved a distance x_0 to the right. At a later time t, the rod has moved an even greater distance x, as part *b* of the drawing indicates. The speed v of the

CONCEPTS AT A GLANCE

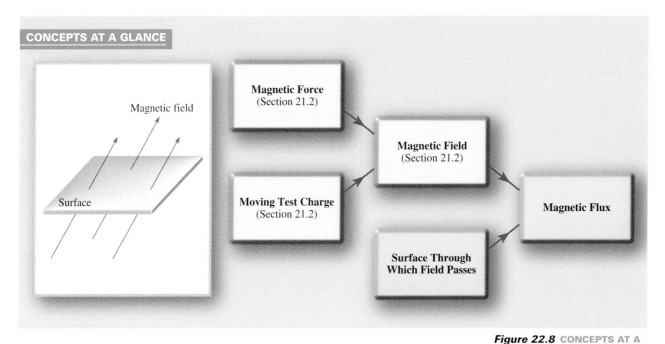

Figure 22.8 CONCEPTS AT A GLANCE Magnetic flux incorporates both the idea of a magnetic field and the surface through which it passes.

rod is the distance traveled divided by the elapsed time: $v = (x - x_0)/(t - t_0)$. Substituting this expression for v into $\mathscr{E} = vBL$ gives

$$\mathscr{E} = \left(\frac{x - x_0}{t - t_0} \right) BL = \left(\frac{xL - x_0 L}{t - t_0} \right) B$$

As the drawing indicates, the term $x_0 L$ is the area A_0 swept out by the rod in moving a distance x_0, while xL is the area A swept out in moving a distance x. In terms of these areas, the emf is

$$\mathscr{E} = \left(\frac{A - A_0}{t - t_0} \right) B = \frac{(BA) - (BA)_0}{t - t_0}$$

Notice how the product BA of the magnetic field strength and the area appears in the numerator of this expression. The quantity BA is given the name **magnetic flux** and is represented by the symbol Φ (Greek capital letter phi); thus $\Phi = BA$. The magnitude of the induced emf is the *change* in flux $\Delta\Phi = \Phi - \Phi_0$ divided by the time interval $\Delta t = t - t_0$ during which the change occurs:

$$\mathscr{E} = \frac{\Phi - \Phi_0}{t - t_0} = \frac{\Delta\Phi}{\Delta t}$$

In other words, the induced emf equals the time rate of change of the magnetic flux.

Figure 22.9 (a) In a time t_0, the moving rod sweeps out an area $A_0 = x_0 L$. (b) The area swept out in a time t is $A = xL$. In both parts of the figure the areas are shaded in color.

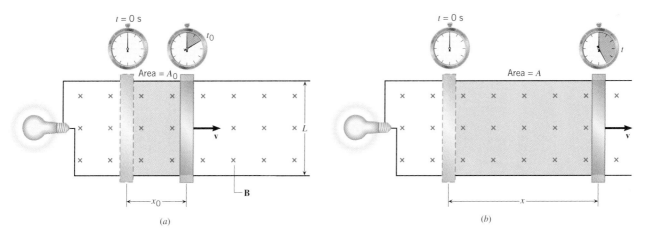

(a)   (b)

You will almost always see the previous equation written with a minus sign—namely, $\mathscr{E} = -\Delta\Phi/\Delta t$. The minus sign is introduced for the following reason: The direction of the current induced in the circuit is such that the magnetic force **F** acts on the rod to *oppose* its motion, thereby tending to slow down the rod (see Figure 22.5). The minus sign is a reminder that the polarity of the induced emf sends the induced current in the proper direction so as to give rise to this opposing magnetic force.

The advantage of writing the induced emf as $\mathscr{E} = -\Delta\Phi/\Delta t$ is that this relation is far more general than our present discussion suggests. In Section 22.4 we will see that $\mathscr{E} = -\Delta\Phi/\Delta t$ can be applied to *all possible ways of generating induced emfs*.

A GENERAL EXPRESSION FOR MAGNETIC FLUX

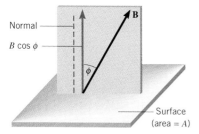

Normal
$B \cos \phi$
ϕ
Surface (area = A)

Figure 22.10 When computing the magnetic flux, the component of the magnetic field that is perpendicular to the surface must be used; this component is $B \cos \phi$.

In Figure 22.9 the direction of the magnetic field **B** is perpendicular to the surface swept out by the moving rod. In general, however, **B** may not be perpendicular to the surface. For instance, in Figure 22.10 the direction perpendicular to the surface is indicated by the normal to the surface, but the magnetic field is inclined at an angle ϕ with respect to this direction. In such a case the flux is computed using only the component of the field that is perpendicular to the surface, $B \cos \phi$. The general expression for magnetic flux is

$$\Phi = (B \cos \phi)A = BA \cos \phi \qquad (22.2)$$

If either the magnitude B of the magnetic field or the angle ϕ is not constant over the surface, an average value of the product $B \cos \phi$ must be used to compute the flux. Equation 22.2 shows that the unit of magnetic flux is the tesla·meter² (T·m²). This unit is called a *weber* (Wb), after the German physicist Wilhelm Weber (1804–1891): 1 Wb = 1 T·m². Example 4 illustrates how to determine the magnetic flux for three different orientations of the surface of a coil relative to the magnetic field.

Example 4 Magnetic Flux

A rectangular coil of wire is situated in a constant magnetic field whose magnitude is 0.50 T. The coil has an area of 2.0 m². Determine the magnetic flux for the three orientations, $\phi = 0°$, 60.0°, and 90.0°, shown in Figure 22.11.

Reasoning The magnetic flux Φ is defined as $\Phi = BA \cos \phi$, where B is the magnitude of the magnetic field, A is the area of the surface through which the magnetic field passes, and ϕ is the angle between the magnetic field and the normal to the surface.

Problem solving insight
The magnetic flux Φ is determined by more than just the magnitude B of the magnetic field and the area A. It also depends on the angle ϕ (see Figure 22.10 and Equation 22.2).

Solution The magnetic flux for the three cases is:

$\phi = 0°$ $\Phi = (0.50 \text{ T})(2.0 \text{ m}^2) \cos 0° = \boxed{1.0 \text{ Wb}}$

$\phi = 60.0°$ $\Phi = (0.50 \text{ T})(2.0 \text{ m}^2) \cos 60.0° = \boxed{0.50 \text{ Wb}}$

$\phi = 90.0°$ $\Phi = (0.50 \text{ T})(2.0 \text{ m}^2) \cos 90.0° = \boxed{0 \text{ Wb}}$

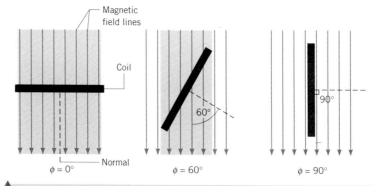

Figure 22.11 Three orientations of a rectangular coil (drawn as an edge view) relative to the magnetic field lines. The magnetic field lines that pass through the coil are those in the regions shaded in blue.

Magnetic field lines

Coil

$\phi = 0°$ Normal $\phi = 60°$ $\phi = 90°$

GRAPHICAL INTERPRETATION OF MAGNETIC FLUX

It is possible to interpret the magnetic flux graphically because the magnitude of the magnetic field **B** is proportional to the number of field lines per unit area that pass through a surface perpendicular to the lines (see Section 21.1). For instance, the magnitude of **B** in Figure 22.12*a* is three times larger than it is in part *b* of the drawing, since the number of field lines drawn through the identical surfaces is in the ratio of 3 : 1. Because Φ is directly proportional to *B* for a given area, the flux in part *a* is also three times larger than that in part *b*. Therefore, *the magnetic flux is proportional to the number of field lines that pass through a surface.*

The graphical interpretation of flux also applies when the surface is oriented at an angle with respect to **B**. For example, as the coil in Figure 22.11 is rotated from *φ* = 0° to 60° to 90°, the number of magnetic field lines passing through the surface (see the field lines in the regions shaded in blue) changes in the ratio of 8 : 4 : 0 or 2 : 1 : 0. The results of Example 4 show that the flux in the three orientations changes by the same ratio. Because the magnetic flux is proportional to the number of field lines passing through a surface, we often use phrases such as "the flux that passes through a surface bounded by a loop of wire."

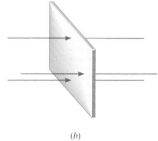

(a)

(b)

Figure 22.12 The magnitude of the magnetic field in (*a*) is three times as great as that in (*b*), because the number of magnetic field lines crossing the surfaces is in the ratio of 3 : 1.

22.4 *Faraday's Law of Electromagnetic Induction*

Two scientists are given credit for the discovery of electromagnetic induction: the Englishman Michael Faraday (1791–1867) and the American Joseph Henry (1797–1878). Although Henry was the first to observe electromagnetic induction, Faraday investigated it in more detail and published his findings first. Consequently, the law that describes the phenomenon bears his name.

▶ CONCEPTS AT A GLANCE Faraday discovered that whenever there is a *change in flux* through a loop of wire, an emf is induced in the loop. In this context, the word "change" refers to a change as time passes. A flux that is constant in time creates no emf. The Concepts-at-a-Glance chart in Figure 22.13, which extends the chart in Figure 22.8,

Figure 22.13 CONCEPTS AT A GLANCE Faraday's law of electromagnetic induction specifies the emf created by a change in the magnetic flux that occurs as time passes. The change in the flux and the time interval over which it occurs both appear in the statement of the law. The device in the photograph is a prototype of a ring-pull mobile telephone. Instead of a battery, the phone contains a generator that provides electric power for over five minutes after the cord is pulled a few times. Generators depend on Faraday's law for their operation, as Section 22.7 discusses. (© Volker Steger/Science Photo Library/Photo Researchers)

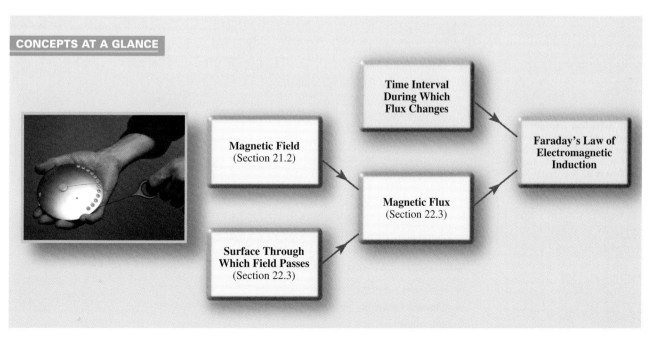

CONCEPTS AT A GLANCE

Magnetic Field (Section 21.2) → Magnetic Flux (Section 22.3)

Surface Through Which Field Passes (Section 22.3) → Magnetic Flux (Section 22.3)

Time Interval During Which Flux Changes → Faraday's Law of Electromagnetic Induction

Magnetic Flux (Section 22.3) → Faraday's Law of Electromagnetic Induction

emphasizes that *Faraday's law of electromagnetic induction* is expressed by bringing together the idea of magnetic flux and the time interval during which it changes. In fact, Faraday found that the magnitude of the induced emf is equal to the time rate of change of the magnetic flux. This is consistent with the relation we obtained in Section 22.3 for the specific case of motional emf: $\mathcal{E} = -\Delta\Phi/\Delta t$. ◄

Often the magnetic flux passes through a coil of wire containing more than one loop (or turn). If the coil consists of N loops, and if the same flux passes through each loop, it is found experimentally that the total induced emf is N times that induced in a single loop. An analogous situation occurs in a flashlight when two 1.5-V batteries are stacked in series on top of one another to give a total emf of 3.0 volts. For the general case of N loops, the total induced emf is described by Faraday's law of electromagnetic induction in the following manner:

■ FARADAY'S LAW OF ELECTROMAGNETIC INDUCTION

The average emf $\mathcal{E}$ induced in a coil of N loops is

$$\mathcal{E} = -N\left(\frac{\Phi - \Phi_0}{t - t_0}\right) = -N\frac{\Delta\Phi}{\Delta t} \qquad (22.3)$$

where $\Delta\Phi$ is the change in magnetic flux through one loop and Δt is the time interval during which the change occurs. The term $\Delta\Phi/\Delta t$ is the average time rate of change of the flux that passes through one loop.

SI Unit of Induced Emf: volt (V)

Faraday's law states that an emf is generated if the magnetic flux changes for any reason. Since the flux is given by Equation 22.2 as $\Phi = BA \cos\phi$, it depends on three factors, B, A, and ϕ, any of which may change. Example 5 considers a change in B.

Example 5 The Emf Induced by a Changing Magnetic Field

A coil of wire consists of 20 turns, or loops, each of which has an area of 1.5×10^{-3} m^2. A magnetic field is perpendicular to the surface of each loop at all times, so that $\phi = \phi_0 = 0°$. At time $t_0 = 0$ s, the magnitude of the field at the location of the coil is $B_0 = 0.050$ T. At a later time $t = 0.10$ s, the magnitude of the field at the coil has increased to $B = 0.060$ T. (a) Find the average emf induced in the coil during this time. (b) What would be the value of the average induced emf if the magnitude of the magnetic field decreased from 0.060 T to 0.050 T in 0.10 s?

Reasoning To find the induced emf, we use Faraday's law of electromagnetic induction (Equation 22.3), combining it with the definition of magnetic flux from Equation 22.2. We note that only the magnitude of the magnetic field changes in time. All other factors remain constant.

Solution

(a) Since $\phi = \phi_0$, the induced emf is

$$\mathcal{E} = -N\left(\frac{\Phi - \Phi_0}{t - t_0}\right) = -N\left(\frac{BA \cos\phi - B_0A \cos\phi_0}{t - t_0}\right)$$

$$= -NA \cos\phi\left(\frac{B - B_0}{t - t_0}\right)$$

$$\mathcal{E} = -(20)(1.5 \times 10^{-3} \text{ m}^2)(\cos 0°)\left(\frac{0.060 \text{ T} - 0.050 \text{ T}}{0.10 \text{ s} - 0 \text{ s}}\right) = \boxed{-3.0 \times 10^{-3} \text{ V}}$$

(b) The calculation here is similar to that in part (a), except the initial and final values of B are interchanged. This interchange reverses the sign of the emf, so $\boxed{\mathcal{E} = +3.0 \times 10^{-3} \text{ V}}$. Because the algebraic sign or polarity of the emf is reversed, the direction of the induced current would be opposite to that in part (a).

▲

The next example demonstrates that an emf can be created when a coil is rotated in a magnetic field.

(sidebar)

Concept Simulation 22.1

When the current in a straight wire is changed, the magnetic field produced by the current also changes. In this simulation the changing field induces an emf in an adjacent loop of wire, and the induced emf leads to an induced current. The current in the straight wire is under your control and, along with the induced current, is recorded graphically as a function of time.

Go to
www.wiley.com/college/cutnell

Problem solving insight
The change in any quantity is the final value minus the initial value: e.g., the change in flux is $\Delta\Phi = \Phi - \Phi_0$ and the change in time is $\Delta t = t - t_0$.

Example 6 The Emf Induced in a Rotating Coil

A flat coil of wire has an area of 0.020 m² and consists of 50 turns. At $t_0 = 0$ s the coil is oriented so the normal to its surface is parallel ($\phi_0 = 0°$) to a constant magnetic field of magnitude 0.18 T. The coil is then rotated through an angle of $\phi = 30.0°$ in a time of 0.10 s (see Figure 22.11). (a) Determine the average induced emf. (b) What would be the induced emf if the coil were returned to its initial orientation in the same time of 0.10 s?

Reasoning As in Example 5 we can determine the induced emf by using Faraday's law of electromagnetic induction, along with the definition of magnetic flux. In the present case, however, only ϕ (the angle between the normal to the surface of the coil and the magnetic field) changes in time. All other factors remain constant.

Solution

(a) Faraday's law yields

$$\mathcal{E} = -N\left(\frac{\Phi - \Phi_0}{t - t_0}\right) = -N\left(\frac{BA \cos \phi - BA \cos \phi_0}{t - t_0}\right)$$

$$= -NBA\left(\frac{\cos \phi - \cos \phi_0}{t - t_0}\right)$$

$$\mathcal{E} = -(50)(0.18 \text{ T})(0.020 \text{ m}^2)\left(\frac{\cos 30.0° - \cos 0°}{0.10 \text{ s} - 0 \text{ s}}\right) = \boxed{+0.24 \text{ V}}$$

(b) When the coil is rotated back to its initial orientation in a time of 0.10 s, the initial and final values of ϕ are interchanged. As a result, the induced emf has the same magnitude, but opposite polarity, so $\boxed{\mathcal{E} = -0.24 \text{ V}}$.

Need more practice?

Interactive LearningWare 22.2
A 75-turn conducting coil has an area of 8.5×10^{-3} m², and the normal to the coil is parallel to a magnetic field **B**. The coil has a resistance of 14 Ω. At what rate (in T/s) must the magnitude of **B** change for an induced current of 7.0 mA to exist in the coil?

Related Homework: Problem 22

Go to
www.wiley.com/college/cutnell
for an interactive solution.

One application of Faraday's law that is found in the home is a safety device called a ground fault interrupter. This device protects against electrical shock from an appliance, such as a clothes dryer. It plugs directly into a wall socket, as in Figure 22.14 or, in new home construction, replaces the socket entirely. The interrupter consists of a circuit breaker that can be triggered to stop the current to the dryer, depending on whether an induced voltage appears across a sensing coil. This coil is wrapped around an iron ring, through which the current-carrying wires pass. In the drawing, the current going to the dryer is shown in red, and the returning current is shown in green. Each of the currents creates a magnetic field that encircles the corresponding wire, according to RHR-2 (see Section 21.7). However, the field lines have opposite directions since the currents have opposite directions. As the drawing shows, the iron ring guides the field lines through the sensing coil. Since the current is ac, the fields from the red and green current are changing, but the red and green field lines always have opposite directions and the opposing fields cancel at all times. As a result, the net flux through the coil remains zero, and no in-

The physics of
a ground fault interrupter.

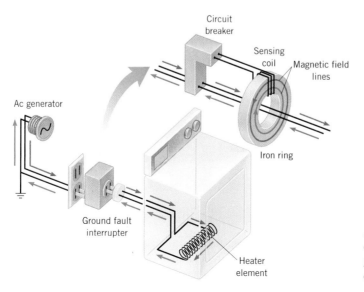

Figure 22.14 The clothes dryer is connected to the wall socket through a ground fault interrupter. The dryer is operating normally.

duced emf appears in the coil. Thus, when the dryer operates normally, the circuit breaker is not triggered and does not shut down the current. The picture changes if the dryer malfunctions, as when a wire inside the unit breaks and accidentally contacts the metal case. When someone touches the case, some of the current begins to pass through the person's body and into the ground, returning to the ac generator *without using the return wire that passes through the ground fault interrupter*. Under this condition, the net magnetic field through the sensing coil is no longer zero and changes with time, since the current is ac. The changing flux causes an induced voltage to appear in the sensing coil, which triggers the circuit breaker to stop the current. Ground fault interrupters work very fast (in less than a millisecond) and turn off the current before it reaches a dangerous level.

Conceptual Example 7 discusses another application of electromagnetic induction—namely, how a stove can cook food without getting hot.

Conceptual Example 7   An Induction Stove

The physics of an induction stove.

Figure 22.15 shows two pots of water that were placed on an induction stove at the same time. There are two interesting features in this drawing. First, the stove itself is cool to the touch. Second, the water in the ferromagnetic metal pot is boiling while that in the glass pot is not. How can such a "cool" stove boil water, and why isn't the water in the glass pot boiling?

Reasoning and Solution The key to this puzzle is related to the fact that one pot is made from a ferromagnetic metal and one from glass. We know that metals are good conductors, while glass is an insulator. Perhaps the stove causes electricity to flow directly in the metal pot. This is exactly what happens. The stove is called an *induction stove* because it operates by using electromagnetic induction. Just beneath the cooking surface is a metal coil that carries an ac current (frequency about 25 kHz). This current produces an alternating magnetic field that extends outward to the location of the metal pot. As the changing field crosses the pot's bottom surface, an emf is induced in it. Because the pot is metallic, an induced current is generated by the induced emf. The metal has a finite resistance to the induced current, however, and heats up as energy is dissipated in this resistance. The fact that the metal is ferromagnetic is important. Ferromagnetic materials contain magnetic domains (see Section 21.9), and the boundaries between them move extremely rapidly in response to the external magnetic field, thus enhancing the induction effect. A normal aluminum cooking pot, in contrast, is not ferromagnetic, so this enhancement is absent and such cookware is not used with induction stoves. An emf is also induced in the glass pot and the cooking surface of the stove. However, these materials are insulators, so very little induced current exists within them. Thus, they do not heat up very much and remain cool to the touch.

Related Homework: *Conceptual Question 8*

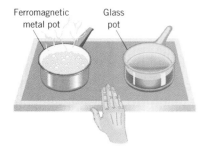

Ferromagnetic metal pot   Glass pot

Figure 22.15 The water in the ferromagnetic metal pot is boiling. Yet the water in the glass pot is not boiling, and the stove top is cool to the touch. The stove operates in this way by using electromagnetic induction.

✔ **Check Your Understanding 2**

A coil is placed in a magnetic field, and the normal to the plane of the coil remains parallel to the field. Which one of the following options causes the average emf induced in the coil to be as large as possible? (a) The magnitude of the field is small, and its rate of change is large. (b) The magnitude of the field is large, and its rate of change is small. (c) The magnitude of the field is large, and it does not change. *(The answer is given at the end of the book.)*

Background: The average emf $\mathscr{E}$ induced in the coil is given by Faraday's law of electromagnetic induction as $\mathscr{E} = -\dfrac{\Delta\Phi}{\Delta t}$, where Φ is the magnetic flux and t is time. The magnetic flux depends on the magnetic field strength B, the area A of the coil, and the angle ϕ between the field and the normal to the coil, according to $\Phi = BA\cos\phi$.

For similar questions (including calculational counterparts), consult Self-Assessment Test 22.1, which is described at the end of Section 22.5.

22.5 Lenz's Law

An induced emf drives current around a circuit just as the emf of a battery does. With a battery, conventional current is directed out of the positive terminal, through the attached device, and into the negative terminal. The same is true for an induced emf, although the locations of the positive and negative terminals are generally not as obvious. Therefore, a method is needed for determining the polarity or algebraic sign of the induced emf, so the terminals can be identified. As we discuss this method, it will be helpful to keep in mind that the net magnetic field penetrating a coil of wire results from two contributions. One is the original magnetic field that produces the changing flux that leads to the induced emf. The other arises because of the induced current, which, like any current, creates its own magnetic field. The field created by the induced current is called the ***induced magnetic field.***

To determine the polarity of the induced emf, we will use a method based on a discovery made by the Russian physicist Heinrich Lenz (1804–1865). This discovery is known as ***Lenz's law.***

■ **LENZ'S LAW**

The induced emf resulting from a changing magnetic flux has a polarity that leads to an induced current whose direction is such that the induced magnetic field opposes the original flux change.

Lenz's law is best illustrated with examples. Each will be worked out according to the following reasoning strategy:

Reasoning Strategy

Determining the Polarity of the Induced Emf

1. Determine whether the magnetic flux that penetrates a coil is increasing or decreasing.

2. Find what the direction of the induced magnetic field must be so that it can ***oppose the change in flux*** by adding to or subtracting from the original field.

3. Having found the direction of the induced magnetic field, use RHR-2 (see Section 21.7) to determine the direction of the induced current. Then the polarity of the induced emf can be assigned because conventional current is directed out of the positive terminal, through the external circuit, and into the negative terminal.

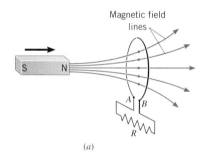

(a)

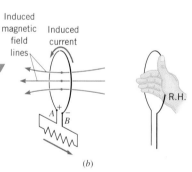

(b)

Figure 22.16 (*a*) As the magnet moves to the right, the magnetic flux through the loop increases. The external circuit attached to the loop has a resistance *R*. (*b*) The polarity of the induced emf is indicated by the + and − symbols.

Conceptual Example 8   The Emf Produced by a Moving Magnet

Figure 22.16*a* shows a permanent magnet approaching a loop of wire. The external circuit attached to the loop consists of the resistance *R*, which could be the resistance of the filament in a light bulb, for instance. Find the direction of the induced current and the polarity of the induced emf.

Reasoning and Solution We apply Lenz's law, the essence of which is that the change in magnetic flux must be opposed by the induced magnetic field. The magnetic flux through the loop is increasing, since the magnitude of the magnetic field at the loop is increasing as the magnet approaches. To oppose the increase in the flux, the direction of the induced magnetic field must be opposite to the field of the bar magnet. Since the field of the bar magnet passes through the loop from left to right in part *a* of the drawing, the induced field must pass through the loop from right to left, as in part *b*. To create such an induced field, ***the induced current must be directed counterclockwise around the loop, when viewed from the side nearest the magnet.*** (See the application of RHR-2 in the drawing.) The loop behaves as a source of emf, just like a battery. Since conventional current is directed into the external circuit from the positive terminal, ***point A in Figure 22.16b must be the positive terminal, and point B must be the negative terminal.***

In Conceptual Example 8 the direction of the induced magnetic field is opposite to the direction of the external field of the bar magnet. However, ***the induced field is not always opposite to the external field, because Lenz's law requires only that it must oppose the change in the flux that generates the emf.*** Conceptual Example 9 illustrates this point.

Problem solving insight

Conceptual Example 9   The Emf Produced by a Moving Copper Ring

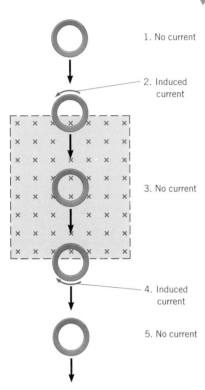

1. No current

2. Induced current

3. No current

4. Induced current

5. No current

Figure 22.17 A copper ring passes through a rectangular region where a constant magnetic field is directed into the page. The picture shows that a current is induced in the ring at locations 2 and 4.

In Figure 22.17 there is a constant magnetic field in a rectangular region of space. This field is directed perpendicularly into the page. Outside this region there is no magnetic field. A copper ring slides through the region, from position 1 to position 5. For each of the five positions, determine whether an induced current exists in the ring and, if so, find its direction.

Reasoning and Solution Lenz's law requires that the induced magnetic field must oppose the change in flux. Sometimes this means that the induced field will be opposite to the external field, as in Example 8, but not always. We will see that the induced field must sometimes have the same direction as the external field in order to oppose the flux change.

Position 1: No flux passes through the ring because the magnetic field is zero outside the rectangular region. Consequently, *there is no change in flux and no induced emf or current.*

Position 2: As the ring moves into the field region, the flux increases and there is an induced emf and current. To determine the direction of the current, we require that the induced field point out of the plane of the paper, opposite to the external field. Only then can the induced field oppose the increase in the flux, in accord with Lenz's law. To produce such an induced field, RHR-2 indicates that *the direction of the induced current must be counterclockwise,* as the drawing shows.

Position 3: Even though a flux passes through the moving ring, *there is no induced emf or current* because the flux remains constant within the rectangular region. To induce an emf, it is not sufficient just to have a flux; the flux must change as time passes.

Position 4: As the ring leaves the field region, the flux decreases, and the induced field must oppose this change. Since the change is a decrease in flux, the induced field must point in the *same* direction as the external field—namely, into the paper. With this orientation, the induced field will increase the net magnetic field through the ring and thereby increase the flux. To produce an induced field pointing into the paper, RHR-2 indicates that *the induced current must be clockwise,* opposite to what it is in position 2. By comparing the results for positions 2 and 4, it should be clear that the induced magnetic field is not always opposite to the external field.

Position 5: As in position 1, *there is no induced current,* since the magnetic field is everywhere zero.

Related Homework: *Problem 31*

Lenz's law should not be thought of as an independent law, because it is a consequence of the law of conservation of energy. The connection between energy conservation and induced emf has already been discussed in Section 22.2 for the specific case of motional emf. However, the connection is valid for any type of induced emf. In fact, the polarity of the induced emf, as specified by Lenz's law, ensures that energy is conserved.

✔ **Check Your Understanding 3**

A circular loop of wire is lying flat on a horizontal table, and you are looking down at it. A magnetic field has a constant direction that is perpendicular to the table, and an induced current flows clockwise in the loop. Is the field directed toward you or away from you, and is its magnitude increasing or decreasing? *(There are two possible answers, and they are given at the end of the book.)*

Background: Lenz's law states that the induced current has a direction such that the induced magnetic field opposes the changing flux that gives rise to the current.

For similar questions (including calculational counterparts), consult Self-Assessment Test 22.1, which is described next.

Self-Assessment Test 22.1

Test your understanding of the material in Sections 22.1–22.5:

• Induced EMF and Induced Current   • Motional EMF   • Magnetic Flux
• Faraday's Law of Electromagnetic Induction   • Lenz's Law

Go to **www.wiley.com/college/cutnell**

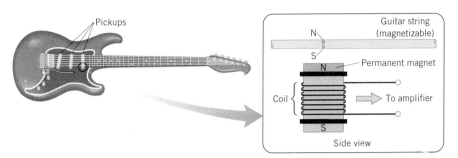

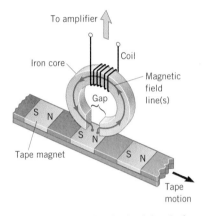

Figure 22.18 When the string of an electric guitar vibrates, an emf is induced in the coil of the pickup. The two ends of the coil are connected to the input of an amplifier.

*22.6 Applications of Electromagnetic Induction to the Reproduction of Sound

Electromagnetic induction plays an important role in the technology used for the reproduction of sound. Virtually all electric guitars, for example, use electromagnetic pickups in which an induced emf is generated in a coil of wire by a vibrating string. Most guitars have at least two pickups located below each string and some, as Figure 22.18 illustrates, have three. Each pickup is sensitive to different harmonics produced by the vibrating string. The string is made from a magnetizable metal, and the pickup consists of a coil of wire within which a permanent magnet is located. The permanent magnet produces a magnetic field that penetrates the guitar string, causing it to become magnetized with north and south poles. When the magnetized string is plucked, it oscillates, thereby changing the magnetic flux that passes through the coil. The changing flux induces an emf in the coil, and the polarity of this emf alternates with the vibratory motion of the string. A string vibrating at 440 Hz, for example, induces a 440-Hz ac emf in the coil. This signal, after being amplified, is sent to the speakers, which produce a 440-Hz sound wave (concert A).

The playback head of a cassette deck uses a moving tape to generate an emf in a coil of wire. Figure 22.19 shows a section of magnetized tape in which a series of "tape magnets" have been created in the magnetic layer of the tape during the recording process (see Section 21.9). The tape moves beneath the playback head, which consists of a coil of wire wrapped around an iron core. The iron core has the approximate shape of a horseshoe with a small gap between the two ends. The drawing shows an instant when a tape magnet is under the gap. Some of the field lines of the tape magnet are routed through the highly magnetizable iron core, and hence through the coil, as they proceed from the north pole to the south pole. Consequently, the flux through the coil changes as the tape moves past the gap. The change in flux leads to an ac emf, which is amplified and sent to the speakers, where the original sound is reproduced.

There are a number of types of microphones, and Figure 22.20 illustrates the one known as a moving coil microphone. When a sound wave strikes the diaphragm of the microphone, the diaphragm vibrates back and forth, and the attached coil moves along with it. Nearby is a stationary magnet. As the coil alternately approaches and recedes from the magnet, the flux through the coil changes. Consequently, an ac emf is induced in the coil. This electrical signal is sent to an amplifier and then to the speakers. In the type of microphone called a moving magnet microphone, the magnet is attached to the diaphragm and moves relative to a stationary coil.

The physics of **the electric guitar pickup.**

Figure 22.19 The playback head of a tape deck. As each tape magnet goes by the gap, some magnetic field lines pass through the core and coil. The changing flux in the coil creates an induced emf. The gap width has been exaggerated.

The physics of **the playback head of a tape deck.**

The physics of **a moving coil and a moving magnet microphone.**

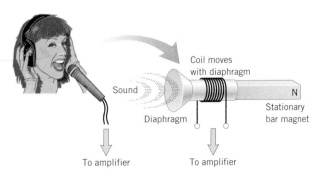

Figure 22.20 A moving coil microphone.

22.7 *The Electric Generator*

HOW A GENERATOR PRODUCES AN EMF

Electric generators, such as those in Figure 22.21, produce virtually all of the world's electrical energy. A generator produces electrical energy from mechanical work, just the opposite of what a motor does. In a motor, an *input* electric current causes a coil to rotate, thereby doing mechanical work on any object attached to the shaft of the motor. In a generator, the shaft is rotated by some mechanical means, such as an engine or a turbine, and an emf is induced in a coil. If the generator is connected to an external circuit, an electric current is the *output* of the generator.

In its simplest form, an ac generator consists of a coil of wire that is rotated in a uniform magnetic field, as Figure 22.22a indicates. Although not shown in the picture, the wire is usually wound around an iron core. As in an electric motor, the coil/core combination is called the *armature*. Each end of the wire forming the coil is connected to the external circuit by means of a metal ring that rotates with the coil. Each ring slides against a stationary carbon brush, to which the external circuit (the lamp in the drawing) is connected.

To see how current is produced by the generator, consider the two vertical sides of the coil in Figure 22.22b. Since each is moving in a magnetic field **B**, the magnetic force exerted on the charges in the wire causes them to flow, thus creating a current. With the aid of RHR-1 (fingers of extended right hand point along **B**, thumb along the velocity **v**, palm pushes in the direction of the force on a positive charge), it can be seen that the direction of the current is from bottom to top in the left side and from top to bottom in the right side. Thus, charge flows around the loop. The upper and lower segments of the loop are also moving. However, these segments can be ignored because the magnetic force on the charges within them points toward the sides of the wire and not along the length.

The magnitude of the motional emf developed in a conductor moving through a magnetic field is given by Equation 22.1. To apply this expression to the left side of the coil, whose length is L (see Figure 22.22c), we need to use the velocity component $v_\perp$ that is perpendicular to **B**. Letting θ be the angle between **v** and **B**, it follows that $v_\perp = v \sin \theta$, and, with the aid of Equation 22.1, the emf can be written as

$$\mathcal{E} = BLv_\perp = BLv \sin \theta$$

The emf induced in the right side has the same magnitude as that in the left side. Since the emfs from both sides drive current in the same direction around the loop, the emf for

Figure 22.21 Electric generators such as this one supply electric power by producing an induced emf according to Faraday's law of electromagnetic induction. (© Peter Bowater/Age Fotostock America, Inc.)

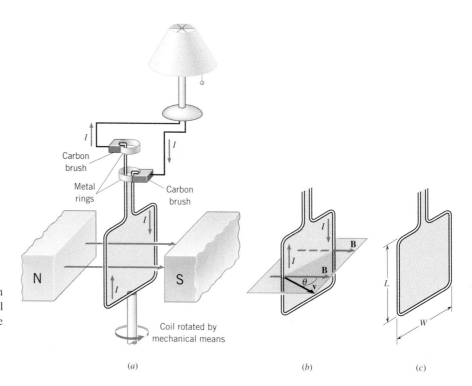

Figure 22.22 (*a*) This electric generator consists of a coil (only one loop is shown) of wire that is rotated in a magnetic field **B** by some mechanical means. (*b*) The current I arises because of the magnetic force exerted on the charges in the moving wire. (*c*) The dimensions of the coil.

Carbon brush

Metal rings

Carbon brush

N

S

Coil rotated by mechanical means

L

W

(*a*)   (*b*)   (*c*)

the complete loop is $\mathscr{E} = 2BLv \sin \theta$. If the coil consists of N loops, the net emf is N times as great as that of one loop, so

$$\mathscr{E} = N(2BLv \sin \theta)$$

It is convenient to express the variables v and θ in terms of the angular speed ω at which the coil rotates. Equation 8.2 shows that the angle θ is the product of the angular speed and the time, $\theta = \omega t$, if it is assumed that $\theta = 0$ rad when $t = 0$ s. Furthermore, any point on each vertical side moves on a circular path of radius $r = W/2$, where W is the width of the coil (see Figure 22.22c). Thus, the tangential speed v of each side is related to the angular speed ω via Equation 8.9 as $v = r\omega = (W/2)\omega$. Substituting these expressions for θ and v in the previous equation for $\mathscr{E}$, and recognizing that the product LW is the area A of the coil, we can write the induced emf as

Emf induced
in a rotating $\mathscr{E} = NAB\omega \sin \omega t = \mathscr{E}_0 \sin \omega t$ where $\omega = 2\pi f$ (22.4)
planar coil

In this result, the angular speed ω is in radians per second and is related to the frequency f [in cycles per second or hertz (Hz)] according to $\omega = 2\pi f$ (Equation 10.6).

Although Equation 22.4 was derived for a rectangular coil, the result is valid for any planar shape of area A (e.g., circular) and shows that the emf varies sinusoidally with time. The peak, or maximum, emf $\mathscr{E}_0$ occurs when $\sin \omega t = 1$ and has the value $\mathscr{E}_0 = NAB\omega$. Figure 22.23 shows a plot of Equation 22.4 and reveals that the emf changes polarity as the coil rotates. This changing polarity is exactly the same as that discussed for an ac voltage in Section 20.5 and illustrated in Figure 20.11. If the external circuit connected to the generator is a closed circuit, an alternating current results that changes direction at the same frequency f as the emf changes polarity. Therefore, this electric generator is also called an *alternating current (ac) generator*. The next two examples show how Equation 22.4 is applied.

Squeezing this hand-held generator produces electric power for portable devices such as cell phones, CD players, and pagers. (© Mario Tama/ Getty Images News and Sport Services)

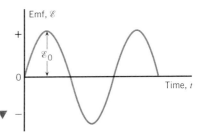

Figure 22.23 An ac generator produces this alternating emf $\mathscr{E}$ according to $\mathscr{E} = \mathscr{E}_0 \sin \omega t$.

Example 10   An AC Generator

In Figure 22.22 the coil of the ac generator rotates at a frequency of $f = 60.0$ Hz and develops an emf of 120 V (rms). The coil has an area of $A = 3.0 \times 10^{-3}$ m^2 and consists of N = 500 turns. Find the magnitude of the magnetic field in which the coil rotates.

Reasoning The magnetic field can be found from the relation $\mathscr{E}_0 = NAB\omega$. However, in using this equation we must remember that $\mathscr{E}_0$ is the peak emf, whereas the given value of 120 V is not a peak value but an rms value. The peak emf is related to the rms emf by $\mathscr{E}_0 = \sqrt{2}\mathscr{E}_{rms}$, according to Equation 20.13.

Solution The peak emf is $\mathscr{E}_0 = \sqrt{2}\mathscr{E}_{rms} = \sqrt{2}(120 \text{ V}) = 170$ V. Solving $\mathscr{E}_0 = NAB\omega$ for B and using the fact that $\omega = 2\pi f$, we find that the magnitude of the magnetic field is

$$B = \frac{\mathscr{E}_0}{NA\omega} = \frac{\mathscr{E}_0}{NA2\pi f} = \frac{170 \text{ V}}{(500)(3.0 \times 10^{-3} \text{ m}^2)2\pi(60.0 \text{ Hz})} = \boxed{0.30 \text{ T}}$$

Problem solving insight
When using the equation $\mathscr{E}_0 = NAB\omega$, remember that the angular frequency ω must be in rad/s and is related to the frequency f (in Hz) according to $\omega = 2\pi f$ (Equation 10.6).

Example 11   A Bike Generator

A generator is mounted on a bicycle to power a headlight. A small wheel on the shaft of the generator is pressed against the bike tire and turns the armature 44 times for each revolution of the tire. The tire has a radius of 0.33 m. The armature has 75 turns, each with an area of 2.6×10^{-3} m^2, and rotates in a 0.10-T magnetic field. When the peak emf being generated is 6.0 V, what is the linear speed of the bike?

Reasoning The relation $\mathscr{E}_0 = NAB\omega$ provides a solution to this problem, if we take advantage of two additional facts. The first is that the angular speed ω of the armature is 44 times larger than the angular speed ω_{tire} of the bike tire. The second is that the tire is rolling, so that ω_{tire} is related to the linear speed v of the bike according to $\omega_{tire} = v/r$ (Equation 8.12), where r is the radius of the tire.

The physics of
a bike generator.

Need more practice?

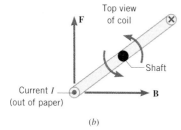

Interactive LearningWare 22.3
An armature of an electric generator consists of a 275-turn circular coil of radius 15 cm. The armature is rotated in a 0.80-T magnetic field, and the polarity changes every 4.2×10^{-3} s. What is the peak emf produced by the generator?

Go to **www.wiley.com/college/cutnell** for an interactive solution.

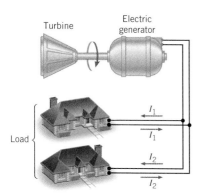

Figure 22.24 The generator supplies a total current of $I = I_1 + I_2$ to the load.

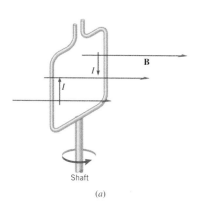

Figure 22.25 (*a*) A current I exists in the rotating coil of the generator. (*b*) A top view of the coil, showing the magnetic force **F** exerted on the left side of the coil.

Solution Using $\omega = 44\omega_{\text{tire}} = 44(v/r)$, we find that the peak emf is $\mathscr{E}_0 = NAB\,[44(v/r)]$. Solving for the linear speed v gives

$$v = \frac{\mathscr{E}_0 r}{44NAB} = \frac{(6.0\ \text{V})(0.33\ \text{m})}{44(75)(2.6 \times 10^{-3}\ \text{m}^2)(0.10\ \text{T})} = \boxed{2.3\ \text{m/s}}$$

THE ELECTRICAL ENERGY DELIVERED BY A GENERATOR AND THE COUNTERTORQUE

Some power-generating stations burn fossil fuel (coal, gas, or oil) to heat water and produce pressurized steam for turning the blades of a turbine, whose shaft is linked to the generator. Others use nuclear fuel or falling water as a source of energy. As the turbine rotates, the generator coil also rotates and mechanical work is transformed into electrical energy.

The devices to which the generator supplies electricity are known collectively as the "load," because they place a burden or load on the generator by taking electrical energy from it. If all the devices are switched off, the generator runs under a no-load condition, because there is no current in the external circuit and the generator does not supply electrical energy. Then, work needs to be done on the turbine only to overcome friction and other mechanical losses within the generator itself, and fuel consumption is at a minimum.

Figure 22.24 illustrates a situation in which a load is connected to a generator. Because there is now a current $I = I_1 + I_2$ in the coil of the generator and the coil is situated in a magnetic field, the current experiences a magnetic force **F**. Figure 22.25 shows the magnetic force acting on the left side of the coil, the direction of **F** being given by RHR-1. A force of equal magnitude but opposite direction acts on the right side of the coil, although this force is not shown in the drawing. The magnetic force **F** gives rise to a *countertorque* that opposes the rotational motion. The greater the current drawn from the generator, the greater the countertorque, and the harder it is for the turbine to turn the coil. To compensate for this countertorque and keep the coil rotating at a constant angular speed, work must be done by the turbine, which means more fuel must be burned. This is another example of the law of conservation of energy, since the electrical energy consumed by the load must ultimately come from the energy source used to drive the turbine.

THE BACK EMF GENERATED BY AN ELECTRIC MOTOR

A generator converts mechanical work into electrical energy; in contrast, an electric motor converts electrical energy into mechanical work. Both devices are similar and consist of a coil of wire that rotates in a magnetic field. In fact, as the armature of a motor rotates, the magnetic flux passing through the coil changes and an emf is induced in the coil. Thus, when a motor is operating, two sources of emf are present: (1) the applied emf V that provides current to drive the motor (e.g., from a 120-V outlet), and (2) the emf $\mathscr{E}$ induced by the generator-like action of the rotating coil. The circuit diagram in Figure 22.26 shows these two emfs.

Consistent with Lenz's law, the induced emf $\mathscr{E}$ acts to oppose the applied emf V and is called the **back emf** or the **counter emf** of the motor. The greater the speed of the motor, the greater the flux change through the coil, and the greater is the back emf. Because V and $\mathscr{E}$ have opposite polarities, the net emf in the circuit is $V - \mathscr{E}$. In Figure 22.26, R is the resistance of the wire in the coil, and the current I drawn by the motor is determined from Ohm's law as the net emf divided by the resistance:

$$I = \frac{V - \mathscr{E}}{R} \tag{22.5}$$

The next example uses this result to illustrate that the current in a motor depends on both the applied emf V and the back emf $\mathscr{E}$.

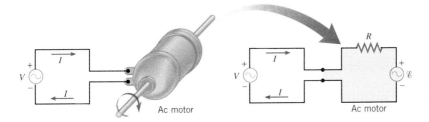

Figure 22.26 The applied emf V supplies the current I to drive the motor. The circuit on the right shows V along with the electrical equivalent of the motor, including the resistance R of its coil and the back emf $\mathscr{E}$.

Example 12 Operating a Motor

The coil of an ac motor has a resistance of $R = 4.1\ \Omega$. The motor is plugged into an outlet where $V = 120.0$ volts (rms), and the coil develops a back emf of $\mathscr{E} = 118.0$ volts (rms) when rotating at normal speed. The motor is turning a wheel. Find (a) the current when the motor first starts up and (b) the current when the motor is operating at normal speed.

Reasoning Once normal operating speed is attained, the motor need only work to compensate for frictional losses. But in bringing the wheel up to speed from rest, the motor must also do work to increase the wheel's rotational kinetic energy. Thus, bringing the wheel up to speed requires more work, and hence more current, than maintaining the normal operating speed. We expect our answers to parts (a) and (b) to reflect this fact.

Solution

(a) When the motor just starts up, the coil is not rotating, so there is no back emf induced in the coil and $\mathscr{E} = 0$ V. The start-up current drawn by the motor is

$$I = \frac{V - \mathscr{E}}{R}$$

$$= \frac{120\ \text{V} - 0\ \text{V}}{4.1\ \Omega} = \boxed{29\ \text{A}} \tag{22.5}$$

(b) At normal speed, the motor develops a back emf of $\mathscr{E} = 118.0$ volts, so the current is

$$I = \frac{V - \mathscr{E}}{R}$$

$$= \frac{120.0\ \text{V} - 118.0\ \text{V}}{4.1\ \Omega} = \boxed{0.49\ \text{A}}$$

Problem solving insight
The current in an electric motor depends on both the applied emf V and any back emf $\mathscr{E}$ developed because the coil of the motor is rotating.

Example 12 illustrates that when a motor is just starting, there is little back emf, and, consequently, a relatively large current exists in the coil. As the motor speeds up, the back emf increases until it reaches a maximum value when the motor is rotating at normal speed. The back emf becomes almost equal to the applied emf, and the current is reduced to a relatively small value, which is sufficient to provide the torque on the coil needed to overcome frictional and other losses in the motor and to drive the load (e.g., a fan).

✔ **Check Your Understanding 4**

You have a fixed length of wire and need to design a generator. You have only two options: use either a one-turn square coil or a two-turn square coil. Which option should you choose so that your generator will produce the greatest peak emf for a given frequency and magnetic field strength, or doesn't it matter? *(The answer is given at the end of the book).*

Background: The peak emf $\mathscr{E}_0$ for a generator is $\mathscr{E}_0 = NAB\omega$, where N is the number of turns in the coil, A is the area of the coil, B is the magnetic field strength, and ω is the angular frequency.

For similar questions (including calculational counterparts), consult Self-Assessment Test 22.2, which is described at the end of Section 22.9.

22.8 *Mutual Inductance and Self-Inductance*

MUTUAL INDUCTANCE

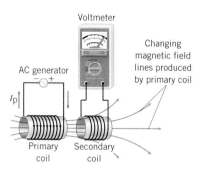

Voltmeter

AC generator

Changing
magnetic field
lines produced
by primary coil

I_p

Primary
coil

Secondary
coil

Figure 22.27 An alternating current I_p in the primary coil creates an alternating magnetic field. This changing field induces an emf in the secondary coil.

We have seen that an emf can be induced in a coil by keeping the coil stationary and moving a magnet nearby, or by moving the coil near a stationary magnet. Figure 22.27 illustrates another important method of inducing an emf. Here, two coils of wire are placed close to each other, the *primary coil* and the *secondary coil*. The primary coil is the one connected to an ac generator, which sends an alternating current I_p through it. The secondary coil is not attached to a generator, although a voltmeter is connected across it to register any induced emf.

The current-carrying primary coil is an electromagnet and creates a magnetic field in the surrounding region. If the two coils are close to each other, a significant fraction of this magnetic field penetrates the secondary coil and produces a magnetic flux. The flux is changing, since the current in the primary coil and its associated magnetic field are changing. Because of the change in flux, an emf is induced in the secondary coil.

The effect in which a changing current in one circuit induces an emf in another circuit is called **mutual induction.** According to Faraday's law of electromagnetic induction, the average emf $\mathscr{E}_s$ induced in the secondary coil is proportional to the change in flux $\Delta\Phi_s$ passing through it. However, $\Delta\Phi_s$ is produced by the change in current ΔI_p in the primary coil. Therefore, it is convenient to recast Faraday's law into a form that relates $\mathscr{E}_s$ to ΔI_p. To see how this recasting is accomplished, note that the net magnetic flux passing through the secondary coil is $N_s\Phi_s$, where N_s is the number of loops in the secondary coil and Φ_s is the flux through one loop (assumed to be the same for all loops). The net flux is proportional to the magnetic field, which, in turn, is proportional to the current I_p in the primary. Thus, we can write $N_s\Phi_s \propto I_p$. This proportionality can be converted into an equation in the usual manner by introducing a proportionality constant M, known as the **mutual inductance:**

$$N_s\Phi_s = MI_p \quad \text{or} \quad M = \frac{N_s\Phi_s}{I_p} \tag{22.6}$$

Substituting this equation into Faraday's law, we find that

$$\mathscr{E}_s = -N_s\frac{\Delta\Phi_s}{\Delta t} = -\frac{\Delta(N_s\Phi_s)}{\Delta t} = -\frac{\Delta(MI_p)}{\Delta t} = -M\frac{\Delta I_p}{\Delta t}$$

***Emf due to
mutual induction***

$$\mathscr{E}_s = -M\frac{\Delta I_p}{\Delta t} \tag{22.7}$$

Writing Faraday's law in this manner makes it clear that the average emf $\mathscr{E}_s$ induced in the secondary coil is due to the change in the current ΔI_p in the primary coil.

Equation 22.7 shows that the measurement unit for the mutual inductance M is $V \cdot s/A$, which is called a henry (H) in honor of Joseph Henry: $1\ V \cdot s/A = 1\ H$. The mutual inductance depends on the geometry of the coils and the nature of any ferromagnetic core material that is present. Although M can be calculated for some highly symmetrical arrangements, it is usually measured experimentally. In most situations, values of M are less than 1 H and are often on the order of millihenries ($1\ mH = 1 \times 10^{-3}\ H$) or microhenries ($1\ \mu H = 1 \times 10^{-6}\ H$).

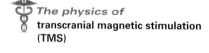

**The physics of
transcranial magnetic stimulation
(TMS)**

A new technique that shows promise for the treatment of psychiatric disorders such as depression is based on mutual induction. This technique is called transcranial magnetic stimulation (TMS) and is a type of indirect and gentler electric shock therapy. In traditional electric shock therapy, electric current is delivered directly through the skull and penetrates the brain, disrupting its electrical circuitry and in the process alleviating the symptoms of the psychiatric disorder. The treatment is not gentle and requires an anesthetic, because relatively large electric currents must be used to penetrate the skull. In contrast, TMS produces its electric current by using a time-varying magnetic field. A primary coil is held over the part of the brain to be treated (see Figure 22.28), and a time-varying current is applied to this coil. The arrangement is analogous to that in Figure 22.27, except that the brain and the electrically conductive pathways within it take the

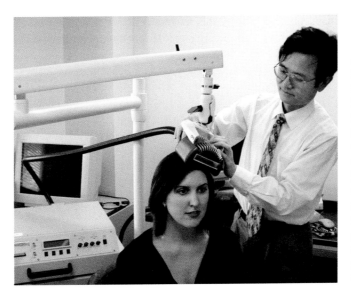

Figure 22.28 In the technique of transcranial magnetic stimulation (TMS), a time-varying electric current is applied to a primary coil, which is held over a region of the brain, as this photograph illustrates. The time-varying magnetic field produced by the coil penetrates the brain and creates an induced emf within it. This induced emf leads to an induced current that disrupts the electric circuits of the brain, thereby relieving some of the symptoms of psychiatric disorders such as depression. (Courtesy Dr. Mark S. George, Medical University of South Carolina.)

place of the secondary coil. The magnetic field produced by the primary coil penetrates the brain and, since the field is changing in time, it induces an emf in the brain. This induced emf causes an electric current to flow in the conductive brain tissue, with therapeutic results similar to those of conventional electric shock treatment. The current delivered to the brain, however, is much smaller than that in the conventional treatment, so that patients receive TMS treatments without anesthetic and severe after-effects such as headaches and memory loss. TMS remains in the experimental stage, however, and the optimal protocol for applying the technique has not yet been determined.

SELF-INDUCTANCE

In all the examples of induced emfs presented so far, the magnetic field has been produced by an external source, such as a permanent magnet or an electromagnet. However, the magnetic field need not arise from an external source. An emf can be induced in a current-carrying coil by a change in the magnetic field that the current itself produces. For instance, Figure 22.29 shows a coil connected to an ac generator. The alternating current creates an alternating magnetic field that, in turn, creates a changing flux through the coil. The change in flux induces an emf in the coil, in accord with Faraday's law. The effect in which a changing current in a circuit induces an emf in the same circuit is referred to as *self-induction.*

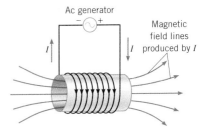

Figure 22.29 The alternating current in the coil generates an alternating magnetic field that induces an emf in the coil.

When dealing with self-induction, as with mutual induction, it is customary to recast Faraday's law into a form in which the induced emf is proportional to the change in current in the coil rather than to the change in flux. If Φ is the magnetic flux that passes through one turn of the coil, then $N\Phi$ is the net flux through a coil of N turns. Since Φ is proportional to the magnetic field, and the magnetic field is proportional to the current I, it follows that $N\Phi \propto I$. By inserting a constant L, called the *self-inductance* or simply the *inductance* of the coil, we can convert this proportionality into Equation 22.8:

$$N\Phi = LI \quad \text{or} \quad L = \frac{N\Phi}{I} \tag{22.8}$$

Faraday's law of induction now gives the average induced emf as

$$\mathcal{E} = -N\frac{\Delta\Phi}{\Delta t} = -\frac{\Delta(N\Phi)}{\Delta t} = -\frac{\Delta(LI)}{\Delta t} = -L\frac{\Delta I}{\Delta t}$$

Emf due to self-induction
$$\mathcal{E} = -L\frac{\Delta I}{\Delta t} \tag{22.9}$$

Like mutual inductance, L is measured in henries. The magnitude of L depends on the geometry of the coil and on the core material. By wrapping the coil around a ferromagnetic (iron) core, the magnetic flux—and therefore the inductance—can be increased substantially relative to that for an air core. Because of their self-inductance, coils are known as **inductors** and are widely used in electronics. Inductors come in all sizes, typically in the range between millihenries and microhenries. Example 13 shows how to determine the inductance of a solenoid.

Example 13 The Self-Inductance of a Long Solenoid

A long solenoid of length $\ell = 8.0 \times 10^{-2}$ m and cross-sectional area $A = 5.0 \times 10^{-5}$ m² contains $n = 6500$ turns per meter. (a) Find the self-inductance of the solenoid, assuming the core is air. (b) Determine the emf induced in the solenoid when the current increases from 0 to 1.5 A in a time of 0.20 s.

Reasoning

(a) The self-inductance can be found by using Equation 22.8 ($L = N\Phi/I$) provided the flux Φ can be determined. The flux is given by Equation 22.2 as $\Phi = BA \cos\phi$. In the case of a solenoid, the interior magnetic field is directed perpendicular to the plane of the loops (see Section 21.7), so $\phi = 0°$ and $\Phi = BA$. The magnetic field inside a long solenoid has the value $B = \mu_0 nI$, according to Equation 21.7, where n is the number of turns per unit length.

(b) The emf induced in the solenoid can be obtained from the relation $\mathscr{E} = -L(\Delta I/\Delta t)$, since L is determined in part (a) and ΔI and Δt are given.

Solution

(a) The self-inductance of the solenoid is

$$L = \frac{N\Phi}{I} = \frac{N(BA)}{I} = \frac{N(\mu_0 nI)A}{I} = \mu_0 nNA = \mu_0 n^2 A\ell$$

where we have replaced N by $n\ell$. Substituting the given values into this result yields

$$L = \mu_0 n^2 A\ell = (4\pi \times 10^{-7}\,\text{T}\cdot\text{m/A})(6500\,\text{turns/m})^2$$
$$\times (5.0 \times 10^{-5}\,\text{m}^2)(8.0 \times 10^{-2}\,\text{m}) = \boxed{2.1 \times 10^{-4}\,\text{H}}$$

(b) The induced emf that results from the increasing current is

$$\mathscr{E} = -L\frac{\Delta I}{\Delta t} = -(2.1 \times 10^{-4}\,\text{H})\left(\frac{1.5\,\text{A} - 0\,\text{A}}{0.20\,\text{s}}\right) = \boxed{-1.6 \times 10^{-3}\,\text{V}} \qquad (22.9)$$

The negative sign reminds us that the induced emf opposes the increasing current.

THE ENERGY STORED IN AN INDUCTOR

An inductor, like a capacitor, can store energy. This stored energy arises because a generator does work to establish a current in an inductor. Suppose an inductor is connected to a generator whose terminal voltage can be varied continuously from zero to some final value. As the voltage is increased, the current I in the circuit rises continuously from zero to its final value. While the current is rising, an induced emf $\mathscr{E} = -L(\Delta I/\Delta t)$ appears across the inductor. Conforming to Lenz's law, the polarity of the induced emf $\mathscr{E}$ is opposite to that of the generator voltage, so as to oppose the increase in the current. Thus, the generator must do work to push the charges through the inductor against this induced emf. The increment of work ΔW done by the generator in moving a small amount of charge ΔQ through the inductor is $\Delta W = -(\Delta Q)\mathscr{E} = -(\Delta Q)[-L(\Delta I/\Delta t)]$, according to Equation 19.4. Since $\Delta Q/\Delta t$ is the current I, the work done is

$$\Delta W = LI(\Delta I)$$

In this expression ΔW represents the work done by the generator to increase the current in the inductor by an amount ΔI. To determine the total work W done while the current is changed from zero to its final value, all the small increments ΔW must be added together. This summation is left as an exercise (see Problem 52). The result is $W = \frac{1}{2}LI^2$, where I

represents the final current in the inductor. This work is stored as energy in the inductor, so that

**Energy stored
in an inductor**

$$\text{Energy} = \tfrac{1}{2}LI^2 \qquad (22.10)$$

It is possible to regard the energy in an inductor as being stored in its magnetic field. For the special case of a long solenoid, Example 13 shows that the self-inductance is $L = \mu_0 n^2 A \ell$, where n is the number of turns per unit length, A is the cross-sectional area, and ℓ is the length of the solenoid. As a result, the energy stored in a long solenoid is

$$\text{Energy} = \tfrac{1}{2}LI^2 = \tfrac{1}{2}\mu_0 n^2 A \ell I^2$$

Since $B = \mu_0 nI$ at the interior of a long solenoid (Equation 21.7), this energy can be expressed as

$$\text{Energy} = \frac{1}{2\mu_0} B^2 A \ell$$

The term $A\ell$ is the volume inside the solenoid where the magnetic field exists, so the energy per unit volume or **energy density** is

$$\text{Energy density} = \frac{\text{Energy}}{\text{Volume}} = \frac{1}{2\mu_0} B^2 \qquad (22.11)$$

Although this result was obtained for the special case of a long solenoid, it is quite general and is valid for any point where a magnetic field exists in air or vacuum or in a non-magnetic material. Thus, energy can be stored in a magnetic field, just as it can in an electric field.

22.9 Transformers

One of the most important applications of mutual induction and self-induction takes place in a transformer. A **transformer** is a device for increasing or decreasing an ac voltage. For instance, whenever cordless appliances (e.g., a hand-held vacuum cleaner) are plugged into a wall receptacle to recharge the batteries, a transformer plays a role in reducing the 120-V ac voltage to a much smaller value. Typically, between 3 and 9 V are needed to energize batteries. In another example, a picture tube in a television set needs about 15 000 V to accelerate the electron beam, and a transformer is used to obtain this high voltage from the 120 V at a wall socket.

The physics of transformers.

Figure 22.30 shows a drawing of a transformer. The transformer consists of an iron core on which two coils are wound: a primary coil with N_p turns and a secondary coil with N_s turns. The primary coil is the one connected to the ac generator. For the moment, suppose the switch in the secondary circuit is open, so there is no current in this circuit.

The alternating current in the primary coil establishes a changing magnetic field in the iron core. Because iron is easily magnetized, it greatly enhances the magnetic field relative to that in an air core and guides the field lines to the secondary coil. In a well-

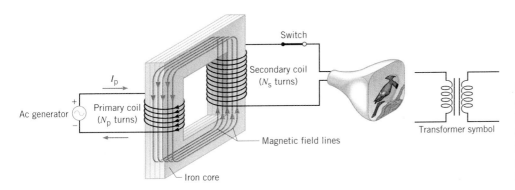

Figure 22.30 A transformer consists of a primary coil and a secondary coil, both wound on an iron core. The changing magnetic flux produced by the current in the primary coil induces an emf in the secondary coil. At the far right is the symbol for a transformer.

designed core, nearly all the magnetic flux Φ that passes through each turn of the primary also goes through each turn of the secondary. Since the magnetic field is changing, the flux through the primary and secondary coils is also changing, and consequently an emf is induced in both coils. In the secondary coil the induced emf $\mathscr{E}_s$ arises from mutual induction and is given by Faraday's law as

$$\mathscr{E}_s = -N_s \frac{\Delta\Phi}{\Delta t}$$

In the primary coil the induced emf $\mathscr{E}_p$ is due to self-induction and is specified by Faraday's law as

$$\mathscr{E}_p = -N_p \frac{\Delta\Phi}{\Delta t}$$

The term $\Delta\Phi/\Delta t$ is the same in both of these equations, since the same flux penetrates each turn of both coils. Dividing the two equations shows that

$$\frac{\mathscr{E}_s}{\mathscr{E}_p} = \frac{N_s}{N_p}$$

In a high-quality transformer the resistances of the coils are negligible, so the magnitudes of the emfs, $\mathscr{E}_s$ and $\mathscr{E}_p$, are nearly equal to the terminal voltages, V_s and V_p, across the coils (see Section 20.9 for a discussion of terminal voltage). The relation $\mathscr{E}_s/\mathscr{E}_p = N_s/N_p$ is called the **transformer equation** and is usually written in terms of the terminal voltages:

Transformer equation
$$\frac{V_s}{V_p} = \frac{N_s}{N_p} \qquad (22.12)$$

According to the transformer equation, if N_s is greater than N_p, the secondary (output) voltage is greater than the primary (input) voltage. In this case we have a *step-up* transformer. On the other hand, if N_s is less than N_p, the secondary voltage is less than the primary voltage, and we have a *step-down* transformer. The ratio N_s/N_p is referred to as the *turns ratio* of the transformer. A turns ratio of 8/1 (often written as 8 : 1) means, for example, that the secondary coil has eight times more turns than does the primary coil. Conversely, a turns ratio of 1 : 8 implies that the secondary coil has one-eighth as many turns as the primary coil.

A transformer operates with ac electricity and not with dc. A steady direct current in the primary coil produces a flux that does not change in time, and thus no emf is induced in the secondary coil. The ease with which transformers can change voltages from one value to another is a principal reason why ac is preferred over dc.

With the switch in the secondary circuit of Figure 22.30 closed, a current I_s exists in the circuit and electrical energy is fed to the TV tube. This energy comes from the ac generator connected to the primary coil. Although the secondary voltage V_s may be larger or smaller than the primary voltage V_p, energy is not being created or destroyed by the transformer. Energy conservation requires that the energy delivered to the secondary coil must be the same as the energy delivered to the primary coil, provided no energy is dissipated in heating these coils or is otherwise lost. In a well-designed transformer, less than 1% of the input energy is lost in the form of heat. Noting that power is energy per unit time, and assuming 100% energy transfer, the average power $\overline{P}_p$ delivered to the primary coil is equal to the average power $\overline{P}_s$ delivered to the secondary coil: $\overline{P}_p = \overline{P}_s$. But $\overline{P} = IV$ (Equation 20.15a), so $I_p V_p = I_s V_s$, or

$$\frac{I_s}{I_p} = \frac{V_p}{V_s} = \frac{N_p}{N_s} \qquad (22.13)$$

Observe that V_s/V_p is equal to the turns ratio N_s/N_p, while I_s/I_p is equal to the inverse turns ratio N_p/N_s. Consequently, *a transformer that steps up the voltage simultaneously steps down the current, and a transformer that steps down the voltage steps up the current.* However, the power is neither stepped up nor stepped down, since $\overline{P}_p = \overline{P}_s$. Example 14 emphasizes this fact.

Power distribution stations use high-voltage transformers similar to these to step-up or step-down voltages. (© Lester Lefkowitz/Taxi/Getty Images)

Problem solving insight

Example 14 A Step-Down Transformer

A step-down transformer inside a stereo receiver has 330 turns in the primary coil and 25 turns in the secondary coil. The plug connects the primary coil to a 120-V wall socket, and there is a current of 0.83 A in the primary coil while the receiver is turned on. Connected to the secondary coil are the transistor circuits of the receiver. Find (a) the voltage across the secondary coil, (b) the current in the secondary coil, and (c) the average electric power delivered to the transistor circuits.

Reasoning The transformer equation, Equation 22.12, states that the secondary voltage V_s is equal to the product of the primary voltage V_p and the turns ratio N_s/N_p. On the other hand, the secondary current I_s is equal to the product of the primary current I_p and the inverse turns ratio N_p/N_s. The average power delivered to the transistor circuits is the product of the secondary current and the secondary voltage.

Solution

(a) The voltage across the secondary coil can be found from the transformer equation:

$$V_s = V_p \frac{N_s}{N_p} = (120 \text{ V})\left(\frac{25}{330}\right) = \boxed{9.1 \text{ V}} \tag{22.12}$$

(b) The current in the secondary coil is

$$I_s = I_p \frac{N_p}{N_s} = (0.83 \text{ A})\left(\frac{330}{25}\right) = \boxed{11 \text{ A}} \tag{22.13}$$

(c) The average power $\overline{P}_s$ delivered to the secondary is the product of I_s and V_s:

$$\overline{P}_s = I_s V_s = (11 \text{ A})(9.1 \text{ V}) = \boxed{1.0 \times 10^2 \text{ W}} \tag{20.15a}$$

As a check on our calculation, we verify that the power delivered to the secondary coil is the same as that sent to the primary coil from the wall receptacle: $\overline{P}_p = I_p V_p = (0.83 \text{ A})(120 \text{ V}) = 1.0 \times 10^2 \text{ W}$.

Transformers play an important role in the transmission of power between electrical generating plants and the communities they serve. Whenever electricity is transmitted, there is always some loss of power in the transmission lines themselves due to resistive heating. Since the resistance of the wires is proportional to their length, the longer the wires the greater is the power loss. Power companies reduce this loss by using transformers that step up the voltage to high levels while reducing the current. A smaller current means less power loss, since $P = I^2 R$, where R is the resistance of the transmission wires (see problem 59). Figure 22.31 shows one possible way of transmitting power. The power plant produces a voltage of 12 000 V. This voltage is then raised to 240 000 V by a 20:1 step-up transformer. The high-voltage power is sent over the long-distance transmission line. Upon arrival at the city, the voltage is reduced to about 8000 V at a substation using

Figure 22.31 Transformers play a key role in the transmission of electric power.

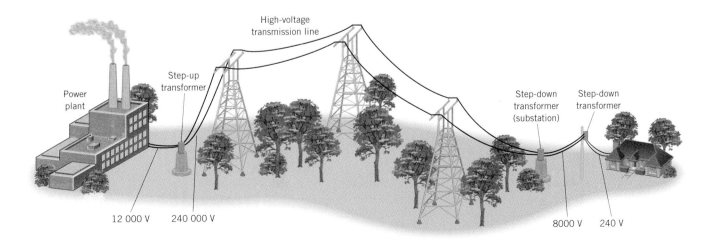

a 1:30 step-down transformer. However, before any domestic use, the voltage is further reduced to 240 V (or possibly 120 V) by another step-down transformer that is often mounted on a utility pole. The power is then distributed to consumers.

 Check Your Understanding 5

A transformer that stepped up the voltage and the current simultaneously would (a) produce less power at the secondary than was supplied at the primary, (b) produce more power at the secondary than was supplied at the primary, (c) produce the same amount of power at the secondary that was supplied at the primary, (d) violate the law of conservation of energy. Choose one or more. *[The answer(s) is (are) given at the end of the book.]*

Background: In a transformer, mutual induction allows the primary and secondary coils to interact. Depending on the turns ratio of the transformer, the voltages across the primary and secondary are not the same, and the currents in the primary and the secondary are not the same. However, for both coils the power is given by the product of current and voltage.

For similar questions (including calculational counterparts), consult Self-Assessment Test 22.2, which is described next.

 Self-Assessment Test 22.2

Test your understanding of the material in Sections 22.7–22.9:
• Generators   • Back Emf   • Mutual Inductance   • Self-Inductance
• Transformers

Go to **www.wiley.com/college/cutnell**

22.10 Concepts & Calculations

In this chapter we have seen that there are three ways to create an induced emf in a coil: by changing the magnitude of a magnetic field, by changing the direction of the field relative to the coil, and by changing the area of the coil. We have already explored the first two methods, and Example 15 now illustrates the third method. In the process, it provides a review of Faraday's law of electromagnetic induction.

Concepts & Calculations Example 15
The Emf Produced by a Changing Area

A circular coil of radius 0.11m contains a single turn and is located in a constant magnetic field of magnitude 0.27 T. The magnetic field is parallel to the normal to the plane of the coil. The radius increases to 0.30 m in a time of 0.080 s. (a) Determine the emf induced in the coil. (b) The coil has a resistance of 0.70 Ω. Find the induced current.

Concept Questions and Answers Why is there an emf induced in the coil?

> *Answer* According to Faraday's law of electromagnetic induction, Equation 22.3, an emf is induced whenever the magnetic flux through the coil changes. The magnetic flux, as expressed by Equation 22.2, depends on the area of the coil, which, in turn, depends on the radius. If the radius changes, the area changes, causing the flux to change and an induced emf to appear.

Does the magnitude of the induced emf depend on whether the area is increasing or decreasing?

> *Answer* No. The magnitude of the induced emf depends only on the magnitude of the rate $\Delta\Phi/\Delta t$ at which the flux changes. It does not matter whether the flux is increasing or decreasing.

What determines the amount of current induced in the coil?

> *Answer* According to Ohm's law (Equation 20.2), the current is equal to the emf induced in the coil divided by the resistance of the coil. Therefore, the larger the induced emf and the smaller the resistance, the larger will be the induced current.

If the coil is cut so it is no longer one continuous piece, are there an induced emf and induced current?

Answer An induced emf is generated in the coil regardless of whether the coil is whole or cut. However, an induced current exists only if the coil is continuous. Cutting the coil is like opening a switch. There is no longer a continuous path for the current to follow, so the current stops.

Solution

(a) The average induced emf is given by Faraday's law of electromagnetic induction.

$$\mathcal{E} = -N\frac{\Delta\Phi}{\Delta t} = -N\left(\frac{\Phi - \Phi_0}{t - t_0}\right) \tag{22.3}$$

where Φ and Φ_0 are the final and initial fluxes, respectively. The definition of magnetic flux is $\Phi = BA\cos\phi$, according to Equation 22.2, where $\phi = 0°$ because the field is parallel to the normal. The area of a circle is $A = \pi r^2$. With these expressions for Φ and A, Faraday's law becomes

$$\mathcal{E} = -N\left(\frac{BA\cos\phi - BA_0\cos\phi}{t - t_0}\right) = -NB(\cos\phi)\left(\frac{\pi r^2 - \pi r_0^2}{t - t_0}\right)$$

$$= -(1)(0.27\text{ T})(\cos 0°)\left[\frac{\pi(0.30\text{ m})^2 - \pi(0.11\text{ m})^2}{0.080\text{ s}}\right] = \boxed{0.83\text{ V}}$$

(b) The induced current I is equal to the induced emf divided by the resistance R of the coil:

$$I = \frac{\mathcal{E}}{R} = \frac{0.83\text{ V}}{0.70\text{ }\Omega} = \boxed{1.2\text{ A}} \tag{20.2}$$

▲

One of the most important applications of Faraday's law of electromagnetic induction is the electric generator, because it is the source of virtually all the electrical energy that we use. A generator produces an emf as a rotating coil changes its orientation relative to a fixed magnetic field. Example 16 reviews the characteristics of this type of induced emf.

Concepts & Calculations Example 16
The Emf Produced by a Generator

▼

Figure 22.32 shows the emf produced by a generator as a function of time t. The coil of the generator has an area of $A = 0.15$ m^2 and consists of $N = 10$ turns. The coil rotates in a magnetic field of magnitude 0.27 T. (a) Determine the period of the motion. (b) What is the angular frequency of the rotating coil? (c) Find the value of the emf when $t = \frac{1}{4}T$, where T denotes the period of the coil motion. (d) What is the emf when $t = 0.025$ s?

Concept Questions and Answers Can the period of the rotating coil be determined from the graph?

Answer Yes. The period is the time for the coil to rotate through one revolution, or cycle. During this time, the emf is positive for one half of the cycle and negative for the other half of the cycle.

The emf produced by a generator depends on its angular frequency. How is the angular frequency of the rotating coil related to the period?

Answer According to Equation 10.6, the angular frequency ω is related to the period T by $\omega = 2\pi/T$.

Starting at $t = 0$ s, how much time is required for the generator to produce its peak emf? Express the answer in terms of the period T of the motion (e.g., $t = \frac{1}{10}T$.)

Answer An examination of the graph shows that the emf reaches its maximum or peak value one-quarter of the way through a cycle. Since the time to complete one cycle is the period T, the time to reach the peak emf is $t = \frac{1}{4}T$.

How often does the polarity of the emf change in one cycle?

Answer The graph shows that in every cycle the emf has an interval of positive polarity followed by an interval of negative polarity. In other words, the polarity changes from positive to negative and then from negative to positive during each cycle. Thus, the polarity changes twice during each cycle.

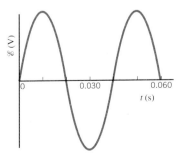

Figure 22.32 A plot of the emf produced by a generator as a function of time.

Solution

(a) The period is the time for the generator to make one complete cycle. From the graph it can be seen that this time is $T = \boxed{0.040 \text{ s}}$.

(b) The angular frequency ω is related to the period by $\omega = 2\pi/T$, so

$$\omega = \frac{2\pi}{T} = \frac{2\pi}{0.040 \text{ s}} = \boxed{160 \text{ rad/s}}$$

(c) When $t = \frac{1}{4}T$, the emf has reached its peak value. According to Equation 22.4, the peak emf is

$$\mathcal{E}_0 = NAB\omega = (10)(0.15 \text{ m}^2)(0.27 \text{ T})(160 \text{ rad/s}) = \boxed{65 \text{ V}}$$

(d) The emf produced by the generator as a function of time is given by Equation 22.4

$$\mathcal{E} = \mathcal{E}_0 \sin \omega t = (65 \text{ V}) \sin [(160 \text{ rad/s})(0.025 \text{ s})] = \boxed{-49 \text{ V}}$$

Problem solving insight

In evaluating sin ωt a calculator must be set to the radian mode (not the degree mode) because ω is expressed in units of rad/s.

At the end of the problem set for this chapter, you will find homework problems that contain both conceptual and quantitative parts. These problems are grouped under the heading *Concepts & Calculations, Group Learning Problems*. They are designed for use by students working alone or in small learning groups. The conceptual part of each problem provides a convenient focus for group discussions.

Concept Summary

This summary presents an abridged version of the chapter, including the important equations and all available learning aids. For convenient reference, the learning aids (including the text's examples) are placed next to or immediately after the relevant equation or discussion. The following learning aids may be found on-line at **www.wiley.com/college/cutnell**:

Interactive LearningWare examples are solved according to a five-step interactive format that is designed to help you develop problem-solving skills.	**Concept Simulations** are animated versions of text figures or animations that illustrate important concepts. You can control parameters that affect the display, and we encourage you to experiment.
Interactive Solutions offer specific models for certain types of problems in the chapter homework. The calculations are carried out interactively.	**Self-Assessment Tests** include both qualitative and quantitative questions. Extensive feedback is provided for both incorrect and correct answers, to help you evaluate your understanding of the material.

Topic	Discussion	Learning Aids
	22.1 Induced Emf and Induced Current	
Electromagnetic induction	Electromagnetic induction is the phenomenon in which an emf is induced in a piece of wire or a coil of wire with the aid of a magnetic field. The emf is called an induced emf, and any current that results from the emf is called an induced current.	
	22.2 Motional Emf	
	An emf $\mathcal{E}$ is induced in a conducting rod of length L when the rod moves with a speed v in a magnetic field of magnitude B, according to	
Motional emf	$$\mathcal{E} = vBL \qquad (22.1)$$	**Example 1** Interactive LearningWare 22.1
	Equation 22.1 applies when the velocity of the rod, the length of the rod, and the magnetic field are mutually perpendicular.	
Conservation of energy	When the motional emf is used to operate an electrical device, such as a light bulb, the energy delivered to the device originates in the work done to move the rod, and the law of conservation of energy applies.	**Examples 2, 3**
	22.3 Magnetic Flux	
	The magnetic flux Φ that passes through a surface is	
Magnetic flux	$$\Phi = BA \cos \phi \qquad (22.2)$$	**Example 4**
	where B is the magnitude of the magnetic field, A is the area of the surface, and ϕ is the angle between the field and the normal to the surface.	

Topic	Discussion	Learning Aids

The magnetic flux is proportional to the number of magnetic field lines that pass through the surface.

22.4 Faraday's Law of Electromagnetic Induction

Faraday's law of electromagnetic induction states that the average emf $\mathscr{E}$ induced in a coil of N loops is

Faraday's law

$$\mathscr{E} = -N\left(\frac{\Phi - \Phi_0}{t - t_0}\right) = -N\frac{\Delta\Phi}{\Delta t} \qquad (22.3)$$

Example 5
Concept Simulation 22.1
Example 6
Interactive LearningWare 22.2
Example 7
Interactive Solution 22.69

where $\Delta\Phi$ is the change in magnetic flux through one loop and Δt is the time interval during which the change occurs. Motional emf is a special case of induced emf.

22.5 Lenz's Law

Lenz's law

Lenz's law provides a way to determine the polarity of an induced emf. Lenz's law is stated as follows: The induced emf resulting from a changing magnetic flux has a polarity that leads to an induced current whose direction is such that the induced magnetic field opposes the original flux change. This statement is a consequence of the law of conservation of energy.

Examples 8, 9
Interactive Solution 22.33

Use *Self-Assessment Test 22.1* to evaluate your understanding of Sections 22.1–22.5.

22.7 The Electric Generator

In its simplest form, an electric generator consists of a coil of N loops that rotates in a uniform magnetic field **B**. The emf produced by this generator is

Emf of a generator

$$\mathscr{E} = NAB\omega \sin \omega t = \mathscr{E}_0 \sin \omega t \qquad (22.4)$$

Examples 10, 11
Interactive LearningWare 22.3
Interactive Solution 22.63

where A is the area of the coil, ω is the angular speed (in rad/s) of the coil, and $\mathscr{E}_0$ is the peak emf. The angular speed in rad/s is related to the frequency f in cycles/s, or Hz, according to $\omega = 2\pi f$.

When an electric motor is running, it exhibits a generator-like behavior by producing an induced emf, called the back emf. The current I needed to keep the motor running at a constant speed is

Back emf of a motor

$$I = \frac{V - \mathscr{E}}{R} \qquad (22.5)$$

Example 12

where V is the emf applied to the motor by an external source, $\mathscr{E}$ is the back emf, and R is the resistance of the motor coil.

22.8 Mutual Inductance and Self-Inductance

Mutual induction is the effect in which a changing current in the primary coil induces an emf in the secondary coil. The average emf $\mathscr{E}_s$ induced in the secondary coil by a change in current ΔI_p in the primary coil is

Emf due to mutual induction

$$\mathscr{E}_s = -M\frac{\Delta I_p}{\Delta t} \qquad (22.7)$$

Mutual inductance

where Δt is the time interval during which the change occurs. The constant M is the mutual inductance between the two coils and is measured in henries (H).

Self-induction is the effect in which a change in current ΔI in a coil induces an average emf $\mathscr{E}$ in the same coil, according to

Emf due to self-induction

$$\mathscr{E} = -L\frac{\Delta I}{\Delta t} \qquad (22.9)$$

Example 13

Self-inductance

The constant L is the self-inductance, or inductance, of the coil and is measured in henries.

To establish a current I in an inductor, work must be done by an external agent. This work is stored as energy in the inductor, the amount being

Energy stored in an inductor

$$\text{Energy} = \tfrac{1}{2}LI^2 \qquad (22.10)$$

Topic	Discussion	Learning Aids

The energy stored in an inductor can be regarded as being stored in its magnetic field. At any point in air or vacuum or in a nonmagnetic material where a magnetic field **B** exists, the energy density, or the energy stored per unit volume, is

Energy density of a magnetic field

$$\text{Energy density} = \frac{1}{2\mu_0} B^2 \qquad (22.11)$$

22.9 Transformers

A transformer consists of a primary coil of N_p turns and a secondary coil of N_s turns. If the resistances of the coils are negligible, the voltage V_p across the primary and the voltage V_s across the secondary are related according to the transformer equation:

Transformer equation

$$\frac{V_s}{V_p} = \frac{N_s}{N_p} \qquad (22.12) \quad \textbf{Example 14}$$

Interactive Solution 22.55

Turns ratio

where the ratio N_s/N_p is called the turns ratio of the transformer.

A transformer functions with ac electricity, not with dc. If the transformer is 100% efficient in transferring power from the primary to the secondary coil, the ratio of the secondary current I_s to the primary current I_p is

$$\frac{I_s}{I_p} = \frac{N_p}{N_s} \qquad (22.13)$$

 Use *Self-Assessment Test 22.2* to evaluate your understanding of Sections 22.7–22.9.

Conceptual Questions

1. Suppose the coil and the magnet in Figure 22.1*a* were each moving with the same velocity relative to the earth. Would there be an induced current in the coil? Explain.

2. In the discussion concerning Figure 22.5, we saw that a force of 0.086 N from an external agent was required to keep the rod moving at a constant speed. Suppose the light bulb in the figure is unscrewed from its socket. How much force would now be needed to keep the rod moving at a constant speed? Justify your answer.

3. Eddy currents are electric currents that can arise in a piece of metal when it moves through a region where the magnetic field is not the same everywhere. The picture shows, for example, a metal sheet moving to the right at a velocity **v** and a magnetic field **B** that is directed perpendicular to the sheet. At the instant represented, the magnetic field only extends over the left half of the sheet. An emf is induced that leads to the eddy current shown. Explain why this current causes the metal sheet to slow down. This action of eddy currents is used in various devices as a brake to damp out unwanted motion.

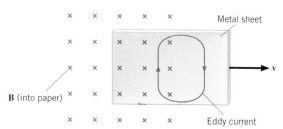

4. A magnetic field is necessary if there is to be a magnetic flux passing through a coil of wire. Yet, just because there is a magnetic field does not mean that a magnetic flux will pass through a coil. Account for this observation.

5. Suppose the magnetic flux through a 1-m² flat surface is known to be 2 Wb. From these data, it is possible to determine certain information about the average magnetic field at the surface, but not the magnitude and direction of the total field. What exactly can be ascertained about the field? Explain.

6. A conducting rod is free to slide along a pair of conducting rails, in a region where a uniform and constant (in time) magnetic field is directed into the plane of the paper, as the drawing illustrates. Initially the rod is at rest. Describe the rod's motion after the switch is closed. Be sure to account for the effect of any motional emf that develops.

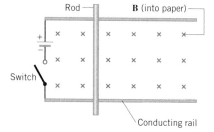

7. Explain how a bolt of lightning can produce a current in the circuit of an electrical appliance, even when the lightning does not directly strike the appliance.

8. Review Conceptual Example 7 before answering this question. A solenoid is connected to an ac source. A copper ring is placed inside the solenoid, with the normal to the ring being parallel to the axis of the solenoid. The copper ring gets hot, yet nothing touches it. Why?

9. A robot is designed to move parallel to a cable hidden under the floor. The cable carries a steady direct current I. A sensor mounted on the robot consists of a coil of wire. The coil is near the floor and parallel to it. As long as the robot moves parallel to the cable, with the coil directly over it, no emf is induced in the coil, since the magnetic flux through the coil does not change. But when the robot deviates from the parallel path, an induced emf appears in the coil. The emf is sent to electronic circuits that bring the robot back to the path. Explain why an emf would be induced in the sensor coil.

10. In a car, the generator-like action of the alternator occurs while the engine is running and keeps the battery fully charged. The headlights would discharge an old and failing battery quickly if it were not for the alternator. Explain why the engine of a parked car runs more quietly with the headlights off than with them on when the battery is in bad shape.

11. In Figure 22.3 a coil of wire is being stretched. (a) Using Lenz's law, verify that the induced current in the coil has the direction shown in the drawing. (b) Deduce the direction of the induced current if the direction of the external magnetic field in the figure were reversed. Explain.

12. (a) When the switch in the circuit in the drawing is closed, a current is established in the coil and the metal ring jumps upward. Explain this behavior.

13. The string of an electric guitar vibrates in a standing wave pattern that consists of nodes and antinodes. (Section 17.5 discusses standing waves.) Where should an electromagnetic pickup be located in the standing wave pattern to produce a maximum emf, at a node or an antinode? Why?

14. An electric motor in a hair dryer is running at normal speed and, thus, is drawing a relatively small current, as in part (b) of Example 12. What happens to the current drawn by the motor if the shaft is prevented from turning, so the back emf is suddenly reduced to zero? Remembering that the wire in the coil of the motor has some resistance, what happens to the temperature of the coil? Justify your answers.

15. One transformer is a step-up device, while another is step-down. These two units have the same voltage across and the same current in their primary coils. Does either one deliver more power to the circuit attached to the secondary coil? If so, which one? Ignore any heat loss within the transformers and account for your answer.

Problems

Section 22.2 Motional Emf

1. ssm A spark can jump between two nontouching conductors if the potential difference between them is sufficiently large. A potential difference of approximately 940 V is required to produce a spark in an air gap of 1.0×10^{-4} m. Suppose the light bulb in Figure 22.4b is replaced by such a gap. How fast would a 1.3-m rod have to be moving in a magnetic field of 4.8 T to cause a spark to jump across the gap?

2. ☤ The drawing shows a type of flow meter that can be used to measure the speed of blood in situations when a blood vessel is sufficiently exposed (e.g., during surgery). Blood is conductive enough that it can be treated as a moving conductor. When it flows perpendicularly with respect to a magnetic field, as in the drawing, electrodes can be used to measure the small voltage that develops across the vessel. Suppose the speed of the blood is 0.30 m/s and the diameter of the vessel is 5.6 mm. In a 0.60-T magnetic field what is the magnitude of the voltage that is measured with the electrodes in the drawing?

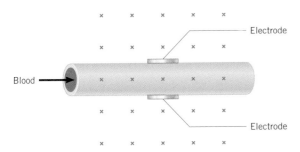

3. The wingspan (tip to tip) of a Boeing 747 jetliner is 59 m. The plane is flying horizontally at a speed of 220 m/s. The vertical component of the earth's magnetic field is 5.0×10^{-6} T. Find the emf induced between the wing tips.

4. In 1996, NASA performed an experiment called the Tethered Satellite experiment. In this experiment a 2.0×10^4-m length of wire was let out by the space shuttle *Atlantis* to generate a motional emf. The shuttle had an orbital speed of 7.6×10^3 m/s, and the magnitude of the earth's magnetic field at the location of the wire was 5.1×10^{-5} T. If the wire had moved perpendicular to the earth's magnetic field, what would have been the motional emf generated between the ends of the wire?

5. ssm www The drawing shows three identical rods (A, B, and C) moving in different planes. A constant magnetic field of magnitude 0.45 T is directed along the +y axis. The length of each rod is $L = 1.3$ m, and the speeds

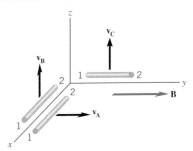

are the same, $v_A = v_B = v_C = 2.7$ m/s. For each rod, find the magnitude of the motional emf, and indicate which end (1 or 2) of the rod is positive.

* **6.** Suppose the light bulb in Figure 22.4b is a 60.0-W bulb with a resistance of 240 Ω. The magnetic field has a magnitude of 0.40 T, and the length of the rod is 0.60 m. The only resistance in the circuit is that due to the bulb. Minimally, how long would the rails on which the moving rod slides have to be, in order that the bulb can remain lit for one-half second?

* **7. ssm** Suppose the light bulb in Figure 22.4b is replaced by a 6.0-Ω electric heater that consumes 15 W of power. The conducting bar moves to the right at a constant speed, the field strength is 2.4 T, and the length of the bar between the rails is 1.2 m. (a) How fast is the bar moving? (b) What force must be applied to the bar to keep it moving to the right at the constant speed found in part (a)?

* **8.** Refer to the drawing that accompanies Conceptual Question 6 (not Problem 6). Suppose that the voltage of the battery in the circuit is 3.0 V, the magnitude of the magnetic field (directed perpendicularly into the plane of the paper) is 0.60 T, and the length of the rod between the rails is 0.20 m. Assuming that the rails are very long and have negligible resistance, find the maximum speed attained by the rod after the switch is closed.

** **9.** Review Conceptual Example 3 and Figure 22.7*b* as an aid in solving this problem. A conducting rod slides down between two frictionless vertical copper tracks at a constant speed of 4.0 m/s perpendicular to a 0.50-T magnetic field. The resistance of the rod and tracks is negligible. The rod maintains electrical contact with the tracks at all times and has a length of 1.3 m. A 0.75-Ω resistor is attached between the tops of the tracks. (a) What is the mass of the rod? (b) Find the change in the gravitational potential energy that occurs in a time of 0.20 s. (c) Find the electrical energy dissipated in the resistor in 0.20 s.

Section 22.3 Magnetic Flux

For problems in this set, assume that the magnetic flux is a positive quantity.

10. The drawing shows two surfaces that have the same area. A uniform magnetic field **B** fills the space occupied by these surfaces and is oriented parallel to the *yz* plane as shown. Find the ratio Φ_{xz}/Φ_{xy} of the magnetic fluxes that pass through the surfaces.

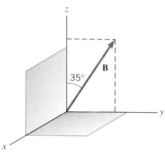

11. ssm A standard door into a house rotates about a vertical axis through one side, as defined by the door's hinges. A uniform magnetic field is parallel to the ground and perpendicular to this axis. Through what angle must the door rotate so that the magnetic flux that passes through it decreases from its maximum value to one-third of its maximum value?

12. A house has a floor area of 112 m² and an outside wall that has an area of 28 m². The earth's magnetic field here has a horizontal component of 2.6×10^{-5} T that points due north and a vertical component of 4.2×10^{-5} T that points straight down, toward the earth. Determine the magnetic flux through the wall if the wall faces (a) north and (b) east. (c) Calculate the magnetic flux that passes through the floor.

13. The drawing shows three square surfaces, one lying in the *xy* plane, one in the *xz* plane, and one in the *yz* plane. The sides of each square have lengths of 2.0×10^{-2} m. A uniform magnetic field exists in this region, and its components are: $B_x = 0.50$ T, $B_y = 0.80$ T, and $B_z = 0.30$ T. Determine the magnetic flux that passes through the surface that lies in (a) the *xy* plane, (b) the *xz* plane, and (c) the *yz* plane.

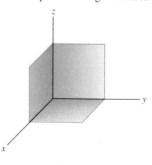

14. A loop of wire has the shape shown in the drawing. The top part of the wire is bent into a semicircle of radius $r = 0.20$ m. The normal to the plane of the loop is parallel to a constant magnetic field ($\phi = 0°$) of magnitude 0.75 T. What is the change $\Delta\Phi$ in the magnetic flux that

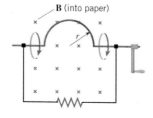
B (into paper)

passes through the loop when, starting with the position shown in the drawing, the semicircle is rotated through half a revolution?

* **15. ssm www** A five-sided object, whose dimensions are shown in the drawing, is placed in a uniform magnetic field. The magnetic field has a magnitude of 0.25 T and points along the positive *y* direction. Determine the magnetic flux through each of the five sides.

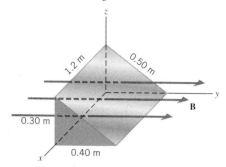

* **16.** A long and narrow rectangular loop of wire is moving toward the bottom of the page with a speed of 0.020 m/s (see the drawing). The loop is leaving a region in which a 2.4-T magnetic field exists; the magnetic field outside this region is zero. During a time of 2.0 s, what is the magnitude of the *change* in the magnetic flux?

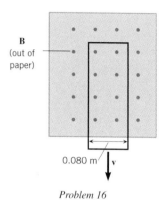

Problem 16

Section 22.4 Faraday's Law of Electromagnetic Induction

17. A planar coil of wire has a single turn. The normal to this coil is parallel to a uniform and constant (in time) magnetic field of 1.7 T. An emf that has a magnitude of 2.6 V is induced in this coil because the coil's area *A* is shrinking. What is the magnitude of $\Delta A/\Delta t$, which is the rate (in m²/s) at which the area changes?

18. In each of two coils the rate of change of the magnetic flux in a single loop is the same. The emf induced in coil 1, which has 184 loops, is 2.82 V. The emf induced in coil 2 is 4.23 V. How many loops does coil 2 have?

19. ssm A circular coil (950 turns, radius = 0.060 m) is rotating in a uniform magnetic field. At $t = 0$ s, the normal to the coil is perpendicular to the magnetic field. At $t = 0.010$ s, the normal makes an angle of $\phi = 45°$ with the field because the coil has made one-eighth of a revolution. An average emf of magnitude 0.065 V is induced in the coil. Find the magnitude of the magnetic field at the location of the coil.

20. A magnetic field is perpendicular to the plane of a single-turn circular coil. The magnitude of the field is changing, so that an emf of 0.80 V and a current of 3.2 A are induced in the coil. The wire is then re-formed into a single-turn square coil, which is used in the same magnetic field (again perpendicular to the plane of the coil and with a magnitude changing at the same rate). What emf and current are induced in the square coil?

21. Magnetic resonance imaging (MRI) is a medical technique for producing pictures of the interior of the body. The patient is placed within a strong magnetic field. One safety concern is what would happen to the positively and negatively charged particles in the body fluids if an equipment failure caused the magnetic field to be shut off suddenly. An induced emf could cause these particles to flow, producing an electric current within the body. Suppose the largest surface of the body through which flux passes has an area of 0.032 m² and a normal that is parallel to a magnetic field of 1.5 T.

Determine the smallest time period during which the field can be allowed to vanish if the magnitude of the average induced emf is to be kept less than 0.010 V.

22. Interactive LearningWare 22.2 at **www.wiley.com/college/cutnell** reviews the fundamental approach in problems such as this. A constant magnetic field passes through a single rectangular loop whose dimensions are 0.35 m × 0.55 m. The magnetic field has a magnitude of 2.1 T and is inclined at an angle of 65° with respect to the normal to the plane of the loop. (a) If the magnetic field decreases to zero in a time of 0.45 s, what is the magnitude of the average emf induced in the loop? (b) If the magnetic field remains constant at its initial value of 2.1 T, what is the magnitude of the rate $\Delta A / \Delta t$ at which the area should change so that the average emf has the same magnitude as in part (a)?

* **23.** A piece of copper wire is formed into a single circular loop of radius 12 cm. A magnetic field is oriented parallel to the normal to the loop, and it increases from 0 to 0.60 T in a time of 0.45 s. The wire has a resistance per unit length of 3.3×10^{-2} Ω/m. What is the average electrical energy dissipated in the resistance of the wire?

* **24.** Parts *a* and *b* of the drawing show the same uniform and constant (in time) magnetic field **B** directed perpendicularly into the paper over a rectangular region. Outside this region, there is no field. Also shown is a rectangular coil (one turn), which lies in the plane of the paper. In part *a* the long side of the coil (length = L) is just at the edge of the field region, while in part *b* the short side (width = W) is just at the edge. It is known that L/W = 3.0. In both parts of the drawing the coil is pushed into the field with the same velocity **v** until it is completely within the field region. The magnitude of the average emf induced in the coil in part *a* is 0.15 V. What is its magnitude in part *b*?

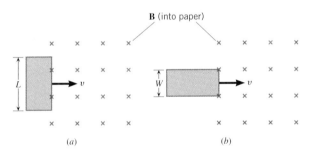

(a)   (b)

* **25.** ssm A magnetic field is passing through a loop of wire whose area is 0.018 m². The direction of the magnetic field is parallel to the normal to the loop, and the magnitude of the field is increasing at the rate of 0.20 T/s. (a) Determine the magnitude of the emf induced in the loop. (b) Suppose the area of the loop can be enlarged or shrunk. If the magnetic field is increasing as in part (a), at what rate (in m²/s) should the area be changed at the instant when B = 1.8 T if the induced emf is to be zero? Explain whether the area is to be enlarged or shrunk.

* **26.** The drawing shows a copper wire (negligible resistance) bent into a circular shape with a radius of 0.50 m. The radial section BC is fixed in place, while the copper bar AC sweeps around at an angular speed of 15 rad/s. The bar makes electrical contact with the wire at all times. The wire and the bar have negligible resistance. A uniform magnetic field exists everywhere, is

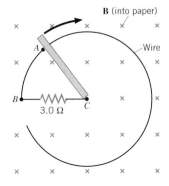

perpendicular to the plane of the circle, and has a magnitude of 3.8×10^{-3} T. Find the magnitude of the current induced in the loop *ABC*.

** **27.** Two 0.68-m-long conducting rods are rotating at the same speed in opposite directions, and both are perpendicular to a 4.7-T magnetic field. As the drawing shows, the ends of these rods come to within 1.0 mm of each other as they rotate. Moreover, the fixed ends about which the rods are rotating are connected by a wire, so these ends are at the same electric potential. If a potential difference of 4.5×10^3 V is required to cause a 1.0-mm spark in air, what is the angular speed (in rad/s) of the rods when a spark jumps across the gap?

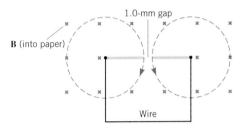

Section 22.5 Lenz's Law

28. The drawing shows that a uniform magnetic field is directed perpendicularly into the plane of the paper and fills the entire region to the left of the y axis. There is no magnetic field to the right of the y axis. A rigid right triangle *ABC* is made of copper wire. The triangle rotates counterclockwise about the origin at point *C*. What is the direction (clockwise or counterclockwise) of the induced current when the triangle is crossing (a) the +y axis, (b) the −x axis, (c) the −y axis, and (d) the +x axis? For each case, justify your answer.

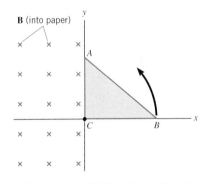

29. ssm Suppose in Figure 22.1 that the bar magnet is held stationary, but the coil of wire is free to move. Which way will current be directed through the ammeter, left to right or right to left, when the coil is moved (a) to the left and (b) to the right? Explain.

30. Review the drawing that accompanies Problem 14. The semicircular piece of wire rotates through half a revolution in the direction shown, starting from the position indicated in the drawing. Which end of the resistor, the left or the right end, is positive? Explain your reasoning.

31. ssm Review Conceptual Example 9 as an aid in understanding this problem. A long, straight wire lies on a table and carries a current *I*. As the drawing shows, a small circular loop of wire is pushed across the top of

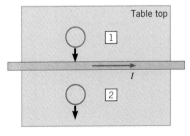

the table from position 1 to position 2. Determine the direction of the induced current, clockwise or counterclockwise, as the loop moves past (a) position 1 and (b) position 2. Justify your answers.

* **32.** Indicate the direction of the electric field between the plates of the parallel plate capacitor shown in the drawing if the magnetic field is decreasing in time. Give your reasoning.

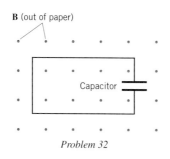

Problem 32

* **33.** Consult **Interactive Solution 22.33** at www.wiley.com/college/cutnell for one approach to this problem. A circular loop of wire rests on a table. A long, straight wire lies on this loop, directly over its center, as the drawing illustrates. The current I in the straight wire is decreasing. In what direction is the induced current, if any, in the loop? Give your reasoning.

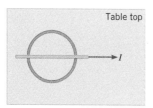

** **34.** A wire loop is suspended from a string that is attached to point P in the drawing. When released, the loop swings downward, from left to right, through a uniform magnetic field, with the plane of the loop remaining perpendicular to the plane of the paper at all times. (a) Determine the direction of the current induced in the loop as it swings past the locations labeled I and II. Specify the direction of the current in terms of the points x, y, and z on the loop (e.g., $x \rightarrow y \rightarrow z$ or $z \rightarrow y \rightarrow x$). The points x, y, and z lie behind the plane of the paper. (b) What is the direction of the induced current at the locations II and I when the loop swings back, from right to left? Provide reasons for your answers.

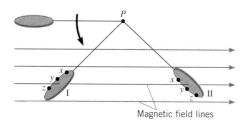

Magnetic field lines

Section 22.7 The Electric Generator

35. ssm The coil of an ac generator has an area per turn of 1.2×10^{-2} m^2 and consists of 500 turns. The coil is situated in a 0.13-T magnetic field and is rotating at an angular speed of 34 rad/s. What is the emf induced in the coil at the instant when the normal to the loop makes an angle of 27° with respect to the direction of the magnetic field?

36. A vacuum cleaner is plugged into a 120.0-V socket and uses 3.0 A of current in normal operation when the back emf generated by the electric motor is 72.0 V. Find the coil resistance of the motor.

37. One generator uses a magnetic field of 0.10 T and has a coil area per turn of 0.045 m^2. A second generator has a coil area per turn of 0.015 m^2. The generator coils have the same number of turns and rotate at the same angular speed. What magnetic field should be used in the second generator so that its peak emf is the same as that of the first generator?

38. A 120.0-volt motor draws a current of 7.00 A when running at normal speed. The resistance of the armature wire is 0.720 Ω. (a) Determine the back emf generated by the motor. (b) What is the current at the instant when the motor is just turned on and has not begun to rotate? (c) What series resistance must be added to limit the starting current to 15.0 A?

39. ssm www The drawing shows a plot of the output emf of a generator as a function of time t. The coil of this device has a cross-sectional area per turn of 0.020 m^2 and contains 150 turns. Find (a) the frequency f of the generator in hertz, (b) the angular speed ω in rad/s, and (c) the magnitude of the magnetic field.

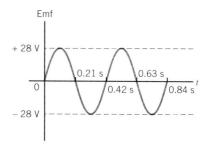

* **40.** A generator uses a coil that has 100 turns and a 0.50-T magnetic field. The frequency of this generator is 60.0 Hz, and its emf has an rms value of 120 V. Assuming that each turn of the coil is a square (an approximation), determine the length of the wire from which the coil is made.

* **41. ssm** At its normal operating speed, an electric fan motor draws only 15.0% of the current it draws when it just begins to turn the fan blade. The fan is plugged into a 120.0-V socket. What back emf does the motor generate at its normal operating speed?

* **42.** The coil of a generator has a radius of 0.14 m. When this coil is unwound, the wire from which it is made has a length of 5.7 m. The magnetic field of the generator is 0.20 T, and the coil rotates at an angular speed of 25 rad/s. What is the peak emf of this generator?

** **43.** The armature of an electric drill motor has a resistance of 15.0 Ω. When connected to a 120.0-V outlet, the motor rotates at its normal speed and develops a back emf of 108 V. (a) What is the current through the motor? (b) If the armature freezes up due to a lack of lubrication in the bearings and can no longer rotate, what is the current in the stationary armature? (c) What is the current when the motor runs at only half speed?

Section 22.8 Mutual Inductance and Self-Inductance

44. The average emf induced in the secondary coil is 0.12 V when the current in the primary coil changes from 3.4 to 1.6 A in 0.14 s. What is the mutual inductance of the coils?

45. ssm The earth's magnetic field, like any magnetic field, stores energy. The maximum strength of the earth's field is about 7.0×10^{-5} T. Find the maximum magnetic energy stored in the space above a city if the space occupies an area of 5.0×10^8 m^2 and has a height of 1500 m.

46. The current through a 3.2-mH inductor varies with time according to the graph shown in the drawing. What is the average induced emf during the time intervals (a) 0–2.0 ms, (b) 2.0–5.0 ms, and (c) 5.0–9.0 ms?

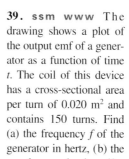

Problem 46

47. Mutual induction can be used as the basis for a metal detector. A typical setup uses two large coils that are parallel to each other and have a common axis. Because of mutual induction, the ac generator connected to the primary coil causes an emf of 0.46 V to be induced in the secondary coil. When someone without metal objects walks through the coils, the mutual inductance and, thus, the induced emf do not change much. But when a person carrying a handgun walks through, the mutual inductance increases. The change in emf can be used to trigger an alarm. If the mutual inductance increases by a factor of three, find the new value of the induced emf.

48. During a 72-ms interval, a change in the current in a primary coil occurs. This change leads to the appearance of a 6.0-mA current in a nearby secondary coil. The secondary coil is part of a circuit in which the resistance is 12 Ω. The mutual inductance between the two coils is 3.2 mH. What is the change in the primary current?

49. ssm Suppose you wish to make a solenoid whose self-inductance is 1.4 mH. The inductor is to have a cross-sectional area of 1.2×10^{-3} m² and a length of 0.052 m. How many turns of wire are needed?

* **50.** A magnetic field has a magnitude of 12 T. What is the magnitude of an electric field that stores the same energy per unit volume as this magnetic field?

* **51.** A long, current-carrying solenoid with an air core has 1750 turns per meter of length and a radius of 0.0180 m. A coil of 125 turns is wrapped tightly around the outside of the solenoid. What is the mutual inductance of this system?

* **52.** The purpose of this problem is to show that the work W needed to establish a final current I_f in an inductor is $W = \frac{1}{2}LI_f^2$ (Equation 22.10). In Section 22.8 we saw that the amount of work ΔW needed to change the current through an inductor by an amount ΔI is $\Delta W = LI(\Delta I)$, where L is the inductance. The drawing shows a graph of LI versus I. Notice that $LI(\Delta I)$ is the area of the shaded vertical rectangle whose height is LI and whose width is ΔI. Use this fact to show that the total work W needed to establish a current I_f is $W = \frac{1}{2}LI_f^2$.

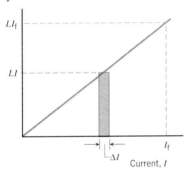

53. Coil 1 is a flat circular coil that has N_1 turns and a radius R_1. At its center is a much smaller flat, circular coil that has N_2 turns and radius R_2. The planes of the coils are parallel. Assume that coil 2 is so small that the magnetic field due to coil 1 has nearly the same value at all points covered by the area of coil 2. Determine an expression for the mutual inductance between these two coils in terms of μ_0, N_1, R_1, N_2, and R_2.

Section 22.9 Transformers

54. A neon sign requires 12 000 V for its operation. It operates from a 220-V receptacle. (a) What type of transformer, step-up or step-down, is needed? (b) What must be the turns ratio N_s/N_p of the transformer?

55. Interactive Solution 22.55 at **www.wiley.com/college/cutnell** offers one approach to problems such as this one. The secondary coil of a step-up transformer provides the voltage that operates an electrostatic air filter. The turns ratio of the transformer is 50:1. The primary coil is plugged into a standard 120-V outlet. The current in the secondary coil is 1.7×10^{-3} A. Find the power consumed by the air filter.

56. The batteries in a portable CD player are recharged by a unit that plugs into a wall socket. Inside the unit is a step-down transformer with a turns ratio of 1:13. The wall socket provides 120 V. What voltage does the secondary coil of the transformer provide?

57. ssm Electric doorbells found in many homes require 10.0 V to operate. To obtain this voltage from the standard 120-V supply, a transformer is used. Is a step-up or a step-down transformer needed, and what is its turns ratio N_s/N_p?

58. In some places, insect "zappers," with their blue lights, are a familiar sight on a summer's night. These devices use a high voltage to electrocute insects. One such device uses an ac voltage of 4320 V, which is obtained from a standard 120.0-V outlet by means of a transformer. If the primary coil has 21 turns, how many turns are in the secondary coil?

* **59. ssm** A generating station is producing 1.2×10^6 W of power that is to be sent to a small town located 7.0 km away. Each of the two wires that comprise the transmission line has a resistance per kilometer of length of 5.0×10^{-2} Ω/km. (a) Find the power lost in heating the wires if the power is transmitted at 1200 V. (b) A 100:1 step-up transformer is used to raise the voltage before the power is transmitted. How much power is now lost in heating the wires?

* **60.** Suppose there are two transformers between your house and the high-voltage transmission line that distributes the power. In addition, assume that your house is the only one using electric power. At a substation the primary of a step-down transformer (turns ratio = 1:29) receives the voltage from the high-voltage transmission line. Because of your usage, a current of 48 mA exists in the primary of this transformer. The secondary is connected to the primary of another step-down transformer (turns ratio = 1:32) somewhere near your house, perhaps up on a telephone pole. The secondary of this transformer delivers a 240-V emf to your house. How much power is your house using? Remember that the current and voltage given in this problem are rms values.

** **61.** A generator is connected across the primary coil (N_p turns) of a transformer, while a resistance R_2 is connected across the secondary coil (N_s turns). This circuit is equivalent to a circuit in which a single resistance R_1 is connected directly across the generator, without the transformer. Show that $R_1 = (N_p/N_s)^2 R_2$, by starting with Ohm's law as applied to the secondary coil.

Additional Problems

62. A magnetic field is perpendicular to a 0.040-m $\times$ 0.060-m rectangular coil of wire that has one hundred turns. In a time of 0.050 s, an average emf of magnitude 1.5 V is induced in the coil. What is the magnitude of the change in the magnetic field?

63. Interactive Solution 22.63 at **www.wiley.com/college/cutnell** provides one model for solving this problem. The maximum strength of the earth's magnetic field is about 6.9×10^{-5} T near the south magnetic pole. In principle, this field could be used with a rotating coil to generate 60.0-Hz ac electricity. What is the minimum number of turns (area per turn = 0.022 m²) that the coil must have to produce an rms voltage of 120 V?

64. A 1.5-m-long aluminum rod is rotating about an axis that is perpendicular to one end. A 0.16-T magnetic field is directed parallel to the axis. The rod rotates through one-fourth of a circle in 0.66 s. What is the magnitude of the average emf generated between the ends of the rod during this time?

65. ssm In Figure 22.1, suppose the north and south poles of the magnet were interchanged. Determine the direction of the current through the ammeter in parts b and c of the picture (left to right or right to left). Give your rationale.

66. The resistances of the primary and secondary coils of a transformer are 56 and 14 Ω, respectively. Both coils are made from lengths of the same copper wire. The circular turns of each coil have the same diameter. Find the turns ratio N_s/N_p.

67. ssm A generator has a square coil consisting of 248 turns. The coil rotates at 79.1 rad/s in a 0.170-T magnetic field. The peak output of the generator is 75.0 V. What is the length of one side of the coil?

*✶**68.** A 3.0-μF capacitor has a voltage of 35 V between its plates. What must be the current in a 5.0-mH inductor, such that the energy stored in the inductor equals the energy stored in the capacitor?

*✶**69. Interactive Solution 22.69** at **www.wiley.com/college/cutnell** offers some help for this problem. A copper rod is sliding on two conducting rails that make an angle of 19° with

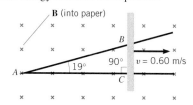

respect to each other, as in the drawing. The rod is moving to the right with a constant speed of 0.60 m/s. A 0.38-T uniform magnetic field is perpendicular to the plane of the paper. Determine the magnitude of the average emf induced in the triangle *ABC* during the 6.0-s period after the rod has passed point *A*.

*✶**70.** The drawing shows a bar magnet falling through a metal ring. In part *a* the ring is solid all the way around, but in part *b* it has been cut through. (a) Explain why the motion of the magnet in part *a* is retarded when the magnet is above the ring and below the ring as well.

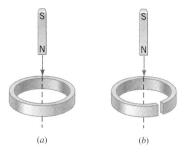

Draw any induced currents that appear in the ring. (b) Explain why the motion of the magnet is unaffected by the ring in part *b*.

*✶**71. ssm www** A conducting coil of 1850 turns is connected to a galvanometer, and the total resistance of the circuit is 45.0 Ω. The area of each turn is 4.70×10^{-4} m². This coil is moved from a region where the magnetic field is zero into a region where it is nonzero, the normal to the coil being kept parallel to the magnetic field. The amount of charge that is induced to flow around the circuit is measured to be 8.87×10^{-3} C. Find the magnitude of the magnetic field. (Such a device can be used to measure the magnetic field strength and is called a *flux meter*.)

*✶**72.** A long solenoid (cross-sectional area = 1.0×10^{-6} m², number of turns per unit length = 2400 turns/m) is bent into a circular shape so it looks like a doughnut. This wire-wound doughnut is called a toroid. Assume that the diameter of the solenoid is small compared to the radius of the toroid, which is 0.050 m. With this assumption, use the results of Example 13 to determine the self-inductance of the toroid.

✶✶73. ssm A solenoid has a cross-sectional area of 6.0×10^{-4} m², consists of 400 turns per meter, and carries a current of 0.40 A. A 10-turn coil is wrapped tightly around the circumference of the solenoid. The ends of the coil are connected to a 1.5-Ω resistor. Suddenly, a switch is opened, and the current in the solenoid dies to zero in a time of 0.050 s. Find the average current induced in the coil.

✶✶74. A motor is designed to operate on 117 V and draws a current of 12.2 A when it first starts up. At its normal operating speed, the motor draws a current of 2.30 A. Obtain (a) the resistance of the armature coil, (b) the back emf developed at normal speed, and (c) the current drawn by the motor at one-third normal speed.

Concepts & Calculations Group Learning Problems

Note: Each of these problems consists of Concept Questions followed by a related quantitative Problem. They are designed for use by students working alone or in small learning groups. The Concept Questions involve little or no mathematics and are intended to stimulate group discussions. They focus on the concepts with which the problems deal. Recognizing the concepts is the essential initial step in any problem-solving techinque.

75. Concept Questions Two circuits contain an emf produced by a moving metal rod, like that in Figure 22.4*b*. The speed of the rod is the same in each circuit, but the bulb in circuit 1 has one-half the resistance of the bulb in circuit 2. The circuits are otherwise identical. Is (a) the motional emf and (b) the current in circuit 1 greater than, the same as, or less than, that in circuit 2? (c) If the speed of the rod in circuit 1 were doubled, how would the power delivered to the light bulb compare to that in circuit 2? Provide a reason for each of your answers.

Problem The resistance of the light bulb in circuit 1 is 55 Ω, and that in circuit 2 is 110 Ω. Determine (a) the ratio $\mathcal{E}_1/\mathcal{E}_2$ of the emfs and (b) the ratio I_1/I_2 of the currents. (c) If the speed of the rod in circuit 1 is twice that in circuit 2, what is the ratio P_1/P_2 of the powers? Check to see that your answers are consistent with your answers to the Concept Questions.

76. Concept Questions A magnetic field passes through a stationary wire loop, and its magnitude changes in time according to the graph in the drawing. The direction of the field remains constant,

however. There are three equal time intervals indicated in the graph: 0–3.0 s, 3.0–6.0 s, and 6.0–9.0 s. (a) Is the induced emf equal to zero during any of the intervals? (b) During which interval is the magnitude of the induced emf the largest? (c) If the direction of the current induced during the first interval is clockwise, what is the direction during the third interval? In all cases, provide a reason for your answer.

Problem The loop consists of 50 turns of wire and has an area of 0.15 m². The magnetic field is oriented parallel to the normal to the loop. (a) For each interval, determine the magnitude of the induced emf. (b) The wire has a resistance of 0.50 Ω. Determine the induced current for the first and third intervals. Make sure your answers are consistent with your answers to the Concept Questions.

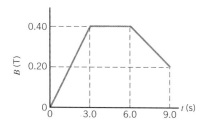

77. Concept Questions The drawing shows a straight wire carrying a current *I*. Above the wire is a rectangular loop that contains a resistor *R*. (a) Does the magnetic field produced by the current *I* penetrate the loop and generate a magnetic flux? (b) When is there an induced current in the loop, if the current *I* is constant or if it is decreasing in time? (c) When there is an induced magnetic field produced by the loop, does it always have a direction that is opposite to the direction of the magnetic field produced by the current I? Provide a reason for each answer.

Problem If the current I is decreasing in time, what is the direction of the induced current through the resistor R—left to right or right to left? Give your reasoning.

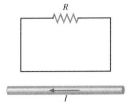

78. Concept Questions A flat coil of wire has an area A, N turns, and a resistance R. It is situated in a magnetic field such that the normal to the coil is parallel to the magnetic field. The coil is then rotated through an angle of 90°, so that the normal becomes perpendicular to the magnetic field. (a) Why is an emf induced in the coil? (b) What determines the amount of induced current in the coil? (c) How is the amount of charge that flows related to the induced current?

Problem The coil has an area of 1.5×10^{-3} m², 50 turns, and a resistance of 140 Ω. During the time it is rotating, a charge of 8.5×10^{-5} C flows in the coil. What is the magnitude of the magnetic field?

79. Concept Questions A constant current I exists in a solenoid whose inductance is L. The current is then reduced to zero in a certain amount of time. (a) If the wire from which the solenoid is made has no resistance, is there a voltage across the solenoid during the time when the current is constant? (b) If the wire from which the solenoid is made has no resistance, is there an emf across the solenoid during the time that the current is being reduced to zero? (c) Does the solenoid store electrical energy when the current is constant? If so, express this energy in terms of the current and the inductance. (d) When the current is reduced from its constant value to zero, what is the rate at which energy is removed from the solenoid? Express your answer in terms of the initial current, the inductance, and the time during which the current goes to zero.

Problem A solenoid has an inductance of $L = 3.1$ H and carries a current of $I = 15$ A. (a) If the current goes from 15 to 0 A in a time of 75 ms, what is the emf induced in the solenoid? (b) How much electrical energy is stored in the solenoid? (c) At what rate must the electrical energy be removed from the solenoid when the current is reduced to zero in 75 ms?

80. Concept Questions The rechargeable batteries for a laptop computer need a much smaller voltage than what a wall socket provides. Therefore, a transformer is plugged into the wall socket and produces the necessary voltage for charging the batteries. (a) Is the transformer a step-up or a step-down transformer? (b) Is the current that goes through the batteries greater than, equal to, or smaller than the current coming from the wall socket? (c) If the transformer has a negligible resistance, is the electric power delivered to the batteries greater than, equal to, or less than the power coming from the wall socket? In all cases, provide a reason for your answer.

Problem The batteries of a laptop computer are rated at 9.0 V, and a current of 225 mA is used to charge them. The wall socket provides a voltage of 120 V. (a) Determine the turns ratio of the transformer. (b) What is the current coming from the wall socket? (c) Find the power delivered by the wall socket and the power sent to the batter-

ies. Be sure your answers are consistent with your answers to the Concept Questions.

* **81. Concept Questions** The drawing shows a coil of copper wire that consists of two semicircles joined by straight sections of wire. In part a the coil is lying flat on a horizontal surface. The dashed line also lies in the plane of the horizontal surface. Starting from the orientation in part a, the smaller semicircle rotates at an angular frequency ω about the dashed line, until its plane becomes perpendicular to the horizontal surface, as shown in part b. A uniform magnetic field is constant in time and is directed upward, perpendicular to the horizontal surface. The field completely fills the region occupied by the coil in either part of the drawing. (a) In which part of the drawing, if either, does a greater magnetic flux pass through the coil? Account for your answer. (b) As the shape of the coil changes from that in part a of the drawing to that in part b, does an induced current flow in the coil, and, if so, in which direction does it flow? Give your reasoning. To describe the flow, imagine that you are above the coil looking down at it. (c) How is the period T of the rotational motion related to the angular frequency ω, and in terms of the period, what is the shortest time interval that elapses between parts a and b of the drawing?

Problem The magnitude of the magnetic field is 0.35 T. The resistance of the coil is 0.025 Ω, and the smaller semicircle has a radius of 0.20 m. The angular frequency at which the small semicircle rotates is 1.5 rad/s. Determine the average current, if any, induced in the coil as the coil changes shape from that in part a of the drawing to that in part b.

(a)   (b)

* **82. Concept Questions** In a television set the power needed to operate the picture tube comes from the secondary of a transformer. The primary of the transformer is connected to a wall receptacle. (a) How is the power delivered by the receptacle to the primary related to the power delivered by the secondary to the picture tube? Give your answer in the form of an equation, and explain what assumptions are implied when this equation is used. (b) How is the turns ratio of the transformer related to the currents in the primary and the secondary? (c) How is the turns ratio of the transformer related to the voltage across the primary and the voltage across the secondary? (d) Given the power used by the picture tube, the voltage across the primary, and the current in the secondary, there are two ways to determine the turns ratio of the transformer. Explain them both.

Problem The primary of the transformer is connected to a 120-V receptacle. The picture tube of a television set uses 91 W, and there is 5.5 mA of current in the secondary coil of the transformer to which the tube is connected. Find the turns ratio N_s/N_p of the transformer according to each of the methods discussed in your answer to Concept Question (d). Verify that you obtain the same answer by either method.

Chapter 23

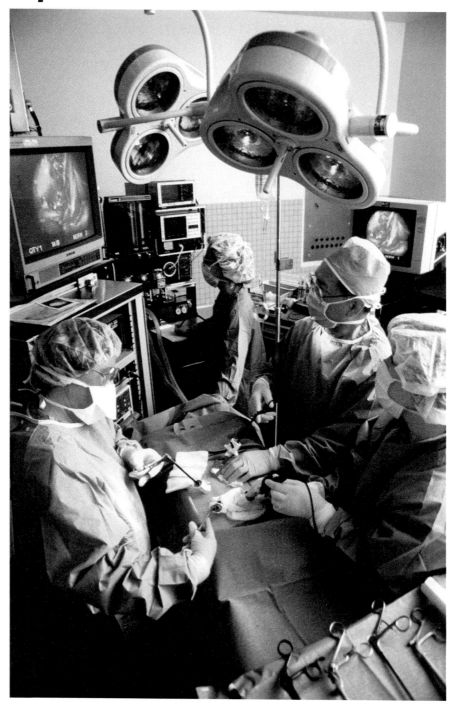

Alternating Current Circuits

All the equipment in this operating room, from the lights to the electronic monitors, use alternating current circuits, the topic of this chapter. (© David Joel/Stone/Getty Images)

23.1 Capacitors and Capacitive Reactance

Our experience with capacitors so far has been in dc circuits. As we have seen in Section 20.13, charge flows in a dc circuit only for the brief period after the battery voltage is applied across the capacitor. In other words, charge flows only while the capacitor is charging up. After the capacitor becomes fully charged, no more charge leaves the battery. However, suppose the battery connections to the fully charged capacitor were suddenly reversed. Then charge would flow again, but in the reverse direction, until the battery recharges the capacitor according to the new connections. What happens in an ac circuit is similar. The polarity of the voltage applied to the capacitor continually switches back and forth, and, in response, charges flow first one way around the circuit and then the other way. This flow of charge, surging back and forth, constitutes an alternating current. Thus, charge flows continuously in an ac circuit containing a capacitor.

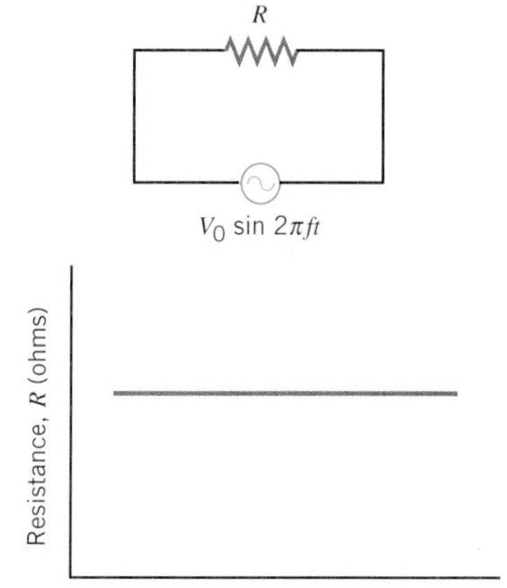

To help set the stage for the present discussion, recall that, for a purely resistive ac circuit, the rms voltage V_{rms} across the resistor is related to the rms current I_{rms} through it by $V_{rms} = I_{rms}R$ (Equation 20.14). The resistance R has the same value for any frequency of the ac voltage or current. Figure 23.1 emphasizes this fact by showing that a graph of resistance versus frequency is a horizontal straight line.

For the rms voltage across a capacitor the following expression applies, which is analogous to $V_{rms} = I_{rms}R$:

$$V_{rms} = I_{rms}X_C \tag{23.1}$$

The term X_C appears in place of the resistance R and is called the *capacitive reactance*. The capacitive reactance, like resistance, is measured in *ohms* and determines how much rms current exists in a capacitor in response to a given rms voltage across the capacitor. It is found experimentally that the capacitive reactance X_C is inversely proportional to both the frequency f and the capacitance C, according to the following equation:

$$X_C = \frac{1}{2\pi f C} \tag{23.2}$$

For a fixed value of the capacitance C, Figure 23.2 gives a plot of X_C versus frequency, according to Equation 23.2. A comparison of this drawing with Figure 23.1 reveals that a capacitor and a resistor behave differently. As the frequency becomes very large, Figure 23.2 shows that X_C approaches zero, signifying that a capacitor offers only a negligibly small opposition to the alternating current. In contrast, in the limit of zero frequency (i.e., direct current), X_C becomes infinitely large, and a capacitor provides so much opposition to the motion of charges that there is no current. Example 1 illustrates how frequency and capacitance determine the amount of current in an ac circuit.

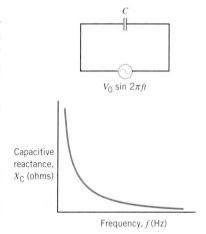

Figure 23.1 The resistance in a purely resistive circuit has the same value at all frequencies. The maximum emf of the generator is V_0.

Figure 23.2 The capacitive reactance X_C is inversely proportional to the frequency f according to $X_C = 1/(2\pi f C)$.

Example 1 A Capacitor in an AC Circuit

For the circuit in Figure 23.2, the capacitance of the capacitor is 1.50 μF, and the rms voltage of the generator is 25.0 V. What is the rms current in the circuit when the frequency of the generator is (a) 1.00×10^2 Hz and (b) 5.00×10^3 Hz?

Reasoning The current can be found from $I_{rms} = V_{rms}/X_C$, once the capacitive reactance X_C is determined. The values for the capacitive reactance will reflect the fact that the capacitor provides more opposition to the current when the frequency is smaller.

Solution

(a) At a frequency of 1.00×10^2 Hz, we find

$$X_C = \frac{1}{2\pi f C} = \frac{1}{2\pi(1.00 \times 10^2 \text{ Hz})(1.50 \times 10^{-6} \text{ F})} = 1060 \ \Omega \tag{23.2}$$

$$I_{rms} = \frac{V_{rms}}{X_C} = \frac{25.0 \text{ V}}{1060 \ \Omega} = \boxed{0.0236 \text{ A}} \tag{23.1}$$

Problem solving insight
The capacitive reactance X_C is inversely proportional to the frequency f of the voltage. If the frequency increases by a factor of 50, for example, the capacitive reactance decreases by a factor of 50.

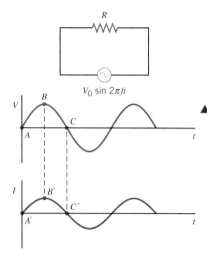

Figure 23.3 The instantaneous voltage V and current I in a purely resistive circuit are *in phase,* which means that they increase and decrease in step with one another.

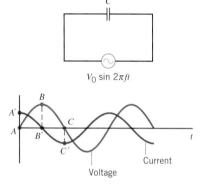

Figure 23.4 In a circuit containing only a capacitor, the instantaneous voltage and current are not in phase. Instead, the current *leads* the voltage by one-quarter of a cycle or by a phase angle of 90°.

Problem solving insight

(b) When the frequency is 5.00×10^3 Hz, the calculations are similar:

$$X_C = \frac{1}{2\pi fC} = \frac{1}{2\pi(5.00 \times 10^3 \text{ Hz})(1.50 \times 10^{-6} \text{ F})} = 21.2 \ \Omega \qquad (23.2)$$

$$I_{\text{rms}} = \frac{V_{\text{rms}}}{X_C} = \frac{25.0 \text{ V}}{21.2 \ \Omega} = \boxed{1.18 \text{ A}} \qquad (23.1)$$

We now consider the behavior of the instantaneous (not rms) voltage and current. For comparison, Figure 23.3 shows graphs of voltage and current versus time in a resistive circuit. These graphs indicate that when only resistance is present, the voltage and current are proportional to each other at every moment. For example, when the voltage increases from A to B on the graph, the current follows along in step, increasing from A' to B' during the same time interval. Likewise, when the voltage decreases from B to C, the current decreases from B' to C'. For this reason, the current in a resistance R is said to be *in phase* with the voltage across the resistance.

For a capacitor, this in-phase relation between instantaneous voltage and current does *not* exist. Figure 23.4 shows graphs of the ac voltage and current versus time for a circuit that contains only a capacitor. As the voltage increases from A to B, the charge on the capacitor increases and reaches its full value at B. The current, however, is not the same thing as the charge. The current is the rate of flow of charge and has a maximum positive value at the start of the charging process at A'. It is a maximum because there is no charge on the capacitor at the start and hence no capacitor voltage to oppose the generator voltage. When the capacitor is fully charged at B, the capacitor voltage has a magnitude equal to that of the generator and completely opposes the generator voltage. The result is that the current decreases to zero at B'. While the capacitor voltage decreases from B to C, the charges flow out of the capacitor in a direction opposite to that of the charging current, as indicated by the negative current from B' to C'. Thus, voltage and current are not in phase but are, in fact, one-quarter wave cycle out of step, or out of phase. More specifically, assuming that the voltage fluctuates as $V_0 \sin (2\pi ft)$, the current varies as $I_0 \sin (2\pi ft + \pi/2) = I_0 \cos (2\pi ft)$. Since $\pi/2$ radians correspond to 90° and since the current reaches its maximum value *before* the voltage does, it is said that the current in a capacitor *leads* the voltage across the capacitor by a phase angle of 90°.

The fact that the current and voltage for a capacitor are 90° out of phase has an important consequence from the point of view of electric power, since power is the product of current and voltage. For the time interval between points A and B (or A' and B') in Figure 23.4, both current and voltage are positive. Therefore, the instantaneous power is also positive, meaning that the generator is delivering energy to the capacitor. However, during the period between B and C (or B' and C'), the current is negative while the voltage remains positive, and the power, being the product of the two, is negative. During this period, the capacitor is returning energy to the generator. Thus, the power alternates between positive and negative values for equal periods of time. In other words, the capacitor alternately absorbs and releases energy. Consequently, ***the average power (and, hence, the average energy) used by a capacitor in an ac circuit is zero.***

It will prove useful later on to use a model for the voltage and current when analyzing ac circuits. In this model, voltage and current are represented by rotating arrows, often called ***phasors,*** whose lengths correspond to the maximum voltage V_0 and maximum current I_0, as Figure 23.5 indicates. These phasors rotate counterclockwise at a frequency f. For a resistor, the phasors are co-linear as they rotate (see part *a* of the drawing) because voltage and current are in phase. For a capacitor (see part *b*), the phasors remain perpendicular while rotating because the phase angle between the current and the voltage is 90°. Since current leads voltage for a capacitor, the current phasor is ahead of the voltage phasor in the direction of rotation.

Note from the two circuit drawings in Figure 23.5 that the instantaneous voltage across the resistor or the capacitor is $V_0 \sin (2\pi ft)$. We can find this instantaneous value of the voltage directly from the phasor diagram. Imagine that the voltage phasor V_0 in this diagram represents the hypotenuse of a right triangle. Then, the *vertical component* of the phasor would be $V_0 \sin (2\pi ft)$. In a similar manner, the instantaneous current can be found as the vertical component of the current phasor.

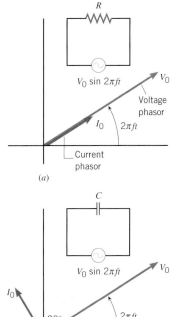

23.2 Inductors and Inductive Reactance

As Section 22.8 discusses, an inductor is usually a coil of wire, and the basis of its operation is Faraday's law of electromagnetic induction. According to Faraday's law, an inductor develops a voltage that opposes a change in the current. This voltage V is given by $V = -L(\Delta I/\Delta t)$ (see Equation 22.9*), where $\Delta I/\Delta t$ is the rate at which the current changes and L is the inductance of the inductor. In an ac circuit the current is always changing, and Faraday's law can be used to show that the rms voltage across an inductor is

$$V_{rms} = I_{rms}X_L \qquad (23.3)$$

Equation 23.3 is analogous to $V_{rms} = I_{rms}R$, with the term X_L appearing in place of the resistance R and being called the ***inductive reactance.*** The inductive reactance is measured in ohms and determines how much rms current exists in an inductor for a given rms voltage across the inductor. It is found experimentally that the inductive reactance X_L is directly proportional to the frequency f and the inductance L, as indicated by the following equation:

$$X_L = 2\pi fL \qquad (23.4)$$

This relation indicates that the larger the inductance, the larger is the inductive reactance. Note that the inductive reactance is directly proportional to the frequency ($X_L \propto f$), whereas the capacitive reactance is inversely proportional to the frequency ($X_C \propto 1/f$).

Figure 23.6 shows a graph of the inductive reactance versus frequency for a fixed value of the inductance, according to Equation 23.4. As the frequency becomes very large, X_L also becomes very large. In such a situation, an inductor provides a large opposition to the alternating current. In the limit of zero frequency (i.e., direct current), X_L becomes zero, indicating that an inductor does not oppose direct current at all. The next example demonstrates the effect of inductive reactance on the current in an ac circuit.

Example 2   An Inductor in an AC Circuit

The circuit in Figure 23.6 contains a 3.60-mH inductor. The rms voltage of the generator is 25.0 V. Find the rms current in the circuit when the generator frequency is (a) 1.00×10^2 Hz and (b) 5.00×10^3 Hz.

Reasoning The current can be calculated from $I_{rms} = V_{rms}/X_L$, provided the inductive reactance is obtained first. The inductor offers more opposition to the changing current when the frequency is larger, and the values for the inductive reactance will reflect this fact.

Solution

(a) At a frequency of 1.00×10^2 Hz, we find

$$X_L = 2\pi fL = 2\pi(1.00 \times 10^2 \text{ Hz})(3.60 \times 10^{-3}\text{H}) = 2.26 \ \Omega \qquad (23.4)$$

$$I_{rms} = \frac{V_{rms}}{X_L} = \frac{25.0 \text{ V}}{2.26 \ \Omega} = \boxed{11.1 \text{ A}} \qquad (23.3)$$

Figure 23.5 These rotating-arrow models represent the voltage and the current in ac circuits that contain (*a*) only a resistor and (*b*) only a capacitor.

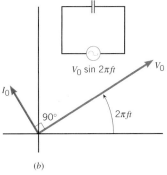

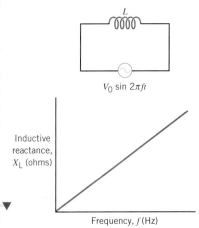

Figure 23.6 In an ac circuit the inductive reactance X_L is directly proportional to the frequency f, according to $X_L = 2\pi fL$.

Problem solving insight
The inductive reactance X_L is directly proportional to the frequency f of the voltage. If the frequency increases by a factor of 50, for example, the inductive reactance also increases by a factor of 50.

* When an inductor is used in a circuit, the notation is simplified if we designate the potential difference across the inductor as the voltage V, rather than the emf $\mathscr{E}$.

(b) The calculation is similar when the frequency is 5.00×10^3 Hz:

$$X_L = 2\pi f L = 2\pi(5.00 \times 10^3 \text{ Hz})(3.60 \times 10^{-3} \text{ H}) = 113 \ \Omega \qquad (23.4)$$

$$I_{rms} = \frac{V_{rms}}{X_L} = \frac{25.0 \text{ V}}{113 \ \Omega} = \boxed{0.221 \text{ A}} \qquad (23.3)$$

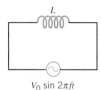

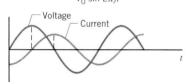

Figure 23.7 The instantaneous voltage and current in a circuit containing only an inductor are not in phase. The current *lags behind* the voltage by one-quarter of a cycle or by a phase angle of 90°.

Concept Simulation 23.1

This simulation is an introduction to ac circuits, and allows you to connect a resistor, a capacitor, or an inductor to an ac generator. You can vary the resistance, the capacitance, the inductance, and the frequency. For each situation, the simulation calculates the maximum current in the circuit and displays plots of the voltage and current as a function of time. For the capacitor and inductor, notice especially how the frequency affects the maximum current, as well as the phase between the voltage and the current.

Go to **www.wiley.com/college/cutnell**

By virtue of its inductive reactance, an inductor affects the amount of current in an ac circuit. The inductor also influences the current in another way, as Figure 23.7 shows. This figure displays graphs of voltage and current versus time for a circuit containing only an inductor. At a maximum or minimum on the current graph, the current does not change much with time, so the voltage generated by the inductor to oppose a change in the current is zero. At the points on the current graph where the current is zero, the graph is at its steepest, and the current has the largest rate of increase or decrease. Correspondingly, the voltage generated by the inductor to oppose a change in the current has the largest positive or negative value. Thus, current and voltage are not in phase but are one-quarter of a wave cycle out of phase. If the voltage varies as $V_0 \sin (2\pi f t)$, the current fluctuates as $I_0 \sin (2\pi f t - \pi/2) = -I_0 \cos (2\pi f t)$. The current reaches its maximum *after* the voltage does, and it is said that the current *lags behind* the voltage by a phase angle of 90° ($\pi/2$ radians). In a purely capacitive circuit, in contrast, the current leads the voltage by 90° (see Figure 23.4).

In an inductor the 90° phase difference between current and voltage leads to the same result for average power that it does in a capacitor. An inductor alternately absorbs and releases energy for equal periods of time, so ***the average power (and, hence, the average energy) used by an inductor in an ac circuit is zero.***

Problem solving insight

As an alternative to the graphs in Figure 23.7, Figure 23.8 uses phasors to describe the instantaneous voltage and current in a circuit containing only an inductor. The voltage and current phasors remain perpendicular as they rotate, because there is a 90° phase angle between them. The current phasor lags behind the voltage phasor, relative to the direction of rotation, in contrast to the equivalent picture in Figure 23.5b for a capacitor. Once again, the instantaneous values are given by the vertical components of the phasors.

✔ Check Your Understanding 2

The drawing shows three ac circuits: one contains a resistor, one a capacitor, and one an inductor. The frequency of each ac generator is reduced to one-half its initial value. Which circuit experiences (a) the greatest increase in current and (b) the least change in current? *(The answers are given at the end of the book.)*

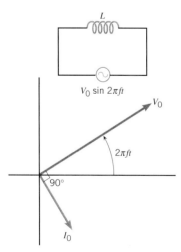

Figure 23.8 This phasor model represents the voltage and current in a circuit that contains only an inductor.

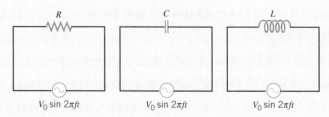

Background: Consider how the inductive reactance X_L, the capacitive reactance X_C, and the resistance R change with frequency. In addition, think about how changes in X_L, X_C, and R alter the current in a circuit.

For similar questions (including calculational counterparts), consult Self-Assessment Test 23.1, which is described next.

Self-Assessment Test 23.1

Test your understanding of the material in Sections 23.1 and 23.2:

• Capacitors and Capacitive Reactance   • Inductors and Inductive Reactance

Go to **www.wiley.com/college/cutnell**

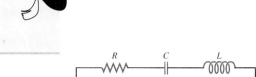

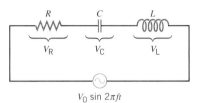

23.3 *Circuits Containing Resistance, Capacitance, and Inductance*

Capacitors and inductors can be combined along with resistors in a single circuit. The simplest combination contains a resistor, a capacitor, and an inductor in series, as Figure 23.9 shows. In a series RCL circuit the total opposition to the flow of charge is called the *impedance* of the circuit and comes partially from (1) the resistance R, (2) the capacitive reactance X_C, and (3) the inductive reactance X_L. It is tempting to follow the analogy of a series combination of resistors and calculate the impedance by simply adding together R, X_C, and X_L. However, such a procedure is not correct. Instead, the phasors shown in Figure 23.10 must be used. The lengths of the voltage phasors in this drawing represent the maximum voltages V_R, V_C, and V_L across the resistor, the capacitor, and the inductor, respectively. The current is the same for each device, since the circuit is wired in series. The length of the current phasor represents the maximum current I_0. Notice that the drawing shows the current phasor to be (1) in phase with the voltage phasor for the resistor, (2) ahead of the voltage phasor for the capacitor by 90°, and (3) behind the voltage phasor for the inductor by 90°. These three facts are consistent with our earlier discussion in Sections 23.1 and 23.2.

The basis for dealing with the voltage phasors in Figure 23.10 is Kirchhoff's loop rule. In an ac circuit this rule applies to the *instantaneous* voltages across each circuit component and the generator. Therefore, it is necessary to take into account the fact that these voltages do not have the same phase; that is, the phasors V_R, V_C, and V_L point in different directions in the drawing. Kirchhoff's loop rule indicates that the phasors add together to give the total voltage V_0 that is supplied to the circuit by the generator. The addition, however, must be like a vector addition, to take into account the different directions. Since V_L and V_C point in opposite directions, they combine to give a resultant phasor of $V_L - V_C$, as Figure 23.11 shows. In this drawing the resultant $V_L - V_C$ is perpendicular to V_R and may be combined with it to give the total voltage V_0. Using the Pythagorean theorem, we find

$$V_0^2 = V_R^2 + (V_L - V_C)^2$$

In this equation each of the symbols stands for a maximum voltage and when divided by $\sqrt{2}$ gives the corresponding rms voltage. Therefore, it is possible to divide both sides of the equation by $(\sqrt{2})^2$ and obtain a result for $V_{rms} = V_0/\sqrt{2}$. This result has exactly the same form as that above, but involves the rms voltages $V_{R\text{-rms}}$, $V_{C\text{-rms}}$, and $V_{L\text{-rms}}$. However, to avoid such awkward symbols, we simply interpret V_R, V_C, and V_L as rms quantities in the following expression:

$$V_{rms}^2 = V_R^2 + (V_L - V_C)^2 \tag{23.5}$$

The last step in determining the impedance of the circuit is to remember that $V_R = I_{rms}R$, $V_C = I_{rms}X_C$, and $V_L = I_{rms}X_L$. With these substitutions Equation 23.5 can be written as

$$V_{rms} = I_{rms}\sqrt{R^2 + (X_L - X_C)^2}$$

Therefore, for the entire RCL circuit, it follows that

$$V_{rms} = I_{rms}Z \tag{23.6}$$

where the impedance Z of the circuit is defined as

Series RCL
combination
$$Z = \sqrt{R^2 + (X_L - X_C)^2} \tag{23.7}$$

Figure 23.9 A series RCL circuit contains a resistor, a capacitor, and an inductor.

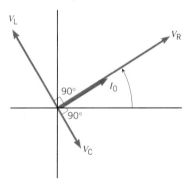

Figure 23.10 The three voltage phasors (V_R, V_C, and V_L) and the current phasor (I_0) for a series RCL circuit.

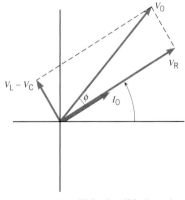

Figure 23.11 This simplified version of Figure 23.10 results when the phasors V_L and V_C, which point in opposite directions, are combined to give a resultant of $V_L - V_C$.

The impedance of the circuit, like R, X_C, and X_L, is measured in ohms. In Equation 23.7, $X_L = 2\pi fL$ and $X_C = 1/(2\pi fC)$.

The phase angle between the current in and the voltage across a series RCL combination is the angle ϕ between the current phasor I_0 and the voltage phasor V_0 in Figure 23.11. According to the drawing, the tangent of this angle is

$$\tan\phi = \frac{V_L - V_C}{V_R} = \frac{I_{rms}X_L - I_{rms}X_C}{I_{rms}R}$$

Series RCL combination

$$\tan\phi = \frac{X_L - X_C}{R} \qquad (23.8)$$

The phase angle ϕ is important because it has a major effect on the average power $\overline{P}$ dissipated by the circuit. Remember that, on the average, only the resistance consumes power; that is, $\overline{P} = I_{rms}^2 R$ (Equation 20.15b). According to Figure 23.11, $\cos\phi = V_R/V_0 = (I_{rms}R)/(I_{rms}Z) = R/Z$, so that $R = Z\cos\phi$. Therefore,

$$\overline{P} = I_{rms}^2 Z\cos\phi = I_{rms}(I_{rms}Z)\cos\phi$$
$$\overline{P} = I_{rms}V_{rms}\cos\phi \qquad (23.9)$$

where V_{rms} is the rms voltage of the generator. The term $\cos\phi$ is called the **power factor** of the circuit. As a check on the validity of Equation 23.9, note that if no resistance is present, $R = 0\ \Omega$, and $\cos\phi = R/Z = 0$. Consequently, $\overline{P} = I_{rms}V_{rms}\cos\phi = 0$, a result that is expected since, on the average, neither a capacitor nor an inductor consumes energy. Conversely, if only resistance is present, $Z = \sqrt{R^2 + (X_L - X_C)^2} = R$, and $\cos\phi = R/Z = 1$. In this case, $\overline{P} = I_{rms}V_{rms}\cos\phi = I_{rms}V_{rms}$, which is the expression for the average power dissipated in a resistor. Example 3 deals with the current, voltages, and power for a series RCL circuit.

Example 3 Current, Voltages, and Power in a Series RCL Circuit

A series RCL circuit contains a 148-Ω resistor, a 1.50-μF capacitor, and a 35.7-mH inductor. The generator has a frequency of 512 Hz and an rms voltage of 35.0 V. Obtain (a) the rms voltage across each circuit element and (b) the average electric power consumed by the circuit.

Reasoning The rms voltages across each circuit element can be determined from $V_R = I_{rms}R$, $V_C = I_{rms}X_C$, and $V_L = I_{rms}X_L$, as soon as the rms current and the reactances X_C and X_L are known. Since the rms current can be found from $I_{rms} = V_{rms}/Z$ and $V_{rms} = 35.0$ V, the first step in the solution is to find the impedance Z from the individual reactances. The average power consumed is given by $\overline{P} = I_{rms}V_{rms}\cos\phi$, where the phase angle ϕ can be obtained from $\tan\phi = (X_L - X_C)/R$.

Solution

(a) The individual reactances are

$$X_C = \frac{1}{2\pi fC} = \frac{1}{2\pi(512\ \text{Hz})(1.50 \times 10^{-6}\ \text{F})} = 207\ \Omega \qquad (23.2)$$
$$X_L = 2\pi fL = 2\pi(512\ \text{Hz})(35.7 \times 10^{-3}\ \text{H}) = 115\ \Omega \qquad (23.4)$$

The impedance of the circuit is

$$Z = \sqrt{R^2 + (X_L - X_C)^2} = \sqrt{(148\ \Omega)^2 + (115\ \Omega - 207\ \Omega)^2} = 174\ \Omega \qquad (23.7)$$

The rms current in each circuit element is

$$I_{rms} = \frac{V_{rms}}{Z} = \frac{35.0\ \text{V}}{174\ \Omega} = 0.201\ \text{A} \qquad (23.6)$$

The rms voltage across each circuit element now follows immediately:

Problem solving insight
In a series RCL circuit, the rms voltages across the resistor, capacitor, and inductor do not add up to equal the rms voltage across the generator.

$$V_R = I_{rms}R = (0.201\ \text{A})(148\ \Omega) = \boxed{29.7\ \text{V}} \qquad (20.14)$$
$$V_C = I_{rms}X_C = (0.201\ \text{A})(207\ \Omega) = \boxed{41.6\ \text{V}} \qquad (23.1)$$
$$V_L = I_{rms}X_L = (0.201\ \text{A})(115\ \Omega) = \boxed{23.1\ \text{V}} \qquad (23.3)$$

Observe that these three rms voltages do not add up to give the generator's rms voltage, which is 35.0 V. Instead, the rms voltages satisfy Equation 23.5. It is the sum of the *instantaneous* voltages across R, C, and L, rather than the sum of the rms voltages, that equals the generator's *instantaneous* voltage, according to Kirchhoff's loop rule.

(b) The average power consumed by the circuit is $\overline{P} = I_{rms}V_{rms} \cos \phi$. Therefore, a value for the phase angle ϕ is needed and can be obtained from $\tan \phi = (X_L - X_C)/R$ as follows:

$$\phi = \tan^{-1}\left(\frac{X_L - X_C}{R}\right) = \tan^{-1}\left(\frac{115\ \Omega - 207\ \Omega}{148\ \Omega}\right) = -32° \qquad (23.8)$$

The phase angle is negative since the circuit is more capacitive than inductive (X_C is greater than X_L), and the current leads the voltage. The average power consumed is

$$\overline{P} = I_{rms}V_{rms} \cos \phi = (0.201\ \text{A})(35.0\ \text{V}) \cos(-32°) = \boxed{6.0\ \text{W}} \qquad (23.9)$$

Concept Simulation 23.2

Use this simulation to investigate the properties of a series RCL circuit. You can adjust the values of the resistance R, the capacitance C, the inductance L, and the frequency f to see how they affect the impedance and the current. For each setting of R, C, L, and f, the simulation calculates the capacitive and inductive reactances, X_C and X_L, the impedance Z of the circuit, the phase angle ϕ between the current and the voltage, and the maximum current. In addition, plots are drawn of the instantaneous voltage and current versus time.

Go to **www.wiley.com/college/cutnell**

In addition to the series RCL circuit, there are many different ways to connect resistors, capacitors, and inductors. In analyzing these additional possibilities, it helps to keep in mind the behavior of capacitors and inductors at the extreme limits of the frequency. When the frequency approaches zero (i.e., dc conditions), the reactance of a capacitor becomes so large that no charge can flow through the capacitor. It is as if the capacitor were cut out of the circuit, leaving an open gap in the connecting wire. In the limit of zero frequency the reactance of an inductor is vanishingly small. The inductor offers no opposition to a dc current. It is as if the inductor were replaced with a wire of zero resistance. In the limit of very large frequency, the behaviors of a capacitor and an inductor are reversed. The capacitor has a very small reactance and offers little opposition to the current, as if it were replaced by a wire with zero resistance. In contrast, the inductor has a very large reactance when the frequency is very large. The inductor offers so much opposition to the current that it might as well be cut out of the circuit, leaving an open gap in the connecting wire. Conceptual Example 4 illustrates how to gain insight into more complicated circuits using the limiting behaviors of capacitors and inductors.

Conceptual Example 4
The Limiting Behavior of Capacitors and Inductors

Figure 23.12*a* shows two circuits. The rms voltage of the generator is the same in each case. The values of the resistance R, the capacitance C, and the inductance L in these circuits are the same. The frequency of the ac generator is very near zero. In which circuit does the generator supply more rms current?

Reasoning and Solution According to Equation 23.6, the rms current is given by $I_{rms} = V_{rms}/Z$. However, the impedance Z is not given by Equation 23.7, since the circuits in Figure 23.12*a* are not series RCL circuits. Since V_{rms} is the same in each case, the greater current is delivered to the circuit with the smaller impedance Z. In the limit of very small frequency, the capacitors behave as if they were cut out of the circuit, leaving gaps in the connecting wires. In this limit, however, the inductors behave as if they were replaced by wires with zero resistance. Figure 23.12*b* shows the circuits as they would appear according to these changes. Circuit I behaves as if it contained only two identical resistors in series, with a total impedance of $Z = 2R$. In contrast, circuit II behaves as if it contained two identical resistors in parallel, in which case the total impedance is given by $1/Z = 1/R + 1/R$, or $Z = R/2$. At a frequency very near zero, then, ***circuit II has the smaller impedance and its generator supplies the greater rms current.***

Related Homework: *Conceptual Questions 6, 7; Problems 22, 38, 44*

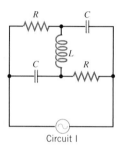

Circuit I

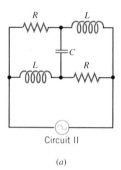

Circuit II

(a)

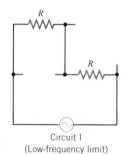

Circuit I
(Low-frequency limit)

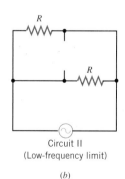

Circuit II
(Low-frequency limit)

(b)

Figure 23.12 (*a*) These circuits are discussed in the limit of very small or low frequency in Conceptual Example 4. (*b*) For a frequency very near zero, the circuits in part *a* behave as if they were as shown here.

✔ Check Your Understanding 3

A long wire of finite resistance is connected to an ac generator. This wire is then wound into a coil of many loops and reconnected to the generator. Is the current in the circuit with the coil greater than, less than, or the same as the current in the circuit with the uncoiled wire? *(The answer is given at the end of the book.)*

Background: The current in a circuit depends on its impedance. The impedance depends not only on the resistance that is present, but also on the capacitance and inductance.

For similar questions (including calculational counterparts), consult Self-Assessment Test 23.2, which is described at the end of Section 23.4.

23.4 *Resonance in Electric Circuits*

▶ CONCEPTS AT A GLANCE   The behavior of the current and voltage in a series RCL circuit can give rise to a condition of ***resonance.*** Resonance occurs when the frequency of a vibrating force exactly matches a natural (resonant) frequency of the object to which the force is applied. When resonance occurs, the force can transmit a large amount of energy to the object, leading to a large-amplitude vibratory motion. As the Concepts-at-a-Glance chart in Figure 23.13 illustrates, we have already encountered several examples of resonance. First, resonance can occur when a vibrating force is applied to an object of mass m that is attached to a spring whose spring constant is k (Section 10.6). In this case there is one natural frequency f_0, whose value is $f_0 = [1/(2\pi)]\sqrt{k/m}$. Second, resonance occurs when standing waves are set up on a string (Section 17.5) or in a tube of air (Section 17.6). As the right-hand side of the concept chart shows, the string and tube of air have many natural frequencies, one for each allowed standing wave. As we will now see, a condition of resonance can also be established in a series RCL circuit. In this case there is only one natural frequency, and the vibrating force is provided by the oscillating electric field that is related to the voltage of the generator. ◀

Figure 23.14 helps us to understand why there is a resonant frequency for an ac circuit. This drawing presents an analogy between the electrical case (ignoring resistance) and the mechanical case of an object attached to a horizontal spring (ignoring friction). Part *a* shows a fully stretched spring that has just been released, with the initial speed v of

Figure 23.13 CONCEPTS AT A GLANCE The phenomenon of resonance can occur when an object vibrates on a spring or when standing waves are established on a string or in a tube of air. Resonance can also occur in a series RCL circuit. The antenna in this photograph is attached to the top of the John Hancock building in Chicago. TV and radio stations use this antenna to broadcast at assigned frequencies, each of which corresponds to a frequency of a resonant circuit. (© Vito Palmisano/Stone/Getty Images)

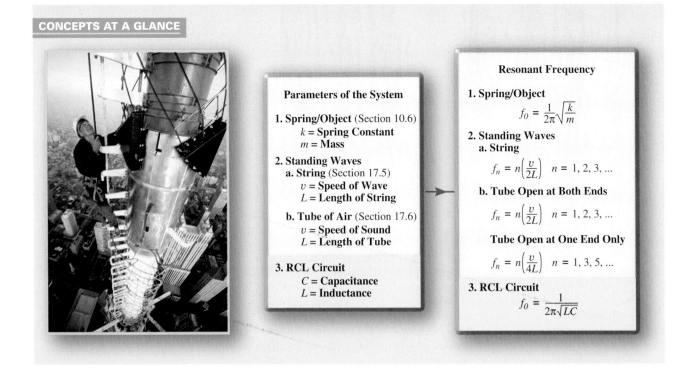

CONCEPTS AT A GLANCE

Parameters of the System

1. **Spring/Object** (Section 10.6)
   k = **Spring Constant**
   m = **Mass**

2. **Standing Waves**
   a. **String** (Section 17.5)
      v = **Speed of Wave**
      L = **Length of String**

   b. **Tube of Air** (Section 17.6)
      v = **Speed of Sound**
      L = **Length of Tube**

3. **RCL Circuit**
   C = **Capacitance**
   L = **Inductance**

Resonant Frequency

1. **Spring/Object**
   $$f_0 = \frac{1}{2\pi}\sqrt{\frac{k}{m}}$$

2. **Standing Waves**
   a. **String**
      $$f_n = n\left(\frac{v}{2L}\right) \quad n = 1, 2, 3, \ldots$$

   b. **Tube Open at Both Ends**
      $$f_n = n\left(\frac{v}{2L}\right) \quad n = 1, 2, 3, \ldots$$

   Tube Open at One End Only
   $$f_n = n\left(\frac{v}{4L}\right) \quad n = 1, 3, 5, \ldots$$

3. **RCL Circuit**
   $$f_0 = \frac{1}{2\pi\sqrt{LC}}$$

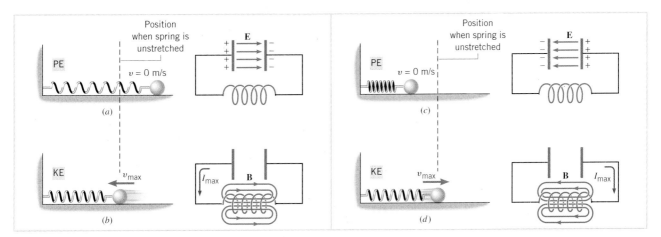

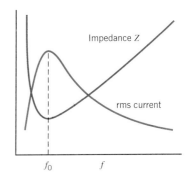

Figure 23.14 The oscillation of an object on a spring is analogous to the oscillation of the electric and magnetic fields that occur, respectively, in a capacitor and in an inductor.

the object being zero. All the energy is stored in the form of elastic potential energy. When the object begins to move, it gradually loses potential energy and picks up kinetic energy. In part b, the object moves with maximum kinetic energy through the position where the spring is unstretched (zero potential energy). Because of its inertia, the moving object coasts through this position and eventually comes to a halt in part c when the spring is fully compressed and all kinetic energy has been converted back into elastic potential energy. Part d of the picture is like part b, except the direction of motion is reversed. The resonant frequency f_0 of the object on the spring is the natural frequency at which the object vibrates and is given as $f_0 = [1/(2\pi)]\sqrt{k/m}$ according to Equations 10.6 and 10.11. In this expression, m is the mass of the object, and k is the spring constant.

In the electrical case, Figure 23.14a begins with a fully charged capacitor that has just been connected to an inductor. At this instant the energy is stored in the electric field between the capacitor plates. As the capacitor discharges, the electric field **E** between the plates decreases, while a magnetic field **B** builds up around the inductor because of the increasing current in the circuit. The maximum current and the maximum magnetic field exist at the instant when the capacitor is completely discharged, as in part b of the figure. Energy is now stored entirely in the magnetic field of the inductor. The voltage induced in the inductor keeps the charges flowing until the capacitor again becomes fully charged, but now with reversed polarity, as in part c. Once again, the energy is stored in the electric field between the plates, and no energy resides in the magnetic field of the inductor. Part d of the cycle repeats part b, but with reversed directions of current and magnetic field. Thus, an ac circuit can have a resonant frequency because there is a natural tendency for energy to shuttle back and forth between the electric field of the capacitor and the magnetic field of the inductor.

To determine the resonant frequency at which energy shuttles back and forth between the capacitor and the inductor, we note that the current in a series RCL circuit is $I_{rms} = V_{rms}/Z$ (Equation 23.6). In this expression Z is the impedance of the circuit and is given by $Z = \sqrt{R^2 + (X_L - X_C)^2}$ (Equation 23.7). As Figure 23.15 illustrates, the rms current is a maximum when the impedance is a minimum, assuming a given generator voltage. The minimum impedance of $Z = R$ occurs when the frequency is f_0, such that $X_L = X_C$ or $2\pi f_0 L = 1/(2\pi f_0 C)$. This result can be solved for f_0, which is the resonant frequency:

$$f_0 = \frac{1}{2\pi\sqrt{LC}} \tag{23.10}$$

The resonant frequency is determined by the inductance and the capacitance, but not the resistance.

The effect of resistance on electrical resonance is to make the "sharpness" of the circuit response less pronounced, as Figure 23.16 indicates. When the resistance is small, the current-versus-frequency graph falls off suddenly on either side of the maximum current. When the resistance is large, the falloff is more gradual, and there is less current at the maximum.

Figure 23.15 In a series RCL circuit the impedance is a minimum, and the current is a maximum, when the frequency f equals the resonant frequency f_0 of the circuit.

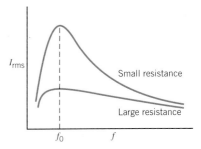

Figure 23.16 The effect of resistance on the current in a series RCL circuit.

The following example deals with one application of resonance in electric circuits. In this example the focus is on the oscillation of energy between a capacitor and an inductor. Once a capacitor/inductor combination is energized, the energy will oscillate indefinitely as in Figure 23.14, provided there is some provision to replace any dissipative losses that occur because of resistance. Circuits that include this type of provision are called oscillator circuits.

Example 5   A Heterodyne Metal Detector

The physics of a heterodyne metal detector.

Figure 23.17 shows a heterodyne metal detector being used. As Figure 23.18 illustrates, this device utilizes two capacitor/inductor oscillator circuits, A and B. Each circuit produces its own resonant frequency, $f_{0A} = 1/(2\pi\sqrt{L_A C})$ and $f_{0B} = 1/(2\pi\sqrt{L_B C})$. Any difference between these two frequencies is detected through earphones as a beat frequency $f_{0B} - f_{0A}$, similar to the beat frequency that two musical tones produce. In the absence of any nearby metal object, the inductances L_A and L_B are the same, and f_{0A} and f_{0B} are identical. There is no beat frequency. When inductor B (the search coil) comes near a piece of metal, the inductance L_B decreases, the corresponding oscillator frequency f_{0B} increases, and a beat frequency is heard. Suppose that initially each inductor is adjusted so $L_B = L_A$, and each oscillator has a resonant frequency of 855.5 kHz. Assuming that the inductance of search coil B decreases by 1.00% due to a nearby piece of metal, determine the beat frequency heard through the earphones.

Reasoning To find the beat frequency $f_{0B} - f_{0A}$, we need to determine the amount by which the resonant frequency f_{0B} changes because of a 1.00% decrease in the inductance L_B.

Solution We begin by obtaining the ratio of f_{0B} to f_{0A}:

$$\frac{f_{0B}}{f_{0A}} = \frac{\dfrac{1}{2\pi\sqrt{L_B C}}}{\dfrac{1}{2\pi\sqrt{L_A C}}} = \sqrt{\frac{L_A}{L_B}}$$

Due to the 1.00% decrease in the inductance of coil B, $L_B = 0.9900 L_A$, so that

$$\frac{f_{0B}}{f_{0A}} = \sqrt{\frac{L_A}{0.9900 L_A}} = 1.005$$

Therefore, the new value for f_{0B} is $f_{0B} = 1.005 f_{0A} = 1.005 \times (855.5 \text{ kHz}) = 859.8 \text{ kHz}$. As a result, the detected beat frequency is

$$f_{0B} - f_{0A} = 859.8 \text{ kHz} - 855.5 \text{ kHz} = \boxed{4.3 \text{ kHz}}$$

Figure 23.17 A heterodyne metal detector can be used to locate buried metal objects, although in this case the objects are just rusty cans. (© American Images, Inc./Taxi/Getty Images)

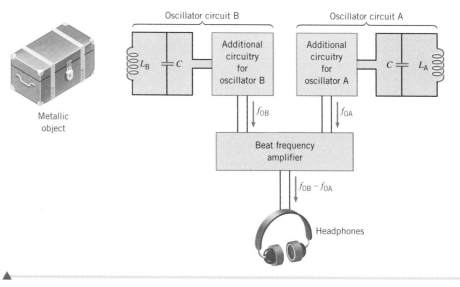

Figure 23.18 A heterodyne metal detector uses two electrical oscillators, A and B, in its operation. When the resonant frequency of oscillator B is changed due to the proximity of a metallic object, a beat frequency, whose value is $f_{0B} - f_{0A}$, is heard in the headphones.

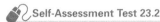

Check Your Understanding 4

The resistance in a series RCL circuit is doubled. (a) Does the resonant frequency increase, decrease, or remain the same? (b) Does the maximum current in the circuit increase, decrease, or remain the same? *(The answers are given at the end of the book.)*

Background: Answering this question requires a knowledge of which circuit elements (resistor, capacitor, inductor) determine the resonant frequency and which control the maximum current in the circuit.

For similar questions (including calculational counterparts), consult Self-Assessment Test 23.2, which is described next.

Self-Assessment Test 23.2

Test your understanding of the material in Sections 23.3 and 23.4:

• Circuits Containing Resistance, Capacitance, and Inductance
• Resonance in Electric Circuits

Go to **www.wiley.com/college/cutnell**

23.5 *Semiconductor Devices*

Semiconductor devices such as diodes and transistors are widely used in modern electronics, and Figure 23.19 illustrates one application. The drawing shows an audio system in which small ac voltages (originating in a compact disc player, an FM tuner, or a cassette deck) are amplified so they can drive the speaker(s). The electric circuits that accomplish the amplification do so with the aid of a dc voltage provided by the power supply. In portable units the power supply is simply a battery. In nonportable units, however, the power supply is a separate electric circuit containing diodes, along with other elements. As we will see, the diodes convert the 60-Hz ac voltage present at a wall outlet into the dc voltage needed by the amplifier, which, in turn, performs its job of amplification with the aid of transistors.

n-TYPE AND *p*-TYPE SEMICONDUCTORS

The materials used in diodes and transistors are semiconductors, such as silicon and germanium. However, they are not pure materials because small amounts of "impurity" atoms (about one part in a million) have been added to them to change their conductive properties. For instance, Figure 23.20*a* shows an array of atoms that symbolizes the crystal structure in pure silicon. Each silicon atom has four outer-shell* electrons, and each

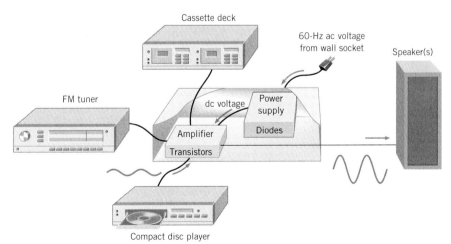

Figure 23.19 In a typical audio system, diodes are used in the power supply to create a dc voltage from the ac voltage present at the wall socket. This dc voltage is necessary so the transistors in the amplifier can perform their task of enlarging the small ac voltages originating in the compact disc player, etc.

* Section 30.6 discusses the electronic structure of the atom in terms of "shells."

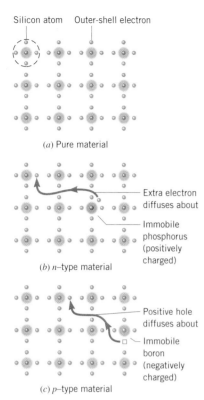

(a) Pure material

Extra electron diffuses about

Immobile phosphorus (positively charged)

(b) n–type material

Positive hole diffuses about

Immobile boron (negatively charged)

(c) p–type material

Figure 23.20 A silicon crystal that is (a) undoped, or pure, (b) doped with phosphorus to produce an n-type material, and (c) doped with boron to produce a p-type material.

electron participates with electrons from neighboring atoms in forming the bonds that hold the crystal together. Since they participate in forming bonds, these electrons generally do not move throughout the crystal. Consequently, pure silicon and germanium are not good conductors of electricity. It is possible, however, to increase their conductivities by adding tiny amounts of impurity atoms, such as phosphorus or arsenic, whose atoms have five outer-shell electrons. For example, when a phosphorus atom replaces a silicon atom in the crystal, only four of the five outer-shell electrons of phosphorus fit into the crystal structure. The extra fifth electron does not fit in and is relatively free to diffuse throughout the crystal, as part b of the drawing suggests. A semiconductor containing small amounts of phosphorus can, therefore, be envisioned as containing immobile, positively charged phosphorus atoms and a pool of electrons that are free to wander throughout the material. These mobile electrons allow the semiconductor to conduct electricity.

The process of adding impurity atoms is called *doping*. A semiconductor doped with an impurity that contributes mobile electrons is called an **n-type semiconductor,** since the mobile charge carriers have a **n**egative charge. Note that an n-type semiconductor is overall electrically neutral, since it contains equal numbers of positive and negative charges.

It is also possible to dope a silicon crystal with an impurity whose atoms have only three outer-shell electrons (e.g., boron or gallium). Because of the missing fourth electron, there is a "hole" in the lattice structure at the boron atom, as Figure 23.20c illustrates. An electron from a neighboring silicon atom can move into this hole, in which event the region around the boron atom, having acquired the electron, becomes negatively charged. Of course, when a nearby electron does move, it leaves behind a hole. This hole is positively charged, since it results from the removal of an electron from the vicinity of a neutral silicon atom. The vast majority of atoms in the lattice are silicon, so the hole is almost always next to another silicon atom. Consequently, an electron from one of these adjacent atoms can move into the hole, with the result that the hole moves to yet another location. In this fashion, a positively charged hole can wander through the crystal. This type of semiconductor can, therefore, be viewed as containing immobile, negatively charged boron atoms and an equal number of positively charged, mobile holes. Because of the mobile holes, the semiconductor can conduct electricity. In this case the charge carriers are positive. A semiconductor doped with an impurity that introduces mobile **p**ositive holes is called a **p-type semiconductor.**

THE SEMICONDUCTOR DIODE

The physics of a semiconductor diode.

A **p-n junction diode** is a device formed from a p-type semiconductor and an n-type semiconductor. The p-n junction between the two materials is of fundamental importance to the operation of diodes and transistors. Figure 23.21 shows separate p-type and n-type semiconductors, each electrically neutral. Figure 23.22a shows them joined together to form a diode. Mobile electrons from the n-type semiconductor and mobile holes from the p-type semiconductor flow across the junction and combine. This process leaves the

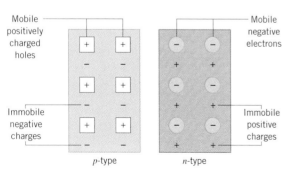

Mobile positively charged holes

Immobile negative charges

p-type   n-type

Mobile negative electrons

Immobile positive charges

Figure 23.21 A p-type semiconductor and an n-type semiconductor.

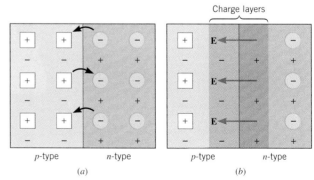

Charge layers

p-type   n-type
(a)

p-type   n-type
(b)

Figure 23.22 (a) At the junction between n and p materials, mobile electrons and holes combine and (b) create positive and negative charge layers. The electric field produced by the charge layers is **E**.

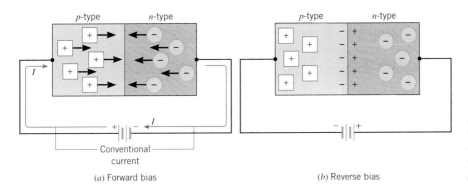

(a) Forward bias   *(b)* Reverse bias

Figure 23.23 (*a*) There is an appreciable current through the diode when the diode is forward biased. (*b*) Under a reverse bias condition, there is almost no current through the diode.

n-type material with a positive charge layer and the *p*-type material with a negative charge layer, as part *b* of the drawing indicates. The positive and negative charge layers on the two sides of the junction set up an electric field **E**, much like that in a parallel plate capacitor. This electric field tends to prevent any further movement of charge across the junction, and all charge flow quickly stops.

Suppose now that a battery is connected across the *p-n* junction, as in Figure 23.23*a*, where the negative terminal of the battery is attached to the *n*-material, and the positive terminal is attached to the *p*-material. In this situation the junction is said to be in a condition of **forward bias,** and, as a result, there is a current in the circuit. The negative terminal of the battery repels the mobile electrons in the *n*-type material, and they move toward the junction. Likewise, the positive terminal repels the positive holes in the *p*-type material, and they also move toward the junction. At the junction the electrons fill the holes. In the meantime, the negative terminal provides a fresh supply of electrons to the *n*-material, and the positive terminal pulls off electrons from the *p*-material, forming new holes in the process. Consequently, a continual flow of charge, and hence a current, is maintained.

In Figure 23.23*b* the battery polarity has been reversed, and the *p-n* junction is in a condition known as **reverse bias.** The battery forces electrons in the *n*-material and holes in the *p*-material away from the junction. As a result, the potential across the junction builds up until it opposes the battery potential, and very little current can be sustained through the diode. The diode, then, is a unidirectional device, for it allows current to pass only in one direction.

The graph in Figure 23.24 shows the dependence of the current on the magnitude and polarity of the voltage applied across a *p-n* junction diode. The exact values of the current depend on the nature of the semiconductor and the extent of the doping. Also shown in the drawing is the symbol used for a diode (—▶︎|—). The direction of the arrowhead in the symbol indicates the direction of the conventional current in the diode under a forward bias condition. In a forward bias condition, the side of the symbol that contains the arrowhead has a positive potential relative to the other side.

A special kind of diode is called an **LED,** which stands for **light-emitting diode.** You can see LEDs in the form of small bright red, green, and yellow lights that appear on many electronic devices, such as computers, TV sets, and stereo systems. These diodes, like others, carry current in only one direction. Imagine a forward biased diode, like that shown in Figure 23.23*a*, in which a current exists. An LED emits light whenever electrons and holes combine, the light coming from the *p-n* junction. Commercial LEDs are often made from gallium, suitably doped with arsenic and phosphorus atoms.

A **fetal oxygen monitor** uses LEDs to measure the level of oxygen in a fetus's blood. A sensor is inserted into the mother's uterus and positioned against the cheek of the fetus, as indicated in Figure 23.25. Two light-emitting diodes are located within the sensor, and each shines light of a different wavelength (or color) into the fetal tissue. The light is reflected by the oxygen-carrying red blood cells and is detected by an adjacent photodetector. Light from one of the LEDs is used to measure the level of oxyhemoglobin in the blood, and light from the other LED is used to measure the level of deoxyhemoglobin. From a comparison of these two levels, the oxygen saturation in the blood is determined.

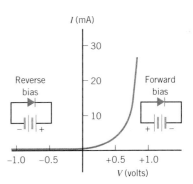

Figure 23.24 The current-versus-voltage characteristics of a typical *p-n* junction diode.

The physics of
light-emitting diodes (LEDs).

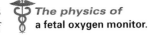
The physics of
a fetal oxygen monitor.

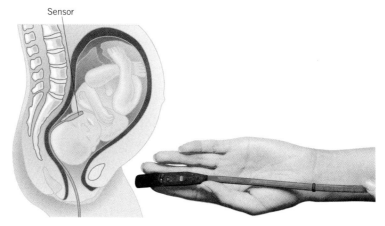

Figure 23.25 A fetal oxygen monitor uses a sensor that contains LEDs to measure the level of oxygen in the fetus's blood. (Reprinted by permission of Nellcor Puritan Bennett, Inc., Pleasanton, California.)

The physics of rectifier circuits.

Because diodes are unidirectional devices, they are commonly used in *rectifier circuits,* which convert an ac voltage into a dc voltage. For instance, Figure 23.26 shows a circuit in which charges flow through the resistance R only while the ac generator biases the diode in the forward direction. Since current occurs only during one-half of every generator voltage cycle, the circuit is called a half-wave rectifier. A plot of the output voltage across the resistor reveals that only the positive halves of each cycle are present. If a capacitor is added in parallel with the resistor, as indicated in the drawing, the capacitor charges up and keeps the voltage from dropping to zero between each positive half-cycle. It is also possible to construct full-wave rectifier circuits, in which both halves of every cycle of the generator voltage drive current through the load resistor in the same direction (see Conceptual Question 11).

When a circuit such as that in Figure 23.26 includes a capacitor and also a transformer to establish the desired voltage level, the circuit is called a power supply. In the audio system in Figure 23.19, the power supply receives the 60-Hz ac voltage from a wall socket and produces a dc output voltage that is used for the transistors within the amplifier. Power supplies using diodes are also found in virtually all electronic appliances, such as televisions and microwave ovens.

SOLAR CELLS

The physics of solar cells.

Solar cells use *p-n* junctions to convert sunlight directly into electricity, as Figure 23.27 illustrates. The solar cell in this drawing consists of a *p*-type semiconductor surrounding an *n*-type semiconductor. As discussed earlier, charge layers form at the junction between the two types of semiconductors, leading to an electric field **E** pointing from the *n*-type toward the *p*-type layer. The outer covering of *p*-type material is so thin that sunlight penetrates into the charge layers and ionizes some of the atoms there. In the process of ionization, the energy of the sunlight causes a negative electron to be ejected from the atom, leaving behind a positive hole. As the drawing indicates, the electric field in the charge layers causes the electron and the hole to move away from the junction. The electron moves into the *n*-type material, and the hole moves into the *p*-type material. As a result, the sunlight causes the solar cell to develop negative and positive terminals, much like the terminals of a battery. The current that a single solar cell can provide is small, so applications of solar cells often use many of them mounted to form large panels, as Figure 23.28 illustrates.

Figure 23.26 A half-wave rectifier circuit, together with a capacitor and a transformer (not shown), constitutes a dc power supply because the rectifier converts an ac voltage into a dc voltage.

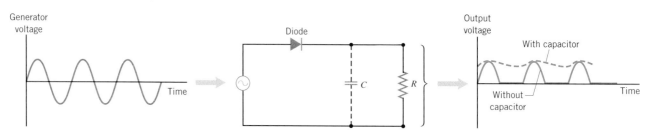

TRANSISTORS

A number of different kinds of transistors are in use today. One type is the **bipolar junction transistor**, which consists of two *p-n* junctions formed by three layers of doped semiconductors. As Figure 23.29 indicates, there are *pnp* and *npn* transistors. In either case, the middle region is made very thin compared to the outer regions.

A transistor is useful because it can be used in circuits that amplify a smaller voltage into a larger one. A transistor plays the same kind of role in an amplifier circuit that a valve does when it controls the flow of water through a pipe. A small change in the valve setting produces a large change in the amount of water per second that flows through the pipe. In other words, a small change in the voltage input to a transistor produces a large change in the output from the transistor.

Figure 23.30 shows a *pnp* transistor connected to two batteries, labeled V_E and V_C. The voltages V_E and V_C are applied in such a way that the *p-n* junction on the left has a forward bias, while the *p-n* junction on the right has a reverse bias. Moreover, the voltage V_C is usually much larger than V_E for a reason to be discussed shortly. The drawing also shows the standard symbol and nomenclature for the three sections of the transistor—namely, the *emitter,* the *base,* and the *collector.* The arrowhead in the symbol points in the direction of the conventional current through the emitter.

The positive terminal of V_E pushes the mobile positive holes in the *p*-type material of the emitter toward the emitter/base junction. Since this junction has a forward bias, the holes enter the base region readily. Once in the base region, the holes come under the strong influence of V_C and are attracted to its negative terminal. Since the base is so thin (about 10^{-6} m or so), approximately 98% of the holes are drawn through the base and into the collector. The remaining 2% of the holes combine with free electrons in the base region, thereby giving rise to a small base current I_B. As the drawing shows, the moving holes in the emitter and collector constitute currents that are labeled I_E and I_C, respectively. From Kirchhoff's junction rule it follows that $I_C = I_E - I_B$.

Because the base current I_B is small, the collector current is determined primarily by current from the emitter ($I_C = I_E - I_B \approx I_E$). This means that a change in I_E will cause a change in I_C of nearly the same amount. Furthermore, a substantial change in I_E can be caused by only a small change in the forward bias voltage V_E. To see that this is the case, look back at Figure 23.24 and notice how steep the current-versus-voltage curve is for a *p-n* junction; small changes in the forward bias voltage give rise to large changes in the current.

With the help of Figure 23.31 we can now appreciate what was meant by the earlier statement that a small change in the voltage input to a transistor leads to a large change in the output. This picture shows an ac generator connected in series with the battery V_E and a resistance R connected in series with the collector. The generator voltage could origi-

The physics of transistors.

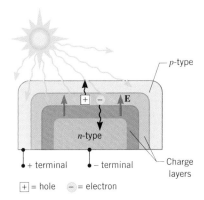

Figure 23.27 A solar cell formed from a *p-n* junction. When sunlight strikes it, the solar cell acts like a battery, with + and − terminals.

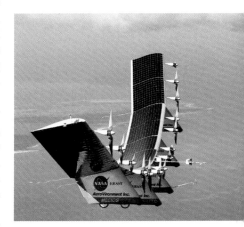

Figure 23.28 The Helios Prototype flying wing is propelled by solar energy. The solar cells are mounted on the top of the wing. (© Gamma Press, Inc.)

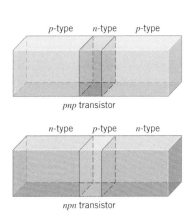

Figure 23.29 There are two kinds of bipolar junction transistors, *pnp* and *npn*.

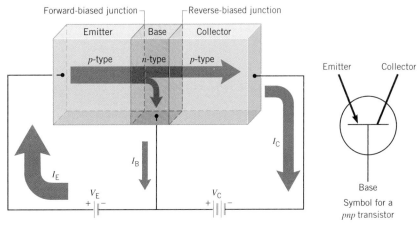

Figure 23.30 A *pnp* transistor, along with its bias voltages V_E and V_C. On the symbol for the *pnp* transistor, the emitter is marked with an arrowhead that denotes the direction of conventional current through the emitter.

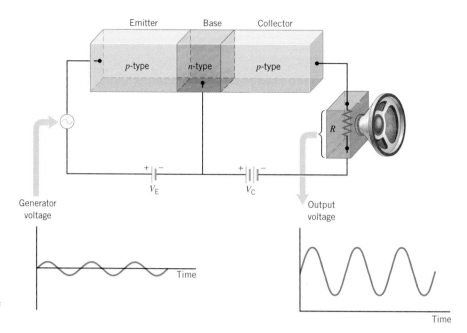

Figure 23.31 The basic *pnp* transistor amplifier in this drawing amplifies a small generator voltage to produce an enlarged voltage across the resistance *R*.

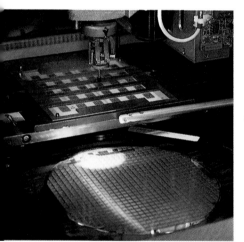

Figure 23.32 Integrated circuit (IC) chips are manufactured on wafers of semiconductor material. Shown here is one wafer containing many chips. Some of the so-called smart cards in which the chips are used are also shown. (Courtesy ORGA Card Systems, Inc.)

nate from many sources, such as an electric guitar pickup or a compact disc player, while the resistance R could represent a loudspeaker. The generator introduces small voltage changes in the forward bias across the emitter/base junction and, thus, causes large corresponding changes in the current I_C leaving the collector and passing through the resistance R. As a result, the output voltage across R is an enlarged or amplified version of the input voltage of the generator. The operation of an *npn* transistor is similar to that of a *pnp* transistor. The main difference is that the bias voltages and current directions are reversed.

It is important to realize that the increased power available at the output of a transistor amplifier does *not* come from the transistor itself. Rather, it comes from the power provided by the voltage source V_C. The transistor, acting like an automatic valve, merely allows the small, weak signals from the input generator to control the power taken from the source V_C and delivered to the resistance R.

Today it is possible to combine arrays of tens of thousands of transistors, diodes, resistors, and capacitors on a tiny chip of silicon that usually measures less than a centimeter on a side. These arrays are called *integrated circuits* (ICs) and can be designed to perform almost any desired electronic function. Integrated circuits, such as the type in Figure 23.32, have revolutionized the electronics industry and lie at the heart of computers, cellular phones, digital watches, and programmable appliances.

23.6 Concepts & Calculations

A capacitor is one of the important elements found in ac circuits. As we have seen in this chapter, the capacitance of a capacitor influences the amount of current in a circuit. The capacitance, in turn, is determined by the geometry of the capacitor and the material that fills the space between its plates, as Section 19.5 discusses. When capacitors are connected together, the equivalent capacitance depends on the nature of the connection—for example, whether it is a series or a parallel connection, as Section 20.12 discusses. The next example provides a review of these issues concerning capacitors.

Concepts & Calculations Example 6   Capacitors in AC Circuits

Two parallel plate capacitors are filled with the same dielectric material and have the same plate area. However, capacitor 1 has a plate separation that is twice that of capacitor 2. When capacitor 1 is connected across the terminals of an ac generator, the generator delivers an rms current of 0.60A. What is the current delivered by the generator when both capacitors are connected in parallel across its terminals? In both cases the generator produces the same frequency and voltage.

Concept Questions and Answers Which of the two capacitors has the greater capacitance?

Answer The capacitance of a capacitor is given by Equation 19.10 as $C = \kappa \epsilon_0 A/d$, where κ is the dielectric constant of the material between the plates, ϵ_0 is the permittivity of free space, A is the area of each plate, and d is the separation between the plates. Since the dielectric constant and the plate area are the same for each capacitor, the capacitance is inversely proportional to the plate separation. Therefore, capacitor 2, with the smaller plate separation, has the greater capacitance.

Is the equivalent capacitance of the parallel combination greater or smaller than the capacitance of capacitor 1?

Answer According to Equation 20.18, the equivalent capacitance of the two capacitors in parallel is $C_P = C_1 + C_2$, where C_1 and C_2 are the individual capacitances. Therefore, the value for C_P is greater than the value for C_1.

Is the capacitive reactance for C_P greater or smaller than for C_1?

Answer The capacitive reactance is given by $X_C = 1/(2\pi fC)$, according to Equation 23.2, where f is the frequency. For a given frequency, the reactance is inversely proportional to the capacitance. Since the capacitance C_P is greater than C_1, the corresponding reactance is smaller.

When both capacitors are connected in parallel across the terminals of the generator, is the current from the generator greater or smaller than when capacitor 1 is connected alone?

Answer According to Equation 23.1, the current is given by $I_{rms} = V_{rms}/X_C$, where V_{rms} is the rms voltage of the generator and X_C is the capacitive reactance. For a given voltage, the current is inversely proportional to the reactance. Since the reactance in the parallel case is smaller than for C_1 alone, the current in the parallel case is greater.

Solution Using Equation 23.1 to express the current as $I_{rms} = V_{rms}/X_C$ and Equation 23.2 to express the reactance as $X_C = 1/(2\pi fC)$, we find that the current is

$$I_{rms} = \frac{V_{rms}}{X_C} = \frac{V_{rms}}{1/(2\pi fC)} = V_{rms}2\pi fC$$

Applying this result to the case where C_1 is connected alone to the generator and to the case where the two capacitors are connected in parallel, we obtain

$$\underbrace{I_{1,\,rms} = V_{rms}2\pi fC_1}_{C_1 \text{ alone}} \quad \text{and} \quad \underbrace{I_{P,\,rms} = V_{rms}2\pi fC_P}_{C_1 \text{ and } C_2 \text{ in parallel}}$$

Dividing the two expressions gives

$$\frac{I_{P,\,rms}}{I_{1,\,rms}} = \frac{V_{rms}2\pi fC_P}{V_{rms}2\pi fC_1} = \frac{C_P}{C_1}$$

According to Equation 20.18, the effective capacitance of the two capacitors in parallel is $C_P = C_1 + C_2$, so that the result for the current ratio becomes

$$\frac{I_{P,\,rms}}{I_{1,\,rms}} = \frac{C_1 + C_2}{C_1} = 1 + \frac{C_2}{C_1}$$

Since the capacitance of a capacitor is given by Equation 19.10 as $C = \kappa \epsilon_0 A/d$, we find that

$$\frac{I_{P,\,rms}}{I_{1,\,rms}} = 1 + \frac{\kappa \epsilon_0 A/d_2}{\kappa \epsilon_0 A/d_1} = 1 + \frac{d_1}{d_2}$$

We know that $d_1 = 2d_2$, so that the current in the parallel case is

$$I_{P,\,rms} = I_{1,\,rms}\left(1 + \frac{d_1}{d_2}\right) = (0.60 \text{ A})\left(1 + \frac{2d_2}{d_2}\right) = \boxed{1.8 \text{ A}}$$

As expected, the current in the parallel case is greater.

In ac circuits that contain capacitance, inductance, and resistance, it is only the resistance that, on average, dissipates power. The average power dissipated in a capacitor or an inductor is zero. However, the presence of a capacitor and/or an inductor does influence the rms current in the circuit. When the current changes for any reason, the power dissipated in a resistor also changes, as Example 7 illustrates.

Concepts & Calculations Example 7
Only Resistance Dissipates Power

An ac generator has a frequency of 1200 Hz and a constant rms voltage. When a 470-Ω resistor is connected between the terminals of the generator, an average power of 0.25 W is dissipated in the resistor. Then, a 0.080-H inductor is connected in series with the resistor, and the combination is connected between the generator terminals. What is the average power dissipated in the series combination?

Concept Questions and Answers In which case does the generator deliver a greater rms current?

Answer When only the resistor is present, the current is given by Equation 20.14 as $I_{rms} = V_{rms}/R$, where V_{rms} is the rms voltage of the generator. When the resistor and the inductor are connected in series, the current is given by $I_{rms} = V_{rms}/Z$, according to Equation 23.6. In this expression Z is the impedance and is given by Equation 23.7 as $Z = \sqrt{R^2 + X_L^2}$, where X_L is the inductive reactance and is given by Equation 23.4 as $X_L = 2\pi fL$. Since Z is greater than R, the current is greater when only the resistance is present.

In which case is a greater average power dissipated in the circuit?

Answer Only the resistor in an ac circuit dissipates power on average, the amount of the average power being $\overline{P} = I_{rms}^2 R$, according to Equation 20.15b. Since the resistance R is the same in both cases, a greater average power is dissipated when the current I_{rms} is greater. Since the current is greater when only the resistor is present, more power is dissipated in that case.

Solution When only the resistor is present, the average power is $\overline{P} = I_{rms}^2 R$, according to Equation 20.15b, and the current is given by $I_{rms} = V_{rms}/R$, according to Equation 20.14. Therefore, we find in this case that

Resistor only
$$\overline{P} = I_{rms}^2 R = \left(\frac{V_{rms}}{R}\right)^2 R = \frac{V_{rms}^2}{R}$$

When the inductor is also present, the average power is still $\overline{P} = I_{rms}^2 R$, but the current is now given by $I_{rms} = V_{rms}/Z$, according to Equation 23.6, in which Z is the impedance. In this case, then, we have

Resistor and inductor
$$\overline{P} = I_{rms}^2 R = \left(\frac{V_{rms}}{Z}\right)^2 R = \frac{V_{rms}^2 R}{Z^2}$$

Dividing this result by the analogous result for the resistor-only case, we obtain

$$\frac{\overline{P}_{\text{Resistor and inductor}}}{\overline{P}_{\text{Resistor only}}} = \frac{V_{rms}^2 R/Z^2}{V_{rms}^2/R} = \frac{R^2}{Z^2}$$

In this expression Z is the impedance and is given by Equation 23.7 as $Z = \sqrt{R^2 + X_L^2}$, where X_L is the inductive reactance and is given by Equation 23.4 as $X_L = 2\pi fL$. With these substitutions, the ratio of the powers becomes

$$\frac{\overline{P}_{\text{Resistor and inductor}}}{\overline{P}_{\text{Resistor only}}} = \frac{R^2}{R^2 + (2\pi fL)^2}$$

The power dissipated in the series combination, then, is

$$\overline{P}_{\text{Resistor and inductor}} = \overline{P}_{\text{Resistor only}} \left[\frac{R^2}{R^2 + (2\pi fL)^2}\right]$$

$$= (0.25 \text{ W}) \frac{(470 \ \Omega)^2}{(470 \ \Omega)^2 + [2\pi(1200 \text{ Hz})(0.080 \text{ H})]^2} = \boxed{0.094 \text{ W}}$$

As expected, more power is dissipated in the resistor-only case.

At the end of the problem set for this chapter, you will find homework problems that contain both conceptual and quantitative parts. These problems are grouped under the heading *Concepts & Calculations, Group Learning Problems*. They are designed for use by students working alone or in small learning groups. The conceptual part of each problem provides a convenient focus for group discussions.

Concept Summary

This summary presents an abridged version of the chapter, including the important equations and all available learning aids. For convenient reference, the learning aids (including the text's examples) are placed next to or immediately after the relevant equation or discussion. The following learning aids may be found on-line at **www.wiley.com/college/cutnell**:

Interactive LearningWare examples are solved according to a five-step interactive format that is designed to help you develop problem-solving skills.	**Concept Simulations** are animated versions of text figures or animations that illustrate important concepts. You can control parameters that affect the display, and we encourage you to experiment.
Interactive Solutions offer specific models for certain types of problems in the chapter homework. The calculations are carried out interactively.	**Self-Assessment Tests** include both qualitative and quantitative questions. Extensive feedback is provided for both incorrect and correct answers, to help you evaluate your understanding of the material.

Topic	*Discussion*	*Learning Aids*
	23.1 Capacitors and Capacitive Reactance	
Relation between voltage and current	In an ac circuit the rms voltage V_{rms} across a capacitor is related to the rms current I_{rms} by $$V_{rms} = I_{rms}X_C \qquad (23.1)$$ where X_C is the capacitive reactance. The capacitive reactance is measured in ohms (Ω) and is given by	
Capacitive reactance	$$X_C = \frac{1}{2\pi fC} \qquad (23.2)$$ where f is the frequency and C is the capacitance.	**Examples 1, 6** Concept Simulation 23.1
Phase angle between current and voltage	The ac current in a capacitor leads the voltage across the capacitor by a phase angle of 90° or $\pi/2$ radians. As a result, a capacitor consumes no power, on average.	
Voltage and current phasors	The phasor model is useful for analyzing the voltage and current in an ac circuit. In this model, the voltage and current are represented by rotating arrows, called phasors.	
	The length of the voltage phasor represents the maximum voltage V_0, and the length of the current phasor represents the maximum current I_0. The phasors rotate in a counterclockwise direction at a frequency f. Since the current leads the voltage by 90° in a capacitor, the current phasor is ahead of the voltage phasor by 90° in the direction of rotation.	
	The instantaneous values of the voltage and current are equal to the vertical components of the corresponding phasors.	
	23.2 Inductors and Inductive Reactance	
Relation between voltage and current	In an ac circuit the rms voltage V_{rms} across an inductor is related to the rms current I_{rms} by $$V_{rms} = I_{rms}X_L \qquad (23.3)$$ where X_L is the inductive reactance. The inductive reactance is measured in ohms (Ω) and is given by	
Inductive reactance	$$X_L = 2\pi fL \qquad (23.4)$$ where f is the frequency and L is the inductance.	**Example 2** Concept Simulation 23.1
Phase angle between current and voltage	The ac current in an inductor lags behind the voltage across the inductor by a phase angle of 90° or $\pi/2$ radians. Consequently, an inductor, like a capacitor, consumes no power, on average.	

Topic	Discussion	Learning Aids
Voltage and current phasors	The voltage and current phasors in a circuit containing only an inductor also rotate in a counterclockwise direction at a frequency f. However, since the current lags the voltage by 90° in an inductor, the current phasor is behind the voltage phasor by 90° in the direction of rotation. The instantaneous values of the voltage and current are equal to the vertical components of the corresponding phasors.	

 Use *Self-Assessment Test 23.1* to evaluate your understanding of Sections 23.1 and 23.2.

23.3 Circuits Containing Resistance, Capacitance, and Inductance

Topic	Discussion	Learning Aids
	When a resistor, a capacitor, and an inductor are connected in series, the rms voltage across the combination is related to the rms current according to	
Relation between voltage and current	$$V_{rms} = I_{rms} Z \qquad (23.6)$$	
	where Z is the impedance of the combination. The impedance is measured in ohms (Ω) and is given by	
Impedance	$$Z = \sqrt{R^2 + (X_L - X_C)^2} \qquad (23.7)$$ where R is the resistance, and X_L and X_C are, respectively, the inductive and capacitive reactances.	**Interactive Solution 23.23** **Concept Simulation 23.2**
	The tangent of the phase angle ϕ between current and voltage in a series RCL circuit is	
Phase angle between current and voltage	$$\tan \phi = \frac{X_L - X_C}{R} \qquad (23.8)$$	
	Only the resistor in the RCL combination dissipates power, on average. The average power $\overline{P}$ dissipated in the circuit is	
Average power dissipated	$$\overline{P} = I_{rms} V_{rms} \cos \phi \qquad (23.9)$$	**Examples 3, 4, 7**
Power factor	where $\cos \phi$ is called the power factor of the circuit.	

23.4 Resonance in Electric Circuits

Topic	Discussion	Learning Aids
	A series RCL circuit has a resonant frequency f_0 that is given by	
Resonant frequency	$$f_0 = \frac{1}{2\pi\sqrt{LC}} \qquad (23.10)$$	**Example 5**
	where L is the inductance and C is the capacitance. At resonance, the impedance of the circuit has a minimum value equal to the resistance R, and the rms current has a maximum value.	

 Use *Self-Assessment Test 23.2* to evaluate your understanding of Sections 23.3 and 23.4.

23.5 Semiconductor Devices

Topic	Discussion	Learning Aids
***n*-type semiconductor**	In an *n*-type semiconductor, mobile negative electrons carry the current. An *n*-type material is produced by doping a semiconductor such as silicon with a small amount of impurity atoms such as phosphorus.	
***p*-type semiconductor**	In a *p*-type semiconductor, mobile positive "holes" in the crystal structure carry the current. A *p*-type material is produced by doping a semiconductor with a small amount of impurity atoms such as boron. These two types of semiconductors are used in *p-n* junction diodes, light-emitting diodes, and solar cells, and in *pnp* and *npn* bipolar junction transistors.	

Conceptual Questions

1. A light bulb and a parallel plate capacitor (containing a dielectric material between the plates) are connected in series to the 60-Hz ac voltage present at a wall outlet. Describe what happens to the brightness of the bulb when the dielectric material is removed from the space between the plates. Explain.

2. The ends of a long, straight wire are connected to the terminals of an ac generator, and the current is measured. The wire is then disconnected, wound into the shape of a multiple-turn coil, and reconnected to the generator. In which case, if either, does the generator deliver a larger current? Account for your answer.

3. An air-core inductor is connected in series with a light bulb, and this circuit is plugged into an electrical outlet. What happens to the brightness of the bulb when a piece of iron is inserted inside the inductor? Give a reason for your answer.

4. Consider the circuit in Figure 23.9. With the capacitor and the inductor present, a certain amount of current is in the circuit. Then the capacitor and the inductor are removed, so that only the resistor remains connected to the generator. Is it possible that the current in the simplified circuit has the same rms value as that in the original circuit? Give your reasoning.

5. Can an ac circuit, such as the one in Figure 23.9, have the same impedance at two different generator frequencies? Justify your answer.

6. Review Conceptual Example 4 as an aid in understanding this question. An inductor and a capacitor are connected in parallel across the terminals of a generator. What happens to the current from the generator as the frequency becomes (a) very large and (b) very small? Give your reasoning.

7. Review Conceptual Example 4. For which of the two circuits discussed there does the generator deliver more current when the frequency is very large? Account for your answer.

8. In a series RCL circuit at resonance, does the current lead or lag behind the voltage across the generator or is it in phase with the voltage? Explain.

9. Is it possible for two series RCL circuits to have the same resonant frequencies and yet have (a) different R values and (b) different C and L values? Justify your answers.

10. Suppose the generator connected to a series RCL circuit has a frequency that is greater than the resonant frequency of the circuit. Suppose, in addition, that it is necessary to match the resonant frequency of the circuit to the frequency of the generator. To accomplish this, should you add a second capacitor in series or in parallel with the one already present? Explain your choice.

11. The drawing shows a full-wave rectifier circuit, in which the direction of the current through the load resistor R is the same for both positive and negative halves of the generator's voltage cycle. Draw the current through the circuit when (a) the top of the generator is positive and the bottom is negative and (b) the top of the generator is negative and the bottom is positive.

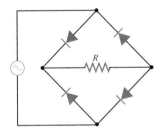

Problems

Note: For problems in this set, the ac current and voltage are rms values, and the power is an average value, unless indicated otherwise.

ssm Solution is in the Student Solutions Manual.   www Solution is available on the World Wide Web at www.wiley.com/college/cutnell
℣ **This icon represents a biomedical application.**

Section 23.1 Capacitors and Capacitive Reactance

1. ssm At what frequency does a 7.50-μF capacitor have a reactance of 168 Ω?

2. What voltage is needed to create a current of 35 mA in a circuit containing only a 0.86-μF capacitor, when the frequency is 3.4 kHz?

3. A capacitor is connected across the terminals of an ac generator that has a frequency of 440 Hz and supplies a voltage of 24 V. When a second capacitor is connected in parallel with the first one, the current from the generator increases by 0.18 A. Find the capacitance of the second capacitor.

4. Two identical capacitors are connected in parallel to an ac generator that has a frequency of 610 Hz and produces a voltage of 24 V. The current in the circuit is 0.16 A. What is the capacitance of each capacitor?

5. ssm A capacitor is attached to a 5.00-Hz generator. The instantaneous current is observed to reach a maximum value at a certain time. What is the least amount of time that passes before the instantaneous voltage across the capacitor reaches its maximum value?

* 6. A capacitor is connected across an ac generator whose frequency is 750 Hz and whose *peak* output voltage is 140 V. The rms current in the circuit is 3.0 A. (a) What is the capacitance of the capacitor? (b) What is the magnitude of the *maximum* charge on one plate of the capacitor?

** 7. A capacitor (capacitance C_1) is connected across the terminals of an ac generator. Without changing the voltage or frequency of the generator, a second capacitor (capacitance C_2) is added in series with the first one. As a result, the current delivered by the generator decreases by a factor of three. Suppose the second capacitor had been added in parallel with the first one, instead of in series. By what factor would the current delivered by the generator have increased?

Section 23.2 Inductors and Inductive Reactance

8. The current in an inductor is 0.20 A, and the frequency is 750 Hz. If the inductance is 0.080 H, what is the voltage across the inductor?

9. ssm At what frequency (in Hz) are the reactances of a 52-mH inductor and a 76-μF capacitor equal?

10. Two ac generators supply the same voltage. However, the first generator has a frequency of 1.5 kHz, and the second has a frequency of 6.0 kHz. When an inductor is connected across the terminals of the first generator, the current delivered is 0.30 A. How much current is delivered when this inductor is connected across the terminals of the second generator?

11. ssm www A 40.0-μF capacitor is connected across a 60.0-Hz generator. An inductor is then connected in parallel with the capacitor. What is the value of the inductance if the rms currents in the inductor and capacitor are equal?

12. A 30.0-mH inductor has a reactance of 2.10 kΩ. (a) What is the frequency of the ac current that passes through the inductor? (b) What is the capacitance of a capacitor that has the same reactance at this frequency? The frequency is tripled, so that the reactances of the inductor and capacitor are no longer equal. What are the new reactances of (c) the inductor and (d) the capacitor?

* **13.** The rms current in a solenoid is 0.036 A when the solenoid is connected to an 18-kHz generator. The solenoid has a cross-sectional area of 3.1×10^{-5} m^2 and a length of 2.5 cm. The solenoid has 135 turns. Determine the *peak voltage* of the generator.

** **14.** Two inductors are connected in parallel across the terminals of a generator. One has an inductance of $L_1 = 0.030$ H, and the other has an inductance of $L_2 = 0.060$ H. A single inductor, with an inductance L, is connected across the terminals of a second generator that has the same frequency and voltage as the first one. The current delivered by the second generator is equal to the *total* current delivered by the first generator. Find L.

Section 23.3 Circuits Containing Resistance, Capacitance, and Inductance

15. ssm A series RCL circuit includes a resistance of 275 Ω, an inductive reactance of 648 Ω, and a capacitive reactance of 415 Ω. The current in the circuit is 0.233 A. What is the voltage of the generator?

16. A light bulb has a resistance of 240 Ω. It is connected to a standard wall socket (120 V, 60.0 Hz). (a) Determine the current in the bulb. (b) Determine the current in the bulb after a 10.0-μF capacitor is added in series in the circuit. (c) It is possible to return the current in the bulb to the value calculated in part (a) by adding an inductor in series with the bulb and the capacitor. What is the value of the inductance of this inductor?

17. Suppose that the inductance is zero ($L = 0$ H) in the series RCL circuit shown in Figure 23.10. The rms voltages across the generator and the resistor are 45 and 24 V, respectively. What is the rms voltage across the capacitor?

18. A 2700-Ω resistor and a 1.1-μF capacitor are connected in series across a generator (60.0 Hz, 120 V). Determine the power dissipated in the circuit.

19. ssm A circuit consists of a 215-Ω resistor and a 0.200-H inductor. These two elements are connected in series across a generator that has a frequency of 106 Hz and a voltage of 234 V. (a) What is the current in the circuit? (b) Determine the phase angle between the current and the voltage of the generator.

20. In a series circuit, a generator (1350 Hz, 15.0 V) is connected to a 16.0-Ω resistor, a 4.10-μF capacitor, and a 5.30-mH inductor. Find the voltage across each circuit element.

* **21. ssm www** A circuit consists of an 85-Ω resistor in series with a 4.0-μF capacitor, the two being connected between the terminals of an ac generator. The voltage of the generator is fixed. At what frequency is the current in the circuit one-half the value that exists when the frequency is very large?

* **22.** Refer to Conceptual Example 4 as background for this problem. For the circuit shown in the drawing, find the current provided by the generator when the frequency is (a) very large and (b) very small.

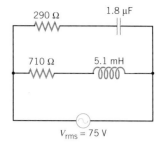

Problem 22

* **23.** Refer to **Interactive Solution 23.23** at **www.wiley.com/college/ cutnell** for help with problems like this one. A series RCL circuit contains only a capacitor ($C = 6.60 \mu$F), an inductor ($L = 7.20$ mH), and a generator (*peak* voltage = 32.0 V, frequency = 1.50×10^3 Hz). When $t = 0$ s, the instantaneous value of the voltage is zero, and it rises to a maximum one-quarter of a period later. (a) Find the *instantaneous* value of the voltage across the capacitor/inductor combination when $t = 1.20 \times 10^{-4}$ s. (b) What is the *instantaneous* value of the current when $t = 1.20 \times 10^{-4}$ s? (*Hint: The instantaneous values of the voltage and current are, respectively, the vertical components of the voltage and current phasors.*)

* **24.** A series circuit contains only a resistor and an inductor. The voltage V of the generator is fixed. If $R = 16$ Ω and $L = 4.0$ mH, find the frequency at which the current is one-half its value at zero frequency.

** **25.** When a resistor is connected by itself to an ac generator, the average power dissipated in the resistor is 1.000 W. When a capacitor is added in series with the resistor, the power dissipated is 0.500 W. When an inductor is added in series with the resistor (without the capacitor), the power dissipated is 0.250 W. Determine the power dissipated when both the capacitor and the inductor are added in series with the resistor.

Section 23.4 Resonance in Electric Circuits

26. A series RCL circuit has a resonant frequency of 690 kHz. If the value of the capacitance is 2.0×10^{-9} F, what is the value of the inductance?

27. ssm A 10.0-Ω resistor, a 12.0-μF capacitor, and a 17.0-mH inductor are connected in series with a 155-V generator. (a) At what frequency is the current a maximum? (b) What is the maximum value of the rms current?

28. A series RCL circuit is at resonance and contains a variable resistor that is set to 175 Ω. The power dissipated in the circuit is 2.6 W. Assuming that the voltage remains constant, how much power is dissipated when the variable resistor is set to 562 Ω?

29. ssm The resonant frequency of a series RCL circuit is 9.3 kHz. The inductance and capacitance of the circuit are each tripled. What is the new resonant frequency?

30. A series RCL circuit has a resonant frequency of 1500 Hz. When operating at a frequency other than 1500 Hz, the circuit has a capacitive reactance of 5.0 Ω and an inductive reactance of 30.0 Ω. What are the values of (a) L and (b) C?

31. The resonant frequency of an RCL circuit is 1.3 kHz, and the value of the inductance is 7.0 mH. What is the resonant frequency (in kHz) when the value of the inductance is 1.5 mH?

* **32.** The ratio of the inductive reactance to the capacitive reactance is observed to be 5.36 in a series RCL circuit. The resonant frequency of the circuit is 225 Hz. What is the frequency (nonresonant) of the generator that is connected to the circuit?

* **33. ssm www** A series RCL circuit contains a 5.10-μF capacitor and a generator whose voltage is 11.0 V. At a resonant frequency of

1.30 kHz the power dissipated in the circuit is 25.0 W. Find the values of (a) the inductance and (b) the resistance. (c) Calculate the power factor when the generator frequency is 2.31 kHz.

** **34.** A 108-Ω resistor, a 0.200-μF capacitor, and a 5.42-mH inductor are connected in series to a generator whose voltage is 26.0 V. The current in the circuit is 0.141 A. Because of the shape of the current-versus-frequency graph (see Figure 23.15), there are two possible

values for the frequency that correspond to this current. Obtain these two values.

** **35.** When the frequency is twice the resonant frequency, the impedance of a series RCL circuit is twice the value of the impedance at resonance. Obtain the ratios of the inductive and capacitive reactances to the resistance; that is, obtain (a) X_L/R and (b) X_C/R when the frequency is twice the resonant frequency.

Additional Problems

36. An ac generator has a frequency of 4.80 kHz and produces a current of 0.0400 A in a series circuit that contains only a 232-Ω resistor and a 0.250-μF capacitor. Obtain (a) the voltage of the generator and (b) the phase angle between the current and the voltage across the resistor/capacitor combination.

37. ssm A circuit consists of a 3.00-μF and a 6.00-μF capacitor connected in series across the terminals of a 510-Hz generator. The voltage of the generator is 120 V. (a) Determine the equivalent capacitance of the two capacitors. (b) Find the current in the circuit.

38. Review Conceptual Example 4 and Figure 23.12. Find the ratio of the current in circuit I to that in circuit II in the *high-frequency limit* for the same generator voltage.

39. The elements in a series RCL circuit are a 106-Ω resistor, a 3.30-μF capacitor, and a 0.0310-H inductor. The frequency is 609 Hz. What are (a) the impedance of the circuit and (b) the phase angle between the current and the voltage of the generator?

40. A 0.047-H inductor is wired across the terminals of a generator that has a voltage of 2.1 V and supplies a current of 0.023 A. Find the frequency of the generator.

41. ssm www An 8.2-mH inductor is connected to an ac generator (10.0 V rms, 620 Hz). Determine the *peak value* of the current supplied by the generator.

42. A circuit consists of a resistor in series with an inductor and an ac generator that supplies a voltage of 115 V. The inductive reactance is 52.0 Ω, and the current in the circuit is 1.75 A. Find the average power dissipated in the circuit.

* **43. ssm** Suppose you have a number of capacitors. Each is identical to the capacitor that is already in a series RCL circuit. How many of these additional capacitors must be inserted in series in the circuit, so the resonant frequency triples?

* **44.** Review Conceptual Example 4 before attempting this problem. Then refer to the drawing, in which the generator delivers four times as much current at very low frequencies as it does at very high frequencies. Find the ratio R_2/R_1 of the resistances.

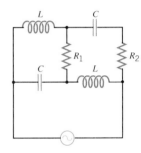

Problem 44

** **45.** In a series RCL circuit the dissipated power drops by a factor of two when the frequency of the generator is changed from the resonant frequency to a nonresonant frequency. The peak voltage is held constant while this change is made. Determine the power factor of the circuit at the nonresonant frequency.

Concepts & Calculations Group Learning Problems

Note: Each of these problems consists of Concept Questions followed by a related quantitative Problem. They are designed for use by students working alone or in small learning groups. The Concept Questions involve little or no mathematics and are intended to stimulate group discussions. They focus on the concepts with which the problems deal. Recognizing the concepts is the essential initial step in any problem-solving technique.

46. Concept Questions (a) Does the capacitance of a parallel plate capacitor increase or decrease when a dielectric material is inserted between the plates? (b) Is the equivalent capacitance of two capacitors in parallel greater or smaller than the capacitance of either capacitor alone? (c) A capacitor is connected between the terminals of an ac generator. When a second capacitor is connected in parallel with the first, does the current supplied by the generator increase or decrease? Explain your answers.

Problem Two parallel plate capacitors are identical, except that one of them is empty and the other contains a material with a dielectric constant of 4.2 in the space between the plates. The empty capacitor is connected between the terminals of an ac generator that has a fixed frequency and rms voltage. The generator delivers a current of 0.22 A. What current does the generator deliver after the other capacitor is connected in parallel with the first one? Verify that your answer is consistent with your answers to the Concept Questions.

47. Concept Questions (a) An inductance L_1 is connected across the terminals of an ac generator, which delivers a current to it. Then a second inductance L_2 is connected in parallel with L_1. Does the presence of L_2 alter the current in L_1? (b) The generator delivers a current to L_2 as well as L_1. Would the removal of L_1 alter the current in L_2? (c) Is the current delivered to the parallel combination greater or smaller than the current to either inductance alone? Give your reasoning.

Problem An ac generator has a frequency of 2.2 kHz and a voltage of 240 V. An inductance L_1 = 6.0 mH is connected across its terminals. Then a second inductance L_2 = 9.0 mH is connected in parallel with L_1. Find the current that the generator delivers to L_1 and to the parallel combination. Check to see that your answers are consistent with your answers to the Concept Questions.

48. Concept Questions (a) An ac circuit contains only a resistor and a capacitor in series. Is the phase angle ϕ between the current and the voltage of the generator positive or negative, and how is the impedance Z of the circuit related to the resistance R and the capacitive reactance X_C? (b) An ac circuit contains only a resistor and an inductor in series. Is the phase angle ϕ positive or negative, and how is the impedance Z of the circuit related to the resistance R and the inductive reactance X_L? Account for your answers.

Problem A series circuit has an impedance of 192 Ω, and the phase angle is $\phi = -75.0°$. The circuit contains a resistor and either a capacitor or an inductor. Find the resistance R and the capacitive reactance X_C or the inductive reactance X_L, whichever is appropriate.

49. Concept Questions (a) A series RCL circuit contains three elements—a resistor, a capacitor, and an inductor. Which of the elements, on average, dissipate(s) power? (b) How is the current in the circuit related to the generator voltage and the impedance of the circuit? (c) What is the impedance of the circuit at resonance? (d) At resonance, how is the average dissipated power related to the generator voltage and the resistance?

Problem The resistor in a series RCL circuit has a resistance of 92 Ω, while the voltage of the generator is 3.0 V. At resonance, what is the average power dissipated in the circuit?

50. Concept Questions The capacitance in a series RCL circuit is C_1. The generator in the circuit has a fixed frequency that is less than the resonant frequency of the circuit. By adding a second capacitance C_2 to the circuit, it is desired to reduce the resonant frequency so that it matches the generator frequency. (a) To reduce the resonant frequency, should the circuit capacitance be increased or decreased? (b) Should C_2 be added in parallel or in series with C_1? Give your reasoning.

Problem The initial circuit capacitance is $C_1 = 2.60$ μF, and the corresponding resonant frequency is $f_{01} = 7.30$ kHz. The generator frequency is 5.60 kHz. What is the value of the capacitance C_2 that should be added so that the circuit will have a resonant frequency that matches the generator frequency?

* 51. Concept Questions Part a of the drawing shows a resistor and a charged capacitor wired in series. When the switch is closed, the capacitor discharges as charge moves from one plate to the other. Part b shows a plot of the amount of charge remaining on each plate of the capacitor as a function of time. (a) What does the time constant τ of this resistor-capacitor circuit physically represent? (b) How is the time constant related to the resistance R and the capacitance C? (c) In part c of the drawing, the switch has been removed and an ac generator has been inserted into the circuit. What

is the impedance Z of this circuit? Express your answer in terms of the resistance R, the time constant τ, and the frequency f of the generator.

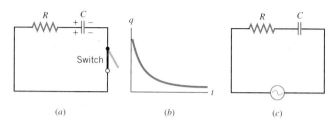

(a)   (b)   (c)

Problem The circuit elements in the drawing have the following values: $R = 18$ Ω, $V_{rms} = 24$ V for the generator, and $f = 380$ Hz. The time constant for the circuit is $\tau = 3.0 \times 10^{-4}$ s. What is the rms current in the circuit?

* 52. Concept Questions A charged capacitor and inductor are connected as shown in the drawing (this circuit is the same as that in Figure 23.14a). There is no resistance in the circuit. As Section 23.4 discusses, the electrical energy initially present in the charged capacitor then oscillates back and forth between the inductor and the capacitor. (a) What is the amount of electrical energy initially stored in the capacitor? Express your answer in terms of its capacitance C and the magnitude q of the charge on each plate. (b) A little later, this energy is transferred completely to the inductor (see Figure 23.14b). Write down an expression for the energy stored in the inductor. Give your answer in terms of its inductance L and the magnitude I of the current in the inductor. (c) If values for q, C, and L are known, how could one obtain a value for the maximum current in the inductor?

Problem The initial charge on the capacitor has a magnitude of $q = 2.90$ μC. The capacitance is $C = 3.60$ μF, and the inductance is $L = 75.0$ mH. (a) What is the electrical energy stored initially in the charged capacitor? (b) Find the maximum current in the inductor.

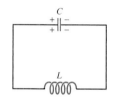

Chapter 24

Electromagnetic Waves

Colorful kites are popular in Malaysia. The colors used by this master kite maker correspond to different wavelengths in the visible region of the spectrum of electromagnetic waves. (© Hugh Sitton/Stone/ Getty Images)

24.1 *The Nature of Electromagnetic Waves*

▶ CONCEPTS AT A GLANCE In Section 13.3 we saw that energy is transported to us from the sun via a class of waves known as electromagnetic waves. This class includes the familiar visible, ultraviolet, and infrared waves. In Sections 18.6, 21.1, and 21.2 we studied the concepts of electric and magnetic fields. It was the great Scottish physicist James Clerk Maxwell (1831–1879) who showed that these two fields fluctuating together can form a propagating *electromagnetic wave.* The Concepts-at-a-Glance chart in Figure 24.1 emphasizes that we will now bring together our knowledge of electric and magnetic fields in order to understand this important type of wave. ◀

Figure 24.2 illustrates one way to create an electromagnetic wave. The setup consists of two straight metal wires that are connected to the terminals of an ac generator and serve as an antenna. The potential difference between the terminals changes sinusoidally with time t and has a period T. Part a shows the instant $t = 0$ s, when there is no charge at the ends of either wire. Since there is no charge, there is no electric field at the point P just to the right of the antenna. As time passes, the top wire becomes positively charged and the bottom wire negatively charged. One-quarter of a cycle later ($t = \frac{1}{4}T$), the charges have attained their maximum values, as part b of the drawing indicates. The corresponding electric field **E** at point P is represented by the red arrow and has increased to its maximum strength in the downward direction.* Part b also shows that the electric field created at earlier times (see the black arrow in the picture) has not disappeared but has moved to the right. Here lies the crux of the matter: At distant points, the electric field of the charges is not felt immediately. Instead, the field is created first near the wires and then, like the effect of a pebble dropped into a pond, moves outward as a wave in all directions. Only the field moving to the right is shown in the picture for the sake of clarity.

Parts c–e of Figure 24.2 show the creation of the electric field at point P (red arrow) at later times during the generator cycle. In each part, the fields produced earlier in the sequence (black arrows) continue propagating toward the right. Part d shows the charges on the wires when the polarity of the generator has reversed, so the top wire is negative and the bottom wire is positive. As a result, the electric field at P has reversed its direction and points upward. In part e of the sequence, a complete sine wave has been drawn through the tips of the electric field vectors to emphasize that the field changes sinusoidally.

Along with the electric field in Figure 24.2, a magnetic field **B** is also created, because the charges flowing in the antenna constitute an electric current, which produces a magnetic field. Figure 24.3 illustrates the field direction at point P at the instant when the current in the antenna wire is upward. With the aid of Right-Hand Rule No. 2 (thumb of

Figure 24.1 CONCEPTS AT A GLANCE Electric and magnetic fields fluctuating together form an electromagnetic wave. The human body emits infrared electromagnetic waves, so called because they have wavelengths that are longer than the wavelength corresponding to the visible color red. The photograph shows a thermogram of a man sunbathing, which was obtained using an infrared camera. The camera registers the intensity of the infrared waves emitted by different parts of the body as different colors that may be taken as an indication of temperature, from 23 °C (purple) to 35 °C (red). (© Dr. Arthur Tucker/Science Photo Library/Photo Researchers)

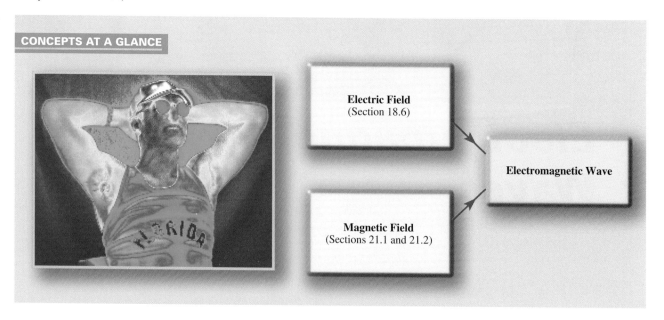

CONCEPTS AT A GLANCE

Electric Field
(Section 18.6)

Magnetic Field
(Sections 21.1 and 21.2)

Electromagnetic Wave

* The direction of the electric field can be obtained by imagining a positive test charge at P and determining the direction in which it would be pushed because of the charges on the wires.

right hand points along the current *I*, fingers curl in the direction of **B**), the magnetic field at *P* can be seen to point into the page. As the oscillating current changes, the magnetic field changes accordingly. The magnetic fields created at earlier times propagate outward as a wave, just as the electric fields do.

Notice that the magnetic field in Figure 24.3 is perpendicular to the page, whereas the electric field in Figure 24.2 lies in the plane of the page. Thus, the electric and magnetic fields created by the antenna are mutually perpendicular and remain so as they move outward. Moreover, both fields are perpendicular to the direction of travel. These perpendicular electric and magnetic fields, moving together, constitute an electromagnetic wave.

The electric and magnetic fields in Figures 24.2 and 24.3 decrease to zero rapidly with increasing distance from the antenna. Therefore, they exist mainly near the antenna and together are called the ***near field.*** Electric and magnetic fields do form a wave at large distances from the antenna, however. These fields arise from an effect that is different from that which produces the near field and are referred to as the ***radiation field.*** Faraday's law of induction provides part of the basis for the radiation field. As Section 22.4 discusses, this law describes the emf or potential difference produced by a changing magnetic field. And, as Section 19.4 explains, a potential difference can be related to an electric field. Thus, a changing magnetic field produces an electric field. Maxwell predicted that the reverse effect also occurs—namely, that a changing electric field produces a magnetic field. The radiation field arises because the changing magnetic field creates an electric field that fluctuates in time and the changing electric field creates the magnetic field.

Figure 24.4 shows the electromagnetic wave of the radiation field far from the antenna. The picture shows only the part of the wave traveling along the +*x* axis. The parts traveling in the other directions have been omitted for clarity. It should be clear from the drawing that ***an electromagnetic wave is a transverse wave*** because the electric and magnetic fields are both perpendicular to the direction in which the wave travels. Moreover, an electromagnetic wave, unlike a wave on a string or a sound wave, does not require a medium in which to propagate. ***Electromagnetic waves can travel through a vacuum or a material substance,*** since electric and magnetic fields can exist in either one.

Electromagnetic waves can be produced in situations that do not involve a wire antenna. In general, any electric charge that is accelerating emits an electromagnetic wave, whether the charge is inside a wire or not. In an alternating current, an electron oscillates in simple harmonic motion along the length of the wire and is one example of an accelerating charge.

All electromagnetic waves move through a vacuum at the same speed, and the symbol *c* is used to denote its value. This speed is called the ***speed of light in a vacuum*** and is $c = 3.00 \times 10^8$ m/s. In air, electromagnetic waves travel at nearly the same speed as they do in a vacuum, but, in general, they move through a substance such as glass at a speed that is substantially less than *c*.

The frequency of an electromagnetic wave is determined by the oscillation frequency of the electric charges at the source of the wave. In Figures 24.2–24.4 the wave frequency would equal the frequency of the ac generator. Suppose, for example, that the antenna is broadcasting electromagnetic waves known as radio waves. The frequencies of AM radio waves lie between 545 and 1605 kHz, these numbers corresponding to the lim-

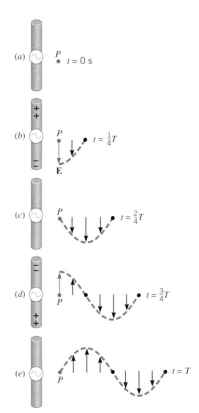

Figure 24.2 In each part of the drawing, the red arrow represents the electric field **E** produced at point *P* by the oscillating charges on the antenna at the indicated time. The black arrows represent the electric fields created at earlier times. For simplicity, only the fields propagating to the right are shown.

Figure 24.3 The oscillating current *I* in the antenna wires creates a magnetic field **B** at point *P* that is tangent to a circle centered on the wires. The field is directed into the page when the current is upward and out of the page when the current is downward.

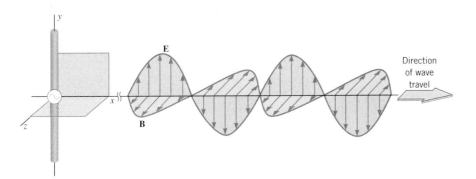

Figure 24.4 This picture shows the wave of the radiation field far from the antenna. Observe that **E** and **B** are perpendicular to each other, and both are perpendicular to the direction of travel.

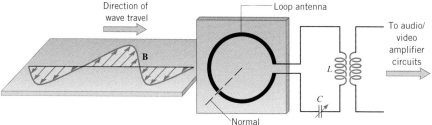

Figure 24.5 A radio wave can be detected with a receiving antenna wire that is parallel to the electric field of the wave. The magnetic field of the radio wave has been omitted for simplicity.

its of the AM broadcast band on the radio dial. The frequencies of FM radio waves lie between 88 and 108 MHz on the dial. Television channels 2–6, on the other hand, utilize electromagnetic waves with frequencies between 54 and 88 MHz, and channels 7–13 use frequencies between 174 and 216 MHz.

The physics of radio and television reception.

Radio and television reception involves a process that is the reverse of that outlined earlier for the creation of electromagnetic waves. When broadcasted waves reach a receiving antenna, they interact with the electric charges in the antenna wires. Either the electric field or the magnetic field of the waves can be used. To take full advantage of the electric field, the wires of the receiving antenna must be parallel to the electric field, as Figure 24.5 indicates. The electric field acts on the electrons in the wire, forcing them to oscillate back and forth along the length of the wire. Consequently, an ac current exists in the antenna and the circuit connected to it. The variable-capacitor C ($\dashv\!\!\!\nearrow\!\!\!\vdash$) and the inductor L in the circuit provide one way to select the frequency of the desired electromagnetic wave. By adjusting the value of the capacitance, it is possible to adjust the corresponding resonant frequency f_0 of the circuit [$f_0 = 1/(2\pi\sqrt{LC})$, Equation 23.10] to match the frequency of the wave. Under the condition of resonance there will be a maximum oscillating current in the inductor. Because of mutual inductance, this current creates a maximum voltage in the second coil in the drawing, and this voltage can then be amplified and processed by the remaining radio or television circuitry.

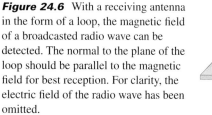

Figure 24.6 With a receiving antenna in the form of a loop, the magnetic field of a broadcasted radio wave can be detected. The normal to the plane of the loop should be parallel to the magnetic field for best reception. For clarity, the electric field of the radio wave has been omitted.

To detect the magnetic field of a broadcasted radio wave, a receiving antenna in the form of a loop can be used, as Figure 24.6 shows. For best reception, the normal to the plane of the wire loop is oriented parallel to the magnetic field. Then, as the wave sweeps by, the magnetic field penetrates the loop, and the changing magnetic flux induces a voltage and a current in the loop, in accord with Faraday's law. Once again, the resonant frequency of a capacitor/inductor combination can be adjusted to match the frequency of the desired electromagnetic wave. Both straight wire and loop antennas can be seen on the ship in Figure 24.7.

Figure 24.7 This cruise ship uses both straight wire and loop antennas to communicate with other vessels and on-shore stations. (© Harvey Lloyd/ Corbis Stock Market)

Cochlear implants use the broadcasting and receiving of radio waves to provide assistance to hearing-impaired people who have auditory nerves that are at least partially intact. These implants utilize radio waves to bypass the damaged part of the hearing mechanism and access the auditory nerve directly, as Figure 24.8 illustrates. An external microphone (often set into an ear mold) detects sound waves and sends a corresponding electrical signal to a speech processor small enough to be carried in a pocket. The speech processor encodes these signals into a radio wave, which is broadcast from an external transmitter coil placed over the site of a miniature receiver (and its receiving antenna) that has been surgically inserted beneath the skin. The receiver acts much like a radio. It detects the broadcasted wave and from the encoded audio information produces electrical

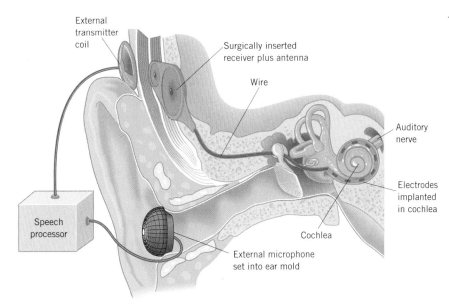

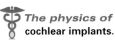
The physics of
cochlear implants.

Figure 24.8 Hearing-impaired people can sometimes recover part of their hearing with the help of a cochlear implant. The broadcasting and receiving of electromagnetic waves lies at the heart of this device.

signals that represent the sound wave. These signals are sent along a wire to electrodes that are implanted in the cochlea of the inner ear. The electrodes stimulate the auditory nerves that feed directly between structures within the cochlea and the brain. To the extent that the nerves are intact, a person can learn to recognize sounds.

The broadcasting and receiving of radio waves is also now being used in the practice of endoscopy. In this medical diagnostic technique a device called an endoscope is used to peer inside the body. For example, to examine the interior of the colon for signs of cancer, a conventional endoscope (known as a colonoscope) is inserted through the rectum. (See Section 26.3.) The wireless capsule endoscope shown in Figure 24.9 bypasses this invasive procedure completely. With a size of about 11×26 mm, this capsule can be swallowed and carried through the gastrointestinal tract by the involuntary contractions of the walls of the intestines (peristalsis). The capsule is self-contained and uses no external wires. A marvel of miniaturization, it contains a radio transmitter and its associated antenna, batteries, a white-light-emitting diode (see Section 23.5) for illumination, and an optical system to capture the digital images. As the capsule moves through the intestine, the transmitter broadcasts the images to an array of small receiving antennas attached to the patient's body. These receiving antennas also are used to determine the position of the capsule within the body. The radio waves that are used lie in the ultrahigh frequency, or UHF band, from 3×10^8 to 3×10^9 Hz.

Radio waves are only one part of the broad spectrum of electromagnetic waves that has been discovered. The next section discusses the entire spectrum.

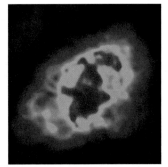

Figure 24.9 This wireless capsule endoscope is designed to be swallowed. As it passes through a patient's intestines, it broadcasts video images of the interior of the intestines. (Courtesy Given Imaging, Ltd.)

The physics of
wireless capsule endoscopy
(see above).

Four views of the Crab Nebula, which is the remnant of a star that underwent a supernova explosion in 1054 AD. It is located 6300 light-years away from the earth. Each view is in a different region of the electromagnetic spectrum, as indicated. [(*a*) Courtesy NASA; (*b*) © Mount Stromlo and Siding Spring Observatories/Photo Researchers; (*c*) Photo by Ken Chambers of UH/IFA. Photo provided courtesy of Keck Observatory; (*d*) © NRAO/AUI/NFS/Science Photo Library/Photo Researchers]

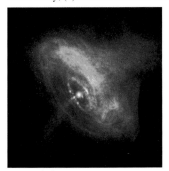
(*a*) X-ray

(*b*) Visible

(*c*) Infrared

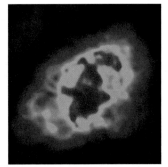
(*d*) Radio wave

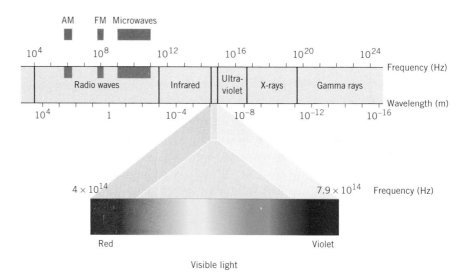

Figure 24.10 The electromagnetic spectrum.

24.2 The Electromagnetic Spectrum

An electromagnetic wave, like any periodic wave, has a frequency f and a wavelength λ that are related to the speed v of the wave by $v = f\lambda$ (Equation 16.1). For electromagnetic waves traveling through a vacuum or, to a good approximation, through air, the speed is $v = c$, so $c = f\lambda$.

As Figure 24.10 shows, electromagnetic waves exist with an enormous range of frequencies, from values less than 10^4 Hz to greater than 10^{24} Hz. Since all these waves travel through a vacuum at the same speed of $c = 3.00 \times 10^8$ m/s, Equation 16.1 can be used to find the correspondingly wide range of wavelengths that the picture also displays. The ordered series of electromagnetic wave frequencies or wavelengths in Figure 24.10 is called the *electromagnetic spectrum.* Historically, regions of the spectrum have been given names such as radio waves and infrared waves. Although the boundary between adjacent regions is shown as a sharp line in the drawing, the boundary is not so well defined in practice, and the regions often overlap.

Beginning on the left in Figure 24.10, we find radio waves. Lower-frequency radio waves are generally produced by electrical oscillator circuits, while higher-frequency radio waves (called microwaves) are usually generated using electron tubes called klystrons. Infrared radiation, sometimes loosely called heat waves, originates with the vibration and rotation of molecules within a material. Visible light is emitted by hot objects, such as the sun, a burning log, or the filament of an incandescent light bulb, when the temperature is high enough to excite the electrons within an atom. Ultraviolet frequencies can be produced from the discharge of an electric arc. X-rays are produced by the sudden deceleration of high-speed electrons. And, finally, gamma rays are radiation from nuclear decay.

The human body, like any object, radiates infrared radiation, and the amount emitted depends on the temperature of the body. Although infrared radiation cannot be seen by the human eye, it can be detected by sensors. An ear thermometer, like the pyroelectric thermometer shown in Figure 24.11, determines the body's temperature by measuring the amount of infrared radiation that emanates from the eardrum and surrounding tissue. The ear is one of the best places to measure body temperature because it is close to the hypothalamus, an area at the bottom of the brain that controls body temperature. The ear is also not cooled or warmed by eating, drinking, or breathing. When the probe of the thermometer is inserted into the ear canal, infrared radiation travels down the barrel of the probe and strikes the sensor. The absorption of infrared radiation warms the sensor, and, as a result, its electrical conductivity changes. The change in electrical conductivity is measured by an electronic circuit. The output from the circuit is sent to a microprocessor, which calculates the body temperature and displays the result on a digital readout.

Of all the frequency ranges in the electromagnetic spectrum, the most familiar is that of visible light, although it is the most narrow (see Figure 24.10). Only waves with fre-

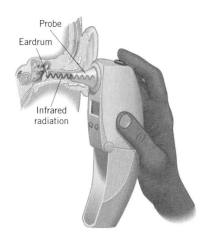

Figure 24.11 A pyroelectric thermometer measures body temperature by determining the amount of infrared radiation emitted by the eardrum and surrounding tissue.

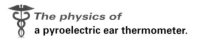

The physics of a pyroelectric ear thermometer.

quencies between about 4.0×10^{14} Hz and 7.9×10^{14} Hz are perceived by the human eye as visible light. Usually visible light is discussed in terms of wavelengths (in vacuum) rather than frequencies. As Example 1 indicates, the wavelengths of visible light are extremely small and, therefore, are normally expressed in *nanometers* (nm); 1 nm $= 10^{-9}$ m. An obsolete (non-SI) unit occasionally used for wavelengths is the *angstrom* (Å); 1 Å $= 10^{-10}$ m.

Example 1 The Wavelengths of Visible Light

Find the range in wavelengths (in vacuum) for visible light in the frequency range between 4.0×10^{14} Hz (red light) and 7.9×10^{14} Hz (violet light). Express the answers in nanometers.

Reasoning According to Equation 16.1, the wavelength (in vacuum) λ of a light wave is equal to the speed of light c in a vacuum divided by the frequency f of the wave, $\lambda = c/f$.

Solution The wavelength corresponding to a frequency of 4.0×10^{14} Hz is

$$\lambda = \frac{c}{f} = \frac{3.00 \times 10^8 \text{ m/s}}{4.0 \times 10^{14} \text{ Hz}} = 7.5 \times 10^{-7} \text{ m}$$

Since 1 nm $= 10^{-9}$ m, it follows that

$$\lambda = (7.5 \times 10^{-7} \text{ m}) \left(\frac{1 \text{ nm}}{10^{-9} \text{ m}} \right) = \boxed{750 \text{ nm}}$$

The calculation for a frequency of 7.9×10^{14} Hz is similar:

$$\lambda = \frac{c}{f} = \frac{3.00 \times 10^8 \text{ m/s}}{7.9 \times 10^{14} \text{ Hz}} = 3.8 \times 10^{-7} \text{ m} \quad \text{or} \quad \boxed{\lambda = 380 \text{ nm}}$$

The eye/brain recognizes light of different wavelengths as different colors. A wavelength of 750 nm (in vacuum) is approximately the longest wavelength of red light, whereas 380 nm (in vacuum) is approximately the shortest wavelength of violet light. Between these limits are found the other familiar colors, as Figure 24.10 indicates.

The association between color and wavelength in the visible part of the electromagnetic spectrum is well known. The wavelength also plays a central role in governing the behavior and use of electromagnetic waves in all regions of the spectrum. For instance, Conceptual Example 2 considers the influence of the wavelength on diffraction.

Conceptual Example 2 The Diffraction of AM and FM Radio Waves

As we have seen in Section 17.3, diffraction is the ability of a wave to bend around an obstacle or the edges of an opening. Based on the discussion in that chapter, would you expect AM or FM radio waves to bend more readily around an obstacle such as a building?

Reasoning and Solution Section 17.3 points out that, other things being equal, sound waves exhibit diffraction to a greater extent when the wavelength is longer than when it is shorter. Based on this information, we expect that longer-wavelength electromagnetic waves will bend more readily around obstacles than will shorter-wavelength waves. Figure 24.10 shows that AM radio waves have considerably longer wavelengths than FM waves. Therefore, *AM waves have a greater ability to bend around buildings than do FM waves.* The inability of FM waves to diffract around obstacles is a major reason why FM stations broadcast their signals essentially in a "line-of-sight" fashion.

The picture of light as a wave is supported by experiments that will be discussed in Chapter 27. However, there are also experiments indicating that light can behave as if it were composed of discrete particles rather than waves. These experiments will be discussed in Chapter 29. Wave theories and particle theories of light have been around for hundreds of years, and it is now widely accepted that light, as well as other electromagnetic radiation, exhibits a dual nature. Either wave-like or particle-like behavior can be observed, depending on the kind of experiment being performed.

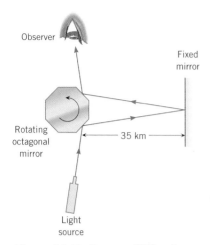

Figure 24.12 Between 1878 and 1931, Michelson used a rotating eight-sided mirror to measure the speed of light. This is a simplified version of the setup.

24.3 The Speed of Light

At a speed of 3.00×10^8 m/s, light travels from the earth to the moon in a little over a second, so the time required for light to travel between two places on earth is very short. Therefore, the earliest attempts at measuring the speed of light had only limited success. One of the first accurate measurements employed a rotating mirror, and Figure 24.12 shows a simplified version of the setup. It was used first by the French scientist Jean Foucault (1819–1868) and later in a more refined version by the American physicist Albert Michelson (1852–1931). If the angular speed of the rotating eight-sided mirror in Figure 24.12 is adjusted correctly, light reflected from one side travels to the fixed mirror, reflects, and can be detected after reflecting from another side that has rotated into place at just the right time. The minimum angular speed must be such that one side of the mirror rotates one-eighth of a revolution during the time it takes for the light to make the round trip between the mirrors. For one of his experiments, Michelson placed mirrors on Mt. San Antonio and Mt. Wilson in California, a distance of 35 km apart. From a value of the minimum angular speed in such experiments, he obtained the value of $c = (2.997\ 96 \pm 0.000\ 04) \times 10^8$ m/s in 1926.

Today, the speed of light has been determined with such high accuracy that it is used to define the meter. As discussed in Section 1.2, the speed of light is now *defined* to be

Speed of light in a vacuum

$$c = 299\ 792\ 458\ \text{m/s}$$

(although a value of 3.00×10^8 m/s is adequate for most calculations). The second is defined in terms of a cesium clock, and the meter is then defined as the distance light travels in a vacuum during a time of $1/(299\ 792\ 458)$ s. Although the speed of light in a vacuum is large, it is finite, so it takes a finite amount of time for light to travel from one place to another. The travel time is especially long for light traveling between astronomical objects, as Conceptual Example 3 discusses.

Conceptual Example 3 Looking Back in Time

A supernova is a violent explosion that occurs at the death of certain stars. For a few days after the explosion, the intensity of the emitted light can become a billion times greater than that of our own sun. But after several years, the intensity usually returns to zero. Supernovae are relatively rare events in the universe, for only six have been observed in our galaxy within the past 400 years. One of them was recorded in 1987. It occurred in a neighboring galaxy, approximately 1.66×10^{21} m away. Figure 24.13 shows a photograph of the sky (*a*) before and (*b*) a

Figure 24.13 A view of the sky (*a*) before and (*b*) a few hours after the 1987 supernova. (© Courtesy Anglo-Australian Telescope Board)

(*a*)

(*b*)

few hours after the explosion. Why do astronomers say that viewing an event like the supernova is like looking back in time?

Reasoning and Solution The light from the supernova traveled to earth at a speed of $c = 3.00 \times 10^8$ m/s. Even at such a speed, the travel time is large because the distance $d = 1.66 \times 10^{21}$ m is enormous. The time t is $t = d/c = (1.66 \times 10^{21}$ m$)/(3.00 \times 10^8$ m/s$) = 5.53 \times 10^{12}$ s. This corresponds to 175 000 years. So *when astronomers saw the explosion in 1987, they were actually seeing the light that left the supernova 175 000 years earlier;* in other words, they were "looking back in time." In fact, whenever we view any celestial object, such as a star, we are seeing it as it was a long time ago. The farther the object is from earth, the longer it takes for the light to reach us, and the further back in time we are looking.

Related Homework: *Problem 14*

In 1865, Maxwell determined theoretically that electromagnetic waves propagate through a vacuum at a speed given by

$$c = \frac{1}{\sqrt{\epsilon_0 \mu_0}} \qquad (24.1)$$

where $\epsilon_0 = 8.85 \times 10^{-12}$ C^2/(N·m^2) is the (electric) permittivity of free space and $\mu_0 = 4\pi \times 10^{-7}$ T·m/A is the (magnetic) permeability of free space. Originally ϵ_0 was introduced in Section 18.5 as an alternative way of writing the proportionality constant k in Coulomb's law [$k = 1/(4\pi\epsilon_0)$] and, hence, plays a basic role in determining the strengths of the electric fields created by point charges. The role of μ_0 is similar for magnetic fields; it was introduced in Section 21.7 as part of a proportionality constant in the expression for the magnetic field created by the current in a long, straight wire. Substituting the values for ϵ_0 and μ_0 into Equation 24.1 shows that

$$c = \frac{1}{\sqrt{[8.85 \times 10^{-12}\ \text{C}^2/(\text{N·m}^2)][4\pi \times 10^{-7}\ \text{T·m/A}]}} = 3.00 \times 10^8\ \text{m/s}$$

The experimental and theoretical values for c agree. Maxwell's success in predicting c provided a basis for inferring that light behaves as a wave consisting of oscillating electric and magnetic fields.

✓ **Check Your Understanding 1**

The frequency of electromagnetic wave A is twice that of electromagnetic wave B. For these two waves, what is the ratio λ_A/λ_B of the wavelengths in a vacuum? (a) $\lambda_A/\lambda_B = 2$, because wave A has twice the speed that wave B has. (b) $\lambda_A/\lambda_B = 2$, because wave A has one-half the speed that wave B has. (c) $\lambda_A/\lambda_B = \frac{1}{2}$, because wave A has one-half the speed that wave B has. (d) $\lambda_A/\lambda_B = \frac{1}{2}$, because wave A has twice the speed that wave B has. (e) $\lambda_A/\lambda_B = \frac{1}{2}$, because both waves have the same speed. *(The answer is given at the end of the book.)*

Background: Electromagnetic waves have a frequency f, wavelength λ, and speed v that are related according to $f\lambda = v$ (Equation 16.1).

For similar questions (including calculational counterparts), consult Self-Assessment Test 24.1, which is described at the end of Section 24.6.

24.4 The Energy Carried by Electromagnetic Waves

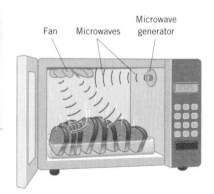

Figure 24.14 A microwave oven. The rotating fan blades reflect the microwaves to all parts of the oven.

Electromagnetic waves, like water waves or sound waves, carry energy. The energy is carried by the electric and magnetic fields that comprise the wave. In a microwave oven, for example, microwaves penetrate food and deliver their energy to it, as Figure 24.14 illustrates. The electric field of the microwaves is largely responsible for delivering the energy, and water molecules in the food absorb it. The absorption occurs because each water molecule has a permanent dipole moment; that is, one end of a molecule has a slight positive charge, and the other end has a negative charge of equal magnitude. As a result,

The physics of a microwave oven.

the positive and negative ends of different molecules can form a bond. However, the electric field of the microwaves exerts forces on the positive and negative ends of a molecule, causing it to spin. Because the field is oscillating rapidly—about 2.4×10^9 times a second—the water molecules are kept spinning at a high rate. In the process, the energy of the microwaves is used to break bonds between neighboring water molecules and ultimately is converted into internal energy. As the internal energy increases, the temperature of the water increases, and the food cooks.

The physics of the greenhouse effect.

The energy carried by electromagnetic waves in the infrared and visible regions of the spectrum plays the key role in the greenhouse effect that is a contributing factor to global warming. The infrared waves from the sun are largely prevented from reaching the earth's surface by carbon dioxide and water in the atmosphere, which reflect them back into space. The visible waves do reach the earth's surface, however, and the energy they carry heats the earth. Heat also flows to the surface from the interior of the earth. The heated surface in turn radiates infrared waves outward, which, if they could, would carry their energy into space. However, the atmospheric carbon dioxide and water reflect these infrared waves back toward the earth, just as they reflect the infrared waves from the sun. Thus, their energy is trapped, and the earth becomes warmer, like plants in a greenhouse. In a greenhouse, however, energy is trapped mainly for a different reason—namely, the lack of effective convection currents to carry warm air past the cold glass walls.

A measure of the energy stored in the electric field **E** of an electromagnetic wave, such as a microwave, is provided by the electric energy density. As we saw in Section 19.5, this density is the electric energy per unit volume of space in which the electric field exists:

$$\frac{\text{Electric energy}}{\text{density}} = \frac{\text{Electric energy}}{\text{Volume}} = \frac{1}{2}\kappa\epsilon_0 E^2 = \frac{1}{2}\epsilon_0 E^2 \qquad (19.12)$$

In this equation, the dielectric constant κ has been set equal to unity, since we are dealing with an electric field in a vacuum (or air). From Section 22.8, the analogous expression for the magnetic energy density is

$$\frac{\text{Magnetic energy}}{\text{density}} = \frac{\text{Magnetic energy}}{\text{Volume}} = \frac{1}{2\mu_0} B^2 \qquad (22.11)$$

The **total energy density** u of an electromagnetic wave in a vacuum is the sum of these two energy densities:

$$u = \frac{\text{Total energy}}{\text{Volume}} = \frac{1}{2}\epsilon_0 E^2 + \frac{1}{2\mu_0} B^2 \qquad (24.2)$$

In an electromagnetic wave propagating through a vacuum or air, the electric field and the magnetic field carry equal amounts of energy per unit volume of space. Since $\frac{1}{2}\epsilon_0 E^2 = \frac{1}{2}(B^2/\mu_0)$, it is possible to rewrite Equation 24.2 for the total energy density in two additional, but equivalent, forms:

$$u = \frac{1}{2}\epsilon_0 E^2 + \frac{1}{2\mu_0} B^2 \qquad (24.2a)$$

$$u = \epsilon_0 E^2 \qquad (24.2b)$$

$$u = \frac{1}{\mu_0} B^2 \qquad (24.2c)$$

The fact that the two energy densities are equal implies that the electric and magnetic fields are related. To see how, we set the electric energy density equal to the magnetic energy density and obtain

$$\frac{1}{2}\epsilon_0 E^2 = \frac{1}{2\mu_0} B^2 \quad \text{or} \quad E^2 = \frac{1}{\epsilon_0\mu_0} B^2$$

But according to Equation 24.1, $c = 1/\sqrt{\epsilon_0\mu_0}$, so it follows that $E^2 = c^2 B^2$. Taking the square root of both sides of this result shows that the relation between the magnitudes of the electric and magnetic fields in an electromagnetic wave is

$$E = cB \qquad (24.3)$$

In an electromagnetic wave, the electric and magnetic fields fluctuate sinusoidally in time, so Equations 24.2a–c give the energy density of the wave at any instant in time. If an average value $\bar{u}$ for the total energy density is desired, average values are needed for E^2 and B^2. In Section 20.5 we faced a similar situation for alternating currents and voltages and introduced rms (root mean square) quantities. Using an analogous procedure here, we find that the rms values for the electric and magnetic fields, E_{rms} and B_{rms}, are related to the maximum values of these fields, E_0 and B_0, by

$$E_{rms} = \frac{1}{\sqrt{2}} E_0 \quad \text{and} \quad B_{rms} = \frac{1}{\sqrt{2}} B_0$$

Equations 24.2a–c can now be interpreted as giving the average energy density $\bar{u}$, provided the symbols E and B are interpreted to mean the rms values given above. The average density of the sunlight reaching the earth is determined in the next example.

Example 4 The Average Energy Density of Sunlight

Sunlight enters the top of the earth's atmosphere with an electric field whose rms value is $E_{rms} = 720$ N/C. Find (a) the average total energy density of this electromagnetic wave and (b) the rms value of the sunlight's magnetic field.

Reasoning The average total energy density $\bar{u}$ of the sunlight can be obtained from Equation 24.2b, provided the rms value is used for the electric field. Since the magnitudes of the magnetic and electric fields are related according to Equation 24.3, the rms value of the magnetic field is $B_{rms} = E_{rms}/c$.

Solution

(a) According to Equation 24.2b, the average total energy density is

$$\bar{u} = \epsilon_0 E_{rms}^2 = [8.85 \times 10^{-12}\ \text{C}^2/(\text{N} \cdot \text{m}^2)](720\ \text{N/C})^2 = \boxed{4.6 \times 10^{-6}\ \text{J/m}^3}$$

(b) Using Equation 24.3, we find that the rms magnetic field is

$$B_{rms} = \frac{E_{rms}}{c} = \frac{720\ \text{N/C}}{3.0 \times 10^8\ \text{m/s}} = \boxed{2.4 \times 10^{-6}\ \text{T}}$$

▶ CONCEPTS AT A GLANCE As an electromagnetic wave moves through space, it carries energy from one region to another. This energy transport is characterized by the *intensity* of the wave. We have encountered the concept of intensity before, in connection with sound waves in Section 16.7. The Concepts-at-a-Glance chart in Figure 16.22 and also that in Figure 24.15 illustrate that the intensity of a wave is formulated by bringing

Figure 24.15 CONCEPTS AT A GLANCE The intensity of an electromagnetic wave is formulated just as it is for a sound wave, by bringing together the power of the wave and the area through which the power passes. The ocean lovers in the photograph are using an umbrella to protect themselves from the intensity of the sun's electromagnetic waves. (© Belinda Banks/Stone/Getty Images)

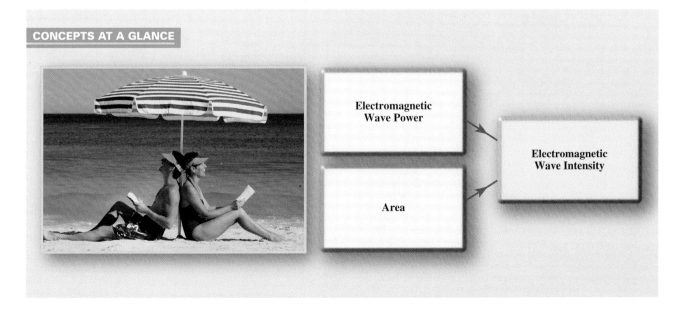

CONCEPTS AT A GLANCE

Electromagnetic Wave Power

Area

Electromagnetic Wave Intensity

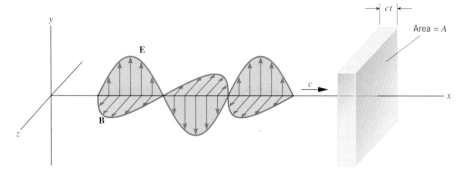

Figure 24.16 In a time t, an electromagnetic wave moves a distance ct along the x axis and passes through a surface of area A.

together the power of the wave and the area through which it passes. The sound intensity is the sound power that passes perpendicularly through a surface divided by the area of the surface. The intensity of an electromagnetic wave is defined similarly. For an electromagnetic wave, the intensity is the electromagnetic power divided by the area of the surface. ◄

Using this definition of intensity, we can show that the electromagnetic intensity S is related to the energy density u. According to Equation 16.8 the intensity is the power P that passes perpendicularly through a surface divided by the area A of that surface, or $S = P/A$. Furthermore, the power is equal to the total energy passing through the surface divided by the elapsed time t (Equation 6.10b), so that $P = $ (Total energy)/t. Combining these two relations gives

$$S = \frac{P}{A} = \frac{\text{Total energy}}{tA}$$

Now, consider Figure 24.16, which shows an electromagnetic wave traveling in a vacuum along the x axis. In a time t the wave travels the distance ct, passing through the surface of area A. Consequently, the volume of space through which the wave passes is ctA. The total (electric and magnetic) energy in this volume is

$$\text{Total energy} = (\text{Total energy density}) \times \text{Volume} = u(ctA)$$

Using this result in the expression for the intensity, we obtain

$$S = \frac{\text{Total energy}}{tA} = \frac{uctA}{tA} = cu \tag{24.4}$$

Thus, the intensity and the energy density are related by the speed of light, c. Substituting Equations 24.2a–c, one at a time, into Equation 24.4 shows that the intensity of an electromagnetic wave depends on the electric and magnetic fields according to the following equivalent relations:

$$S = cu = \frac{1}{2}c\epsilon_0 E^2 + \frac{c}{2\mu_0}B^2 \tag{24.5a}$$

$$S = c\epsilon_0 E^2 \tag{24.5b}$$

$$S = \frac{c}{\mu_0}B^2 \tag{24.5c}$$

If the rms values for the electric and magnetic fields are used in Equations 24.5a–c, the intensity becomes an average intensity, $\overline{S}$, as Example 5 illustrates.

Example 5   A Neodymium–Glass Laser

A neodymium–glass laser emits short pulses of high-intensity electromagnetic waves. The electric field has an rms value of $E_{\text{rms}} = 2.0 \times 10^9$ N/C. Find the average power of each pulse that passes through a 1.6×10^{-5}-m² surface that is perpendicular to the laser beam.

Reasoning Since the intensity of a wave is the power per unit area that passes perpendicularly through a surface, the average power $\overline{P}$ of the wave is the product of the average intensity $\overline{S}$ and the area A.

Solution The average power is $\overline{P} = \overline{S}A$. But from Equation 24.5b, $\overline{S} = c\epsilon_0 E_{rms}^2$ so that

$$\overline{P} = c\epsilon_0 E_{rms}^2 A$$
$$\overline{P} = (3.0 \times 10^8 \text{ m/s})[8.85 \times 10^{-12} \text{ C}^2/(\text{N}\cdot\text{m}^2)](2.0 \times 10^9 \text{ N/C})^2(1.6 \times 10^{-5} \text{ m}^2)$$
$$= \boxed{1.7 \times 10^{11} \text{ W}}$$

Problem solving insight
The concepts of power and intensity are similar, but they are not the same. Intensity is the power that passes perpendicularly through a surface divided by the area of the surface.

Check Your Understanding 2

If both the electric and magnetic fields of an electromagnetic wave double in magnitude, how does the intensity of the wave change? (a) The intensity decreases by a factor of four. (b) The intensity decreases by a factor of two. (c) The intensity increases by a factor of two. (d) The intensity increases by a factor of four. (e) The intensity increases by a factor of eight. *(The answer is given at the end of the book.)*

Background: The intensity of an electromagnetic wave is the energy per second per square meter that the wave carries through a surface that is perpendicular to the direction of travel of the wave. It depends on the magnitudes of both the electric and magnetic fields.

For similar questions (including calculational counterparts), consult Self-Assessment Test 24.1, which is described at the end of Section 24.6.

24.5 The Doppler Effect and Electromagnetic Waves

Section 16.9 presents a discussion of the Doppler effect that sound waves exhibit when either the source of a sound wave, the observer of the wave, or both are moving with respect to the medium of propagation (e.g., air). This effect is one in which the observed sound frequency is greater or smaller than the frequency emitted by the source. A different Doppler effect arises when the source moves than when the observer moves.

Electromagnetic waves also can exhibit a Doppler effect, but it differs from that for sound waves for two reasons. First, sound waves require a medium such as air in which to propagate. In the Doppler effect for sound, it is the motion (of the source, the observer, and the waves themselves) relative to this medium that is important. In the Doppler effect for electromagnetic waves, motion relative to a medium plays no role, because the waves do not require a medium in which to propagate. They can travel in a vacuum. Second, in the equations for the Doppler effect in Section 16.9, the speed of sound plays an important role, and it depends on the reference frame relative to which it is measured. For example, the speed of sound with respect to moving air is different than it is with respect to stationary air. As we will see in Section 28.2, electromagnetic waves behave in a different way. The speed at which they travel has the same value, whether it is measured relative to a stationary observer or relative to one moving at a constant velocity. For these two reasons, the same Doppler effect arises for electromagnetic waves when either the source or the observer of the waves moves; only the relative motion of the source and the observer with respect to one another is important.

When electromagnetic waves and the source and the observer of the waves all travel along the same line in a vacuum (or in air, to a good degree of approximation), the single equation that specifies the Doppler effect is

$$f_o = f_s \left(1 \pm \frac{v_{rel}}{c}\right) \qquad \text{if } v_{rel} \ll c \qquad (24.6)$$

In this expression, f_o is the observed frequency, and f_s is the frequency emitted by the source. The symbol v_{rel} stands for the speed of the source and the observer relative to one another, and c is the speed of light in a vacuum. Equation 24.6 applies only if v_{rel} is very small compared to c—that is, if $v_{rel} \ll c$. The plus sign in Equation 24.6 applies when the source and observer are moving toward one another, and the minus sign applies when they are moving apart.

It is essential to realize that v_{rel} is the ***relative*** speed of the source and the observer. Thus, if the source is moving due east at a speed of 28 m/s with respect to the earth, while the observer is moving due east at a speed of 22 m/s, the value for v_{rel} is 28 m/s − 22 m/s = 6 m/s. Because v_{rel} is the relative ***speed,*** it has no algebraic sign. The direction of the relative motion is taken into account by choosing the plus or minus sign in Equation 24.6. The plus sign is used when the source and the observer come together, and the minus sign is used when they move apart. Example 6 illustrates one familiar use of the Doppler effect for electromagnetic waves.

Example 6   Radar Guns and Speed Traps

The physics of
radar speed traps.

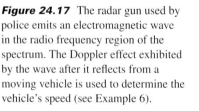

Police use radar guns and the Doppler effect to catch speeders. One such gun emits an electromagnetic wave with a frequency of $f_s = 8.0 \times 10^9$ Hz, as Figure 24.17 illustrates. In this picture, a car is approaching a police car parked on the side of the road. The direction of approach is essentially head-on. The wave from the radar gun reflects from the speeding car and returns to the police car, where on-board equipment measures its frequency to be greater than that of the emitted wave by 2100 Hz. Find the speed of the car with respect to the highway.

Reasoning The Doppler effect depends on the relative speed v_{rel} between the speeding car and the police car. We will find this speed and then relate it to the speed of the moving car with respect to the highway by using the fact that the police car is at rest. There are two Doppler frequency changes in this situation. First, the speeder's car "observes" the wave frequency coming from the radar gun to have a frequency f_o that is different from the emitted frequency f_s. According to Equation 24.6 (with the plus sign, since the two cars are coming together), $f_o - f_s = f_s(v_{rel}/c)$. Then, the wave reflects and returns to the police car, where it is observed to have a frequency f_o' that is different than its frequency f_o at the instant of reflection. Again using Equation 24.6, we find that $f_o' - f_o = f_o(v_{rel}/c)$. Adding the two previous equations gives the following result for the total Doppler change in frequency:

Problem solving insight
The Doppler effect for electromagnetic waves depends on the speed v_{rel} of the observer and the source of the waves relative to one another. In general, do not use the speed of the observer or of the source with respect to the ground in Equation 24.6.

$$(f_o' - f_o) + (f_o - f_s) = f_o' - f_s = f_o\left(\frac{v_{rel}}{c}\right) + f_s\left(\frac{v_{rel}}{c}\right) \approx 2f_s\left(\frac{v_{rel}}{c}\right)$$

where we have assumed that f_o and f_s differ by only a negligibly small amount, since v_{rel} is small compared to the speed of light c. Solving for the relative speed v_{rel} gives

$$v_{rel} \approx \left(\frac{f_o' - f_s}{2f_s}\right)c$$

Solution The speed of the moving car relative to the police car is

$$v_{rel} \approx \left(\frac{f_o' - f_s}{2f_s}\right)c = \left[\frac{2100 \text{ Hz}}{2(8.0 \times 10^9 \text{ Hz})}\right](3.0 \times 10^8 \text{ m/s}) = 39 \text{ m/s}$$

Since the police car is at rest, its speed is $v_P = 0$ m/s, and the speeder has a speed of $v_{rel} = v - v_P = v = \boxed{39 \text{ m/s}}$.

Figure 24.17 The radar gun used by police emits an electromagnetic wave in the radio frequency region of the spectrum. The Doppler effect exhibited by the wave after it reflects from a moving vehicle is used to determine the vehicle's speed (see Example 6).

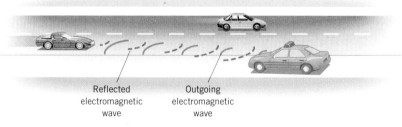

Reflected
electromagnetic
wave

Outgoing
electromagnetic
wave

The Doppler effect of electromagnetic waves provides a powerful tool for astronomers. For instance, Example 10 in Chapter 5 discusses how astronomers have identified a supermassive black hole at the center of galaxy M87 by using the Hubble space telescope. They focused the telescope on regions to either side of the center of the galaxy (see Figure 5.15). From the light emitted by these two regions, they were able to use the Doppler effect to determine that one side is moving away from the earth, while the other side is moving toward the earth. In other words, the galaxy is rotating. The speeds of recession and approach enabled astronomers to determine the rotational speed of the galaxy, and Example 10 in Chapter 5 shows how the value for this speed leads to the identification of the black hole. Astronomers routinely study the Doppler effect of the light that reaches the earth from distant parts of the universe. From such studies, they have determined the speeds at which distant light-emitting objects are receding from the earth.

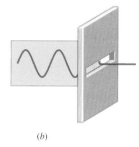

(a)

✔ **Check Your Understanding 3**

The drawing shows three situations A, B, and C in which an observer and a source of electromagnetic waves are moving along the same line. In each case the source emits a wave of the same frequency. The arrows in each situation denote velocity vectors relative to the ground and have magnitudes of either v or $2v$. Rank the frequencies of the observed waves in descending order (largest first) according to magnitude. *(The answer is given at the end of the book.)*

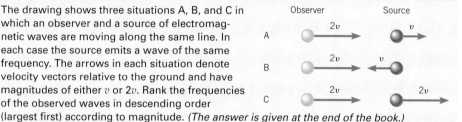

Background: The electromagnetic wave frequency that is observed is not necessarily the same as the frequency emitted by the source because of the Doppler effect. The observed frequency depends on the speed of the source and the observer relative to one another.

For similar questions (including calculational counterparts), consult Self-Assessment Test 24.1, which is described at the end of Section 24.6.

(b)

Figure 24.18 A transverse wave is linearly polarized when its vibrations always occur along one direction. (*a*) A linearly polarized wave on a rope can pass through a slit that is parallel to the direction of the rope vibrations, but (*b*) cannot pass through a slit that is perpendicular to the vibrations.

24.6 *Polarization*

POLARIZED ELECTROMAGNETIC WAVES

One of the essential features of electromagnetic waves is that they are transverse waves, and because of this feature they can be polarized. Figure 24.18 illustrates the idea of polarization by showing a transverse wave as it travels along a rope toward a slit. The wave is said to be ***linearly polarized,*** which means that its vibrations always occur along one direction. This direction is called the direction of polarization. In part *a* of the picture, the direction of polarization is vertical, parallel to the slit. Consequently, the wave passes through easily. However, when the slit is turned perpendicular to the direction of polarization, as in part *b*, the wave cannot pass, because the slit prevents the rope from oscillating. For longitudinal waves, such as sound waves, the notion of polarization has no meaning. In a longitudinal wave the direction of vibration is along the direction of travel, and the orientation of the slit would have no effect on the wave.

In an electromagnetic wave such as that in Figure 24.4, the electric field oscillates along the *y* axis. Similarly, the magnetic field oscillates along the *z* axis. Therefore, the wave is linearly polarized, with the direction of polarization taken arbitrarily to be that along which the electric field oscillates. If the wave is a radio wave generated by a straight-wire antenna, the direction of polarization is determined by the orientation of the antenna. In comparison, the visible light given off by an incandescent light bulb consists of electromagnetic waves that are completely unpolarized. In this case the waves are emitted by a large number of atoms in the hot filament of the bulb. When an electron in an atom oscillates, the atom behaves as a miniature antenna that broadcasts light for brief periods of time, about 10^{-8} seconds. However, the directions of these atomic antennas change randomly as a result of collisions. Unpolarized light, then, consists of many individual waves, emitted in short bursts by many "atomic antennas," each with its own direction of polarization. Figure 24.19 compares polarized and unpolarized light. In the unpolarized case, the arrows shown around the direction of wave travel symbolize the random directions of polarization of the individual waves that comprise the light.

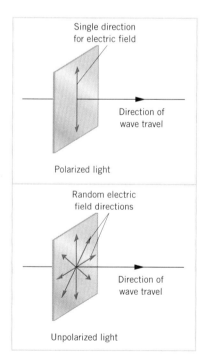

Figure 24.19 In polarized light, the electric field of the electromagnetic wave fluctuates along a single direction. Unpolarized light consists of short bursts of electromagnetic waves emitted by many different atoms. The electric field directions of these bursts are perpendicular to the direction of wave travel but are distributed randomly about it.

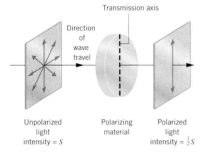

Figure 24.20 With the aid of a piece of polarizing material, polarized light may be produced from unpolarized light. The transmission axis of the material is the direction of polarization of the light that passes through the material.

Linearly polarized light can be produced from unpolarized light with the aid of certain materials. One commercially available material goes under the name of Polaroid. Such materials allow only the component of the electric field along one direction to pass through, while absorbing the field component perpendicular to this direction. As Figure 24.20 indicates, the direction of polarization that a polarizing material allows through is called the ***transmission axis.*** No matter how this axis is oriented, the intensity of the transmitted polarized light is one-half that of the incident unpolarized light. The reason for this is that the unpolarized light contains all polarization directions to an equal extent. Moreover, the electric field for each direction can be resolved into components perpendicular and parallel to the transmission axis, with the result that the average components perpendicular and parallel to the axis are equal. As a result, the polarizing material absorbs as much of the electric (and magnetic) field strength as it transmits.

MALUS' LAW

Once polarized light has been produced with a piece of polarizing material, it is possible to use a second piece to change the polarization direction and simultaneously adjust the intensity of the light. Figure 24.21 shows how. As in this picture the first piece of polarizing material is called the ***polarizer,*** and the second piece is referred to as the ***analyzer.*** The transmission axis of the analyzer is oriented at an angle θ relative to the transmission axis of the polarizer. If the electric field strength of the polarized light incident on the analyzer is E, the field strength passing through is the component parallel to the transmission axis, or $E \cos \theta$. According to Equation 24.5b, the intensity is proportional to the square of the electric field strength. Consequently, the average intensity of polarized light passing through the analyzer is proportional to $\cos^2 \theta$. Thus, both the polarization direction and the intensity of the light can be adjusted by rotating the transmission axis of the analyzer relative to that of the polarizer. The average intensity $\overline{S}$ of the light leaving the analyzer, then, is

Malus' law
$$\overline{S} = \overline{S}_0 \cos^2 \theta \qquad (24.7)$$

where $\overline{S}_0$ is the average intensity of the light entering the analyzer. Equation 24.7 is sometimes called ***Malus' law,*** for it was discovered by the French engineer Etienne-Louis Malus (1775–1812). Example 7 illustrates the use of Malus' law.

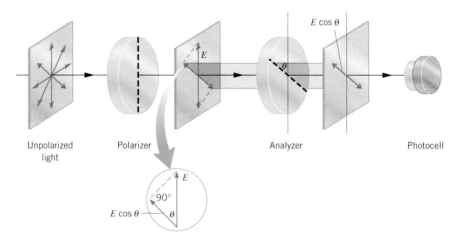

Figure 24.21 Two sheets of polarizing material, called the polarizer and the analyzer, may be used to adjust the polarization direction and intensity of the light reaching the photocell. This can be done by changing the angle θ between the transmission axes of the polarizer and analyzer.

Example 7 Using Polarizers and Analyzers

What value of θ should be used in Figure 24.21, so the average intensity of the polarized light reaching the photocell is one-tenth the average intensity of the unpolarized light?

Reasoning Both the polarizer and the analyzer reduce the intensity of the light. The polarizer reduces the intensity by a factor of one-half, as discussed earlier. Therefore, if the average intensity of the unpolarized light is $\overline{I}$, the average intensity of the polarized light leaving the

Problem solving insight
Remember that when unpolarized light strikes a polarizer, only one half of the incident light is transmitted, the other half being absorbed by the polarizer.

polarizer and striking the analyzer is $\bar{S}_0 = \bar{I}/2$. The angle θ must now be selected so the average intensity of the light leaving the analyzer is $\bar{S} = \bar{I}/10$. Malus' law provides the solution.

Solution Using $\bar{S}_0 = \bar{I}/2$ and $\bar{S} = \bar{I}/10$ in Malus' law, we find

$$\tfrac{1}{10}\bar{I} = \tfrac{1}{2}\bar{I} \cos^2 \theta$$

$$\tfrac{1}{5} = \cos^2 \theta \quad \text{or} \quad \theta = \cos^{-1}\left(\frac{1}{\sqrt{5}}\right) = \boxed{63.4^\circ}$$

Figure 24.22 When Polaroid sunglasses are uncrossed (left photograph), the transmitted light is dimmed due to the extra thickness of tinted plastic. However, when they are crossed (right photograph), the intensity of the transmitted light is reduced to zero because of the effects of polarization. (© Diane Schiumo/ Fundamental Photographs)

When $\theta = 90^\circ$ in Figure 24.21, the polarizer and analyzer are said to be **crossed,** and no light is transmitted by the polarizer/analyzer combination. As an illustration of this effect, Figure 24.22 shows two pairs of Polaroid sunglasses in uncrossed and crossed configurations.

An exciting application of crossed polarizers is used in viewing IMAX 3-D movies. These movies are recorded on two separate rolls of film, using a camera that provides images from the two different perspectives that correspond to what is observed by human eyes and allow us to see in three dimensions. The camera has two apertures or openings located at roughly the spacing between our eyes. The films are projected using a projector with two lenses, as Figure 24.23 indicates. Each lens has its own polarizer, and the two polarizers are crossed (see the drawing). In one type of theater, viewers watch the action on-screen using glasses with corresponding polarizers for the left and right eyes, as the drawing shows. Because of the crossed polarizers the left eye sees only the image from the left lens of the projector, and the right eye sees only the image from the right lens. Since the two images have the approximate perspectives that the left and right eyes would see in reality, the brain combines the images to produce a realistic 3-D effect.

The physics of
IMAX 3-D films.

Projector

Figure 24.23 In an IMAX 3-D film, two separate rolls of film are projected using a projector with two lenses, each with its own polarizer. The two polarizers are crossed. Viewers watch the action on-screen through glasses that have corresponding crossed polarizers for each eye. The result is a 3-D moving picture, as the text discusses.

Conceptual Example 8 illustrates an interesting result that occurs when a piece of polarizing material is inserted between a crossed polarizer and analyzer.

Conceptual Example 8
How Can a Crossed Polarizer and Analyzer Transmit Light?

As explained earlier, no light reaches the photocell in Figure 24.21 when the polarizer and analyzer are crossed. Suppose that a third piece of polarizing material is inserted between the polarizer and analyzer, as in Figure 24.24*a*. Does light now reach the photocell?

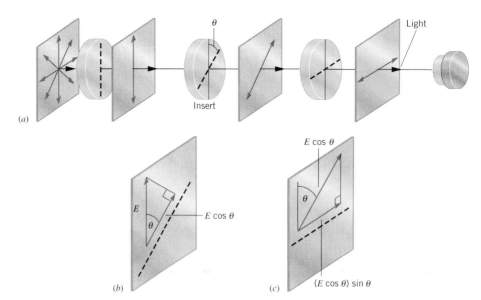

Figure 24.24 (a) Light reaches the photocell when a piece of polarizing material is inserted between a crossed polarizer and analyzer. (b) The component of the electric field parallel to the transmission axis of the insert is $E \cos \theta$. (c) Light incident on the analyzer has a component $(E \cos \theta) \sin \theta$ parallel to its transmission axis.

Reasoning and Solution Surprisingly, the answer is yes, even though the polarizer and analyzer are crossed. To show why, we note that if any light is to pass through the analyzer, it must have an electric field component parallel to the transmission axis of the analyzer. Without the insert, there is no such component. But with the insert, there is, as parts b and c of the drawing illustrate. Part b shows that the electric field E of the light leaving the polarizer makes an angle θ with respect to the transmission axis of the insert. The component of the field that is parallel to the insert's transmission axis is $E \cos \theta$, and this component passes through the insert. Part c of the drawing illustrates that the field $(E \cos \theta)$ incident on the analyzer has a component parallel to the transmission axis of the analyzer—namely, $(E \cos \theta) \sin \theta$. This component passes through the analyzer, so that light now reaches the photocell.

Light will reach the photocell as long as the angle θ is between 0 and 90°. If the angle is 0 or 90°, however, no light reaches the photocell. Can you explain, without using any equations, why no light reaches the photocell under either of these conditions?

Related Homework: *Problems 37, 50*

Need More Practice?

Interactive LearningWare 24.1
The drawing shows three sheets of polarizing material. The orientation of the transmission axis for each is labeled relative to the vertical direction. A beam of unpolarized light, whose intensity is 1550 W/m², is incident on the first sheet. What is the intensity of the beam transmitted through the three sheets when $\theta_1 = 30.0°$, $\theta_2 = 45.0°$, and $\theta_3 = 70.0°$?

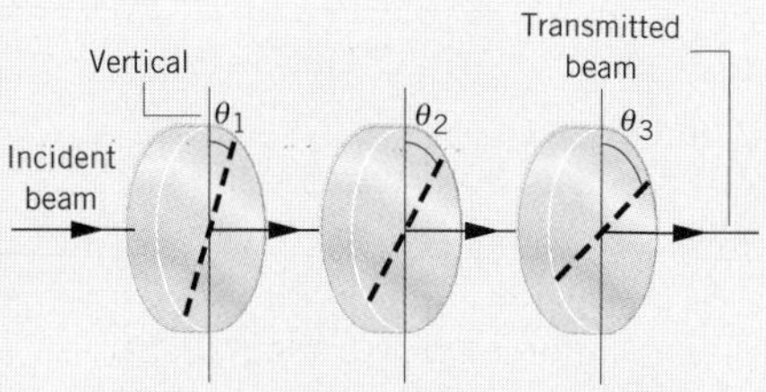

Related Homework: *Problem 36*

Go to **www.wiley.com/college/cutnell** for an interactive solution.

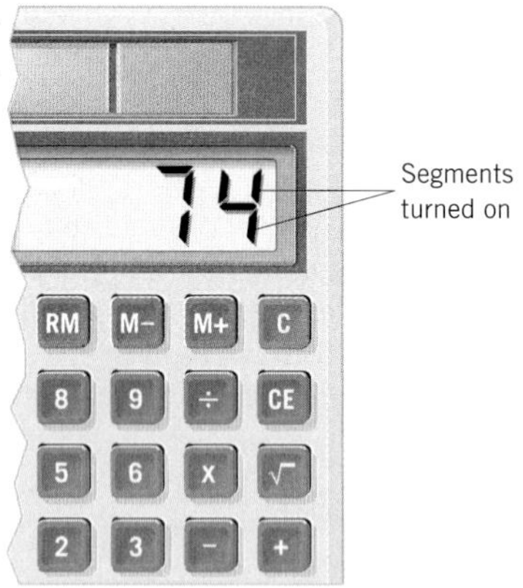

Figure 24.25 Liquid crystal displays use liquid crystal segments to form the numbers.

The physics of
a liquid crystal display.

An application of a crossed polarizer/analyzer combination occurs in one kind of liquid crystal display (LCD). LCDs are widely used in pocket calculators and digital watches. The display usually consists of blackened numbers and letters set against a light gray background. As Figure 24.25 indicates, each number or letter is formed from a combination of liquid crystal segments that have been turned on and appear black. The liquid crystal part of an LCD segment consists of the liquid crystal material sandwiched between two transparent electrodes, as in Figure 24.26. When a voltage is applied between the electrodes, the liquid crystal is said to be "on." Part a of the picture shows that linearly polarized incident light passes through the "on" material without having its direction of polarization affected. When the voltage is removed, as in part b, the liquid crystal is said to be

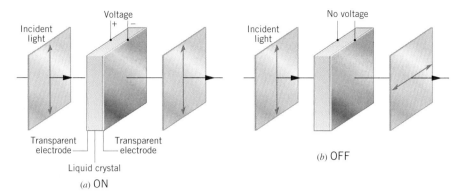

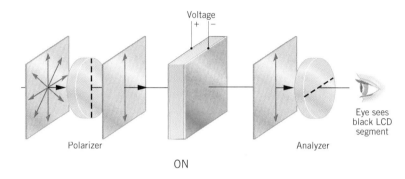

(a) ON

(b) OFF

Figure 24.26 A liquid crystal in its *(a)* "on" state and *(b)* "off" state.

Figure 24.27 A liquid crystal display (LCD) incorporates a crossed polarizer/analyzer combination. When the LCD segment is turned on (voltage applied), no light is transmitted through the analyzer, and the observer sees a black segment.

"off" and now rotates the direction of polarization by 90°. A complete LCD segment also includes a crossed polarizer/analyzer combination, as Figure 24.27 illustrates. The polarizer, analyzer, electrodes, and liquid crystal material are packaged as a single unit. The polarizer produces polarized light from incident unpolarized light. With the display segment turned on, as in Figure 24.27, the polarized light emerges from the liquid crystal only to be absorbed by the analyzer, since the light is polarized perpendicular to the transmission axis of the analyzer. Since no light emerges from the analyzer, an observer sees a black segment against a light gray background, as in Figure 24.25. On the other hand, the segment is turned off when the voltage is removed, in which case the liquid crystal rotates the direction of polarization by 90° to coincide with the axis of the analyzer. The light now passes through the analyzer and enters the eye of the observer. However, the light coming from the segment has been designed to have the same color and shade (light gray) as the background of the display, so the segment becomes indistinguishable from the background.

Color LCD display screens and computer monitors are popular because they occupy less space and weigh less than traditional units do. An LCD display screen, such as that in Figure 24.28, uses thousands of LCD segments arranged like the squares on graph paper. To produce color, three segments are grouped together to form a tiny picture element (or "pixel"). Color filters are used to enable one segment in the pixel to produce red light, one to produce green, and one to produce blue. The eye blends the colors from each pixel into a composite color. By varying the intensity of the red, green, and blue colors, the pixel can generate an entire spectrum of colors.

THE OCCURRENCE OF POLARIZED LIGHT IN NATURE

Polaroid is a familiar material because of its widespread use in sunglasses. Such sunglasses are designed so that the transmission axis of the Polaroid is oriented vertically when the glasses are worn in the usual fashion. Thus, the glasses prevent any light that is polarized horizontally from reaching the eye. Light from the sun is unpolarized, but a considerable amount of horizontally polarized sunlight originates by reflection from horizontal surfaces such as that of a lake. Section 26.4 discusses this effect. Polaroid sunglasses reduce glare by preventing the horizontally polarized reflected light from reaching the eyes.

Polarized sunlight also originates from the scattering of light by molecules in the atmosphere. Figure 24.29 shows light being scattered by a single atmospheric molecule.

Figure 24.28 This personal entertainment organizer utilizes a color LCD display screen because it is lightweight and space-efficient. (© AP/Wide World Photos)

The physics of
Polaroid sunglasses.

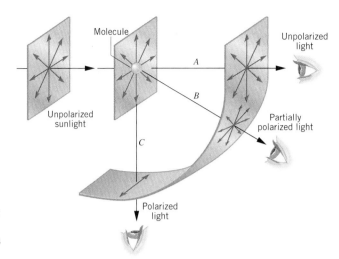

Figure 24.29 In the process of being scattered from atmospheric molecules, unpolarized light from the sun becomes partially polarized.

The electric fields in the unpolarized sunlight cause the electrons in the molecule to vibrate perpendicular to the direction in which the light is traveling. The electrons, in turn, reradiate the electromagnetic waves in different directions, as the drawing illustrates. The light radiated straight ahead in direction *A* is unpolarized, just like the incident light. But light radiated perpendicular to the incident light in direction *C* is polarized. Light radiated in the intermediate direction *B* is partially polarized. There is experimental evidence that some bird species use such polarized light as a navigational aid.

✔ Check Your Understanding 4

The drawing shows two sheets of polarizing material. The transmission axis of one is vertical, and that of the other makes an angle of 45° with the vertical. Unpolarized light shines on this arrangement from the left and from the right. From which direction does at least some of the light pass through both sheets: (a) from the left, (b) from the right, (c) from either direction, or (d) from neither direction? What is the answer when the light is horizontally polarized? *(The answers are given at the end of the book.)*

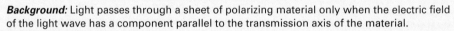

Background: Light passes through a sheet of polarizing material only when the electric field of the light wave has a component parallel to the transmission axis of the material.

For similar questions (including calculational counterparts), consult Self-Assessment Test 24.1, which is described next.

🖱 Self-Assessment Test 24.1

Test your understanding of the material in Sections 24.1–24.6:

• The Nature of Electromagnetic Waves   • The Electromagnetic Spectrum
• The Speed of Light   • The Energy Carried by Electromagnetic Waves
• The Doppler Effect and Electromagnetic Waves   • Polarization

Go to **www.wiley.com/college/cutnell**

24.7 *Concepts & Calculations*

One of the central ideas of this chapter is that electromagnetic waves carry energy. Two concepts are used to describe this energy—the wave's intensity and its energy density. The next example reviews these important ideas.

Concepts & Calculations Example 9
Intensity and Energy Density of a Wave

Figure 24.30 shows the popular dish antenna that receives digital TV signals from a satellite. The average intensity of the electromagnetic wave that carries a particular TV program is

$\overline{S} = 7.5 \times 10^{-14}$ W/m², and the circular aperture of the antenna has a radius of $r = 15$ cm. (a) Determine the electromagnetic energy delivered to the dish during a one-hour program. (b) What is the average energy density of the electromagnetic wave?

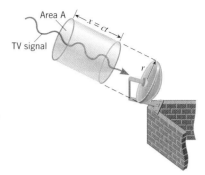

Concept Questions and Answers How is the average power passing through the circular aperture of the antenna related to the average intensity of the TV signal?

Answer According to Equation 16.8, the average intensity $\overline{S}$ of a wave is equal to the average power $\overline{P}$ that passes perpendicularly through a surface divided by the area A of the surface. Thus, the average power is $\overline{P} = \overline{S}A$, where $A = \pi r^2$.

How much energy does the antenna receive in a time t?

Answer Since the average power is the energy per unit time according to Equation 6.10b, the energy received by the antenna is the product of the average power and the time, or Energy $= \overline{P}t$.

Figure 24.30 A dish antenna picks up the TV signal from a satellite. The signal extends over the dish's entire aperture (radius r).

What is the average energy density, or average energy per unit volume, of the electromagnetic wave?

Answer The energy delivered to the dish is carried by the electromagnetic wave. The wave passes perpendicularly through an imaginary surface that has an area A that matches the circular area of the dish's aperture. In a time t, all the energy that passes through the surface is contained, therefore, in a cylinder of length x (see Figure 24.30). Since the energy is being carried by an electromagnetic wave, it travels at the speed of light c. Thus, the length of the cylinder is $x = ct$. The energy density u of the wave, or energy per unit volume, is the energy contained within the cylinder divided by its volume.

Solution

(a) According to Equation 6.10b, the energy received by the antenna during a time t is equal to the average power multiplied by the time, Energy $= \overline{P}t$. On the other hand, Equation 16.8 gives the average power as the product of the average intensity and the area, $\overline{P} = \overline{S}A$, so that

$$\text{Energy} = \underbrace{\overline{S}A}_{\substack{\text{Average} \\ \text{power}}} t = \overline{S}(\pi r^2)t$$

$$= (7.5 \times 10^{-14} \text{ W/m}^2)\pi(0.15 \text{ m})^2(3600 \text{ s}) = \boxed{1.9 \times 10^{-11} \text{ J}}$$

(b) The energy density is equal to the energy divided by the volume of the cylinder in Figure 24.30. The volume of the cylinder is equal to the cross-sectional area πr^2 times its length ct.

$$u = \frac{\text{Energy}}{\text{Volume}}$$

$$= \frac{\text{Energy}}{(\pi r^2)(ct)}$$

$$= \frac{1.9 \times 10^{-11} \text{ J}}{\pi(0.15 \text{ m})^2(3.00 \times 10^8 \text{ m/s})(3600 \text{ s})} = \boxed{2.5 \times 10^{-22} \text{ J/m}^3}$$

This calculation is equivalent to using Equation 24.4, which gives the energy density as

$$u = \frac{\text{Energy}}{\text{Volume}} = \frac{\overline{S}}{c} = \frac{7.5 \times 10^{-14} \text{ W/m}^2}{3.00 \times 10^8 \text{ m/s}} = 2.5 \times 10^{-22} \text{ J/m}^3$$

We have seen how the intensities of completely polarized or completely unpolarized light beams can change as they pass through a polarizer. But what about light that is partially polarized and partially unpolarized? Can the concepts that we discussed in Section 24.6 be applied to such light? The answer is "yes," and Example 10 illustrates how.

Concepts & Calculations Example 10
Partially Polarized and Partially Unpolarized Light

The light beam in Figure 24.31 passes through a polarizer whose transmission axis makes an angle ϕ with the vertical. The beam is partially polarized and partially unpolarized, and the

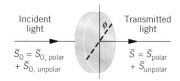

Incident
light
$$\overline{S}_0 = \overline{S}_{0,\,\text{polar}}$$
$$+ \overline{S}_{0,\,\text{unpolar}}$$

Transmitted
light
$$\overline{S} = \overline{S}_{\text{polar}}$$
$$+ \overline{S}_{\text{unpolar}}$$

Figure 24.31 Light that is partially polarized and partially unpolarized is incident on a sheet of polarizing material.

average intensity $\overline{S}_0$ of the incident light is the sum of the average intensity $\overline{S}_{0,\,\text{polar}}$ of the polarized light and the average intensity $\overline{S}_{0,\,\text{unpolar}}$ of the unpolarized light; $\overline{S}_0 = \overline{S}_{0,\,\text{polar}} + \overline{S}_{0,\,\text{unpolar}}$. The intensity $\overline{S}$ of the transmitted light is also the sum of two parts: $\overline{S} = \overline{S}_{\text{polar}} + \overline{S}_{\text{unpolar}}$. As the polarizer is rotated clockwise, the intensity of the transmitted light has a minimum value of $\overline{S} = 2.0$ W/m^2 when $\phi = 20.0°$ and has a maximum value of $\overline{S} = 8.0$ W/m^2 when the angle is $\phi = \phi_{\text{max}}$. (a) What is the intensity $\overline{S}_{0,\,\text{unpolar}}$ of the unpolarized incident light? (b) What is the intensity $\overline{S}_{0,\,\text{polar}}$ of the polarized incident light?

Concept Questions and Answers How is $\overline{S}_{\text{unpolar}}$ related to $\overline{S}_{0,\,\text{unpolar}}$?

Answer When unpolarized light strikes a polarizer, one half is absorbed and the other half is transmitted, so that $\overline{S}_{\text{unpolar}} = \frac{1}{2}\overline{S}_{0,\,\text{unpolar}}$. This is true for any angle ϕ of the transmission axis.

How is $\overline{S}_{\text{polar}}$ related to $\overline{S}_{0,\,\text{polar}}$?

Answer When polarized light strikes a polarizer, the transmitted intensity is given by Malus' law (Equation 24.7), $\overline{S}_{\text{polar}} = \overline{S}_{0,\,\text{polar}} \cos^2 \theta$, where θ is the angle between the direction of polarization of the incident light and the transmission axis. Note that θ is not the same as ϕ because the direction of polarization of the incident beam is not given and may not be in the vertical direction.

The minimum transmitted intensity is 2.0 W/m^2. Why isn't it 0 W/m^2?

Answer When the polarized part of the incident beam passes through the polarizer, the intensity of its transmitted portion changes as the angle ϕ changes, ranging between some maximum value and 0 W/m^2, in accord with Malus' law. However, when the unpolarized part of the beam passes through the polarizer, its transmitted intensity does not change as the angle ϕ changes. Thus, the minimum intensity of the transmitted beam is 2.0 W/m^2, rather than 0 W/m^2, because of the unpolarized light.

At what angle ϕ_{max} is the intensity of the light leaving the polarizing material a maximum?

Answer When the polarized part of the beam passes through the polarizer, the intensity of its transmitted portion is zero when $\phi = 20.0°$. For this angle, the direction of polarization of the polarized light is perpendicular to the transmission axis of the polarizer. The intensity rises to a maximum when the polarizer is rotated by 90.0° relative to this position, at which orientation the direction of polarization is aligned with the transmission axis. Thus, the intensity is a maximum when $\phi = \phi_{\text{max}} = 20.0° + 90.0° = 110.0°$.

Solution

(a) The average intensity $\overline{S}$ of the transmitted light is (see Figure 24.31) $\overline{S} = \overline{S}_{\text{polar}} + \overline{S}_{\text{unpolar}}$. We are given that the minimum intensity is $\overline{S} = 2.0$ W/m^2 when $\phi = 20.0°$, and we know that $\overline{S}_{\text{polar}} = 0$ W/m^2 for this angle. Thus, $\overline{S}_{\text{unpolar}} = \overline{S} - \overline{S}_{\text{polar}} = 2.0$ W/m^2. The intensity $\overline{S}_{0,\,\text{unpolar}}$ of the unpolarized incident light is twice this amount because the polarizer absorbs one-half the incident light:

$$\overline{S}_{0,\,\text{unpolar}} = 2\overline{S}_{\text{unpolar}} = 2(2.0 \text{ W/m}^2) = \boxed{4.0 \text{ W/m}^2}$$

(b) When $\phi = \phi_{\text{max}} = 110°$, the intensity of the transmitted beam is at its maximum value of $\overline{S} = 8.0$ W/m^2. Writing the transmitted intensity as the sum of two parts, we have

$$\overline{S} = 8.0 \text{ W/m}^2 = \underbrace{\overline{S}_{0,\,\text{polar}} \cos^2 \theta}_{\overline{S}_{\text{polar}}} + \underbrace{2.0 \text{ W/m}^2}_{\overline{S}_{\text{unpolar}}}$$

The angle θ is the angle between the direction of polarization of the incident polarized light and the transmission axis of the polarizer. Since $\overline{S}_{\text{polar}}$ is a maximum, we know that $\theta = 0°$. Solving this equation for $\overline{S}_{0,\,\text{polar}}$ yields

$$\overline{S}_{0,\,\text{polar}} = 8.0 \text{ W/m}^2 - 2.0 \text{ W/m}^2 = \boxed{6.0 \text{ W/m}^2}$$

At the end of the problem set for this chapter, you will find homework problems that contain both conceptual and quantitative parts. These problems are grouped under the heading *Concepts & Calculations, Group Learning Problems*. They are designed for use by students working alone or in small learning groups. The conceptual part of each problem provides a convenient focus for group discussions.

Concept Summary

This summary presents an abridged version of the chapter, including the important equations and all available learning aids. For convenient reference, the learning aids (including the text's examples) are placed next to or immediately after the relevant equation or discussion. The following learning aids may be found on-line at **www.wiley.com/college/cutnell:**

Interactive LearningWare examples are solved according to a five-step interactive format that is designed to help you develop problem-solving skills.	**Concept Simulations** are animated versions of text figures or animations that illustrate important concepts. You can control parameters that affect the display, and we encourage you to experiment.
Interactive Solutions offer specific models for certain types of problems in the chapter homework. The calculations are carried out interactively.	**Self-Assessment Tests** include both qualitative and quantitative questions. Extensive feedback is provided for both incorrect and correct answers, to help you evaluate your understanding of the material.

Topic	Discussion	Learning Aids

24.1 The Nature of Electromagnetic Waves

Electromagnetic wave

An electromagnetic wave consists of mutually perpendicular and oscillating electric and magnetic fields. The wave is a transverse wave, since the fields are perpendicular to the direction in which the wave travels. Electromagnetic waves can travel through a vacuum or a material substance. All electromagnetic waves travel through a vacuum at the same speed, which is known as the speed of light c ($c = 3.00 \times 10^8$ m/s).

Speed of light

24.2 The Electromagnetic Spectrum

Relation between frequency, wavelength, and speed of light in a vacuum

The frequency f and wavelength λ of an electromagnetic wave in a vacuum are related to its speed c through the relation

$$c = f\lambda$$

Examples 1, 2

Electromagnetic spectrum

The series of electromagnetic waves, arranged in order of their frquencies or wavelengths, is called the electromagnetic spectrum. In increasing order of frequency (decreasing order of wavelength), the spectrum includes radio waves, infrared radiation, visible light, ultraviolet radiation, X-rays, and gamma rays. Visible light has

Visible light

frequencies between about 4.0×10^{14} and 7.9×10^{14} Hz. The human eye and brain perceive different frequencies or wavelengths as different colors.

24.3 The Speed of Light

James Clerk Maxwell showed that the speed of light in a vacuum is

Speed of light

$$c = \frac{1}{\sqrt{\epsilon_0 \mu_0}}$$

(24.1) **Example 3**

where ϵ_0 is the (electric) permittivity of free space and μ_0 is the (magnetic) permeability of free space.

24.4 The Energy Carried by Electromagnetic Waves

The total energy density u of an electromagnetic wave is the total energy per unit volume of the wave and, in a vacuum, is given by

Total energy density

$$u = \frac{1}{2}\epsilon_0 E^2 + \frac{1}{2\mu_0} B^2$$

(24.2a)

where E and B, respectively, are the magnitudes of the electric and magnetic fields of the wave. Since the electric and magnetic parts of the total energy density are equal, the following two equations are equivalent to Equation 24.2a:

$$u = \epsilon_0 E^2$$

(24.2b)

$$u = \frac{1}{\mu_0} B^2$$

(24.2c)

In a vacuum, E and B are related according to

Relation between electric and magnetic fields

$$E = cB$$

(24.3)

Equations 24.2a–c can be used to determine the average total energy density, if the rms average values E_{rms} and B_{rms} are used in place of the symbols E and B. **Example 4** The rms values are related to the peak values E_0 and B_0 in the usual way:

Topic	Discussion	Learning Aids
Root mean square fields	$$E_{\mathrm{rms}} = \frac{1}{\sqrt{2}} E_0 \quad \text{and} \quad B_{\mathrm{rms}} = \frac{1}{\sqrt{2}} B_0$$	
Intensity	The intensity of an electromagnetic wave is the power that the wave carries perpendicularly through a surface divided by the area of the surface. In a vacuum, the intensity S is related to the total energy density u according to	Example 5 Interactive Solution 24.27
Relation between intensity and total energy density	$$S = cu \qquad (24.4)$$	Example 9

24.5 The Doppler Effect and Electromagnetic Waves

When electromagnetic waves and the source and observer of the waves all travel along the same line in a vacuum, the Doppler effect is given by

Topic	Discussion	Learning Aids
Doppler effect	$$f_{\mathrm{o}} = f_{\mathrm{s}}\left(1 \pm \frac{v_{\mathrm{rel}}}{c}\right) \qquad \text{if } v_{\mathrm{rel}} \ll c \qquad (24.6)$$	Example 6 Interactive Solution 24.31

where f_{o} and f_{s} are, respectively, the observed and emitted wave frequencies and v_{rel} is the relative speed of the source and the observer. The plus sign is used when the source and the observer come together, and the minus sign is used when they move apart.

24.6 Polarization

Topic	Discussion	Learning Aids
Linear polarized electromagnetic wave Unpolarized electromagnetic wave	A linearly polarized electromagnetic wave is one in which the oscillation of the electric field occurs only along one direction, which is taken to be the direction of polarization. The magnetic field also oscillates along only one direction, which is perpendicular to the electric field direction. In an unpolarized wave such as the light from an incandescent bulb, the direction of polarization does not remain fixed, but fluctuates randomly in time.	
Transmission axis	Polarizing materials allow only the component of the wave's electric field along one direction (and the associated magnetic field component) to pass through them. The preferred transmission direction for the electric field is called the transmission axis of the material. When unpolarized light is incident on a piece of polarizing material, the transmitted polarized light has an intensity that is one-half that of the incident light.	
Polarizer and analyzer	When two pieces of polarizing material are used one after the other, the first is called the polarizer, and the second is referred to as the analyzer. If the average intensity of polarized light falling on the analyzer is $\overline{S}_0$, the average intensity $\overline{S}$ of the light leaving the analyzer is given by Malus' law as	
Malus' law	$$\overline{S} = \overline{S}_0 \cos^2 \theta \qquad (24.7)$$	Examples 7, 8 Interactive LearningWare 24.1 Example 10
Crossed polarizer and analyzer	where θ is the angle between the transmission axes of the polarizer and analyzer. When $\theta = 90°$, the polarizer and the analyzer are said to be "crossed," and no light passes through the analyzer.	

 Use Self-Assessment Test 24.1 to evaluate your understanding of Sections 24.1–24.6.

Conceptual Questions

1. Which of the following concepts applies to both sound waves and electromagnetic waves: (a) intensity or (b) polarization? Account for your answer.

2. Refer to Figure 24.2. Between the times indicated in parts *c* and *d* in the drawing, what is the direction of the magnetic field at the point *P* for the electromagnetic wave being generated? Is it directed into or out of the plane of the paper? Justify your answer.

3. A transmitting antenna is located at the origin of an *x*, *y*, *z* axis system and broadcasts an electromagnetic wave whose electric field oscillates along the *y* axis. The wave travels along the +*x* axis. Three possible wire loops are available for use with an LC-tuned circuit to detect this wave: One loop lies in the *xy* plane, another in the *xz* plane, and the third in the *yz* plane. Which of the loops will detect the wave? Why?

4. Why does the peak value of the emf induced in a loop antenna (see Figure 24.6) depend on the frequency of the electromagnetic wave, whereas the peak value of the emf induced in a straight-wire antenna (see Figure 24.5) does not?

5. Suppose that the electric field of an electromagnetic wave decreases in magnitude. Does the magnetic field increase, decrease, or remain the same? Account for your answer.

6. An astronomer measures the Doppler change in frequency for the light reaching the earth from a distant star. From this measurement, can the astronomer tell whether the star is moving away from the earth or whether the earth is moving away from the star? Explain.

7. Is there any real difference between a polarizer and an analyzer? In other words, can a polarizer be used as an analyzer, and vice versa?

8. Malus' law applies to the setup in Figure 24.21, which shows the analyzer rotated through an angle θ and the polarizer held fixed. Does Malus' law apply when the analyzer is held fixed and the polarizer is rotated? Give your reasoning.

9. In Example 7, we saw that when the angle between the polarizer and analyzer is $63.4°$, the intensity of the transmitted light drops to one-tenth of that of the incident unpolarized light. What happens to the light intensity that is not transmitted?

10. Light is incident from the left on two pieces of polarizing material, 1 and 2. As part *a* of the drawing illustrates, the transmission axis of material 1 is along the vertical direction, and that of material 2 makes an angle of θ with respect to the vertical. In part *b* of the drawing the two polarizing materials are interchanged. (a) Assume that the incident light is unpolarized and determine whether the intensity of the transmitted light in part *a* is greater than, equal to, or less than that in part *b*. (b) Repeat part (a), assuming that the incident light is linearly polarized along the vertical direction. Justify your answers to both parts (a) and (b).

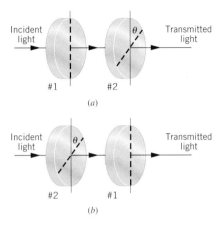

11. You are sitting upright on the beach near a lake on a sunny day, wearing Polaroid sunglasses. When you lie down on your side, facing the lake, the sunglasses don't work as well as they did while you were sitting upright. Why not?

Problems

Section 24.1 The Nature of Electromagnetic Waves

1. ssm www The distance between earth and the moon can be determined from the time it takes for a laser beam to travel from earth to a reflector on the moon and back. If the round-trip time can be measured to an accuracy of one-tenth of a nanosecond (1 ns = 10^{-9} s), what is the corresponding error in the earth–moon distance?

2. (a) Neil A. Armstrong was the first person to walk on the moon. The distance between the earth and the moon is 3.85×10^8 m. Find the time it took for his voice to reach earth via radio waves. (b) Someday a person will walk on Mars, which is 5.6×10^{10} m from earth at the point of closest approach. Determine the minimum time that will be required for that person's voice to reach earth.

3. An AM station is broadcasting a radio wave whose frequency is 1400 kHz. The value of the capacitance in Figure 24.5 is 8.4×10^{-11} F. What must be the value of the inductance in order that this station can be tuned in by the radio?

4. FM radio stations use radio waves with frequencies from 88.0 to 108 MHz to broadcast their signals. Assuming that the inductance in Figure 24.5 has a value of 6.00×10^{-7} H, determine the range of capacitance values that are needed so the antenna can pick up all the radio waves broadcasted by FM stations.

* **5. ssm** Equation 16.3, $y = A \sin (2\pi ft - 2\pi x/\lambda)$, gives the mathematical representation of a wave oscillating in the y direction and traveling in the positive x direction. Let y in this equation equal the electric field of an electromagnetic wave traveling in a vacuum. The maximum electric field is $A = 156$ N/C, and the frequency is $f = 1.50 \times 10^8$ Hz. Plot a graph of the electric field strength versus position, using for x the following values: 0, 0.50, 1.00, 1.50, and 2.00 m. Plot this graph for (a) a time $t = 0$ s and (b) a time t that is one-fourth of the wave's period.

** **6.** A flat coil of wire is used with an LC-tuned circuit as a receiving antenna. The coil has a radius of 0.25 m and consists of 450 turns. The transmitted radio wave has a frequency of 1.2 MHz. The magnetic field of the wave is parallel to the normal to the coil and has a maximum value of 2.0×10^{-13} T. Using Faraday's law of electromagnetic induction and the fact that the magnetic field changes from zero to its maximum value in one-quarter of a wave period, find the magnitude of the average emf induced in the antenna during this time.

Section 24.2 The Electromagnetic Spectrum

7. Some of the X-rays produced in an X-ray machine have a wavelength of 2.1 nm. What is the frequency of these electromagnetic waves?

8. TV channel 3 (VHF) broadcasts at a frequency of 63.0 MHz. TV channel 23 (UHF) broadcasts at a frequency of 527 MHz. Find the ratio (VHF/UHF) of the wavelengths for these channels.

9. ssm The human eye is most sensitive to light having a frequency of about 5.5×10^{14} Hz, which is in the yellow-green region of the electromagnetic spectrum. How many wavelengths of this light can fit across the width of your thumb, a distance of about 2.0 cm?

10. Magnetic resonance imaging, or MRI (see Section 21.7), and positron emission tomography, or PET scanning (see Section 32.6), are two medical diagnostic techniques. Both employ electromagnetic waves. For these waves, find the ratio of the MRI wavelength (frequency = 6.38×10^7 Hz) to the PET scanning wavelength (frequency = 1.23×10^{20} Hz).

11. ssm www At one time television sets used "rabbit-ears" antennas. Such an antenna consists of a pair of metal rods. The length of each rod can be adjusted to be one-quarter of a wavelength of an electromagnetic wave whose frequency is 60.0 MHz. How long is each rod?

12. Two radio waves are used in the operation of a cellular telephone. To receive a call, the phone detects the wave emitted at one frequency by the transmitter station or base unit. To send your message to the base unit, your phone emits its own wave at a different frequency. The difference between these two frequencies is fixed for all channels of cell phone operation. Suppose the wavelength of the wave emitted by the base unit is 0.34339 m and the wavelength of the wave emitted by the phone is 0.36205 m. Using a value of 2.9979×10^8 m/s for the speed of light, determine the difference between the two frequencies used in the operation of a cell phone.

* **13.** Section 17.5 deals with transverse standing waves on a string. Electromagnetic waves also can form standing waves. In a standing wave pattern formed from microwaves, the distance between a node and an adjacent antinode is 0.50 cm. What is the microwave frequency?

Section 24.3 The Speed of Light

14. Review Conceptual Example 3 before attempting this problem. The brightest star in the night sky is Sirius, which is at a distance of 8.3×10^{16} m. When we look at this star, how far back in time are we seeing it? Express your answer in years. (There are $365\frac{1}{4}$ days in one year.)

15. ssm Two astronauts are 1.5 m apart in their spaceship. One speaks to the other. The conversation is transmitted to earth via electromagnetic waves. The time it takes for sound waves to travel at 343 m/s through the air between the astronauts equals the time it takes for the electromagnetic waves to travel to the earth. How far away from the earth is the spaceship?

16. A communications satellite is in a synchronous orbit that is 3.6×10^7 m directly above the equator. The satellite is located midway between Quito, Equador, and Belém, Brazil, two cities almost on the equator that are separated by a distance of 3.5×10^6 m. Find the time it takes for a telephone call to go by way of satellite between these cities. Ignore the curvature of the earth.

17. Figure 24.12 illustrates Michelson's setup for measuring the speed of light with the mirrors placed on Mt. San Antonio and Mt. Wilson in California, which are 35 km apart. Using a value of 3.00×10^8 m/s for the speed of light, find the minimum angular speed (in rev/s) for the rotating mirror.

18. A lidar (laser radar) gun is an alternative to the standard radar gun that uses the Doppler effect to catch speeders. A lidar gun uses an infrared laser and emits a precisely timed series of pulses of infrared electromagnetic waves. The time for each pulse to travel to the speeding vehicle and return to the gun is measured. In one situation a lidar gun in a stationary police car observes a difference of 1.27×10^{-7} s in round-trip travel times for two pulses that are emitted 0.450 s apart. Assuming that the speeding vehicle is approaching the police car essentially head-on, determine the speed of the vehicle.

* **19. ssm** A mirror faces a cliff located some distance away. Mounted on the cliff is a second mirror, directly opposite the first mirror and facing toward it. A gun is fired very close to the first mirror. The speed of sound is 343 m/s. How many times does the flash of the gunshot travel the round-trip distance between the mirrors before the echo of the gunshot is heard?

* **20.** A celebrity holds a press conference, which is televised live. A television viewer hears the sound picked up by a microphone directly in front of the celebrity. This viewer is seated 2.3 m from the television set. A reporter at the press conference is located 4.1 m from the microphone and hears the words directly *at the very same instant* that the television viewer hears them. Using a value of 343 m/s for the speed of sound, determine the maximum distance between the television viewer and the celebrity.

Section 24.4 The Energy Carried by Electromagnetic Waves

21. A laser emits a narrow beam of light. The radius of the beam is 1.0×10^{-3} m, and the power is 1.2×10^{-3} W. What is the intensity of the laser beam?

22. The maximum strength of the magnetic field in an electromagnetic wave is 3.3×10^{-6} T. What is the maximum strength of the wave's electric field?

23. ssm The microwave radiation left over from the Big Bang explosion of the universe has an average energy density of 4×10^{-14} J/m^3. What is the rms value of the electric field of this radiation?

24. On a cloudless day, the sunlight that reaches the surface of the earth has an average intensity of about 1.0×10^3 W/m^2. What is the average electromagnetic energy contained in 5.5 m^3 of space just above the earth's surface?

25. ssm A future space station in orbit about the earth is being powered by an electromagnetic beam from the earth. The beam has a cross-sectional area of 135 m^2 and transmits an average power of 1.20×10^4 W. What are the rms values of the (a) electric and (b) magnetic fields?

* **26.** The intensity of sunlight at the top of the earth's atmosphere is about 1390 W/m^2. The distance between the sun and earth is 1.50×10^{11} m, while that between the sun and Mars is 2.28×10^{11} m. What is the intensity of sunlight at the surface of Mars?

* **27. Interactive Solution 24.27** at **www. wiley.com/college/ cutnell** provides one model for problems like this one. The drawing shows an edge-on view of the solar panels on a communications satellite. The dashed line specifies the normal to the panels. Sunlight strikes the panels at an angle θ with respect to the normal. If the solar power impinging on the panels is 2600 W when $\theta = 65°$, what is it when $\theta = 25°$?

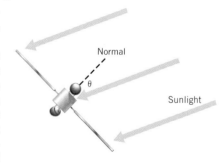

28. A heat lamp emits infrared radiation whose rms electric field is $E_{rms} = 2800$ N/C. (a) What is the average intensity of the radiation? (b) The radiation is focused on a person's leg over a circular area of radius 4.0 cm. What is the average power delivered to the leg? (c) The portion of the leg being radiated has a mass of 0.28 kg and a specific heat capacity of 3500 J/(kg·C°). How long does it take to raise its temperature by 2.0 C°? Assume that there is no other heat transfer into or out of the portion of the leg being heated.

* **29. ssm** The mean distance between earth and the sun is 1.50×10^{11} m. The average intensity of solar radiation incident on the upper atmosphere of the earth is 1390 W/m^2. Assuming the sun emits radiation uniformly in all directions, determine the total power radiated by the sun.

** **30.** The average intensity of sunlight reaching the earth is 1390 W/m^2. A charge of 2.6×10^{-8} C is placed in the path of this electromagnetic wave. (a) What is the magnitude of the maximum electric force that the charge experiences? (b) If the charge is moving at a speed of 3.7×10^4 m/s, what is the magnitude of the maximum magnetic force that the charge could experience?

Section 24.5 The Doppler Effect and Electromagnetic Waves

31. Review **Interactive Solution 24.31** at **www.wiley.com/college/cutnell** to see one model for solving this problem. A distant galaxy emits light that has a wavelength of 434.1 nm. On earth, the wavelength of this light is measured to be 438.6 nm. (a) Decide whether this galaxy is approaching or receding from the earth. Give your reasoning. (b) Find the speed of the galaxy relative to the earth.

32. Suppose that the police car in Example 6 is moving to the right at 27 m/s, while the speeder is coming up from behind at a speed of 39 m/s, both speeds being with respect to the ground. Assume that the electromagnetic wave emitted by the radar gun has a frequency of 8.0×10^9 Hz. (a) Find the magnitude of the difference between the frequency of the emitted wave and the wave that returns to the police car after reflecting from the speeder's car. (b) Which wave has the greater frequency? Why?

* **33.** **ssm** A distant galaxy is simultaneously rotating and receding from the earth. As the drawing shows, the galactic center is receding from the earth at a relative speed of $u_G = 1.6 \times 10^6$ m/s. Relative to the center, the tangential speed is $v_T = 0.4 \times 10^6$ m/s for locations A and B, which are equidistant from the center. When the frequencies of the light coming from regions A and B are measured on earth, they are not the same and each is different from the emitted frequency of 6.200×10^{14} Hz. Find the measured frequency for the light from (a) region A and (b) region B.

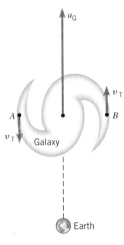

Section 24.6 Polarization

34. Unpolarized light whose intensity is 1.10 W/m² is incident on the polarizer in Figure 24.21. (a) What is the intensity of the light leaving the polarizer? (b) If the analyzer is set at an angle of $\theta = 75°$ with respect to the polarizer, what is the intensity of the light that reaches the photocell?

35. **ssm** In the polarizer/analyzer combination in Figure 24.21, 90.0% of the light intensity falling on the analyzer is absorbed. Determine the angle between the transmission axes of the polarizer and the analyzer.

36. For one approach to this problem, consult **Interactive Learning-Ware 24.1** at **www.wiley.com/college/cutnell.** For each of the three sheets of polarizing material shown in the drawing, the orientation of the transmission axis is labeled relative to the vertical. The incident beam of light is unpolarized and has an intensity of 1260.0 W/m². What is the intensity of the beam transmitted through the three sheets when $\theta_1 = 19.0°$, $\theta_2 = 55.0°$, and $\theta_3 = 100.0°$?

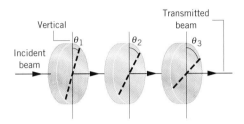

37. Review Conceptual Example 8 before solving this problem. Suppose unpolarized light of intensity 150 W/m² falls on the polarizer in Figure 24.24a, and the angle θ in the drawing is 30.0°. What is the light intensity reaching the photocell?

38. Light that is polarized along the vertical direction is incident on a sheet of polarizing material. Only 94% of the intensity of the light passes through the sheet and strikes a second sheet of polarizing material. No light passes through the second sheet. What angle does the transmission axis of the second sheet make with the vertical?

* **39.** **ssm www** More than one analyzer can be used in a setup like that in Figure 24.21, each analyzer following the previous one. Suppose that the transmission axis of the first analyzer is rotated 27° relative to the transmission axis of the polarizer, and that the transmission axis of each additional analyzer is rotated 27° relative to the transmission axis of the previous one. What is the minimum number of analyzers needed, so the light reaching the photocell has an intensity that is reduced by at least a factor of one hundred relative to that striking the first analyzer?

* **40.** A beam of polarized light has an average intensity of 15 W/m² and is sent through a polarizer. The transmission axis makes an angle of 25° with respect to the direction of polarization. Determine the rms value of the electric field of the transmitted beam.

Additional Problems

41. A truck driver is broadcasting at a frequency of 26.965 MHz with a CB (citizen's band) radio. Determine the wavelength of the electromagnetic wave being used.

42. An industrial laser is used to burn a hole through a piece of metal. The average intensity of the light is $\overline{S} = 1.23 \times 10^9$ W/m². What is the rms value of (a) the electric field and (b) the magnetic field in the electromagnetic wave emitted by the laser?

43. **ssm** In astronomy, distances are often expressed in light-years. One light-year is the distance traveled by light in one year. The distance to Alpha Centauri, the closest star other than our own sun that can be seen by the naked eye, is 4.3 light-years. Express this distance in meters.

44. The electromagnetic wave that delivers a cellular phone call to a car has a magnetic field with an rms value of 1.5×10^{-10} T. The wave passes perpendicularly through an open window, the area of

which is 0.20 m². How much energy does this wave carry through the window during a 45-s phone call?

45. The average intensity of sunlight at the top of the earth's atmosphere is 1390 W/m². What is the maximum energy that a 25-m × 45-m solar panel could collect in one hour in this sunlight?

46. A radio station broadcasts a radio wave whose wavelength is 274 m. (a) What is the frequency of the wave? (b) Is this radio wave AM or FM? (See Figure 24.10.)

47. **ssm** Linearly polarized light is incident on a piece of polarizing material. What is the ratio of the transmitted light intensity to the incident light intensity when the angle between the transmission axis and the incident electric field is (a) 25° and (b) 65°?

48. A speeder is pulling directly away and increasing his distance from a police car that is moving at 25 m/s with respect to the ground. The radar gun in the police car emits an electromagnetic wave with a frequency of 7.0×10^9 Hz. The wave reflects from the speeder's car and returns to the police car, where its frequency is measured to be 320 Hz less than the emitted frequency. Find the speeder's speed with respect to the ground.

*49. In a traveling electromagnetic wave, the electric field is represented mathematically as

$$E = E_0 \sin [(1.5 \times 10^{10} \text{ s}^{-1})t - (5.0 \times 10^1 \text{ m}^{-1})x]$$

where E_0 is the maximum field strength. (a) What is the frequency of the wave? (b) This wave and the wave that results from its reflection can form a standing wave, in a way similar to that in which standing waves can arise on a string (see Section 17.5). What is the separation between adjacent nodes in the standing wave?

*50. Before attempting this problem, review Conceptual Example 8. The intensity of the light that reaches the photocell in Figure 24.24a is 110 W/m², when $\theta = 23°$. What would be the intensity reaching the photocell if the *analyzer* were removed from the setup, everything else remaining the same?

*51. **ssm www** The power radiated by the sun is 3.9×10^{26} W. The earth orbits the sun in a nearly circular orbit of radius 1.5×10^{11} m. The earth's axis of rotation is tilted by 27° relative to the plane of the orbit (see the drawing), so sunlight does not strike the

equator perpendicularly. What power strikes a 0.75-m² patch of flat land at the equator at point Q?

*52. The rms electric field strength at a spot is 19.0 N/C. This field is created by a tiny source of light located 2.50 m away that emits light uniformly in all directions. Assuming that the light does not reflect from anything in the environment, determine the average power of the light emitted by the source.

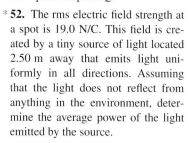
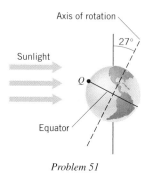

Problem 51

*53. An argon-ion laser produces a cylindrical beam of light whose average power is 0.750 W. How much energy is contained in a 2.50-m length of the beam?

**54. Suppose that the light falling on the polarizer in Figure 24.21 is partially polarized (average intensity = $\overline{S}_P$) and partially unpolarized (average intensity = $\overline{S}_U$). The total incident intensity is $\overline{S}_P + \overline{S}_U$, and the percentage polarization is $100 \, \overline{S}_P/(\overline{S}_P + \overline{S}_U)$. When the polarizer is rotated in such a situation, the intensity reaching the photocell varies between a minimum value of $\overline{S}_{\min}$ and a maximum value of $\overline{S}_{\max}$. Show that the percentage polarization can be expressed as $100 \, (\overline{S}_{\max} - \overline{S}_{\min})/(\overline{S}_{\max} + \overline{S}_{\min})$.

Concepts & Calculations Group Learning Problems

Note: Each of these problems consists of Concept Questions followed by a related quantitative Problem. They are designed for use by students working alone or in small learning groups. The Concept Questions involve little or no mathematics and are intended to stimulate group discussions. They focus on the concepts with which the problems deal. Recognizing the concepts is the essential initial step in any problem-solving technique.

55. Concept Questions A certain type of laser puts out light of known frequency. The light, however, occurs as a series of short pulses, each lasting for a time t_0. (a) How is the wavelength of the light related to its frequency? (b) How is the length (in meters) of each pulse related to the time t_0?

Problem A laser puts out a pulse of light that lasts for 2.7×10^{-11} s. The frequency of the light is 5.2×10^{14} Hz. (a) How many wavelengths are there in one pulse? (b) The light enters a pool of water. Its frequency remains the same, but the light slows down to a speed of 2.3×10^8 m/s. How many wavelengths are there now in one pulse?

56. Concept Questions (a) Suppose that the magnitude E of the electric field in an electromagnetic wave triples. By what factor does the intensity S of the wave change? (b) The magnitude B of the magnetic field is much smaller than E because, according to Equation 24.3, $B = E/c$, where c is the speed of light in a vacuum. If B triples, by what factor does the intensity change? Account for your answers.

Problem The magnitude of the electric field of an electromagnetic field increases from 315 to 945 N/C. (a) Determine the intensities for the two values of the electric field. (b) What is the magnitude of the magnetic field associated with each electric field? (c) Determine the intensity for each value of the magnetic field. Make sure your answers are consistent with your answers to the Concept Questions.

57. Concept Questions A source is radiating light waves uniformly in all directions. At a certain distance r from the source a

person measures the average intensity of the waves. (a) Does the average intensity increase, decrease, or remain the same as r increases? (b) If the magnitude of the electric field is determined from the average intensity, is the electric field the rms value or the peak value? In both cases, justify your answers.

Problem A light bulb emits light uniformly in all directions. The emitted power is 150.0 W. At a distance of 5.00 m from the bulb, what are (a) the average intensity and the magnitudes of the (b) rms and (c) peak electric fields?

58. Concept Questions An electric charge is placed in a laser beam. Does a stationary charge experience a force due to (a) the electric field and (b) the magnetic field of the electromagnetic wave? Now suppose that the charge is moving perpendicular to the magnetic field of the beam. Does it experience (c) an electric force and (d) a magnetic force? Account for your answers.

Problem A stationary particle of charge $q = 2.6 \times 10^{-8}$ C is placed in a laser beam whose intensity is 2.5×10^3 W/m². Determine the magnitude of the (a) electric and (b) magnetic forces exerted on the charge. If the charge is moving perpendicular to the magnetic field with a speed of 3.7×10^4 m/s, find the magnitudes of the (c) electric and (d) magnetic forces exerted on it. Verify that your answers are consistent with your answers to the Concept Questions.

59. Concept Questions The drawing shows light incident on a polarizer whose transmission axis is parallel to the z axis. The polarizer is rotated clockwise through an angle α between 0 and 90°. While the polarizer is being rotated, does the intensity of the transmitted light increase, decrease, or remain the same if the incident light is (a) unpolarized, (b) polarized parallel to the z axis, and (c) polarized parallel to the y axis? Provide a reason for each of your answers.

Problem The intensity of the incident light is 7.0 W/m². Determine the intensity of the transmitted light for each of the six cases shown in the table.

Incident Light	Intensity of Transmitted Light	
	$\alpha = 0°$	$\alpha = 35°$
Unpolarized		
Polarized parallel to z axis		
Polarized parallel to y axis		

Be sure that your answers are consistent with your answers to the Concept Questions.

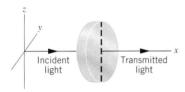

60. Concept Question The drawing shows three polarizer/analyzer pairs. The incident light on each pair is unpolarized and has the same intensity. Rank the pairs according to the intensity of the transmitted light, largest first. Provide reasons for your answers.

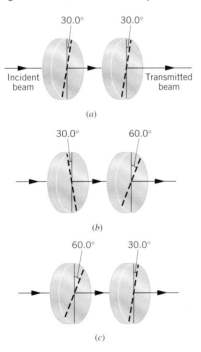

Problem The intensity of the unpolarized incident beam is 48 W/m². Find the intensity of the transmitted beams for each of the three cases shown in the drawing. Be sure your answers are consistent with your answer to the Concept Question.

* **61. Concept Question** The drawing shows three situations A, B, and C in which an observer and a source of electromagnetic waves are moving along the same line. In each case the source emits a wave of the same frequency. The arrows in each situation denote velocity vectors relative to the ground and have the magnitudes indicated, either v or $2v$. Rank the frequencies of the observed electromagnetic waves in descending order (largest first) according to magnitude. Explain your reasoning.

Problem Each of the sources in the drawing emits a frequency of 4.57×10^{14} Hz, and the speed v is 1.50×10^6 m/s. Calculate the observed frequency in each of the three cases. Verify that your answers are consistent with your answer to the Concept Question.

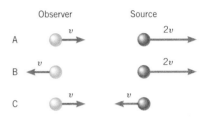

* **62. Concept Questions** The drawing shows four sheets of polarizing material, each with its transmission axis oriented differently. Light that is polarized in the vertical direction is incident from the left. One of the sheets is to be removed, with the goal of having some light still pass through the remaining three sheets and emerge on the right. (a) There are a number of possibilities for the sheet that is removed. What are they? Account for your answer. (b) Of the possibilities identified in Concept Question (a), which one allows the greatest light intensity to pass through, and which one allows the least to pass through? Justify your answer.

Problem The light incident from the left in the drawing has an average intensity of 27 W/m². For each of the possibilities identified in Concept Question (a), determine the average intensity of the light that emerges on the right in the drawing. Be sure that your answer is consistent with your answer to Concept Question (b).

Chapter 25

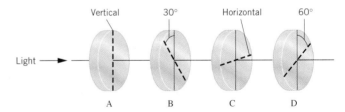

The Reflection of Light: Mirrors

The reflection of light from the plane surface of the water acts to double the presence of this bullfrog. This chapter discusses the images formed by the reflection of light from plane and spherical mirrors. (© Planet Earth/ Chris Howes/Taxi/Getty Images)

25.1 Wave Fronts and Rays

Mirrors are usually close at hand. It is difficult, for example, to put on makeup, shave, or drive a car without them. We see images in mirrors because some of the light that strikes them is reflected into our eyes. To discuss reflection, it is necessary to introduce the concepts of a wave front and a ray of light, and we can do so by taking advantage of the familiar topic of sound waves (see Chapter 16). Both sound and light are kinds of waves. Sound is a pressure wave, whereas light is electromagnetic in nature. However, the ideas of a wave front and a ray apply to both.

Consider a small spherical object whose surface is pulsating in simple harmonic motion. A sound wave is emitted that moves spherically outward from the object at a constant speed. To represent this wave, we draw surfaces through all points of the wave that are in the same phase of motion. These surfaces of constant phase are called *wave fronts.* Figure 25.1 shows a hemispherical view of the wave fronts. In this view they appear as concentric spherical shells about the vibrating object. If the wave fronts are drawn through the condensations, or crests, of the sound wave, as they are in the picture, the distance between adjacent wave fronts equals the wavelength λ. The radial lines pointing outward from the source and perpendicular to the wave fronts are called *rays.* The rays point in the direction of the velocity of the wave.

Figure 25.2*a* shows small sections of two adjacent spherical wave fronts. At large distances from the source, the wave fronts become less and less curved and approach the shape of flat surfaces, as in part *b* of the drawing. Waves whose wave fronts are flat surfaces (i.e., planes) are known as *plane waves* and are important in understanding the properties of mirrors and lenses. Since rays are perpendicular to the wave fronts, the rays for a plane wave are parallel to each other.

The concepts of wave fronts and rays can also be used to describe light waves. For light waves, the ray concept is particularly convenient when showing the path taken by the light. We will make frequent use of light rays, and they can be regarded essentially as narrow beams of light much like those that lasers produce.

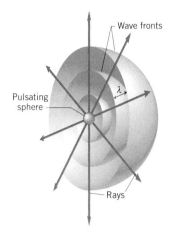

Figure 25.1 A hemispherical view of a sound wave emitted by a pulsating sphere. The wave fronts are drawn through the condensations of the wave, so the distance between two successive wave fronts is the wavelength λ. The rays are perpendicular to the wave fronts and point in the direction of the velocity of the wave.

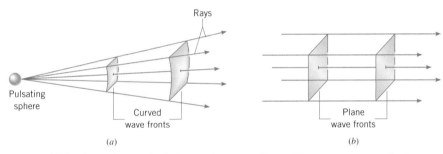

Figure 25.2 (*a*) Portions of two spherical wave fronts are shown. The rays are perpendicular to the wave fronts and diverge. (*b*) For a plane wave, the wave fronts are flat surfaces, and the rays are parallel to each other.

25.2 The Reflection of Light

Most objects reflect a certain portion of the light falling on them. Suppose a ray of light is incident on a flat, shiny surface, such as the mirror in Figure 25.3. As the drawing shows, the *angle of incidence* θ_i is the angle that the incident ray makes with respect to the normal, which is a line drawn perpendicular to the surface at the point of incidence. The *angle of reflection* θ_r is the angle that the reflected ray makes with the normal. The *law of reflection* describes the behavior of the incident and reflected rays.

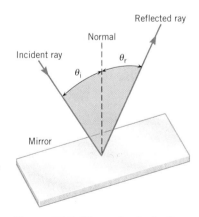

Figure 25.3 The angle of reflection θ_r equals the angle of incidence θ_i. These angles are measured with respect to the normal, which is a line drawn perpendicular to the surface of the mirror at the point of incidence.

■ **LAW OF REFLECTION**

The incident ray, the reflected ray, and the normal to the surface all lie in the same plane, and the angle of reflection θ_r equals the angle of incidence θ_i:

$$\theta_r = \theta_i$$

Figure 25.4 (*a*) The drawing shows specular reflection from a polished plane surface, such as a mirror. The reflected rays are parallel to each other. (*b*) A rough surface reflects the light rays in all directions; this type of reflection is known as diffuse reflection.

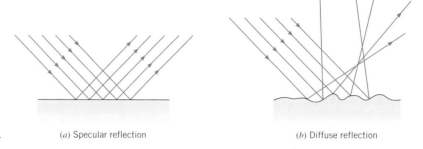

(*a*) Specular reflection   (*b*) Diffuse reflection

When parallel light rays strike a smooth, plane surface, such as that in Figure 25.4*a*, the reflected rays are parallel to each other. This type of reflection is one example of what is known as *specular reflection* and is important in determining the properties of mirrors. Most surfaces, however, are not perfectly smooth, because they contain irregularities the sizes of which are equal to or greater than the wavelength of the light. The law of reflection applies to each ray, but the irregular surface reflects the light rays in various directions, as part *b* of the drawing suggests. This type of reflection is known as *diffuse reflection*. Common surfaces that give rise to diffuse reflection are most papers, wood, nonpolished metals, and walls covered with a "flat" (nongloss) paint.

A revolution in digital technology is occurring in the movie industry, where digital techniques are now being used to produce films. Until recently, films have been viewed primarily by using projectors that shine light directly through a strip of film containing the images. Now, however, projectors are available that allow a movie produced using digital techniques to be viewed completely without film by using a digital representation (zeros and ones) of the images. These digital projectors depend on the law of reflection and tiny mirrors called micromirrors, each about the size of one-fourth the diameter of a human hair. Each micromirror creates a tiny portion of an individual movie frame on the screen and serves as a pixel, like one of the glowing spots that comprise the picture on a TV screen or computer monitor. This pixel action is possible because a micromirror pivots in one direction or the reverse in response to the "zero" or "one" in the digital representation of the frame. One of the directions puts a portion of the light from a powerful xenon lamp on the screen, and the other does not. The pivoting action can occur as fast as 1000 times per second, leading to a series of light pulses for each pixel that the eye and the brain combine and interpret as a continuously changing image. Present-generation digital projectors use about 800 000 micromirrors to reproduce each of the three primary colors (red, green, and blue) that comprise a color image.

The physics of digital movie projectors and micromirrors.

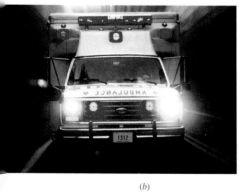

Right hand   Left hand of image

(*a*)

(*b*)

Figure 25.5 (*a*) The person's right hand becomes the image's left hand when viewed in a plane mirror. (*b*) Many emergency vehicles are reverse-lettered so the lettering appears normal when viewed through the rearview mirror of a car. (© Mug Shots/Corbis Images)

25.3 The Formation of Images by a Plane Mirror

When you look into a plane (flat) mirror, you see an image of yourself that has three properties:

1. The image is upright.
2. The image is the same size as you are.
3. The image is located as far behind the mirror as you are in front of it.

As Figure 25.5*a* illustrates, the image of yourself in the mirror is also reversed right to left and left to right. If you wave your *right* hand, it is the *left* hand of the image that waves back. Similarly, letters and words held up to a mirror are reversed. Ambulances and other emergency vehicles are often lettered in reverse, as in Figure 25.5*b*, so that the letters will appear normal when seen in the rearview mirror of a car.

To illustrate why an image appears to originate from behind a plane mirror, Figure 25.6*a* shows a light ray leaving the top of an object. This ray reflects from the mirror (an-

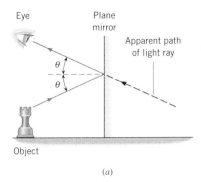

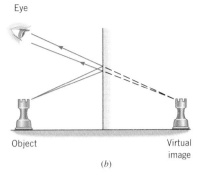

Figure 25.6 (*a*) A ray of light from the top of the chess piece reflects from the mirror. To the eye, the ray seems to come from behind the mirror. (*b*) The bundle of rays from the top of the object appears to originate from the image behind the mirror.

gle of reflection equals angle of incidence) and enters the eye. To the eye, it appears that the ray originates from behind the mirror, somewhere back along the dashed line. Actually, rays going in all directions leave each point on the object, but only a small bundle of such rays is intercepted by the eye. Part *b* of the figure shows a bundle of two rays leaving the top of the object. All the rays that leave a given point on the object, no matter what angle θ they have when they strike the mirror, appear to originate from a corresponding point on the image behind the mirror (see the dashed lines in part *b*). For each point on the object, there is a single corresponding point on the image, and it is this fact that makes the image in a plane mirror a sharp and undistorted one.

Although rays of light *seem* to come from the image, it is evident from Figure 25.6*b* that they do not originate from behind the plane mirror where the image appears to be. Because none of the light rays actually emanate from the image, it is called a ***virtual image.*** In this text the parts of the light rays that appear to come from a virtual image are represented by dashed lines. *Curved* mirrors, on the other hand, can produce images from which all the light rays actually do emanate. Such images are known as ***real images*** and are discussed later.

With the aid of the law of reflection, it is possible to show that the image is located as far behind a plane mirror as the object is in front of it. In Figure 25.7 the object distance is d_o and the image distance is d_i. A ray of light leaves the base of the object, strikes the mirror at an angle of incidence θ, and is reflected at the same angle. To the eye, this ray appears to come from the base of the image. For the angles β_1 and β_2 in the drawing it follows that $\theta + \beta_1 = 90°$ and $\alpha + \beta_2 = 90°$. But the angle α is equal to the angle of reflection θ, since the two are opposite angles formed by intersecting lines. Therefore, $\beta_1 = \beta_2$. As a result, triangles *ABC* and *DBC* are identical (congruent) because they share a common side *BC* and have equal angles ($\beta_1 = \beta_2$) at the top and equal angles (90°) at the base. Thus, the magnitude of the object distance d_o equals the magnitude of the image distance d_i.

By starting with a light ray from the top of the object, rather than the bottom, we can extend the line of reasoning given above to show also that the height of the image equals the height of the object.

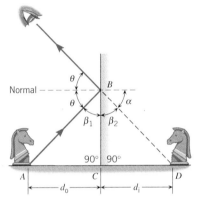

Figure 25.7 This drawing illustrates the geometry used with a plane mirror to show that the image distance d_i equals the object distance d_o.

Concept Simulation 25.1

By clicking on the "Plane mirror" button in this simulation, you can explore the image produced by a plane mirror. The user can change the object distance and the object height, and the program displays numerical values for the image distance and the image height. The simulation also uses a ray diagram to locate the image, and it can be seen whether the image is real or virtual, upright or inverted, enlarged or reduced (or neither).

Go to **www.wiley.com/college/cutnell**

Conceptual Examples 1 and 2 discuss some interesting features of plane mirrors.

Conceptual Example 1 Full-Length Versus Half-Length Mirrors

In Figure 25.8 a woman is standing in front of a plane mirror. What is the minimum mirror height necessary for her to see her full image?

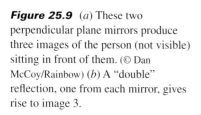

Figure 25.8 For the woman to see her full-sized image, only a half-sized mirror is needed.

Reasoning and Solution The mirror is labeled *ABCD* in the drawing and is the same height as the woman. Light emanating from her body is reflected by the mirror, and some of this light enters her eyes. Consider a ray of light from her foot *F*. This ray strikes the mirror at *B* and enters her eyes at *E*. According to the law of reflection, the angles of incidence and reflection are both *θ*. Any light from her foot that strikes the mirror below *B* is reflected toward a point on her body that is below her eyes. Since light striking the mirror below *B* does not enter her eyes, the part of the mirror between *B* and *A* may be removed. The section *BC* of the mirror that produces the image is one-half the woman's height between *F* and *E*. This follows because the right triangles *FBM* and *EBM* are identical. They are identical because they share a common side *BM* and have two angles, *θ* and 90°, that are the same.

The blowup in Figure 25.8 illustrates a similar line of reasoning, starting with a ray from the woman's head at *H*. This ray is reflected from the mirror at *P* and enters her eyes. The top mirror section *PD* can be removed without disturbing this reflection. The necessary section *CP* is one-half the woman's height between her head at *H* and her eyes at *E*. We find, then, that only the sections *BC* and *CP* are needed for the woman to see her full height. The height of section *BC* plus section *CP* is exactly one-half the woman's height. Thus, *to view one's full length in a mirror, only a half-length mirror is needed.* The conclusions here are valid regardless of how far the person stands from the mirror.

Related Homework: *Problems 2, 32*

Conceptual Example 2 Multiple Reflections

A person is sitting in front of two mirrors that intersect at a 90° angle. As Figure 25.9*a* illustrates, the person sees three images of himself. (The person himself is not shown; only the images are present.) Why are there three images rather than two?

Reasoning and Solution Figure 25.9*b* shows a top view of the person in front of the mirrors. It is a straightforward matter to understand two of the images that he sees. These are the images that are normally seen when one stands in front of a mirror. Standing in front of mirror 1, he sees image 1, which is located as far behind that mirror as he is in front of it. He also sees image 2 behind mirror 2, at a distance that matches his distance in front of that mirror. Each of these images arises from light emanating from his body and reflecting from a single mirror. However, it is also possible for light to undergo two reflections in sequence, first from one mirror and then from the other. *When a double reflection occurs, an additional image becomes possible.* Figure 25.9*b* shows two rays of light that strike mirror 1. Each one, according to the law of reflection, has an angle of reflection that equals the angle of incidence. The rays then strike mirror 2, where they again are reflected according to the law of reflection. When the outgoing rays are extended backward (see the dashed lines in the drawing), they intersect and appear to originate from image 3.

Related Homework: *Problem 4*

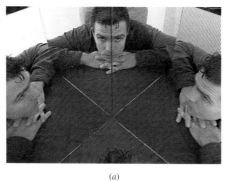

(*a*)

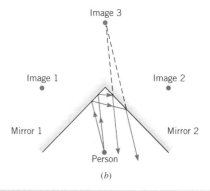

(*b*)

Figure 25.9 (*a*) These two perpendicular plane mirrors produce three images of the person (not visible) sitting in front of them. (© Dan McCoy/Rainbow) (*b*) A "double" reflection, one from each mirror, gives rise to image 3.

✔ **Check Your Understanding 1**

The drawing shows a light ray undergoing multiple reflections from a mirrored corridor. The walls of the corridor are either parallel or perpendicular to one another. If the initial angle of incidence is 35°, what is the angle of reflection when the ray makes its last reflection? *(The answer is given at the end of the book.)*

(Continues)

Background: Understanding the law of reflection is the key to answering this question. To use this law, one must know how the angles of incidence and reflection are measured.

For similar questions (including calculational counterparts), consult Self-Assessment Test 25.1, which is described next.

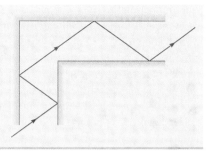

Self-Assessment Test 25.1

Test your understanding of the material in Sections 25.1–25.3:

• Wave Fronts and Rays • The Reflection of Light
• Formation of Images by a Plane Mirror

Go to **www.wiley.com/college/cutnell**

25.4 *Spherical Mirrors*

The most common type of curved mirror is a spherical mirror. As Figure 25.10 shows, a spherical mirror has the shape of a section from the surface of a sphere. If the inside surface of the mirror is polished, it is a *concave mirror.* If the outside surface is polished, it is a *convex mirror.* The drawing shows both types of mirrors, with a light ray reflecting from the polished surface. The law of reflection applies, just as it does for a plane mirror. For either type of spherical mirror, the normal is drawn perpendicular to the mirror at the point of incidence. For each type, the center of curvature is located at point C, and the radius of curvature is R. The *principal axis* of the mirror is a straight line drawn through C and the midpoint of the mirror.

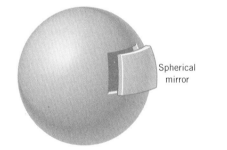

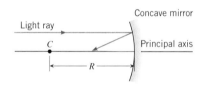

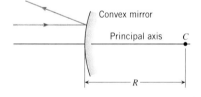

Figure 25.11 shows a tree in front of a concave mirror. A point on this tree lies on the principal axis of the mirror and is beyond the center of curvature C. Light rays emanate from this point and reflect from the mirror, consistent with the law of reflection. If the rays are near the principal axis, they cross it at a common point after reflection. This point is called the *image point.* The rays continue to diverge from the image point as if there were an object there. Since light rays actually come from the image point, the image is a real image.

If the tree in Figure 25.11 is infinitely far from the mirror, the rays are parallel to each other and to the principal axis as they approach the mirror. Figure 25.12 shows rays

Figure 25.10 A spherical mirror has the shape of a segment of a spherical surface. The center of curvature is point C and the radius is R. For a concave mirror, the reflecting surface is the inner one; for a convex mirror it is the outer one.

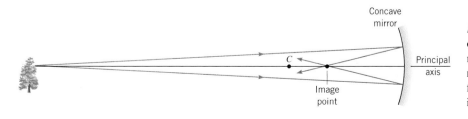

Figure 25.11 A point on the tree lies on the principal axis of the concave mirror. Rays from this point that are near the principal axis are reflected from the mirror and cross the axis at the image point.

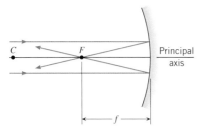

Figure 25.12 Light rays near and parallel to the principal axis are reflected from a concave mirror and converge at the focal point F. The focal length f is the distance between F and the mirror.

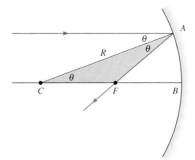

Figure 25.13 This drawing is used to show that the focal point F of a concave mirror is halfway between the center of curvature C and the mirror at point B.

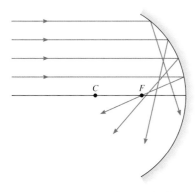

Figure 25.14 Rays that are farthest from the principal axis have the greatest angle of incidence and miss the focal point F after reflection from the mirror.

Figure 25.15 This long row of parabolic mirrors focuses the sun's rays to heat an oil-filled pipe located at the focal point of each mirror. It is one of many that are used by a solar-thermal electric plant in the Mojave Desert. (© Mastrorillo/Corbis Stock Market)

near and parallel to the principal axis, as they reflect from the mirror and pass through an image point. In this special case the image point is referred to as the ***focal point F*** of the mirror. Therefore, an object infinitely far away on the principal axis gives rise to an image at the focal point of the mirror. The distance between the focal point and the middle of the mirror is the ***focal length f*** of the mirror.

We can show that the focal point F lies halfway between the center of curvature C and the middle of a concave mirror. In Figure 25.13, a light ray parallel to the principal axis strikes the mirror at point A. The line CA is the radius of the mirror and, therefore, is the normal to the spherical surface at the point of incidence. The ray reflects from the mirror such that the angle of reflection θ equals the angle of incidence. Furthermore, the angle ACF is also θ because the radial line CA is a transversal of two parallel lines. Since two of its angles are equal, the colored triangle CAF is an isosceles triangle; thus, sides CF and FA are equal. But when the incoming ray lies close to the principal axis, the angle of incidence θ is small, and the distance FA does not differ appreciably from the distance FB. Therefore, in the limit that θ is small, $CF = FA = FB$, and so the focal point F lies halfway between the center of curvature and the mirror. In other words, the focal length f is one-half of the radius R:

Focal length of a concave mirror
$$f = \tfrac{1}{2}R \qquad (25.1)$$

Rays that lie close to the principal axis are known as ***paraxial rays,**** and Equation 25.1 is valid only for such rays. Rays that are far from the principal axis do not converge to a single point after reflection from the mirror, as Figure 25.14 shows. The result is a blurred image. The fact that a spherical mirror does not bring all rays parallel to the principal axis to a single image point is known as ***spherical aberration.*** Spherical aberration can be minimized by using a mirror whose height is small compared to the radius of curvature.

A sharp image point can be obtained with a large mirror, if the mirror is parabolic in shape instead of spherical. The shape of a parabolic mirror is such that all light rays parallel to the principal axis, regardless of their distance from the axis, are reflected through a single image point. However, parabolic mirrors are costly to manufacture and are used where the sharpest images are required, as in research-quality telescopes. Parabolic mirrors are also used in one method of capturing solar energy for commercial purposes. Figure 25.15 shows a long row of concave parabolic mirrors that reflect the sun's rays to the focal point. Located at the focal point and running the length of the row is an oil-filled pipe. The focused rays of the sun heat the oil. In a solar-thermal electric plant, the heat from many such rows is used to generate steam. The steam, in turn, drives a turbine con-

* Paraxial rays are close to the principal axis but not necessarily parallel to it.

nected to an electric generator. Another application of parabolic mirrors is in automobile headlights. Here, however, the situation is reversed from the operation of a solar collector. In a headlight, a high-intensity light bulb is placed at the focal point of the mirror, and light emerges parallel to the principal axis.

A convex mirror also has a focal point, and Figure 25.16 illustrates its meaning. In this picture, parallel rays are incident on a convex mirror. Clearly, the rays diverge after being reflected. If the incident parallel rays are paraxial, the reflected rays seem to come from a single point *F* behind the mirror. This point is the focal point of the convex mirror, and its distance from the midpoint of the mirror is the focal length *f*. The focal length of a convex mirror is also one-half of the radius of curvature, just as it is for a concave mirror. However, we assign the focal length of a convex mirror a negative value because it will be convenient later on:

Focal length of a convex mirror

$$f = -\tfrac{1}{2}R \qquad (25.2)$$

The physics of **capturing solar energy with mirrors; automobile headlights.**

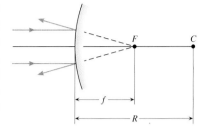

Figure 25.16 When paraxial light rays that are parallel to the principal axis strike a convex mirror, the reflected rays appear to originate from the focal point *F*. The radius of curvature is *R* and the focal length is *f*.

 **Check Your Understanding 2**

A section of a sphere has a radius of curvature of 0.60 m, and both the inside and outside surfaces have a mirror-like polish. What are the focal lengths of the inside and outside surfaces? *(The answers are given at the end of the book.)*

Background: The two types of spherical mirrors, concave and convex, are made from a section of a sphere. The radius of curvature determines the focal lengths of these types of mirrors.

For similar questions (including calculational counterparts), consult Self-Assessment Test 25.2, which is described at the end of Section 25.6.

25.5 *The Formation of Images by Spherical Mirrors*

▶ CONCEPTS AT A GLANCE As we have seen, some of the light rays emitted from an object in front of a mirror strike the mirror, reflect from it, and form an image. We can analyze the image produced by either concave or convex mirrors by using a graphical method called *ray tracing.* As the Concepts-at-a-Glance chart in Figure 25.17 shows, this method is based on the law of reflection and the notion that a spherical mirror has a center of curvature *C* and a focal point *F*. Ray tracing enables us to find the location of the image, as well as its size, by taking advantage of this fact: paraxial rays leave from a point on the object and intersect at a corresponding point on the image after reflection. ◀

Figure 25.17 CONCEPTS AT A GLANCE The method of ray tracing is used to predict the location and size of an image produced by a spherical mirror. Ray tracing is based on the law of reflection and two points associated with the mirror, its center of curvature *C* and its focal point *F*. The red and blue rays in the photograph, after reflecting from the mirror, pass through the focal point. (© Richard Megna/ Fundamental Photographs)

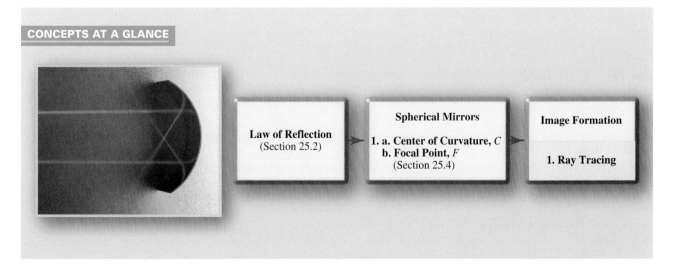

CONCEPTS AT A GLANCE

Law of Reflection (Section 25.2) → **Spherical Mirrors** 1. a. Center of Curvature, *C* b. Focal Point, *F* (Section 25.4) → **Image Formation** 1. Ray Tracing

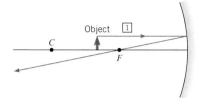

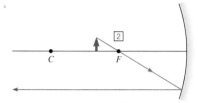

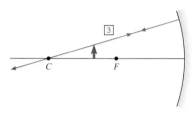

Figure 25.18 The rays labeled 1, 2, and 3 are useful in locating the image of an object placed in front of a concave spherical mirror. The object is represented as a vertical arrow.

Problem solving insight

CONCAVE MIRRORS

Three specific paraxial rays are especially convenient to use. Figure 25.18 shows an object in front of a concave mirror, and these three rays leave from a point on the top of the object. The rays are labeled 1, 2, and 3, and when tracing their paths, we use the following reasoning strategy.

Reasoning Strategy

Ray Tracing for a Concave Mirror

Ray 1. This ray is initially parallel to the principal axis and, therefore, passes through the focal point *F* after reflection from the mirror.

Ray 2. This ray initially passes through the focal point *F* and is reflected parallel to the principal axis. Ray 2 is analogous to ray 1 except that the reflected, rather than the incident, ray is parallel to the principal axis.

Ray 3. This ray travels along a line that passes through the center of curvature *C* and follows a radius of the spherical mirror; as a result, the ray strikes the mirror perpendicularly and reflects back on itself.

If rays 1, 2, and 3 are superimposed on a scale drawing, they converge at a point on the top of the image, as can be seen in Figure 25.19a.* Although three rays have been used here to locate the image, only two are really needed; the third ray is usually drawn to serve as a check. In a similar fashion, rays from all other points on the object locate corresponding points on the image, and the mirror forms a complete image of the object. If you were to place your eye as shown in the drawing, you would see an image that is *larger* and *inverted* relative to the object. The image is *real* because the light rays actually pass through the image.

If the locations of the object and image in Figure 25.19a are interchanged, the situation in part *b* of the drawing results. The three rays in part *b* are the same as those in part *a*, except the directions are reversed. These drawings illustrate the **principle of reversibility,** which states that **if the direction of a light ray is reversed, the light retraces its original path.** This principle is quite general and is not restricted to reflection from mirrors. The image is *real,* and it is *smaller* and *inverted* relative to the object.

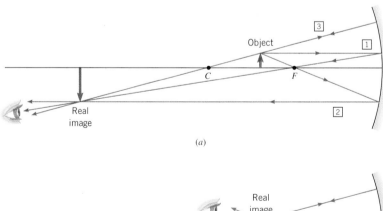

(a)

Figure 25.19 (a) When an object is placed between the focal point *F* and the center of curvature *C* of a concave mirror, a real image is formed. The image is enlarged and inverted relative to the object. (b) When the object is located beyond the center of curvature *C*, a real image is created that is reduced in size and inverted relative to the object.

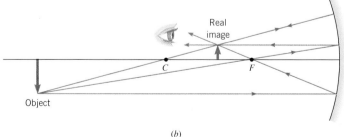

(b)

* In the drawings that follow, we assume that the rays are paraxial, although the distance between the rays and the principal axis is often exaggerated for clarity.

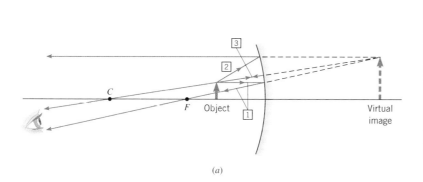

(a)

(b)

Figure 25.20 (a) When an object is located between a concave mirror and its focal point F, an enlarged, upright, and virtual image is produced. (b) A shaving mirror (or makeup mirror) is a concave mirror that can form an enlarged virtual image, as this photograph shows. (© Laurence Monneret/Stone/Getty Images)

When the object is placed between the focal point F and a concave mirror, as in Figure 25.20a, three rays can again be drawn to find the image. But now ray 2 does not go through the focal point on its way to the mirror, since the object is inside the focal point. However, when projected backward, ray 2 appears to come from the focal point. Therefore, after reflection, ray 2 is directed parallel to the principal axis. In this case the three reflected rays diverge from each other and do not converge to a common point. However, when projected behind the mirror, the three rays appear to come from a common point; thus, a *virtual* image is formed. This virtual image is *larger* than the object and *upright*. Makeup and shaving mirrors are concave mirrors. When you place your face between the mirror and its focal point, you see an enlarged virtual image of yourself, as part b of the drawing shows.

The physics of makeup and shaving mirrors.

Concave mirrors are also used in one method for displaying the speed of a car. The method presents a digital readout (e.g., "55 mph") that the driver sees when looking directly through the windshield, as in Figure 25.21a. The advantage of the method, which is called a head-up display (HUD), is that the driver does not need to take his or her eyes off the road to monitor the speed. Figure 25.21b shows how a HUD works. Located below the windshield is a readout device that displays the speed in digital form. This device is located in front of a concave mirror but within its focal point. The arrangement is similar to that in Figure 25.20a and produces a virtual, upright, and enlarged image of the speed

The physics of a head-up display for automobiles.

Figure 25.21 (a) A head-up display (HUD) presents the driver with a digital readout of the car's speed in the field of view seen through the windshield. (Copyright GM Corporation. All rights reserved.) (b) One version of a HUD uses a concave mirror. (See text for explanation.)

(a)

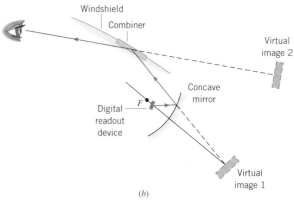

(b)

readout (see virtual image 1 in Figure 25.21*b*). Light rays that appear to come from this image strike the windshield at a place where a so-called combiner is located. The purpose of the combiner is to combine the digital readout information with the field of view that the driver sees through the windshield. The combiner is virtually undetectable by the driver because it allows all colors except one to pass through it unaffected. The one exception is the color produced by the digital readout device. For this color, the combiner behaves as a plane mirror and reflects the light that appears to originate from image 1. Thus, the combiner produces image 2, which is what the driver sees. The location of image 2 is near the front bumper. The driver can then read the speed with his eyes focused just as they are to see the road.

CONVEX MIRRORS

The procedure for determining the location and size of an image in a convex mirror is similar to that for a concave mirror. The same three rays are used. However, the focal point and center of curvature of a convex mirror lie behind the mirror, not in front of it. Figure 25.22*a* shows the rays. When tracing their paths, we use the following reasoning strategy, which takes into account the different locations of the focal point and center of curvature.

Reasoning Strategy

Ray Tracing for a Convex Mirror

Ray 1. This ray is initially parallel to the principal axis and, therefore, appears to originate from the focal point *F* after reflection from the mirror.

Ray 2. This ray heads toward *F*, emerging parallel to the principal axis after reflection. Ray 2 is analogous to ray 1, except that the reflected, rather than the incident, ray is parallel to the principal axis.

Ray 3. This ray travels toward the center of curvature *C*; as a result, the ray strikes the mirror perpendicularly and reflects back on itself.

The three rays in Figure 25.22*a* appear to come from a point on a *virtual* image that is behind the mirror. The virtual image is *diminished in size* and *upright*, relative to the object. A convex mirror *always* forms a virtual image of the object, no matter where in front of the mirror the object is placed. Figure 25.22*b* shows an example of such an image.

Figure 25.22 (*a*) An object placed in front of a convex mirror produces a virtual image behind the mirror. The virtual image is reduced in size and upright. (*b*) The sun shield in this pilot's helmet acts as a convex mirror and reflects an image of his plane. (© Chad Slattery/Stone/Getty Images)

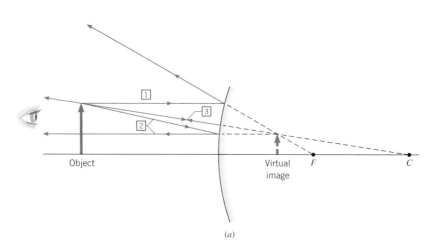

Object · Virtual image · *F* · *C*

(*a*)

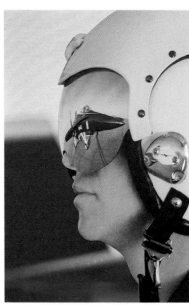

(*b*)

Because of its shape, a convex mirror gives a wider field of view than do other types of mirrors. Therefore, they are often used in stores for security purposes. A mirror with a wide field of view is also needed to give a driver a good rear view. Thus, the outside mirror on the passenger side is often a convex mirror. Printed on such a mirror is usually the warning "VEHICLES IN MIRROR ARE CLOSER THAN THEY APPEAR." The reason for the warning is that the virtual image in Figure 25.22a is reduced in size and therefore looks smaller, just as a distant object would appear in a plane mirror. An unwary driver, thinking that the side-view mirror is a plane mirror, might incorrectly deduce from the small size of the image that the car behind is far enough away to ignore.

The physics of
passenger-side automobile mirrors.

✔ **Check Your Understanding 3**

An object is placed between the focal point and the center of curvature of a concave mirror. The object is then moved closer to the mirror, but still remains between the focal point and the center of curvature. What happens to the magnitudes of the image distance and the image height? Larger or smaller? *(The answers are given at the end of the book.)*

Background: The location and height of an image can be found by constructing a ray diagram and using the three rays discussed in this section.

For similar questions (including calculational counterparts), consult Self-Assessment Test 25.2, which is described at the end of Section 25.6.

25.6 The Mirror Equation and the Magnification Equation

▶ CONCEPTS AT A GLANCE Ray diagrams drawn to scale are useful for determining the location and size of the image formed by a mirror. However, for an accurate description of the image, a more analytical technique is needed. We will now derive two equations, known as the *mirror equation* and the *magnification equation,* that will provide a complete description of the image. The Concepts-at-a-Glance chart in Figure 25.23, which is an extension of the chart in Figure 25.17, emphasizes that these equations are based on the law of reflection and provide relationships between:

f = the focal length of the mirror
d_o = the object distance, which is the distance between the object and the mirror
d_i = the image distance, which is the distance between the image and the mirror
m = the magnification of the mirror, which is the ratio of the height of the image to the height of the object. ◀

🖱 Concept Simulation 25.3

Selecting the "Convex mirror" option in this simulation allows you to investigate the image produced by a convex mirror. The object distance and/or the object height can be changed by clicking and dragging the object. The user can also alter the focal length of the mirror. The simulation uses a ray diagram to locate the image. In the case of a convex mirror, you will see that the image is always virtual, upright, and smaller than the object, regardless of where the object is located. The program also displays numerical values for the image distance and the image height.

Related Homework: Conceptual Questions 11, 12, Problem 30

Go to
www.wiley.com/college/cutnell

Figure 25.23 CONCEPTS AT A GLANCE The law of reflection gives rise to the mirror equation and the magnification equation, which give the location and relative size of an image. The photograph shows an image of an eye chart, the light from which has been reflected from the curved surface of the eye. (© Art Montes de Oca/Taxi/ Getty Images)

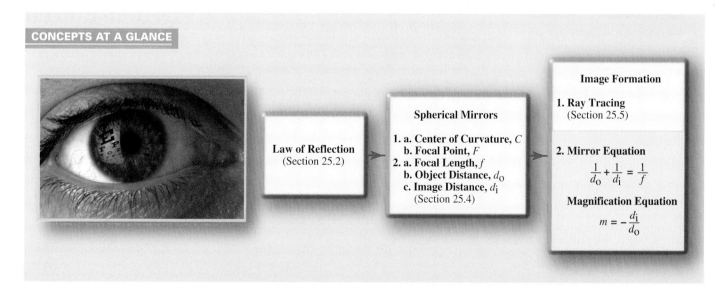

CONCEPTS AT A GLANCE

Law of Reflection
(Section 25.2)

Spherical Mirrors

1. a. Center of Curvature, C
   b. Focal Point, F
2. a. Focal Length, f
   b. Object Distance, d_o
   c. Image Distance, d_i
   (Section 25.4)

Image Formation

1. **Ray Tracing**
   (Section 25.5)

2. **Mirror Equation**

$$\frac{1}{d_o} + \frac{1}{d_i} = \frac{1}{f}$$

Magnification Equation

$$m = -\frac{d_i}{d_o}$$

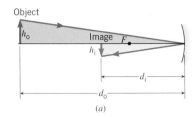

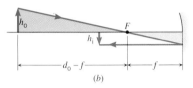

Figure 25.24 These diagrams are used to derive the mirror equation and the magnification equation. (*a*) The two colored triangles are similar triangles. (*b*) If the ray is close to the principal axis, the two colored regions are almost similar triangles.

The reflections of the boy and girl in this fun-house mirror are distorted because the mirror is curved and not flat. It is not a spherical mirror, however. (© Jeremy Homer/Stone/Getty Images)

Problem solving insight
According to the mirror equation, the image distance d_i has a reciprocal given by $d_i^{-1} = f^{-1} - d_o^{-1}$. After combining the reciprocals f^{-1} and d_o^{-1}, do not forget to take the reciprocal of the result to find d_i.

CONCAVE MIRRORS

We begin our derivation of the mirror equation by referring to Figure 25.24*a*, which shows a ray leaving the top of the object and striking a concave mirror at the point where the principal axis intersects the mirror. Since the principal axis is perpendicular to the mirror, it is also the normal at this point of incidence. Therefore, the ray reflects at an equal angle and passes through the image. The two colored triangles are similar triangles because they have equal angles, so

$$\frac{h_o}{-h_i} = \frac{d_o}{d_i}$$

where h_o is the height of the object and h_i is the height of the image. The minus sign appears on the left in this equation because the image is inverted in Figure 25.24*a*. In part *b* another ray leaves the top of the object, this one passing through the focal point *F*, reflecting parallel to the principal axis, and then passing through the image. Provided the ray remains close to the axis, the two colored areas can be considered to be similar triangles, with the result that

$$\frac{h_o}{-h_i} = \frac{d_o - f}{f}$$

Setting the two equations above equal to each other yields $d_o/d_i = (d_o - f)/f$. Rearranging this result gives the ***mirror equation:***

Mirror equation $$\frac{1}{d_o} + \frac{1}{d_i} = \frac{1}{f} \qquad (25.3)$$

We have derived this equation for a real image formed in front of a concave mirror. In this case, the image distance is a positive quantity, as are the object distance and the focal length. However, we have seen in the last section that a concave mirror can also form a virtual image, if the object is located between the focal point and the mirror. Equation 25.3 can also be applied to such a situation, provided that we adopt the convention that d_i is negative for an image behind the mirror, as it is for a virtual image.

In deriving the magnification equation, we remember that the ***magnification*** *m* of a mirror is the ratio of the image height to the object height: $m = h_i/h_o$. If the image height is less than the object height, the magnitude of *m* is less than one. Conversely, if the image is larger than the object, the magnitude of *m* is greater than one. We have already shown that $h_o/(-h_i) = d_o/d_i$, so it follows that

Magnification equation $$m = \frac{\text{Image height, } h_i}{\text{Object height, } h_o} = -\frac{d_i}{d_o} \qquad (25.4)$$

As Examples 3 and 4 show, the value of *m* is negative if the image is inverted and positive if the image is upright.

Example 3   A Real Image Formed by a Concave Mirror

A 2.0-cm-high object is placed 7.10 cm from a concave mirror whose radius of curvature is 10.20 cm. Find (a) the location of the image and (b) its size.

Reasoning Since $f = \frac{1}{2}R = \frac{1}{2}(10.20 \text{ cm}) = 5.10 \text{ cm}$, the object is located between the focal point *F* and the center of curvature *C* of the mirror, as in Figure 25.19*a*. Based on this figure, we expect that the image is real and that, relative to the object, it is farther away from the mirror, inverted, and larger.

Solution

(a) With $d_o = 7.10$ cm and $f = 5.10$ cm, the mirror equation (Equation 25.3) can be used to find the image distance:

$$\frac{1}{d_i} = \frac{1}{f} - \frac{1}{d_o} = \frac{1}{5.10 \text{ cm}} - \frac{1}{7.10 \text{ cm}} = 0.055 \text{ cm}^{-1} \quad \text{or} \quad \boxed{d_i = 18 \text{ cm}}$$

In this calculation, *f* and d_o are positive numbers, indicating that the focal point and the object are in front of the mirror. The positive answer for d_i means that the image is also in front of the mirror, and the reflected rays actually pass through the image, as Figure 25.19a shows. In other words, the positive value for d_i indicates that the image is a real image.

(b) The height of the image can be determined once the magnification m of the mirror is known. The magnification equation (Equation 25.4) can be used to find m:

$$m = -\frac{d_i}{d_o} = -\frac{18 \text{ cm}}{7.10 \text{ cm}} = -2.5$$

The image height is $h_i = mh_o = (-2.5)(2.0 \text{ cm}) = \boxed{-5.0 \text{ cm}}$. The image is 2.5 times larger than the object, the negative values for m and h_i indicating that the image is inverted with respect to the object, as in Figure 25.19a.

Example 4 A Virtual Image Formed by a Concave Mirror

An object is placed 6.00 cm in front of a concave mirror that has a 10.0-cm focal length. (a) Determine the location of the image. (b) The object is 1.2 cm high. Find the image height.

Reasoning The object is located between the focal point and the mirror, as in Figure 25.20a. The setup is analogous to a person using a makeup or shaving mirror. Therefore, we expect that the image is virtual and that, relative to the object, it is upright and larger.

Solution

(a) Using the mirror equation with $d_o = 6.00$ cm and $f = 10.0$ cm, we have

$$\frac{1}{d_i} = \frac{1}{f} - \frac{1}{d_o} = \frac{1}{10.0 \text{ cm}} - \frac{1}{6.00 \text{ cm}} = -0.067 \text{ cm}^{-1} \quad \text{or} \quad \boxed{d_i = -15 \text{ cm}}$$

The answer for d_i is negative, indicating that the image is *behind* the mirror. Thus, as expected, the image is virtual.

(b) The image height h_i can be found from the magnification m and the object height h_o:

$$m = -\frac{d_i}{d_o} = -\frac{(-15 \text{ cm})}{6.00 \text{ cm}} = 2.5$$

The image height is $h_i = mh_o = (2.5)(1.2 \text{ cm}) = \boxed{3.0 \text{ cm}}$. The image is larger than the object, and the positive values for m and h_i indicate that the image is upright (see Figure 25.20a).

Need more practice?

Interactive LearningWare 25.1
A gemstone is placed 20.0 cm in front of a concave mirror and is within the focal point. When the concave mirror is replaced by a plane mirror, the image moves 15.0 cm toward the mirror. Find the focal length of the concave mirror.

Go to **www.wiley.com/college/cutnell** for an interactive solution.

CONVEX MIRRORS

The mirror equation and the magnification equation can also be used with convex mirrors, provided the focal length f is taken to be a *negative number*, as indicated earlier in Equation 25.2. One way to remember this is to recall that the focal point of a convex mirror lies *behind* the mirror. Example 5 deals with a convex mirror.

Example 5 A Virtual Image Formed by a Convex Mirror

A convex mirror is used to reflect light from an object placed 66 cm in front of the mirror. The focal length of the mirror is $f = -46$ cm (note the minus sign). Find (a) the location of the image and (b) the magnification.

Reasoning We have seen that a convex mirror always forms a virtual image, as in Figure 25.22a, where the image is upright and smaller than the object. These characteristics should also be indicated by the results of our analysis here.

Problem solving insight
When using the mirror equation, it is useful to construct a ray diagram to guide your thinking and to check your calculation.

Solution

(a) With $d_o = 66$ cm and $f = -46$ cm, the mirror equation gives

$$\frac{1}{d_i} = \frac{1}{f} - \frac{1}{d_o} = \frac{1}{-46 \text{ cm}} - \frac{1}{66 \text{ cm}} = -0.037 \text{ cm}^{-1} \quad \text{or} \quad \boxed{d_i = -27 \text{ cm}}$$

The negative sign for d_i indicates that the image is behind the mirror and, therefore, is a virtual image.

(b) According to the magnification equation, the magnification is

$$m = -\frac{d_i}{d_o} = -\frac{(-27 \text{ cm})}{66 \text{ cm}} = \boxed{0.41}$$

The image is smaller (m is less than one) and upright (m is positive) with respect to the object.

Convex mirrors, like plane (flat) mirrors, always produce virtual images behind the mirror. However, the virtual image in a convex mirror is closer to the mirror than it would be if the mirror were planar, as Example 6 illustrates.

Example 6 A Convex Versus a Plane Mirror

An object is placed 9.00 cm in front of a mirror. The image is 3.00 cm closer to the mirror when the mirror is convex than when it is planar. Find the focal length of the convex mirror.

Reasoning For a plane mirror, the image and the object are the same distance on either side of the mirror. Thus, the image would be 9.00 cm behind a plane mirror. If the image in a convex mirror is 3.00 cm closer than this, the image must be located 6.00 cm behind the convex mirror. In other words, when the object distance is $d_o = 9.00$ cm, the image distance for the convex mirror is $d_i = -6.00$ cm (negative because the image is virtual). The mirror equation can be used to find the focal length of the mirror.

Solution According to the mirror equation, the reciprocal of the focal length is

$$\frac{1}{f} = \frac{1}{d_o} + \frac{1}{d_i} = \frac{1}{9.00 \text{ cm}} + \frac{1}{-6.00 \text{ cm}} = -0.056 \text{ cm}^{-1} \quad \text{or} \quad \boxed{f = -18 \text{ cm}}$$

The Reasoning Strategy below summarizes the sign conventions that are used with the mirror equation and the magnification equation. These conventions apply to both concave and convex mirrors.

Reasoning Strategy

Summary of Sign Conventions for Spherical Mirrors

Focal length

  f is $+$ for a concave mirror.

  f is $-$ for a convex mirror.

Object distance

  d_o is $+$ if the object is in front of the mirror (real object).

  d_o is $-$ if the object is behind the mirror (virtual object).*

Image distance

  d_i is $+$ if the image is in front of the mirror (real image).

  d_i is $-$ if the image is behind the mirror (virtual image).

Magnification

  m is $+$ for an image that is upright with respect to the object.

  m is $-$ for an image that is inverted with respect to the object.

---

* Sometimes optical systems use two (or more) mirrors, and the image formed by the first mirror serves as the object for the second mirror. Occasionally, such an object falls *behind* the second mirror. In this case the object distance is negative, and the object is said to be a virtual object.

✓ **Check Your Understanding 4**

An object is placed in front of a spherical mirror, and the magnification of the system is $m = -6$. What does this number tell you about the image? Select one or more of the choices below. *(The answer is given at the end of the book.)*

    A. The image is larger than the object.

    B. The image is smaller than the object.

    C. The image is upright relative to the object.

    D. The image is inverted relative to the object.

    E. The image is a real image.

    F. The image is a virtual image.

Background: The magnification equation is the key here; it contains a wealth of information about the image.

For similar questions (including calculational counterparts), consult Self-Assessment Test 25.2, which is described next.

Self-Assessment Test 25.2

Test your understanding of the material in Sections 25.4–25.6:

• Spherical Mirrors   • The Formation of Images by Spherical Mirrors
• The Mirror Equation and the Magnification Equation

Go to **www.wiley.com/college/cutnell**

25.7 Concepts & Calculations

Relative to the object in front of a spherical mirror, the image can differ in a number of respects. The image can be real (in front of the mirror) or virtual (behind the mirror). It can be larger or smaller than the object, and it can be upright or inverted. As you solve problems dealing with spherical mirrors, keep these image characteristics in mind. They can help guide you to the correct answer, as Examples 7 and 8 illustrate.

Concepts & Calculations Example 7   Finding the Focal Length

An object is located 7.0 cm in front of a mirror. The virtual image is located 4.5 cm away from the mirror and is smaller than the object. Find the focal length of the mirror.

Concept Questions and Answers Based solely on the fact that the image is virtual, is the mirror concave or convex, or is either type possible?

    Answer Either type is possible. A concave mirror can form a virtual image if the object is within the focal point of the mirror, as Figure 25.20a illustrates. A convex mirror always forms a virtual image, as Figure 25.22a shows.

The image is smaller than the object, as well as virtual. Do these characteristics together indicate a concave or convex mirror, or do they indicate either type?

    Answer They indicate a convex mirror. A concave mirror can produce an image that is smaller than the object if the object is located beyond the center of curvature of the mirror, as in Figure 25.19b. However, the image in Figure 25.19b is real, not virtual. A convex mirror, in contrast, always produces an image that is virtual and smaller than the object, as Figure 25.22a illustrates.

Is the focal length positive or negative?

    Answer The focal length is negative because the mirror is convex. A concave mirror has a positive focal length.

Solution The virtual image is located behind the mirror and, therefore, has a negative image distance, $d_i = -4.5$ cm. Using this value together with the object distance of $d_o = 7.0$ cm, we can apply the mirror equation to find the focal length:

$$\frac{1}{f} = \frac{1}{d_o} + \frac{1}{d_i} = \frac{1}{7.0 \text{ cm}} + \frac{1}{-4.5 \text{ cm}} = -0.079 \text{ cm}^{-1} \quad \text{or} \quad \boxed{f = -13 \text{ cm}}$$

As expected, the focal length is negative.

Concepts & Calculations Example 8   Two Choices

The radius of curvature of a mirror is 24 cm. A diamond ring is placed in front of this mirror. The image is twice the size of the ring. Find the object distance of the ring.

Concept Questions and Answers Is the mirror concave or convex, or is either type possible?

Answer   A convex mirror always forms an image that is smaller than the object, as Figure 25.22a shows. Therefore, the mirror must be concave.

How many places are there in front of a concave mirror where the ring can be placed and produce an image that is twice the size of the object?

Answer   There are two places. Figure 25.19a illustrates that one of the places is between the center of curvature and the focal point. The enlarged image is real and inverted. Figure 25.20a shows that another possibility is between the focal point and the mirror, in which case the enlarged image is virtual and upright.

What are the possible values for the magnification of the image of the ring?

Answer   Since the image is inverted in Figure 25.19a, the magnification for this possibility is $m = -2$. In Figure 25.20a, however, the image is upright, so the magnification is $m = +2$ for this option. In either case, the image is twice the size of the ring.

Solution According to the mirror equation and the magnification equation, we have

$$\underbrace{\frac{1}{d_o} + \frac{1}{d_i} = \frac{1}{f}}_{\text{Mirror equation}} \quad \text{and} \quad \underbrace{m = -\frac{d_i}{d_o}}_{\text{Magnification equation}}$$

We can solve the magnification equation for the image distance and obtain $d_i = -md_o$. Substituting this expression for d_i into the mirror equation gives

$$\frac{1}{d_o} + \frac{1}{(-md_o)} = \frac{1}{f} \quad \text{or} \quad d_o = \frac{f(m - 1)}{m}$$

Applying this result with the two magnifications (and noting that $f = \frac{1}{2}R = 12$ cm), we obtain

$$m = -2 \qquad d_o = \frac{f(m - 1)}{m} = \frac{(12 \text{ cm})(-2 - 1)}{-2} = \boxed{18 \text{ cm}}$$

$$m = +2 \qquad d_o = \frac{f(m - 1)}{m} = \frac{(12 \text{ cm})(+2 - 1)}{+2} = \boxed{6.0 \text{ cm}}$$

At the end of the problem set for this chapter, you will find homework problems that contain both conceptual and quantitative parts. These problems are grouped under the heading *Concepts & Calculations, Group Learning Problems.* They are designed for use by students working alone or in small learning groups. The conceptual part of each problem provides a convenient focus for group discussions.

Concept Summary

This summary presents an abridged version of the chapter, including the important equations and all available learning aids. For convenient reference, the learning aids (including the text's examples) are placed next to or immediately after the relevant equation or discussion. The following learning aids may be found on-line at **www.wiley.com/college/cutnell**:

Interactive LearningWare examples are solved according to a five-step interactive format that is designed to help you develop problem-solving skills.	**Concept Simulations** are animated versions of text figures or animations that illustrate important concepts. You can control parameters that affect the display, and we encourage you to experiment.
Interactive Solutions offer specific models for certain types of problems in the chapter homework. The calculations are carried out interactively.	**Self-Assessment Tests** include both qualitative and quantitative questions. Extensive feedback is provided for both incorrect and correct answers, to help you evaluate your understanding of the material.

Topic	Discussion	Learning Aids
	25.1 Wave Fronts and Rays	
Wave fronts **Plane waves**	Wave fronts are surfaces on which all points of a wave are in the same phase of motion. Waves whose wave fronts are flat surfaces are known as plane waves.	
Rays	Rays are lines that are perpendicular to the wave fronts and point in the direction of the velocity of the wave.	
	25.2 The Reflection of Light	
Law of reflection	When light reflects from a smooth surface, the reflected light obeys the law of reflection: a. The incident ray, the reflected ray, and the normal to the surface all lie in the same plane. b. The angle of reflection θ_r equals the angle of incidence θ_i; $\theta_r = \theta_i$.	
	25.3 The Formation of Images by a Plane Mirror	
Virtual image	A virtual image is one from which all the rays of light do not actually come, but only appear to do so.	
Real image	A real image is one from which all the rays of light actually do emanate.	
Plane mirror	A plane mirror forms an upright, virtual image that is located as far behind the mirror as the object is in front of it. In addition, the heights of the image and the object are equal.	Concept Simulation 25.1 Examples 1, 2

 Use *Self Assessment Test 25.1* to evaluate your understanding of Sections 25.1–25.3.

Topic	Discussion	Learning Aids
	25.4 Spherical Mirrors	
Concave and convex mirrors	A spherical mirror has the shape of a section from the surface of a sphere. If the inside surface of the mirror is polished, it is a concave mirror. If the outside surface is polished, it is a convex mirror.	
Principal axis **Paraxial rays**	The principal axis of a mirror is a straight line drawn through the center of curvature and the middle of the mirror's surface. Rays that are close to the principal axis are known as paraxial rays. Paraxial rays are not necessarily parallel to the principal axis.	
Radius of curvature	The radius of curvature R of a mirror is the distance from the center of curvature to the mirror.	
Focal point of a concave mirror	The focal point of a concave spherical mirror is a point on the principal axis, in front of the mirror. Incident paraxial rays that are parallel to the principal axis converge to the focal point after being reflected from the concave mirror.	
Spherical aberration	The fact that a spherical mirror does not bring all rays parallel to the principal axis to a single image point after reflection is known as spherical aberration.	
Focal point of a convex mirror	The focal point of a convex spherical mirror is a point on the principal axis, behind the mirror. For a convex mirror, incident paraxial rays that are parallel to the principal axis diverge after reflecting from the mirror. These rays seem to originate from the focal point.	

Topic	Discussion	Learning Aids
Focal length	The focal length f of a mirror is the distance along the principal axis between the focal point and the mirror. The focal length and the radius of curvature R are related by	
Concave mirror	$$f = \tfrac{1}{2}R \qquad (25.1)$$	
Convex mirror	$$f = -\tfrac{1}{2}R \qquad (25.2)$$	

25.5 The Formation of Images by Spherical Mirrors

Ray tracing	The image produced by a mirror can be located by a graphical method known as ray tracing.	
	For a concave mirror, the following paraxial rays are especially useful for ray tracing (see Figure 25.18):	Concept Simulation 25.2
Three rays for a concave mirror	**Ray 1.** This ray leaves the object traveling parallel to the principal axis. The ray reflects from the mirror and passes through the focal point.	
	Ray 2. This ray leaves the object and passes through the focal point. The ray reflects from the mirror and travels parallel to the principal axis.	
	Ray 3. This ray leaves the object and travels along a line that passes through the center of curvature. The ray strikes the mirror perpendicularly and reflects back on itself.	
	For a convex mirror, these paraxial rays are useful for ray tracing (see Figure 25.22a):	Concept Simulation 25.3
Three rays for a convex mirror	**Ray 1.** This ray leaves the object traveling parallel to the principal axis. After reflection from the mirror, the ray appears to originate from the focal point of the mirror.	
	Ray 2. This ray leaves the object and heads toward the focal point. After reflection, the ray travels parallel to the principal axis.	
	Ray 3. This ray leaves the object and travels toward the center of curvature. The ray strikes the mirror perpendicularly and reflects back on itself.	

25.6 The Mirror Equation and the Magnification Equation

	The mirror equation specifies the relation between the object distance d_o, the image distance d_i, and the focal length f of the mirror:	Examples 3, 4
Mirror equation	$$\frac{1}{d_o} + \frac{1}{d_i} = \frac{1}{f} \qquad (25.3)$$	Interactive LearningWare 25.1
	The mirror equation can be used with either concave or convex mirrors.	Examples 5–8
	The magnification m of a mirror is the ratio of the image height h_i to the object height h_o:	Interactive Solution 25.19
Magnification	$$m = \frac{h_i}{h_o}$$	Interactive Solution 25.25
	The magnification is also related to d_i and d_o by the magnification equation:	
Magnification equation	$$m = -\frac{d_i}{d_o} \qquad (25.4)$$	
	The algebraic sign conventions for the variables appearing in these equations are summarized in the Reasoning Strategy at the end of Section 25.6.	

 Use Self-Assessment Test 25.2 to evaluate your understanding of Sections 25.4–25.6.

Conceptual Questions

1. A sign painted on a store window is reversed when viewed from inside the store. If a person inside the store views the reversed sign in a plane mirror, does the sign appear as it would when viewed from outside the store? (Try it by writing some letters on a transpar-

ent sheet of paper and then holding the back side of the paper up to a mirror.) Explain.

2. If a clock is held in front of a mirror, its image is reversed left to right. From the point of view of a person looking into the mirror, does the image of the second hand rotate in the reverse (counter-clockwise) direction? Justify your answer.

3. (a) Which kind of spherical mirror, concave or convex, can be used to start a fire with sunlight? (b) For the best results, how far from the mirror should the paper to be ignited be placed? Explain your answers.

4. The photograph shows an experimental device at Sandia National Laboratories in New Mexico. This device is a mirror that focuses sunlight to heat sodium to a boil, which then heats helium gas in an engine. The engine does the work of driving a generator to produce electricity. The sodium unit and the engine are labeled in the photo. (a) What kind of mirror is being used, and (b) where is the sodium unit located relative to the mirror? Express your answer in terms of the focal length of the mirror. Give your reasoning.

(Courtesy Sandia National Laboratories)

5. Refer to Figure 25.14 and the related discussion about spherical aberration. To bring the top ray closer to the focal point F after reflection, describe how you would change the shape of the mirror. Would you open it up to produce a more gently curving shape or bring the top and bottom edges closer to the principal axis? Account for your answer using the law of reflection.

6. (a) Can the image formed by a concave mirror ever be projected directly onto a screen, without the help of other mirrors or lenses? If so, specify where the object should be placed relative to the mirror. (b) Repeat part (a) assuming that the mirror is convex.

7. (a) When you look at the back side of a shiny teaspoon, held at arm's length, you see yourself upright. Why? (b) When you look at the other side of the spoon, you see yourself upside down. Why?

8. If you stand between two parallel plane mirrors, you see an infinite number of images of yourself. This occurs because an image in one mirror is reflected in the other mirror to produce another image, which is then re-reflected, and so forth. The multiple images are equally spaced. Suppose that you are facing a convex mirror, with a plane mirror behind you. Describe what you would see and comment about the spacing between any multiple images. Explain your reasoning.

9. Sometimes news personnel covering an event use a microphone arrangement that is designed to increase the ability of the mike to pick up weak sounds. The drawing shows that the arrangement consists of a hollowed-out shell behind the mike. The shell acts like a mirror for sound waves. Explain how this arrangement enables the mike to detect weak sounds.

Microphone

10. When you see the image of yourself formed by a mirror, it is because (1) light rays actually coming from a real image enter your eyes or (2) light rays appearing to come from a virtual image enter your eyes. If light rays from the image do not enter your eyes, you do not see yourself. Are there any places on the principal axis where you cannot see yourself when you are in front of a mirror that is (a) convex and (b) concave? If so, where are these places?

11. Concept Simulation 25.3 at **www.wiley.com/college/cutnell** reviews the concepts that are important in this question. Plane mirrors and convex mirrors form virtual images. With a plane mirror, the image may be infinitely far behind the mirror, depending on where the object is located in front of the mirror. For an object in front of a single convex mirror, what is the greatest distance behind the mirror at which the image can be found? Justify your answer.

12. Concept Simulation 25.3 at **www.wiley.com/college/cutnell** allows you to explore the concepts to which this question relates. Is it possible to use a convex mirror to produce an image that is larger than the object? Provide a reason for your answer.

13. Suppose you stand in front of a spherical mirror (concave or convex). Is it possible for your image to be (a) real and upright or (b) virtual and inverted? Justify your answers.

Problems

ssm Solution is in the Student Solutions Manual.   **www** Solution is available on the World Wide Web at www.wiley.com/college/cutnell
This icon represents a biomedical application.

Section 25.2 The Reflection of Light,
Section 25.3 The Formation of Images by a Plane Mirror

1. ssm Two plane mirrors are separated by 120°, as the drawing illustrates. If a ray strikes mirror M_1 at a 65° angle of incidence, at what angle θ does it leave mirror M_2?

Problem 1

2. Review Conceptual Example 1 before attempting this problem. A person whose eyes are 1.70 m above the floor stands in front of a plane mirror. The top of her head is 0.12 m above her eyes. (a) What is the height of the shortest mirror in which she can see her entire image? (b) How far above the floor should the bottom edge of the mirror be placed?

3. A person stands 3.6 m in front of a wall that is covered floor-to-ceiling with a plane mirror. His eyes are 1.8 m above the floor. He holds a flashlight between his feet and manages to point it at the mirror. At what angle of incidence must the light strike the mirror so the light will reach his eyes?

4. Review Conceptual Example 2. Suppose that in Figure 25.9b the two perpendicular plane mirrors are represented by the $-x$ and $-y$ axes of an x, y coordinate system. An object is in front of these mirrors at a point whose coordinates are $x = -2.0$ m and $y = -1.0$ m. Find the coordinates that locate each of the three images.

5. ssm www Two diverging light rays, originating from the same point, have an angle of 10° between them. After the rays reflect from a plane mirror, what is the angle between them? Construct one possible ray diagram that supports your answer.

6. You are trying to photograph a bird sitting on a tree branch, but a tall hedge is blocking your view. However, as the drawing shows, a plane mirror reflects light from the bird into your camera. For what distance must you set the focus of the camera lens in order to snap a sharp picture of the bird's image?

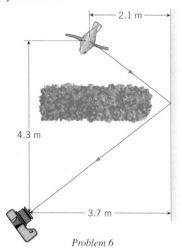

Problem 6

* **7.** The drawing shows a top view of a square room. One wall is missing, and the other three are each mirrors. From point *P* in the center of the open side, a laser is fired, with the intent of hitting a small target located at the center of one wall. Identify six directions in which the laser can be fired and score a hit, assuming that the light does not strike any mirror more than once. Draw the rays to confirm your choices.

* **8.** A ray of light strikes a plane mirror at a 45° angle of incidence. The mirror is then rotated by 15° into the position shown in red in the drawing, while the incident ray is kept fixed. (a) Through what angle ϕ does the reflected ray rotate? (b) What is the answer to part (a) if the angle of incidence is 60° instead of 45°?

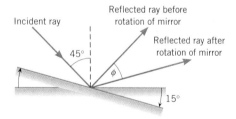

** **9.** In the drawing for problem 7, a laser is fired from point *P* in the center of the open side of the square room. The laser is pointed at the mirrored wall on the right. At what angle of incidence must the light strike the right-hand wall so that, after being reflected, the light hits the left corner of the back wall?

Section 25.4 Spherical Mirrors,
Section 25.5 The Formation of Images by Spherical Mirrors

10. Concept Simulation 25.2 at **www.wiley.com/college/cutnell** illustrates the concepts pertinent to this problem. A 2.0-cm-high object is situated 15.0 cm in front of a concave mirror that has a radius of curvature of 10.0 cm. Using a ray diagram drawn to scale, measure (a) the location and (b) the height of the image. The mirror must be drawn to scale.

11. Repeat problem 10 for a concave mirror with a focal length of 20.0 cm, an object distance of 12.0 cm, and a 2.0-cm-high object.

12. The image of a very distant car is located 12 cm behind a convex mirror. (a) What is the radius of curvature of the mirror? (b) Draw a ray diagram to scale showing this situation.

13. ssm Repeat problem 10 for a convex mirror with a radius of curvature of 1.00×10^2 cm, an object distance of 25.0 cm, and a 10.0-cm-high object.

14. An object is placed in front of a convex mirror. Draw a convex mirror (radius of curvature = 15 cm) to scale, and place an object 25 cm in front of it. Make the object height 4 cm. Using a ray diagram, locate the image and measure its height. Now move the object closer to the mirror, so the object distance is 5 cm. Again, locate its image using a ray diagram. As the object moves closer to the mirror, (a) does the magnitude of the image distance become larger or smaller, and (b) does the magnitude of the image height become larger or smaller? (c) What is the ratio of the image height when the object distance is 5 cm to that when the object distance is 25 cm? Give your answer to one significant figure.

* **15.** A plane mirror and a concave mirror ($f = 8.0$ cm) are facing each other and are separated by a distance of 20.0 cm. An object is placed 10.0 cm in front of the plane mirror. Consider the light from the object that reflects first from the plane mirror and then from the concave mirror. Using a ray diagram drawn to scale, find the location of the image that this light produces in the concave mirror. Specify this distance relative to the concave mirror.

Section 25.6 The Mirror Equation and the Magnification Equation

16. A coin is placed 8.0 cm in front of a concave mirror. The mirror produces a real image that has a diameter 4.0 times larger than that of the coin. What is the image distance?

17. ssm A concave mirror has a focal length of 42 cm. The image formed by this mirror is 97 cm in front of the mirror. What is the object distance?

18. The focal length of a concave mirror is 17 cm. An object is located 38 cm in front of this mirror. Where is the image located?

19. Refer to **Interactive Solution 25.19** at **www.wiley.com/college/cutnell** for help with problems like this one. A concave mirror ($R = 56.0$ cm) is used to project a transparent slide onto a wall. The slide is located at a distance of 31.0 cm from the mirror, and a small flashlight shines light through the slide and onto the mirror. The setup is similar to that in Figure 25.19a. (a) How far from the wall should the mirror be located? (b) The height of the object on the slide is 0.95 cm. What is the height of the image? (c) How should the slide be oriented, so that the picture on the wall looks normal?

20. An object that is 25 cm in front of a convex mirror has an image located 17 cm behind the mirror. How far behind the mirror is the image located when the object is 19 cm in front of the mirror?

21. ssm The image produced by a concave mirror is located 26 cm in front of the mirror. The focal length of the mirror is 12 cm. How far in front of the mirror is the object located?

22. A concave mirror ($f = 45$ cm) produces an image whose distance from the mirror is one-third the object distance. Determine (a) the object distance and (b) the (positive) image distance.

23. ssm www When viewed in a spherical mirror, the image of a setting sun is a virtual image. The image lies 12.0 cm behind the mirror. (a) Is the mirror concave or convex? Why? (b) What is the radius of curvature of the mirror?

* **24.** A candle is placed 15.0 cm in front of a convex mirror. When the convex mirror is replaced with a plane mirror, the image moves 7.0 cm farther away from the mirror. Find the focal length of the convex mirror.

* **25.** Consult **Interactive Solution 25.25** at **www.wiley.com/college/cutnell** for insight into this problem. An object is placed in front of a convex mirror, and the size of the image is one-fourth that of the object. What is the ratio d_o/f of the object distance to the focal length of the mirror?

* **26.** The same object is located at the same distance from two spherical mirrors, A and B. The magnifications produced by the mirrors are $m_A = 4.0$ and $m_B = 2.0$. Find the ratio f_A/f_B of the focal lengths of the mirrors.

* **27. ssm** An object is located 14.0 cm in front of a convex mirror, the image being 7.00 cm behind the mirror. A second object, twice as tall as the first one, is placed in front of the mirror, but at a different location. The image of this second object has the same height as the other image. How far in front of the mirror is the second object located?

** **28.** A spherical mirror is polished on both sides. When the convex side is used as a mirror, the magnification is $+1/4$. What is the magnification when the concave side is used as a mirror, the object remaining the same distance from the mirror?

** **29.** Using the mirror equation and the magnification equation, show that for a convex mirror the image is always (a) virtual (i.e., d_i is always negative) and (b) upright and smaller, relative to the object (i.e., m is positive and less than one).

Additional Problems

30. Concept Simulation 25.3 at **www.wiley.com/college/cutnell** illustrates the concepts pertinent to this problem. A convex mirror has a focal length of -40.0 cm. A 12.0-cm-tall object is located 40.0 cm in front of this mirror. Using a ray diagram drawn to scale, determine the (a) location and (b) size of the image. Note that the mirror must be drawn to scale.

31. ssm www A small postage stamp is placed in front of a concave mirror (radius = R), such that the image distance equals the object distance. (a) In terms of R, what is the object distance? (b) What is the magnification of the mirror? (c) State whether the image is upright or inverted relative to the object. Draw a ray diagram to guide your thinking.

32. Review Conceptual Example 1 as an aid in understanding this problem. The drawings show two arrows, A and B, that are located in front of a plane mirror. A person at point P is viewing the image of each arrow. Which images can be seen in their entirety? Determine your answers by drawing a ray from the head and foot of each arrow that reflects from the mirror according to the law of reflection and reaches point P. Only if *both* rays reach point P after reflection can the image of that arrow be seen in its entirety.

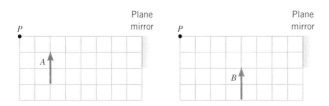

33. A clown is using a concave makeup mirror to get ready for a show and is 27 cm in front of the mirror. The image is 65 cm *behind* the mirror. Find (a) the focal length of the mirror and (b) the magnification.

34. Convex mirrors are being used to monitor the aisles in a store. The mirrors have a radius of curvature of 4.0 m. (a) What is the image distance if a customer is 15 m in front of the mirror? (b) Is the image real or virtual? (c) If a customer is 1.6 m tall, how tall is the image?

35. ssm The image behind a convex mirror (radius of curvature = 68 cm) is located 22 cm from the mirror. (a) Where is the object located and (b) what is the magnification of the mirror? Determine whether the image is (c) upright or inverted and (d) larger or smaller than the object.

* **36.** Two plane mirrors are facing each other. They are parallel, 3.00 cm apart, and 17.0 cm in length, as the drawing indicates. A laser beam is directed at the top mirror from the left edge of the bottom mirror. What is the smallest angle of incidence with respect to the top mirror, such that the laser beam (a) hits only one of the mirrors and (b) hits each mirror only once?

* **37.** In Figure 25.21*b* the head-up display is designed so that the distance between the digital readout device and virtual image 1 is 2.00 m. The magnification of virtual image 1 is 4.00. Find the focal length of the concave mirror. *(Hint: Remember that the image distance for virtual image 1 is a negative quantity.)*

* **38.** A concave makeup mirror is designed so the virtual image it produces is twice the size of the object when the distance between the object and the mirror is 14 cm. What is the radius of curvature of the mirror?

* **39. ssm www** An image formed by a convex mirror ($f = -24.0$ cm) has a magnification of 0.150. Which way and by how much should the object be moved to double the size of the image?

** **40.** A concave mirror has a focal length of 30.0 cm. The distance between an object and its image is 45.0 cm. Find the object and image distances, assuming that (a) the object lies beyond the center of curvature and (b) the object lies within the focal point.

** **41. ssm** A lamp is twice as far in front of a plane mirror as a person is. Light from the lamp reaches the person via two paths. It strikes the mirror at a 30.0° angle of incidence and reflects from it before reaching the person. It also travels directly to the person without reflecting. Find the ratio of the travel time along the reflected path to the travel time along the direct path.

Concepts & Calculations   Group Learning Problems

Note: Each of these problems consists of Concept Questions followed by a related quantitative Problem. They are designed for use by students working alone or in small learning groups. The Concept Questions involve little or no mathematics and are intended to stimulate group discussions. They focus on the concepts with which the problems deal. Recognizing the concepts is the essential initial step in any problem-solving technique.

42. Concept Questions A small mirror is attached to a vertical wall, and it hangs a distance y above the floor. A ray of sunlight strikes the mirror, and the reflected ray forms a spot on the floor. (a) From a knowledge of y and the horizontal distance x from the base of the wall to the spot, describe how one can determine the angle of incidence of the ray striking the mirror. If it is morning and the mirror is facing due east, would (b) the angle of incidence and (c) the distance x increase or decrease in time? Why?

Problem Suppose the mirror is 1.80 m above the floor. The reflected ray of sunlight strikes the floor at a distance of 3.86 m from the base of the wall. Later in the morning, the ray is observed to strike the floor at a distance of 1.26 m from the wall. The earth rotates at a rate of 15.0° per hour. How much time (in hours) has elapsed between the two observations?

43. Concept Questions (a) Suppose that you are walking toward a stationary plane mirror. Following the method discussed in Section 3.4, express your image's velocity v_{IY} relative to you in terms of the image's velocity v_{IM} relative to the mirror and the mirror's velocity v_{MY} relative to you. (b) How is the mirror's velocity v_{MY} relative to you related to your velocity v_{YM} relative to the mirror? Explain. (c) Consider both velocities v_{YM} and v_{IM}. Do they have the same magnitudes and the same directions? Explain.

Problem When you walk perpendicularly with a velocity of $+0.90$ m/s toward a stationary plane mirror, what is the velocity of your image relative to you? The direction in which you walk is the positive direction.

44. Concept Questions (a) For an image that is in front of a mirror, is the image distance positive or negative? (b) Given the image distance, what additional information is needed to determine the focal length? Explain. (c) For an inverted image is the image height positive or negative? (d) Given the object and image heights and a statement as to whether the image is upright or inverted, what additional information is needed to determine the object distance?

Problem A small statue has a height of 3.5 cm and is placed in front of a concave mirror. The image of the statue is inverted, 1.5 cm tall, and is located 13 cm in front of the mirror. Find the focal length of the mirror.

45. Concept Questions These questions refer to Figure 25.22a. (a) As the object distance increases, does reflected ray 1 change? (b) As the object distance increases, does reflected ray 3 make a greater or smaller angle with respect to the principal axis? (c) Extending the reflected rays 1 and 3 behind the mirror allows us to locate the top of the image. As the object distance increases, does the image height increase or decrease?

Problem A convex mirror has a focal length of -27.0 cm. Find the magnification produced by the mirror when the object distance is 9.0 cm and 18.0 cm. Verify that your answers are consistent with your answers to the Concept Questions.

* **46. Concept Questions** (a) Suppose that you are walking toward a plane mirror as in the drawing. The view is from above. Following the method discussed in Section 3.4, express your image's velocity v_{IY} relative to you in terms of the image's velocity v_{IM} relative to the mirror and the mirror's velocity v_{MY} relative to you. (b) How is the mirror's velocity v_{MY} relative to you related to your velocity v_{YM} relative to the mirror? Explain. (c) Consider both velocities v_{YM} and v_{IM}. Do they have the same x and y components? Explain.

Problem You walk at an angle of $\theta = 50.0°$ toward a plane mirror, as in the drawing. Your walking velocity has a magnitude of 0.90 m/s. What is the velocity of your image relative to you (magnitude and direction)?

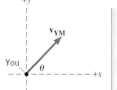

* **47. Concept Questions** A tall tree is growing across a river from you. You would like to know the distance between yourself and the tree, as well as its height, but are unable to make the measurements directly. However, by using only a mirror to form an image of the tree, and then measuring the image distance and the image height, you can calculate the distance to the tree, as well as its height. (a) What kind of mirror, concave or convex, must you use? Why? (b) You will need to know the focal length of the mirror. The sun is shining. You aim the mirror at the sun and form an image of it. How is the image distance of the sun related to the focal length of the mirror? (c) Having measured the image distance d_i and the image height h_i of the tree, as well as the image distance of the sun, describe how you would use these numbers to determine the distance and height of the tree.

Problem A mirror produces an image of the sun, and the image is located 0.9000 m from the mirror. The same mirror is then used to produce an image of the tree. The image of the tree is 0.9100 m from the mirror. (a) How far away is the tree? (b) The image height of the tree has a magnitude of 0.12 m. How tall is the tree?

Chapter 26

The Refraction of Light: Lenses and Optical Instruments

Light rays emanating from the bee's body enter the plastic tube and, in part, are guided around the 90° turn via total internal reflection, eventually exiting through the front surface and allowing you to see the bee. Total internal reflection is one of the refraction phenomena that this chapter discusses. (Courtesy Schott Fiber Optics, Inc., Southbridge, MA)

26.1 The Index of Refraction

Table 26.1   *Index of Refraction[a]*
for Various Substances

Substance	Index of Refraction, n
Solids at 20 °C	
Diamond	2.419
Glass, crown	1.523
Ice (0 °C)	1.309
Sodium chloride	1.544
Quartz	
Crystalline	1.544
Fused	1.458
Liquids at 20 °C	
Benzene	1.501
Carbon disulfide	1.632
Carbon tetrachloride	1.461
Ethyl alcohol	1.362
Water	1.333
Gases at 0 °C, 1 atm	
Air	1.000 293
Carbon dioxide	1.000 45
Oxygen, O_2	1.000 271
Hydrogen, H_2	1.000 139

[a] Measured with light whose wavelength in a vacuum is 589 nm.

As Section 24.3 discusses, light travels through a vacuum at a speed $c = 3.00 \times 10^8$ m/s. It can also travel through many materials, such as air, water, and glass. Atoms in the material absorb, reemit, and scatter the light, however. Therefore, light travels through the material at a speed that is less than c, the actual speed depending on the nature of the material. In general, we will see that the change in speed as a ray of light goes from one material to another causes the ray to deviate from its incident direction. This change in direction is called *refraction,* and it is governed by Snell's law of refraction, which will be discussed in the next section.

▶ CONCEPTS AT A GLANCE   To describe the extent to which the speed of light in a material medium differs from that in a vacuum, we use a parameter called the *index of refraction* (or *refractive index*). As the Concepts-at-a-Glance chart in Figure 26.1 illustrates, the index of refraction incorporates both the speed of light c in a vacuum and the speed v in the material. The index of refraction is an important parameter because it appears in Snell's law of refraction, the basis of all the phenomena discussed in this chapter. ◀

■ DEFINITION OF THE INDEX OF REFRACTION

The index of refraction n of a material is the ratio of the speed c of light in a vacuum to the speed v of light in the material:

$$n = \frac{\text{Speed of light in a vacuum}}{\text{Speed of light in the material}} = \frac{c}{v} \tag{26.1}$$

Table 26.1 lists the refractive indices for some common substances. The values of n are greater than unity because the speed of light in a material medium is less than it is in a vacuum. For example, the index of refraction for diamond is $n = 2.419$, so the speed of light in diamond is $v = c/n = (3.00 \times 10^8 \text{ m/s})/2.419 = 1.24 \times 10^8$ m/s. In contrast, the index of refraction for air (and also for other gases) is so close to unity that $n_{\text{air}} = 1$ for most purposes. The index of refraction depends slightly on the wavelength of the light, and the values in Table 26.1 correspond to a wavelength of $\lambda = 589$ nm in a vacuum.

Figure 26.1 CONCEPTS AT A GLANCE The index of refraction of a material incorporates both the speed of light in a vacuum and the speed in the material. The index of refraction appears in Snell's law of refraction, which describes how light changes its direction of travel in passing from one material to another. This change in direction is evident in the photograph, which shows light traveling through air and being bent sharply downward upon entering a tank of water. (© Richard Megna/Fundamental Photographs)

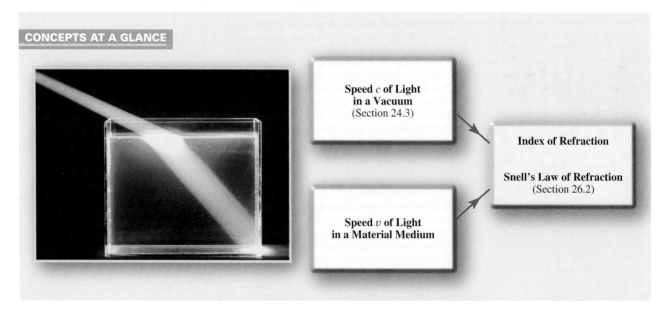

CONCEPTS AT A GLANCE

Speed c of Light
in a Vacuum
(Section 24.3)

Speed v of Light
in a Material Medium

Index of Refraction

Snell's Law of Refraction
(Section 26.2)

26.2 Snell's Law and the Refraction of Light

SNELL'S LAW

When light strikes the interface between two transparent materials, such as air and water, the light generally divides into two parts, as Figure 26.2a illustrates. Part of the light is reflected, with the angle of reflection equaling the angle of incidence. The remainder is transmitted across the interface. If the incident ray does not strike the interface at normal incidence, the transmitted ray has a different direction than the incident ray. The ray that enters the second material is said to be refracted.

In Figure 26.2a the light travels from a medium where the refractive index is smaller (air) into a medium where it is larger (water), and the refracted ray is bent *toward* the normal. Both the incident and refracted rays obey the principle of reversibility, so their directions can be reversed to give a situation like that in part b of the drawing. Here light travels from a material with a greater refractive index (water) into one with a smaller refractive index (air), and the refracted ray is bent *away* from the normal. In this case the reflected ray lies in the water rather than in the air. In both parts of the drawing the angles of incidence, refraction, and reflection are measured relative to the normal. Note that the index of refraction of air is labeled n_1 in part a, whereas it is n_2 in part b, because *we label all variables associated with the incident (and reflected) ray with subscript 1 and all variables associated with the refracted ray with subscript 2.*

The angle of refraction θ_2 depends on the angle of incidence θ_1 and on the indices of refraction, n_2 and n_1, of the two media. The relation between these quantities is known as **Snell's law of refraction,** after the Dutch mathematician Willebrord Snell (1591–1626) who discovered it experimentally. At the end of this section is a proof of Snell's law.

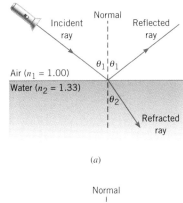

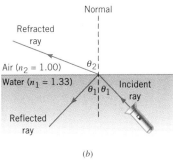

Figure 26.2 (a) When a ray of light is directed from air into water, part of the light is reflected at the interface and the remainder is refracted into the water. The refracted ray is bent *toward* the normal ($\theta_2 < \theta_1$). (b) When a ray of light is directed from water into air, the refracted ray in air is bent *away* from the normal ($\theta_2 > \theta_1$).

■ SNELL'S LAW OF REFRACTION

When light travels from a material with refractive index n_1 into a material with refractive index n_2, the refracted ray, the incident ray, and the normal to the interface between the materials all lie in the same plane. The angle of refraction θ_2 is related to the angle of incidence θ_1 by

$$n_1 \sin \theta_1 = n_2 \sin \theta_2 \qquad (26.2)$$

Example 1 illustrates the use of Snell's law.

Example 1 Determining the Angle of Refraction

A light ray strikes an air/water surface at an angle of 46° with respect to the normal. The refractive index for water is 1.33. Find the angle of refraction when the direction of the ray is (a) from air to water and (b) from water to air.

Reasoning Snell's law of refraction applies to both part (a) and part (b). However, in part (a) the incident ray is in air, whereas in part (b) it is in water. We keep track of this difference by always labeling the variables associated with the incident ray with a subscript 1 and the variables associated with the refracted ray with a subscript 2.

Solution

(a) The incident ray is in air, so $\theta_1 = 46°$ and $n_1 = 1.00$. The refracted ray is in water, so $n_2 = 1.33$. Snell's law can be used to find the angle of refraction θ_2:

$$\sin \theta_2 = \frac{n_1 \sin \theta_1}{n_2} = \frac{(1.00) \sin 46°}{1.33} = 0.54 \qquad (26.2)$$

$$\theta_2 = \sin^{-1}(0.54) = \boxed{33°}$$

Since θ_2 is less than θ_1, the refracted ray is bent *toward* the normal, as Figure 26.2a shows.

(b) With the incident ray in water, we find that

$$\sin \theta_2 = \frac{n_1 \sin \theta_1}{n_2} = \frac{(1.33) \sin 46°}{1.00} = 0.96$$

$$\theta_2 = \sin^{-1}(0.96) = \boxed{74°}$$

Since θ_2 is greater than θ_1, the refracted ray is bent *away* from the normal, as in Figure 26.2b.

Problem solving insight
The angle of incidence θ_1 and the angle of refraction θ_2 that appear in Snell's law are measured with respect to the normal to the surface, and not with respect to the surface itself.

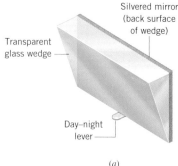

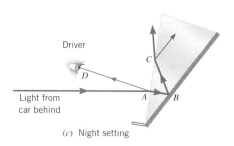

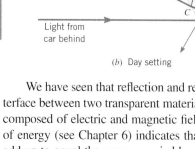

(a)   (b) Day setting   (c) Night setting

Figure 26.3 An interior rearview mirror with a day–night adjustment lever.

The physics of rearview mirrors that have a day–night adjustment.

We have seen that reflection and refraction of light waves occur simultaneously at the interface between two transparent materials. It is important to keep in mind that light waves are composed of electric and magnetic fields, which carry energy. The principle of conservation of energy (see Chapter 6) indicates that the energy reflected plus the energy refracted must add up to equal the energy carried by the incident light, provided that none of the energy is absorbed by the materials. The percentage of incident energy that appears as reflected versus refracted light depends on the angle of incidence and the refractive indices of the materials on either side of the interface. For instance, when light travels from air toward water at perpendicular incidence, most of its energy is refracted and little is reflected. But when the angle of incidence is nearly 90° and the light barely grazes the water surface, most of its energy is reflected, with only a small amount refracted into the water. On a rainy night, you probably have experienced the annoying glare that results when light from an oncoming car just grazes the wet road. Under such conditions, most of the light energy reflects into your eyes.

The simultaneous reflection and refraction of light have applications in a number of devices. For instance, interior rearview mirrors in cars often have adjustment levers. One position of the lever is for day driving, while the other is for night driving and reduces glare from the headlights of the car behind. As Figure 26.3a indicates, this kind of mirror is a glass wedge, the back side of which is silvered and highly reflecting. Part b of the picture shows the day setting. Light from the car behind follows path ABCD in reaching the driver's eye. At points A and C, where the light strikes the air–glass surface, there are both reflected and refracted rays. The reflected rays are drawn as thin lines, the thinness denoting that only a small percentage (about 10%) of the light during the day is reflected at A and C. The weak reflected rays at A and C do not reach the driver's eye. In contrast, almost all the light reaching the silvered back surface at B is reflected toward the driver. Since most of the light follows path ABCD, the driver sees a bright image of the car behind. During the night, the adjustment lever can be used to rotate the top of the mirror away from the driver (see part c of the drawing). Now, most of the light from the headlights behind follows path ABC and does not reach the driver. Only the light that is weakly reflected from the front surface along path AD is seen, and the result is significantly less glare.

✔ **Check Your Understanding 1**

The drawing shows three layers of liquids, A, B, and C, each with a different index of refraction. Light begins in liquid A, passes into B, and eventually into C, as the ray of light in the drawing shows. The dashed lines denote the normals to the interfaces between the layers. Which liquid has the smallest index of refraction? *(The answer is given at the end of the book.)*

Background: The index of refraction is the ratio of the speed of light in a vacuum to the speed of light in a material. As light travels from one material into another, its direction of travel can change, depending on the indices of refraction of the materials. Snell's law of refraction describes the change in direction.

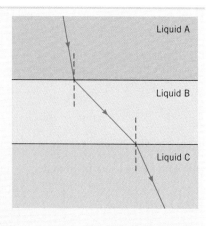

For similar questions (including calculational counterparts), consult Self-Assessment Test 26.1, which is described at the end of Section 26.5.

APPARENT DEPTH

One interesting consequence of refraction is that an object lying under water appears to be closer to the surface than it actually is. Example 2 sets the stage for explaining why, by showing what must be done to shine a light on such an object.

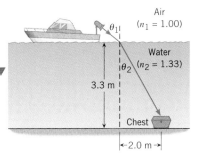

Example 2 Finding a Sunken Chest

A searchlight on a yacht is being used at night to illuminate a sunken chest, as in Figure 26.4. At what angle of incidence θ_1 should the light be aimed?

Reasoning The angle of incidence θ_1 must be such that, after refraction, the light strikes the chest. The angle of incidence can be obtained from Snell's law, once the angle of refraction θ_2 is determined. This angle can be found using the data in Figure 26.4 and trigonometry. The light travels from a region of lower into a region of higher refractive index, so the light is bent toward the normal and we expect θ_1 to be greater than θ_2.

Figure 26.4 The beam from the searchlight is refracted when it enters the water.

Solution From the data in the drawing it follows that $\tan \theta_2 = (2.0 \text{ m})/(3.3 \text{ m})$, so $\theta_2 = 31°$. With $n_1 = 1.00$ for air and $n_2 = 1.33$ for water, Snell's law gives

$$\sin \theta_1 = \frac{n_2 \sin \theta_2}{n_1} = \frac{(1.33) \sin 31°}{1.00} = 0.69$$

$$\theta_1 = \sin^{-1}(0.69) = \boxed{44°}$$

As expected, θ_1 is greater than θ_2.

Problem solving insight
Remember that the refractive indices are written as n_1 for the medium in which the incident light travels and n_2 for the medium in which the refracted light travels.

When the sunken chest in Example 2 is viewed from the boat (Figure 26.5*a*), light rays from the chest pass upward through the water, refract away from the normal when they enter the air, and then travel to the observer. This picture is similar to Figure 26.4, except the direction of the rays is reversed and the searchlight is replaced by an observer. When the rays entering the air are extended back into the water (see the dashed lines), they indicate that the observer sees a virtual image of the chest at an *apparent depth* that is less than the actual depth. The image is virtual because light rays do not actually pass through it. For the situation shown in Figure 26.5*a*, it is difficult to determine the apparent depth. A case that is much simpler is shown in part *b* of the drawing, where the observer is *directly above* the submerged object, and the apparent depth d' is related to the actual depth d by

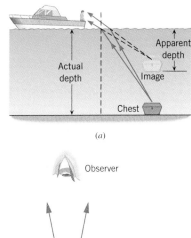

Apparent depth, observer directly above object

$$d' = d\left(\frac{n_2}{n_1}\right) \tag{26.3}$$

In this result, n_1 is the refractive index of the medium associated with the incident ray (the medium in which the object is located), and n_2 refers to the medium associated with the refracted ray (the medium in which the observer is situated). The proof of Equation 26.3 is the focus of problem 19 at the end of the chapter. Example 3 illustrates that the effect of apparent depth is quite noticeable in water.

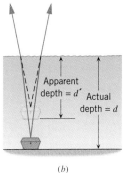

Example 3 The Apparent Depth of a Swimming Pool

A swimmer is treading water (with her head above the water) at the surface of a 3.00-m-deep pool. She sees a coin on the bottom directly below. How deep does the coin appear to be?

Reasoning Equation 26.3 may be used to find the apparent depth, provided we remember that the light rays travel from the coin to the swimmer. Therefore, the incident ray is coming from the coin under the water ($n_1 = 1.33$), while the refracted ray is in the air ($n_2 = 1.00$).

Figure 26.5 (*a*) Because light from the chest is refracted away from the normal when the light enters the air, the apparent depth of the image is less than the actual depth. (*b*) The observer is viewing the submerged object from directly overhead.

Solution The apparent depth d' of the coin is

$$d' = d\left(\frac{n_2}{n_1}\right) = (3.00 \text{ m})\left(\frac{1.00}{1.33}\right) = \boxed{2.26 \text{ m}} \tag{26.3}$$

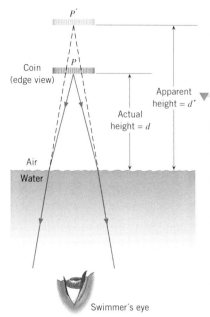

Figure 26.6 Rays from point P on a coin in the air above the water refract toward the normal as they enter the water. An underwater swimmer perceives the rays as originating from a point P' that is farther above the surface than the actual point P.

In Example 3, a person sees a coin on the bottom of a pool at an apparent depth that is less than the actual depth. Conceptual Example 4 considers the reverse situation—namely, a person looking from under the water at a coin in the air.

Conceptual Example 4   On the Inside Looking Out

A swimmer is under water and looking up at the surface. Someone holds a coin in the air, directly above the swimmer's eyes. To the swimmer, the coin appears to be at a certain height above the water. Is the apparent height of the coin greater than, less than, or the same as its actual height?

Reasoning and Solution Figure 26.6 shows that the apparent height of the coin above the water surface is greater than the actual height. Consider two rays of light leaving a point P on the coin. When the rays enter the water, they are refracted toward the normal because water has a larger index of refraction than air has. By extending the refracted rays backward (see the dashed lines in the drawing), we find that the rays appear to the swimmer as if they originate from a point P' that is farther from the water than point P is. Thus, the swimmer sees the coin at an apparent height d' that is *greater* than the actual height d. Equation 26.3 [$d' = d(n_2/n_1)$] reveals the same result, because n_1 represents the medium (air) associated with the incident ray and n_2 represents the medium (water) associated with the refracted ray. Since n_2 for water is greater than n_1 for air, the ratio n_2/n_1 is greater than one and d' is larger than d. This situation is the opposite of that in Figure 26.5b, where an object beneath the water appears to be closer to the surface than it actually is.

Related Homework: *Problems 11, 18*

THE DISPLACEMENT OF LIGHT BY A TRANSPARENT SLAB OF MATERIAL

A window pane is an example of a transparent slab of material. It consists of a plate of glass with parallel surfaces. When a ray of light passes through the glass, the emergent ray is parallel to the incident ray but displaced from it, as Figure 26.7 shows. This result can be verified by applying Snell's law to each of the two glass surfaces, with the result that $n_1 \sin\theta_1 = n_2 \sin\theta_2 = n_3 \sin\theta_3$. Since air surrounds the glass, $n_1 = n_3$, and it follows that $\sin\theta_1 = \sin\theta_3$. Therefore, $\theta_1 = \theta_3$, and the emergent and incident rays are parallel. However, as the drawing shows, the emergent ray is displaced laterally relative to the incident ray. The extent of the displacement depends on the angle of incidence, the thickness of the slab, and its refractive index.

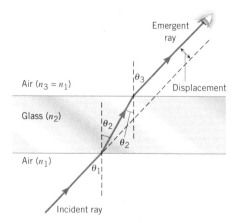

Figure 26.7 When a ray of light passes through a pane of glass that has parallel surfaces and is surrounded by air, the emergent ray is parallel to the incident ray ($\theta_3 = \theta_1$), but is displaced from it.

DERIVATION OF SNELL'S LAW

Snell's law can be derived by considering what happens to the wave fronts when the light passes from one medium into another. Figure 26.8a shows light propagating from

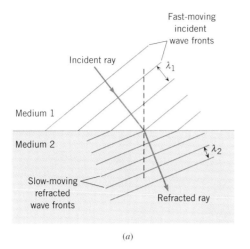

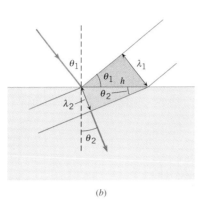

Figure 26.8 (*a*) The wave fronts are refracted as the light passes from medium 1 into medium 2. (*b*) An enlarged view of the incident and refracted wave fronts at the surface.

medium 1, where the speed is relatively large, into medium 2, where the speed is smaller; therefore, n_1 is less than n_2. The plane wave fronts in this picture are drawn perpendicular to the incident and refracted rays. Since the part of each wave front that penetrates medium 2 slows down, the wave fronts in medium 2 are rotated clockwise relative to those in medium 1. Correspondingly, the refracted ray in medium 2 is bent toward the normal, as the drawing shows.

Although the incident and refracted waves have different speeds, *they have the same frequency f.* The fact that the frequency does not change can be understood in terms of the atomic mechanism underlying the generation of the refracted wave. When the electromagnetic wave strikes the surface, the oscillating electric field forces the electrons in the molecules of medium 2 to oscillate at the same frequency as the wave. The accelerating electrons behave like atomic antennas that radiate "extra" electromagnetic waves, which combine with the original wave. The net electromagnetic wave within medium 2 is a superposition of the original wave plus the extra radiated waves, and it is this superposition that constitutes the refracted wave. Since the extra waves are radiated at the same frequency as the original wave, the refracted wave also has the same frequency as the original wave.

The distance between successive wave fronts in Figure 26.8*a* has been chosen to be the wavelength λ. Since the frequencies are the same in both media but the speeds are different, it follows from Equation 16.1 that the wavelengths are different: $\lambda_1 = v_1/f$ and $\lambda_2 = v_2/f$. Since v_1 is assumed to be larger than v_2, λ_1 is larger than λ_2, and the wave fronts are farther apart in medium 1.

Figure 26.8*b* shows an enlarged view of the incident and refracted wave fronts at the surface. The angles θ_1 and θ_2 within the colored right triangles are, respectively, the angles of incidence and refraction. In addition, the triangles share the same hypotenuse h. Therefore,

$$\sin \theta_1 = \frac{\lambda_1}{h} = \frac{v_1/f}{h} = \frac{v_1}{hf}$$

and

$$\sin \theta_2 = \frac{\lambda_2}{h} = \frac{v_2/f}{h} = \frac{v_2}{hf}$$

Combining these two equations into a single equation by eliminating the common term hf gives

$$\frac{\sin \theta_1}{v_1} = \frac{\sin \theta_2}{v_2}$$

By multiplying each side of this result by c, the speed of light in a vacuum, and recognizing that the ratio c/v is the index of refraction n, we arrive at Snell's law of refraction: $n_1 \sin \theta_1 = n_2 \sin \theta_2$.

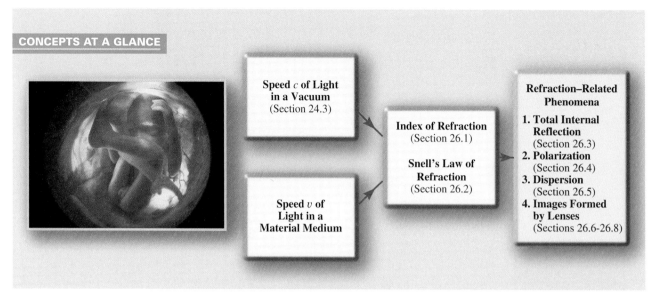

CONCEPTS AT A GLANCE

Figure 26.9 CONCEPTS AT A GLANCE Four of the phenomena that we will discuss in this chapter are listed at the right of this chart. The spectacular intrauterine photograph of a fetus was taken with the aid of optical fibers, which utilize the phenomenon of total internal reflection. (© Lennart Nilsson, from *A Child Is Born.*)

▶ CONCEPTS AT A GLANCE A number of important phenomena are related to refraction and Snell's law. The Concepts-at-a-Glance chart in Figure 26.9 is an expansion of that in Figure 26.1 and lists four of these phenomena, each of which will be discussed in the remainder of this chapter. In the next section, we turn our attention to the first of them, total internal reflection. ◀

26.3 Total Internal Reflection

When light passes from a medium of larger refractive index into one of smaller refractive index—for example, from water to air—the refracted ray bends *away* from the normal, as in Figure 26.10a. As the angle of incidence increases, the angle of refraction also increases. When the angle of incidence reaches a certain value, called the ***critical angle, θ_c,*** the angle of refraction is 90°. Then the refracted ray points along the surface; part *b* illustrates what happens at the critical angle. When the angle of incidence exceeds the critical angle, as in part *c* of the drawing, there is no refracted light. All the incident light is reflected back into the medium from which it came, a phenomenon called ***total internal reflection.*** Total internal reflection occurs only when light travels from a higher-index medium toward a lower-index medium. It does not occur when light propagates in the reverse direction—for example, from air to water.

An expression for the critical angle θ_c can be obtained from Snell's law by setting $\theta_1 = \theta_c$ and $\theta_2 = 90°$ (see Figure 26.10b):

$$\sin \theta_c = \frac{n_2 \sin 90°}{n_1}$$

Critical angle $$\sin \theta_c = \frac{n_2}{n_1} \qquad (n_1 > n_2) \qquad (26.4)$$

For instance, the critical angle for light traveling from water ($n_1 = 1.33$) to air ($n_2 = 1.00$) is $\theta_c = \sin^{-1}(1.00/1.33) = 48.8°$. For incident angles greater than 48.8°, Snell's law predicts that $\sin \theta_2$ is greater than unity, a value that is not possible. Thus, light rays with incident angles exceeding 48.8° yield no refracted light, and the light is totally reflected back into the water, as Figure 26.10c indicates.

Concept Simulation 26.1

This simulation illustrates the refraction and total internal reflection that can occur at the interface between two materials. The indices of refraction of both materials and the angle of incidence of the light are under your control. The angle of refraction and the critical angle are also displayed, so that you can see what happens when the angle of incidence is made larger than the critical angle.

Related Homework: Problem 99

Go to
www.wiley.com/college/cutnell

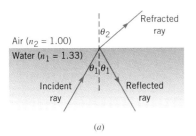 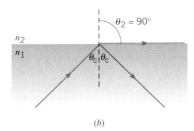 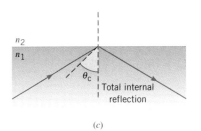

(a) (b) (c)

Figure 26.10 (a) When light travels from a higher-index medium (water) into a lower-index medium (air), the refracted ray is bent away from the normal. (b) When the angle of incidence is equal to the critical angle θ_c, the angle of refraction is 90°. (c) If θ_1 is greater than θ_c, there is no refracted ray, and total internal reflection occurs.

The next example illustrates how the critical angle changes when the indices of refraction change.

Example 5 Total Internal Reflection

A beam of light is propagating through diamond ($n_1 = 2.42$) and strikes a diamond–air interface at an angle of incidence of 28°. (a) Will part of the beam enter the air ($n_2 = 1.00$) or will the beam be totally reflected at the interface? (b) Repeat part (a), assuming that the diamond is surrounded by water ($n_2 = 1.33$).

Reasoning Total internal reflection occurs only when the beam of light has an angle of incidence that is greater than the critical angle θ_c. The critical angle is different in parts (a) and (b), since it depends on the ratio n_2/n_1 of the refractive indices of the incident (n_1) and refracting (n_2) media.

Solution

(a) The critical angle θ_c for total internal reflection at the diamond–air interface is given by Equation 26.4 as

$$\theta_c = \sin^{-1}\left(\frac{n_2}{n_1}\right)$$

$$= \sin^{-1}\left(\frac{1.00}{2.42}\right) = 24.4°$$

Because the angle of incidence of 28° is greater than the critical angle, there is no refraction, and the light is totally reflected back into the diamond.

(b) If water, rather than air, surrounds the diamond, the critical angle for total internal reflection becomes larger:

$$\theta_c = \sin^{-1}\left(\frac{n_2}{n_1}\right)$$

$$= \sin^{-1}\left(\frac{1.33}{2.42}\right) = 33.3°$$

Now a beam of light that has an angle of incidence of 28° (less than the critical angle of 33.3°) at the diamond–water interface is refracted into the water.

The critical angle plays an important role in why a diamond sparkles, as Conceptual Example 6 discusses.

Conceptual Example 6 The Sparkle of a Diamond

A diamond gemstone is famous for its sparkle because the light coming from it glitters as it is moved about. Why does a diamond exhibit such brilliance? And why does a diamond lose much of its brilliance when placed under water?

The physics of why a diamond sparkles.

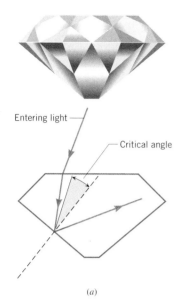

Entering light

Critical angle

(a)

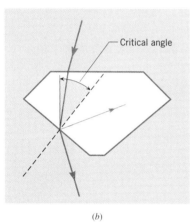

Critical angle

(b)

Figure 26.11 (a) Near the bottom of the diamond, light is totally internally reflected, because the incident angle exceeds the critical angle for diamond and air. (b) When the diamond is in water, the same light is partially reflected and partially refracted, since the incident angle is less than the critical angle for diamond and water.

Reasoning and Solution A diamond sparkles because when it is held a certain way, the intensity of the light coming from it is greatly enhanced. Figure 26.11 helps to explain that this enhancement is related to total internal reflection. Part a of the drawing shows a ray of light striking a bottom facet of the diamond at an angle of incidence that is greater than the critical angle. This ray, and all other rays whose angles of incidence exceed the critical angle, are totally reflected back into the diamond, eventually exiting the top surface to give the diamond its sparkle. Many of the rays striking a bottom facet behave in this fashion because diamond has a relatively small critical angle in air. As part (a) of Example 5 shows, the critical angle is 24.4°. The value is so small because the index of refraction of diamond ($n = 2.42$) is large compared to that of air ($n = 1.00$).

Now consider what happens to the same ray of light within the diamond when the diamond is placed in water. Because water has a larger index of refraction than air, the critical angle increases to 33.3°, as part (b) of Example 5 shows. Therefore, this particular ray is no longer totally internally reflected. As Figure 26.11b illustrates, only some of the light is reflected back into the diamond, and the remainder escapes into the water. Consequently, less light exits from the top of the diamond, causing it to lose much of its brilliance.

Need more practice?

Interactive LearningWare 26.1
The drawing shows a ray of light whose angle of incidence is $\theta_1 = 68.3°$, traveling through two solid materials and then undergoing total internal reflection at a solid–liquid interface. What is the largest possible index of refraction for the liquid?

Related Homework: Problem 30

Go to
www.wiley.com/college/cutnell for an interactive solution.

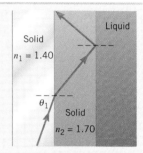

Liquid

Solid
$n_1 = 1.40$

θ_1
Solid
$n_2 = 1.70$

Many optical instruments, such as binoculars, periscopes, and telescopes, use glass prisms and total internal reflection to turn a beam of light through 90° or 180°. Figure 26.12a shows a light ray entering a 45°–45°–90° glass prism ($n_1 = 1.5$) and striking the hypotenuse of the prism at an angle of incidence of $\theta_1 = 45°$. The critical angle for a glass–air interface is $\theta_c = \sin^{-1}(n_2/n_1) = \sin^{-1}(1.0/1.5) = 42°$. Since the angle of incidence is greater than the critical angle, the light is totally reflected at the hypotenuse and is directed vertically upward in the drawing, having been turned through an angle of 90°. Part b of the picture shows how the same prism can turn the beam through 180° when total internal reflection occurs twice. Prisms can also be used in tandem to produce a lateral displacement of a light ray, while leaving its initial direction unaltered. Figure 26.12c illustrates such an application in binoculars.

An important application of total internal reflection occurs in fiber optics, where hair-thin threads of glass or plastic, called optical fibers, "pipe" light from one place to another. Figure 26.13a shows that an optical fiber consists of a cylindrical inner *core* that carries the light and an outer concentric shell, the *cladding*. The core is made from transparent glass or plastic that has a relatively high index of refraction. The cladding is also made of glass, but of a type that has a relatively low index of refraction. Light enters one

Figure 26.12 Total internal reflection at a glass–air interface can be used to turn a ray of light through an angle of (a) 90° or (b) 180°. (c) Two prisms, each reflecting the light twice by total internal reflection, are sometimes used in binoculars to produce a lateral displacement of a light ray.

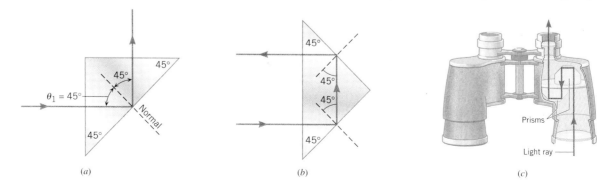

(a)

(b)

(c)

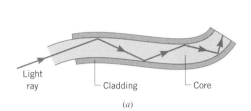

Figure 26.13 (*a*) Light can travel with little loss in a curved optical fiber because the light is totally reflected whenever it strikes the core–cladding interface and because the absorption of light by the core itself is small. (*b*) Light is being transmitted by a bundle of optical fibers. (© Masashiro Sano/Corbis Stock Market)

end of the core, strikes the core/cladding interface at an angle of incidence greater than the critical angle, and, therefore, is reflected back into the core. Light thus travels inside the optical fiber along a zigzag path. In a well-designed fiber, little light is lost as a result of absorption by the core, so light can travel many kilometers before its intensity diminishes appreciably. Fibers are often bundled together to produce cables. Because the fibers themselves are so thin, the cables are relatively small and flexible and can fit into places inaccessible to larger metal wires.

Optical fiber cables are the medium of choice for high-quality telecommunications because the cables are relatively immune to external electrical interference and because a light beam can carry information through an optical fiber just as electricity carries information through copper wires. The information-carrying capacity of light, however, is thousands of times greater than that of electricity. A laser beam traveling through a single optical fiber can carry tens of thousands of telephone conversations and several TV programs simultaneously.

In the field of medicine, optical fiber cables have had extraordinary impact. In the practice of endoscopy, for instance, a device called an endoscope is used to peer inside the body. Figure 26.14 shows a bronchoscope being used, which is a kind of endoscope that is inserted through the nose or mouth, down the bronchial tubes, and into the lungs. It consists of two optical fiber cables. One provides light to illuminate interior body parts, while the other sends back an image for viewing. A bronchoscope greatly simplifies the diagnosis of pulmonary disease. Tissue samples can even be collected with some bronchoscopes. A colonoscope is another kind of endoscope, and its design is similar to that of the bronchoscope. It is inserted through the rectum and used to examine the interior of the colon (see Figure 26.15). The colonoscope currently offers the best hope for diagnosing colon cancer in its early stages, when it can be treated.

The physics of **fiber optics.**

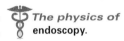 *The physics of* **endoscopy.**

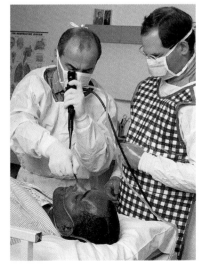

Figure 26.14 A bronchoscope is being used to look for signs of pulmonary disease. (© Len Lessin/Peter Arnold, Inc.)

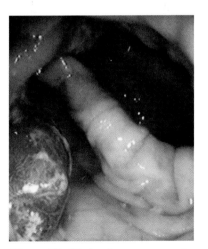

Figure 26.15 A colonoscope reveals a polyp (red) attached to the wall of the colon. Polyps that turn cancerous or grow large enough to obstruct the colon are surgically removed. (© Dr. Larpent/CNRI/Photo Researchers)

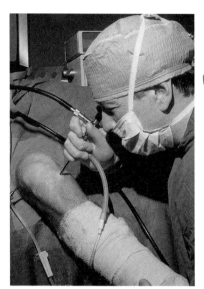

The physics of **arthroscopic surgery.**

The use of optical fibers has also revolutionized surgical techniques. In arthroscopic surgery, a small surgical instrument, several millimeters in diameter, is mounted at the end of an optical fiber cable. The surgeon can insert the instrument and cable into a joint, such as the knee, with only a tiny incision and minimal damage to the surrounding tissue (see Figure 26.16). Consequently, recovery from the procedure is relatively rapid compared to traditional surgical techniques.

✔ **Check Your Understanding 2**

The drawing shows a 30°–60°–90° prism and two light rays, A and B, both of which strike the prism perpendicularly. The prism is surrounded by an unknown liquid. When ray A reaches the hypotenuse in the drawing, it is totally internally reflected. Which one of the following statements applies to ray B when it reaches the hypotenuse? (a) It may or may not be totally internally reflected, depending on what the surrounding liquid is. (b) It is not totally internally reflected, no matter what the surrounding liquid is. (c) It is totally internally reflected, no matter what the surrounding liquid is. *(The answer is given at the end of the book.)*

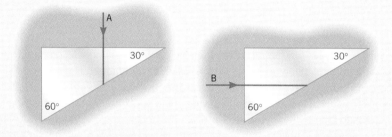

Figure 26.16 Optical fibers have made arthroscopic surgery possible, such as the repair of a damaged knee shown in this photograph. (© Margaret Rose Orthopaedic Hospital/Photo Researchers)

Background: When light travels from a higher toward a lower index of refraction, total internal reflection occurs when the light strikes the interface between the two materials at an angle greater than the critical angle. The critical angle depends on the values of the two refractive indices.

For similar questions (including calculational counterparts), consult Self-Assessment Test 26.1, which is described at the end of Section 26.5.

26.4 *Polarization and the Reflection and Refraction of Light*

For incident angles other than 0°, unpolarized light becomes partially polarized in reflecting from a nonmetallic surface, such as water. To demonstrate this fact, rotate a pair of Polaroid sunglasses in the sunlight reflected from a lake. You will see that the light intensity transmitted through the glasses is a minimum when the glasses are oriented as they are normally worn. Since the transmission axis of the glasses is aligned vertically, it follows that the light reflected from the lake is partially polarized in the horizontal direction.

There is one special angle of incidence at which the reflected light is completely polarized parallel to the surface, the refracted ray being only partially polarized. This angle is called the ***Brewster angle*** θ_B. Figure 26.17 summarizes what happens when unpolarized light strikes a nonmetallic surface at the Brewster angle. The value of θ_B is given by ***Brewster's law***, in which n_1 and n_2 are, respectively, the refractive indices of the materials in which the incident and refracted rays propagate:

Brewster's law
$$\tan \theta_B = \frac{n_2}{n_1} \tag{26.5}$$

This relation is named after the Scotsman David Brewster (1781–1868), who discovered it. Figure 26.17 also indicates that the reflected and refracted rays are perpendicular to each other when light strikes the surface at the Brewster angle (see problem 37 at the end of the chapter).

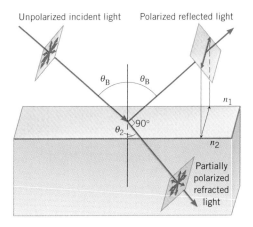

Figure 26.17 When unpolarized light is incident on a nonmetallic surface at the Brewster angle θ_B, the reflected light is 100% polarized in a direction parallel to the surface. The angle between the reflected and refracted rays is 90°.

26.5 The Dispersion of Light: Prisms and Rainbows

Figure 26.18a shows a ray of monochromatic light passing through a glass prism surrounded by air. When the light enters the prism at the left face, the refracted ray is bent toward the normal, because the refractive index of glass is greater than that of air. When the light leaves the prism at the right face, it is refracted away from the normal. Thus, the net effect of the prism is to change the direction of the ray, causing it to bend downward upon entering the prism, and downward again upon leaving. Because the refractive index of the glass depends on wavelength (see Table 26.2), rays corresponding to different colors are bent by different amounts by the prism and depart traveling in different directions. The greater the index of refraction for a given color, the greater the bending, and part b of the drawing shows the refractions for the colors red and violet, which are at opposite ends of the visible spectrum. If a beam of sunlight, which contains all colors, is sent through the prism, the sunlight is separated into a spectrum of colors, as part c shows. The spreading of light into its color components is called **dispersion.**

Table 26.2 Indices of Refraction n of Crown Glass at Various Wavelengths

Approximate Color	Wavelength in Vacuum (nm)	Index of Refraction, n
Red	660	1.520
Orange	610	1.522
Yellow	580	1.523
Green	550	1.526
Blue	470	1.531
Violet	410	1.538

Concept Simulation 26.2

This simulation is essentially an interactive version of Figure 26.18, in which you can select the color of the light incident on the prism. Since the index of refraction is different for different colors or wavelengths (see Table 26.2), the exiting ray departs in a different direction for different colors.

Go to **www.wiley.com/college/cutnell**

In Figure 26.18a the ray of light is refracted twice by a glass prism surrounded by air. Conceptual Example 7 explores what happens to the light when the prism is surrounded by materials other than air.

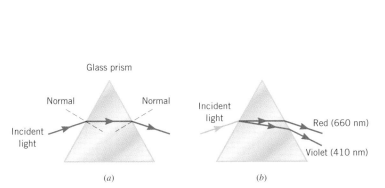

(a)   (b)   (c)

Figure 26.18 (a) A ray of light is refracted as it passes through the prism. The prism is surrounded by air. (b) Two different colors are refracted by different amounts. For clarity, the amount of refraction has been exaggerated. (c) Sunlight is dispersed into its color components by this prism. (© David Parker/Photo Researchers)

(a)

(b)

Figure 26.19 A ray of light passes through identical prisms, each surrounded by a different fluid. The ray of light is (a) refracted upward and (b) not refracted at all.

Conceptual Example 7
The Refraction of Light Depends on Two Refractive Indices

In Figure 26.18a the glass prism is surrounded by air and bends the ray of light downward. It is also possible for the prism to bend the ray upward, as in Figure 26.19a, or to not bend the ray at all, as in part b of the drawing. How can the situations illustrated in Figure 26.19 arise?

Reasoning and Solution Snell's law of refraction includes the refractive indices of *both* materials on either side of an interface. With this in mind, we note that the ray bends upward, or away from the normal, as it enters the prism in Figure 26.19a. A ray bends away from the normal when it travels from a medium with a larger refractive index into a medium with a smaller refractive index. When the ray leaves the prism, it again bends upward, which is toward the normal at the point of exit. A ray bends toward the normal when traveling from a smaller toward a larger refractive index. Thus, **the situation in Figure 26.19a could arise if the prism were immersed in a fluid, such as carbon disulfide, which has a larger refractive index than does glass** (see Table 26.1).

We have seen in Figures 26.18a and 26.19a that a glass prism can bend a ray of light either downward or upward, depending on whether the surrounding fluid has a smaller or larger index of refraction than the glass. It seems logical to conclude, then, that **a prism will not bend a ray at all, neither up nor down, if the surrounding fluid has the same index of refraction as the glass**—a condition known as *index matching*. This is exactly what is happening in Figure 26.19b, where the ray proceeds straight through the prism as if it were not even there. If the index of refraction of the surrounding fluid equals that of the glass prism, then $n_1 = n_2$, and Snell's law ($n_1 \sin \theta_1 = n_2 \sin \theta_2$) reduces to $\sin \theta_1 = \sin \theta_2$. Therefore, the angle of refraction equals the angle of incidence, and no bending of the light occurs.

Related Homework: *Conceptual Questions 17, 20, Problems 42, 108*

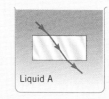 **Check Your Understanding 3**

Two blocks, made from the same transparent material, are immersed in different liquids. A ray of light strikes each block at the same angle of incidence. From the drawing, determine which liquid has the greater index of refraction. (*The answer is given at the end of the book.*)

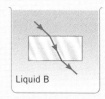

Liquid A   Liquid B

Background: When a ray of light is refracted at the interface between two materials, the change in direction of the ray depends on the refractive indices of both materials, according to Snell's law of refraction.

For similar questions (including calculational counterparts), consult Self-Assessment Test 26.1, which is described at the end of this section.

The physics of rainbows.

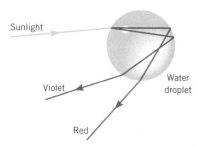

Figure 26.20 When sunlight emerges from a water droplet, the light is dispersed into its constituent colors, of which only two are shown.

Another example of dispersion occurs in rainbows, in which refraction by water droplets gives rise to the colors. You can often see a rainbow just as a storm is leaving, if you look at the departing rain with the sun at your back. When light from the sun enters a spherical raindrop, as in Figure 26.20, light of each color is refracted or bent by an amount that depends on the refractive index of water for that wavelength. After reflection from the back surface of the droplet, the different colors are again refracted as they reenter the air. Although each droplet disperses the light into its full spectrum of colors, the observer in Figure 26.21a sees only one color of light coming from any given droplet, since only one color travels in the right direction to reach the observer's eyes. However, all colors are visible in a rainbow (see Figure 26.21b) because each color originates from different droplets at different angles of elevation.

 Self-Assessment Test 26.1

Test your understanding of the material in Sections 26.1–26.5:

• The Index of Refraction   • Snell's Law and the Refraction of Light
• Total Internal Reflection   • Polarization and the Reflection and Refraction of Light
• The Dispersion of Light

Go to **www.wiley.com/college/cutnell**

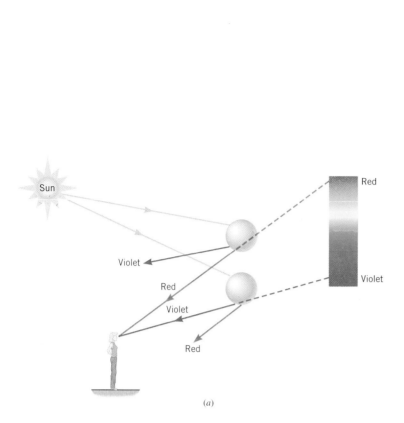

(a)

(b)

Figure 26.21 The different colors seen in a rainbow originate from water droplets at different angles of elevation. (© David Woodfall/Stone/Getty Images)

26.6 Lenses

The lenses used in optical instruments, such as eyeglasses, cameras, and telescopes, are made from transparent materials that refract light. They refract the light in such a way that an image of the source of the light is formed. Figure 26.22a shows a crude lens formed from two glass prisms. Suppose an object centered on the principal axis is infinitely far from the lens so the rays from the object are parallel to the principal axis. In passing through the prisms, these rays are bent toward the axis because of refraction. Unfortunately, the rays do not all cross the axis at the same place, and, therefore, such a crude lens gives rise to a blurred image of the object.

A better lens can be constructed from a single piece of transparent material with properly curved surfaces, often spherical, as in part *b* of the drawing. With this improved lens, rays that are near the principal axis (paraxial rays) and parallel to it converge to a single point on the axis after emerging from the lens. This point is called the *focal point F* of the lens. Thus, an object located infinitely far away on the principal axis leads to an image at the focal point of the lens. The distance between the focal point and the lens is the *focal length f.* In what follows, we assume the lens is so thin compared to f that it makes no difference whether f is measured between the focal point and either surface of the lens or the center of the lens. The type of lens in Figure 26.22b is known as a *converging lens* because it causes incident parallel rays to converge at the focal point.

Figure 26.22 (a) These two prisms cause rays of light that are parallel to the principal axis to change direction and cross the axis at different points. (b) With a converging lens, paraxial rays that are parallel to the principal axis converge to the focal point F after passing through the lens.

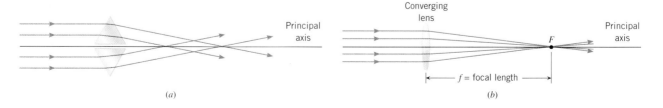

(a)   (b)

Figure 26.23 (*a*) These two prisms cause parallel rays to diverge. (*b*) With a diverging lens, paraxial rays that are parallel to the principal axis appear to originate from the focal point *F* after passing through the lens.

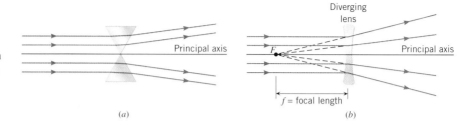

(*a*)   (*b*)

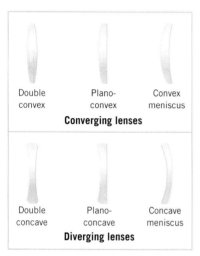

Double convex   Plano-convex   Convex meniscus
Converging lenses

Double concave   Plano-concave   Concave meniscus
Diverging lenses

Figure 26.24 Converging and diverging lenses come in a variety of shapes.

Another type of lens found in optical instruments is a *diverging lens,* which causes incident parallel rays to diverge after exiting the lens. Two prisms can also be used to form a crude diverging lens, as in Figure 26.23*a*. In a properly designed diverging lens, such as that in part *b* of the picture, paraxial rays that are parallel to the principal axis appear to originate from a single point on the axis after passing through the lens. This point is the focal point *F*, and its distance *f* from the lens is the focal length. Again, we assume that the lens is thin compared to the focal length.

Converging and diverging lenses come in a variety of shapes, as Figure 26.24 illustrates. Observe that converging lenses are thicker at the center than at the edges, whereas diverging lenses are thinner at the center.

26.7 The Formation of Images by Lenses
RAY DIAGRAMS

Each point on an object emits light rays in all directions, and when some of these rays pass through a lens, they form an image. As with mirrors, ray diagrams can be drawn to determine the location and size of the image. Lenses differ from mirrors, however, in that light can pass through a lens from left to right or from right to left. Therefore, when constructing ray diagrams, begin by locating a focal point *F* on *each side of the lens;* each point lies on the principal axis at the same distance *f* from the lens. The lens is assumed to be thin, in that its thickness is small compared with the focal length and the distances of the object and the image from the lens. For convenience, it is also assumed that the object is located to the left of the lens and is oriented perpendicular to the principal axis. There are three paraxial rays that leave a point on the top of the object and are especially helpful in drawing ray diagrams. They are labeled 1, 2, and 3 in Figure 26.25. When tracing their paths, we use the following reasoning strategy.

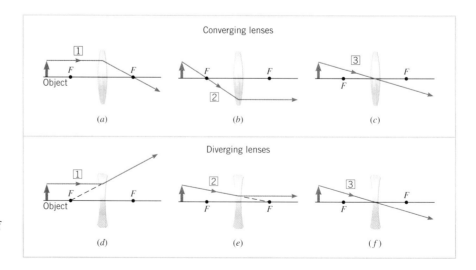

Figure 26.25 The rays shown here are useful in determining the nature of the images formed by converging and diverging lenses.

Reasoning Strategy

Ray Tracing for Converging and Diverging Lenses

Converging Lens	Diverging Lens

Ray 1

This ray initially travels parallel to the principal axis. In passing through a converging lens, the ray is refracted toward the axis and travels through the focal point on the right side of the lens, as Figure 26.25*a* shows.	This ray initially travels parallel to the principal axis. In passing through a diverging lens, the ray is refracted away from the axis, and *appears* to have originated from the focal point on the left of the lens. The dashed line in Figure 26.25*d* represents the apparent path of the ray.

Ray 2

This ray first passes through the focal point on the left and then is refracted by the lens in such a way that it leaves traveling parallel to the axis, as in Figure 26.25*b*.	This ray leaves the object and moves toward the focal point on the right of the lens. Before reaching the focal point, however, the ray is refracted by the lens so as to exit parallel to the axis. See Figure 26.25*e*, where the dashed line indicates the ray's path in the absence of the lens.

*Ray 3**

This ray travels directly through the center of the thin lens without any appreciable bending, as in Figure 26.25*c*.	This ray travels directly through the center of the thin lens without any appreciable bending, as in Figure 26.25*f*.

* Ray 3 does not bend as it proceeds through the lens because the left and right surfaces of each type of lens are nearly parallel at the center. Thus, in either case, the lens behaves as a transparent slab. As Figure 26.7 shows, the rays incident on and exiting from a slab travel in the same direction with only a lateral displacement. If the lens is sufficiently thin, the displacement is negligibly small.

IMAGE FORMATION BY A CONVERGING LENS

Figure 26.26*a* illustrates the formation of a real image by a converging lens. Here the object is located at a distance from the lens that is greater than twice the focal length (beyond the point labeled 2*F*). To locate the image, any two of the three special rays can be drawn from the tip of the object, although all three are shown in the drawing. The point on the right side of the lens where these rays intersect locates the tip of the image. The ray diagram indicates that the image is real, inverted, and smaller than the object. This optical arrangement is similar to that used in a camera, where a piece of film records the image (see part *b* of the drawing).

The physics of a camera.

When the object is placed between 2*F* and *F*, as in Figure 26.27*a*, the image is still real and inverted; however, the image is now larger than the object. This optical system is

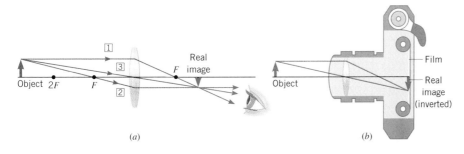

(a) *(b)*

Figure 26.26 (*a*) When the object is placed to the left of the point labeled 2*F*, a real, inverted, and smaller image is formed. (*b*) The arrangement in part *a* is like that used in a camera.

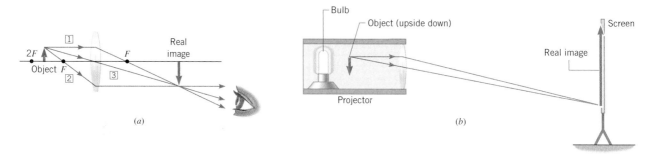

Figure 26.27 (*a*) When the object is placed between 2*F* and *F*, the image is real, inverted, and larger than the object. (*b*) This arrangement is found in projectors.

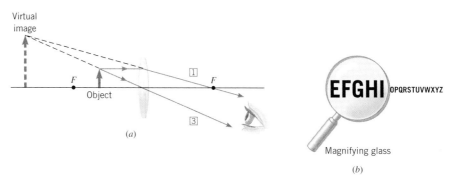

Figure 26.28 (*a*) When an object is placed within the focal point *F* of a converging lens, an upright, enlarged, and virtual image is created. (*b*) Such an image is seen when looking through a magnifying glass.

The physics of a slide or film projector.

used in a slide or film projector in which a small piece of film is the object and the enlarged image falls on a screen. However, to obtain an image that is right-side up, the film must be placed in the projector upside down.

When the object is located between the focal point and the lens, as in Figure 26.28, the rays diverge after leaving the lens. To a person viewing the diverging rays, they appear to come from an image behind (to the left of) the lens. Because none of the rays actually come from the image, it is a virtual image. The ray diagram shows that the virtual image is upright and enlarged. A magnifying glass uses this arrangement, as can be seen in part *b* of the drawing.

Concept Simulation 26.3

By selecting the "Converging lens" option in this simulation, the user can investigate the nature of the image produced by a converging lens. The object distance and/or the object height can be changed by clicking and dragging the object. The user can also change the focal length of the lens. The simulation uses a ray diagram to locate the image. Depending on where the object is located, the image can be real or virtual. The real image is always inverted and can be larger or smaller than the object. The virtual image is always upright and larger than the object. The program also displays numerical values for the image distance and the image height.

Related Homework: Conceptual Question 29, Problem 48

Go to **www.wiley.com/college/cutnell**

IMAGE FORMATION BY A DIVERGING LENS

Light rays diverge upon leaving a diverging lens, as Figure 26.29 shows, and the ray diagram indicates that a virtual image is formed on the left side of the lens. In fact, regardless of the position of a real object, a diverging lens always forms a virtual image that is upright and smaller relative to the object.

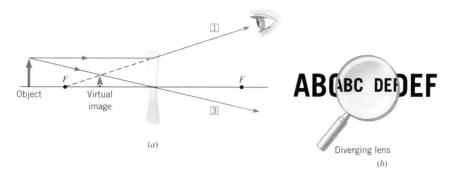

Figure 26.29 (*a*) A diverging lens always forms a virtual image of a real object. The image is upright and smaller relative to the object. (*b*) The image seen through a diverging lens.

Concept Simulation 26.4

Selecting the "Diverging lens" option in this simulation allows you to investigate the nature of the image produced by a diverging lens. The object distance and/or the object height can be changed by clicking and dragging the object. The user can also change the focal length of the lens. The simulation uses a ray diagram to locate the image. In the case of a diverging lens, you will see that the image is always virtual, upright, and smaller than the object, regardless of where the object is located. The program also displays numerical values for the image distance and the image height.

Related Homework: Conceptual Question 29, Problem 52

Go to **www.wiley.com/college/cutnell**

26.8 The Thin-Lens Equation and the Magnification Equation

▶ CONCEPTS AT A GLANCE When an object is placed in front of a spherical mirror, we can determine the location, size, and nature of its image by using the technique of ray tracing or the mirror and magnification equations. Both options are based on the law of reflection. The mirror and magnification equations relate the distances d_o and d_i of the object and image from the mirror to the focal length f and magnification m. For an object placed in front of a lens, Snell's law of refraction leads to the technique of ray tracing and to equations that are identical to the mirror and magnification equations. The similarity between mirrors and lenses can be seen by comparing the Concepts-at-a-Glance chart in Figure 26.30 with that for mirrors in Figure 25.23. ◀

Figure 26.30 CONCEPTS AT A GLANCE The thin-lens equation and the magnification equation describe how thin lenses form images and are based on Snell's law of refraction. They are used to design lenses that correct vision, such as the contact lens that the woman in the photograph is putting in her eye. (© Med. Illus. SBHA/Stone/Getty Images)

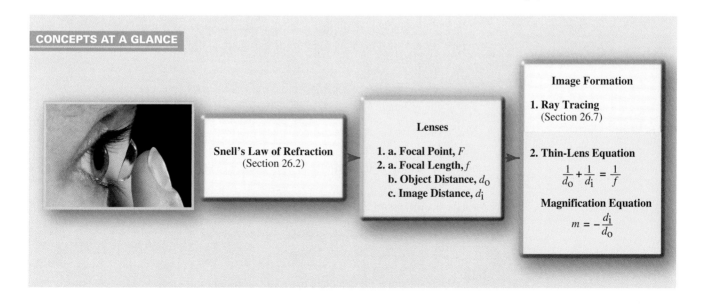

CONCEPTS AT A GLANCE

Snell's Law of Refraction
(Section 26.2)

Lenses
1. a. Focal Point, F
2. a. Focal Length, f
   b. Object Distance, d_o
   c. Image Distance, d_i

Image Formation

1. **Ray Tracing**
   (Section 26.7)

2. **Thin-Lens Equation**
$$\frac{1}{d_o} + \frac{1}{d_i} = \frac{1}{f}$$

Magnification Equation
$$m = -\frac{d_i}{d_o}$$

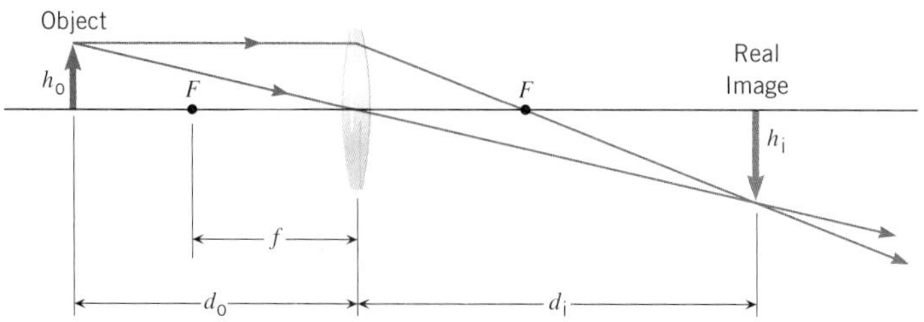

Figure 26.31 The drawing shows the focal length f, the object distance d_o, and the image distance d_i for a converging lens. The object and image heights are, respectively, h_o and h_i.

The chart in Figure 26.30 shows that Snell's law leads to equations that are referred to as the thin-lens equation and the magnification equation:

Thin-lens equation
$$\frac{1}{d_o} + \frac{1}{d_i} = \frac{1}{f}$$
(26.6)

Magnification equation
$$m = \frac{\text{Image height}}{\text{Object height}} = \frac{h_i}{h_o} = -\frac{d_i}{d_o}$$
(26.7)

Figure 26.31 defines the symbols in these expressions with the aid of a thin converging lens, but the expressions also apply to a diverging lens, if it is thin. The derivations of these equations are presented at the end of this section.

Certain sign conventions accompany the use of the thin-lens and magnification equations, and the conventions are similar to those used with mirrors in Section 25.6. The issue of real versus virtual images, however, is slightly different with lenses than with mirrors. With a mirror, a real image is formed on the *same side* of the mirror as the object (see Figure 25.19), in which case the image distance d_i is a positive number. With a lens, a positive value for d_i also means the image is real. But, starting with an actual object, a real image is formed on the side of the lens *opposite to* the object (see Figure 26.31). The *sign conventions* listed in the following Reasoning Strategy apply to light rays traveling from left to right from a real object.

Reasoning Strategy

Summary of Sign Conventions for Lenses

Focal length

f is + for a converging lens.

f is − for a diverging lens.

Object distance

d_o is + if the object is to the left of the lens (real object), as is usual.

d_o is − if the object is to the right of the lens (virtual object).*

Image distance

d_i is + for an image (real) formed to the right of the lens by a real object.

d_i is − for an image (virtual) formed to the left of the lens by a real object.

Magnification

m is + for an image that is upright with respect to the object.

m is − for an image that is inverted with respect to the object.

* This situation arises in systems containing more than one lens, where the image formed by the first lens becomes the object for the second lens. In such a case, the object of the second lens may lie to the right of that lens, in which event d_o is assigned a negative value and the object is called a virtual object.

Examples 8 and 9 illustrate the use of the thin-lens and magnification equations.

Example 8 The Real Image Formed by a Camera Lens

A 1.70-m-tall person is standing 2.50 m in front of a camera. The camera uses a converging lens whose focal length is 0.0500 m. (a) Find the image distance (the distance between the lens and the film) and determine whether the image is real or virtual. (b) Find the magnification and the height of the image on the film.

Reasoning This optical arrangement is similar to that in Figure 26.26a, where the object distance is greater than twice the focal length of the lens. Therefore, we expect the image to be real, inverted, and smaller than the object.

Solution

(a) To find the image distance d_i we use the thin-lens equation with $d_o = 2.50$ m and $f = 0.0500$ m:

$$\frac{1}{d_i} = \frac{1}{f} - \frac{1}{d_o} = \frac{1}{0.0500 \text{ m}} - \frac{1}{2.50 \text{ m}}$$

$$= 19.6 \text{ m}^{-1} \quad \text{or} \quad \boxed{d_i = 0.0510 \text{ m}}$$

Since the image distance is a positive number, a $\boxed{\text{real image}}$ is formed on the film.

(b) The magnification follows from the magnification equation:

$$m = -\frac{d_i}{d_o} = -\frac{0.0510 \text{ m}}{2.50 \text{ m}} = \boxed{-0.0204}$$

The image is 0.0204 times as large as the object, and it is inverted since m is negative. Since the object height is $h_o = 1.70$ m, the image height is

$$h_i = mh_o = (-0.0204)(1.70 \text{ m}) = \boxed{-0.0347 \text{ m}}$$

Problem solving insight
In the thin-lens equation, the reciprocal of the image distance d_i is given by $d_i^{-1} = f^{-1} - d_o^{-1}$, where f is the focal length and d_o is the object distance. After combining the reciprocals f^{-1} and d_o^{-1}, do not forget to take the reciprocal of the result to find d_i.

Example 9 The Virtual Image Formed by a Diverging Lens

An object is placed 7.10 cm to the left of a diverging lens whose focal length is $f = -5.08$ cm (a diverging lens has a negative focal length). (a) Find the image distance and determine whether the image is real or virtual. (b) Obtain the magnification.

Reasoning This situation is similar to that in Figure 26.29a. The ray diagram shows that the image is virtual, erect, and smaller than the object.

Solution

(a) The thin-lens equation can be used to find the image distance d_i:

$$\frac{1}{d_i} = \frac{1}{f} - \frac{1}{d_o} = \frac{1}{-5.08 \text{ cm}} - \frac{1}{7.10 \text{ cm}}$$

$$= -0.338 \text{ cm}^{-1} \quad \text{or} \quad \boxed{d_i = -2.96 \text{ cm}}$$

The image distance is negative, indicating that the image is $\boxed{\text{virtual}}$ and located to the left of the lens.

(b) Since d_i and d_o are known, the magnification equation shows that

$$m = -\frac{d_i}{d_o} = -\frac{-2.96 \text{ cm}}{7.10 \text{ cm}} = \boxed{0.417}$$

The image is upright (m is +) and smaller ($m < 1$) than the object.

The thin-lens and magnification equations can be derived by considering rays 1 and 3 in Figure 26.32a. Ray 1 is shown separately in part b of the drawing, where the angle θ is the same in each of the two colored triangles. Thus, $\tan \theta$ is the same for each triangle:

$$\tan \theta = \frac{h_o}{f} = \frac{-h_i}{d_i - f}$$

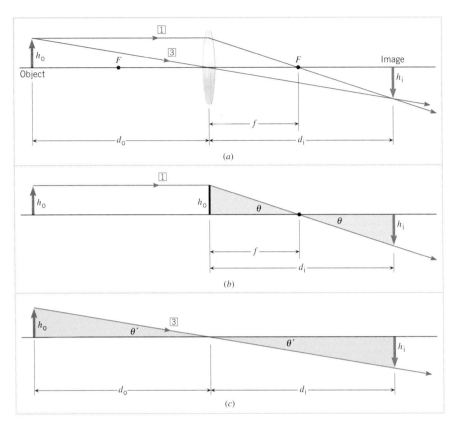

Figure 26.32 These ray diagrams are used for deriving the thin-lens and magnification equations.

A minus sign has been inserted in the numerator of the ratio $h_i/(d_i - f)$ for the following reason. The angle θ in Figure 26.32b is assumed to be positive. Since the image is inverted relative to the object, the image height h_i is a negative number. The insertion of the minus sign ensures that the term $-h_i/(d_i - f)$, and hence $\tan \theta$, is a positive quantity.

Ray 3 is shown separately in part c of the drawing, where the angles θ' are the same. Therefore,

$$\tan \theta' = \frac{h_o}{d_o} = \frac{-h_i}{d_i}$$

A minus sign has been inserted in the numerator of the term h_i/d_i for the same reason that a minus sign was inserted earlier—namely, to ensure that $\tan \theta'$ is a positive quantity. The first equation gives $h_i/h_o = -(d_i - f)/f$, while the second equation yields $h_i/h_o = -d_i/d_o$. Equating these two expressions for h_i/h_o and rearranging the result produces the thin-lens equation, $1/d_o + 1/d_i = 1/f$. The magnification equation follows directly from the equation $h_i/h_o = -d_i/d_o$, if we recognize that h_i/h_o is the magnification m of the lens.

✓ **Check Your Understanding 4**

An object is located at a distance d_o in front of a lens. The lens has a focal length f and produces an upright image that is twice as tall as the object. What kind of lens is it? What is the object distance? Express your answer as a fraction or multiple of the focal length. *(The answer is given at the end of the book.)*

Background: The characteristics of the possible images for converging and diverging lenses allow you to determine the type of lens here. The thin-lens equation and the magnification equation provide the basis for determining the object distance in terms of the focal length.

For similar questions (including calculational counterparts), consult Self-Assessment Test 26.2, which is described at the end of Section 26.14.

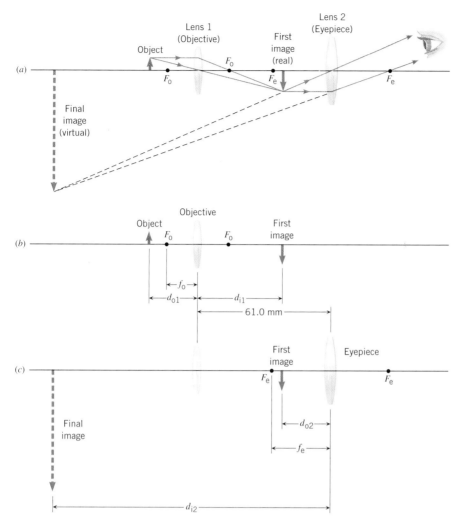

Figure 26.33 (*a*) This two-lens system can be used as a compound microscope to produce a virtual, enlarged, and inverted final image. (*b*) The objective forms the first image and (*c*) the eyepiece forms the final image.

26.9 *Lenses in Combination*

Many optical instruments, such as microscopes and telescopes, use a number of lenses together to produce an image. Among other things, a multiple-lens system can produce an image that is magnified more than is possible with a single lens. For instance, Figure 26.33*a* shows a two-lens system used in a microscope. The first lens, the lens closest to the object, is referred to as the *objective*. The second lens is known as the *eyepiece* (or *ocular*). The object is placed just outside the focal point F_o of the objective. The image formed by the objective—called the "first image" in the drawing—is real, inverted, and enlarged compared to the object. This first image then serves as the object for the eyepiece. Since the first image falls between the eyepiece and its focal point F_e, the eyepiece forms an enlarged, virtual, final image, which is what the observer sees.

The location of the final image in a multiple-lens system can be determined by applying the thin-lens equation to each lens separately. The key point to remember in such situations is that **the image produced by one lens serves as the object for the next lens,** as the next example illustrates.

Problem solving insight

Example 10 A Microscope—Two Lenses in Combination

The objective and eyepiece of the compound microscope in Figure 26.33 are both converging lenses and have focal lengths of $f_o = 15.0$ mm and $f_e = 25.5$ mm. A distance of 61.0 mm separates the lenses. The microscope is being used to examine an object placed $d_{o1} = 24.1$ mm in front of the objective. Find the final image distance.

Reasoning The thin-lens equation can be used to locate the final image produced by the eyepiece. We know the focal length of the eyepiece, but to determine the final image distance from the thin-lens equation we also need to know the object distance, which is not given. To obtain this distance, we recall that the image produced by one lens (the objective) is the object for the next lens (the eyepiece). We can use the thin-lens equation to locate the image produced by the objective, since the focal length and the object distance for this lens are given. The location of this image relative to the eyepiece will tell us the object distance for the eyepiece.

Solution The final image distance relative to the eyepiece is d_{i2}, and we can determine it by using the thin-lens equation:

$$\frac{1}{d_{i2}} = \frac{1}{f_e} - \frac{1}{d_{o2}}$$

The focal length f_e of the eyepiece is known, but to obtain a value for the object distance d_{o2} we must locate the first image produced by the objective. The first image distance d_{i1} (see Figure 26.33b) can be determined using the thin-lens equation with $d_{o1} = 24.1$ mm and $f_o = 15.0$ mm.

$$\frac{1}{d_{i1}} = \frac{1}{f_o} - \frac{1}{d_{o1}}$$
$$= \frac{1}{15.0 \text{ mm}} - \frac{1}{24.1 \text{ mm}}$$
$$= 0.0252 \text{ mm}^{-1} \quad \text{or} \quad d_{i1} = 39.7 \text{ mm}$$

The first image now becomes the object for the eyepiece (see part c of the drawing). Since the distance between the lenses is 61.0 mm, the object distance for the eyepiece is $d_{o2} = 61.0$ mm $- d_{i1} = 61.0$ mm $- 39.7$ mm $= 21.3$ mm. Noting that the focal length of the eyepiece is $f_e = 25.5$ mm, we can determine the final image distance with the aid of the thin-lens equation:

$$\frac{1}{d_{i2}} = \frac{1}{f_e} - \frac{1}{d_{o2}}$$
$$= \frac{1}{25.5 \text{ mm}} - \frac{1}{21.3 \text{ mm}}$$
$$= -0.0077 \text{ mm}^{-1} \quad \text{or} \quad \boxed{d_{i2} = -130 \text{ mm}}$$

The fact that d_{i2} is negative indicates that the final image is virtual. It lies to the left of the eyepiece, as the drawing shows.

Need more practice?

Interactive LearningWare 26.2
A diverging lens ($f = -14.0$ cm) is 30.0 cm to the left of a converging lens ($f = 12.0$ cm). An object is placed 16.0 cm to the left of the diverging lens. Find the image distance of the first image, measured from the diverging lens, and the image distance of the final image, measured from the converging lens.

Related Homework: Problem 62

Go to
www.wiley.com/college/cutnell
for an interactive solution.

26.10   *The Human Eye*

ANATOMY

Without doubt, the human eye is the most remarkable of all optical devices. Figure 26.34 shows some of its main anatomical features. The eyeball is approximately spherical with a diameter of about 25 mm. Light enters the eye through a transparent membrane (the *cornea*). This membrane covers a clear liquid region (the *aqueous humor*), behind which are a diaphragm (the *iris*), the *lens*, a region filled with a jelly-like substance (the *vitreous humor*), and, finally, the *retina*. The retina is the light-sensitive part of the eye, consisting of millions of structures called *rods* and *cones*. When stimulated by light, these structures send electrical impulses via the *optic nerve* to the brain, which interprets the image on the retina.

The iris is the colored portion of the eye and controls the amount of light reaching the retina. The iris acts as a controller because it is a muscular diaphragm with a variable opening at its center, through which the light passes. The opening is called the *pupil*. The diameter of the pupil varies from about 2 to 7 mm, decreasing in bright light and increasing (dilating) in dim light.

Of prime importance to the operation of the eye is the fact that the lens is flexible, and its shape can be altered by the action of the *ciliary muscle*. The lens is connected to the ciliary muscle by the *suspensory ligaments* (see the drawing). We will see shortly how the shape-changing ability of the lens affects the focusing ability of the eye.

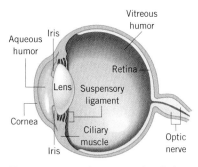

Figure 26.34 A cross-sectional view of the human eye.

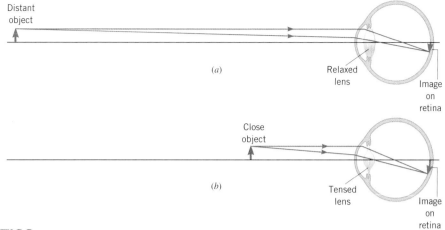

Figure 26.35 (*a*) When fully relaxed, the lens of the eye has its longest focal length, and an image of a very distant object is formed on the retina. (*b*) When the ciliary muscle is tensed, the lens has a shorter focal length. Consequently, an image of a closer object is also formed on the retina.

OPTICS

Optically, the eye and the camera are similar; both have a lens system and a diaphragm with a variable opening or aperture at its center. Moreover, the retina of the eye and the film in the camera serve similar functions, for both record the image formed by the lens system. In the eye, the image formed on the retina is real, inverted, and smaller than the object, just as it is in a camera. Although the image on the retina is inverted, it is interpreted by the brain as being right-side up.

The physics of the human eye.

For clear vision, the eye must refract the incoming light rays, so as to form a sharp image on the retina. In reaching the retina, the light travels through five different media, each with a different index of refraction n: air ($n = 1.00$), the cornea ($n = 1.38$), the aqueous humor ($n = 1.33$), the lens ($n = 1.40$, on the average), and the vitreous humor ($n = 1.34$). Each time light passes from one medium into another, it is refracted at the boundary. The greatest amount of refraction, about 70% or so, occurs at the air/cornea boundary. According to Snell's law, the large refraction at this interface occurs primarily because the refractive index of air ($n = 1.00$) is so different from that of the cornea ($n = 1.38$). The refraction at all the other boundaries is relatively small because the indices of refraction on either side of these boundaries are nearly equal. The lens itself contributes only about 20–25% of the total refraction, since the surrounding aqueous and vitreous humors have indices of refraction that are nearly the same as that of the lens.

Even though the lens contributes only a quarter of the total refraction or less, its function is an important one. The eye has a fixed image distance; that is, the distance between the lens and the retina is constant. Therefore, the only way that objects located at different distances can produce images on the retina is for the focal length of the lens to be adjustable. And it is the ciliary muscle that adjusts the focal length. When the eye looks at a very distant object, the ciliary muscle is not tensed. The lens has its least curvature and, consequently, its longest focal length. Under this condition the eye is said to be "fully relaxed," and the rays form a sharp image on the retina, as in Figure 26.35*a*. When the object moves closer to the eye, the ciliary muscle tenses automatically, thereby increasing the curvature of the lens, shortening the focal length, and permitting a sharp image to form again on the retina (Figure 26.35*b*). When a sharp image of an object is formed on the retina, we say the eye is "focused" on the object. The process in which the lens changes its focal length to focus on objects at different distances is called ***accommodation.***

When you hold a book too close, the print is blurred because the lens cannot adjust enough to bring the book into focus. The point nearest the eye at which an object can be placed and still produce a sharp image on the retina is called the ***near point*** of the eye. The ciliary muscle is fully tensed when an object is placed at the near point. For people in their early twenties with normal vision, the near point is located about 25 cm from the eye. It increases to about 50 cm at age 40 and to roughly 500 cm at age 60. Since most reading material is held at a distance of 25–45 cm from the eye, older adults typically need eyeglasses to overcome the loss of accommodation. The ***far point*** of the eye is the location of the farthest object on which the fully relaxed eye can focus. A person with normal eyesight can see objects very far away, such as the planets and stars, and thus has a far point located nearly at infinity.

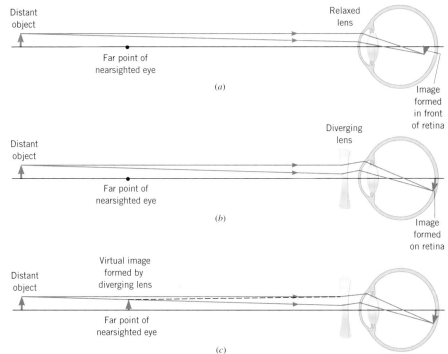

Figure 26.36 (*a*) When a nearsighted person views a distant object, the image is formed in front of the retina. The result is blurred vision. (*b*) With a diverging lens in front of the eye, the image is moved onto the retina and clear vision results. (*c*) The diverging lens is designed to form a virtual image at the far point of the nearsighted eye.

NEARSIGHTEDNESS

The physics of nearsightedness.

A person who is ***nearsighted (myopic)*** can focus on nearby objects but cannot clearly see objects far away. For such a person, the far point of the eye is not at infinity and may even be as close to the eye as three or four meters. When a nearsighted eye tries to focus on a distant object, the eye is fully relaxed, like a normal eye. However, the nearsighted eye has a focal length that is shorter than it should be, so rays from the distant object form a sharp image in front of the retina, as Figure 26.36*a* shows, and blurred vision results.

The nearsighted eye can be corrected with glasses or contacts that use *diverging* lenses, as Figure 26.36*b* suggests. The rays from the object diverge after leaving the eyeglass lens. Therefore, when they are subsequently refracted toward the principal axis by the eye, a sharp image is formed farther back and falls on the retina. Since the relaxed (but nearsighted) eye can focus on an object at the eye's far point—but not on objects farther away—the diverging lens is designed to transform a very distant object into an image located at the far point. Part *c* of the drawing shows this transformation, and the next example illustrates how to determine the focal length of the diverging lens that accomplishes it.

Example 11 Eyeglasses for the Nearsighted Person

A nearsighted person has a far point located only 521 cm from the eye. Assuming that eyeglasses are to be worn 2 cm in front of the eye, find the focal length needed for the diverging lenses of the glasses so the person can see distant objects.

Problem solving insight
Eyeglasses are worn about 2 cm from the eyes. Be sure, if necessary, to take this 2 cm into account when determining the object and image distances (d_o and d_i) that are used in the thin-lens equation.

Reasoning In Figure 26.36*c* the far point is 521 cm away from the eye. Since the glasses are worn 2 cm from the eye, the far point is 519 cm to the left of the diverging lens. The image distance, then, is −519 cm, the negative sign indicating that the image is a virtual image formed to the left of the lens. The object is assumed to be infinitely far from the diverging lens. The thin-lens equation can be used to find the focal length of the eyeglasses. We expect the focal length to be negative, since the lens is a diverging lens.

Solution With $d_i = -519$ cm and $d_o = \infty$, the focal length can be found as follows:

$$\frac{1}{f} = \frac{1}{d_o} + \frac{1}{d_i} = \frac{1}{\infty} + \frac{1}{-519 \text{ cm}} \quad \text{or} \quad \boxed{f = -519 \text{ cm}} \qquad (26.6)$$

The value for *f* is negative, as expected, for a diverging lens.

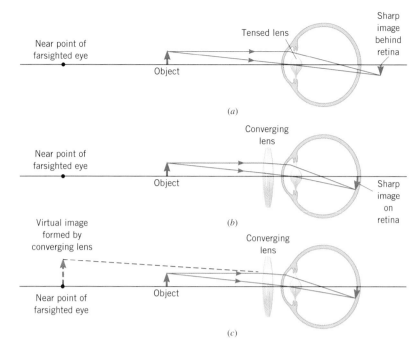

(a)

(b)

(c)

Figure 26.37 (a) When a farsighted person views an object located inside the near point, a sharp image would be formed behind the retina if light could pass through it. Only a blurred image forms on the retina. (b) With a converging lens in front of the eye, the sharp image is moved onto the retina and clear vision results. (c) The converging lens is designed to form a virtual image at the near point of the farsighted eye.

FARSIGHTEDNESS

A *farsighted (hyperopic)* person can usually see distant objects clearly, but cannot focus on those nearby. Whereas the near point of a young and normal eye is located about 25 cm from the eye, the near point of a farsighted eye may be considerably farther away than that, perhaps as far as several hundred centimeters. When a farsighted eye tries to focus on a book held closer than the near point, it accommodates and shortens its focal length as much as it can. However, even at its shortest, the focal length is longer than it should be. Therefore, the light rays from the book would form a sharp image behind the retina if they could do so, as Figure 26.37a suggests. In reality, no light passes through the retina, but a blurred image does form on it.

Figure 26.37b shows that farsightedness can be corrected by placing a *converging* lens in front of the eye. The lens refracts the light rays more toward the principal axis before they enter the eye. Consequently, when the rays are refracted even more by the eye, they converge to form an image on the retina. Part *c* of the figure illustrates what the eye sees when it looks through the converging lens. The lens is designed so that the eye perceives the light to be coming from a virtual image located at the near point. Example 12 shows how the focal length of the converging lens is determined to correct for farsightedness.

The physics of **farsightedness.**

Example 12 Contact Lenses for the Farsighted Person

A farsighted person has a near point located 210 cm from the eyes. Obtain the focal length of the converging lenses in a pair of contacts that can be used to read a book held 25.0 cm from the eyes.

Reasoning A contact lens is placed directly against the eye. Thus, the object distance, which is the distance from the book to the lens, is 25.0 cm. The lens forms an image of the book at the near point of the eye, so the image distance is -210 cm. The minus sign indicates that the image is a virtual image formed to the left of the lens, as in Figure 26.37c. The focal length can be obtained from the thin-lens equation.

Solution With $d_o = 25.0$ cm and $d_i = -210$ cm, the focal length can be determined from the thin-lens equation as follows:

$$\frac{1}{f} = \frac{1}{d_o} + \frac{1}{d_i} = \frac{1}{25.0 \text{ cm}} + \frac{1}{-210 \text{ cm}} = 0.0352 \text{ cm}^{-1} \quad \text{or} \quad \boxed{f = 28.4 \text{ cm}}$$

Object

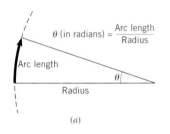

Figure 26.38 The angle θ is the angular size of both the image and the object.

THE REFRACTIVE POWER OF A LENS—THE DIOPTER

The extent to which rays of light are refracted by a lens depends on its focal length. However, the optometrists who prescribe correctional lenses and the opticians who make the lenses do not specify the focal length directly in prescriptions. Instead, they use the concept of *refractive power* to describe the extent to which a lens refracts light:

$$\begin{array}{c} \text{Refractive power} \\ \text{of a lens (in diopters)} \end{array} = \frac{1}{f\,(\text{in meters})} \qquad (26.8)$$

The refractive power is measured in units of *diopters*. One diopter is $1\ \text{m}^{-1}$.

Equation 26.8 shows that a converging lens has a refractive power of 1 diopter if it focuses parallel light rays to a focal point 1 m beyond the lens. If a lens refracts parallel rays even more and converges them to a focal point only 0.25 m beyond the lens, the lens has four times more refractive power, or 4 diopters. Since a converging lens has a positive focal length and a diverging lens has a negative focal length, the refractive power of a converging lens is positive while that of a diverging lens is negative. For instance, the eyeglasses in Example 11 would be described in a prescription from an optometrist in the following way: Refractive power = $1/(-5.19\ \text{m})$ = -0.193 diopters. The contact lenses in Example 12 would be described in a similar fashion: Refractive power = $1/(0.284\ \text{m})$ = 3.52 diopters.

26.11 *Angular Magnification and the Magnifying Glass*

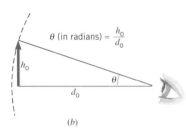

Figure 26.39 (a) The angle θ, measured in radians, is the arc length divided by the radius. (b) For small angles (less than $9°$), θ is approximately equal to h_o/d_o, where h_o and d_o are the object height and distance.

If you hold a penny at arm's length, the penny looks larger than the moon. The reason is that the penny, being so close, forms a larger image on the retina of the eye than does the more distant moon. The brain interprets the larger image of the penny as arising from a larger object. The size of the image on the retina determines how large an object appears to be. However, the size of the image on the retina is difficult to measure. Alternatively, the angle θ subtended by the image can be used as an indication of the image size. Figure 26.38 shows this alternative, which has the advantage that θ is also the angle subtended by the object and, hence, can be measured more easily. The angle θ is called the *angular size* of both the image and the object. The larger the angular size, the larger the image on the retina, and the larger the object appears to be.

According to Equation 8.1, the angle θ (measured in radians) is the length of the circular arc that is subtended by the angle divided by the radius of the arc, as Figure 26.39a indicates. Part b of the drawing shows the situation for an object of height h_o viewed at a distance d_o from the eye. When θ is small, h_o is approximately equal to the arc length and d_o is nearly equal to the radius, so that

$$\theta\ (\text{in radians}) = \text{Angular size} \approx \frac{h_o}{d_o}$$

This approximation is good to within one percent for angles of $9°$ or smaller. In the next example the angular size of a penny is compared with that of the moon.

Example 13   A Penny and the Moon

Compare the angular size of a penny (diameter = h_o = 1.9 cm) held at arm's length (d_o = 71 cm) with that of the moon (diameter = h_o = 3.5×10^6 m, and d_o = 3.9×10^8 m).

Reasoning The angular size θ of an object is given approximately by its height h_o divided by its distance d_o from the eye, $\theta \approx h_o/d_o$, provided that the angle involved is less than roughly $9°$; this approximation applies here. The "heights" of the penny and the moon are their diameters.

Solution The angular sizes of the penny and moon are

Penny
$$\theta \approx \frac{h_o}{d_o} = \frac{1.9\ \text{cm}}{71\ \text{cm}} = \boxed{0.027\ \text{rad}\ (1.5°)}$$

Moon
$$\theta \approx \frac{h_o}{d_o} = \frac{3.5 \times 10^6\ \text{m}}{3.9 \times 10^8\ \text{m}} = \boxed{0.0090\ \text{rad}\ (0.52°)}$$

The penny thus appears to be about three times as large as the moon.

An optical instrument, such as a magnifying glass, allows us to view small or distant objects because it produces a larger image on the retina than would be possible otherwise. In other words, an optical instrument magnifies the angular size of the object. The *angular magnification* (or *magnifying power*) M is the angular size θ' of the final image produced by the instrument divided by a reference angular size θ. The reference angular size is the angular size of the object when seen without the instrument.

Angular magnification
$$M = \frac{\text{Angular size of final image produced by optical instrument}}{\text{Reference angular size of object seen without optical instrument}} = \frac{\theta'}{\theta} \qquad (26.9)$$

A magnifying glass is the simplest device that provides angular magnification. In this case, the reference angular size θ is chosen to be the angular size of the object when placed at the near point of the eye and seen without the magnifying glass. Since an object cannot be brought closer than the near point and still produce a sharp image on the retina, θ represents the largest angular size obtainable without the magnifying glass. Figure 26.40*a* indicates that the reference angular size is $\theta \approx h_o/N$, where N is the distance from the eye to the near point. To compute θ', recall from Section 26.7 and Figure 26.28 that a magnifying glass is usually a single converging lens, with the object located inside the focal point. In this situation, Figure 26.40*b* indicates that the lens produces a virtual image that is enlarged and upright with respect to the object. Assuming the eye is next to the magnifying glass, the angular size θ' seen by the eye is $\theta' \approx h_o/d_o$, where d_o is the object distance. The angular magnification is

$$M = \frac{\theta'}{\theta} \approx \frac{h_o/d_o}{h_o/N} = \frac{N}{d_o}$$

According to the thin-lens equation, d_o is related to the image distance d_i and the focal length f of the lens by

$$\frac{1}{d_o} = \frac{1}{f} - \frac{1}{d_i}$$

Substituting this expression for $1/d_o$ into the previous expression for M leads to the following result:

Angular magnification of a magnifying glass
$$M = \frac{\theta'}{\theta} \approx \left(\frac{1}{f} - \frac{1}{d_i} \right) N \qquad (26.10)$$

Two special cases of this result are of interest, depending on whether the image is located as close to the eye as possible or as far away as possible. To be seen clearly, the closest the image can be relative to the eye is at the near point, or $d_i = -N$. The minus sign indicates that the image lies to the left of the lens and is virtual. In this event, Equation 26.10 becomes $M \approx (N/f) + 1$. The farthest the image can be from the eye is at infinity ($d_i = -\infty$); this occurs when the object is placed at the focal point of the lens. When the image is at infinity, Equation 26.10 simplifies to $M \approx N/f$. Clearly, the angular magnification is greater when the image is at the near point of the eye rather than at infinity. In either case, however, the greatest magnification is achieved by using a magnifying glass with the shortest possible focal length. Example 14 illustrates how to determine the angular magnification of a magnifying glass that is used in these two ways.

The physics of a magnifying glass.

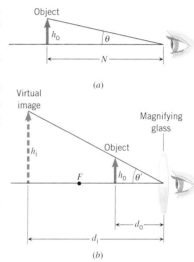

Figure 26.40 (*a*) Without a magnifying glass, the largest angular size θ occurs when the object is placed at the near point, a distance N from the eye. (*b*) A magnifying glass produces an enlarged, virtual image of an object placed inside the focal point F of the lens. The angular size of both the image and the object is θ'.

Example 14   Examining a Diamond with a Magnifying Glass

A jeweler, whose near point is 40.0 cm from his eye and whose far point is at infinity, is using a small magnifying glass (called a loupe) to examine a diamond. The lens of the magnifying glass has a focal length of 5.00 cm, and the image of the gem is -185 cm from the lens. The image distance is negative because the image is virtual and is formed on the same side of the lens as the object. (a) Determine the angular magnification of the magnifying glass. (b) Where should the image be located so the jeweler's eye is fully relaxed and has the least strain? What is the angular magnification under this "least strain" condition?

A gem cutter is using a magnifying glass to examine a piece of amber containing ancient insects. (© Steven L. Raymer/National Geographic/Getty Images)

Reasoning The angular magnification of the magnifying glass can be determined from Equation 26.10. In part (a) the image distance is -185 cm. In part (b) the ciliary muscle of the jeweler's eye is fully relaxed, so the image must be located infinitely far from the eye, at its far point, as Section 26.10 discusses.

Solution

(a) With $f = 5.00$ cm, $d_i = -185$ cm, and $N = 40.0$ cm, the angular magnification is

$$M = \left(\frac{1}{f} - \frac{1}{d_i}\right)N = \left(\frac{1}{5.00 \text{ cm}} - \frac{1}{-185 \text{ cm}}\right)(40.0 \text{ cm}) = \boxed{8.22}$$

(b) With $f = 5.00$ cm, $d_i = -\infty$, and $N = 40.0$ cm, the angular magnification is

$$M = \left(\frac{1}{f} - \frac{1}{d_i}\right)N = \left(\frac{1}{5.00 \text{ cm}} - \frac{1}{-\infty}\right)(40.0 \text{ cm}) = \boxed{8.00}$$

Jewelers often prefer to minimize eyestrain when viewing objects, even though it means a slight reduction in angular magnification.

✔ Check Your Understanding 5

A person who has a near point of 25.0 cm is looking with unaided eyes at an object that is located at the near point. The object has an angular size of 0.012 rad. Then, holding a magnifying glass ($f = 10.0$ cm) next to her eye, she views the image of this object, the image being located at the near point. What is the angular size of the image? *(The answer is given at the end of the book.)*

Background: The angular magnification of a magnifying glass is the angular size of the image seen with the magnifying glass divided by the reference angular size. The reference angular size is that of the object when placed at the near point and seen without the magnifying glass.

For similar questions (including conceptual counterparts), consult Self-Assessment Test 26.2, which is described at the end of Section 26.14.

The physics of the compound microscope.

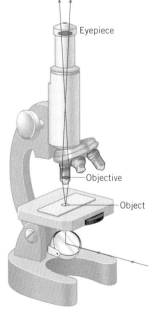

Figure 26.41 A compound microscope.

26.12 The Compound Microscope

To increase the angular magnification beyond that possible with a magnifying glass, an additional converging lens can be included to "premagnify" the object before the magnifying glass comes into play. The result is an optical instrument known as the **compound microscope** (Figure 26.41). As discussed in Section 26.9, the magnifying glass is called the eyepiece, and the additional lens is called the objective.

The angular magnification of the compound microscope is $M = \theta'/\theta$ (Equation 26.9), where θ' is the angular size of the final image and θ is the reference angular size. As with the magnifying glass in Figure 26.40, the reference angular size is determined by the height h_o of the object when the object is located at the near point of the unaided eye: $\theta \approx h_o/N$, where N is the distance between the eye and the near point. Assuming that the object is placed just outside the focal point F_o of the objective (see Figure 26.33a) and that the final image is very far from the eyepiece (i.e., near infinity, see Figure 26.33c), it can be shown that

Angular magnification of a compound microscope

$$M \approx -\frac{(L - f_e)N}{f_o f_e} \qquad (L > f_o + f_e) \qquad (26.11)$$

In Equation 26.11, f_o and f_e are, respectively, the focal lengths of the objective and the eyepiece. The angular magnification is greatest when f_o and f_e are as small as possible (since they are in the denominator in Equation 26.11) and when the distance L between the lenses is as large as possible. Furthermore, L must be greater than the sum of f_o and f_e for this equation to be valid. Example 15 deals with the angular magnification of a compound microscope.

Example 15 The Angular Magnification of a Compound Microscope

The focal length of the objective of a compound microscope is $f_o = 0.40$ cm, and that of the eyepiece is $f_e = 3.0$ cm. The two lenses are separated by a distance of $L = 20.0$ cm. A person with a near point distance of $N = 25$ cm is using the microscope. (a) Determine the angular magnification of the microscope. (b) Compare the answer in part (a) with the largest angular magnification obtainable by using the eyepiece alone as a magnifying glass.

Reasoning The angular magnification of the compound microscope can be obtained directly from Equation 26.11, since all the variables are known. When the eyepiece is used alone as a magnifying glass, as in Figure 26.40b, the largest angular magnification occurs when the image seen through the eyepiece is as close as possible to the eye. The image in this case is at the near point, and according to Equation 26.10, the angular magnification is $M \approx (N/f_e) + 1$.

Solution

(a) The angular magnification of the compound microscope is

$$M \approx -\frac{(L - f_e)N}{f_o f_e} = -\frac{(20.0 \text{ cm} - 3.0 \text{ cm})(25 \text{ cm})}{(0.40 \text{ cm})(3.0 \text{ cm})} = \boxed{-350}$$

The minus sign indicates that the final image is inverted relative to the initial object.

(b) The maximum angular magnification of the eyepiece by itself is

$$M \approx \frac{N}{f_e} + 1 = \frac{25 \text{ cm}}{3.0 \text{ cm}} + 1 = \boxed{9.3}$$

The effect of the objective is to increase the angular magnification of the compound microscope by a factor of $350/9.3 = 38$ compared to that of a magnifying glass.

26.13 The Telescope

A telescope is an instrument for magnifying distant objects, such as stars and planets. Like a microscope, a telescope consists of an objective and an eyepiece (also called the ocular). Since the object is usually far away, the light rays striking the telescope are nearly parallel, and the "first image" is formed just beyond the focal point F_o of the objective, as Figure 26.42a illustrates. The first image is real and inverted. Unlike that in the compound microscope, however, this image is *smaller* than the object. If, as in part b of the drawing, the telescope is constructed so the first image lies just inside the focal point F_e of the eyepiece, the eyepiece acts like a magnifying glass. It forms a final image that is greatly enlarged, virtual, and located near infinity. This final image can then be viewed with a fully relaxed eye.

The angular magnification M of a telescope, like that of a magnifying glass or a microscope, is the angular size θ' subtended by the final image of the telescope divided by the reference angular size θ of the object. For an astronomical object, such as a planet, it is convenient to use as a reference the angular size of the object seen in the sky with the unaided eye. Since the object is far away, the angular size seen by the unaided eye is nearly the same as the angle θ subtended at the objective of the telescope in Figure

The physics of the telescope.

Figure 26.42 (a) An astronomical telescope is used to view distant objects. (Note the "break" in the principal axis, between the object and the objective.) The objective produces a real, inverted first image. (b) The eyepiece magnifies the first image to produce the final image near infinity.

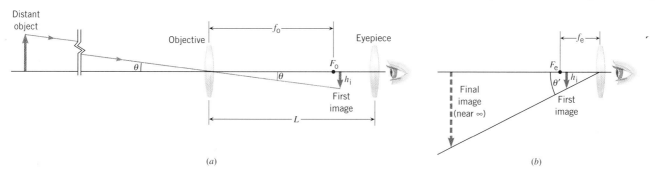

(a)

(b)

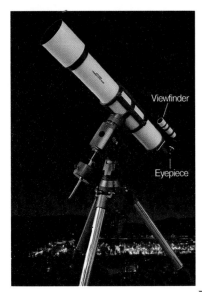

Figure 26.43 An astronomical telescope. The viewfinder is a separate small telescope with low magnification and serves as an aid in locating the object. Once the object has been found, the viewer looks through the eyepiece to obtain the full magnification of the telescope. (© Tony Freeman/PhotoEdit)

26.42*a*. Moreover, θ is also the angle subtended by the first image, so $\theta \approx -h_i/f_o$, where h_i is the height of the first image and f_o is the focal length of the objective. A minus sign has been inserted into this equation because the first image is inverted relative to the object and the image height h_i is a negative number. The insertion of the minus sign ensures that the term $-h_i/f_o$, and hence θ, is a positive quantity. To obtain an expression for θ', refer to Figure 26.42*b* and note that the first image is located very near the focal point F_e of the eyepiece, which has a focal length f_e. Therefore, $\theta' \approx h_i/f_e$. The angular magnification of the telescope is approximately

Angular magnification of an astronomical telescope
$$M = \frac{\theta'}{\theta} \approx \frac{h_i/f_e}{-h_i/f_o} \approx -\frac{f_o}{f_e} \qquad (26.12)$$

The angular magnification is determined by the ratio of the focal length of the objective to the focal length of the eyepiece. For large angular magnifications, the objective should have a long focal length and the eyepiece a short one. Some of the design features of a telescope are the topic of the next example.

Example 16   The Angular Magnification of an Astronomical Telescope

The telescope shown in Figure 26.43 has the following specifications: $f_o = 985$ mm and $f_e = 5.00$ mm. From these data, find (a) the angular magnification of the telescope and (b) the approximate length of the telescope.

Reasoning The angular magnification of the telescope follows directly from Equation 26.12, since the focal lengths of the objective and eyepiece are known. We can find the length of the telescope by noting that it is approximately equal to the distance L between the objective and eyepiece. Figure 26.42 shows that the first image is located just beyond the focal point F_o of the objective and just inside the focal point F_e of the eyepiece. These two focal points are, therefore, very close together, so the distance L is approximately the sum of the two focal lengths: $L \approx f_o + f_e$.

Solution

(a) The angular magnification is approximately

$$M \approx -\frac{f_o}{f_e} = -\frac{985 \text{ mm}}{5.00 \text{ mm}} = \boxed{-197} \qquad (26.12)$$

(b) The approximate length of the telescope is

$$L \approx f_o + f_e = 985 \text{ mm} + 5.00 \text{ mm} = \boxed{990 \text{ mm}}$$

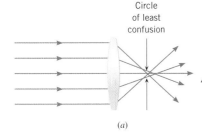

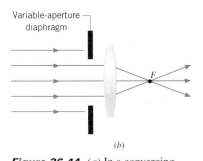

Figure 26.44 (*a*) In a converging lens, spherical aberration prevents light rays parallel to the principal axis from converging to a common point. (*b*) Spherical aberration can be reduced by allowing only rays near the principal axis to pass through the lens. The refracted rays now converge more nearly to a single focal point *F*.

26.14 *Lens Aberrations*

Rather than forming a sharp image, a single lens typically forms an image that is slightly out of focus. This lack of sharpness arises because the rays originating from a single point on the object are not focused to a single point on the image. As a result, each point on the image becomes a small blur. The lack of point-to-point correspondence between object and image is called an *aberration.*

One common type of aberration is ***spherical aberration,*** and it occurs with converging and diverging lenses made with spherical surfaces. Figure 26.44*a* shows how spherical aberration arises with a converging lens. Ideally, all rays traveling parallel to the principal axis are refracted so they cross the axis at the same point after passing through the lens. However, rays far from the principal axis are refracted more by the lens than those closer in. Consequently, the outer rays cross the axis closer to the lens than do the inner rays, so a lens with spherical aberration does not have a unique focal point. Instead, as the drawing suggests, there is a location along the principal axis where the light converges to the smallest cross-sectional area. This area is circular and is known as the ***circle of least confusion.*** The circle of least confusion is where the most satisfactory image can be formed by the lens.

Spherical aberration can be reduced substantially by using a variable-aperture diaphragm to allow only those rays close to the principal axis to pass through the lens. Fig-

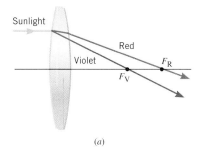
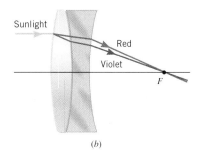

Figure 26.45 (*a*) Chromatic aberration arises when different colors are focused at different points along the principal axis: F_V = focal point for violet light, F_R = focal point for red light. (*b*) A converging and a diverging lens in tandem can be designed to bring different colors more nearly to the same focal point F.

ure 26.44*b* indicates that a reasonably sharp focal point can be achieved by this method, although less light now passes through the lens. Lenses with parabolic surfaces are also used to reduce this type of aberration, but they are difficult and expensive to make.

Chromatic aberration also causes blurred images. It arises because the index of refraction of the material from which the lens is made varies with wavelength. Section 26.5 discusses how this variation leads to the phenomenon of dispersion, in which different colors refract by different amounts. Figure 26.45*a* shows sunlight incident on a converging lens, in which the light spreads into its color spectrum because of dispersion. For clarity, however, the picture shows only the colors at the opposite ends of the visible spectrum—red and violet. Violet is refracted more than red, so the violet ray crosses the principal axis closer to the lens than does the red ray. Thus, the focal length of the lens is shorter for violet than for red, with intermediate values of the focal length corresponding to the colors in between. As a result of chromatic aberration, an undesirable color fringe surrounds the image.

Chromatic aberration can be greatly reduced by using a compound lens, such as the combination of a converging lens and a diverging lens shown in Figure 26.45*b*. Each lens is made from a different type of glass. With this lens combination the red and violet rays almost come to a common focus and, thus, chromatic aberration is reduced. A lens combination designed to reduce chromatic aberration is called an *achromatic lens* (from the Greek "achromatos," meaning "without color"). All high-quality cameras use achromatic lenses.

 **Self-Assessment Test 26.2**

Test your understanding of the material in Sections 26.6–26.14:

- Lenses • The Formation of Images by Lenses
- The Thin-Lens Equation and the Magnification Equation • Lenses in Combination
- The Human Eye • Angular Magnification and the Magnifying Glass
- The Compound Microscope • The Telescope • Lens Aberrations

Go to **www.wiley.com/college/cutnell**

26.15 *Concepts & Calculations*

One important phenomenon discussed in this chapter is how a ray of light is refracted when it goes from one medium into another. Example 17 reviews some of the important aspects of refraction, including Snell's law, the concept of a critical angle, and the notion of index matching.

Concepts & Calculations Example 17 Refraction

A ray of light is incident on a glass–water interface at the critical angle θ_c, as Figure 26.46 illustrates. The reflected light then passes through a liquid (immiscible with water) and into air. The indices of refraction for the four substances are given in the drawing. Determine the angle of refraction θ_5 for the ray as it passes into the air.

Concept Questions and Answers What determines the critical angle when the ray strikes the glass–water interface?

Answer According to Equation 26.4, the critical angle θ_c is determined by the indices of refraction n_{glass} and n_{water}; $\theta_c = \sin^{-1}(n_{water}/n_{glass})$.

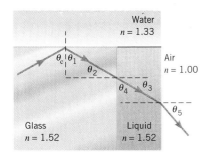

Figure 26.46 A ray of light strikes the glass–water interface at the critical angle θ_c. The reflected ray passes through a liquid and then into air.

When light is incident at the glass–water interface at the critical angle, what is the angle of refraction, and how is the angle of reflection θ_1 related to the critical angle?

Answer When light is incident at the critical angle, the angle of refraction is 90°, so the refracted ray skims along the surface. (For simplicity, this ray is not shown in the drawing.) We know from our work with mirrors that the angle of reflection is equal to the angle of incidence. Therefore, the angle of reflection is equal to the critical angle, so $\theta_1 = \theta_c$.

When the reflected ray strikes the glass–liquid interface, how is the angle of refraction θ_3 related to the angle of incidence θ_2? Note that the two materials have the same indices of refraction.

Answer According to Snell's law, Equation 26.2, the angle of refraction θ_3 is related to the angle of incidence θ_2 by $\sin \theta_3 = (n_{\text{glass}} \sin \theta_2)/n_{\text{liquid}}$. Since the two indices of refraction are equal, a condition known as "index matching," $\sin \theta_3 = \sin \theta_2$, and the two angles are equal. Therefore, the ray crosses the glass–liquid interface without being refracted.

When the ray passes from the liquid into the air, is the ray refracted?

Answer Yes. Because the two materials have different indices of refraction, the angle of refraction θ_5 is different from the angle of incidence θ_4. The angle of refraction can be found from Snell's law.

Solution Let us follow the ray of light as it progresses from left to right in Figure 26.46. It strikes the glass–water interface at the critical angle of

$$\theta_c = \sin^{-1}\left(\frac{n_{\text{water}}}{n_{\text{glass}}}\right) = \sin^{-1}\left(\frac{1.33}{1.52}\right) = 61.0°$$

The angle of reflection θ_1 from the glass–water interface is the same as the critical angle, so $\theta_1 = 61.0°$.

The ray then strikes the glass–liquid interface with an angle of incidence of $\theta_2 = 90.0° - 61.0° = 29.0°$. Note, as usual, that the angle of incidence is measured relative to the normal. At the glass–liquid interface, the angle of refraction θ_3 is the same as the angle of incidence θ_2, since the indices of refraction of the two materials are the same ($n_{\text{glass}} = n_{\text{liquid}} = 1.52$). Thus, $\theta_3 = \theta_2 = 29.0°$.

At the liquid–air interface we use Snell's law, Equation 26.2, to determine the angle θ_5 at which the refracted ray enters the air: $n_{\text{liquid}} \sin \theta_4 = n_{\text{air}} \sin \theta_5$. Note from the drawing that the angle of incidence θ_4 is the same as θ_3, so $\theta_4 = 29.0°$. The angle of refraction is

$$\theta_5 = \sin^{-1}\left(\frac{1.52 \sin 29.0°}{1.00}\right) = \boxed{47.5°}$$

One of the most important uses of refraction is in lenses, the behavior of which is governed by the thin-lens and magnification equations. Example 18 discusses how these equations are applied to a two-lens system and reviews the all-important sign conventions that must be followed (see the Reasoning Strategy in Section 26.8). In particular, the example shows how to account for a virtual object, in which the object lies to the right (rather than to the left) of a lens.

Concepts & Calculations Example 18   A Two-Lens System

Figure 26.47 illustrates a converging lens ($f_1 = +20.0$ cm) and a diverging lens ($f_2 = -15.0$ cm) that are separated by a distance of 10.0 cm. An object with a height of $h_{o1} = 5.00$ mm is placed $d_{o1} = 45.0$ cm to the left of the first (converging) lens. What is (a) the image distance d_{i1} and (b) the height h_{i1} of the image produced by the first lens? (c) What is the object distance for the second (diverging) lens? Find (d) the image distance d_{i2} and (e) the height h_{i2} of the image produced by the second lens.

Concept Questions and Answers Is the image produced by the first (converging) lens real or virtual?

Answer A converging lens can form either a real or a virtual image, depending on where the object is located relative to the focal point. If the object is to the left of the focal point, as it is in this example, the image is real and falls to the right of the lens. If, on the other hand, the object had been located between the focal point and the lens, the image would have been a virtual image located to the left of the converging lens.

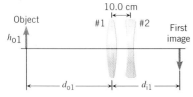

Figure 26.47 The image produced by the first lens, the "first image," falls to the right of the second lens, so the image becomes a virtual object for the second lens.

As far as the second lens is concerned, what role does the image produced by the first lens play?

  Answer   The image produced by the first lens acts as the object for the second lens.

Note in Figure 26.47 that the image produced by the first lens is called the "first image," and it falls to the right of the second lens. This image acts as the object for the second lens. Normally, however, an object would lie to the left of a lens. How do we take into account that this object lies to the right of the diverging lens?

  Answer   According to the Reasoning Strategy in Section 26.8, an object that is located to the right of a lens is called a virtual object and is assigned a negative object distance.

How do we find the location of the image produced by the second lens when its object is a virtual object?

  Answer   Once the object distance for the second lens has been assigned a negative number, we can use the thin-lens equation in the usual manner to find the image distance.

Solution

(a) The distance d_{i1} of the image from the first lens can be found from the thin-lens equation, Equation 26.6. The focal length, $f_1 = +20.0$ cm, is positive because the lens is a converging lens, and the object distance, $d_{o1} = +45.0$ cm, is positive because it lies to the left of the lens:

$$\frac{1}{d_{i1}} = \frac{1}{f_1} - \frac{1}{d_{o1}} = \frac{1}{20.0 \text{ cm}} - \frac{1}{45.0 \text{ cm}} = 0.0278 \text{ cm}^{-1} \quad \text{or} \quad d_{i1} = \boxed{36.0 \text{ cm}}$$

This image distance is positive, indicating that the image is real.

(b) The height h_{i1} of the image produced by the first lens can be obtained from the magnification equation, Equation 26.7:

$$h_{i1} = h_{o1}\left(-\frac{d_{i1}}{d_{o1}}\right) = (5.00 \text{ mm})\left(-\frac{36.0 \text{ cm}}{45.0 \text{ cm}}\right) = \boxed{-4.00 \text{ mm}}$$

The minus sign indicates that the first image is inverted with respect to the object (see Figure 26.47).

(c) The first image falls 36.0 cm to the right of the first lens. However, the second lens is located 10.0 cm to the right of the first lens, so the first image is located 36.0 cm − 10.0 cm = 26.0 cm to the right of the second lens. This image acts as the object for the second lens. Since the object for the second lens lies to the right of it, the object is a virtual object and assigned a negative number; $d_{o2} = \boxed{-26.0 \text{ cm}}$.

(d) The distance d_{i2} of the image from the second lens can be found from the thin-lens equation. The focal length, $f_2 = -15.0$ cm, is negative because the lens is a diverging lens, and the object distance, $d_{o2} = -26.0$ cm, is negative because it lies to the right of the lens:

$$\frac{1}{d_{i2}} = \frac{1}{f_2} - \frac{1}{d_{o2}} = \frac{1}{-15.0 \text{ cm}} - \frac{1}{-26.0 \text{ cm}} = -0.0282 \text{ cm}^{-1}$$

or

$$d_{i2} = \boxed{-35.5 \text{ cm}}$$

The negative sign for d_{i2} means that the image is formed to the left of the diverging lens and, hence, is a virtual image.

(e) The image height h_{i1} produced by the first lens becomes the object height h_{o2} for the second lens; $h_{o2} = h_{i1} = -4.00$ mm. The height h_{i2} of the image produced by the second lens follows from the magnification equation:

$$h_{i2} = h_{o2}\left(-\frac{d_{i2}}{d_{o2}}\right) = (-4.00 \text{ mm})\left(-\frac{-35.5 \text{ cm}}{-26.0 \text{ cm}}\right) = \boxed{5.46 \text{ mm}}$$

The positive value for h_{i2} means that the final image has the same orientation as the first image — that is, inverted.

  ▲

  At the end of the problem set for this chapter, you will find homework problems that contain both conceptual and quantitative parts. These problems are grouped under the heading *Concepts & Calculations, Group Learning Problems*. They are designed for use by students working alone or in small learning groups. The conceptual part of each problem provides a convenient focus for group discussions.

Concept Summary

This summary presents an abridged version of the chapter, including the important equations and all available learning aids. For convenient reference, the learning aids (including the text's examples) are placed next to or immediately after the relevant equation or discussion. The following learning aids may be found on-line at **www.wiley.com/college/cutnell**:

Interactive LearningWare examples are solved according to a five-step interactive format that is designed to help you develop problem-solving skills.	**Concept Simulations** are animated versions of text figures or animations that illustrate important concepts. You can control parameters that affect the display, and we encourage you to experiment.
Interactive Solutions offer specific models for certain types of problems in the chapter homework. The calculations are carried out interactively.	**Self-Assessment Tests** include both qualitative and quantitative questions. Extensive feedback is provided for both incorrect and correct answers, to help you evaluate your understanding of the material.

Topic	Discussion	Learning Aids
	26.1 The Index of Refraction	
Refraction	The change in speed as a ray of light goes from one material to another causes the ray to deviate from its incident direction. This change in direction is called refraction. The index of refraction n of a material is the ratio of the speed c of light in a vacuum to the speed v of light in the material:	
Index of refraction	$$n = \frac{c}{v}$$ (26.1)	Interactive Solution 26.7
	The values for n are less than unity, because the speed of light in a material medium is less than it is in a vacuum.	
	26.2 Snell's Law and the Refraction of Light	
	The refraction that occurs at the interface between two materials obeys Snell's law of refraction. This law states that (1) the refracted ray, the incident ray, and the normal to the interface all lie in the same plane, and (2) the angle of refraction θ_2 is related to the angle of incidence θ_1 according to	
Snell's law of refraction	$$n_1 \sin \theta_1 = n_2 \sin \theta_2$$ (26.2)	Example 1
	where n_1 and n_2 are the indices of refraction of the incident and refracting media, respectively. The angles are measured relative to the normal.	
	Because of refraction, a submerged object has an apparent depth that is different from its actual depth. If the observer is directly above (or below) the object, the apparent depth (or height) d' is related to the actual depth (or height) d according to	Example 2
Apparent depth	$$d' = d\left(\frac{n_2}{n_1}\right)$$ (26.3)	Examples 3, 4
	where n_1 and n_2 are the refractive indices of the materials in which the object and the observer, respectively, are located.	
	26.3 Total Internal Reflection	
	When light passes from a material with a larger refractive index n_1 into a material with a smaller refractive index n_2, the refracted ray is bent away from the normal. If the incident ray is at the critical angle θ_c, the angle of refraction is 90°. The critical angle is determined from Snell's law and is given by	Concept Simulation 26.1
		Examples 5, 6, 17
Critical angle	$$\sin \theta_c = \frac{n_2}{n_1} \quad (n_1 > n_2)$$ (26.4)	Interactive LearningWare 26.1
		Interactive Solution 26.25
Total internal reflection	When the angle of incidence exceeds the critical angle, all the incident light is reflected back into the material from which it came, a phenomenon known as total internal reflection.	
	26.4 Polarization and the Reflection and Refraction of Light	
Brewster angle	When light is incident on a nonmetallic surface at the Brewster angle θ_B, the reflected light is completely polarized parallel to the surface. The Brewster angle is given by	
Brewster's law	$$\tan \theta_B = \frac{n_2}{n_1}$$ (26.5)	

Topic	Discussion	Learning Aids
	where n_1 and n_2 are the refractive indices of the incident and refracting media, respectively. When light is incident at the Brewster angle, the reflected and refracted rays are perpendicular to each other.	

26.5 The Dispersion of Light: Prisms and Rainbows

Dispersion	A glass prism can spread a beam of sunlight into a spectrum of colors because the index of refraction of the glass depends on the wavelength of the light. Thus, a prism bends the refracted rays corresponding to different colors by different amounts. The spreading of light into its color components is known as dispersion. The dispersion of light by water droplets in the air leads to the formation of rainbows.	Concept Simulation 26.2
Index matching	A prism will not bend a light ray at all, neither up nor down, if the surrounding fluid has the same refractive index as the glass, a condition known as index matching.	**Example 7**

 Use *Self-Assessment Test 26.1* **to evaluate your understanding of Sections 26.1–26.5.**

26.6 Lenses

26.7 The Formation of Images by Lenses

Focal point and focal length of a converging lens **Focal point and focal length of a diverging lens** **Ray tracing**	Converging lenses and diverging lenses depend on the phenomenon of refraction in forming an image. With a converging lens, paraxial rays that are parallel to the principal axis are focused to a point on the axis by the lens. This point is called the focal point of the lens, and its distance from the lens is the focal length f. Paraxial light rays that are parallel to the principal axis of a diverging lens appear to originate from its focal point after passing through the lens. The distance of this point from the lens is the focal length f. The image produced by a converging or a diverging lens can be located via a technique known as ray tracing, which utilizes the three rays outlined in the Reasoning Strategy given in Section 26.7.	
Image formed by a converging lens	The nature of the image formed by a converging lens depends on where the object is situated relative to the lens. When the object is located at a distance from the lens that is greater than twice the focal length, the image is real, inverted, and smaller than the object. When the object is located at a distance from the lens that is between the focal length and twice the focal length, the image is real, inverted, and larger than the object. When the object is located within the focal length, the image is virtual, upright, and larger than the object.	Concept Simulation 26.3
Image formed by a diverging lens	Regardless of the position of a real object, a diverging lens always produces an image that is virtual, upright, and smaller than the object.	Concept Simulation 26.4

26.8 The Thin-Lens Equation and the Magnification Equation

	The thin-lens equation can be used with either converging or diverging lenses that are thin, and it relates the object distance d_o, the image distance d_i, and the focal length f of the lens:	
Thin-lens equation	$$\frac{1}{d_o} + \frac{1}{d_i} = \frac{1}{f} \qquad (26.6)$$	Interactive Solution 26.51
Magnification	The magnification m of a lens is the ratio of the image height h_i to the object height h_o and is also related to d_o and d_i by the magnification equation:	**Examples 8, 9**
Magnification equation	$$m = \frac{h_i}{h_o} = -\frac{d_i}{d_o} \qquad (26.7)$$	
	The algebraic sign conventions for the variables appearing in the thin-lens and magnification equations are summarized in the Reasoning Strategy given in Section 26.8.	

26.9 Lenses in Combination

	When two or more lenses are used in combination, the image produced by one lens serves as the object for the next lens.	**Examples 10, 18** Interactive LearningWare 26.2

Topic	Discussion	Learning Aids

26.10 The Human Eye

Accommodation

Near point

Far point

In the human eye, a real, inverted image is formed on a light-sensitive surface, called the retina. Accommodation is the process by which the focal length of the eye is automatically adjusted, so that objects at different distances produce sharp images on the retina. The near point of the eye is the point nearest the eye at which an object can be placed and still have a sharp image produced on the retina. The far point of the eye is the location of the farthest object on which the fully relaxed eye can focus. For a young and normal eye, the near point is located 25 cm from the eye, and the far point is located at infinity.

Nearsightedness

Farsightedness

A nearsighted (myopic) eye is one that can focus on nearby objects, but not on distant objects. Nearsightedness can be corrected with eyeglasses or contacts made from diverging lenses. A farsighted (hyperopic) eye can see distant objects clearly, but not objects close up. Farsightedness can be corrected with converging lenses.

Example 11

Example 12

Interactive Solution 26.67

The refractive power of a lens is measured in diopters and is given by

Refractive power

$$\text{Refractive power (in diopters)} = \frac{1}{f \text{ (in meters)}} \qquad (26.8)$$

where f is the focal length of the lens and must be expressed in meters. A converging lens has a positive refractive power, and a diverging lens has a negative refractive power.

26.11 Angular Magnification and the Magnifying Glass

The angular size of an object is the angle that it subtends at the eye of the viewer. For small angles, the angular size θ in radians is

Angular size

$$\theta \text{ (in radians)} \approx \frac{h_o}{d_o}$$

Example 13

where h_o is the height of the object and d_o is the object distance. The angular magnification M of an optical instrument is the angular size θ' of the final image produced by the instrument divided by the reference angular size θ of the object, which is that seen without the instrument:

Angular magnification

$$M = \frac{\theta'}{\theta} \qquad (26.9)$$

A magnifying glass is usually a single converging lens that forms an enlarged, upright, and virtual image of an object placed at or inside the focal point of the lens. For a magnifying glass held close to the eye, the angular magnification M is approximately

Angular magnification of a magnifying glass

$$M \approx \left(\frac{1}{f} - \frac{1}{d_i} \right) N \qquad (26.10)$$

Example 14

where f is the focal length of the lens, d_i is the image distance, and N is the distance of the viewer's near point from the eye.

26.12 The Compound Microscope

A compound microscope usually consists of two lenses, an objective and an eyepiece. The final image is enlarged, inverted, and virtual. The angular magnification M of such a microscope is approximately

Angular magnification of a compound microscope

$$M \approx -\frac{(L - f_e)N}{f_o f_e} \qquad (L > f_o + f_e) \qquad (26.11)$$

Example 15

where f_o and f_e are, respectively, the focal lengths of the objective and eyepiece, L is the distance between the two lenses, and N is the distance of the viewer's near point from his or her eye.

26.13 The Telescope

An astronomical telescope magnifies distant objects with the aid of an objective and an eyepiece, and it produces a final image that is inverted and virtual. The

Topic	Discussion	Learning Aids
Angular magnification of an astronomical telescope	angular magnification M of a telescope is approximately $$M \approx -\frac{f_o}{f_e}$$ where f_o and f_e are, respectively, the focal lengths of the objective and eyepiece.	(26.12) **Example 16**

26.14 Lens Aberrations

Spherical aberration	Lens aberrations limit the formation of perfectly focused or sharp images by optical instruments. Spherical aberration occurs because rays that pass through the outer edge of a lens with spherical surfaces are not focused at the same point as those that pass through near the center of the lens.	
Chromatic aberration	Chromatic aberration arises because a lens focuses different colors at slightly different points.	

   Use *Self-Assessment Test 24.2* **to evaluate your understanding of Sections 26.6–26.14.**

Conceptual Questions

1. In Figure 26.2a, suppose that a layer of oil were added on top of the water. The angle θ_1 at which the incident light travels through the air remains the same. Assuming that light still enters the water, does the angle of refraction at which it does so change because of the presence of the oil? Explain.

2. Two slabs with parallel faces are made from different types of glass. A ray of light travels through air and enters each slab at the same angle of incidence, as the drawing shows. Which slab has the greater index of refraction? Why?

Question 2

3. When an observer peers over the edge of a deep empty bowl, he does not see the entire bottom surface, so a small object lying on the bottom is hidden from view. However, when the bowl is filled with water, the object can be seen. Explain this effect.

4. Two identical containers, one filled with water ($n = 1.33$) and the other filled with ethyl alcohol ($n = 1.36$), are viewed from directly above. Which container (if either) appears to have a greater depth of fluid? Why?

5. When you look through an aquarium window at a fish, is the fish as close as it appears? Explain.

6. At night, when it's dark outside and you are standing in a brightly lit room, it is easy to see your reflection in a window. During the day it is not so easy. Account for these facts.

7. A man is fishing from a dock. (a) If he is using a bow and arrow, should he aim above the fish, at the fish, or below the fish, to strike it? (b) How would he aim if he were using a laser gun? Give your reasoning.

8. Two rays of light converge to a point on a screen. A plane-parallel plate of glass is placed in the path of this converging light, and the glass plate is parallel to the screen. Will the point of convergence remain on the screen? If not, will the point move toward the glass or away from it? Justify your answer by drawing a diagram and showing how the rays are affected by the glass.

9. A person sitting at the beach is wearing a pair of Polaroid sunglasses and notices little discomfort due to the glare from the water on a bright sunny day. When she lies on her side, however, she notices that the glare increases. Why?

10. You are sitting by the shore of a lake on a sunny and windless day. Your Polaroid sunglasses are not equally effective at all times of the day in reducing the glare of the sunlight reflected from the lake. Account for this observation.

11. The dispersion of sunlight by a prism is discussed in connection with Figure 26.18. Is it possible to direct a mixture of colors into a prism and have sunlight emerge from it? If so, explain how this could be done.

12. Refer to Figure 26.7. Note that the ray within the glass slab is traveling from a medium with a larger refractive index toward a medium with a smaller refractive index. Is it possible, for θ_1 less than 90°, that the ray within the glass will experience total internal reflection at the glass–air interface? Account for your answer.

13. Suppose you want to make a rainbow by spraying water from a garden hose into the air. (a) Where must you stand relative to the water and the sun to see the rainbow? (b) Why can't you ever walk under the rainbow?

14. A person is floating on an air mattress in the middle of a swimming pool. His friend is sitting on the side of the pool. The person on the air mattress claims that there is a light shining up from the bottom of the pool directly beneath him. His friend insists, however, that she cannot see any light from where she sits on the side. Can both individuals be correct? Give your reasoning.

15. A beam of blue light is propagating in glass. When the light reaches the boundary between the glass and the surrounding air, the beam is totally reflected back into the glass. However, red light with the same angle of incidence is not totally reflected, and some of the light is refracted into the air. Why do these two colors behave differently?

16. A beacon in a lighthouse is to produce a parallel beam of light. The beacon consists of a light bulb and a converging lens. Should the bulb be placed outside the focal point, at the focal point, or inside the focal point of the lens? State your reason.

17. Review Conceptual Example 7 as an aid in answering this question. Is it possible that a converging lens (in air) behaves as a diverging lens when surrounded by another medium? Give a reason for your answer.

18. A spherical mirror and a lens are immersed in water. Compared to the way they work in air, which one do you expect will be more affected by the water? Why?

19. A converging lens is used to project a real image onto a screen, as in Figure 26.27b. A piece of black tape is then placed over the upper half of the lens. Will only the lower half of the image be visible on the screen? Justify your answer by drawing rays from various points on the object to the corresponding points on the image.

20. Review Conceptual Example 7 as an aid in answering this question. A converging lens is made from glass whose index of refraction is *n*. The lens is surrounded by a fluid whose index of refraction is also *n*. Can this lens still form an image, either real or virtual, of an object? Why?

21. In a TV mystery program, a photographic negative is introduced as evidence in a court trial. The negative shows an image of a house (now burned down) that was the scene of the crime. At the trial the defendant's acquittal depends on knowing exactly how far above the ground a window was. An expert called by the defense claims that this height can be calculated from only two pieces of information: (1) the measured height on the film, and (2) the focal length of the camera lens. Explain whether the expert is making sense, using the thin-lens and magnification equations to guide your thinking.

22. Two people who wear glasses are camping. One is nearsighted and the other is farsighted. Whose glasses may be useful in starting a fire with the sun's rays? Give your reasoning.

23. Suppose that a 21-year-old with normal vision (near point = 25 cm) is standing in front of a plane mirror. How close can he stand to the mirror and still see himself in focus? Explain.

24. ☤ If we read for a long time, our eyes become "tired." When this happens, it helps to stop reading and look at a distant object. From the point of view of the ciliary muscle, why does this refresh the eyes?

25. To a swimmer under water, objects look blurred and out of focus. However, when the swimmer wears goggles that keep the water away from the eyes, the objects appear sharp and in focus. Why do goggles improve a swimmer's underwater vision?

26. ☤ The refractive power of the lens of the eye is 15 diopters when surrounded by the aqueous and vitreous humors. If this lens is removed from the eye and surrounded by air, its refractive power increases to about 150 diopters. Why is the refractive power of the lens so much greater outside the eye?

27. The light shining through a full glass of wine forms an irregularly shaped bright spot on the table, but does not do so when the glass is empty. Explain.

28. Jupiter is the largest planet in our solar system. Yet, to the naked eye, it looks smaller than Venus. Why?

29. Try using **Concept Simulations 26.3** and **26.4** at **www.wiley. com/college/cutnell** to guide your reasoning here. By means of a ray diagram, show that the eyes of a person wearing glasses appear to be (a) smaller when the glasses use diverging lenses to correct for nearsightedness and (b) larger when the glasses use converging lenses to correct for farsightedness.

30. Can a diverging lens be used as a magnifying glass? Justify your answer with a ray diagram.

31. Who benefits more from using a magnifying glass, a person whose near point is located 25 cm away from the eyes or a person whose near point is located 75 cm away from the eyes? Provide a reason for your answer.

32. Two lenses, whose focal lengths are 3.0 and 45 cm, are used to build a telescope. Which lens should be the objective? Why?

33. Two refracting telescopes have identical eyepieces, although one telescope is twice as long as the other. Which has the greater angular magnification? Provide a reason for your answer.

34. Suppose a well-designed optical instrument is composed of two converging lenses separated by 14 cm. The focal lengths of the lenses are 0.60 and 4.5 cm. Is the instrument a microscope or a telescope? Why?

35. It is often thought that virtual images are somehow less important than real images. To show that this is not true, identify which of the following instruments normally produce final images that are virtual: (a) a projector, (b) a camera, (c) a magnifying glass, (d) eyeglasses, (e) a compound microscope, and (f) an astronomical telescope.

36. Why does chromatic aberration occur in lenses, but not in mirrors?

Problems

Unless specified otherwise, use the values given in Table 26.1 for the refractive indices.

ssm Solution is in the Student Solutions Manual.   **www** Solution is available on the World Wide Web at www.wiley.com/college/cutnell
☤ This icon represents a biomedical application.

Section 26.1 The Index of Refraction

1. ssm What is the speed of light in benzene?

2. The refractive indices of materials *A* and *B* have a ratio of $n_A/n_B = 1.33$. The speed of light in material *A* is 1.25×10^8 m/s. What is the speed of light in material *B*?

3. The frequency of a light wave is the same when the light travels in ethyl alcohol as it is when it travels in carbon disulfide. Find the ratio of the wavelength of the light in ethyl alcohol to that in carbon disulfide.

4. The speed of light is 1.25 times as large in material *A* as in material *B*. Determine the ratio n_A/n_B of the refractive indices of these materials.

5. ssm www A plate glass window ($n = 1.5$) has a thickness of 4.0×10^{-3} m. How long does it take light to pass perpendicularly through the plate?

6. Light has a wavelength of 340.0 nm and a frequency of 5.403×10^{14} Hz when traveling through a certain substance. What substance from Table 26.1 could this be?

* 7. **Interactive Solution 26.7** at **www.wiley.com/college/cutnell** offers one model for problems like this one. In a certain time, light travels 6.20 km in a vacuum. During the same time, light travels only 3.40 km in a liquid. What is the refractive index of the liquid?

* 8. A flat sheet of ice has a thickness of 2.0 cm. It is on top of a flat sheet of crystalline quartz that has a thickness of 1.1 cm. Light strikes the ice perpendicularly and travels through it and then through the quartz. In the time it takes the light to travel through the two sheets, how far (in cm) would it have traveled in a vacuum?

Section 26.2 Snell's Law and the Refraction of Light

9. ssm A light ray in air is incident on a water surface at a 43° angle of incidence. Find (a) the angle of reflection and (b) the angle of refraction.

10. A ray of light traveling in material A strikes the interface between materials A and B at an angle of incidence of 72°. The angle of refraction is 56°. Find the ratio n_A/n_B of the refractive indices of the two materials.

11. As an aid in understanding this problem, refer to Conceptual Example 4. A swimmer, who is looking up from under the water, sees a diving board directly above at an apparent height of 4.0 m above the water. What is the actual height of the diving board?

12. A person working on the transmission of a car accidentally drops a bolt into a tray of oil. The oil is 5.00 cm deep. The bolt appears to be 3.40 cm beneath the surface of the oil, when viewed from directly above. What is the index of refraction of the oil?

13. ssm The drawing shows a coin resting on the bottom of a beaker filled with an unknown liquid. A ray of light from the coin travels to the surface of the liquid and is refracted as it enters into the air. A person sees the ray as it skims just above the surface of the liquid. How fast is the light traveling in the liquid?

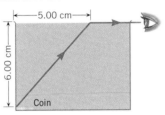

14. A ray of sunlight hits a frozen lake at a 45° angle of incidence. At what angle of refraction does the ray penetrate (a) the ice and (b) the water beneath the ice?

15. Light in a vacuum is incident on a transparent glass slab. The angle of incidence is 35.0°. The slab is then immersed in a pool of liquid. When the angle of incidence for the light striking the slab is 20.3°, the angle of refraction for the light entering the slab is the same as when the slab was in a vacuum. What is the index of refraction of the liquid?

* 16. A silver medallion is sealed within a transparent block of plastic. An observer in air, viewing the medallion from directly above, sees the medallion at an apparent depth of 1.6 cm beneath the top surface of the block. How far below the top surface would the medallion appear if the observer (not wearing goggles) and the block were under water?

* 17. ssm www In Figure 26.7, suppose that the angle of incidence is $\theta_1 = 30.0°$, the thickness of the glass pane is 6.00 mm, and the refractive index of the glass is $n_2 = 1.52$. Find the amount (in mm) by which the emergent ray is displaced relative to the incident ray.

* 18. Review Conceptual Example 4 as background for this problem. A man in a boat is looking straight down at a fish in the water directly beneath him. The fish is looking straight up at the man. They are equidistant from the air–water interface. To the man, the fish appears to be 2.0 m beneath his eyes. To the fish, how far above its eyes does the man appear to be?

* 19. Refer to Figure 26.5a and assume the observer is nearly above the submerged object. For this situation, derive the expression for the apparent depth: $d' = d(n_2/n_1)$, Equation 26.3. (*Hint: Use Snell's law of refraction and the fact that the angles of incidence and refraction are small, so tan $\theta \approx$ sin θ.*)

* 20. The drawing shows a rectangular block of glass ($n = 1.52$) surrounded by liquid carbon disulfide ($n = 1.63$). A ray of light is incident on the glass at point A with a 30.0° angle of incidence. At what angle of refraction does the ray leave the glass at point B?

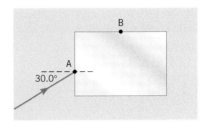

** 21. ssm A small logo is embedded in a thick block of crown glass ($n = 1.52$), 3.20 cm beneath the top surface of the glass. The block is put under water, so there is 1.50 cm of water above the top surface of the block. The logo is viewed from directly above by an observer in air. How far beneath the top surface of the water does the logo appear to be?

** 22. A beaker has a height of 30.0 cm. The lower half of the beaker is filled with water, and the upper half is filled with oil ($n = 1.48$). To a person looking down into the beaker from above, what is the apparent depth of the bottom?

Section 26.3 Total Internal Reflection

23. ssm One method of determining the refractive index of a transparent solid is to measure the critical angle when the solid is in air. If θ_c is found to be 40.5°, what is the index of refraction of the solid?

24. What is the critical angle for light emerging from carbon disulfide into air?

25. Interactive Solution 26.25 at **www.wiley.com/college/cutnell** provides one model for solving problems such as this. A glass block ($n = 1.56$) is immersed in a liquid. A ray of light within the glass hits a glass–liquid surface at a 75.0° angle of incidence. Some of the light enters the liquid. What is the smallest possible refractive index for the liquid?

26. A point source of light is submerged 2.2 m below the surface of a lake and emits rays in all directions. On the surface of the lake, directly above the source, the area illuminated is a circle. What is the maximum radius that this circle could have?

27. ssm The drawing shows a crown glass slab with a rectangular cross section. As illustrated, a laser beam strikes the upper surface at an angle of 60.0°. After reflecting from the upper surface, the beam reflects from the side and bottom surfaces. (a) If the glass is surrounded by air, determine where part of the beam first exits the glass, at point A, B, or C. (b) Repeat part (a), assuming that the glass is surrounded by water.

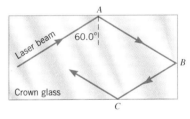

28. A layer of liquid B floats on liquid A. A ray of light begins in liquid A and undergoes total internal reflection at the interface between the liquids when the angle of incidence exceeds 36.5°. When liquid B is replaced with liquid C, total internal reflection occurs for angles of incidence greater than 47.0°. Find the ratio n_B/n_C of the refractive indices of liquids B and C.

* 29. The drawing shows an optical fiber that consists of a core made of flint glass ($n_{flint} = 1.667$) surrounded by a cladding made of crown glass ($n_{crown} = 1.523$). A beam of light enters

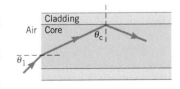

the fiber from air at an angle θ_1 with respect to the normal. What is θ_1 if the light strikes the core–cladding interface at the critical angle θ_c?

30. Interactive Learning Ware 26.1 at **www.wiley. com/college/cutnell** provides some helpful background for this problem. The drawing shows a crystalline quartz slab with a rectangular cross section. A ray of light

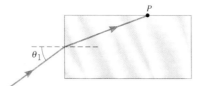

strikes the slab at an incident angle of $\theta_1 = 34°$, enters the quartz, and travels to point P. This slab is surrounded by a fluid with a refractive index n. What is the maximum value of n such that total internal reflection occurs at point P?

Section 26.4 Polarization and the Reflection and Refraction of Light

31. ssm www When light strikes the surface between two materials from above, the Brewster angle is 65.0°. What is the Brewster angle when the light encounters the same surface from below?

32. Sunlight strikes a diamond surface. At what angle of incidence is the reflected light completely polarized?

33. Light is incident from air onto a beaker of carbon tetrachloride. If the reflected light is 100% polarized, what is the angle of refraction of the light that penetrates into the carbon tetrachloride?

34. For light that originates within a liquid and strikes the liquid–air interface, the critical angle is 39°. What is Brewster's angle for this light?

35. ssm Light is reflected from a glass coffee table. When the angle of incidence is 56.7°, the reflected light is completely polarized parallel to the surface of the glass. What is the index of refraction of the glass?

36. When red light in a vacuum is incident at the Brewster angle on a certain type of glass, the angle of refraction is 29.9°. What are (a) the Brewster angle and (b) the index of refraction of the glass?

37. In Section 26.4 it is mentioned that the reflected and refracted rays are perpendicular to each other when light strikes the surface at the Brewster angle. This is equivalent to saying that the angle of reflection plus the angle of refraction is 90°. Using Snell's law and Brewster's law, prove that the angle of reflection plus the angle of refraction is 90°.

Section 26.5 The Dispersion of Light: Prisms and Rainbows

38. A ray of sunlight is passing from diamond into crown glass; the angle of incidence is 35.00°. The indices of refraction for the blue and red components of the ray are: blue ($n_{diamond} = 2.444$, $n_{crown\ glass} = 1.531$), and red ($n_{diamond} = 2.410$, $n_{crown\ glass} = 1.520$). Determine the angle between the refracted blue and red rays in the crown glass.

39. ssm A beam of sunlight encounters a plate of crown glass at a 45.00° angle of incidence. Using the data in Table 26.2, find the angle between the violet ray and the red ray in the glass.

40. Red light ($n = 1.520$) and violet light ($n = 1.538$) traveling in air are incident on a slab of crown glass. Both colors enter the glass at the same angle of refraction. The red light has an angle of incidence of 30.00°. What is the angle of incidence of the violet light?

41. Horizontal rays of red light ($\lambda = 660$ nm, in vacuum) and violet light ($\lambda = 410$ nm, in vacuum) are incident on the flint-glass prism shown in the drawing. The indices of refraction for the red and violet light are 1.662 and 1.698, respectively. What is the angle of refraction for each ray as it emerges from the prism?

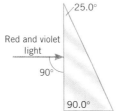

42. Refer to Conceptual Example 7 as background material for this problem. The drawing shows a horizontal beam of light that is incident on a prism made from ice. The base of the prism is also horizontal.

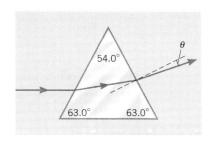

The prism ($n = 1.31$) is surrounded by oil whose index of refraction is 1.48. Determine the angle θ that the exiting light makes with the normal to the right face of the prism.

43. ssm This problem relates to Figure 26.18 which illustrates the dispersion of light by a prism. The prism is made from glass, and its cross section is an equilateral triangle. The indices of refraction for the red and violet light are 1.662 and 1.698, respectively. The angle of incidence for both the red and violet light is 60.0°. Find the angles of refraction at which the red and violet rays emerge into the air from the prism.

Section 26.6 Lenses,
Section 26.7 The Formation of Images by Lenses,
Section 26.8 The Thin-Lens Equation and the Magnification Equation

(Note: When drawing ray diagrams, be sure that the object height h_o is much smaller than the focal length f of the lens or mirror.)

44. A diverging lens has a focal length of -32 cm. An object is placed 19 cm in front of this lens. Calculate (a) the image distance and (b) the magnification. Is the image (c) real or virtual, (d) upright or inverted, and (e) enlarged or reduced in size?

45. ssm An object is located 9.0 cm in front of a converging lens ($f = 6.0$ cm). Using an accurately drawn ray diagram, determine where the image is located.

46. When a diverging lens is held 13 cm above a line of print, as in Figure 26.29, the image is 5.0 cm beneath the lens. What is the focal length of the lens?

47. A tourist takes a picture of a mountain 14 km away using a camera that has a lens with a focal length of 50 mm. She then takes a second picture when she is only 5.0 km away. What is the ratio of the height of the mountain's image on the film for the second picture to its height on the film for the first picture?

48. Concept Simulation 26.3 at **www.wiley.com/college/cutnell** reviews the concepts that play a role in this problem. A converging lens has a focal length of 88.00 cm. A 13.0-cm-tall object is located 155.0 cm in front of this lens. (a) What is the image distance? (b) Is the image real or virtual? (c) What is the image height? Be sure to include the proper algebraic sign.

49. ssm An object is located 30.0 cm to the left of a converging lens whose focal length is 50.0 cm. (a) Draw a ray diagram to scale and from it determine the image distance and the magnification. (b) Use the thin-lens and magnification equations to verify your answers to part (a).

50. A slide projector has a converging lens whose focal length is 105.00 mm. (a) How far (in meters) from the lens must the screen be located if a slide is placed 108.00 mm from the lens? (b) If the slide measures 24.0 mm × 36.0 mm, what are the dimensions (in mm) of its image?

51. Consult **Interactive Solution 26.51** at **www.wiley.com/college/ cutnell** to review the concepts on which this problem depends. A camera is supplied with two interchangeable lenses, whose focal lengths are 35.0 and 150.0 mm. A woman whose height is 1.60 m stands 9.00 m in front of the camera. What is the height (including sign) of her image on the film, as produced by (a) the 35.0-mm lens and (b) the 150.0-mm lens?

* **52. Concept Simulation 26.4** at **www.wiley.com/college/cutnell** provides the option of exploring the ray diagram that applies to this problem. The distance between an object and its image formed by a diverging lens is 49.0 cm. The focal length of the lens is −233.0 cm. Find (a) the image distance and (b) the object distance.

* **53. ssm** An office copier uses a lens to place an image of a document onto a rotating drum. The copy is made from this image. (a) What kind of lens is used? If the document and its copy are to have the same size, but are inverted with respect to one another, (b) how far from the document is the lens located and (c) how far from the lens is the image located? Express your answers in terms of the focal length f of the lens.

* **54.** An object is in front of a converging lens ($f = 0.30$ m). The magnification of the lens is $m = 4.0$. (a) Relative to the lens, in what direction should the object be moved so that the magnification changes to $m = -4.0$? (b) Through what distance should the object be moved?

* **55. ssm** An object is 18 cm in front of a diverging lens that has a focal length of −12 cm. How far in front of the lens should the object be placed so that the size of its image is reduced by a factor of 2.0?

* **56.** From a distance of 72 m, a photographer uses a telephoto lens ($f = 300.0$ mm) to take a picture of a charging rhinoceros. How far from the rhinoceros would the photographer have to be to record an image of the same size using a lens whose focal length is 50.0 mm?

** **57.** A converging lens ($f = 25.0$ cm) is used to project an image of an object onto a screen. The object and the screen are 125 cm apart, and between them the lens can be placed at either of two locations. Find the two object distances.

** **58.** The equation

$$\frac{1}{d_o} + \frac{1}{d_i} = \frac{1}{f}$$

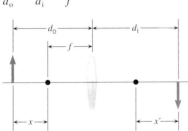

is called the *Gaussian* form of the thin-lens equation. The drawing shows the variables d_o, d_i, and f. The drawing also shows the distances x and x', which are, respectively, the distance from the object to the focal point on the left of the lens and the distance from the focal point on the right of the lens to the image. An equivalent form of the thin-lens equation, involving x, x', and f, is called the *Newtonian* form. Show that the Newtonian form of the thin-lens equation can be written as $xx' = f^2$.

Section 26.9 Lenses in Combination

59. A converging lens has a focal length of 0.080 m. An object is located 0.040 m to the left of this lens. A second converging lens has the same focal length as the first one and is located 0.120 m to the right of it. Relative to the second lens, where is the final image located?

60. Two identical diverging lenses are separated by 16 cm. The focal length of each lens is −8.0 cm. An object is located 4.0 cm to the left of the lens that is on the left. Determine the final image distance relative to the lens on the right.

61. ssm A converging lens ($f = 12.0$ cm) is located 30.0 cm to the left of a diverging lens ($f = -6.00$ cm). A postage stamp is placed 36.0 cm to the left of the converging lens. (a) Locate the final image of the stamp relative to the diverging lens. (b) Find the overall magnification. (c) Is the final image real or virtual? With respect to the original object, is the final image (d) upright or inverted, and is it (e) larger or smaller?

62. Interactive LearningWare 26.2 at **www.wiley.com/college/cutnell** offers a review of the concepts that play roles in this problem. A diverging lens ($f = -10.0$ cm) is located 20.0 cm to the left of a converging lens ($f = 30.0$ cm). A 3.00-cm-tall object stands to the left of the diverging lens, exactly at its focal point. (a) Determine the distance of the final image relative to the converging lens. (b) What is the height of the final image (including the proper algebraic sign)?

63. A converging lens ($f_1 = 24.0$ cm) is located 56.0 cm to the left of a diverging lens ($f_2 = -28.0$ cm). An object is placed to the left of the converging lens, and the final image produced by the two-lens combination lies 20.7 cm to the left of the diverging lens. How far is the object from the converging lens?

* **64.** A coin is located 20.0 cm to the left of a converging lens ($f = 16.0$ cm). A second, identical lens is placed to the right of the first lens, such that the image formed by the combination has the same size and orientation as the original coin. Find the separation between the lenses.

* **65. ssm** An object is placed 20.0 cm to the left of a diverging lens ($f = -8.00$ cm). A concave mirror ($f = 12.0$ cm) is placed 30.0 cm to the right of the lens. (a) Find the final image distance, measured relative to the mirror. (b) Is the final image real or virtual? (c) Is the final image upright or inverted with respect to the original object?

** **66.** Two converging lenses ($f_1 = 9.00$ cm and $f_2 = 6.00$ cm) are separated by 18.0 cm. The lens on the left has the longer focal length. An object stands 12.0 cm to the left of the left-hand lens in the combination. (a) Locate the final image relative to the lens on the right. (b) Obtain the overall magnification. (c) Is the final image real or virtual? With respect to the original object, is the final image (d) upright or inverted and is it (e) larger or smaller?

Section 26.10 The Human Eye

67. Interactive Solution 26.67 at **www.wiley.com/college/cutnell** illustrates one approach to solving problems such as this one. A farsighted person has a near point that is 67.0 cm from her eyes. She wears eyeglasses that are designed to enable her to read a newspaper held at a distance of 25.0 cm from her eyes. Find the focal length of the eyeglasses, assuming that they are worn (a) 2.2 cm from the eyes and (b) 3.3 cm from the eyes.

68. A person holds a book 25 cm in front of the effective lens of her eye; the print in the book is 2.0 mm high. If the effective lens of the eye is located 1.7 cm from the retina, what is the size (including the sign) of the print image on the retina?

69. ssm A nearsighted person has a far point located only 220 cm from his eyes. Determine the focal length of contact lenses that will enable him to see distant objects clearly.

70. A student is reading a lecture written on a blackboard. The lenses in her eyes have a refractive power of 57.50 diopters, and the lens-to-retina distance is 1.750 cm. (a) How far (in meters) is the blackboard from her eyes? (b) If the writing on the blackboard is 5.00 cm high, what is the size of the image on her retina?

71. A nearsighted person cannot read a sign that is more than 5.2 m from his eyes. To deal with this problem, he wears contact lenses that do not correct his vision completely, but do allow him to read signs located up to distances of 12.0 m from his eyes. What is the focal length of the contacts?

72. A farsighted woman breaks her current eyeglasses and is using an old pair whose refractive power is 1.660 diopters. Since these eyeglasses do not completely correct her vision, she must hold a newspaper 42.00 cm from her eyes in order to read it. She wears the eyeglasses 2.00 cm from her eyes. How far is her near point from her eyes?

* **73.** ssm At age forty, a man requires contact lenses ($f = 65.0$ cm) to read a book held 25.0 cm from his eyes. At age forty-five, he finds that while wearing these contacts he must now hold a book 29.0 cm from his eyes. (a) By what distance has his near point *changed?* (b) What focal length lenses does he require at age forty-five to read a book at 25.0 cm?

** **74.** The contacts worn by a farsighted person allow her to see objects clearly that are as close as 25.0 cm, even though her uncorrected near point is 79.0 cm from her eyes. When she is looking at a poster, the contacts form an image of the poster at a distance of 217 cm from her eyes. (a) How far away is the poster actually located? (b) If the poster is 0.350 m tall, how tall is the image formed by the contacts?

** **75.** The far point of a nearsighted person is 6.0 m from her eyes, and she wears contacts that enable her to see distant objects clearly. A tree is 18.0 m away and 2.0 m high. (a) When she looks through the contacts at the tree, what is its image distance? (b) How high is the image formed by the contacts?

Section 26.11 Angular Magnification and the Magnifying Glass

76. A jeweler whose near point is 72 cm from his eye uses a magnifying glass as in Figure 26.40b to examine a watch. The watch is held 4.0 cm from the magnifying glass. Find the angular magnification of the magnifying glass.

77. ssm A quarter (diameter = 2.4 cm) is held at arm's length (70.0 cm). The sun has a diameter of 1.39×10^9 m and is 1.50×10^{11} m from the earth. What is the ratio of the angular size of the quarter to that of the sun?

78. An object has an angular size of 0.0150 rad when placed at the near point (21.0 cm) of an eye. When the eye views this object using a magnifying glass, the largest possible angular size of the image is 0.0380 rad. What is the focal length of the magnifying glass?

79. A magnifying glass is held next to the eye, above a magazine. The image formed by the magnifying glass is located at the near point of the eye. The near point is 0.30 m away from the eye, and the angular magnification is 3.4. Find the focal length of the magnifying glass.

80. The near point of a naked eye is 32 cm. When an object is placed at the near point and viewed by the naked eye, it has an angular size of 0.060 rad. A magnifying glass has a focal length of 16 cm, and is held next to the eye. The enlarged image that is seen is located 64 cm from the magnifying glass. Determine the angular size of the image.

* **81.** ssm A person using a magnifying glass as in Figure 26.40b observes that for clear vision its maximum angular magnification is 1.25 times as large as its minimum angular magnification. Assuming that the person has a near point located 25 cm from her eye, what is the focal length of the magnifying glass?

** **82.** A farsighted person can read printing as close as 25.0 cm when she wears contacts that have a focal length of 45.4 cm. One day,

however, she forgets her contacts and uses a magnifying glass, as in Figure 26.40b. It has a maximum angular magnification of 7.50 for a young person with a normal near point of 25.0 cm. What is the maximum angular magnification that the magnifying glass can provide for her?

Section 26.12 The Compound Microscope

83. An insect subtends an angle of only 4.0×10^{-3} rad at the unaided eye when placed at the near point. What is the angular size (magnitude only) when the insect is viewed through a microscope whose angular magnification has a magnitude of 160?

84. An anatomist is viewing heart muscle cells with a microscope that has two selectable objectives with refracting powers of 100 and 300 diopters. When she uses the 100-diopter objective, the image of a cell subtends an angle of 3×10^{-3} rad with the eye. What angle is subtended when she uses the 300-diopter objective?

85. ssm A compound microscope has a barrel whose length is 16.0 cm and an eyepiece whose focal length is 1.4 cm. The viewer has a near point located 25 cm from his eyes. What focal length must the objective have so that the angular magnification of the microscope will be -320?

86. The near point of a naked eye is 25 cm. When placed at the near point and viewed by the naked eye, a tiny object would have an angular size of 5.2×10^{-5} rad. When viewed through a compound microscope, however, it has an angular size of -8.8×10^{-3} rad. (The minus sign indicates that the image produced by the microscope is inverted.) The objective of the microscope has a focal length of 2.6 cm, and the distance between the objective and the eyepiece is 16 cm. Find the focal length of the eyepiece.

* **87.** The maximum angular magnification of a magnifying glass is 12.0 when a person uses it who has a near point that is 25.0 cm from his eyes. The same person finds that a microscope, using this magnifying glass as the eyepiece, has an angular magnification of -525. The separation between the eyepiece and the objective of the microscope is 23.0 cm. Obtain the focal length of the objective.

* **88.** In a compound microscope, the focal length of the objective is 3.50 cm and that of the eyepiece is 6.50 cm. The distance between the lenses is 26.0 cm. (a) What is the angular magnification of the microscope if the person using it has a near point of 35.0 cm? (b) If, as usual, the first image lies just inside the focal point of the eyepiece (see Figure 26.33), how far is the object from the objective? (c) What is the magnification (not the angular magnification) of the objective?

Section 26.13 The Telescope

89. ssm An astronomical telescope has an angular magnification of -184 and uses an objective with a focal length of 48.0 cm. What is the focal length of the eyepiece?

90. An astronomical telescope has an angular magnification of -132. Its objective has a refractive power of 1.50 diopters. What is the refractive power of its eyepiece?

91. Mars subtends an angle of 8.0×10^{-5} rad at the unaided eye. An astronomical telescope has an eyepiece with a focal length of 0.032 m. When Mars is viewed using this telescope, it subtends an angle of 2.8×10^{-3} rad. Find the focal length of the telescope's objective lens.

92. An astronomical telescope for hobbyists has an angular magnification of -155. The eyepiece has a focal length of 5.00 mm. (a) Determine the focal length of the objective. (b) About how long is the telescope?

93. ssm An amateur astronomer decides to build a telescope from a discarded pair of eyeglasses. One of the lenses has a refractive power of 11 diopters, and the other has a refractive power of 1.3 diopters. (a) Which lens should be the objective? (b) How far apart should the lenses be separated? (c) What is the angular magnification of the telescope?

* **94.** The telescope at Yerkes Observatory in Wisconsin has an objective whose focal length is 19.4 m. Its eyepiece has a focal length of 10.0 cm. (a) What is the angular magnification of the telescope? (b) If the telescope is used to look at a lunar crater (diameter = 1500 m), what is the size of the first image, assuming the surface of the moon is 3.77×10^8 m from the surface of the earth? (c) How close does the crater appear to be when seen through the telescope?

* **95.** The objective and eyepiece of an astronomical telescope are 1.25 m apart, and the eyepiece has a focal length of 5.0 cm. What is the angular magnification of the telescope?

** **96.** An astronomical telescope is being used to examine a relatively close object that is only 114.00 m away from the objective of the telescope. The objective and eyepiece have focal lengths of 1.500 and 0.070 m, respectively. Noting that the expression $M \approx -f_o/f_e$ is no longer applicable because the object is so close, use the thin-lens and magnification equations to find the angular magnification of this telescope. *(Hint: See Figure 26.42 and note that the focal points F_o and F_e are so close together that the distance between them may be ignored.)*

Additional Problems

97. ssm A macroscopic (or macro) lens for a camera is usually a converging lens of normal focal length built into a lens barrel that can be adjusted to provide the additional lens-to-film distance needed when focusing at very close range. Suppose that a macro lens ($f = 50.0$ mm) has a maximum lens-to-film distance of 275 mm. How close can the object be located in front of the lens?

98. A microscope for viewing blood cells has an objective with a focal length of 0.50 cm and an eyepiece with a focal length of 2.5 cm. The distance between the objective and eyepiece is 14.0 cm. If a blood cell subtends an angle of 2.1×10^{-5} rad when viewed with the naked eye at a near point of 25.0 cm, what angle (magnitude only) does it subtend when viewed through the microscope?

99. Concept Simulation 26.1 at www.wiley.com/college/cutnell illustrates the concepts that are pertinent to this problem. A ray of light is traveling in glass and strikes a glass–liquid interface. The angle of incidence is 58.0°, and the index of refraction of glass is $n = 1.50$. (a) What must be the index of refraction of the liquid such that the direction of the light entering the liquid is not changed? (b) What is the largest index of refraction that the liquid can have, such that none of the light is transmitted into the liquid and all of it is reflected back into the glass?

100. A spotlight on a boat is 2.5 m above the water, and the light strikes the water at a point that is 8.0 m horizontally displaced from the spotlight (see the drawing). The depth of the water is 4.0 m. Determine the distance d, which locates the point where the light strikes the bottom.

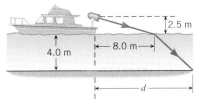

101. ssm A person has far points of 5.0 m from the right eye and 6.5 m from the left eye. Write a prescription for the refractive power of each corrective contact lens.

102. Amber ($n = 1.546$) is a transparent brown-yellow fossil resin. An insect, trapped and preserved within the amber, appears to be 2.5 cm beneath the surface when viewed directly from above. How far below the surface is the insect actually located?

103. A diverging lens has a focal length of -25 cm. (a) Find the image distance when an object is placed 38 cm from the lens. (b) Is the image real or virtual?

104. To focus a camera on objects at different distances, the converging lens is moved toward or away from the film, so a sharp image always falls on the film. A camera with a telephoto lens

($f = 200.0$ mm) is to be focused on an object located first at a distance of 3.5 m and then at 50.0 m. Over what distance must the lens be movable?

105. ssm A beam of light is traveling in air and strikes a material. The angles of incidence and refraction are 63.0° and 47.0°, respectively. Obtain the speed of light in the material.

106. A spectator, seated in the left field stands, is watching a 1.9-m-tall baseball player who is 75 m away. On a TV screen, located 3.0 m from a person watching the game at home, the same player has a 0.12-m image. Find the angular size of the player as seen by (a) the spectator watching the game live and (b) the TV viewer. (c) To whom does the player appear to be larger?

107. ssm An optometrist prescribes contact lenses that have a focal length of 55.0 cm. (a) Are the lenses converging or diverging, and (b) is the person who wears them nearsighted or farsighted? (c) Where is the unaided near point of the person located, if the lenses are designed so that objects no closer than 35.0 cm can be seen clearly?

108. Refer to Conceptual Example 7 as an aid in understanding this problem. The drawing shows a ray of light traveling through a gas ($n = 1.00$), a solid ($n = 1.55$), and a liquid ($n = 1.55$). At what angle θ does the light enter the liquid?

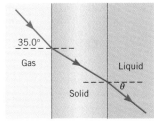

109. A camper is trying to start a fire by focusing sunlight onto a piece of paper. The diameter of the sun is 1.39×10^9 m, and its mean distance from the earth is 1.50×10^{11} m. The camper is using a converging lens whose focal length is 10.0 cm. (a) What is the area of the sun's image on the paper? (b) If 0.530 W of sunlight pass through the lens, what is the intensity of the sunlight at the paper?

110. (a) For a diverging lens ($f = -20.0$ cm), construct a ray diagram to scale and find the image distance for an object that is 20.0 cm from the lens. (b) Determine the magnification of the lens from the diagram.

* **111. ssm www** The moon's diameter is 3.48×10^6 m, and its mean distance from the earth is 3.85×10^8 m. The moon is being photographed by a camera whose lens has a focal length of 50.0 mm. (a) Find the diameter of the moon's image on the slide film. (b) When the slide is projected onto a screen that is 15.0 m from the lens of the projector ($f = 110.0$ mm), what is the diameter of the moon's image on the screen?

* **112.** A stamp collector is viewing a stamp with a magnifying glass held next to her eye. Her near point is 25 cm from her eye. (a) What is the refractive power of a magnifying glass that has an angular magnification of 6.0 when the image of the stamp is located at the near point? (b) What is the angular magnification when the image of the stamp is 45 cm from the eye?

* **113.** A nearsighted person wears contacts to correct for a far point that is only 3.62 m from his eyes. The near point of his unaided eyes is 25.0 cm from his eyes. If he does not remove the lenses when reading, how close can he hold a book and see it clearly?

* **114.** When a converging lens is used in a camera (as in Figure 26.26b), the film must be placed at a distance of 0.210 m from the lens to record an image of an object that is 4.00 m from the lens. The same lens is then used in a projector (see Figure 26.27b), with the screen 0.500 m from the lens. How far from the projector lens should the film be placed?

** **115. ssm** The angular magnification of a telescope is 32 800 times as large when you look through the correct end of the telescope as when you look through the wrong end. What is the angular magnification of the telescope?

** **116.** An object is 20.0 cm from a converging lens, and the image falls on a screen. When the object is moved 4.00 cm closer to the lens, the screen must be moved 2.70 cm farther away from the lens to register a sharp image. Determine the focal length of the lens.

** **117.** Bill is farsighted and has a near point located 125 cm from his eyes. Anne is also farsighted, but her near point is 75.0 cm from her eyes. Both have glasses that correct their vision to a normal near point (25.0 cm from the eyes), and both wear the glasses 2.0 cm from the eyes. Relative to the eyes, what is the closest object that can be seen clearly (a) by Anne when she wears Bill's glasses and (b) by Bill when he wears Anne's glasses?

Concepts & Calculations *Group Learning Problems*

Note: Each of these problems consists of Concept Questions followed by a related quantitative Problem. They are designed for use by students working alone or in small learning groups. The Concept Questions involve little or no mathematics and are intended to stimulate group discussions. They focus on the concepts with which the problems deal. Recognizing the concepts is the essential initial step in any problem-solving technique.

118. Concept Question The drawing shows four different situations in which a light ray is traveling from one medium into another. Without doing any calculations, but taking note of the relative sizes of the angles of incidence and refraction, decide which situations (if any) show a refraction that is physically possible. Provide a reason as to why the refraction is possible or impossible.

Problem For the first three cases, the angle of incidence is 55°; for the fourth case, the angle of incidence is 0°. For each case, determine the angle of refraction. Check to be sure that your answers are consistent with your answers to the Concept Question.

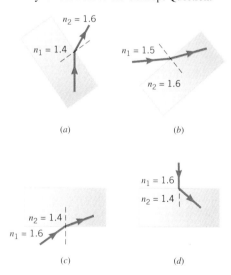

119. Concept Question The drawing shows a ray of light traveling through three materials whose surfaces are parallel to each other. The refracted rays (but not the reflected rays) are shown as the light passes through each material. Taking into account the relative sizes

of the angles of incidence and refraction, rank the materials according to their indices of refraction, greatest first. Provide reasons for your ranking.

Problem A ray of light strikes the *a–b* interface at a 50.0° angle of incidence. The index of refraction of material *a* is $n_a = 1.20$. The angles of refraction in materials *b* and *c* are, respectively, 45.0° and 56.7°. Find the indices of refraction in these two media. Verify that your answers are consistent with your answers to the Concept Question.

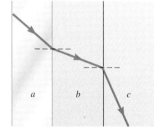

120. Concept Question The drawing shows three materials, *a*, *b*, and *c*. A ray of light strikes the *a–b* interface at an angle of incidence that just barely exceeds its critical angle. The reflected ray then strikes the *a–c* interface at an angle of incidence that just barely exceeds its critical angle. Rank the three materials according to their indices of refraction, largest first.

Problem A ray of light is incident at the *a–b* interface with an angle of incidence that just barely exceeds the critical angle of $\theta_c = 40.0°$. The index of refraction of material *a* is $n_a = 1.80$. Find the indices of refraction for the two other materials. Be sure your ranking of the indices is consistent with that determined in the Concept Question.

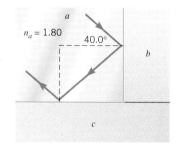

121. Concept Question The drawing shows three layers of different materials, with air above and below the layers. The interfaces between the layers are parallel. The index of refraction of each layer is given in the drawing. Identical rays of light are sent into the layers, and each ray zigzags through the layer, reflecting from the top and bottom surfaces. Fill in the table on page 819, specifying a "yes" or "no" as to whether total internal reflection is possible from the top and bottom surfaces of each layer. Provide a reason for each of your answers.

	Is total internal reflection possible?	
Layer	Top surface of layer	Bottom surface of layer
a		
b		
c		

Problem For each layer, the ray of light has an angle of incidence of 75.0°. For the cases in which total internal reflection is possible from either the top or bottom surface of a layer, determine the amount by which the angle of incidence exceeds the critical angle.

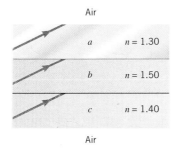

Air

a $n = 1.30$

b $n = 1.50$

c $n = 1.40$

Air

122. Concept Questions An object is placed to the left of a lens, and a real image is formed to the right of the lens. The image is inverted relative to the object and is one-half the size of the object. (a) What kind of lens, converging or diverging, is used to produce this image? (b) How is the height h_i of the image related to the height h_o of the object? Don't forget to take into account the fact that the image is inverted relative to the object. (c) What is the ratio d_i/d_o of the image distance to the object distance?

Problem For the situation described in the Concept Questions, the distance between the object and the image is 90.0 cm. (a) How far from the lens is the object? (b) What is the focal length of the lens?

123. Concept Question Two systems are formed from a converging lens and a diverging lens, as shown in parts *a* and *b* of the drawing. (The point labeled "F" is the focal point of the converging lens.) An object is placed inside the focal point of lens 1. Without doing any

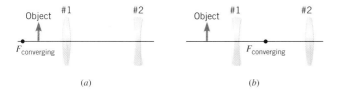

(a)   (b)

calculations, determine for each system whether the final image lies to the left or to the right of lens 2. Provide a reason for each answer.

Problem The focal lengths of the converging and diverging lenses are 15.00 and −20.0 cm, respectively. The distance between the lenses is 50.0 cm, and an object is placed 10.00 cm to the left of lens 1. Determine the final image distance for each system, measured with respect to lens 2. Check to be sure your answers are consistent with your answers to the Concept Question.

* **124. Concept Questions** The back wall of a home aquarium is a mirror that is a distance L away from the front wall. The walls of the tank are negligibly thin. A fish, swimming midway between the front and back walls, is being viewed by a person looking through the front wall. (a) Does the fish appear to be at a distance greater than, less than, or equal to $\frac{1}{2}L$ from the front wall? Express your answer in terms of L. (b) The mirror forms an image of the fish. How far from the front wall is this image located? (c) Assume that your answer to Question (b) is a distance D. Does the image of the fish appear to be at a distance greater than, less than, or equal to D from the front wall? (d) Could the image of the fish appear to be in front of the mirror if the index of refraction of water were different than it actually is, and, if so, would the index of refraction have to be greater than or less than its actual value? Explain each of your answers.

Problem The distance between the back and front walls of the aquarium is 40.0 cm. (a) Calculate the apparent distance between the fish and the front wall. (b) Calculate the apparent distance between the image of the fish and the front wall. Verify that your answers are consistent with your answers to the Concept Questions.

* **125. Concept Questions** (a) For astronomical telescopes that have large angular magnifications, which lens has the greater focal length, the objective or the eyepiece? (b) How is the length L of the telescope related to the focal length f_o of the objective and the focal length f_e of the eyepiece? (c) Three astronomical telescopes have different lengths L, such that $L_A < L_B < L_C$. They have identical eyepieces. Rank the angular magnifications (magnitudes only) of these telescopes in descending order (largest first). For each answer, give your reasoning.

Problem The lengths of the telescopes are $L_A = 455$ mm, $L_B = 615$ mm, and $L_C = 824$ mm. The focal length of the eyepieces is 3.00 mm for each telescope. Find the angular magnification of each telescope. Be sure that your answers are consistent with your answers to the Concept Questions.

Chapter 27

Interference and the Wave Nature of Light

Owls, such as this long-eared owl, have eyes that can distinguish between two distant objects that are close together. This ability is related, in part, to the large pupils of the eyes and the wave nature of light, as this chapter discusses. (© George F. Mobley/National Geographic/Getty Images)

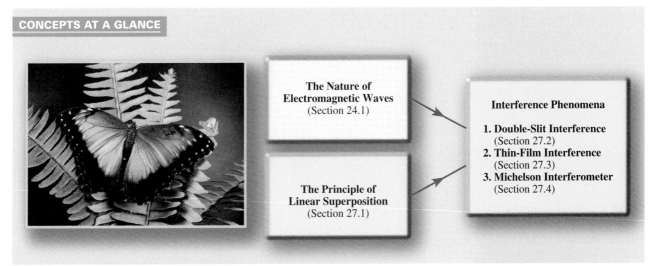

CONCEPTS AT A GLANCE

The Nature of
Electromagnetic Waves
(Section 24.1)

The Principle of
Linear Superposition
(Section 27.1)

Interference Phenomena

1. **Double-Slit Interference**
(Section 27.2)
2. **Thin-Film Interference**
(Section 27.3)
3. **Michelson Interferometer**
(Section 27.4)

27.1 THE PRINCIPLE OF LINEAR SUPERPOSITION

▶ CONCEPTS AT A GLANCE Chapter 17 examines what happens when several sound waves are present at the same place at the same time. The pressure disturbance that results is governed by the principle of linear superposition, which states that the resultant disturbance is the sum of the disturbances from the individual waves. Light is also a wave, an electromagnetic wave, and it too obeys the superposition principle. The Concepts-at-a-Glance chart in Figure 27.1 illustrates that this principle is used to explain interference phenomena associated with light, including Young's double-slit interference, thin-film interference, and the interference effects that occur in the Michelson interferometer. This chart is analogous to the one in Figure 17.17 for the interference of sound waves. When two or more light waves pass through a given point, their electric fields combine according to the principle of linear superposition and produce a resultant electric field. According to Equation 24.5b, the square of the electric field strength is proportional to the intensity of the light, which, in turn, is related to its brightness. Thus, interference can and does alter the brightness of light, just as it affects the loudness of sound. ◀

Figure 27.2 illustrates what happens when two identical waves (same wavelength λ and same amplitude) arrive at the point P in phase—that is, crest-to-crest and trough-to-trough. According to the principle of linear superposition, the waves reinforce each other and *constructive interference* occurs. The resulting total wave at P has an amplitude that is twice the amplitude of either individual wave, and in the case of light waves, the brightness at P is greater than that due to either wave alone. The waves start out in phase and are in phase at P because the distances ℓ_1 and ℓ_2 between this spot and the sources of the

Figure 27.1 CONCEPTS AT A GLANCE The principle of linear superposition explains various interference phenomena associated with electromagnetic waves. The beautiful iridescent blue color of the wings of this butterfly are due to the thin-film interference of light. A transparent cuticlelike material on the upper wing surface acts like a thin film. (© Gail Shumway/Taxi/Getty Images)

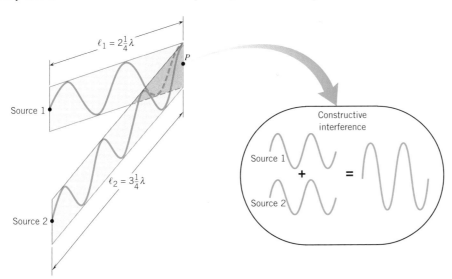

$\ell_1 = 2\frac{1}{4}\lambda$

P

Source 1

$\ell_2 = 3\frac{1}{4}\lambda$

Source 2

Constructive
interference

Source 1

$+$

$=$

Source 2

Figure 27.2 The waves emitted by source 1 and source 2 start out in phase and arrive at point P in phase, leading to constructive interference at that point.

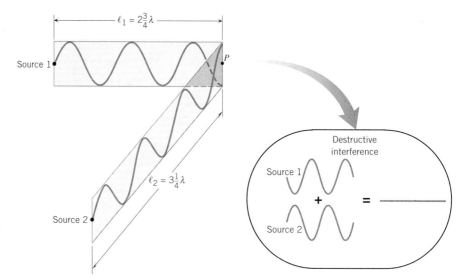

Figure 27.3 The waves emitted by the two sources have the same phase to begin with, but they arrive at point P out of phase. As a result, destructive interference occurs at P.

waves differ by one wavelength λ. In Figure 27.2, these distances are $\ell_1 = 2\frac{1}{4}$ wavelengths and $\ell_2 = 3\frac{1}{4}$ wavelengths. In general, when the waves start out in phase, constructive interference will result at P whenever the distances are the same or differ by any integer number of wavelengths—in other words, assuming ℓ_2 is the larger distance, whenever $\ell_2 - \ell_1 = m\lambda$, where $m = 0, 1, 2, 3, \ldots$.

Figure 27.3 shows what occurs when two identical waves arrive at the point P out of phase with one another, or crest-to-trough. Now the waves mutually cancel, according to the principle of linear superposition, and ***destructive interference*** results. With light waves this would mean that there is no brightness. The waves begin with the same phase but are out of phase at P because the distances through which they travel in reaching this spot differ by one-half of a wavelength ($\ell_1 = 2\frac{3}{4}\lambda$ and $\ell_2 = 3\frac{1}{4}\lambda$ in the drawing). In general, for waves that start out in phase, destructive interference will take place at P whenever the distances differ by any odd integer number of half-wavelengths—that is, whenever $\ell_2 - \ell_1 = \frac{1}{2}\lambda, \frac{3}{2}\lambda, \frac{5}{2}\lambda \ldots$, where ℓ_2 is the larger distance. This is equivalent to $\ell_2 - \ell_1 = (m + \frac{1}{2})\lambda$, where $m = 0, 1, 2, 3, \ldots$.

If constructive or destructive interference is to continue occurring at a point, the sources of the waves must be ***coherent sources.*** Two sources are coherent if the waves they emit maintain a constant phase relation. Effectively, this means that the waves do not shift relative to one another as time passes. For instance, suppose that the wave pattern of source 1 in Figure 27.3 shifted forward or backward by random amounts at random moments. Then, on average, neither constructive nor destructive interference would be observed at point P because there would be no stable relation between the two wave patterns. Lasers are coherent sources of light, whereas incandescent light bulbs and fluorescent lamps are incoherent sources.

✔ **Check Your Understanding 1**

Two separate coherent sources produce waves whose wavelengths are 0.10 m. The two waves spread out and intersect at a certain point. Does constructive or destructive interference occur at this point when (a) one wave travels 3.20 m and the other travels 3.00 m, (b) one wave travels 3.20 m and the other travels 3.05 m, and (c) one wave travels 3.20 m and the other travels 2.95 m? *(The answers are given at the end of the book.)*

Background: Whether constructive or destructive interference occurs at a given point depends on how the difference in path lengths is related to the wavelength of the waves.

For similar questions (including calculational counterparts), consult Self-Assessment Test 27.1, which is described at the end of Section 27.4.

27.2 *Young's Double-Slit Experiment*

In 1801 the English scientist Thomas Young (1773–1829) performed a historic experiment that demonstrated the wave nature of light by showing that two overlapping light waves interfered with each other. His experiment was particularly important because he was also able to determine the wavelength of the light from his measurements, the first such determination of this important property. Figure 27.4 shows one arrangement of Young's experiment, in which light of a single wavelength (monochromatic light) passes through a single narrow slit and falls on two closely spaced, narrow slits S_1 and S_2. These two slits act as coherent sources of light waves that interfere constructively and destructively at different points on the screen to produce a pattern of alternating bright and dark fringes. The purpose of the single slit is to ensure that only light from one direction falls on the double slit. Without it, light coming from different points on the light source would strike the double slit from different directions and cause the pattern on the screen to be washed out. The slits S_1 and S_2 act as coherent sources of light waves because the light from each originates from the same primary source—namely, the single slit.

To help explain the origin of the bright and dark fringes, Figure 27.5 presents three top views of the double slit and the screen. Part *a* illustrates how a bright fringe arises directly opposite the midpoint between the two slits. In this part of the drawing the waves (identical) from each slit travel to the midpoint on the screen. At this location, the distances ℓ_1 and ℓ_2 to the slits are equal, each containing the same number of wavelengths. Therefore, constructive interference results, leading to the bright fringe. Part *b* indicates that constructive interference produces another bright fringe on one side of the midpoint when the distance ℓ_2 is larger than ℓ_1 by exactly one wavelength. A bright fringe also occurs symmetrically on the other side of the midpoint when the distance ℓ_1 exceeds ℓ_2 by one wavelength; for clarity, however, this bright fringe is not shown. Constructive interference produces additional bright fringes (not shown) on both sides of the middle wherever the difference between ℓ_1 and ℓ_2 is an integer number of wavelengths: λ, 2λ, 3λ, and so on. Part *c* shows how the first dark fringe arises. Here the distance ℓ_2 is larger than ℓ_1 by exactly one-half a wavelength, so the waves interfere destructively, giving rise to the dark fringe. Destructive interference creates additional dark fringes on both sides of the center wherever the difference between ℓ_1 and ℓ_2 equals an odd integer number of half-wavelengths: $1(\frac{\lambda}{2})$, $3(\frac{\lambda}{2})$, $5(\frac{\lambda}{2})$, and so on.

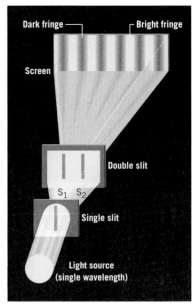

Figure 27.4 In Young's double-slit experiment, two slits S_1 and S_2 act as coherent sources of light. Light waves from these slits interfere constructively and destructively on the screen to produce, respectively, the bright and dark fringes. The slit widths and the distance between the slits have been exaggerated for clarity.

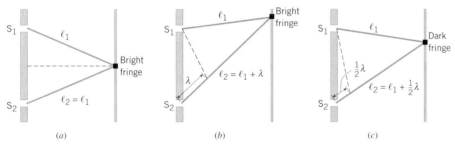

Figure 27.5 The waves from slits S_1 and S_2 interfere constructively (parts *a* and *b*) or destructively (part *c*) on the screen, depending on the difference in distances between the slits and the screen. The slit widths and the distance between the slits have been exaggerated for clarity.

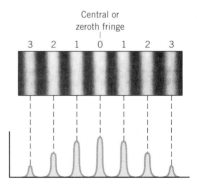

Figure 27.6 The results of Young's double-slit experiment, showing a photograph of the bright and dark fringes formed on the screen and a graph of the light intensity. The central or zeroth fringe is the brightest fringe (greatest intensity). (From Michel Cagnet, et al., *Atlas of Optical Phenomena*, Springer-Verlag, Berlin.)

The brightness of the fringes in Young's experiment varies, as the photograph in Figure 27.6 shows. Below the photograph is a graph to suggest the way in which the intensity varies for the fringe pattern. The central fringe is labeled with a zero, and the other bright fringes are numbered in ascending order on either side of the center. It can be seen that the central fringe has the greatest intensity. To either side of the center, the intensities of the other fringes decrease symmetrically in a way that depends on how small the slit widths are relative to the wavelength of the light.

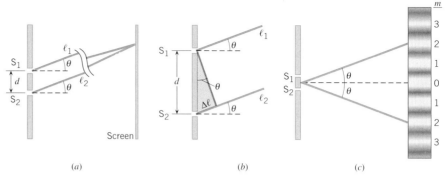

Figure 27.7 (*a*) Rays from slits S_1 and S_2, which make approximately the same angle θ with the horizontal, strike a distant screen at the same spot. (*b*) The difference in the path lengths of the two rays is $\Delta\ell = d \sin \theta$. (*c*) The angle θ is the angle at which a bright fringe ($m = 2$, here) occurs on either side of the central bright fringe ($m = 0$).

The position of the fringes observed on the screen in Young's experiment can be calculated with the aid of Figure 27.7. If the screen is located far away compared with the separation d of the slits, then the lines labeled ℓ_1 and ℓ_2 in part *a* are nearly parallel. Being nearly parallel, these lines make approximately equal angles θ with the horizontal. The distances ℓ_1 and ℓ_2 differ by an amount $\Delta\ell$, which is the length of the short side of the colored triangle in part *b* of the drawing. Since the triangle is a right triangle, it follows that $\Delta\ell = d \sin \theta$. Constructive interference occurs when the distances differ by an integer number m of wavelengths λ, or $\Delta\ell = d \sin \theta = m\lambda$. Therefore, the angle θ for the interference maxima can be determined from the following expression:

Bright fringes
of a double slit
$$\sin \theta = m\frac{\lambda}{d} \qquad m = 0, 1, 2, 3, \quad \ldots \tag{27.1}$$

The value of m specifies the *order* of the fringe. Thus, $m = 2$ identifies the "second-order" bright fringe. Part *c* of the drawing stresses that the angle θ given by Equation 27.1 locates bright fringes on either side of the midpoint between the slits. A similar line of reasoning leads to the conclusion that the dark fringes, which lie between the bright fringes, are located according to

Dark fringes
of a double slit
$$\sin \theta = (m + \tfrac{1}{2})\frac{\lambda}{d} \qquad m = 0, 1, 2, 3, \quad \ldots \tag{27.2}$$

Example 1 illustrates how to determine the distance of a higher-order bright fringe from the central bright fringe with the aid of these expressions.

Example 1   Young's Double-Slit Experiment

Red light ($\lambda = 664$ nm in vacuum) is used in Young's experiment with the slits separated by a distance $d = 1.20 \times 10^{-4}$ m. The screen in Figure 27.8 is located at a distance of $L = 2.75$ m from the slits. Find the distance y on the screen between the central bright fringe and the third-order bright fringe.

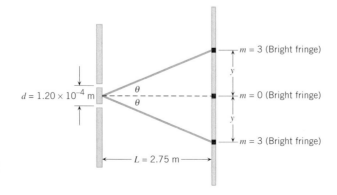

Figure 27.8 The third-order bright fringe ($m = 3$) is observed on the screen at a distance y from the central bright fringe ($m = 0$).

Reasoning This problem can be solved by first using Equation 27.1 to determine the value of θ that locates the third-order ($m = 3$) bright fringe. Then trigonometry can be used to obtain the distance y.

Solution According to Equation 27.1, we find

$$\theta = \sin^{-1}\left(\frac{m\lambda}{d}\right) = \sin^{-1}\left[\frac{3(664 \times 10^{-9}\ \text{m})}{1.20 \times 10^{-4}\ \text{m}}\right] = 0.951°$$

According to Figure 27.8, the distance y can be calculated from $\tan \theta = y/L$:

$$y = L \tan \theta = (2.75\ \text{m}) \tan 0.951° = \boxed{0.0456\ \text{m}}$$

▲

Need more practice?

Interactive LearningWare 27.1
In Young's double-slit experiment, blue light (λ = 440 nm) gives a second-order bright fringe at a certain location on a flat screen. What wavelength of visible light would produce a dark fringe at the same location? Assume that the range of visible wavelengths extends from 380 to 750 nm.

Related Homework: Problem 50

Go to **www.wiley.com/college/cutnell** for an interactive solution.

In the preceding version of Young's experiment, monochromatic light has been used. Light that contains a mixture of wavelengths can also be used. Conceptual Example 2 deals with some of the interesting features of the resulting interference pattern.

Conceptual Example 2 White Light and Young's Experiment

Figure 27.9 shows a photograph that illustrates the kind of interference fringes that can result when white light, which is a mixture of all colors, is used in Young's experiment. Except for the central fringe, which is white, the bright fringes are a rainbow of colors. Why does Young's experiment separate white light into its constituent colors? In any group of colored fringes, such as the two singled out in Figure 27.9, why is red farther out from the central fringe than green is? And finally, why is the central fringe white rather than colored?

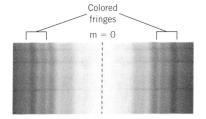

Figure 27.9 This photograph shows the results observed on the screen in one version of Young's experiment in which white light (a mixture of all colors) is used. (© Andy Washnik)

Reasoning and Solution To understand how the color separation arises, we need to remember that each color corresponds to a different wavelength λ and that constructive and destructive interference depend on the wavelength. According to Equation 27.1 ($\sin \theta = m\lambda/d$), there is a different angle that locates a bright fringe for each value of λ, and thus for each color. These different angles lead to the separation of colors on the observation screen. In fact, on either side of the central fringe, there is one group of colored fringes for $m = 1$ and another for each value of m.

Now, consider what it means that, within any single group of colored fringes, red is farther out from the central fringe than green is. It means that, in the equation $\sin \theta = m\lambda/d$, red light has a larger angle θ than green light does. Does this make sense? Yes, because red has the larger wavelength (see Table 26.2, where $\lambda_{\text{red}} = 660$ nm and $\lambda_{\text{green}} = 550$ nm).

In Figure 27.9, the central fringe is distinguished from all the other colored fringes by being white. In Equation 27.1, the central fringe is different from the other fringes because it is the only one for which $m = 0$. In Equation 27.1, a value of $m = 0$ means that $\sin \theta = m\lambda/d = 0$, which reveals that $\theta = 0°$, no matter what the wavelength λ is. In other words, all wavelengths have a zeroth-order bright fringe located at the same place on the screen, so that all colors strike the screen there and mix together to produce the white central fringe.

Related Homework: *Problem 6*

▲

Historically, Young's experiment provided strong evidence that light has a wave-like character. If light behaved only as a stream of "tiny particles," as others believed at the time,* then the two slits would deliver the light energy into only two bright fringes located directly opposite the slits on the screen. Instead, Young's experiment shows that wave interference redistributes the energy from the two slits into many bright fringes.

*It is now known that the particle or corpuscular theory of light, which Isaac Newton promoted, does indeed explain some experiments that the wave theory cannot explain. Today, light is regarded as having both particle and wave characteristics. Chapter 29 discusses this dual nature of light.

✔ Check Your Understanding 2

The drawing shows two double-slits that have slit separations of d_1 and d_2. Light whose wavelength is either λ_1 or λ_2 passes through the slits. For comparison, the wavelengths are also illustrated in the drawing. For which combination of slit separation and wavelength would the pattern of bright and dark fringes on the observation screen be (a) the most spread out and (b) the least spread out? *(The answers are given at the end of the book.)*

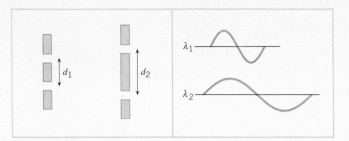

Background: The amount by which the fringe pattern on the observation screen would be spread out depends on how large the wavelength is relative to the slit separation.

For similar questions (including calculational counterparts), consult Self-Assessment Test 27.1, which is described at the end of Section 27.4.

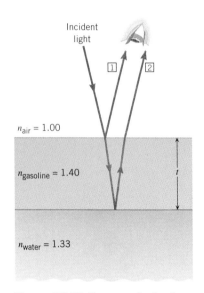

Figure 27.10 Because of reflection and refraction, two light waves, represented by rays 1 and 2, enter the eye when light shines on a thin film of gasoline floating on a thick layer of water.

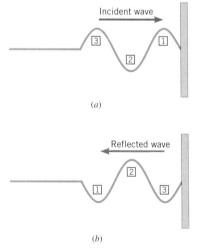

Figure 27.11 When a wave on a string reflects from a wall, the wave undergoes a phase change. Thus, an upward-pointing half-cycle of the wave becomes, after reflection, a downward-pointing half-cycle, and vice versa, as the numbered labels in the drawing indicate.

27.3 *Thin-Film Interference*

Young's double-slit experiment is one example of interference between light waves. Interference also occurs in more common circumstances. For instance, Figure 27.10 shows a thin film, such as gasoline floating on water. To begin with, let us assume that the film has a constant thickness. Consider what happens when monochromatic light (a single wavelength) strikes the film nearly perpendicularly. At the top surface of the film reflection occurs and produces the light wave represented by ray 1. However, refraction also occurs, and some light enters the film. Part of this light reflects from the bottom surface of the film and passes back up through the film, eventually reentering the air. Thus, a second light wave, which is represented by ray 2, also exists. Moreover, this wave, having traversed the film twice, has traveled farther than wave 1. Because of the extra travel distance, there can be interference between the two waves. If constructive interference occurs, an observer, whose eyes detect the superposition of waves 1 and 2, would see a uniformly bright film. If destructive interference occurs, an observer would see a uniformly dark film. The controlling factor is whether the extra distance for wave 2 is an integer number of whole wavelengths (for constructive interference) or an odd integer number of half-wavelengths (for destructive interference).

In Figure 27.10 the difference in path lengths between waves 1 and 2 occurs inside the thin film. Therefore, *the wavelength that is important for thin-film interference is the wavelength within the film,* not the wavelength in vacuum. The wavelength within the film can be calculated from the wavelength in vacuum by using the index of refraction n for the film. According to Equations 26.1 and 16.1, $n = c/v = (c/f)/(v/f) = \lambda_{\text{vacuum}}/\lambda_{\text{film}}$. In other words,

$$\lambda_{\text{film}} = \frac{\lambda_{\text{vacuum}}}{n} \tag{27.3}$$

In explaining the interference that can occur in Figure 27.10, we need to add one other important part to the story. Whenever waves reflect at a boundary, it is possible for them to change phase. Figure 27.11, for example, shows that a wave on a string is inverted when it reflects from the end that is tied to a wall (see also Figure 17.19). This inversion is equivalent to a half-cycle of the wave, as if the wave had traveled an additional distance of one-half of a wavelength. In contrast, a phase change does not occur when a wave on a string reflects from the end of a string that is hanging free. When light waves undergo reflection, similar phase changes occur as follows:

1. When light travels through a material with a smaller refractive index toward a material with a larger refractive index (e.g., air to gasoline), reflection at the boundary occurs along with a phase change that is equivalent to one-half of a wavelength in the film.

2. When light travels from a larger toward a smaller refractive index, there is no phase change upon reflection at the boundary.

The next example indicates how the phase change that can accompany reflection is taken into account when dealing with thin-film interference.

Example 3 A Colored Thin Film of Gasoline

A thin film of gasoline floats on a puddle of water. Sunlight falls almost perpendicularly on the film and reflects into your eyes. Although sunlight is white since it contains all colors, the film has a yellow hue because destructive interference eliminates the color of blue (λ_{vacuum} = 469 nm) from the reflected light. The refractive indices of the blue light in gasoline and water are 1.40 and 1.33, respectively. Determine the minimum nonzero thickness t of the film.

Reasoning To solve this problem, we must express the condition for destructive interference in terms of the film thickness t and the wavelength in the gasoline film λ_{film}. We must also take into account any phase changes that occur upon reflection.

Solution Equation 27.3, with n = 1.40, gives the wavelength of blue light in the film as λ_{film} = (469 nm)/1.40 = 335 nm. In Figure 27.10, the phase change for wave 1 is equivalent to one-half of a wavelength, since this light travels from a smaller refractive index (n_{air} = 1.00) toward a larger refractive index ($n_{gasoline}$ = 1.40). In contrast, there is no phase change when wave 2 reflects from the bottom surface of the film, since this light travels from a larger refractive index ($n_{gasoline}$ = 1.40) toward a smaller one (n_{water} = 1.33). The net phase change between waves 1 and 2 due to reflection is, thus, equivalent to one-half of a wavelength, $\frac{1}{2}\lambda_{film}$. This half-wavelength must be combined with the extra travel distance for wave 2, to determine the condition for destructive interference. For destructive interference, the combined total must be an odd integer number of half-wavelengths. Since wave 2 travels back and forth through the film and since light strikes the film nearly perpendicularly, the extra travel distance is twice the film thickness, or $2t$. Thus, the condition for destructive interference is

Problem solving insight
When analyzing thin-film interference effects, remember to use the wavelength of the light in the film (λ_{film}) instead of the wavelength in a vacuum (λ_{vacuum}).

$$\underbrace{2t}_{\substack{\text{Extra distance} \\ \text{traveled by} \\ \text{wave 2}}} + \underbrace{\tfrac{1}{2}\lambda_{film}}_{\substack{\text{Half-wavelength} \\ \text{net phase change} \\ \text{due to reflection}}} = \underbrace{\tfrac{1}{2}\lambda_{film}, \tfrac{3}{2}\lambda_{film}, \tfrac{5}{2}\lambda_{film}, \dots}_{\substack{\text{Condition for} \\ \text{destructive interference}}}$$

After subtracting the term $\frac{1}{2}\lambda_{film}$ from the left-hand side of this equation and from each term on the right-hand side, we can solve for the thickness t of the film that yields destructive interference:

$$t = \frac{m\lambda_{film}}{2} \qquad m = 0, 1, 2, 3, \dots$$

With m = 1, the expression above gives the minimum nonzero film thickness for which the blue color is missing in the reflected light:

$$t = \tfrac{1}{2}\lambda_{film} = \tfrac{1}{2}(335 \text{ nm}) = \boxed{168 \text{ nm}}$$

Example 3 deals with a thin film that has the same yellow color everywhere. In nature, such a uniformly colored thin film would be unusual, and the next example deals with a more realistic situation.

Conceptual Example 4 Multicolored Thin Films

Under natural conditions, thin films, like gasoline on water or like the soap bubble in Figure 27.12 on page 828, have a multicolored appearance that often changes while you are watching them. Why are such films multicolored, and what can be inferred from the fact that the colors change in time?

Figure 27.12 A soap bubble is multicolored when viewed in sunlight because of the effects of thin-film interference. (© Paul A. Souders/Corbis Images)

Problem solving insight

Reasoning and Solution In Example 3 we have seen that a thin film can appear yellow if destructive interference removes blue light from the reflected sunlight. The thickness of the film is the key. If the thickness were different, so that destructive interference removed green light from the reflected sunlight, the film would appear magenta. Constructive interference can also cause certain colors to appear brighter than others in the reflected light and give the film a colored appearance. The colors that are enhanced by constructive interference, like those removed by destructive interference, depend on the thickness of the film. Thus, we conclude that *the different colors in a thin film of gasoline on water or in a soap bubble arise because the thickness is different in different places on the film. Moreover, the fact that the colors change as you watch them indicates that the thickness is changing.* A number of factors can cause the thickness to change, including air currents, temperature fluctuations, and the pull of gravity, which tends to make a vertical film sag, leading to thicker regions at the bottom than at the top.

Related Homework: *Problems 12, 14, 16*

Need more practice?

Interactive LearningWare 27.2
A thin film of gasoline ($n = 1.40$) floats on a surface of glass ($n = 1.52$). Sunlight falls almost perpendicularly on the film and reflects into your eyes. Although sunlight is white since it contains all colors, the film has a yellow hue because destructive interference eliminates the color of blue ($\lambda_{vacuum} = 469$ nm) from the reflected light. Determine the minimum nonzero thickness t of the film.

Related Homework: *Problems 48, 58*

Go to **www.wiley.com/college/cutnell** for an interactive solution.

The colors that you see when sunlight is reflected from a thin film also depend on your viewing angle. At an oblique angle, the light corresponding to ray 2 in Figure 27.10 would travel a greater distance within the film than it does at nearly perpendicular incidence. The greater distance would lead to destructive interference for a different wavelength.

Thin-film interference can be beneficial in optical instruments. For example, some cameras contain six or more lenses. Reflections from all the lens surfaces can reduce considerably the amount of light directly reaching the film. In addition, multiple reflections

The physics of
nonreflecting lens coatings.

from the lenses often reach the film indirectly and degrade the quality of the image. To minimize such unwanted reflections, high-quality lenses are often covered with a thin nonreflective coating of magnesium fluoride ($n = 1.38$). The thickness of the coating is usually chosen to ensure that destructive interference eliminates the reflection of green light, which is in the middle of the visible spectrum. It should be pointed out that the absence of any reflected light does not mean that it has been destroyed by the nonreflective coating. Rather, the "missing" light has been transmitted into the coating and the lens.

Another interesting illustration of thin-film interference is the air wedge. As Figure 27.13a shows, an air wedge is formed when two flat plates of glass are separated along one side, perhaps by a thin sheet of paper. The thickness of this film of air varies between zero, where the plates touch, and the thickness of the paper. When monochromatic light reflects from this arrangement, alternate bright and dark fringes are formed by constructive and destructive interference, as the drawing indicates and Example 5 discusses.

Example 5   An Air Wedge

(a) Assuming that green light ($\lambda_{\text{vacuum}} = 552$ nm) strikes the glass plates nearly perpendicularly in Figure 27.13, determine the number of bright fringes that occur between the place where the plates touch and the edge of the sheet of paper (thickness $= 4.10 \times 10^{-5}$ m). (b) Explain why there is a dark fringe where the plates touch.

Reasoning A bright fringe occurs wherever there is constructive interference, as determined by any phase changes due to reflection and the effects of the thickness of the air wedge. We examine the phase changes and the effects of the thickness separately. There is no phase change upon reflection for wave 1, since this light travels from a larger (glass) toward a smaller (air) refractive index. In contrast, there is a half-wavelength phase change for wave 2, since the ordering of the refractive indices is reversed at the lower air/glass boundary where reflection occurs. The net phase change due to reflection for waves 1 and 2, then, is equivalent to a half wavelength. Now we combine any extra distance traveled by ray 2 with this half wavelength and determine the condition for the constructive interference that creates the bright fringes. Constructive interference occurs whenever the *combination* yields an integer number of wavelengths. At nearly perpendicular incidence, the extra travel distance for wave 2 is approximately twice the thickness t of the wedge at any point, so the condition for constructive interference is

$$\underbrace{2t}_{\substack{\text{Extra distance} \\ \text{traveled by} \\ \text{wave 2}}} + \underbrace{\tfrac{1}{2}\lambda_{\text{film}}}_{\substack{\text{Half-wavelength} \\ \text{net phase change} \\ \text{due to reflection}}} = \underbrace{\lambda_{\text{film}}, 2\lambda_{\text{film}}, 3\lambda_{\text{film}}, \dots}_{\substack{\text{Condition for} \\ \text{constructive interference}}}$$

Subtracting the term $\tfrac{1}{2}\lambda_{\text{film}}$ from the left-hand side of this equation and from each term on the right-hand side yields

$$2t = \tfrac{1}{2}\lambda_{\text{film}}, \tfrac{3}{2}\lambda_{\text{film}}, \tfrac{5}{2}\lambda_{\text{film}}, \dots$$
$$(m + \tfrac{1}{2})\lambda_{\text{film}} \qquad m = 0, 1, 2, 3, \dots$$

Therefore,

$$t = \frac{(m + \tfrac{1}{2})\lambda_{\text{film}}}{2} \qquad m = 0, 1, 2, 3, \dots$$

In this expression, note that the "film" is a film of air. Since the refractive index of air is nearly one, λ_{film} is virtually the same as that in a vacuum, so $\lambda_{\text{film}} = 552$ nm.

Solution

(a) When t equals the thickness of the paper holding the plates apart, the corresponding value of m can be obtained from the equation above:

$$m = \frac{2t}{\lambda_{\text{film}}} - \frac{1}{2} = \frac{2(4.10 \times 10^{-5} \text{ m})}{552 \times 10^{-9} \text{ m}} - \frac{1}{2} = 148$$

Since the first bright fringe occurs when $m = 0$, the number of bright fringes is $m + 1 = \boxed{149}$.

(b) Where the plates touch, there is a dark fringe because of destructive interference between the light waves represented by rays 1 and 2. Destructive interference occurs because the thickness of the wedge is zero here and the only difference between the rays is the half-wavelength phase change due to reflection from the lower plate.

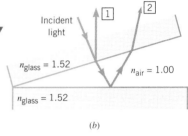

Figure 27.13 (a) The wedge of air formed between two flat glass plates causes an interference pattern of alternating dark and bright fringes to appear in reflected light. (b) A side view of the glass plates and the air wedge.

Problem solving insight
Reflected light will experience a phase change only if the light travels from a material with a smaller refractive index toward a material with a larger refractive index. Be sure to take such a phase change into account when analyzing thin-film interference.

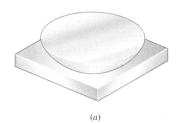

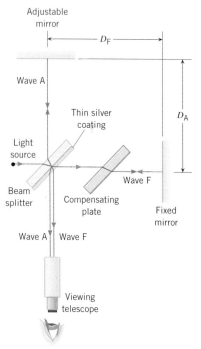

Figure 27.14 (*a*) The air wedge between a convex spherical glass surface and an optically flat plate leads to (*b*) a pattern of circular interference fringes that is known as Newton's rings. (Courtesy Bausch & Lomb)

Another type of air wedge can also be used to determine the degree to which the surface of a lens or mirror is spherical. When an accurate spherical surface is put in contact with an optically flat plate, as in Figure 27.14*a*, the circular interference fringes shown in part *b* of the figure can be observed. The circular fringes are called *Newton's rings*. They arise in the same way that the straight fringes arise in Figure 27.13*a*.

✔ **Check Your Understanding 3**

The figure shows three situations in which light reflects almost perpendicularly from the top and bottom surfaces of a thin film, with the indices of refraction as shown. (a) For which situation(s) is there a net phase shift (due to reflection) between waves 1 and 2 that is equivalent to either zero wavelengths or one wavelength (λ_{film}), where λ_{film} is the wavelength of the light in the film? (b) For which situation(s) will the film appear dark when the thickness of the film is equal to $\frac{1}{2}\lambda_{film}$? (*The answers are given at the end of the book.*)

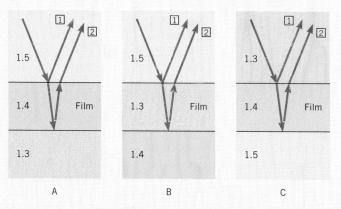

Background: For a film to appear dark, the waves represented by rays 1 and 2 must be out of phase. This can arise from phase shifts associated with the reflection at the top and bottom surfaces of the film and the extra distance traveled by ray 2 as it moves through the film.

For similar questions (including calculational counterparts), consult Self-Assessment Test 27.1, which is described at the end of Section 27.4.

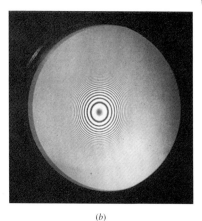

Figure 27.15 A schematic drawing of a Michelson interferometer.

27.4 *The Michelson Interferometer*

An interferometer is an apparatus that can be used to measure the wavelength of light by utilizing interference between two light waves. One particularly famous interferometer is that developed by Albert A. Michelson (1852–1931). The Michelson interferometer uses reflection to set up conditions where two light waves interfere. Figure 27.15 presents a schematic drawing of the instrument. Waves emitted by the monochromatic light source strike a *beam splitter*, so called because it splits the beam of light into two parts. The beam splitter is a glass plate, the far side of which is coated with a thin layer of silver that reflects part of the beam upward as wave A in the drawing. The coating is so thin, however, that it also allows the remainder of the beam to pass directly through as wave F. Wave A strikes an adjustable mirror and reflects back on itself. It again crosses the beam splitter and then enters the viewing telescope. Wave F strikes a fixed mirror and returns, to be partly reflected into the viewing telescope by the beam splitter. Note that wave A passes through the glass plate of the beam splitter three times in reaching the viewing scope, while wave F passes through it only once. The compensating plate in the path of wave F has the same thickness as the beam splitter plate and ensures that wave F also passes three times through the same thickness of glass on the way to the viewing scope. Thus, an observer who views the superposition of waves A and F through the telescope sees constructive or destructive interference, depending only on the difference in path lengths D_A and D_F traveled by the two waves.

Now suppose the mirrors are perpendicular to each other, the beam splitter makes a 45° angle with each, and the distances D_A and D_F are equal. Waves A and F travel the same distance, and the field of view in the telescope is uniformly bright due to constructive interference. However, if the adjustable mirror is moved away from the telescope by a distance of $\frac{1}{4}\lambda$, wave A travels back and forth by an amount that is twice this value, leading to an extra distance of $\frac{1}{2}\lambda$. Then, the waves are out of phase when they reach the viewing scope, destructive inteference occurs, and the viewer sees a dark field. If the adjustable mirror is moved farther, full brightness returns as soon as the waves are in phase and interfere constructively. The in-phase condition occurs when wave A travels a total extra distance of λ relative to wave F. Thus, as the mirror is continuously moved, the viewer sees the field of view change from bright to dark, then back to bright, and so on. The amount by which D_A has been changed can be measured and related to the wavelength of the light, since a bright field changes into a dark field and back again each time D_A is changed by a half wavelength. (The back-and-forth change in distance is λ.) If a sufficiently large number of wavelengths are counted in this manner, the Michelson interferometer can be used to obtain a very accurate value for the wavelength from the measured changes in D_A.

The physics of
the Michelson interferometer.

 Self-Assessment Test 27.1

Test your understanding of the material in Sections 27.1–27.4:

- The Principle of Linear Superposition   • Young's Double-Slit Experiment
- Thin-Film Interference   • The Michelson Interferometer

Go to **www.wiley.com/college/cutnell**

27.5 *Diffraction*

▶ CONCEPTS AT A GLANCE In the previous sections of this chapter we have used the principle of linear superposition to investigate some interference phenomena associated with light waves. As the Concepts-at-a-Glance chart in Figure 27.16 shows, we will now use the principle of linear superposition to explore three more topics—diffraction, resolving power, and the diffraction grating—all of which involve interference phenomena. This chart is an extension of the earlier version in Figure 27.1. ◀

As we have seen in Section 17.3, *diffraction* is the bending of waves around obstacles or the edges of an opening. In Figure 27.17 on page 832, which is similar to Figure 17.11, sound waves are leaving a room through an open doorway. Because the exiting sound waves bend, or diffract, around the edges of the opening, a listener outside the room can hear the sound even when standing around the corner from the doorway.

Figure 27.16 The principle of linear superposition can be used to describe the topics highlighted in gold. The grooved surface of a CD behaves like a diffraction grating and separates white light into a spectrum of colors.
(© Bruno/Corbis Stock Market)

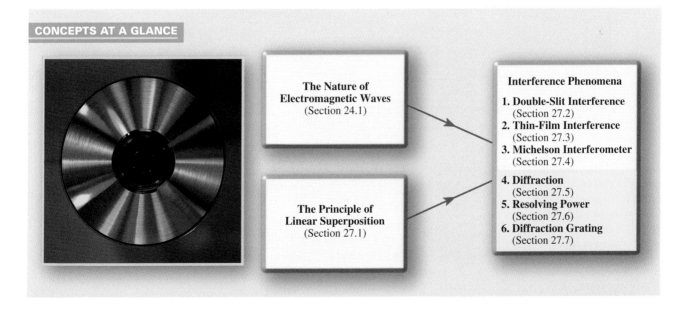

CONCEPTS AT A GLANCE

The Nature of
Electromagnetic Waves
(Section 24.1)

The Principle of
Linear Superposition
(Section 27.1)

Interference Phenomena

1. **Double-Slit Interference**
   (Section 27.2)
2. **Thin-Film Interference**
   (Section 27.3)
3. **Michelson Interferometer**
   (Section 27.4)
4. **Diffraction**
   (Section 27.5)
5. **Resolving Power**
   (Section 27.6)
6. **Diffraction Grating**
   (Section 27.7)

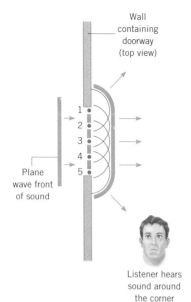

Wall
containing
doorway
(top view)

Plane
wave front
of sound

Listener hears
sound around
the corner

Figure 27.17 Sound bends or diffracts around the edges of a doorway, so even a person who is not standing directly in front of the opening can hear the sound. The five red points within the doorway act as sources and emit the five Huygens wavelets shown in red.

Diffraction is an interference effect, and the Dutch scientist Christian Huygens (1629–1695) developed a principle that is useful in explaining why diffraction arises. *Huygens' principle* describes how a wave front that exists at one instant gives rise to the wave front that exists later on. This principle states that:

> *Every point on a wave front acts as a source of tiny wavelets that move forward with the same speed as the wave; the wave front at a later instant is the surface that is tangent to the wavelets.*

We begin by using Huygens' principle to explain the diffraction of sound waves in Figure 27.17. The drawing shows the top view of a plane wave front of sound approaching a doorway and identifies five points on the wave front just as it is leaving the opening. According to Huygens' principle, each of these points acts as a source of wavelets, which are shown as red circular arcs at some moment after they are emitted. The tangent to the wavelets from points 2, 3, and 4 indicates that in front of the doorway the wave front is flat and moving straight ahead. But at the edges, points 1 and 5 are the last points that produce wavelets. Huygens' principle suggests that in conforming to the curved shape of the wavelets near the edges, the new wave front moves into regions that it would not reach otherwise. The sound wave, then, bends or diffracts around the edges of the doorway.

Huygens' principle applies not just to sound waves, but to all kinds of waves. For instance, light has a wave-like nature and, consequently, exhibits diffraction. Therefore, you may ask, "Since I can hear around the edges of a doorway, why can't I also see around them?" As a matter of fact, light waves do bend around the edges of a doorway. However, the degree of bending is extremely small, so the diffraction of light is not enough to allow you to see around the corner.

As we will learn, the extent to which a wave bends around the edges of an opening is determined by the ratio λ/W, where λ is the wavelength of the wave and W is the width of the opening. The photographs in Figure 27.18 illustrate the effect of this ratio on the diffraction of water waves. The degree to which the waves are diffracted or bent is indicated by the two red arrows in each photograph. In part *a*, the ratio λ/W is small because the wavelength (as indicated by the distance between the wave fronts) is small relative to the width of the opening. The wave fronts move through the opening with little bending or diffraction into the regions around the edges. In part *b*, the wavelength is larger and the width of the opening is smaller. As a result, the ratio λ/W is larger, and the degree of bending becomes more pronounced, with the wave fronts penetrating more into the regions around the edges of the opening.

Based on the pictures in Figure 27.18, we might expect that light waves of wavelength λ will bend or diffract appreciably when they pass through an opening whose width W is small enough to make the ratio λ/W sufficiently large. This is indeed the case, as Figure 27.19 illustrates. In this picture, it is assumed that parallel rays (or plane wave fronts) of light fall on a very narrow slit and illuminate a viewing screen that is located far

Figure 27.18 These photographs show water waves (horizontal lines) approaching an opening whose width W is greater in (*a*) than in (*b*). In addition, the wavelength λ of the waves is smaller in (*a*) than in (*b*). Therefore, the ratio λ/W increases from (*a*) to (*b*) and so does the extent of the diffraction, as the red arrows indicate. (Courtesy Education Development Center)

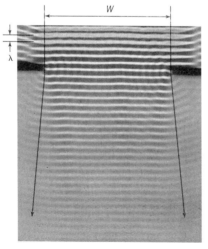

(*a*) Smaller value for λ/W, less diffraction.

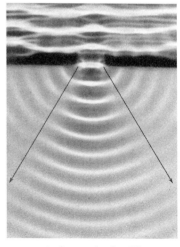

(*b*) Larger value for λ/W, more diffraction.

Figure 27.19 (*a*) If light were to pass through a very narrow slit *without* being diffracted, only the region on the screen directly opposite the slit would be illuminated. (*b*) Diffraction causes the light to bend around the edges of the slit into regions it would not otherwise reach, forming a pattern of alternating bright and dark fringes on the screen. The slit width has been exaggerated for clarity.

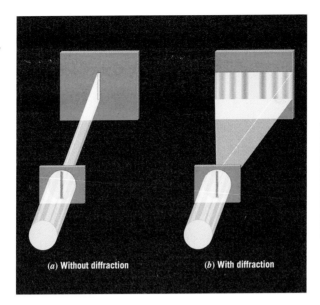

(*a*) Without diffraction   (*b*) With diffraction

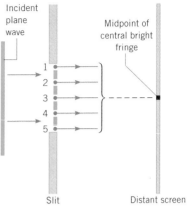

Figure 27.20 A plane wave front is incident on a single slit. This top view of the slit shows five sources of Huygens wavelets. The wavelets travel toward the midpoint of the central bright fringe on the screen, as the red rays indicate. The screen is very far from the slit.

from the slit. Part *a* of the drawing shows what would happen if light were *not* diffracted: it would pass through the slit without bending around the edges and produce an image of the slit on the screen. Part *b* shows what actually happens. The light diffracts around the edges of the slit and brightens regions on the screen that are not directly opposite the slit. The diffraction pattern on the screen consists of a bright central band, accompanied by a series of narrower faint fringes that are parallel to the slit itself.

To help explain how the pattern of diffraction fringes arises, Figure 27.20 shows a top view of a plane wave front approaching the slit and singles out five sources of Huygens wavelets. Consider how the light from these five sources reaches the midpoint on the screen. To simplify things, the screen is assumed to be so far from the slit that the rays from each Huygens source are nearly parallel.* Then, all the wavelets travel virtually the same distance to the midpoint, arriving there in phase. As a result, constructive interference creates a bright central fringe on the screen, directly opposite the slit.

The wavelets emitted by the Huygens sources in the slit can also interfere destructively on the screen, as Figure 27.21 illustrates. Part *a* shows light rays directed from each source toward the first dark fringe. The angle θ gives the position of this dark fringe relative to the line between the midpoint of the slit and the midpoint of the central bright fringe. Since the screen is very far from the slit, the rays from each Huygens source are nearly parallel and oriented at nearly the same angle θ, as in part *b* of the drawing. The wavelet from source 1 travels the shortest distance to the screen, while that from source 5 travels the farthest. Destructive interference creates the first dark fringe when the extra distance traveled by the wavelet from source 5 is exactly one wavelength, as the colored right triangle in the drawing indicates. Under this condition, the extra distance traveled by the wavelet from source 3 at the center of the slit is exactly one-half of a wavelength.

Therefore, wavelets from sources 1 and 3 in Figure 27.21*b* are exactly out of phase and interfere destructively when they reach the screen. Similarly, a wavelet that originates slightly below source 1 cancels a wavelet that originates the same distance below source 3. Thus, each wavelet from the upper half of the slit cancels a corresponding wavelet from the lower half, and no light reaches the screen. As can be seen from the colored right triangle, the angle θ locating the first dark fringe is given by sin $\theta = \lambda/W$, where W is the width of the slit.

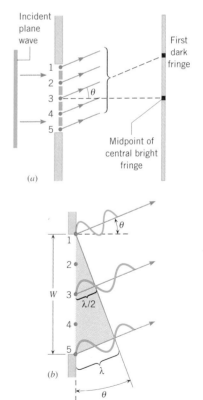

Figure 27.21 These drawings pertain to single-slit diffraction and show how destructive interference leads to the first dark fringe on either side of the central bright fringe. For clarity, only one of the dark fringes is shown. The screen is very far from the slit.

*When the rays are parallel, the diffraction is called Fraunhofer diffraction in tribute to the German optician Joseph von Fraunhofer (1787–1826). When the rays are not parallel, the diffraction is referred to as Fresnel diffraction, named for the French physicist Augustin Jean Fresnel (1788–1827).

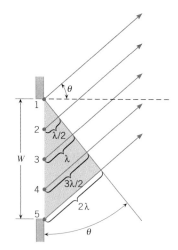

Figure 27.22 In a single-slit diffraction pattern, multiple dark fringes occur on either side of the central bright fringe. This drawing shows how destructive interference creates the second dark fringe on a very distant screen.

Figure 27.22 shows the condition that leads to destructive interference at the second dark fringe on either side of the midpoint on the screen. In reaching the screen, the light from source 5 now travels a distance of two wavelengths farther than the light from source 1. Under this condition, the wavelet from source 5 travels one wavelength farther than the wavelet from source 3, and the wavelet from source 3 travels one wavelength farther than the wavelet from source 1. Therefore, each half of the slit can be treated as the entire slit was in the previous paragraph; all the wavelets from the top half interfere destructively with each other, and all the wavelets from the bottom half do likewise. As a result, no light from either half reaches the screen, and another dark fringe occurs. The colored triangle in the drawing shows that this second dark fringe occurs when $\sin \theta = 2\lambda/W$. Similar arguments hold for the third- and higher-order dark fringes, with the general result being

Dark fringes for single-slit diffraction
$$\sin \theta = m \frac{\lambda}{W} \qquad m = 1, 2, 3, \ldots \qquad (27.4)$$

Between each pair of dark fringes there is a bright fringe due to constructive interference. The brightness of the fringes is related to the light intensity, just as loudness is related to sound intensity. The intensity of the light at any location on the screen is the amount of light energy per second per unit area that strikes the screen there. Figure 27.23 gives a graph of the light intensity along with a photograph of a single-slit diffraction pattern. The central bright fringe, which is approximately twice as wide as the other bright fringes, has by far the greatest intensity.

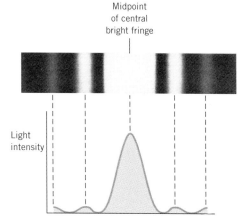

Figure 27.23 The photograph shows a single-slit diffraction pattern, with a bright and wide central fringe. The higher-order bright fringes are much less intense than the central fringe, as the graph indicates. (From Michel Cagnet, et al., *Atlas of Optical Phenomena*, Springer-Verlag, Berlin.)

Concept Simulation 27.1

By clicking on the "Single-slit mode" button in this simulation, you can explore the diffraction pattern produced by a single slit. The wavelength of the light striking the slit, the slit width, and the distance between the slit and the screen are under your control. Make predictions on how you think each of these variables affects the diffraction pattern, and then use the simulation to check your predictions.

Go to **www.wiley.com/college/cutnell**

The width of the central fringe provides some indication of the extent of the diffraction, as Example 6 illustrates.

Example 6   Single-Slit Diffraction

▼

Light passes through a slit and shines on a flat screen that is located $L = 0.40$ m away (see Figure 27.24). The width of the slit is $W = 4.0 \times 10^{-6}$ m. The distance between the middle of the central bright fringe and the first dark fringe is y. Determine the width $2y$ of the central bright fringe when the wavelength of the light in a vacuum is (a) $\lambda = 690$ nm (red) and (b) $\lambda = 410$ nm (violet).

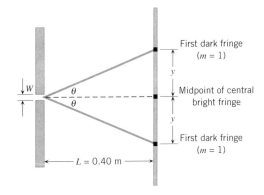

Figure 27.24 The distance $2y$ is the width of the central bright fringe.

Reasoning The width of the central bright fringe is determined by two factors. One is the angle θ that locates the first dark fringe on either side of the midpoint. The other is the distance L between the screen and the slit. Larger values for θ and L lead to a wider central bright fringe. Larger values of θ mean greater diffraction and occur when the ratio λ/W is larger. Thus, we expect the width of the central bright fringe to be greater when λ is greater.

Solution

(a) The angle θ in Equation 27.4 locates the first dark fringe when $m = 1$; $\sin\theta = (1)\lambda/W$. Therefore,

$$\theta = \sin^{-1}\left(\frac{\lambda}{W}\right) = \sin^{-1}\left(\frac{690 \times 10^{-9}\text{ m}}{4.0 \times 10^{-6}\text{ m}}\right) = 9.9°$$

According to Figure 27.24, $\tan\theta = y/L$, so

$$y = L\tan\theta = (0.40\text{ m})\tan 9.9° = 0.070\text{ m}$$

The width of the central fringe, then, is $\boxed{2y = 0.14\text{ m}}$.

(b) Repeating the calculation above with $\lambda = 410$ nm shows that $\theta = 5.9°$ and $\boxed{2y = 0.083\text{ m}}$. As expected, the width $2y$ is greater in part (a), where the wavelength λ is greater.

▲

Need more practice?

Interactive LearningWare 27.3

A single slit is illuminated by monochromatic light ($\lambda_{\text{vacuum}} = 645$ nm). The width of the slit is 4.85×10^{-6} m. Determine the angular width 2θ of the central bright fringe when the region between the slit and the screen is filled with (a) water ($n = 1.33$) and (b) air ($n = 1.00$).

Go to www.wiley.com/college/cutnell for an interactive solution.

In the production of computer chips it is important to minimize the effects of diffraction. As Figure 23.32 illustrates, such chips are very small and yet contain enormous numbers of electronic components. Such miniaturization is achieved using the techniques of photolithography. The patterns on the chip are created first on a "mask," which is similar to a photographic slide. Light is then directed through the mask onto silicon wafers that have been coated with a photosensitive material. The light-activated parts of the coating can be removed chemically, to leave the ultrathin lines that correspond to the miniature patterns on the chip. As the light passes through the narrow slit-like patterns on the mask, the light spreads out due to diffraction. If excessive diffraction occurs, the light spreads out so much that sharp patterns are not formed on the photosensitive material coating the silicon wafer. Ultraminiaturization of the patterns requires the absolute minimum of diffraction, and currently this is achieved by using ultraviolet light, which has a wavelength shorter than that of visible light. The shorter the wavelength λ, the smaller the ratio λ/W, and the less the diffraction, as illustrated in Example 6. Intensive research programs are under way to develop the technique of X-ray lithography. The wavelengths of X-rays are much shorter than those of ultraviolet light and, thus, will reduce diffraction even more, allowing further miniaturization.

The physics of **producing computer chips using photolithography.**

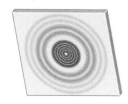

Opaque disk

Light

Figure 27.25 The diffraction pattern formed by an opaque disk consists of a small bright spot in the center of the dark shadow, circular bright fringes within the shadow, and concentric bright and dark fringes surrounding the shadow.

Another example of diffraction can be seen when light from a point source falls on an opaque disk, such as a coin (Figure 27.25). The effects of diffraction modify the dark shadow cast by the disk in several ways. First, the light waves diffracted around the circular edge of the disk interfere constructively at the center of the shadow to produce a small bright spot. There are also circular bright fringes in the shadow area. In addition, the boundary between the circular shadow and the lighted screen is not sharply defined but consists of concentric bright and dark fringes. The various fringes are analogous to those produced by a single slit and are due to interference between Huygens wavelets that originate from different points near the edge of the disk.

✔ **Check Your Understanding 4**

A diffraction pattern is produced on a viewing screen by using a single slit with blue light. Does the pattern broaden or contract (become narrower) when (a) the blue light is replaced by red light and (b) the slit width is increased? *(The answers are given at the end of the book.)*

Background: When light passes through a single slit, the extent to which it is bent, or diffracted, depends on two parameters: the wavelength of the light and the width of the slit. This question involves both of these parameters.

For similar questions (including calculational counterparts), consult Self-Assessment Test 27.2, which is described at the end of Section 27.9.

27.6 Resolving Power

Figure 27.26 shows three photographs of an automobile's headlights taken at progressively greater distances from the camera. In parts *a* and *b*, the two separate headlights can be seen clearly. In part *c*, however, the car is so far away that the headlights are barely distinguishable and appear almost as a single light. The ***resolving power*** of an optical instrument, such as a camera, is its ability to distinguish between two closely spaced objects. If a camera with a higher resolving power had taken these pictures, the photograph in part *c* would have shown two distinct and separate headlights. Any instrument used for viewing objects that are close together must have a high resolving power. This is true, for example, for a telescope used to view distant stars or for a microscope used to view tiny organisms. We will now see that diffraction occurs when light passes through the circular, or nearly circular, openings that admit light into cameras, telescopes, microscopes, and human eyes. The resulting diffraction pattern places a natural limit on the resolving power of these instruments.

Figure 27.27 shows the diffraction pattern created by a small circular opening when the viewing screen is far from the opening. The pattern consists of a central bright circular region, surrounded by alternating bright and dark circular fringes. These fringes are analogous to the rectangular fringes that a single slit produces. The angle θ in the picture locates the first circular dark fringe relative to the central bright region and is given by

$$\sin \theta = 1.22 \frac{\lambda}{D} \tag{27.5}$$

Figure 27.26 These automobile headlights were photographed at various distances from the camera, closest in part (*a*) and farthest in part (*c*). In part (*c*), the headlights are so far away that they are barely distinguishable. (© Truax/The Image Finders)

(*a*)

(*b*)

(*c*)

where λ is the wavelength of the light and D is the diameter of the opening. This expression is similar to Equation 27.4 for a slit ($\sin \theta = \lambda/W$, when $m = 1$) and is valid when the distance to the screen is much larger than the diameter D.

An optical instrument with the ability to resolve two closely spaced objects can produce images of them that can be identified separately. Think about the images on the film when light from two widely separated point objects passes through the circular aperture of a camera. As Figure 27.28 illustrates, each image is a circular diffraction pattern, but the two patterns do not overlap and are completely resolved. On the other hand, if the objects are sufficiently close together, the intensity patterns created by the diffraction overlap, as Figure 27.29a suggests. In fact, if the overlap is extensive, it may no longer be possible to distinguish the patterns separately. In such a case, the picture from a camera would show a single blurred object instead of two separate objects. In Figure 27.29b the diffraction patterns overlap, but not enough to prevent us from seeing that two objects are present. Ultimately, then, diffraction limits the ability of an optical instrument to produce distinguishable images of objects that are close together.

It is useful to have a criterion for judging whether two closely spaced objects will be resolved by an optical instrument. Figure 27.29a presents the ***Rayleigh criterion*** for resolution, first proposed by Lord Rayleigh (1842–1919):

> ***Two point objects are just resolved when the first dark fringe in the diffraction pattern of one falls directly on the central bright fringe in the diffraction pattern of the other.***

The minimum angle θ_{min} between the two objects in the drawing is that given by Equation 27.5. If θ_{min} is small and is expressed in radians, $\sin \theta_{min} \approx \theta_{min}$. Then, Equation 27.5 can be rewritten as

$$\theta_{min} \approx 1.22 \frac{\lambda}{D} \qquad (\theta_{min} \text{ in radians}) \qquad (27.6)$$

For a given wavelength λ and aperture diameter D, this result specifies the smallest angle that two point objects can subtend at the aperture and still be resolved. According to Equation 27.6, optical instruments designed to resolve closely spaced objects (small values of θ_{min}) must utilize the smallest possible wavelength and the largest possible aperture diameter. For example, when short-wavelength ultraviolet light is collected by its large 2.4-m-diameter mirror, the Hubble Space Telescope is capable of resolving two closely spaced stars that have an angular separation of about $\theta_{min} = 1 \times 10^{-7}$ rad. This angle is equivalent to resolving two objects only 1 cm apart when they are 1×10^5 m (about 62 miles) from the telescope. Example 7 deals with the resolving power of the human eye.

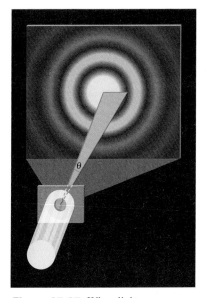

Figure 27.27 When light passes through a small circular opening, a circular diffraction pattern is formed on a screen. The angle θ locates the first dark fringe relative to the central bright region. The intensities of the bright fringes and the diameter of the opening have been exaggerated for clarity.

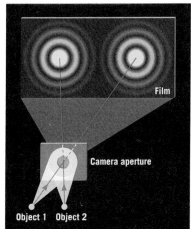

Figure 27.28 When light from two point objects passes through the circular aperture of a camera, two circular diffraction patterns are formed as images on the film. The images here are completely separated or resolved because the objects are widely separated.

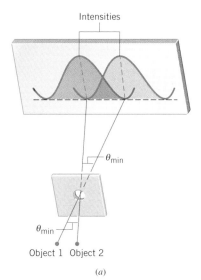

(a)

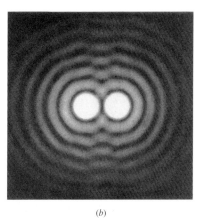

(b)

Figure 27.29 (a) According to the Rayleigh criterion, two point objects are just resolved when the first dark fringe (zero intensity) of one of the images falls on the central bright fringe (maximum intensity) of the other image. (b) This photograph shows two overlapping but still resolvable diffraction patterns. (From Michel Cagnet et al., *Atlas of Optical Phenomena*, Springer-Verlag, Berlin.)

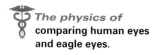

The physics of
**comparing human eyes
and eagle eyes**.

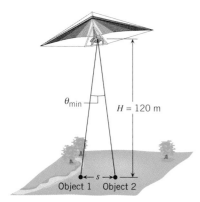

Figure 27.30 The Rayleigh criterion can be used to estimate the smallest distance s that can separate two objects on the ground, if a person on a hang glider is to have any hope of distinguishing one object from the other.

Problem solving insight

The minimum angle θ_{min} between two objects that are just resolved must be expressed in radians, not degrees, when using Equation 27.6 ($\theta_{min} \approx 1.22 \; \lambda/D$).

Example 7 The Human Eye Versus the Eagle's Eye

(a) A hang glider is flying at an altitude of $H = 120$ m. Green light (wavelength = 555 nm in vacuum) enters the pilot's eye through a pupil that has a diameter of $D = 2.5$ mm. Determine how far apart two point objects must be on the ground if the pilot is to have any hope of distinguishing between them (see Figure 27.30). (b) An eagle's eye has a pupil with a diameter of $D = 6.2$ mm. Repeat part (a) for an eagle flying at the same altitude as the glider.

Reasoning According to the Rayleigh criterion, the two objects must be separated by a distance s sufficient to subtend an angle $\theta_{min} \approx 1.22\lambda/D$ at the pupil of the pilot's eye.* Remembering to use radians for the angle, we will relate θ_{min} to the altitude H and the desired distance s.

Solution

(a) Using the Rayleigh criterion, we find

$$\theta_{min} \approx 1.22 \frac{\lambda}{D} = 1.22 \left(\frac{555 \times 10^{-9} \text{ m}}{2.5 \times 10^{-3} \text{ m}} \right) = 2.7 \times 10^{-4} \text{ rad} \qquad (27.6)$$

According to Equation 8.1, θ_{min} in radians is $\theta_{min} \approx s/H$, so

$$s \approx \theta_{min} H = (2.7 \times 10^{-4} \text{ rad})(120 \text{ m}) = \boxed{0.032 \text{ m}}$$

(b) Since the pupil of an eagle's eye is larger than that of a human eye, diffraction creates less of a limitation for the eagle. A calculation like that above, using $D = 6.2$ mm, reveals that the diffraction limit for the eagle is $\boxed{s = 0.013 \text{ m}}$.

Many optical instruments have a resolving power that exceeds that of the human eye. The typical camera does, for instance. Conceptual Example 8 compares the abilities of the human eye and a camera to resolve two closely spaced objects.

Conceptual Example 8 What You See Is Not What You Get

The French postimpressionist artist Georges Seurat developed a technique of painting in which dots of color are placed close together on the canvas. From sufficiently far away the in-

Figure 27.31 This person is about to take a photograph of a famous painting by Georges Seurat, who developed the technique of using tiny dots of color to construct his images. Conceptual Example 8 discusses what the person sees when the film is developed. (Georges Seurat, *A Sunday on La Grande Jatte.* © 2003. All rights reserved. Reproduced with permission of The Art Institute of Chicago.)

*In applying the Rayleigh criterion, we use the given vacuum wavelength because it is nearly identical to the wavelength in air. We use the wavelength in air or vacuum, even though the diffraction occurs within the eye, where the index of refraction is about $n = 1.36$ and the wavelength is $\lambda_{eye} = \lambda_{vacuum}/n$, according to Equation 27.3. The reason is that in entering the eye, the light is refracted according to Snell's law, which also includes an effect due to the index of refraction. If the angle of incidence is small, the effect of n in Snell's law cancels the effect of n in Equation 27.3, to a good degree of approximation.

dividual dots are not distinguishable, and the images in the picture take on a more normal appearance. Figure 27.31 shows a person in an art museum, looking at one of Seurat's paintings. Suppose that this person stands close to the painting, then backs up until the dots just become indistinguishable to his eyes and takes a picture from this position. When he gets home and has the film developed, however, he can see the individual dots in the enlarged photograph. Why does the camera resolve the individual dots, while his eyes do not? Assume that light enters his eyes through pupils that have diameters of 2.5 mm and enters the camera through an aperture or opening with a diameter of 25 mm.

Reasoning and Solution To answer this question, we turn to the Rayleigh criterion for resolving two point objects—namely, $\theta_{min} \approx 1.22 \, \lambda/D$. According to the Rayleigh criterion, a larger value of the diameter D of the opening means that the value for θ_{min} is smaller, so that an optical instrument can be farther away and still resolve the objects. In other words, a larger value of D means a greater resolving power. For the eye and the camera the diameters are $D_{eye} = 2.5$ mm and $D_{camera} = 25$ mm, so the diameter for the camera is ten times larger than that for the eye. Thus, at the distance at which the eye loses its ability to resolve the individual dots in the painting, the camera can easily resolve the dots. As discussed in the footnote to Example 7, we can ignore the effect on the wavelength of the index of refraction of the material from which the eye is made. We conclude, then, that *the camera succeeds in resolving the dots in the painting while the eye fails because the camera gathers light through a much larger opening.*

Related Homework: *Conceptual Question 14, Problem 32*

27.7 The Diffraction Grating

Diffraction patterns of bright and dark fringes occur when monochromatic light passes through a single or double slit. Fringe patterns also result when light passes through more than two slits, and an arrangement consisting of a large number of parallel, closely spaced slits called a *diffraction grating,* has proved very useful. Gratings with as many as 40 000 slits per centimeter can be made, depending on the production method. In one method a diamond-tipped cutting tool is used to inscribe closely spaced parallel lines on a glass plate, the spaces between the lines serving as the slits. In fact, the number of slits per centimeter is often quoted as the number of lines per centimeter.

The physics of **the diffraction grating.**

Figure 27.32 illustrates how light travels to a distant viewing screen from each of five slits in a grating and forms the central bright fringe and the first-order bright fringes on either side. Higher-order bright fringes are also formed but are not shown in the drawing. Each bright fringe is located by an angle θ relative to the central fringe. These bright fringes are sometimes called the *principal fringes* or *principal maxima,* since they are places where the light intensity is a maximum. The term "principal" distinguishes them from other, much less bright fringes that are referred to as secondary fringes or secondary maxima.

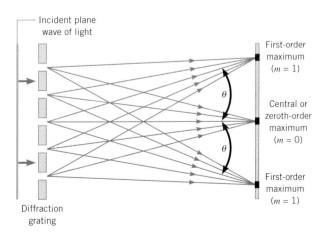

Figure 27.32 When light passes through a diffraction grating, a central bright fringe ($m = 0$) and higher-order bright fringes ($m = 1, 2, \ldots$) form when the light falls on a distant viewing screen.

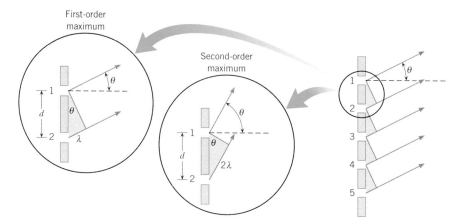

Figure 27.33 The conditions shown here lead to the first- and second-order intensity maxima in the diffraction pattern produced by the diffraction grating on the right.

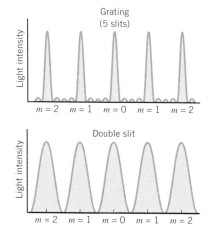

Figure 27.34 The bright fringes produced by a diffraction grating are much narrower than those produced by a double slit. Note the three small secondary bright fringes between the principal bright fringes of the grating. For a large number of slits, these secondary fringes become very small.

Constructive interference creates the principal fringes. To show how, we assume the screen is far from the grating, so that the rays remain nearly parallel while the light travels toward the screen, as in Figure 27.33. In reaching the place on the screen where the first-order maximum is located, light from slit 2 travels a distance of one wavelength farther than light from slit 1. Similarly, light from slit 3 travels one wavelength farther than light from slit 2, and so forth, as emphasized by the four colored right triangles on the right-hand side of the drawing. For the first-order maximum, the enlarged view of one of these right triangles shows that constructive interference occurs when $\sin\theta = \lambda/d$, where d is the separation between the slits. The second-order maximum forms when the extra distance traveled by light from adjacent slits is two wavelengths, so that $\sin\theta = 2\lambda/d$. The general result is

Principal maxima of a diffraction grating
$$\sin\theta = m\frac{\lambda}{d} \qquad m = 0, 1, 2, 3, \ldots \qquad (27.7)$$

The separation d between the slits can be calculated from the number of slits per centimeter of grating; for instance, a grating with 2500 slits per centimeter has a slit separation of $d = (1/2500)$ cm $= 4.0 \times 10^{-4}$ cm. Equation 27.7 is identical to Equation 27.1 for the double slit. A grating, however, produces bright fringes that are much *narrower* or *sharper* than those from a double slit, as the intensity patterns in Figure 27.34 reveal. Note also from the drawing that between the principal maxima of a diffraction grating there are secondary maxima with much smaller intensities.

The next example illustrates the ability of a grating to separate the components in a mixture of colors.

Example 9   Separating Colors with a Diffraction Grating

A mixture of violet light ($\lambda = 410$ nm in vacuum) and red light ($\lambda = 660$ nm in vacuum) falls on a grating that contains 1.0×10^4 lines/cm. For each wavelength, find the angle θ that locates the first-order maxima.

Reasoning Before Equation 27.7 can be used here, a value for the separation d between the slits is needed: $d = 1/(1.0 \times 10^4$ lines/cm$) = 1.0 \times 10^{-4}$ cm, or 1.0×10^{-6} m. For violet light, the angle θ_{violet} for the first-order maxima ($m = 1$) is given by $\sin\theta_{violet} = m\lambda_{violet}/d$, with an analogous equation applying for the red light.

Solution For violet light, the angle locating the first-order maxima is

$$\theta_{violet} = \sin^{-1}\frac{\lambda_{violet}}{d} = \sin^{-1}\left(\frac{410 \times 10^{-9} \text{ m}}{1.0 \times 10^{-6} \text{ m}}\right) = \boxed{24°}$$

For red light, a similar calculation with $\lambda_{red} = 660 \times 10^{-9}$ m shows that $\boxed{\theta_{red} = 41°}$. Because θ_{violet} and θ_{red} are different, separate first-order bright fringes are seen for violet and red light on a viewing screen.

If the light in Example 9 had been sunlight, the angles for the first-order maxima would cover all values in the range between 24° and 41°, since sunlight contains all colors or wavelengths between violet and red. Consequently, a rainbow-like dispersion of the colors would be observed to either side of the central fringe on a screen, as can be seen in Figure 27.35. This drawing shows that the spectrum of colors associated with the *m* = 2 order is completely separate from that of the *m* = 1 order. For higher orders, however, the spectra from adjacent orders may overlap (see problems 43 and 55). The central maximum (*m* = 0) is white because all the colors overlap there.

An instrument designed to measure the angles at which the principal maxima of a grating occur is called a grating spectroscope. With a measured value of the angle, calculations such as those in Example 9 can be turned around to provide the corresponding value of the wavelength. As we will point out in Chapter 30, the atoms in a hot gas emit discrete wavelengths, and determining the values of these wavelengths is one important technique used to identify the atoms. Figure 27.36 shows the principle of a grating spectroscope. The slit that admits light from the source (e.g., a hot gas) is located at the focal point of the collimating lens, so the light rays striking the grating are parallel. The telescope is used to detect the bright fringes and, hence, to measure the angle θ.

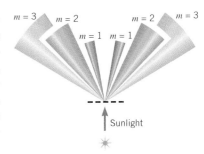

Figure 27.35 When sunlight falls on a diffraction grating, a rainbow of colors is produced at each principal maximum (*m* = 1, 2, . . .). The central maximum (*m* = 0), however, is white but is not shown in the drawing.

The physics of **a grating spectroscope.**

Check Your Understanding 5

The drawing shows a top view of a diffraction grating and the *m*th-order principal maxima that are obtained with red and blue light. Red light has the longer wavelength. (a) Which principal maximum is associated with blue light? (b) If the number of slits per centimeter in the grating were increased, would these two principal maxima move away from the central maximum, move toward the central maximum, or remain in the same place? *(The answers are given at the end of the book.)*

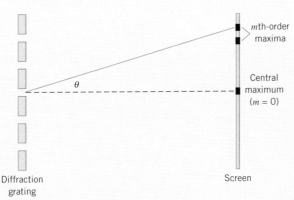

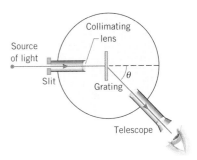

Figure 27.36 A grating spectroscope.

Background: The angle θ that specifies the *m*th principal maximum of a diffraction grating depends on two parameters—the wavelength λ of the light and the separation *d* between the slits. This question deals with both of these parameters.

For similar questions (including calculational counterparts), consult Self-Assessment Test 27.2, which is described at the end of Section 27.9.

*27.8 Compact Discs, Digital Video Discs, and the Use of Interference

The compact disc (CD) and the digital video disc (DVD) have revolutionized how text, graphics, music, and movies are stored for use in computers, stereo sound systems, and televisions. The operation of these discs uses interference effects in some interesting ways. Each contains a spiral track that holds the information, which is detected using a

The physics of
retrieving information from compact
discs and digital video discs.

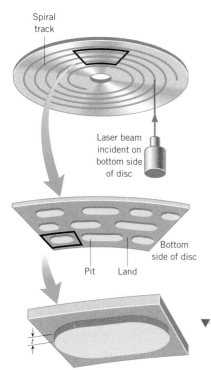

Spiral
track

Laser beam
incident on
bottom side
of disc

Bottom
side of disc

Pit   Land

Figure 27.37 The bottom surface of a compact disc (CD) or digital video disc (DVD) carries information in the form of raised areas ("pits") and flat areas ("land") along a spiral track. A CD or DVD is played by using a laser beam that strikes the bottom surface and reflects from it.

The physics of
the three-beam tracking method
for compact discs.

laser beam that reflects from the bottom of the disc, as Figure 27.37 illustrates. The information is encoded in the form of raised areas on the bottom of the disc. Under a microscope these raised areas appear as "pits" when viewed from the *top* or labeled side of the disc. They are separated by flat areas called "land." The pits and land are covered with a transparent plastic coating, which has been omitted from the drawing for simplicity.

As the disc rotates, the laser beam reflects off the bottom and into a detector. The reflected light intensity fluctuates as the pits and land areas pass by, and the fluctuations convey the information as a series of binary numbers (zeros and ones). To make the fluctuations easier to detect, the pit thickness t (see Figure 27.37) is chosen with destructive interference in mind. As the laser beam overlaps the edges of a pit, part of the beam is reflected from the raised pit surface and part from the land. The part that reflects from the land travels an additional distance of $2t$. The thickness of the pits is chosen so that $2t$ is one-half of a wavelength of the laser beam in the plastic coating. With this choice, destructive interference occurs when the two parts of the reflected beam combine. As a result, there is markedly less reflected intensity when the laser beam passes over a pit edge than when it passes over other places on the surface. Thus, the fluctuations in reflected light that occur while the disc rotates are large enough to detect because of the effects of destructive interference. Example 10 determines the theoretical thickness of the pits on a compact disc. In reality, a value slightly less than that obtained in the example is used for technical reasons that are not pertinent here.

Example 10   Pit Thickness on a Compact Disc

The laser in a CD player has a wavelength of 780 nm in a vacuum. The plastic coating over the pits has an index of refraction of $n = 1.5$. Find the thickness of the pits on a CD.

Reasoning As we have discussed, the thickness t is chosen so that $2t = \frac{1}{2}\lambda_{\text{coating}}$ in order to achieve destructive interference. Equation 27.3 gives the wavelength in the plastic coating as $\lambda_{\text{coating}} = \lambda_{\text{vacuum}}/n$.

Solution The thickness of the pits is

$$t = \frac{\lambda_{\text{coating}}}{4} = \frac{\lambda_{\text{vacuum}}}{4n} = \frac{780 \times 10^{-9}\text{ m}}{4(1.5)} = \boxed{1.3 \times 10^{-7}\text{ m}}$$

The pit thickness calculated in Example 10 applies only to a CD. The pit thickness (and other dimensions as well) are smaller on a DVD because the lasers used for DVDs have smaller wavelengths (635 nm, for example). The fact that the pit dimensions are smaller is one of the reasons that a DVD has more information storage capacity than a CD, from seven to twenty-six times more, depending on the type of DVD.

As a CD or DVD rotates, the laser beam must accurately follow the pits and land along the spiral track. One type of CD player utilizes a three-beam tracking method to ensure that the laser beam follows the spiral properly. Figure 27.38 shows that a diffraction grating is the key element in this method. Before the laser beam strikes the CD, the beam passes through a grating that produces a central maximum and two first-order maxima, one on either side. As the picture indicates, the central maximum beam falls directly on the spiral track. This beam reflects into a detector, and the reflected light intensity fluctuates as the pits and land areas pass by, the fluctuations conveying the information. The two first-order maxima beams are called *tracking beams*. They hit the CD between the arms of the spiral and also reflect into detectors of their own. Under perfect conditions, the intensities of the two reflected tracking beams do not fluctuate, since they originate from the smooth surface between the arms of the spiral where there are no pits. As a result, each tracking-beam detector puts out the same constant electrical signal. However, if the tracking drifts to either side, the reflected intensity of each tracking beam changes because of the pits. In response, the tracking-beam detectors produce different electrical signals. The difference between the signals is used in a "feedback" circuit to correct for the drift and put the three beams back into their proper positions.

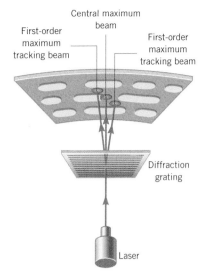

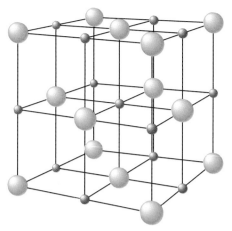

Figure 27.38 A three-beam tracking method is sometimes used in CD players to ensure that the laser follows the spiral track correctly. The three beams are derived from a single laser beam by using a diffraction grating.

Figure 27.39 In this drawing of the crystalline structure of sodium chloride, the small red spheres represent positive sodium ions, and the large blue spheres represent negative chloride ions.

27.9 X-Ray Diffraction

Not all diffraction gratings are commercially made. Nature also creates diffraction gratings, although these gratings do not look like an array of closely spaced slits. Instead, nature's gratings are the arrays of regularly spaced atoms that exist in crystalline solids. For example, Figure 27.39 shows the structure of a crystal of ordinary salt (NaCl). Typically, the atoms in a crystalline solid are separated by distances of about 1.0×10^{-10} m, so we might expect a crystalline array of atoms to act like a grating with roughly this "slit" spacing for electromagnetic waves of the appropriate wavelength. Assuming that $\sin \theta = 0.5$ and that $m = 1$ in Equation 27.7, then $0.5 = \lambda/d$. A value of $d = 1.0 \times 10^{-10}$ m in this equation gives a wavelength of $\lambda = 0.5 \times 10^{-10}$ m. This wavelength is much shorter than that of visible light and falls in the X-ray region of the electromagnetic spectrum. (See Figure 24.10.)

A diffraction pattern does indeed result when X-rays are directed onto a crystalline material, as Figure 27.40*a* illustrates for a crystal of NaCl. The pattern consists of a complicated arrangement of spots because a crystal has a complex three-dimensional structure. It is from patterns such as this that the spacing between atoms and the nature of the crystal structure can be determined. X-ray diffraction has also been applied with great success toward understanding the structure of biologically important molecules, such as proteins and nucleic acids. One of the most famous results was the discovery in 1953 by Watson and Crick that the structure of the nucleic acid DNA is a double helix. X-ray diffraction patterns such as that in Figure 27.40*b* played the pivotal role in their research.

The physics of **X-ray diffraction**.

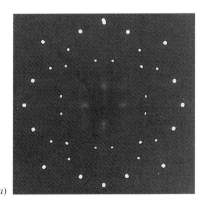

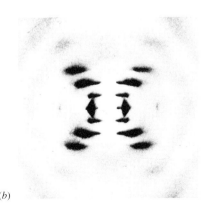

(*a*) (*b*)

Figure 27.40 The X-ray diffraction patterns from (*a*) crystalline NaCl and (*b*) crystalline DNA. This image of DNA was obtained by Rosalind Franklin in 1953, the year in which Watson and Crick discovered DNA's structure. (*a*: Courtesy Edwin Jones, University of South Carolina; *b*: © Omikron/ Photo Researchers)

Self-Assessment Test 27.2

Test your understanding of the material in Sections 27.5–27.9:

• Diffraction   • Resolving Power   • The Diffraction Grating
• Compact Discs, Digital Video Discs, and the Use of Interference   • X-Ray Diffraction

Go to **www.wiley.com/college/cutnell**

27.10 Concepts & Calculations

The ability to exhibit interference effects is a fundamental characteristic of any kind of wave. Our understanding of these effects depends on the principle of linear superposition, which we first encountered in Chapter 17. Only by means of this principle can we understand the constructive and destructive interference of light waves that lie at the heart of every topic in this chapter. Therefore, it is fitting that we review the essence of the principle in the next example.

Concepts & Calculations Example 11
The Principle of Linear Superposition

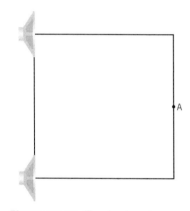

Figure 27.41 Two loudspeakers are located at the corners of a square and produce identical sound waves. As Example 11 discusses, constructive interference occurs at point A, which is at the midpoint of the opposite side of the square, and destructive interference occurs at either empty corner.

A square is 3.5 m on a side, and point A is the midpoint of one of its sides. On the side opposite this spot, two in-phase loudspeakers are located at adjacent corners, as Figure 27.41 indicates. Standing at point A, you hear a loud sound because constructive interference occurs between the identical sound waves coming from the speakers. As you walk along the side of the square toward either empty corner, the loudness diminishes gradually but does not entirely disappear until you reach either empty corner, where you hear no sound at all. Thus, at each empty corner destructive interference occurs. Find the wavelength of the sound waves.

Concept Questions and Answers Why does constructive interference occur at point A?

Answer The loudspeakers are in-phase sources of identical sound waves. The waves from one speaker travel a distance ℓ_1 in reaching point A and the waves from the second speaker travel a distance ℓ_2. The condition that leads to constructive interference is $\ell_2 - \ell_1 = m\lambda$, where λ is the wavelength of the waves and $m = 0, 1, 2, 3. \ldots$. Thus, the two distances are the same or differ by an integer number of wavelengths. Point A is the midpoint of a side of the square, so that ℓ_1 and ℓ_2 are the same, and constructive interference occurs.

What is the general condition that leads to destructive interference?

Answer The waves start out in phase, so the general condition that leads to destructive interference is $\ell_2 - \ell_1 = (m + \frac{1}{2})\lambda$, where $m = 0, 1, 2, 3. \ldots$.

The general condition that leads to destructive interference entails a number of possibilities. Which one of them, if any, applies at either empty corner of the square?

Answer The general condition that leads to destructive interference is $\ell_2 - \ell_1 = (m + \frac{1}{2})\lambda$, where $m = 0, 1, 2, 3. \ldots$. There are many possible values for m, and we are being asked what the specific value is. As you walk from point A toward either empty corner, one of the distances decreases and the other increases; the loudness diminishes gradually and disappears at either empty corner. Therefore, destructive interference occurs for the first time at the corner, where the difference between the distances has attained the smallest possible value of $\ell_2 - \ell_1 = \frac{1}{2}\lambda$. This means that we are dealing with the specific case in which $m = 0$.

Solution Consider the destructive interference that occurs at either empty corner. Using L to denote the length of a side of the square and taking advantage of the Pythagorean theorem, we have

$$\ell_1 = L \quad \text{and} \quad \ell_2 = \sqrt{L^2 + L^2} = \sqrt{2}\,L$$

The specific condition for the destructive interference at the empty corner is

$$\ell_2 - \ell_1 = \sqrt{2}\,L - L = \tfrac{1}{2}\lambda$$

Solving for the wavelength of the waves gives

$$\lambda = 2L\,(\sqrt{2} - 1) = 2(3.5\ \text{m})\,(\sqrt{2} - 1) = \boxed{2.9\ \text{m}}$$

The next example deals with thin-film interference and serves as a review of the factors that must be considered in such cases.

Concepts & Calculations Example 12   A Soap Film

▼

A soap film ($n = 1.33$) is 375 nm thick and is surrounded on both sides by air. Sunlight, whose wavelengths (in vacuum) extend from 380 to 750 nm, strikes the film nearly perpendicularly. For which wavelength(s) in this range does constructive interference cause the film to look bright in reflected light?

Concept Questions and Answers What, if any, phase change occurs when light, traveling in air, reflects from the air–film interface?

Answer   In air the index of refraction is nearly $n = 1$, whereas in the film it is $n = 1.33$. A phase change occurs whenever light travels through a material with a smaller refractive index toward a material with a larger refractive index and reflects from the boundary between the two. The phase change is equivalent to $\frac{1}{2}\lambda_{film}$, where λ_{film} is the wavelength in the film.

What, if any, phase change occurs when light, traveling in the film, reflects from the film–air interface?

Answer   In the film the index of refraction is $n = 1.33$, whereas in air it is nearly $n = 1$. A phase change does *not* occur when light travels through a material with a larger refractive index toward a material with a smaller refractive index and reflects from the boundary between the two.

Is the wavelength of the light in the film greater than, smaller than, or equal to the wavelength in a vacuum?

Answer   The wavelength of the light in the film (refractive index $= n$) is given by Equation 27.3 as $\lambda_{film} = \lambda_{vacuum}/n$. Since the refractive index of the film is $n = 1.33$, the wavelength in the film is less than the wavelength in a vacuum.

Solution Figure 27.42 shows the soap film and the two rays of light that represent the interfering light waves. At nearly perpendicular incidence, ray 2 travels a distance of $2t$ farther than ray 1, where t is the thickness of the film. In addition, as discussed in the Concept Questions, ray 2 experiences no phase change upon reflection at the bottom film surface, while ray 1 does experience a phase shift of $\frac{1}{2}\lambda_{film}$ at the upper film surface. Therefore, the net phase change for the two reflected rays is $\frac{1}{2}\lambda_{film}$. We must combine this amount with the extra travel distance to determine the condition for constructive interference. For constructive interference, the combined total must be an integer number of wavelengths in the film:

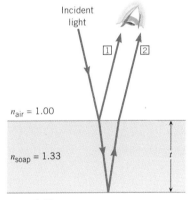

$n_{air} = 1.00$

$n_{soap} = 1.33$

$n_{air} = 1.00$

Figure 27.42 The constructive interference that occurs with this thin soap film is analyzed in Example 12.

$$
\underbrace{2t}_{\substack{\text{Extra distance} \\ \text{traveled by} \\ \text{ray 2}}} + \underbrace{\tfrac{1}{2}\lambda_{film}}_{\substack{\text{Half-wavelength} \\ \text{net phase change} \\ \text{due to reflection}}} = \underbrace{\lambda_{film},\, 2\lambda_{film},\, 3\lambda_{film},\, \cdots}_{\substack{\text{Condition for} \\ \text{constructive interference}}}
$$

Subtracting the term $\frac{1}{2}\lambda_{film}$ from the left side of this equation and from each term on the right side yields

$$
2t = \underbrace{\tfrac{1}{2}\lambda_{film},\, \tfrac{3}{2}\lambda_{film},\, \tfrac{5}{2}\lambda_{film},\, \cdots}_{(m + \frac{1}{2})\lambda_{film} \quad m = 0, 1, 2, 3, \ldots}
$$

Solving for the wavelength gives

$$
\lambda_{film} = \frac{2t}{m + \frac{1}{2}} \qquad m = 0, 1, 2, 3, \ldots
$$

According to Equation 27.3, we have

$$
\lambda_{film} = \lambda_{vacuum}/n
$$

Using this substitution in our result for λ_{film}, we obtain

$$
\lambda_{vacuum} = \frac{2nt}{m + \frac{1}{2}} \qquad m = 0, 1, 2, 3, \ldots
$$

For the first four values of m and the given values for n and t, we find

$$m = 0 \qquad \lambda_{\text{vacuum}} = \frac{2nt}{m + \frac{1}{2}} = \frac{2(1.33)(375 \text{ nm})}{0 + \frac{1}{2}} = 2000 \text{ nm}$$

$m = 1 \qquad \lambda_{\text{vacuum}} = 665 \text{ nm}$

$m = 2 \qquad \lambda_{\text{vacuum}} = 399 \text{ nm}$

$m = 3 \qquad \lambda_{\text{vacuum}} = 285 \text{ nm}$

The range of visible wavelengths (in vacuum) extends from 380 to 750 nm. Therefore, the visible wavelengths in which the film appears bright due to constructive interference are $\boxed{399 \text{ and } 665 \text{ nm}}$.

At the end of the problem set for this chapter, you will find homework problems that contain both conceptual and quantitative parts. These problems are grouped under the heading *Concepts & Calculations, Group Learning Problems.* They are designed for use by students working alone or in small learning groups. The conceptual part of each problem provides a convenient focus for group discussions.

Concept Summary

This summary presents an abridged version of the chapter, including the important equations and all available learning aids. For convenient reference, the learning aids (including the text's examples) are placed next to or immediately after the relevant equation or discussion. The following learning aids may be found on-line at **www.wiley.com/college/cutnell**:

Interactive LearningWare examples are solved according to a five-step interactive format that is designed to help you develop problem-solving skills.	**Concept Simulations** are animated versions of text figures or animations that illustrate important concepts. You can control parameters that affect the display, and we encourage you to experiment.
Interactive Solutions offer specific models for certain types of problems in the chapter homework. The calculations are carried out interactively.	**Self-Assessment Tests** include both qualitative and quantitative questions. Extensive feedback is provided for both incorrect and correct answers, to help you evaluate your understanding of the material.

Topic	Discussion	Learning Aids
	27.1 The Principle of Linear Superposition	
Principle of linear superposition	The principle of linear superposition states that when two or more waves are present simultaneously in the same region of space, the resultant disturbance is the sum of the disturbances from the individual waves.	
Constructive interference	Constructive interference occurs at a point when two waves meet there crest-to-crest and trough-to-trough, thus reinforcing each other. When two waves that start out in phase and have traveled some distance meet at a point, constructive interference occurs whenever the travel distances are the same or differ by any integer number of wavelengths: $\ell_2 - \ell_1 = m\lambda$, where ℓ_1 and ℓ_2 are the distances traveled by the waves, and $m = 0, 1, 2, 3, \ldots$.	**Example 11**
Destructive interference	Destructive interference occurs at a point when two waves meet there crest-to-trough, thus mutually canceling each other. When two waves that start out in phase and have traveled some distance meet at a point, destructive interference occurs whenever the travel distances differ by any odd integer number of half-wavelengths: $\ell_2 - \ell_1 = (m + \frac{1}{2})\lambda$, where ℓ_1 and ℓ_2 are the distances traveled by the waves, and $m = 0, 1, 2, 3, \ldots$.	

Topic	Discussion	Learning Aids

Coherent sources

Two sources are coherent if the waves they emit maintain a constant phase relation. In other words, the waves do not shift relative to one another as time passes. If constructive and destructive interference are to be observed, coherent sources are necessary.

27.2 Young's Double-Slit Experiment

Double-slit experiment

In Young's double-slit experiment, light passes through a pair of closely spaced narrow slits and produces a pattern of alternating bright and dark fringes on a viewing screen. The fringes arise because of constructive and destructive interference. The angle θ that locates the mth-order bright fringe is given by

Bright fringes of a double slit

$$\sin \theta = \frac{m\lambda}{d} \qquad m = 0, 1, 2, 3, \ldots \qquad (27.1)$$

Examples 1, 2
Interactive LearningWare 27.1
Interactive Solution 27.7

where λ is the wavelength of the light, and d is the spacing between the slits. The angle that locates the mth dark fringe is given by

Dark fringes of a double slit

$$\sin \theta = \frac{(m + \frac{1}{2})\lambda}{d} \qquad m = 0, 1, 2, 3, \ldots \qquad (27.2)$$

27.3 Thin-Film Interference

Constructive and destructive interference of light waves can occur with thin films of transparent materials. The interference occurs between light waves that reflect from the top and bottom surfaces of the film.

One important factor in thin-film interference is the thickness of a film relative to the wavelength of the light within the film. The wavelength λ_{film} within a film is

Wavelength within a film

$$\lambda_{\text{film}} = \frac{\lambda_{\text{vacuum}}}{n} \qquad (27.3)$$

Examples 3, 4, 5, 12
Interactive LearningWare 27.2
Interactive Solution 27.17

where λ_{vacuum} is the wavelength in a vacuum, and n is the index of refraction of the film.

Phase change due to reflection

A second important factor is the phase change that can occur when light reflects at each surface of the film:

1. When light travels through a material with a smaller index of refraction toward a material with a larger index of refraction, reflection at the boundary occurs along with a phase change that is equivalent to one-half a wavelength in the film.

2. When light travels through a material with a larger index of refraction toward a material with a smaller index of refraction, there is no phase change upon reflection at the boundary.

27.4 The Michelson Interferometer

Michelson interferometer

An interferometer is an instrument that can be used to measure the wavelength of light by employing interference between two light waves. The Michelson interferometer splits the light into two beams, one of which travels to a fixed mirror, reflects from it, and returns. The other beam travels to a movable mirror, reflects from it, and returns. When the two returning beams are combined, interference is observed, the amount of which depends on the travel distances.

 Use *Self-Assessment Test 27.1* to evaluate your understanding of Sections 27.1–27.4.

Topic	Discussion	Learning Aids
	27.5 Diffraction	
Diffraction	Diffraction is the bending of waves around obstacles or around the edges of an opening. Diffraction is an interference effect that can be explained with the aid of Huygens' principle. This principle states that every point on a wave front acts as a source of tiny wavelets that move forward with the same speed as the wave; the wave front at a later instant is the surface that is tangent to the wavelets.	
Huygens' principle		
	When light passes through a single narrow slit and falls on a viewing screen, a pattern of bright and dark fringes is formed because of the superposition of Huygens wavelets. The angle θ that specifies the mth dark fringe on either side of the central bright fringe is given by	Concept Simulation 27.1
Dark fringes for single-slit diffraction	$$\sin\theta = m\frac{\lambda}{W} \qquad m = 1, 2, 3, \ldots \qquad (27.4)$$ where λ is the wavelength of the light and W is the width of the slit.	Example 6 Interactive LearningWare 27.3
	27.6 Resolving Power	
Resolving power	The resolving power of an optical instrument is the ability of the instrument to distinguish between two closely spaced objects. Resolving power is limited by the diffraction that occurs when light waves enter an instrument, often through a circular opening.	
Rayleigh criterion	The Rayleigh criterion specifies that two point objects are just resolved when the first dark fringe in the diffraction pattern of one falls directly on the central bright fringe in the diffraction pattern of the other. According to this specification, the minimum angle (in radians) that two point objects can subtend at a circular aperture of diameter D and still be resolved as separate objects is	
Minimum angle for resolving two point objects	$$\theta_{\min} \approx 1.22\frac{\lambda}{D} \qquad (\theta_{\min} \text{ in radians}) \qquad (27.6)$$ where λ is the wavelength of the light.	Examples 7, 8 Interaction Solution 27.35
	27.7 The Diffraction Grating	
Diffraction grating	A diffraction grating consists of a large number of parallel, closely spaced slits. When light passes through a diffraction grating and falls on a viewing screen, the light forms a pattern of bright and dark fringes. The bright fringes are referred to as principal maxima and are found at an angle θ such that	
Principal maxima of a diffraction grating	$$\sin\theta = m\frac{\lambda}{d} \qquad m = 0, 1, 2, 3, \ldots \qquad (27.7)$$ where λ is the wavelength of the light and d is the separation between two adjacent slits.	Example 9
	27.8 Compact Discs, Digital Video Discs, and the Use of Interference	
	Compact discs and digital video discs depend on intereference for their operation.	Example 10
	27.9 X-Ray Diffraction	
	A diffraction pattern forms when X-rays are directed onto a crystalline material. The pattern arises because the regularly spaced atoms in a crystal act like a diffraction grating. Because the spacing is extremely small, on the order of 1×10^{-10} m, the wavelength of the electromagnetic waves must also be very small—hence, the use of X-rays. The crystal structure of a material can be determined from its X-ray diffraction pattern.	

    Use Self-Assessment Test 27.2 **to evaluate your understanding of Sections 27.5–27.9.**

Conceptual Questions

1. Suppose that a radio station broadcasts simultaneously from two transmitting antennas at two different locations. Is it clear that your radio will have better reception with two transmitting antennas rather than one? Justify your answer.

2. Two sources of waves are in phase and produce identical waves. These sources are mounted at the corners of a square. At the center of the square waves from the sources produce constructive interference, no matter which two corners of the square are occupied by the sources. Explain why.

3. (a) How would the pattern of bright and dark fringes produced in a Young's double-slit experiment change if the light waves coming from *both* slits had their phases shifted by an amount equivalent to a half wavelength? (b) How would the pattern change if the light coming from *only one* of the slits had its phase shifted by an amount equivalent to a half wavelength?

4. Replace slits S_1 and S_2 in Figure 27.4 with identical in-phase loudspeakers and use the same ac electrical signal to drive them. The two sound waves produced will then be identical, and you will have the audio equivalent of Young's double-slit experiment. In terms of loudness and softness, describe what you would hear as you walked along the screen, starting from the center and going to either end.

5. In Young's double-slit experiment, is it possible to see interference fringes when the wavelength of the light is greater than the distance between the slits? Provide a reason for your answer.

6. A camera lens is covered with a nonreflective coating that eliminates the reflection of perpendicularly incident green light. Recalling Snell's law of refraction (see Section 26.2), would you expect the reflected green light to be eliminated if it were incident on the nonreflective coating at an angle of 45° rather than perpendicularly? Justify your answer.

7. When sunlight reflects from a thin film of soapy water, the film appears multicolored, in part because destructive interference removes different wavelengths from the light reflected at different places, depending on the thickness of the film. As the film becomes thinner and thinner, it looks darker and darker in reflected light, appearing black just before it breaks. The blackness means that destructive interference removes *all* wavelengths from the reflected light when the film is very thin. Explain why.

8. Two pieces of the same glass are covered with thin films of different materials. The thickness of each film is the same. In reflected sunlight, however, the films have different colors. Give a reason that accounts for the difference in colors.

9. In Figure 27.14b there is a dark spot at the center of the pattern of Newton's rings. By considering the phase changes that occur when light reflects from the upper curved surface and the lower flat surface, account for the dark spot.

10. A thin film of a material is floating on water ($n = 1.33$). When the material has a refractive index of $n = 1.20$, the film looks bright in reflected light as its thickness approaches zero. But when the material has a refractive index of $n = 1.45$, the film looks black in reflected light as its thickness approaches zero. Explain these observations in terms of constructive and destructive interference and the phase changes that occur when light waves undergo reflection.

11. A transparent coating is deposited on a glass plate and has a refractive index that is *larger than that of the glass,* not smaller, as it is for a typical nonreflective coating. For a certain wavelength within the coating, the thickness of the coating is a quarter wavelength. The coating *enhances* the reflection of the light corresponding to this wavelength. Explain why, referring to Example 3 in the text to guide your thinking.

12. On most cameras one can select the *f*-number setting, or *f*-stop. The *f*-number gives the ratio of the focal length of the camera lens to the diameter of the aperture through which light enters the camera. If one wishes to resolve two closely spaced objects in a picture, should a small or a large *f*-number setting be used? Account for your answer.

13. Explain why a sound wave diffracts much more than a light wave does when the two pass through the same doorway.

14. Review Conceptual Example 8 before answering this question. A person is viewing one of Seurat's paintings that consists of dots of color. She is so close to the painting that the dots are distinguishable. Without moving, however, she can squint, which makes the painting take on a more normal appearance. In terms of the Rayleigh criterion, why does squinting make the painting look more normal?

15. Four light bulbs are arranged at the corners of a rectangle that is three times longer than it is wide. You look at this arrangement perpendicular to the plane of the rectangle. From very far away, your eyes cannot resolve the individual bulbs and you see a single smear of light. From close in, you see the individual bulbs. Between these two extremes, what do you see? Draw two pictures to illustrate the possibilities that exist, depending on how far away you are. Explain your drawings.

16. Suppose the pupil of your eye were elliptical instead of circular in shape, with the long axis of the ellipse oriented in the vertical direction. (a) Would the resolving power of your eye be the same in the horizontal and vertical directions? (b) In which direction would the resolving power be greatest? Justify your answers by discussing how the diffraction of light waves would differ in the two directions.

17. Suppose you were designing an eye and could select the size of the pupil and the wavelengths of the electromagnetic waves to which the eye is sensitive. As far as the limitation created by diffraction is concerned, rank the following design choices in order of decreasing resolving power (greatest first): (a) large pupil and ultraviolet wavelengths, (b) small pupil and infrared wavelengths, and (c) small pupil and ultraviolet wavelengths. Justify your answer.

18. In our discussion of single-slit diffraction, we considered the ratio of the wavelength λ to the width W of the slit. We ignored the height of the slit, in effect assuming that the height was much larger than the width. Suppose the height and width were the same size, so that diffraction in both dimensions occurred. How would the diffraction pattern in Figure 27.19b be altered? Give your reasoning.

19. What would happen to the distance between the bright fringes produced by a diffraction grating if the entire interference apparatus (light source, grating, and screen) were immersed in water? Why?

Problems

ssm Solution is in the Student Solutions Manual.   **www** Solution is available on the World Wide Web at www.wiley.com/college/cutnell

⚕ This icon represents a biomedical application.

Section 27.1 The Principle of Linear Superposition,
Section 27.2 Young's Double-Slit Experiment

1. ssm A flat observation screen is placed at a distance of 4.5 m from a pair of slits. The separation on the screen between the central bright fringe and the first-order bright fringe is 0.037 m. The light illuminating the slits has a wavelength of 490 nm. Determine the slit separation.

2. In a Young's double-slit experiment, the seventh dark fringe is located 0.025 m to the side of the central bright fringe on a flat screen, which is 1.1 m away from the slits. The separation between the slits is 1.4×10^{-4} m. What is the wavelength of the light being used?

3. A rock concert is being held in an open field. Two loudspeakers are separated by 7.00 m. As an aid in arranging the seating, a test is conducted in which both speakers vibrate in phase and produce an 80.0-Hz bass tone simultaneously. The speed of sound is 343 m/s. A reference line is marked out in front of the speakers, perpendicular to the midpoint of the line between the speakers. Relative to either side of this reference line, what is the smallest angle that locates the places where destructive interference occurs? People seated in these places would have trouble hearing the 80.0-Hz bass tone.

4. Two in-phase sources of waves are separated by a distance of 4.00 m. These sources produce identical waves that have a wavelength of 5.00 m. On the line between them, there are two places at which the same type of interference occurs. (a) Is it constructive or destructive interference, and (b) where are the places located?

5. ssm In a Young's double-slit experiment, the angle that locates the second-order bright fringe is 2.0°. The slit separation is 3.8×10^{-5} m. What is the wavelength of the light?

* 6. Review Conceptual Example 2 before attempting this problem. Two slits are 0.158 mm apart. A mixture of red light (wavelength = 665 nm) and yellow-green light (wavelength = 565 nm) falls on the slits. A flat observation screen is located 2.24 m away. What is the distance on the screen between the third-order red fringe and the third-order yellow-green fringe?

* 7. Refer to **Interactive Solution 27.7** at **www.wiley.com/college/cutnell** for help in solving this problem. In a Young's double-slit experiment the separation y between the second-order bright fringe and the central bright fringe on a flat screen is 0.0180 m, when the light has a wavelength of 425 nm. Assume that the angles that locate the fringes on the screen are small enough so that $\sin \theta \approx \tan \theta$. Find the separation y when the light has a wavelength of 585 nm.

** 8. In Young's experiment a mixture of orange light (611 nm) and blue light (471 nm) shines on the double slit. The centers of the first-order bright blue fringes lie at the outer edges of a screen that is located 0.500 m away from the slits. However, the first-order bright orange fringes fall off the screen. By how much and in which direction (toward or away from the slits) should the screen be moved, so that the centers of the first-order bright orange fringes just appear on the screen? It may be assumed that θ is small, so that $\sin \theta \approx \tan \theta$.

** 9. ssm www A sheet of plastic ($n = 1.60$) covers *one slit* of a double slit (see the drawing). When the double slit is illuminated by monochromatic light ($\lambda_{\text{vacuum}} = 586$ nm), the center of the screen appears dark rather than bright. What is the minimum thickness of the plastic?

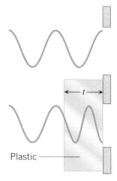

Plastic

Problem 9

Section 27.3 Thin-Film Interference

10. Light of wavelength 691 nm (in vacuum) is incident perpendicularly on a soap film ($n = 1.33$) suspended in air. What are the two smallest nonzero film thicknesses (in nm) for which the reflected light undergoes constructive interference?

11. ssm A nonreflective coating of magnesium fluoride ($n = 1.38$) covers the glass ($n = 1.52$) of a camera lens. Assuming that the coating prevents reflection of yellow-green light (wavelength in vacuum = 565 nm), determine the minimum nonzero thickness that the coating can have.

12. For background on this problem, review Conceptual Example 4. A mixture of yellow light (wavelength = 580 nm in vacuum) and violet light (wavelength = 410 nm in vacuum) falls perpendicularly on a film of gasoline that is floating on a puddle of water. For both wavelengths, the refractive index of gasoline is $n = 1.40$ and that of water is $n = 1.33$. What is the minimum nonzero thickness of the film in a spot that looks (a) yellow and (b) violet because of destructive interference?

13. A mixture of red light ($\lambda_{\text{vacuum}} = 661$ nm) and green light ($\lambda_{\text{vacuum}} = 551$ nm) shines perpendicularly on a soap film ($n = 1.33$) that has air on either side. What is the minimum nonzero thickness of the film, so that destructive interference causes it to look red in reflected light?

14. Review Conceptual Example 4 before beginning this problem. A soap film with different thicknesses at different places has an unknown refractive index n and air on both sides. In reflected light it looks multicolored. One region looks yellow because destructive interference has removed blue ($\lambda_{\text{vacuum}} = 469$ nm) from the reflected light, while another looks magenta because destructive interference has removed green ($\lambda_{\text{vacuum}} = 555$ nm). In these regions the film has the minimum nonzero thickness t required for the destructive interference to occur. Find the ratio $t_{\text{magenta}}/t_{\text{yellow}}$.

* 15. ssm www Orange light ($\lambda_{\text{vacuum}} = 611$ nm) shines on a soap film ($n = 1.33$) that has air on either side of it. The light strikes the film perpendicularly. What is the minimum thickness of the film for which constructive interference causes it to look bright in reflected light?

* 16. Review Conceptual Example 4 before attempting this problem. A film of gasoline ($n = 1.40$) floats on water ($n = 1.33$). Yellow light (wavelength = 580 nm in vacuum) shines perpendicularly on this film. (a) Determine the minimum nonzero thickness of the film, such that the film appears bright yellow due to constructive interference. (b) Repeat part (a), assuming that the gasoline film is on glass ($n = 1.52$) instead of water.

* 17. Consult **Interactive Solution 27.17** at **www.wiley.com/college/cutnell** to review a model for solving this problem. A film of oil lies on wet pavement. The refractive index of the oil exceeds that of the water. The film has the minimum nonzero thickness such that it appears dark due to destructive interference when viewed in red light (wavelength = 640.0 nm in vacuum). Assuming that the visible spectrum extends from 380 to 750 nm, for which visible wavelength(s) (in vacuum) will the film appear bright due to constructive interference?

** **18.** A piece of curved glass has a radius of curvature of 10.0 m and is used to form Newton's rings, as in Figure 27.14. Not counting the dark spot at the center of the pattern, there are one hundred dark fringes, the last one being at the outer edge of the curved piece of glass. The light being used has a wavelength of 654 nm in vacuum. What is the radius of the outermost dark ring in the pattern?

Section 27.5 Diffraction

19. ssm A diffraction pattern forms when light passes through a single slit. The wavelength of the light is 675 nm. Determine the angle that locates the first dark fringe when the width of the slit is (a) 1.8×10^{-4} m and (b) 1.8×10^{-6} m.

20. A slit whose width is 4.30×10^{-5} m is located 1.32 m from a flat screen. Light shines through the slit and falls on the screen. Find the width of the central fringe of the diffraction pattern when the wavelength of the light is 635 nm.

21. A single slit has a width of 2.1×10^{-6} m and is used to form a diffraction pattern. Find the angle that locates the second dark fringe when the wavelength of the light is (a) 430 nm and (b) 660 nm.

22. Light of wavelength 668 nm passes through a slit 6.73×10^{-6} m wide and falls on a screen that is 1.85 m away. What is the distance on the screen from the center of the central bright fringe to the third dark fringe on either side?

23. ssm www Light shines through a single slit whose width is 5.6×10^{-4} m. A diffraction pattern is formed on a flat screen located 4.0 m away. The distance between the middle of the central bright fringe and the first dark fringe is 3.5 mm. What is the wavelength of the light?

* **24.** How many dark fringes will be produced on either side of the central maximum if light ($\lambda = 651$ nm) is incident on a single slit that is 5.47×10^{-6} m wide?

* **25.** ssm The central bright fringe in a single-slit diffraction pattern has a width that equals the distance between the screen and the slit. Find the ratio λ/W of the wavelength of the light to the width of the slit.

* **26.** The width of a slit is 2.0×10^{-5} m. Light with a wavelength of 480 nm passes through this slit and falls on a screen that is located 0.50 m away. In the diffraction pattern, find the width of the bright fringe that is next to the central bright fringe.

** **27.** In a single-slit diffraction pattern, the central fringe is 450 times as wide as the slit. The screen is 18 000 times farther from the slit than the slit is wide. What is the ratio λ/W, where λ is the wavelength of the light shining through the slit and W is the width of the slit? Assume that the angle that locates a dark fringe on the screen is small, so that $\sin \theta \approx \tan \theta$.

Section 27.6 Resolving Power

28. You are looking down at the earth from inside a jetliner flying at an altitude of 8690 m. The pupil of your eye has a diameter of 2.00 mm. Determine how far apart two cars must be on the ground if you are to have any hope of distinguishing between them in (a) red light (wavelength = 665 nm in vacuum) and (b) violet light (wavelength = 405 nm in vacuum).

29. It is claimed that some professional baseball players can see which way the ball is spinning as it travels toward home plate. One way to judge this claim is to estimate the distance at which a batter can first hope to resolve two points on opposite sides of a baseball, which has a diameter of 0.0738 m. (a) Estimate this distance, assuming that the pupil of the eye has a diameter of 2.0 mm and the wavelength of the light is 550 nm in vacuum. (b) Considering that the distance between the pitcher's mound and home plate is 18.4 m, can you rule out the claim based on your answer to part (a)?

30. Two stars are 3.7×10^{11} m apart and are equally distant from the earth. A telescope has an objective lens with a diameter of 1.02 m and just detects these stars as separate objects. Assume that light of wavelength 550 nm is being observed. Also assume that diffraction effects, rather than atmospheric turbulence, limit the resolving power of the telescope. Find the maximum distance that these stars could be from the earth.

31. ssm Late one night on a highway, a car speeds by you and fades into the distance. Under these conditions the pupils of your eyes have diameters of about 7.0 mm. The taillights of this car are separated by a distance of 1.2 m and emit red light (wavelength = 660 nm in vacuum). How far away from you is this car when its taillights appear to merge into a single spot of light because of the effects of diffraction?

32. Review Conceptual Example 8 as background for this problem. In addition to the data given there, assume that the dots in the painting are separated by 1.5 mm and that the wavelength of the light is $\lambda_{\text{vacuum}} = 550$ nm. Find the distance at which the dots can just be resolved by (a) the eye and (b) the camera.

33. Astronomers have discovered a planetary system orbiting the star Upsilon Andromedae, which is at a distance of 4.2×10^{17} m from the earth. One planet is believed to be located at a distance of 1.2×10^{11} m from the star. Using visible light with a vacuum wavelength of 550 nm, what is the minimum necessary aperture diameter that a telescope must have so that it can resolve the planet and the star?

* **34.** The pupil of an eagle's eye has a diameter of 6.0 mm. Two field mice are separated by 0.010 m. From a distance of 176 m, the eagle sees them as one unresolved object and dives toward them at a speed of 17 m/s. Assume that the eagle's eye detects light that has a wavelength of 550 nm in a vacuum. How much time passes until the eagle sees the mice as separate objects?

* **35.** Refer to **Interactive Solution 27.35** at **www.wiley.com/college/ cutnell** to review a method by which this problem can be solved. You are using a microscope to examine a blood sample. Recall from Section 26.12 that the sample should be placed just outside the focal point of the objective lens of the microscope. (a) If the specimen is being illuminated with light of wavelength λ and the diameter of the objective equals its focal length, determine the closest distance between two blood cells that can just be resolved. Express your answer in terms of λ. (b) Based on your answer to (a), should you use light with a longer wavelength or a shorter wavelength if you wish to resolve two blood cells that are even closer together?

** **36.** Two concentric circles of light emit light whose wavelength is 555 nm. The larger circle has a radius of 4.0 cm, and the smaller circle has a radius of 1.0 cm. When taking a picture of these lighted circles, a camera admits light through an aperture whose diameter is 12.5 mm. What is the maximum distance at which the camera can (a) distinguish one circle from the other and (b) reveal that the inner circle is a circle of light rather than a solid disk of light?

Section 27.7 The Diffraction Grating,
Section 27.8 Compact Discs, Digital Video Discs,
and the Use of Interference

37. ssm The diffraction gratings discussed in the text are transmission gratings because light *passes through* them. There are also gratings in which the light *reflects from* the grating to form a pattern of fringes. Equation 27.7 also applies to a reflection grating with straight parallel lines when the incident light shines perpendicularly on the grating. The surface of a compact disc (CD) has a multicolored appearance because it acts like a reflection grating and spreads sunlight into its colors. The arms of the spiral track on the CD are separated by 1.1×10^{-6} m. Using Equation 27.7, estimate the angle that corresponds to the first-order maximum for a wavelength of (a) 660 nm (red) and (b) 410 nm (violet).

38. A diffraction grating produces a first-order bright fringe that is 0.0894 m away from the central bright fringe on a flat screen. The separation between the slits of the grating is 4.17×10^{-6} m, and the distance between the grating and the screen is 0.625 m. What is the wavelength of the light shining on the grating?

39. For a wavelength of 420 nm, a diffraction grating produces a bright fringe at an angle of 26°. For an unknown wavelength, the same grating produces a bright fringe at an angle of 41°. In both cases the bright fringes are of the same order m. What is the unknown wavelength?

40. A diffraction grating is 1.50 cm wide and contains 2400 lines. When used with light of a certain wavelength, a third-order maximum is formed at an angle of 18.0°. What is the wavelength (in nm)?

41. ssm The wavelength of the laser beam used in a compact disc player is 780 nm. Suppose that a diffraction grating produces first-order tracking beams that are 1.2 mm apart at a distance of 3.0 mm from the grating. Estimate the spacing between the slits of the grating.

* **42.** A diffraction grating has 2604 lines per centimeter, and it produces a principal maximum at $\theta = 30.0°$. The grating is used with light that contains all wavelengths between 410 and 660 nm. What is (are) the wavelength(s) of the incident light that could have produced this maximum?

* **43. ssm** The same diffraction grating is used with two different wavelengths of light, λ_A and λ_B. The fourth-order principal maximum of light A exactly overlaps the third-order principal maximum of light B. Find the ratio λ_A/λ_B.

* **44.** Three, and only three, bright fringes can be seen on either side of the central maximum when a grating is illuminated with light ($\lambda = 510$ nm). What is the maximum number of lines/cm for the grating?

** **45.** There are 5620 lines per centimeter in a grating that is used with light whose wavelength is 471 nm. A flat observation screen is located at a distance of 0.750 m from the grating. What is the minimum width that the screen must have so the *centers* of all the principal maxima formed on either side of the central maximum fall on the screen?

Additional Problems

46. A Young's double-slit experiment is performed using light that has a wavelength of 630 nm. The separation between the slits is 5.3×10^{-5} m. Find the angles that locate the (a) first-, (b) second-, and (c) third-order bright fringes on the screen.

47. ssm The transmitting antenna for a radio station is 7.00 km from your house. The frequency of the electromagnetic wave broadcast by this station is 536 kHz. The station builds a second transmitting antenna that broadcasts an identical electromagnetic wave in phase with the original one. The new antenna is 8.12 km from your house. Does constructive or destructive interference occur at the receiving antenna of your radio? Show your calculations.

48. Interactive LearningWare 27.2 at **www.wiley.com/college/ cutnell** provides some pertinent background for this problem. A transparent film ($n = 1.43$) is deposited on a glass plate ($n = 1.52$) to form a nonreflecting coating. The film has a thickness of 1.07×10^{-7} m. What is the longest possible wavelength (in vacuum) of light for which this film has been designed?

49. A doorway is 0.91 m wide. (a) Obtain the angle that locates the first dark fringe in the Fraunhofer diffraction pattern formed when red light (wavelength = 660 nm) passes through the doorway. (b) Repeat part (a) for a 440-Hz sound wave (concert A), assuming that the speed of sound is 343 m/s.

50. Interactive LearningWare 27.1 at **www.wiley.com/college/ cutnell** explores the approach taken in problems such as this one. Two parallel slits are illuminated by light composed of two wavelengths, one of which is 645 nm. On a viewing screen, the light whose wavelength is known produces its third dark fringe at the same place where the light whose wavelength is unknown produces its fourth-order bright fringe. The fringes are counted relative to the central or zeroth-order bright fringe. What is the unknown wavelength?

51. ssm The largest refracting telescope in the world is at the Yerkes Observatory in Williams Bay, Wisconsin. The objective of the telescope has a diameter of 1.02 m. Two objects are 3.75×10^4 m from the telescope. With light of wavelength 565 nm, how close can the objects be to each other so that they are just resolved by the telescope?

* **52.** At most, how many bright fringes can be formed on either side of the central bright fringe when light of wavelength 625 nm falls on a double slit whose slit separation is 3.76×10^{-6} m?

* **53. ssm www** In an experiment, red light (wavelength = 694.3 nm) is sent to the moon. At the surface of the moon, which is 3.77×10^8 m away, the light strikes a reflector left there by astronauts. The reflected light returns to the earth, where it is detected. When it leaves the spotlight, the circular beam of light has a diameter of about 0.20 m, and diffraction causes the beam to spread as the light travels to the moon. In effect, the first circular dark fringe in the diffraction pattern defines the size of the central bright spot on the moon. Determine the diameter (not the radius) of the central bright spot on the moon.

* **54.** In a single-slit diffraction pattern on a flat screen, the central bright fringe is 1.2 cm wide when the slit width is 3.2×10^{-5} m. When the slit is replaced by a second slit, the wavelength of the light and the distance to the screen remaining unchanged, the central bright fringe broadens to a width of 1.9 cm. What is the width of the second slit? It may be assumed that θ is so small that $\sin \theta \approx \tan \theta$.

* **55. ssm** Violet light (wavelength = 410 nm) and red light (wavelength = 660 nm) lie at opposite ends of the visible spectrum. (a) For each wavelength, find the angle θ that locates the first-order maximum produced by a grating with 3300 lines/cm. This grating converts a mixture of all colors between violet and red into a rainbow-like dispersion between the two angles. Repeat the calculation above for (b) the second-order maximum and (c) the third-order maximum. (d) From your results, decide whether there is an overlap between any of the "rainbows" and, if so, specify which orders overlap.

* **56.** The drawing shows a cross section of a plano-concave lens resting on a flat glass plate. (A plano-concave lens has one surface that is a plane and the other that is concave spherical.) The thickness t is 1.37×10^{-5} m. The lens is illuminated with monochromatic light ($\lambda_{\text{vacuum}} = 550$ nm), and a series of concentric bright and dark rings is formed, much like Newton's rings. How many bright rings are there?

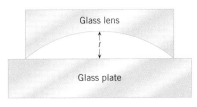

Glass lens

t

Glass plate

** **57.** Two gratings A and B have slit separations d_A and d_B, respectively. They are used with the same light and the same observation screen. When grating A is replaced with grating B, it is observed that the first-order maximum of A is exactly replaced by the second-order maximum of B. (a) Determine the ratio d_B/d_A of the spacings between the slits of the gratings. (b) Find the next two principal maxima of grating A and the principal maxima of B that exactly replace them when the gratings are switched. Identify these maxima by their order numbers.

** **58. Interactive LearningWare 27.2** at **www.wiley.com/college/cutnell** reviews the concepts that are important in this problem. A uniform layer of water ($n = 1.33$) lies on a glass plate ($n = 1.52$). Light shines perpendicularly on the layer. Because of constructive interference, the layer looks maximally bright when the wavelength of the light is 432 nm in vacuum and *also* when it is 648 nm in vacuum. (a) Obtain the minimum thickness of the film. (b) Assuming that the film has the minimum thickness and that the visible spectrum extends from 380 to 750 nm, determine the visible wavelength(s) (in vacuum) for which the film appears completely dark.

Concepts & Calculations Group Learning Problems

Note: Each of these problems consists of Concept Questions followed by a related quantitative Problem. They are designed for use by students working alone or in small learning groups. The Concept Questions involve little or no mathematics and are intended to stimulate group discussions. They focus on the concepts with which the problems deal. Recognizing the concepts is the essential initial step in any problem-solving technique.

59. Concept Questions (a) Point A is the midpoint of one of the sides of a square. On the side opposite this spot, two in-phase loudspeakers are located at adjacent corners, as in Figure 27.41. Standing at point A, you hear a loud sound because constructive interference occurs between the identical sound waves coming from the speakers. Why does constructive interference occur at point A? (b) As you walk along the side of the square toward either empty corner, the loudness diminishes gradually to nothing and then increases again until you hear a maximally loud sound at the corner. Explain these observations in terms of destructive and constructive interference. (c) The general condition that leads to constructive interference entails a number of possibilities for the variable m, which designates the order of the interference maxima. Which one of those possibilities, if any, applies at either empty corner of the square? Explain.

Problem The square referred to in the Concept Questions is 4.6 m on a side. Find the wavelength of the sound waves.

60. Concept Questions (a) The screen in Figure 27.8 has a fixed width and is centered on the midpoint between the two slits. Should the screen be moved to the left or to the right to ensure that the third-order bright fringe does *not* lie on the screen? (b) Do fewer fringes lie on the screen when the distance L between the double-slit and the screen is smaller or larger? Account for your answers.

Problem In a setup like that in Figure 27.8, a wavelength of 625 nm is used in a Young's double-slit experiment. The separation between the slits is $d = 1.4 \times 10^{-5}$ m. The total width of the screen is 0.20 m. In one version of the setup, the separation between the double slit and the screen is $L_A = 0.35$ m, whereas in another version it is $L_B = 0.50$ m. On one side of the central bright fringe, how many bright fringes lie on the screen in the two versions of the setup? Do not include the central bright fringe in your counting. Verify that your answer is consistent with your answers to the Concept Questions.

61. Concept Questions (a) What, if any, phase change occurs when light, traveling in air, reflects from the interface between the air and a soap film ($n = 1.33$)? (b) What, if any, phase change occurs when light, traveling in a soap film, reflects from the interface between the soap film and a glass plate ($n = 1.52$)? (c) Is the wavelength of the light in a soap film greater than, smaller than, or equal to the wavelength in a vacuum?

Problem A soap film ($n = 1.33$) is 465 nm thick and lies on a glass plate ($n = 1.52$). Sunlight, whose wavelengths (in vacuum) extend from 380 to 750 nm, travels through the air and strikes the film perpendicularly. For which wavelength(s) in this range does destructive interference cause the film to look dark in reflected light?

62. Concept Questions (a) In a single-slit diffraction pattern the width of the central bright fringe is defined by the location of the first dark fringe that lies on either side of it. For a given slit width, does the width of the central bright fringe increase, decrease, or remain the same as the wavelength of the light increases? (b) For a given wavelength, does the width of the central bright fringe increase, decrease, or remain the same as the slit width increases? (c) When both the wavelength and the slit width change, it is possible for the width of the central bright fringe to remain the same. What condition must be satisfied for this to happen? In each case, give your reasoning.

Problem A slit has a width of $W_1 = 2.3 \times 10^{-6}$ m. When light with a wavelength of $\lambda_1 = 510$ nm passes through this slit, the width of the central bright fringe on a flat observation screen has a certain value. With the screen kept in the same place, this slit is replaced with a second slit (width W_2) and a wavelength of $\lambda_2 = 740$ nm is used. The width of the central bright fringe on the screen is observed to be unchanged. Find W_2.

63. Concept Questions An inkjet color printer uses tiny dots of red, green, and blue ink to produce an image. At normal viewing distances, the eye does not resolve the individual dots, so that the image has a normal look. The angle θ_{min} is the minimum angle that two dots can subtend at the eye and still be resolved separately. (a) For which color does θ_{min} have the largest value? (b) For which color does θ_{min} have the smallest value? (c) Corresponding to each value of θ_{min}, there is a value for the separation distance s between the dots. How is θ_{min} related to s and the viewing distance L? (d) Assume that the dot separation on the printed page is the same for all colors and is chosen so that none of the colored dots can be seen as separate objects. Should the maximum allowable dot separation be s_{red}, s_{green}, or s_{blue}? For each answer, give your reasoning.

Problem The wavelengths for red, green, and blue are $\lambda_{red} = 660$ nm, $\lambda_{green} = 550$ nm, and $\lambda_{blue} = 470$ nm. The diameter of the pupil through which light enters the eye is 2.0 mm. For a viewing distance of 0.40 m, find the maximum separation distance that the dots can have and not be resolved separately. Check to see that your answer is consistent with your answers to the Concept Questions.

64. Concept Questions (a) Two diffraction gratings are located at the same distance from observation screens. Light with the same wavelength λ is used for each. The principal maxima of grating A are observed to be closer together on the screen than the principal maxima of grating B. Which grating diffracts the light to a greater extent? (b) Which grating has the smaller slit separation d? (c) Which grating has the greater number of lines per meter? Justify each of your answers.

Problem The separation between adjacent principal maxima for grating A is 2.7 cm, and for grating B it is 3.2 cm. Grating A has 2000 lines per meter. How many lines per meter does grating B have? The diffraction angles are small enough that $\sin \theta \approx \tan \theta$. Be sure that your answer is consistent with your answers to the Concept Questions.

* **65. Concept Questions** A beam of light is sent directly down onto a glass plate ($n = 1.5$) and a plastic plate ($n = 1.2$) that form a thin wedge of air (see the drawing). An observer looking down through the glass plate sees the fringe pattern shown in the lower part of the drawing, with dark fringes at the ends A and B. (a) What, if any, phase change occurs when the light, traveling in the glass, reflects from the interface between the glass and the air wedge? (b) What, if any, phase change occurs when the light, traveling in the air wedge, reflects from the interface between the air and the top surface of the plastic plate? (c) Why is there a dark fringe at A? (d) How many wavelengths of light fit into the air between the

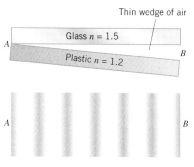

glass and plastic plates at B? Account for your answers to each of the questions.

Problem The wavelength of the light is 520 nm. Using the fringe pattern shown in the drawing, determine the thickness (in nm) of the air wedge at B. Be sure your answer is consistent with part (d) of the Concept Questions.

* **66. Concept Questions** Light shines on a diffraction grating, and a diffraction pattern is produced on a viewing screen that consists of a central bright fringe and higher-order bright fringes (see the drawing). (a) From trigonometry, how is the distance y from the central bright fringe to the second-order bright fringe related to the diffraction angle θ and the distance L between the grating and the screen? (b) From physics, how is θ related to the order m of the bright fringe, the wavelength λ of the light, and the separation d between the slits? (c) In this problem, the angle θ is small (less than a few degrees). When the angle is small, $\tan \theta$ is approximately equal to $\sin \theta$, or $\tan \theta \approx \sin \theta$. Using this approximation, obtain an expression for y in terms of L, m, λ, and d. (d) If the entire apparatus in the drawing is submerged in water, would you expect the distance y to increase, decrease, or remain unchanged. Why?

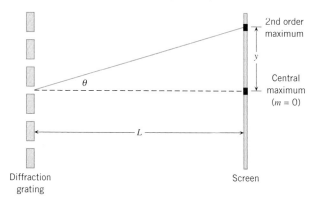

Problem Light of wavelength 480 nm (in vacuum) is incident on a diffraction grating that has a slit separation of 5.0×10^{-7} m. The distance between the grating and the viewing screen is 0.15 m. (a) Determine the distance y from the central bright fringe to the second-order bright fringe. (b) If the entire apparatus is submerged in water ($n_{water} = 1.33$), what is the distance y? Be sure your answer is consistent with part (d) of the Concept Questions.

Chapter 28

Special Relativity

Albert Einstein (1879–1955), the author of the theory of special relativity, is one of the most famous scientists of the twentieth century. The top photograph shows the Einstein Memorial Sculpture by Robert Berks, which is located in Washington, DC, at the National Academy of Sciences. Shown beneath this photograph is a portion of a working copy (1912) of the manuscript in which Einstein outlines his theory. The equation shown is presented in this chapter as Equation 28.4, and from it, the famous $E_0 = mc^2$ equation is obtained, as Section 28.6 discusses. (*Top photo:* The Einstein Memorial Sculpture © Robert Berks, 1978, at the National Academy of Sciences. Photograph by Diana H. Walker, courtesy National Academy of Sciences; *bottom photo:* © AP/Wide World Photos)

28.1 *Events and Inertial Reference Frames*

In the theory of special relativity, an *event*, such as the launching of the space shuttle in Figure 28.1, is a physical "happening" that occurs at a certain place and time. In this drawing two observers are watching the lift-off, one standing on the earth and one in an airplane that is flying at a constant velocity relative to the earth. To record the event, each observer uses a *reference frame* that consists of a set of *x, y, z* axes (called a *coordinate system*) and a clock. The coordinate systems are used to establish where the event occurs, and the clocks to specify when. Each observer is at rest relative to his own reference frame. However, the earth-based observer and the airborne observer are moving relative to each other and so are their respective reference frames.

The theory of special relativity deals with a "special" kind of reference frame, called an *inertial reference frame*. As Section 4.2 discusses, an inertial reference frame is one in which Newton's law of inertia is valid. That is, if the net force acting on a body is zero, the body either remains at rest or moves at a constant velocity. In other words, the acceleration of such a body is zero when measured in an inertial reference frame. Rotating and otherwise accelerating reference frames are not inertial reference frames. The earth-based reference frame in Figure 28.1 is not quite an inertial frame because it is subjected to centripetal accelerations as the earth spins on its axis and revolves around the sun. In most situations, however, the effects of these accelerations are small, and we can neglect them. To the extent that the earth-based reference frame is an inertial frame, so is the plane-based reference frame, because the plane moves at a constant velocity relative to the earth. The next section discusses why inertial reference frames are important in relativity.

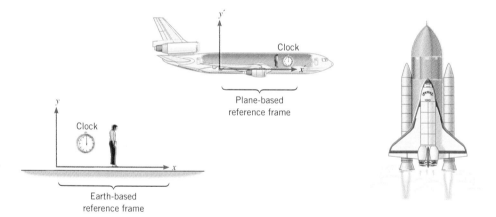

Figure 28.1 Using an earth-based reference frame, an observer standing on the earth records the location and time of an event (the space shuttle lift-off). Likewise, an observer in the airplane uses a plane-based reference frame to describe the event.

28.2 *The Postulates of Special Relativity*

Einstein built his theory of special relativity on two fundamental assumptions or postulates about the way nature behaves.

■ **THE POSTULATES OF SPECIAL RELATIVITY**

1. *The Relativity Postulate.* The laws of physics are the same in every inertial reference frame.
2. *The Speed of Light Postulate.* The speed of light in a vacuum, measured in any inertial reference frame, always has the same value of *c*, no matter how fast the source of light and the observer are moving relative to each other.

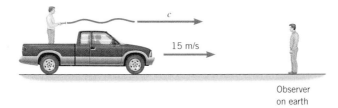

Figure 28.2 Both the person on the truck and the observer on the earth measure the speed of the light to be c, regardless of the speed of the truck.

It is not too difficult to accept the relativity postulate. For instance, in Figure 28.1 each observer, using his own inertial reference frame, can make measurements on the motion of the space shuttle. The relativity postulate asserts that both observers find their data to be consistent with Newton's laws of motion. Similarly, both observers find that the behavior of the electronics on the space shuttle is described by the laws of electromagnetism. According to the relativity postulate, *any inertial reference frame is as good as any other for expressing the laws of physics.* As far as inertial reference frames are concerned, nature does not play favorites.

Since the laws of physics are the same in all inertial reference frames, there is no experiment that can distinguish between an inertial frame that is at rest and one that is moving at a constant velocity. When you are seated on the aircraft in Figure 28.1, for instance, it is just as valid to say that you are at rest and the earth is moving as it is to say the converse. It is not possible to single out one particular inertial reference frame as being at "absolute rest." Consequently, it is meaningless to talk about the "absolute velocity" of an object—that is, its velocity measured relative to a reference frame at "absolute rest." Thus, the earth moves relative to the sun, which itself moves relative to the center of our galaxy. And the galaxy moves relative to other galaxies, and so on. According to Einstein, only the relative velocity between objects, not their absolute velocities, can be measured and is physically meaningful.

Whereas the relativity postulate seems plausible, the speed of light postulate defies common sense. For instance, Figure 28.2 illustrates a person standing on the bed of a truck that is moving at a constant speed of 15 m/s relative to the ground. Now, suppose you are standing on the ground and the person on the truck shines a flashlight at you. The person on the truck observes the speed of light to be c. What do you measure for the speed of light? You might guess that the speed of light would be c + 15 m/s. However, this guess is inconsistent with the speed of light postulate, which states that all observers in inertial reference frames measure the speed of light to be c—nothing more, nothing less. Therefore, you must also measure the speed of light to be c, the same as that measured by the person on the truck. According to the speed of light postulate, the fact that the flashlight is moving has no influence whatsoever on the speed of the light approaching you. This property of light, although surprising, has been verified many times by experiment.

Since waves, such as water waves and sound waves, require a medium through which to propagate, it was natural for scientists before Einstein to assume that light did too. This hypothetical medium was called the *luminiferous ether* and was assumed to fill all of space. Furthermore, it was believed that light traveled at the speed c only when measured with respect to the ether. According to this view, an observer moving relative to the ether would measure a speed for light that was slower or faster than c, depending on whether the observer moved with or against the light, respectively. During the years 1883–1887, however, the American scientists A. A. Michelson and E. W. Morley carried out a series of famous experiments whose results were not consistent with the ether theory. Their results indicated that the speed of light is indeed the same in all inertial reference frames and does not depend on the motion of the observer. These experiments, and others, led eventually to the demise of the ether theory and the acceptance of the theory of special relativity.

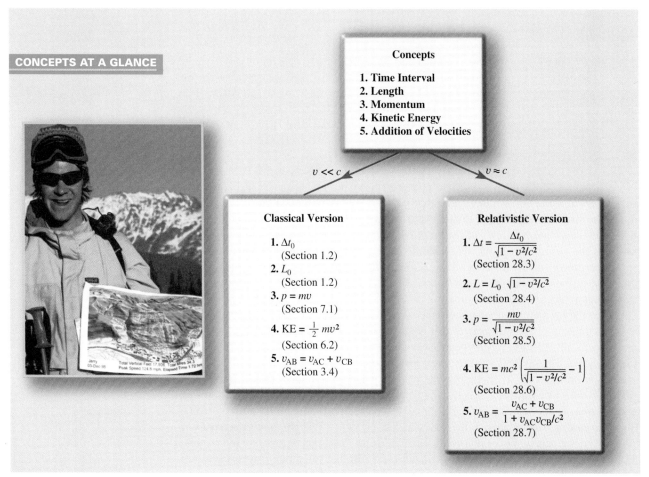

CONCEPTS AT A GLANCE

Concepts

1. Time Interval
2. Length
3. Momentum
4. Kinetic Energy
5. Addition of Velocities

$v \ll c$ $v \approx c$

Classical Version

1. Δt_0
   (Section 1.2)
2. L_0
   (Section 1.2)
3. $p = mv$
   (Section 7.1)
4. $KE = \frac{1}{2} mv^2$
   (Section 6.2)
5. $v_{AB} = v_{AC} + v_{CB}$
   (Section 3.4)

Relativistic Version

1. $\Delta t = \dfrac{\Delta t_0}{\sqrt{1 - v^2/c^2}}$
   (Section 28.3)
2. $L = L_0 \sqrt{1 - v^2/c^2}$
   (Section 28.4)
3. $p = \dfrac{mv}{\sqrt{1 - v^2/c^2}}$
   (Section 28.5)
4. $KE = mc^2 \left(\dfrac{1}{\sqrt{1 - v^2/c^2}} - 1 \right)$
   (Section 28.6)
5. $v_{AB} = \dfrac{v_{AC} + v_{CB}}{1 + v_{AC}v_{CB}/c^2}$
   (Section 28.7)

Figure 28.3 CONCEPTS AT A GLANCE This chart emphasizes that it is the speed v of a moving object, as compared to the speed c of light in a vacuum, that determines whether the effects of special relativity are measurably important. As Section 28.3 discusses, these effects must be taken into account by the Global Positioning System (GPS), which can locate objects on the earth with remarkable accuracy. The device on the skier's left shoulder is known as a "Sports Tracker" and uses the GPS to provide a map showing exactly where he has skied. (© AP/Wide World Photos)

▶ CONCEPTS AT A GLANCE The remainder of this chapter reexamines, from the viewpoint of special relativity, a number of fundamental concepts that have been discussed in earlier chapters from the viewpoint of classical physics. These concepts are time, length, momentum, kinetic energy, and the addition of velocities. We will see that each is modified by special relativity in a way that depends on the speed v of a moving object relative to the speed c of light in a vacuum. Figure 28.3 illustrates that when the object moves slowly (v is much smaller than c [$v \ll c$]), the modification is negligibly small, and the classical version of each concept provides an accurate description of reality. However, when the object moves so rapidly that v is an appreciable fraction of the speed of light, the effects of special relativity must be considered. The gold panel in Figure 28.3 lists the various equations that convey the modifications imposed by special relativity. Each of these equations will be discussed in later sections of this chapter. ◀

It is important to realize that the modifications imposed by special relativity do not imply that the classical concepts of time, length, momentum, kinetic energy, and the addition of velocities, as developed by Newton and others, are wrong. They are just limited to speeds that are very small compared to the speed of light. In contrast, the relativistic view of the concepts applies to all speeds between zero and the speed of light.

28.3 The Relativity of Time: Time Dilation

TIME DILATION

Common experience indicates that time passes just as quickly for a person standing on the ground as it does for an astronaut in a spacecraft. In contrast, special relativity reveals that the person on the ground measures time passing more slowly for the astronaut than for himself. We can see how this curious effect arises with the help of the clock illustrated in Figure 28.4, which uses a pulse of light to mark time. A short pulse of light is emitted by a light source, reflects from a mirror, and then strikes a detector that is situated next to the source. Each time a pulse reaches the detector, a "tick" registers on the chart recorder, another short pulse of light is emitted, and the cycle repeats. Thus, the time interval between successive "ticks" is marked by a beginning event (the firing of the light source) and an ending event (the pulse striking the detector). The source and detector are so close to each other that the two events can be considered to occur at the same location.

Suppose two identical clocks are built. One is kept on earth, and the other is placed aboard a spacecraft that travels at a constant velocity relative to the earth. The astronaut is at rest with respect to the clock on the spacecraft and, therefore, sees the light pulse move along the up/down path shown in Figure 28.5a. According to the astronaut, the time interval Δt_0 required for the light to follow this path is the distance $2D$ divided by the speed of light c; $\Delta t_0 = 2D/c$. To the astronaut, Δt_0 is the time interval between the "ticks" of the spacecraft clock—that is, the time interval between the beginning and ending events of the clock. An earth-based observer, however, does *not* measure Δt_0 as the time interval between these two events. Since the spacecraft is moving, the earth-based observer sees the light pulse follow the diagonal path shown in red in part b of the drawing. This path is longer than the up/down path seen by the astronaut. But light travels at the *same speed* c for both observers, in accord with the speed of light postulate. Therefore, the earth-based observer measures a time interval Δt between the two events that is *greater* than the time interval Δt_0 measured by the astronaut. In other words, the earth-based observer, using his own earth-based clock to measure the performance of the astronaut's clock, finds that the astronaut's clock runs slowly. This result of special relativity is known as **time dilation.** (To *dilate* means to expand, and the time interval Δt is "expanded" relative to Δt_0.)

The time interval Δt that the earth-based observer measures in Figure 28.5b can be determined as follows. While the light pulse travels from the source to the detector, the spacecraft moves a distance $2L = v\Delta t$ to the right, where v is the speed of the spacecraft relative to the earth. From the drawing it can be seen that the light pulse travels a total diagonal distance of $2s$ during the time interval Δt. Applying the Pythagorean theorem, we find that

$$2s = 2\sqrt{D^2 + L^2} = 2\sqrt{D^2 + \left(\frac{v\,\Delta t}{2}\right)^2}$$

But the distance $2s$ is also equal to the speed of light times the time interval Δt, so $2s = c\,\Delta t$. Therefore,

$$c\,\Delta t = 2\sqrt{D^2 + \left(\frac{v\,\Delta t}{2}\right)^2}$$

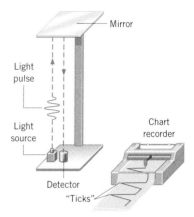

Figure 28.4 A light clock.

Concept Simulation 28.1

This simulation is based on Figure 28.5 and gives you control over the speed at which the spaceship moves. Thus, you can experiment and see the effect that the speed has on the clocks.

Go to
www.wiley.com/college/cutnell

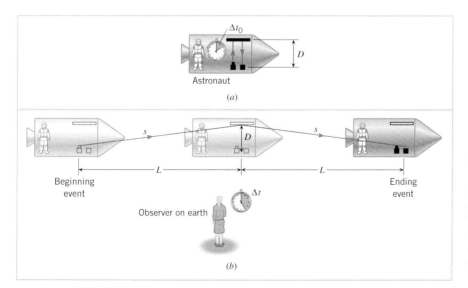

(a)

(b)

Figure 28.5 (a) The astronaut measures the time interval Δt_0 between successive "ticks" of his light clock. (b) An observer on earth watches the astronaut's clock and sees the light pulse travel a greater distance between "ticks" than it does in part a. Consequently, the earth-based observer measures a time interval Δt between "ticks" that is greater than Δt_0.

Squaring this result and solving for Δt gives

$$\Delta t = \frac{2D}{c} \frac{1}{\sqrt{1 - \dfrac{v^2}{c^2}}}$$

But $2D/c = \Delta t_0$, the time interval between successive "ticks" of the spacecraft's clock as measured by the astronaut. With this substitution, the equation for Δt can be expressed as

Time dilation

$$\Delta t = \frac{\Delta t_0}{\sqrt{1 - \dfrac{v^2}{c^2}}} \qquad (28.1)$$

The symbols in this formula are defined as follows:

Δt_0 = proper time interval, which is the interval between two events as measured by an observer who is at rest with respect to the events and who views them as occurring *at the same place*

Δt = dilated time interval, which is the interval measured by an observer who is in motion with respect to the events and who views them as occurring at *different places*

v = relative speed between the two observers

c = speed of light in a vacuum

For a speed v that is less than c, the term $\sqrt{1 - v^2/c^2}$ in Equation 28.1 is less than 1, and the dilated time interval Δt is greater than Δt_0. Example 1 shows this time dilation effect.

Example 1 Time Dilation

The spacecraft in Figure 28.5 is moving past the earth at a constant speed v that is 0.92 times the speed of light. Thus, $v = (0.92)(3.0 \times 10^8 \text{ m/s})$, which is often written as $v = 0.92c$. The astronaut measures the time interval between successive "ticks" of the spacecraft clock to be $\Delta t_0 = 1.0$ s. What is the time interval Δt that an earth observer measures between "ticks" of the astronaut's clock?

Reasoning Since the clock on the spacecraft is moving relative to the earth observer, the earth observer measures a greater time interval Δt between "ticks" than does the astronaut, who is at rest relative to the clock. The dilated time interval Δt can be determined from the time dilation relation, Equation 28.1.

Solution The dilated time interval is

$$\Delta t = \frac{\Delta t_0}{\sqrt{1 - \dfrac{v^2}{c^2}}} = \frac{1.0 \text{ s}}{\sqrt{1 - \left(\dfrac{0.92c}{c}\right)^2}} = \boxed{2.6 \text{ s}}$$

From the point of view of the earth-based observer, the astronaut is using a clock that is running slowly, because the earth-based observer measures a time between "ticks" that is longer (2.6 s) than what the astronaut measures (1.0 s).

Present-day spacecrafts fly nowhere near as fast as the craft in Example 1. Yet circumstances exist in which time dilation can create appreciable errors if not accounted for. The Global Positioning System (GPS), for instance, uses highly accurate and stable atomic clocks on board each of 24 satellites orbiting the earth at speeds of about 4000 m/s. These clocks make it possible to measure the time it takes for electromagnetic waves to travel from a satellite to a ground-based GPS receiver. From the speed of light and the times measured for signals from three or more of the satellites, it is possible to locate the position of the receiver (see Section 5.5). The stability of the clocks must be better than one part in 10^{13} to ensure the positional accuracy demanded of the GPS. Using Equation 28.1 and the speed of the GPS satellites, we can calculate the difference between the dilated time interval and the proper time interval as a fraction of the proper time interval and compare the result to the stability of the GPS clocks:

The physics of the Global Positioning System and special relativity.

$$\frac{\Delta t - \Delta t_0}{\Delta t_0} = \frac{1}{\sqrt{1 - v^2/c^2}} - 1 = \frac{1}{\sqrt{1 - (4000 \text{ m/s})^2/(3.00 \times 10^8 \text{ m/s})^2}} - 1$$

$$= \frac{1}{1.1 \times 10^{10}}$$

This result is approximately one thousand times greater than the GPS-clock stability of one part in 10^{13}. Thus, if not taken into account, time dilation would cause an error in the measured position of the earth-based GPS receiver roughly equivalent to that caused by a thousand-fold degradation in the stability of the atomic clocks.

PROPER TIME INTERVAL

In Figure 28.5 both the astronaut and the person standing on the earth are measuring the time interval between a beginning event (the firing of the light source) and an ending event (the light pulse striking the detector). For the astronaut, who is at rest with respect to the light clock, the two events occur at the same location. (Remember, we are assuming that the light source and detector are so close together that they are considered to be at the same place.) Being at rest with respect to a clock is the usual or "proper" situation, so the time interval Δt_0 measured by the astronaut is called the ***proper time interval.*** In general, the proper time interval Δt_0 between two events is the time interval measured by an observer who is at rest relative to the events and sees them at the *same location* in space. On the other hand, the earth-based observer does not see the two events occurring at the same location in space, since the spacecraft is in motion. The time interval Δt that the earth-based observer measures is, therefore, not a proper time interval in the sense that we have defined it.

To understand situations involving time dilation, it is essential to distinguish between Δt_0 and Δt. It is helpful if one first identifies the two events that define the time interval. These may be something other than the firing of a light source and the light pulse striking a detector. Then determine the reference frame in which the two events occur at the same place. An observer at rest in this reference frame measures the proper time interval Δt_0.

SPACE TRAVEL

One of the intriguing aspects of time dilation occurs in conjunction with space travel. Since enormous distances are involved, travel to even the closest star outside our solar system would take a long time. However, as the following example shows, the travel time can be considerably less for the passengers than one might guess.

Shown here in orbit is astronaut Lee M. E. Morin on April 13, 2002, as he works on the International Space Station. His feet are secured to the end of the station's robotic arm. (© AP/Wide World Photos)

Example 2 Space Travel

▼

Alpha Centauri, a nearby star in our galaxy, is 4.3 light-years away. This means that, as measured by a person on earth, it would take light 4.3 years to reach this star. If a rocket leaves for Alpha Centauri and travels at a speed of $v = 0.95c$ relative to the earth, by how much will the passengers have aged, according to their own clock, when they reach their destination? Assume that the earth and Alpha Centauri are stationary with respect to one another.

The physics of **space travel and special relativity.**

Reasoning The two events in this problem are the departure from earth and the arrival at Alpha Centauri. At departure, earth is just outside the spaceship. Upon arrival at the destination, Alpha Centauri is just outside. Therefore, relative to the passengers, the two events occur at the same place—namely, just outside the spaceship. Thus, the passengers measure the proper time interval Δt_0 on their clock, and it is this interval that we must find. For a person left behind on earth, the events occur at *different places,* so such a person measures the dilated time interval Δt rather than the proper time interval. To find Δt we note that the time to travel a given distance is inversely proportional to the speed. Since it takes 4.3 years to traverse the distance between earth and Alpha Centauri at the speed of light, it would take even longer at the slower speed of $v = 0.95c$. Thus, a person on earth measures the dilated time interval to be $\Delta t = (4.3 \text{ years})/0.95 = 4.5$ years. This value can be used with the time-dilation equation to find the proper time interval Δt_0.

Problem solving insight
In dealing with time dilation, decide which interval is the proper time interval as follows: (1) Identify the two events that define the interval. (2) Determine the reference frame in which the events occur at the same place; an observer at rest in this frame measures the proper time interval Δt_0.

Solution Using the time-dilation equation, we find that the proper time interval by which the passengers judge their own aging is

$$\Delta t_0 = \Delta t \sqrt{1 - \frac{v^2}{c^2}} = (4.5 \text{ years}) \sqrt{1 - \left(\frac{0.95c}{c}\right)^2} = \boxed{1.4 \text{ years}}$$

Thus, the people aboard the rocket will have aged by only 1.4 years when they reach Alpha Centauri, and not the 4.5 years an earthbound observer has calculated.

▲

VERIFICATION OF TIME DILATION

A striking confirmation of time dilation was achieved in 1971 by an experiment carried out by J. C. Hafele and R. E. Keating.* They transported very precise cesium-beam atomic clocks around the world on commercial jets. Since the speed of a jet plane is considerably less than c, the time-dilation effect is extremely small. However, the atomic clocks were accurate to about $\pm 10^{-9}$ s, so the effect could be measured. The clocks were in the air for 45 hours, and their times were compared to reference atomic clocks kept on earth. The experimental results revealed that, within experimental error, the readings on the clocks on the planes were different from those on earth by an amount that agreed with the prediction of relativity.

The behavior of subatomic particles called *muons* provides additional confirmation of time dilation. These particles are created high in the atmosphere, at altitudes of about 10 000 m. When at rest, muons exist only for about 2.2×10^{-6} s before disintegrating. With such a short lifetime, these particles could never make it down to the earth's surface, even traveling at nearly the speed of light. However, *a large number of muons do reach the earth.* The only way they can do so is to live longer because of time dilation, as Example 3 illustrates.

Example 3   The Lifetime of a Muon

The average lifetime of a muon at rest is 2.2×10^{-6} s. A muon created in the upper atmosphere, thousands of meters above sea level, travels toward the earth at a speed of $v = 0.998c$. Find, on the average, (a) how long a muon lives according to an observer on earth, and (b) how far the muon travels before disintegrating.

Reasoning The two events of interest are the generation and subsequent disintegration of the muon. When the muon is at rest, these events occur at the same place, so the muon's average (at rest) lifetime of 2.2×10^{-6} s is a proper time interval Δt_0. When the muon moves at a speed $v = 0.998c$ relative to the earth, an observer on the earth measures a dilated lifetime Δt that is given by Equation 28.1. The average distance x traveled by a muon, as measured by an earth observer, is equal to the muon's speed times the dilated time interval.

Solution

(a) The observer on earth measures a dilated lifetime given by

Problem solving insight
The proper time interval Δt_0 is always shorter than the dilated time interval Δt.

$$\Delta t = \frac{\Delta t_0}{\sqrt{1 - \dfrac{v^2}{c^2}}} = \frac{2.2 \times 10^{-6}\text{ s}}{\sqrt{1 - \left(\dfrac{0.998c}{c}\right)^2}} = \boxed{35 \times 10^{-6}\text{ s}} \qquad (28.1)$$

(b) The distance traveled by the muon before it disintegrates is

$$x = v\,\Delta t = (0.998)(3.00 \times 10^8\text{ m/s})(35 \times 10^{-6}\text{ s}) = \boxed{1.0 \times 10^4\text{ m}}$$

Thus, the dilated, or extended, lifetime provides sufficient time for the muon to reach the surface of the earth. If its lifetime were only 2.2×10^{-6} s, a muon would travel only 660 m before disintegrating and could never reach the earth.

✔ Check Your Understanding 1

One of the following statements concerning the dilated time interval between two events is false. Which is it? (a) The dilated time interval is always greater than the proper time interval. (b) The dilated time interval depends on the relative speed between the observers who measure the proper and dilated time intervals. (c) The dilated time interval depends on the speed of light in a vacuum. (d) The dilated time interval is measured by an observer who is at rest with respect to the events. *(The answer is given at the end of the book)*

Background: In the theory of special relativity the time interval between two events can be either a proper or a dilated time interval. Which type of interval it is depends on who measures it—an observer at rest with respect to the events or an observer moving with respect to the events.

For similar questions (including calculational counterparts), consult Self-Assessment Test 28.1, which is described at the end of Section 28.4.

* J. C. Hafele and R. E. Keating, "Around-the-World Atomic Clocks: Observed Relativistic Time Gains," *Science*, Vol. 177, July 14, 1972, p. 168.

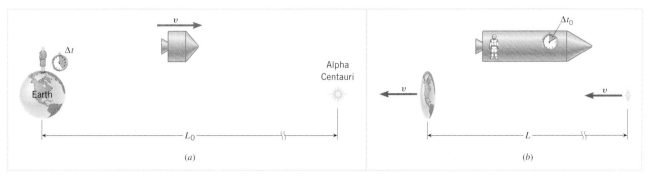

28.4 The Relativity of Length: Length Contraction

Because of time dilation, observers moving at a constant velocity relative to each other measure different time intervals between two events. For instance, Example 2 in the previous section illustrates that a trip from earth to Alpha Centauri at a speed of $v = 0.95c$ takes 4.5 years according to a clock on earth, but only 1.4 years according to a clock in the rocket. These two times differ by the factor $\sqrt{1 - v^2/c^2}$. Since the times for the trip are different, one might ask whether the observers measure different distances between earth and Alpha Centauri. The answer, according to special relativity, is yes. After all, both the earth-based observer and the rocket passenger agree that the relative speed between the rocket and earth is $v = 0.95c$. Since speed is distance divided by time and the time is different for the two observers, it follows that the distances must also be different, if the relative speed is to be the same for both individuals. Thus, the earth observer determines the distance to Alpha Centauri to be $L_0 = v\Delta t = (0.95c)(4.5 \text{ years}) = 4.3$ light-years. On the other hand, a passenger aboard the rocket finds the distance is only $L = v\Delta t_0 = (0.95c)(1.4 \text{ years}) = 1.3$ light-years. The passenger, measuring the shorter time, also measures the shorter distance. This shortening of the distance between two points is one example of a phenomenon known as *length contraction.*

The relation between the distances measured by two observers in relative motion at a constant velocity can be obtained with the aid of Figure 28.6. Part *a* of the drawing shows the situation from the point of view of the earth-based observer. This person measures the time of the trip to be Δt, the distance to be L_0, and the relative speed of the rocket to be $v = L_0/\Delta t$. Part *b* of the drawing presents the point of view of the passenger, for whom the rocket is at rest, and the earth and Alpha Centauri appear to move by at a speed v. The passenger determines the distance of the trip to be L, the time to be Δt_0, and the relative speed to be $v = L/\Delta t_0$. Since the relative speed computed by the passenger equals that computed by the earth-based observer, it follows that $v = L/\Delta t_0 = L_0/\Delta t$. Using this result and the time-dilation equation, Equation 28.1, we obtain the following relation between L and L_0:

Length contraction
$$L = L_0 \sqrt{1 - \frac{v^2}{c^2}} \tag{28.2}$$

The length L_0 is called the **proper length;** it is the length (or distance) between two points *as measured by an observer at rest with respect to them.* Since v is less than c, the term $\sqrt{1 - v^2/c^2}$ is less than 1, and L is less than L_0. It is important to note that this length contraction occurs only along the direction of the motion. Those dimensions that are perpendicular to the motion are not shortened, as the next example discusses.

Example 4 The Contraction of a Spacecraft

An astronaut, using a meter stick that is at rest relative to a cylindrical spacecraft, measures the length and diameter of the spacecraft to be 82 and 21 m, respectively. The spacecraft moves with a constant speed of $v = 0.95c$ relative to the earth, as in Figure 28.6. What are the dimensions of the spacecraft, as measured by an observer on earth?

Figure 28.6 (*a*) As measured by an observer on the earth, the distance to Alpha Centauri is L_0, and the time required to make the trip is Δt. (*b*) According to the passenger on the spacecraft, the earth and Alpha Centauri move with speed v relative to the craft. The passenger measures the distance and time of the trip to be L and Δt_0, respectively, both quantities being less than those in part *a*.

Reasoning The length of 82 m is a proper length L_0, since it is measured using a meter stick that is at rest relative to the spacecraft. The length L measured by the observer on earth can be determined from the length-contraction formula, Equation 28.2. On the other hand, the diameter of the spacecraft is perpendicular to the motion, so the earth observer does not measure any change in the diameter.

Solution The length L of the spacecraft, as measured by the observer on earth, is

$$L = L_0 \sqrt{1 - \frac{v^2}{c^2}} = (82 \text{ m})\sqrt{1 - \left(\frac{0.95c}{c}\right)^2} = \boxed{26 \text{ m}}$$

Both the astronaut and the observer on earth measure the same value for the diameter of the spacecraft: $\boxed{\text{Diameter} = 21 \text{ m}}$. Figure 28.6*a* shows the size of the spacecraft as measured by the earth observer, and part *b* shows the size measured by the astronaut.

Problem solving insight
The proper length L_0 is always larger than the contracted length L.

Need more practice?

Interactive LearningWare 28.1
After a long journey, an astronaut is heading home at a speed of 0.850*c* relative to earth. Her instruments tell her that she has 3.20×10^{13} m to go. (a) Based on her clock, how much time (in hours) will it take her to travel this distance? (b) According to a stationary observer on earth, how far is she from the planet? (c) What does this observer calculate for the time (in hours) it will take for her to arrive?

Related Homework: *Problem 10*

Go to
www.wiley.com/college/cutnell
for an interactive solution.

When dealing with relativistic effects we need to distinguish carefully between the criteria for the proper time interval and the proper length. The proper time interval Δt_0 between two events is the time interval measured by an observer who is at rest relative to the events and sees them occurring at the *same place*. All other moving inertial observers will measure a larger value for this time interval. The proper length L_0 of an object is the length measured by an observer who is *at rest* with respect to the object. All other moving inertial observers will measure a shorter value for this length. The observer who measures the proper time interval may not be the same one who measures the proper length. For instance, Figure 28.6 shows that the astronaut measures the proper time interval Δt_0 for the trip between earth and Alpha Centauri, whereas the earth-based observer measures the proper length (or distance) L_0 for the trip.

It should be emphasized that the word "proper" in the phrases proper time and proper length does *not* mean that these quantities are the correct or preferred quantities in any absolute sense. If this were so, the observer measuring these quantities would be using a preferred reference frame for making the measurement, a situation that is prohibited by the relativity postulate. According to this postulate, there is no preferred inertial reference frame. When two observers are moving relative to each other at a constant velocity, each measures the other person's clock to run more slowly than his own, and each measures the other person's length, along that person's motion, to be contracted.

Check Your Understanding 2

The drawing shows an object that has the shape of a square when it is at rest in an inertial frame. When this square moves, length contraction changes its shape. In each of the following three situations the speed is the same, and the velocity is in the plane of the square, directed parallel to (a) side AB, (b) diagonal AC, or (c) side AD. In which situation is the area of the object the least? *(The answer is given at the end of the book.)*

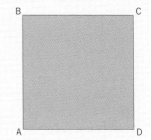

Background: Length contraction occurs only along the direction of the motion. Dimensions perpendicular to the motion are not contracted.

For similar questions (including calculational couterparts), consult Self-Assessment Test 28.1, which is described next.

Self-Assessment Test 28.1

Test your understanding of the material in Sections 28.1–28.4:

• Events and Inertial Reference Frames   • The Postulates of Special Relativity
• The Relativity of Time: Time Dilation   • The Relativity of Length: Length Contraction

Go to **www.wiley.com/college/cutnell**

28.5 Relativistic Momentum

Thus far we have discussed how time intervals and distances between two events are measured by observers moving at a constant velocity relative to each other. Special relativity also alters our ideas about momentum and energy.

Recall from Chapter 7 that when two or more objects interact, the principle of conservation of linear momentum applies if the system of objects is isolated. This principle states that the total linear momentum of an isolated system remains constant at all times. (An isolated system is one in which the sum of the external forces acting on the objects is zero.) The conservation of linear momentum is a law of physics and, in accord with the relativity postulate, is valid in all inertial reference frames. That is, when the total linear momentum is conserved in one inertial reference frame, it is conserved in all inertial reference frames.

As an example of momentum conservation, suppose several people are watching two billiard balls collide on a frictionless pool table. One person is standing next to the pool table, and the other is moving past the table with a constant velocity. Since the two balls constitute an isolated system, the relativity postulate requires that both observers must find the total linear momentum of the two-ball system to be the same before, during, and after the collision. For this kind of situation, Section 7.1 defines the classical linear momentum $\mathbf{p}$ of an object to be the product of its mass m and velocity $\mathbf{v}$. As a result, the magnitude of the classical momentum is $p = mv$. As long as the speed of an object is considerably smaller than the speed of light, this definition is adequate. However, when the speed approaches the speed of light, an analysis of the collision shows that the total linear momentum is not conserved in all inertial reference frames if one defines linear momentum simply as the product of mass and velocity. In order to preserve the conservation of linear momentum, it is necessary to modify this definition. The theory of special relativity reveals that the magnitude of the ***relativistic momentum*** must be defined as in Equation 28.3:

Magnitude of the
relativistic momentum

$$p = \frac{mv}{\sqrt{1 - \dfrac{v^2}{c^2}}} \qquad (28.3)$$

The total relativistic momentum of an isolated system is conserved in all inertial reference frames.

From Equation 28.3, we can see that the magnitudes of the relativistic and nonrelativistic momenta differ by the same factor of $\sqrt{1 - v^2/c^2}$ that occurs in the time-dilation and length-contraction equations. Since this factor is always less than 1 and occurs in the denominator in Equation 28.3, the relativistic momentum is always larger than the nonrelativistic momentum. To illustrate how the two quantities differ as the speed v increases, Figure 28.7 shows a plot of the ratio of the momentum magnitudes (relativistic/nonrelativistic) as a function of v. According to Equation 28.3, this ratio is just $1/\sqrt{1 - v^2/c^2}$. The graph shows that for speeds attained by ordinary objects, such as cars and planes, the relativistic and nonrelativistic momenta are almost equal because their ratio is nearly 1. Thus, at speeds much less than the speed of light, either the nonrelativistic momentum or the relativistic momentum can be used to describe collisions. On the other hand, when the speed of the object becomes comparable to the speed of light, the relativistic momentum becomes significantly greater than the nonrelativistic momentum and must be used. Example 5 deals with the relativistic momentum of an electron traveling close to the speed of light.

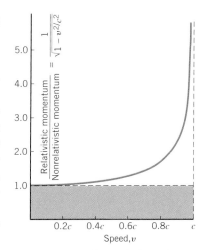

Figure 28.7 This graph shows how the ratio of the magnitude of the relativistic momentum to the magnitude of the nonrelativistic momentum increases as the speed of an object approaches the speed of light.

Example 5 The Relativistic Momentum of a High-Speed Electron

The particle accelerator at Stanford University (Figure 28.8) is three kilometers long and accelerates electrons to a speed of 0.999 999 999 7c, which is very nearly equal to the speed of light. Find the magnitude of the relativistic momentum of an electron that emerges from the accelerator, and compare it with the nonrelativistic value.

Reasoning and Solution The magnitude of the electron's relativistic momentum can be obtained from Equation 28.3 if we recall that the mass of an electron is $m = 9.11 \times 10^{-31}$ kg:

$$p = \frac{mv}{\sqrt{1 - \dfrac{v^2}{c^2}}} = \frac{(9.11 \times 10^{-31} \text{ kg})(0.999\ 999\ 999\ 7c)}{\sqrt{1 - \dfrac{(0.999\ 999\ 999\ 7c)^2}{c^2}}} = \boxed{1 \times 10^{-17} \text{ kg} \cdot \text{m/s}}$$

This value for the magnitude of the momentum agrees with that measured experimentally. The relativistic momentum is greater than the nonrelativistic momentum by a factor of

$$\frac{1}{\sqrt{1 - \dfrac{v^2}{c^2}}} = \frac{1}{\sqrt{1 - \dfrac{(0.999\ 999\ 999\ 7c)^2}{c^2}}} = \boxed{4 \times 10^4}$$

Figure 28.8 The Stanford three-kilometer linear accelerator accelerates electrons almost to the speed of light. (© Bill Marsh/Photo Researchers)

28.6 *The Equivalence of Mass and Energy*

THE TOTAL ENERGY OF AN OBJECT

One of the most astonishing results of special relativity is that mass and energy are equivalent, in the sense that a gain or loss of mass can be regarded equally well as a gain or loss of energy. Consider, for example, an object of mass m traveling at a speed v. Einstein showed that the *total energy* E of the moving object is related to its mass and speed by the following relation:

Total energy
of an object
$$E = \frac{mc^2}{\sqrt{1 - \dfrac{v^2}{c^2}}} \qquad (28.4)$$

To gain some understanding of Equation 28.4, consider the special case in which the object is at rest. When $v = 0$ m/s, the total energy is called the *rest energy* E_0, and Equation 28.4 reduces to Einstein's now-famous equation:

Rest energy
of an object
$$E_0 = mc^2 \qquad (28.5)$$

The rest energy represents the energy equivalent of the mass of an object at rest. As Example 6 shows, even a small mass is equivalent to an enormous amount of energy.

Example 6 The Energy Equivalent of a Golf Ball

A 0.046-kg golf ball is lying on the green. (a) Find the rest energy of the golf ball. (b) If this rest energy were used to operate a 75-W light bulb, for how many years could the bulb stay on?

Reasoning The rest energy E_0 that is equivalent to the mass m of the golf ball is found from the relation $E_0 = mc^2$. The power used by the bulb is 75 W, which means that it consumes 75 J of energy per second. If the entire rest energy of the ball were available for use, the bulb could stay on for a time equal to the rest energy divided by the power.

Solution

(a) The rest energy of the ball is

$$E_0 = mc^2 = (0.046 \text{ kg})(3.0 \times 10^8 \text{ m/s})^2 = \boxed{4.1 \times 10^{15} \text{ J}} \qquad (28.5)$$

(b) This rest energy can keep the bulb burning for a time t given by

$$t = \frac{\text{Rest energy}}{\text{Power}} = \frac{4.1 \times 10^{15} \text{ J}}{75 \text{ W}} = 5.5 \times 10^{13} \text{ s} \qquad (6.10b)$$

Since one year contains 3.2×10^7 s, we find $\boxed{t = 1.7 \times 10^6 \text{ yr}}$, or 1.7 million years!

When an object is accelerated from rest to a speed v, the object acquires kinetic energy in addition to its rest energy. The total energy E is the sum of the rest energy E_0 and the kinetic energy KE, or $E = E_0 + \text{KE}$. Therefore, the kinetic energy is the difference

between the object's total energy and its rest energy. Using Equations 28.4 and 28.5, we can write the kinetic energy as

$$KE = E - E_0 = mc^2 \left(\frac{1}{\sqrt{1 - \dfrac{v^2}{c^2}}} - 1 \right) \tag{28.6}$$

This equation is the relativistically correct expression for the kinetic energy of an object of mass m moving at speed v.

Equation 28.6 looks nothing like the kinetic energy expression introduced in Chapter 6—namely, $KE = \frac{1}{2}mv^2$. However, for speeds much less than the speed of light ($v \ll c$), the relativistic equation for the kinetic energy reduces to $KE = \frac{1}{2}mv^2$, as can be seen by using the binomial expansion* to represent the square root term in Equation 28.6:

$$\frac{1}{\sqrt{1 - \dfrac{v^2}{c^2}}} = 1 + \frac{1}{2}\left(\frac{v^2}{c^2}\right) + \frac{3}{8}\left(\frac{v^2}{c^2}\right)^2 + \cdots$$

Suppose v is much smaller than c—say $v = 0.01c$. The second term in the expansion has the value $\frac{1}{2}(v^2/c^2) = 5.0 \times 10^{-5}$, while the third term has the much smaller value $\frac{3}{8}(v^2/c^2)^2 = 3.8 \times 10^{-9}$. The additional terms are smaller still, so if $v \ll c$, we can neglect the third and additional terms in comparison with the first and second terms. Substituting the first two terms into Equation 28.6 gives

$$KE \approx mc^2 \left(1 + \frac{1}{2}\frac{v^2}{c^2} - 1 \right) = \frac{1}{2}mv^2$$

which is the familiar form for the kinetic energy. However, Equation 28.6 gives the correct kinetic energy for all speeds and must be used for speeds near the speed of light, as in Example 7.

Example 7 A High-Speed Electron

An electron ($m = 9.109 \times 10^{-31}$ kg) is accelerated from rest to a speed of $v = 0.9995c$ in a particle accelerator. Determine the electron's (a) rest energy, (b) total energy, and (c) kinetic energy in millions of electron volts or MeV.

Reasoning and Solution

(a) The electron's rest energy is

$$E_0 = mc^2 = (9.109 \times 10^{-31} \text{ kg})(2.998 \times 10^8 \text{ m/s})^2 = 8.187 \times 10^{-14} \text{ J} \tag{28.5}$$

Since 1 eV = 1.602×10^{-19} J, the electron's rest energy is

$$(8.187 \times 10^{-14} \text{ J}) \left(\frac{1 \text{ eV}}{1.602 \times 10^{-19} \text{ J}} \right) = \boxed{5.11 \times 10^5 \text{ eV} \quad \text{or} \quad 0.511 \text{ MeV}}$$

(b) The total energy of an electron traveling at a speed of $v = 0.9995c$ is

$$E = \frac{mc^2}{\sqrt{1 - \dfrac{v^2}{c^2}}} = \frac{(9.109 \times 10^{-31} \text{ kg})(2.998 \times 10^8 \text{ m/s})^2}{\sqrt{1 - \left(\dfrac{0.9995c}{c}\right)^2}} \tag{28.4}$$

$$= \boxed{2.59 \times 10^{-12} \text{ J} \quad \text{or} \quad 16.2 \text{ MeV}}$$

(c) The kinetic energy is the difference between the total energy and the rest energy:

$$KE = E - E_0 = 2.59 \times 10^{-12} \text{ J} - 8.2 \times 10^{-14} \text{ J} \tag{28.6}$$

$$= \boxed{2.51 \times 10^{-12} \text{ J} \quad \text{or} \quad 15.7 \text{ MeV}}$$

For comparison, if the kinetic energy of the electron had been calculated from $\frac{1}{2}mv^2$, a value of only 0.26 MeV would have been obtained.

---

* The binomial expansion states that $(1 - x)^n = 1 - nx + n(n - 1)x^2/2 + \cdots$. In our case, $x = v^2/c^2$ and $n = -1/2$.

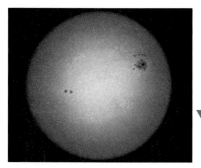

Visible light image.

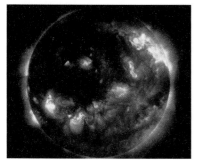

X-ray image.

Figure 28.9 The sun emits electromagnetic energy over a broad portion of the electromagnetic spectrum. These photographs were obtained using that energy in the indicated regions of the spectrum. (*Top photo:* © Mark Marten/NASA/Photo Researchers; *bottom photo:* Dr. Leon Golub/Photo Researchers)

Since mass and energy are equivalent, any change in one is accompanied by a corresponding change in the other. For instance, life on earth is dependent on electromagnetic energy (light) from the sun. Because this energy is leaving the sun (see Figure 28.9), there is a decrease in the sun's mass. Example 8 illustrates how to determine this decrease.

Example 8   The Sun Is Losing Mass

The sun radiates electromagnetic energy at the rate of 3.92×10^{26} W. (a) What is the change in the sun's mass during each second that it is radiating energy? (b) The mass of the sun is 1.99×10^{30} kg. What fraction of the sun's mass is lost during a human lifetime of 75 years?

Reasoning Since 1 W = 1 J/s, the amount of electromagnetic energy radiated during each second is 3.92×10^{26} J. Thus, during each second, the sun's rest energy decreases by this amount. The change ΔE_0 in the sun's rest energy is related to the change Δm in its mass by $\Delta E_0 = (\Delta m)c^2$, according to Equation 28.5.

Solution

(a) For each second that the sun radiates energy, the change in its mass is

$$\Delta m = \frac{\Delta E_0}{c^2} = \frac{3.92 \times 10^{26} \text{ J}}{(3.00 \times 10^8 \text{ m/s})^2} = \boxed{4.36 \times 10^9 \text{ kg}}$$

Over 4 billion kilograms of mass are lost by the sun during each second.

(b) The amount of mass lost by the sun in 75 years is

$$\Delta m = (4.36 \times 10^9 \text{ kg/s}) \left(\frac{3.16 \times 10^7 \text{ s}}{1 \text{ year}} \right) (75 \text{ years}) = 1.0 \times 10^{19} \text{ kg}$$

Although this is an enormous amount of mass, it represents only a tiny fraction of the sun's total mass:

$$\frac{\Delta m}{m_{\text{sun}}} = \frac{1.0 \times 10^{19} \text{ kg}}{1.99 \times 10^{30} \text{ kg}} = \boxed{5.0 \times 10^{-12}}$$

Any change in the energy of a system causes a change in the mass of the system according to $\Delta E_0 = (\Delta m)c^2$. It does not matter whether the change in energy is due to a change in electromagnetic energy, potential energy, thermal energy, or so on. Although any change in energy gives rise to a change in mass, in most instances the change in mass is too small to be detected. For instance, when 4186 J of heat is used to raise the temperature of 1 kg of water by 1 C°, the mass changes by only $\Delta m = \Delta E_0/c^2 = (4186 \text{ J})/(3.00 \times 10^8 \text{ m/s})^2 = 4.7 \times 10^{-14}$ kg. Conceptual Example 9 illustrates further how a change in the energy of an object leads to an equivalent change in its mass.

Conceptual Example 9   When Is a Massless Spring Not Massless?

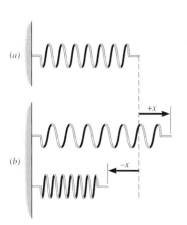

Figure 28.10 (*a*) This spring is unstrained and assumed to have no mass. (*b*) When the spring is either stretched or compressed by an amount *x*, it gains elastic potential energy and, hence, mass.

Figure 28.10*a* shows the top view of a spring lying on a horizontal table. The spring is initially unstrained and assumed to be massless. Suppose that the spring is either stretched or compressed by an amount *x* from its unstrained length, as part *b* of the drawing shows. Is the mass of the spring still zero, or has it changed? And, if the mass has changed, is the change greater, smaller, or the same when the spring is stretched rather than compressed?

Reasoning and Solution Whenever a spring is stretched or compressed, its elastic potential energy changes. In Section 10.3 we showed that the elastic potential energy of an ideal spring is equal to $\frac{1}{2}kx^2$, where *k* is the spring constant and *x* is the amount of stretch or compression. Consistent with the theory of special relativity, any change in the total energy of a system, including a change in the elastic potential energy, is equivalent to a change in the mass of the system. Thus, the mass of a strained spring is greater than that of an unstrained (and massless) spring. Furthermore, since the elastic potential energy depends on x^2, the increase in mass of the spring is the same whether it is compressed or stretched, provided the magnitude of *x* is the same in both cases.

The increase is exceedingly small because it is equal to the elastic potential energy divided by c^2 and c^2 is so large.

Related Homework: *Conceptual Questions 8, 9, 10*

It is also possible to transform matter itself into other forms of energy. For example, the positron (see Section 31.4) has the same mass as an electron but an opposite electrical charge. If these two particles of matter collide, they are completely annihilated, and a burst of high-energy electromagnetic waves is produced. Thus, matter is transformed into electromagnetic waves, the energy of the electromagnetic waves being equal to the total energies of the two colliding particles. The medical diagnostic technique known as positron emission tomography or PET scanning depends on the electromagnetic energy produced when a positron and an electron are annihilated (see Section 32.6).

The transformation of electromagnetic waves into matter also happens. In one experiment, an extremely high-energy electromagnetic wave, called a gamma ray (see Section 31.4), passes close to the nucleus of an atom. If the gamma ray has sufficient energy, it can create an electron and a positron. The gamma ray disappears, and the two particles of matter appear in its place. Except for picking up some momentum, the nearby nucleus remains unchanged. The process in which the gamma ray is transformed into the two particles is known as *pair production.*

THE RELATION BETWEEN TOTAL ENERGY AND MOMENTUM

It is possible to derive a useful relation between the total relativistic energy E and the relativistic momentum p. We begin by rearranging Equation 28.3 for the momentum, to obtain

$$\frac{m}{\sqrt{1 - v^2/c^2}} = \frac{p}{v}$$

With this substitution, Equation 28.4 for the total energy becomes

$$E = \frac{mc^2}{\sqrt{1 - v^2/c^2}} = \frac{pc^2}{v} \quad \text{or} \quad \frac{v}{c} = \frac{pc}{E}$$

Using this expression to replace v/c in Equation 28.4 gives

$$E = \frac{mc^2}{\sqrt{1 - v^2/c^2}} = \frac{mc^2}{\sqrt{1 - p^2c^2/E^2}} \quad \text{or} \quad E^2 = \frac{m^2c^4}{1 - p^2c^2/E^2}$$

Solving this expression for E^2 shows that

$$E^2 = p^2c^2 + m^2c^4 \tag{28.7}$$

THE SPEED OF LIGHT IS THE ULTIMATE SPEED

One of the important consequences of the theory of special relativity is that objects with mass cannot reach the speed of light in a vacuum. Thus, the speed of light represents the ultimate speed. To see that this speed limitation is a consequence of special relativity, consider Equation 28.6, which gives the kinetic energy of a moving object. As v approaches the speed of light c, the $\sqrt{1 - v^2/c^2}$ term in the denominator approaches zero. Hence, the kinetic energy becomes infinitely large. However, the work–energy theorem (Chapter 6) tells us that an infinite amount of work would have to be done to give the object an infinite kinetic energy. Since an infinite amount of work is not available, we are left with the conclusion that objects with mass cannot attain the speed of light c.

✔ **Check Your Understanding 3**

Consider the same cup of coffee in the following four situations. (a) It is hot (95 °C) and sits on a table at sea level. (b) It is cold (60 °C) and sits on the same table at sea level. (c) It is hot (95 °C) and sits on a table in a cottage in the mountains. (d) It is cold (60 °C) and sits on a table in a cottage in the mountains. In which situation does the cup of coffee have the greatest and the smallest mass? *(The answers are given at the end of the book.)*

Background: Mass and energy are equivalent. A gain or loss of energy in any form is equivalent to a gain or loss of mass.

For similar questions (including calculational counterparts), consult Self-Assessment Test 28.2, which is described at the end of Section 28.7.

28.7 The Relativistic Addition of Velocities

The velocity of an object relative to an observer plays a central role in special relativity, and to determine this velocity, it is sometimes necessary to add two or more velocities together. We first encountered relative velocity in Section 3.4, so we will begin by reviewing some of the ideas presented there. Figure 28.11 illustrates a truck moving at a constant velocity of $v_{TG} = +15$ m/s relative to an observer standing on the ground, where the plus sign denotes a direction to the right. Suppose someone on the truck throws a baseball toward the observer at a velocity of $v_{BT} = +8.0$ m/s relative to the truck. We might conclude that the observer on the ground sees the ball approaching at a velocity of $v_{BG} = v_{BT} + v_{TG} = 8.0$ m/s $+$ 15 m/s $= +23$ m/s. These symbols are similar to those used in Section 3.4 and have the following meaning:

$$v_{\boxed{BG}} = \text{velocity of the } \boxed{\text{Baseball}} \text{ relative to the } \boxed{\text{Ground}} = +23 \text{ m/s}$$

$$v_{\boxed{BT}} = \text{velocity of the } \boxed{\text{Baseball}} \text{ relative to the } \boxed{\text{Truck}} = +8.0 \text{ m/s}$$

$$v_{\boxed{TG}} = \text{velocity of the } \boxed{\text{Truck}} \text{ relative to the } \boxed{\text{Ground}} = +15.0 \text{ m/s}$$

Although the result that $v_{BG} = +23$ m/s seems reasonable, careful measurements would show that it is not quite right. According to special relativity, the equation $v_{BG} = v_{BT} + v_{TG}$ is not valid for the following reason. If the velocity of the truck had a magnitude sufficiently close to the speed of light, the equation would predict that the observer on the earth could see the baseball moving faster than the speed of light. This is not possible, since no object with a finite mass can move faster than the speed of light.

For the case where the truck and ball are moving along the same straight line, the theory of special relativity reveals that the velocities are related according to

$$v_{BG} = \frac{v_{BT} + v_{TG}}{1 + \dfrac{v_{BT}v_{TG}}{c^2}}$$

The subscripts in this equation have been chosen for the specific situation shown in Figure 28.11. For the general situation, the relative velocities are related by the *velocity-addition formula:*

Velocity addition

$$v_{AB} = \frac{v_{AC} + v_{CB}}{1 + \dfrac{v_{AC}v_{CB}}{c^2}} \tag{28.8}$$

where all the velocities are assumed to be constant and the symbols have the following meanings:

$$v_{\boxed{AB}} = \text{velocity of } \boxed{\text{object A}} \text{ relative to } \boxed{\text{object B}}$$

$$v_{\boxed{AC}} = \text{velocity of } \boxed{\text{object A}} \text{ relative to } \boxed{\text{object C}}$$

$$v_{\boxed{CB}} = \text{velocity of } \boxed{\text{object C}} \text{ relative to } \boxed{\text{object B}}$$

The ordering of the subscripts in Equation 28.8 follows the discussion in Section 3.4. For motion along a straight line, the velocities can have either positive or negative values, depending on whether they are directed along the positive or negative direction. Furthermore, switching the order of the subscripts changes the sign of the velocity, so, for example, $v_{BA} = -v_{AB}$ (see Example 11 in Chapter 3).

Equation 28.8 differs from the nonrelativistic formula ($v_{AB} = v_{AC} + v_{CB}$) by the presence of the $v_{AC}v_{CB}/c^2$ term in the denominator. This term arises because of the effects

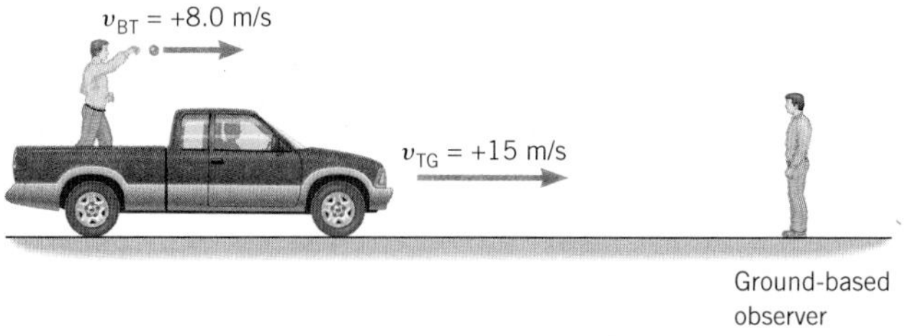

Figure 28.11 The truck is approaching the ground-based observer at a relative velocity of $v_{TG} = +15$ m/s. The velocity of the baseball relative to the truck is $v_{BT} = +8.0$ m/s.

$v_{BT} = +8.0$ m/s

$v_{TG} = +15$ m/s

Ground-based observer

of time dilation and length contraction that occur in special relativity. When v_{AC} and v_{CB} are small compared to c, the $v_{AC}v_{CB}/c^2$ term is small compared to 1, so the velocity-addition formula reduces to $v_{AB} \approx v_{AC} + v_{CB}$. However, when either v_{AC} or v_{CB} is comparable to c, the results can be quite different, as Example 10 illustrates.

Example 10 The Relativistic Addition of Velocities

Imagine a hypothetical situation in which the truck in Figure 28.11 is moving relative to the ground with a velocity of $v_{TG} = +0.8c$. A person riding on the truck throws a baseball at a velocity relative to the truck of $v_{BT} = +0.5c$. What is the velocity v_{BG} of the baseball relative to a person standing on the ground?

Reasoning The observer on the ground does *not* see the baseball approaching at $v_{BG} = 0.5c + 0.8c = 1.3c$. This cannot be because the speed of the ball would then exceed the speed of light. The velocity-addition formula gives the correct velocity, which has a magnitude less than the speed of light.

Solution The ground-based observer sees the ball approaching with a velocity of

$$v_{BG} = \frac{v_{BT} + v_{TG}}{1 + \dfrac{v_{BT}v_{TG}}{c^2}} = \frac{0.5c + 0.8c}{1 + \dfrac{(0.5c)(0.8c)}{c^2}} = \boxed{0.93c} \tag{28.8}$$

Example 10 discusses how the speed of a baseball is viewed by observers in different inertial reference frames. The next example deals with a similar situation, except that the baseball is replaced by the light of a laser beam.

Conceptual Example 11 The Speed of a Laser Beam

Figure 28.12 shows an intergalactic cruiser approaching a hostile spacecraft. The velocity of the cruiser relative to the spacecraft is $v_{CS} = +0.7c$. Both vehicles are moving at a constant velocity. The cruiser fires a beam of laser light at the enemy. The velocity of the laser beam relative to the cruiser is $v_{LC} = +c$. (a) What is the velocity of the laser beam v_{LS} relative to the renegades aboard the spacecraft? (b) At what velocity do the renegades aboard the spacecraft see the laser beam move away from the cruiser?

Reasoning and Solution

(a) Since both vehicles move at a constant velocity, each constitutes an inertial reference frame. According to the speed of light postulate, *all* observers in inertial reference frames measure the speed of light in a vacuum to be c. Thus, the renegades aboard the hostile spacecraft see the laser beam travel toward them at the speed of light, even though the beam is emitted from the cruiser, which itself is moving at seven-tenths the speed of light.

(b) The renegades aboard the spacecraft see the cruiser approach them at a relative velocity of $v_{CS} = +0.7c$, and they also see the laser beam approach them at a relative velocity of $v_{LS} = +c$. Both these velocities are measured relative to the *same* inertial reference frame—namely, that of the spacecraft. Therefore, the renegades aboard the spacecraft see the laser beam move away from the cruiser at a velocity that is the difference between these two velocities, or $+c - (+0.7c) = +0.3c$. The velocity-addition formula, Equation 28.8, is not applicable here because both velocities are measured relative to the *same* inertial reference frame (the spacecraft's reference frame). The velocity-addition formula can be used only when the velocities are measured relative to different inertial reference frames.

Related Homework: *Conceptual Question 12, Problem 34*

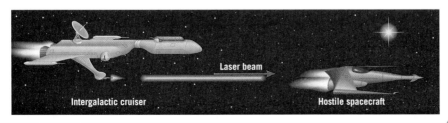

Laser beam

Intergalactic cruiser

Hostile spacecraft

Figure 28.12 An intergalactic cruiser, closing in on a hostile spacecraft, fires a beam of laser light.

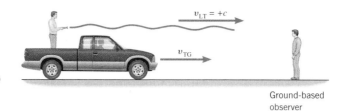

Figure 28.13 The speed of the light emitted by the flashlight is c relative to both the truck and the observer on the ground.

Ground-based observer

It is a straightforward matter to show that the velocity-addition formula is consistent with the speed of light postulate. Consider Figure 28.13, which shows a person riding on a truck and holding a flashlight. The velocity of the light, relative to the person on the truck, is $v_{LT} = +c$. The velocity v_{LG} of the light relative to the observer standing on the ground is given by the velocity-addition formula as

$$v_{LG} = \frac{v_{LT} + v_{TG}}{1 + \dfrac{v_{LT}v_{TG}}{c^2}} = \frac{c + v_{TG}}{1 + \dfrac{cv_{TG}}{c^2}} = \frac{(c + v_{TG})c}{(c + v_{TG})} = c$$

Thus, the velocity-addition formula indicates that the observer on the ground and the person on the truck both measure the speed of light to be c, independent of the relative velocity v_{TG} between them. This is exactly what the speed of light postulate states.

✔ **Check Your Understanding 4**

Car A and car B are both traveling due east on a straight section of an interstate highway. The speed of car A relative to the ground is 10 m/s faster than the speed of car B relative to the ground. According to special relativity, is the speed of car A relative to car B (a) 10 m/s, (b) less than 10 m/s, or (c) greater than 10 m/s? *(The answer is given at the end of the book.)*

Background: The relativistic addition of velocities obeys the velocity-addition formula. However, the method of using subscripts to identify the moving objects, as introduced in Section 3.4, still applies.

For similar questions (including calculational counterparts), consult Self-Assessment Test 28.2, which is described next.

 Self-Assessment Test 28.2

Test your understanding of the material in Sections 28.5–28.7:

• Relativistic Momentum   • The Equivalence of Mass and Energy
• The Relativistic Addition of Velocities

Go to **www.wiley.com/college/cutnell**

28.8 *Concepts & Calculations*

There are many astonishing consequences of special relativity, two of which are time dilation and length contraction. Example 12 reviews these important concepts in the context of a golf game in a world where the speed of light is a little faster than that of a golf cart.

 Concepts & Calculations Example 12   **Golf and Special Relativity**

Imagine playing golf in a world where the speed of light is only $c = 3.40$ m/s. Golfer A drives a ball down a flat horizontal fairway for a distance that he measures as 75.0 m. Golfer B, riding in a cart, happens to pass by just as the ball is hit. Golfer A stands at the tee and watches while golfer B moves down the fairway toward the ball at a constant speed of 2.80 m/s. (a) How far is the ball hit according to a measurement made by golfer B? (b) According to each golfer, how much time does it take for golfer B to reach the ball?

Concept Questions and Answers Who measures the proper length of the drive, and who measures the contracted length?

Answer Consider two coordinate systems, one attached to the earth and the other to the golf cart. The proper length L_0 is the distance between two points as measured by an observer *who is at rest with respect to them.* Since golfer A is standing at the tee and is at rest relative to the earth, golfer A measures the proper length of the drive, which is $L_0 = 75.0$ m. Golfer B is not at rest with respect to the two points, however, and measures a contracted length for the drive that is less than 75.0 m.

Who measures the proper time interval, and who measures the dilated time interval?

Answer The proper time interval Δt_0 is the time interval measured by an observer who is at rest in his or her coordinate system and *who views the beginning and ending events as occurring at the same place.* The beginning event is when the ball is hit, and the ending event is when golfer B arrives at the ball. When the ball is hit, it is alongside the origin of golfer B's coordinate system. When golfer B arrives at the ball down the fairway, it is again alongside the origin of his coordinate system. Thus, golfer B measures the proper time interval. Golfer A, who does not see the beginning and ending events occurring at the same place, measures a dilated, or longer, time interval.

Solution

(a) Golfer A, being at rest with respect to the beginning and ending points, measures the proper length of the drive, so $L_0 = 75.0$ m. Golfer B, who is moving, measures a contracted length L given by Equation 28.2:

$$L = L_0 \sqrt{1 - \frac{v^2}{c^2}} = (75.0 \text{ m}) \sqrt{1 - \frac{(2.80 \text{ m/s})^2}{(3.40 \text{ m/s})^2}} = \boxed{42.5 \text{ m}}$$

Thus, the moving golfer measures the length of the drive to be a shortened 42.5 m rather than the 75.0 m measured by the stationary golfer.

(b) According to golfer B, the time interval Δt_0 it takes to reach the ball is equal to the contracted length L that he measures divided by the speed of the ground with respect to him. The speed of the ground with respect to golfer B is the same as the speed v of the cart with respect to the ground. Thus, we find

$$\Delta t_0 = \frac{L}{v} = \frac{42.5 \text{ m}}{2.80 \text{ m/s}} = \boxed{15.2 \text{ s}}$$

Golfer A, standing at the tee, measures a dilated time interval Δt, which is related to the proper time interval by Equation 28.1:

$$\Delta t = \frac{\Delta t_0}{\sqrt{1 - \frac{v^2}{c^2}}} = \frac{15.2 \text{ s}}{\sqrt{1 - \frac{(2.80 \text{ m/s})^2}{(3.40 \text{ m/s})^2}}} = \boxed{26.8 \text{ s}}$$

In summary, golfer A measures the proper length (75.0 m) and a dilated time interval (26.8 s), and golfer B measures a shortened length (42.5 m) and a proper time interval (15.2 s).

Other important consequences of special relativity are the equivalence of mass and energy, and how kinetic energy depends on the total and rest energies. Example 13 compares these properties for three different particles.

Concepts & Calculations Example 13 Mass and Energy

The rest energy E_0 and the total energy E of three particles, expressed in terms of a basic amount of energy $E' = 5.98 \times 10^{-10}$ J, are listed in the table below. The speeds of these particles are large, in some cases approaching the speed of light. For each particle, determine its mass and kinetic energy.

Particle	Rest Energy	Total Energy
a	E'	$2E'$
b	E'	$4E'$
c	$5E'$	$6E'$

Concept Questions and Answers Given the rest energies specified in the table, what is the ranking (largest first) of the masses of the particles?

Answer The rest energy is the energy that an object has when its speed is zero. According to special relativity, the rest energy E_0 and the mass m are equivalent. The relation between the two is given by Equation 28.5 as $E_0 = mc^2$, where c is the speed of light in a vacuum. Thus, the rest energy is directly proportional to the mass. From the table it can be seen that particles a and b have identical rest energies, so they have identical masses. Particle c has the greatest rest energy, so it has the greatest mass. The ranking of the masses, largest first, is c, then a and b (a tie).

Is the kinetic energy KE given by the expression $\text{KE} = \frac{1}{2}mv^2$, and what is the ranking (largest first) of the kinetic energies of the particles?

Answer No. The expression $\text{KE} = \frac{1}{2}mv^2$ applies only when the speed of the object is much, much less than the speed of light. According to special relativity, the kinetic energy is the difference between the total energy E and the rest energy E_0, so $\text{KE} = E - E_0$. Therefore, we can examine the table and determine the kinetic energy of each particle in terms of E'. The kinetic energies of particles a, b, and c are, respectively, $2E' - E' = E'$, $4E' - E' = 3E'$, and $6E' - 5E' = E'$. The ranking of the kinetic energies, largest first, is b, then a and c (a tie).

Solution

(a) The mass of particle a can be found from its rest energy $E_0 = mc^2$. Since $E_0 = E'$ (see the table), its mass is

$$m_a = \frac{E'}{c^2} = \frac{5.98 \times 10^{-10}\ \text{J}}{(3.00 \times 10^8\ \text{m/s})^2} = \boxed{6.64 \times 10^{-27}\ \text{kg}}$$

In a similar manner, we find that the masses of particles b and c are

$$\boxed{m_b = 6.64 \times 10^{-27}\ \text{kg}} \quad \text{and} \quad \boxed{m_c = 33.2 \times 10^{-27}\ \text{kg}}$$

As expected, the ranking is $m_c > m_a = m_b$.

(b) According to Equation 28.6, the kinetic energy KE of a particle is equal to its total energy E minus its rest energy E_0; $\text{KE} = E - E_0$. For particle a, its total energy is $E = 2E'$ and its rest energy is $E_0 = E'$, so its kinetic energy is

$$\text{KE}_a = 2E' - E' = E' = \boxed{5.98 \times 10^{-10}\ \text{J}}$$

The kinetic energies of particles b and c can be determined in a similar fashion:

$$\boxed{\text{KE}_b = 17.9 \times 10^{-10}\ \text{J}} \quad \text{and} \quad \boxed{\text{KE}_c = 5.98 \times 10^{-10}\ \text{J}}$$

As anticipated, the ranking is $\text{KE}_b > \text{KE}_a = \text{KE}_c$.

At the end of the problem set for this chapter, you will find homework problems that contain both conceptual and quantitative parts. These problems are grouped under the heading *Concepts & Calculations, Group Learning Problems*. They are designed for use by students working alone or in small learning groups. The conceptual part of each problem provides a convenient focus for group discussions.

Concept Summary

This summary presents an abridged version of the chapter, including the important equations and all available learning aids. For convenient reference, the learning aids (including the text's examples) are placed next to or immediately after the relevant equation or discussion. The following learning aids may be found on-line at **www.wiley.com/college/cutnell**:

Interactive LearningWare examples are solved according to a five-step interactive format that is designed to help you develop problem-solving skills.	**Concept Simulations** are animated versions of text figures or animations that illustrate important concepts. You can control parameters that affect the display, and we encourage you to experiment.
Interactive Solutions offer specific models for certain types of problems in the chapter homework. The calculations are carried out interactively.	**Self-Assessment Tests** include both qualitative and quantitative questions. Extensive feedback is provided for both incorrect and correct answers, to help you evaluate your understanding of the material.

Topic	Discussion	Learning Aids

28.1 Events and Inertial Reference Frames

Reference frame

An event is a physical "happening" that occurs at a certain place and time. To record the event an observer uses a reference frame that consists of a coordinate system and a clock. Different observers may use different reference frames.

Inertial reference frame

The theory of special relativity deals with inertial reference frames. An inertial reference frame is one in which Newton's law of inertia is valid. Accelerating reference frames are not inertial reference frames.

28.2 The Postulates of Special Relativity

Relativity postulate

Speed of light postulate

The theory of special relativity is based on two postulates. The relativity postulate states that the laws of physics are the same in every inertial reference frame. The speed of light postulate says that the speed of light in a vacuum, measured in any inertial reference frame, always has the same value of c, no matter how fast the source of the light and the observer are moving relative to each other.

28.3 The Relativity of Time: Time Dilation

Proper time interval

Dilated time interval

The proper time interval Δt_0 between two events is the time interval measured by an observer who is at rest relative to the events and views them occurring at the same place. An observer who is in motion with respect to the events and who views them as occurring at different places measures a dilated time interval Δt. The dilated time interval is greater than the proper time interval, according to the time-dilation equation:

Concept Simulation 28.1

Time-dilation equation

$$\Delta t = \frac{\Delta t_0}{\sqrt{1 - \dfrac{v^2}{c^2}}} \qquad (28.1)$$

Examples 1, 2, 3

Interactive Solution 28.5

In this expression, v is the relative speed between the observer who measures Δt_0 and the observer who measures Δt.

28.4 The Relativity of Length: Length Contraction

Proper length

Contracted length

Length-contraction formula

The proper length L_0 between two points is the length measured by an observer who is at rest relative to the points. An observer moving with a relative speed v parallel to the line between the two points does not measure the proper length. Instead, such an observer measures a contracted length L given by the length-contraction formula:

Examples 4, 12

$$L = L_0 \sqrt{1 - \frac{v^2}{c^2}} \qquad (28.2)$$

Interactive LearningWare 28.1

Interactive Solution 28.13

Length contraction occurs only along the direction of the motion. Those dimensions that are perpendicular to the motion are not shortened.

The observer who measures the proper length may not be the observer who measures the proper time interval.

 Use *Self-Assessment Test 28.1* to evaluate your understanding of Sections 28.1–28.4.

28.5 Relativistic Momentum

An object of mass m, moving with speed v, has a relativistic momentum whose magnitude p is given by

Relativistic momentum

$$p = \frac{mv}{\sqrt{1 - \dfrac{v^2}{c^2}}} \qquad (28.3)$$

Example 5

28.6 The Equivalence of Mass and Energy

Energy and mass are equivalent. The total energy E of an object of mass m, moving at speed v, is

Total energy

$$E = \frac{mc^2}{\sqrt{1 - \dfrac{v^2}{c^2}}} \qquad (28.4)$$

Topic	Discussion	Learning Aids
Rest energy	The rest energy E_0 is the total energy of an object at rest ($v = 0$ m/s): $$E_0 = mc^2 \quad (28.5)$$	Examples 6, 8, 9
Kinetic energy	The total energy of an object is the sum of its rest energy and its kinetic energy KE, or $E = E_0 + KE$. Therefore, the kinetic energy is $$KE = E - E_0 = mc^2 \left(\frac{1}{\sqrt{1 - \frac{v^2}{c^2}}} - 1 \right) \quad (28.6)$$	Examples 7, 13
Relation between total energy and momentum	The relativistic total energy and momentum are related according to $$E^2 = p^2c^2 + m^2c^4 \quad (28.7)$$	
The ultimate speed	Objects with mass cannot attain the speed of light, which is the ultimate speed for such objects.	

28.7 The Relativistic Addition of Velocities

According to special relativity, the velocity addition formula specifies how the relative velocities of moving objects are related. For objects that move along the same straight line, this formula is

Velocity-addition formula	$$v_{AB} = \frac{v_{AC} + v_{CB}}{1 + \frac{v_{AC}v_{CB}}{c^2}} \quad (28.8)$$	Examples 10, 11

where v_{AB} is the velocity of object A relative to object B, v_{AC} is the velocity of object A relative to object C, and v_{CB} is the velocity of object C relative to object B. The velocities can have positive or negative values, depending on whether they are directed along the positive or negative direction. Furthermore, switching the order of the subscripts changes the sign of the velocity, so that, for example, $v_{BA} = -v_{AB}$.

 Use *Self-Assessment Test 28.2* to evaluate your understanding of Sections 28.5–28.7.

Conceptual Questions

1. The speed of light in water is c/n, where $n = 1.33$ is the index of refraction of water. Thus, the speed of light in water is less than c. Why doesn't this violate the speed of light postulate?

2. A baseball player at home plate hits a pop fly straight up (the beginning event) that is caught by the catcher at home plate (the ending event). Which of the following observers record the proper time interval between the two events: (a) a spectator sitting in the stands, (b) a spectator sitting on the couch and watching the game on TV, and (c) the third baseman running in to cover the play? Explain your answers.

3. The earth spins on its axis once each day. To a person viewing the earth from an inertial reference frame in space, which clock runs slower, a clock at the north pole or one at the equator? Why? (Ignore the orbital motion of the earth about the sun.)

4. Suppose you are standing at a railroad crossing, watching a train go by. (a) Both you and a passenger in the train are looking at a clock on the train. Which of you measures the proper time interval? (b) Who measures the proper length of the train car? (c) Who measures the proper distance between the railroad ties under the track? Justify your answers.

5. There are tables that list data for the various particles of matter that physicists have discovered. Often, such tables list the masses of

the particles in units of energy, such as in MeV (million electron volts), rather than in kilograms. Why is this possible?

6. Why is it easier to accelerate an electron to a speed that is close to the speed of light, compared to accelerating a proton to the same speed?

7. Light travels in water at a speed of 2.26×10^8 m/s. It is possible, in principle, for a particle that has mass to travel through water at a faster speed than this. In what sense, then, does the speed of light represent the ultimate speed attainable by objects that have mass?

8. Review Conceptual Example 9 for background pertinent to this question. Do two positive electric charges separated by a finite distance have more mass than when they are infinitely far apart (assume that the charges remain stationary)? Provide a reason for your answer.

9. Review Conceptual Example 9 for background relevant to this question. One system consists of two stationary electrons, separated by a distance r. Another consists of a positron and an electron, both stationary and separated by the same distance r. A positron has the same mass as an electron, but it has a positive electric charge $+e$. Which system of charges, if either, has the greater mass? Provide a reason for your answer.

10. Review Conceptual Example 9 for background pertinent to this question. Suppose that a parallel plate capacitor is initially uncharged. The capacitor is charged up by removing electrons from one plate and placing them on the other plate. Does the uncharged or the charged capacitor, if either, have the greater mass? Why?

11. The speed limit on many interstate highways is 65 miles per hour. If the speed of light were 65 miles per hour, would you be able to drive at the speed limit? Give your reasoning.

12. Review Conceptual Example 11 as an aid in answering this question. A person is approaching you in a truck that is traveling very close to the speed of light. This person throws a baseball toward you. Relative to the truck, the ball is thrown with a speed nearly equal to the speed of light, so the person on the truck sees the baseball move away from the truck at a very high speed. Yet you see the baseball move away from the truck very slowly. Why? Use the velocity-addition formula to guide your thinking.

13. Which of the following quantities will two observers always measure to be the *same*, regardless of the relative velocity between the observers: (a) the time interval between two events; (b) the length of an object; (c) the speed of light in a vacuum; (d) the relative speed between the observers? In each case, give a reason for your answer.

14. If the speed of light were infinitely large instead of 3.0×10^8 m/s, would the effects of time dilation and length contraction be observable? Explain, using the equations presented in the text to support your reasoning.

Problems

Before doing any calculations involving time dilation or length contraction, it is useful to identify which observer measures the proper time interval Δt_0 or the proper length L_0.

ssm Solution is in the Student Solutions Manual.   **www Solution is available on the World Wide Web at www.wiley.com/college/cutnell**

 This icon represents a biomedical application.

Section 28.3 The Relativity of Time: Time Dilation

1. ssm A radar antenna is rotating at an angular speed of 0.25 rad/s, as measured on earth. To an observer moving past the antenna at a speed of $0.80c$, what is its angular speed?

2. A Klingon spacecraft has a speed of $0.75c$ with respect to the earth. The Klingons measure 37.0 h for the time interval between two events on the earth. What value for the time interval would they measure if their ship had a speed of $0.94c$ with respect to the earth?

3. ssm A law enforcement officer in an intergalactic "police car" turns on a red flashing light and sees it generate a flash every 1.5 s. A person on earth measures that the time between flashes is 2.5 s. How fast is the "police car" moving relative to the earth?

4. Suppose that you are traveling on board a spacecraft that is moving with respect to the earth at a speed of $0.975c$. You are breathing at a rate of 8.0 breaths per minute. As monitored on earth, what is your breathing rate?

*5. Interactive Solution 28.5 at **www.wiley.com/college/cutnell** illustrates one way to model this problem. A 6.00-kg object oscillates back and forth at the end of a spring whose spring constant is 76.0 N/m. An observer is traveling at a speed of 1.90×10^8 m/s relative to the fixed end of the spring. What does this observer measure for the period of oscillation?

*6. An astronaut travels at a speed of 7800 m/s relative to the earth, a speed that is very small compared to c. According to a clock on the earth, the trip lasts 15 days. Determine the *difference* (in seconds) between the time recorded by the earth clock and the astronaut's clock. [*Hint: When $v \ll c$, the following approximation is valid:* $\sqrt{1 - v^2/c^2} \approx 1 - \frac{1}{2}(v^2/c^2)$.]

**7. As observed on earth, a certain type of bacterium is known to double in number every 24.0 hours. Two cultures of these bacteria are prepared, each consisting initially of one bacterium. One culture is left on earth and the other placed on a rocket that travels at a speed of $0.866c$ relative to the earth. At a time when the earthbound culture has grown to 256 bacteria, how many bacteria are in the culture on the rocket, according to an earth-based observer?

Section 28.4 The Relativity of Length: Length Contraction

8. A tourist is walking at a speed of 1.3 m/s along a 9.0-km path that follows an old canal. If the speed of light in a vacuum were 3.0 m/s, how long would the path be, according to the tourist?

9. ssm www How fast must a meter stick be moving if its length is observed to shrink to one-half of a meter?

10. Interactive LearningWare 28.1 at **www.wiley.com/college/cutnell** reviews the concepts that play roles in this problem. The distance from earth to the center of our galaxy is about 23 000 ly (1 ly = 1 light-year = 9.47×10^{15} m), as measured by an earth-based observer. A spaceship is to make this journey at a speed of $0.9990c$. According to a clock on board the spaceship, how long will it take to make the trip? Express your answer in years (1 yr = 3.16×10^7 s).

11. ssm A UFO streaks across the sky at a speed of $0.90c$ relative to the earth. A person on earth determines the length of the UFO to be 230 m along the direction of its motion. What length does the person measure for the UFO when it lands?

12. Suppose you are traveling in space and pass a rectangular landing pad on a planet. Your spacecraft has a speed of $0.85c$ relative to the planet and moves in a direction parallel to the length of the pad. While moving, you measure the length to be 1800 m and the width to be 1500 m. What are the dimensions of the landing pad according to the engineer who built it?

13. Interactive Solution 28.13 at **www.wiley/com/college/cutnell** illustrates one approach to solving this problem. A space traveler moving at a speed of $0.70c$ with respect to the earth makes a trip to a distant star that is stationary relative to the earth. He measures the length of this trip to be 6.5 light-years. What would be the length of this same trip (in light-years) as measured by a traveler moving at a speed of $0.90c$ with respect to the earth?

* **14.** As the drawing shows, a carpenter on a space station has constructed a 30.0° ramp. A rocket moves past the space station with a relative speed of 0.730c in a direction parallel to side x. What does a person aboard the rocket measure for the angle of the ramp?

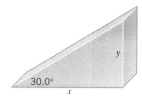

** **15. ssm www** A rectangle has the dimensions of 3.0 m × 2.0 m when viewed by someone at rest with respect to it. When you move past the rectangle along one of its sides, the rectangle looks like a square. What dimensions do you observe when you move at the same speed along the adjacent side of the rectangle?

** **16.** A woman and a man are on separate rockets, which are flying parallel to each other and have a relative speed of 0.940c. The woman measures the same value for the length of her own rocket and for the length of the man's rocket. What is the ratio of the value that the man measures for the length of his own rocket to the value he measures for the length of the woman's rocket?

Section 28.5 Relativistic Momentum

17. At what speed is the magnitude of the relativistic momentum of a particle three times the magnitude of the nonrelativistic momentum?

18. A jetliner has a mass of 1.2×10^5 kg and flies at a speed of 140 m/s. (a) Find the magnitude of its momentum. (b) If the speed of light in a vacuum had the hypothetical value of 170 m/s, what would be the magnitude of the jetliner's momentum?

19. ssm A rocket of mass 1.40×10^5 kg has a relativistic momentum the magnitude of which is 3.15×10^{13} kg·m/s. How fast is the rocket traveling?

20. A woman is 1.6 m tall and has a mass of 55 kg. She moves past an observer with the direction of the motion parallel to her height. The observer measures her relativistic momentum to have a magnitude of 2.0×10^{10} kg·m/s. What does the observer measure for her height?

* **21.** Starting from rest, two skaters "push off" against each other on smooth level ice, where friction is negligible. One is a woman and one is a man. The woman moves away with a velocity of +2.5 m/s relative to the ice. The mass of the woman is 54 kg, and the mass of the man is 88 kg. Assuming that the speed of light is 3.0 m/s, so that the relativistic momentum must be used, find the recoil velocity of the man relative to the ice. (*Hint: This problem is similar to Example 6 in Chapter 7.*)

Section 28.6 The Equivalence of Mass and Energy

22. An electron and a positron each have a mass of 9.11×10^{-31} kg. They collide and both vanish, with only electromagnetic radiation appearing after the collision. If each particle is moving at a speed of 0.20c relative to the laboratory before the collision, determine the energy of the electromagnetic radiation.

23. ssm Determine the ratio of the relativistic kinetic energy to the nonrelativistic kinetic energy ($\frac{1}{2}mv^2$) when a particle has a speed of (a) $1.00 \times 10^{-3}c$ and (b) 0.970c.

24. Suppose one gallon of gasoline produces 1.1×10^8 J of energy, and this energy is sufficient to operate a car for twenty miles. An aspirin tablet has a mass of 325 mg. If the aspirin could be converted completely into thermal energy, how many miles could the car go on a single tablet?

25. A nuclear power reactor generates 3.0×10^9 W of power. In one year, what is the change in the mass of the nuclear fuel due to the energy being taken from the reactor?

26. Four kilograms of water are heated from 20.0 °C to 60.0 °C. (a) How much heat is required to produce this change in temperature? [The specific heat capacity of water is 4186 J/(kg·C°).] (b) By how much does the mass of the water increase?

27. ssm How much work must be done on an electron to accelerate it from rest to a speed of 0.990c?

* **28.** An object has a total energy of 5.0×10^{15} J and a kinetic energy of 2.0×10^{15} J. What is the magnitude of the object's relativistic momentum?

* **29.** How close would two stationary electrons have to be positioned so that their total mass is twice what it is when the electrons are very far apart?

28.7 The Relativistic Addition of Velocities

30. Spaceship Y is between spaceship X and spaceship Z. Spaceship Y is moving toward spaceship Z at a speed of 0.68c. Spaceship Z is moving toward spaceship X at a speed of 0.42c. Assuming that all of the spaceships are moving at constant velocities, so they are inertial reference frames, find the speed of spaceship Y with respect to spaceship X.

31. ssm A spacecraft approaching the earth launches an exploration vehicle. After the launch, an observer on earth sees the spacecraft approaching at a speed of 0.50c and the exploration vehicle approaching at a speed of 0.70c. What is the speed of the exploration vehicle relative to the spacecraft?

32. Galaxy A is moving away from us with a speed of 0.75c relative to the earth. Galaxy B is moving away from us in the opposite direction with a relative speed of 0.55c. Assume that the earth and the galaxies are moving at constant velocities, so they are inertial reference frames. How fast is galaxy A moving according to an observer in galaxy B?

* **33. ssm www** The crew of a rocket that is moving away from the earth launches an escape pod, which they measure to be 45 m long. The pod is launched toward the earth with a speed of 0.55c relative to the rocket. After the launch, the rocket's speed relative to the earth is 0.75c. What is the length of the escape pod as determined by an observer on earth?

* **34.** Refer to Conceptual Example 11 as an aid in solving this problem. An intergalactic cruiser has two types of guns: a photon cannon that fires a beam of laser light and an ion gun that shoots ions at a velocity of 0.950c relative to the cruiser. The cruiser closes in on an alien spacecraft at a velocity of 0.800c relative to this spacecraft. The captain fires both types of guns. At what velocity do the aliens see (a) the laser light and (b) the ions approach them? At what velocity do the aliens see (c) the laser light and (d) the ions move away from the cruiser?

** **35.** Two atomic particles approach each other in a head-on collision. Each particle has a mass of 2.16×10^{-25} kg. The speed of each particle is 2.10×10^8 m/s when measured by an observer standing in the laboratory. (a) What is the speed of one particle as seen by the other particle? (b) Determine the relativistic momentum of one particle, as it would be observed by the other.

Additional Problems

36. Radium is a radioactive element whose nucleus emits an α particle (a helium nucleus) that has a kinetic energy of about 7.8×10^{-13} J (4.9 MeV). To what amount of mass is this energy equivalent?

37. ssm www Two spaceships A and B are exploring a new planet. Relative to this planet, spaceship A has a speed of $0.60c$, and spaceship B has a speed of $0.80c$. What is the ratio D_A/D_B of the values for the planet's diameter that each spaceship measures in a direction that is parallel to its motion?

38. Suppose you are planning a trip in which a spacecraft is to travel at a constant velocity for exactly six months, as measured by a clock on board the spacecraft, and then return home at the same speed. Upon return, the people on earth will have advanced exactly one hundred years into the future. According to special relativity, how fast must you travel? Express your answer to 5 significant figures as a multiple of c—for example, $0.955\,85c$.

39. A spaceship is approaching the earth at a relative speed of $0.85c$. The mass of the ship is 2.0×10^7 kg. Find the magnitude of (a) the classical momentum and (b) the relativistic momentum of the ship.

* **40.** A person on earth notices a rocket approaching from the right at a speed of $0.75c$ and another rocket approaching from the left at $0.65c$. What is the relative speed between the two rockets, as measured by a passenger on one of them?

* **41. ssm** An unstable particle is at rest and suddenly breaks up into two fragments. No external forces act on the particle or its fragments. One of the fragments has a velocity of $+0.800c$ and a mass of 1.67×10^{-27} kg, and the other has a mass of 5.01×10^{-27} kg. What is the velocity of the more massive fragment? *(Hint: This problem is similar to Example 6 in Chapter 7.)*

* **42.** An electron is accelerated from rest through a potential difference of 2.40×10^7 V. (a) What is the relativistic kinetic energy (in joules) of the electron? (b) What is the speed of the electron?

** **43.** Twins who are 19.0 years of age leave the earth and travel to a distant planet 12.0 light-years away. Assume that the planet and earth are at rest with respect to each other. The twins depart at the same time on different spaceships. One twin travels at a speed of $0.900c$, and the other twin travels at $0.500c$. (a) According to the theory of special relativity, what is the difference between their ages when they meet again at the earliest possible time? (b) Which twin is older?

Concepts & Calculations Group Learning Problems

Note: Each of these problems consists of Concept Questions followed by a related quantitative Problem. They are designed for use by students working alone or in small learning groups. The Concept Questions involve little or no mathematics and are intended to stimulate group discussions. They focus on the concepts with which the problems deal. Recognizing the concepts is the essential initial step in any problem-solving technique.

44. Concept Questions A Martian leaves Mars in a spaceship that is heading to Venus. On the way, the spaceship passes earth with a speed v relative to it. Assume that the three planets do not move relative to each other during the trip. (a) Who, if anyone, measures the proper time interval Δt_0 for the trip—the Martian or a person on earth? (b) Who, if anyone, measures the proper length between Mars and Venus? (c) As measured by the Martian, how is the time of the trip related to the length of the trip and the speed v of the spaceship?

Problem The spaceship has a speed of $0.80c$ relative to earth, and the distance between Mars and Venus is 1.20×10^{11} m, as measured by a person on earth. (a) What does the Martian measure for the distance between Mars and Venus? (b) What is the time of the trip (in seconds) as measured by the Martian?

45. Concept Questions Examine Equation 28.3 for the magnitude p of the relativistic momentum. Decide whether doubling the speed v causes p to double or more than double. Decide whether halving the mass m causes p to decrease by a factor of two or more than a factor of two. With this in mind, consider the three particles in the table and rank them according to the magnitude of their relativistic momenta, largest first. Provide a reason for your ranking.

Particle	Mass	Speed
a	m	v
b	$\frac{1}{2}m$	$2v$
c	$\frac{1}{4}m$	$4v$

Problem The mass and speed in the table are $m = 1.20 \times 10^{-8}$ kg and $v = 0.200c$. Determine the momentum for each particle according to special relativity. Check to see that your answers are consistent with your answer to the Concept Questions.

46. Concept Questions (a) Suppose that a certain mass of ice at $0\,°C$ is melted into liquid water at $0\,°C$. According to special relativity, which has the greater mass, the ice or the water? (b) The same mass of liquid water at $100\,°C$ is boiled into steam at $100\,°C$. According to special relativity, which has the greater mass, the water or the steam? (c) Is the change in mass in question (b) greater than, less than, or the same as the change in mass in question (a)? Explain your answers.

Problem Two kilograms of water are changed (a) from ice at $0\,°C$ into liquid water at $0\,°C$ and (b) from liquid water at $100\,°C$ into steam at $100\,°C$. For each situation, determine the change in mass of the water. Verify that your answers are consistent with your answers to the Concept Questions.

47. Concept Questions Suppose you're driving down the highway at a speed of 25 m/s. A truck in the next lane is traveling toward you with a speed of 35 m/s relative to the ground. Relative to you, the truck is approaching at a speed of 25 m/s + 35 m/s = 60 m/s. (a) Since the speeds are much less than the speed of light, would the relative speed be the same if you were going 5.0 m/s relative to the ground and the truck were moving at 55 m/s relative to the ground? (b) Suppose the speed of light were $c = 65$ m/s. Would the relative speeds be the same for the two situations? Give a reason for each of your answers.

Problem Using the data presented in the Concept Questions and assuming that the speed of light is $c = 65$ m/s, determine the speed of the truck relative to you for the two situations in which your speed and the truck's speed are, respectively, (a) 25 m/s and 35 m/s, and (b) 5.0 m/s and 55 m/s. Check to see that your answers are consistent with your answer to Concept Question (b).

*48. Concept Questions** (a) How is the total relativistic energy of an object related to its rest energy and its speed? (b) If one object has a greater total energy than another, does it necessarily follow that their speeds are different? Explain, taking into account the role of the rest energy. (c) Consider the following table, which gives the total energy and the rest energy for three objects. The energies are given in terms of an energy increment ε. Rank the speeds of the objects in descending order (largest first). Justify your ranking.

Object	Total Energy (E)	Rest Energy (E_0)
A	2ε	ε
B	3ε	ε
C	3ε	2ε

Problem For each of the objects in Concept Question (c), determine the speed as a multiple of the speed c of light in a vacuum. Verify that your answers are consistent with your answers to the Concept Questions.

*49. Concept Questions** Two identical spaceships are under construction. After being launched, spaceship A moves away from the earth at a constant velocity with respect to the earth. Spaceship B follows in the same direction at a different constant velocity with respect to the earth. (a) Passengers on board each spaceship measure a length for the other spaceship that does not match that measured during construction. How is this possible? (b) What kind of length do the passengers measure, a proper length or a contracted length? Explain. (c) Describe the information needed to calculate the length that the passengers measure and explain how the calculation can be carried out.

Problem The constructed length of each spaceship is 1.50 km. Spaceship A moves away from the earth at a speed of $0.850c$ with respect to the earth. Spaceship B moves away at a speed of $0.500c$. Determine the length that a passenger on one spaceship measures for the other spaceship.

Chapter 29

Newly born star
sitting on surface
of EGG

Particles and Waves

This is the central portion of the Eagle Nebula located about 7000 light-years from earth, as seen by the Hubble Space Telescope. Among the topics in this chapter are particle-like entities called photons. As we will see, they play the central role in the process of photo-evaporation, which allows astronomers to peer into the dense star-forming regions of the nebula. (Courtesy NASA)

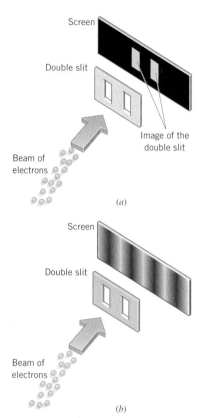

Figure 29.1 (*a*) If electrons behaved as discrete particles with no wave properties, they would pass through one or the other of the two slits and strike the screen, causing it to glow and produce exact images of the slits. (*b*) In reality, the screen reveals a pattern of bright and dark fringes, similar to the pattern produced when a beam of light is used and interference occurs between the light waves coming from each slit.

29.1 The Wave–Particle Duality

The ability to exhibit interference effects is an essential characteristic of waves. For instance, Section 27.2 discusses Young's famous experiment in which light passes through two closely spaced slits and produces a pattern of bright and dark fringes on a screen (see Figure 27.4). The fringe pattern is a direct indication that interference is occurring between the light waves coming from each slit.

One of the most incredible discoveries of twentieth-century physics is that particles can also behave like waves and exhibit interference effects. For instance, Figure 29.1 shows a version of Young's experiment performed by directing *a beam of electrons* onto a double slit. In this experiment, the screen is like a television screen and glows wherever an electron strikes it. Part *a* of the drawing indicates the pattern that would be seen on the screen if each electron, behaving strictly as a particle, were to pass through one slit or the other and strike the screen. The pattern would consist of an image of each slit. Part *b* shows the pattern actually observed, which consists of bright and dark fringes, reminiscent of the pattern obtained when light waves pass through a double slit. The fringe pattern indicates that the electrons are exhibiting interference effects that are associated with waves.

But how can electrons behave like waves in the experiment shown in Figure 29.1*b*? And what kind of waves are they? The answers to these profound questions will be discussed later in this chapter. For the moment, we intend only to emphasize that the concept of an electron as a tiny discrete particle of matter does not account for the fact that the electron can behave as a wave in some circumstances. In other words, the electron exhibits a dual nature, with both particle-like characteristics and wave-like characteristics.

Here is another interesting question: If a particle can exhibit wave-like properties, can waves exhibit particle-like behavior? As the next three sections reveal, the answer is yes. In fact, experiments that demonstrated the particle-like behavior of waves were performed near the beginning of the twentieth century, before the experiments that demonstrated the wave-like properties of the electrons. Scientists now accept the ***wave–particle duality*** as an essential part of nature:

> ***Waves can exhibit particle-like characteristics, and particles can exhibit wave-like characteristics.***

Section 29.2 begins the remarkable story of the wave–particle duality by discussing the electromagnetic waves that are radiated by a perfect blackbody. It is appropriate to begin with blackbody radiation, because it provided the first link in the chain of experimental evidence leading to our present understanding of the wave–particle duality.

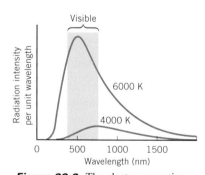

Figure 29.2 The electromagnetic radiation emitted by a perfect blackbody has an intensity per unit wavelength that varies from wavelength to wavelength, as each curve indicates. At the higher temperature, the intensity per unit wavelength is greater, and the maximum occurs at a shorter wavelength.

29.2 Blackbody Radiation and Planck's Constant

All bodies, no matter how hot or cold, continuously radiate electromagnetic waves. For instance, we see the glow of very hot objects because they emit electromagnetic waves in the visible region of the spectrum. Our sun, which has a surface temperature of about 6000 K, appears yellow, while the cooler star Betelgeuse has a red-orange appearance due to its lower surface temperature of 2900 K. However, at relatively low temperatures, cooler objects emit visible light waves only weakly and, as a result, do not appear to be glowing. Certainly the human body, at only 310 K, does not emit enough visible light to be seen in the dark with the unaided eye. But the body does emit electromagnetic waves in the infrared region of the spectrum, and these can be detected with infrared-sensitive devices.

At a given temperature, the intensities of the electromagnetic waves emitted by an object vary from wavelength to wavelength throughout the visible, infrared, and other regions of the spectrum. Figure 29.2 illustrates how the intensity per unit wavelength depends on wavelength for a perfect blackbody emitter. As Section 13.3 discusses, a perfect blackbody at a constant temperature absorbs and reemits all the electromagnetic radiation that falls on it. The two curves in the drawing show that at a higher temperature the maximum emitted intensity per unit wavelength increases and shifts to shorter wavelengths, toward the visible region of the spectrum. In accounting for the shape of these curves, the German physicist Max Planck (1858–1947) took the first step toward our present understanding of the wave–particle duality.

In 1900 Planck calculated the blackbody radiation curves, using a model that represents a blackbody as a large number of atomic oscillators, each of which emits and absorbs electromagnetic waves. To obtain agreement between the theoretical and experimental curves, Planck assumed that the energy E of an atomic oscillator could have only the discrete values of $E = 0$, hf, $2hf$, $3hf$, and so on. In other words, he assumed that

$$E = nhf \qquad n = 0, 1, 2, 3, \ldots \qquad (29.1)$$

where n is either zero or a positive integer, f is the frequency of vibration (in hertz), and h is a constant now called ***Planck's constant.*** Planck's constant has a value of

$$h = 6.626\ 068\ 76 \times 10^{-34}\ \text{J} \cdot \text{s}$$

The radical feature of Planck's assumption was that the energy of an atomic oscillator could have only discrete values (hf, $2hf$, $3hf$, etc.), with energies in between these values being forbidden. Whenever the energy of a system can have only certain definite values, and nothing in between, the energy is said to be *quantized*. This quantization of the energy was unexpected on the basis of the traditional physics of the time. However, it was soon realized that energy quantization had wide-ranging implications.

Conservation of energy requires that the energy carried off by the radiated electromagnetic waves must equal the energy lost by the atomic oscillators in Planck's model. Suppose, for example, that an oscillator with an energy of $3hf$ emits an electromagnetic wave. According to Equation 29.1, the next smallest allowed value for the energy of the oscillator is $2hf$. In such a case, the energy carried off by the electromagnetic wave would have the value of hf, equaling the amount of energy lost by the oscillator. Thus, Planck's model for blackbody radiation sets the stage for the idea that electromagnetic energy occurs as a collection of discrete amounts or packets of energy, the energy of a packet being equal to hf. Einstein made the proposal that light consists of such energy packets.

29.3 Photons and the Photoelectric Effect

▶ CONCEPTS AT A GLANCE The total energy E and the linear momentum $\mathbf{p}$ are fundamental concepts in physics. We have seen in Chapters 6 and 7 how they apply to moving particles, such as electrons and protons. The total energy of a (nonrelativistic) particle is the sum of its kinetic energy (KE) and potential energy (PE), or $E = \text{KE} + \text{PE}$. The magnitude p of the particle's momentum is the product of its mass m and speed v, or $p = mv$. These particle concepts are listed in the upper-right portion of the Concepts-at-a-Glance chart in Figure 29.3. We will now discuss the fact that electromagnetic waves are composed of particle-like entities called ***photons,*** and the lower-right portion of the chart shows that the ideas of energy and momentum also apply to them. However, as we will see, the equations defining photon energy ($E = hf$) and momentum ($p = h/\lambda$) are different from those for a particle, as the chart indicates. ◀

Figure 29.3 CONCEPTS AT A GLANCE A moving particle has energy E and momentum p. An electromagnetic wave is composed of particle-like entities called photons, each of which has energy and momentum. Although the laser beams in this photograph look like continuous beams of light, each is composed of discrete photons. (© AP/Wide World Photos)

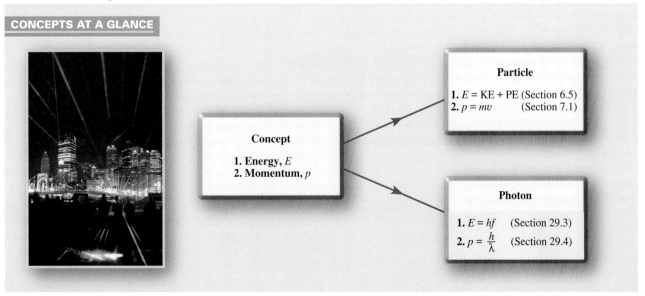

CONCEPTS AT A GLANCE

Concept

1. Energy, E
2. Momentum, p

Particle

1. $E = \text{KE} + \text{PE}$ (Section 6.5)
2. $p = mv$ (Section 7.1)

Photon

1. $E = hf$ (Section 29.3)
2. $p = \dfrac{h}{\lambda}$ (Section 29.4)

* It is now known that the energy of a harmonic oscillator is $E = (n + \frac{1}{2})hf$, the extra term of $\frac{1}{2}$ being unimportant to the present discussion.

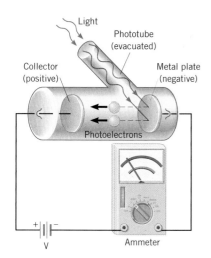

Figure 29.4 In the photoelectric effect, light with a sufficiently high frequency ejects electrons from a metal surface. These photoelectrons, as they are called, are drawn to the positive collector, thus producing a current.

Experimental evidence that light consists of photons comes from a phenomenon called the **photoelectric effect,** in which electrons are emitted from a metal surface when light shines on it. Figure 29.4 illustrates the effect. The electrons are emitted if the light being used has a sufficiently high frequency. The ejected electrons move toward a positive electrode called the *collector* and cause a current to register on the ammeter. Because the electrons are ejected with the aid of light, they are called **photoelectrons.** As will be discussed shortly, a number of features of the photoelectric effect could not be explained solely with the ideas of classical physics.

In 1905 Einstein presented an explanation of the photoelectric effect that took advantage of Planck's work concerning blackbody radiation. It was primarily for his theory of the photoelectric effect that he was awarded the Nobel Prize in physics in 1921. In his photoelectric theory, Einstein proposed that light of frequency f could be regarded as a collection of discrete packets of energy (photons), each packet containing an amount of energy E given by

Energy of a photon $\qquad\qquad\qquad E = hf \qquad\qquad\qquad$ (29.2)

where h is Planck's constant. The light energy given off by a light bulb, for instance, is carried by photons. The brighter the bulb, the greater is the number of photons emitted per second. Example 1 estimates the number of photons emitted per second by a typical light bulb.

Example 1 Photons from a Light Bulb

In converting electrical energy into light energy, a sixty-watt incandescent light bulb operates at about 2.1% efficiency. Assuming that all the light is green light (vacuum wavelength = 555 nm), determine the number of photons per second given off by the bulb.

Reasoning The number of photons emitted per second can be found by dividing the amount of light energy emitted per second by the energy E of one photon. The energy of a single photon is $E = hf$, according to Equation 29.2. The frequency f of the photon is related to its wavelength λ by Equation 16.1 as $f = c/\lambda$.

Solution At an efficiency of 2.1%, the light energy emitted per second by a sixty-watt bulb is $(0.021)(60.0 \text{ J/s}) = 1.3$ J/s. The energy of a single photon is

$$E = hf = \frac{hc}{\lambda} = \frac{(6.63 \times 10^{-34}\ \text{J} \cdot \text{s})(3.00 \times 10^{8}\ \text{m/s})}{555 \times 10^{-9}\ \text{m}} = 3.58 \times 10^{-19}\ \text{J}$$

Therefore,

$$\begin{matrix} \text{Number of} \\ \text{photons emitted} \\ \text{per second} \end{matrix} = \frac{1.3\ \text{J/s}}{3.58 \times 10^{-19}\ \text{J/photon}} = \boxed{3.6 \times 10^{18}\ \text{photons/s}}$$

Need more practice?

Interactive LearningWare 29.1
Example 1 calculates the number of photons per second given off by a sixty-watt incandescent light bulb. The photons are emitted uniformly in all directions. From a distance of 3.1 m you glance at this bulb for 0.10 s. The light from the bulb travels directly to your eye and does not reflect from anything. The pupil of the eye has a diameter of 2.0 mm. How many photons enter your eye?

Related Homework: Problem 9

Go to **www.wiley.com/college/cutnell** for an interactive solution.

According to Einstein, when light shines on a metal, a photon can give up its energy to an electron in the metal. If the photon has enough energy to do the work of removing the electron from the metal, the electron can be ejected. The work required depends on how strongly the electron is held. For the *least strongly* held electrons, the necessary work has a minimum value W_0 and is called the **work function** of the metal. If a photon has energy in excess of the work needed to remove an electron, the excess appears as kinetic energy of the ejected electron. Thus, the least strongly held electrons are ejected with the maximum kinetic energy KE_{max}. Einstein applied the conservation-of-energy principle and proposed the following relation to describe the photoelectric effect:

$$\underbrace{hf}_{\substack{\text{Photon}\\\text{energy}}} = \underbrace{KE_{max}}_{\substack{\text{Maximum}\\\text{kinetic energy}\\\text{of ejected}\\\text{electron}}} + \underbrace{W_0}_{\substack{\text{Minimum}\\\text{work needed to}\\\text{eject electron}}} \qquad (29.3)$$

According to this equation, $KE_{max} = hf - W_0$, which is plotted in Figure 29.5, with KE_{max} along the y axis and f along the x axis. The graph is a straight line that crosses the x axis at $f = f_0$. At this frequency, the electron departs from the metal with no kinetic energy ($KE_{max} = 0$ J). According to Equation 29.3, when $KE_{max} = 0$ J the energy hf_0 of the incident photon is equal to the work function W_0 of the metal: $hf_0 = W_0$.

The photon concept provides an explanation for a number of features of the photoelectric experiment that are difficult to explain without photons. It is observed, for instance, that only light with a frequency above a certain minimum value f_0 will eject electrons. If the frequency of the light is below this value, no electrons are ejected, regardless of how intense the light is. The next example determines the minimum frequency value for a silver surface.

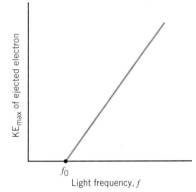

Figure 29.5 Photons can eject electrons from a metal when the light frequency is above a minimum value f_0. For frequencies above this value, ejected electrons have a maximum kinetic energy KE_{max} that is linearly related to the frequency, as the graph shows.

Example 2 The Photoelectric Effect for a Silver Surface

The work function for a silver surface is $W_0 = 4.73$ eV. Find the minimum frequency that light must have to eject electrons from this surface.

Reasoning The minimum frequency f_0 is that frequency at which the photon energy equals the work function W_0 of the metal, so the electron is ejected with zero kinetic energy. Since 1 eV $= 1.60 \times 10^{-19}$ J, the work function expressed in joules is $W_0 = (4.73$ eV$)[(1.60 \times 10^{-19}$ J$)/(1$ eV$)] = 7.57 \times 10^{-19}$ J. Using Equation 29.3, we find

$$hf_0 = \underbrace{KE_{max}}_{= 0 \text{ J}} + W_0 \quad \text{or} \quad f_0 = \frac{W_0}{h}$$

Solution The minimum frequency f_0 is

$$f_0 = \frac{W_0}{h} = \frac{7.57 \times 10^{-19} \text{ J}}{6.63 \times 10^{-34} \text{ J} \cdot \text{s}} = \boxed{1.14 \times 10^{15} \text{ Hz}}$$

Photons with frequencies less than f_0 do not have enough energy to eject electrons from a silver surface. Since $\lambda_0 = c/f_0$, the wavelength of this light is $\lambda_0 = 263$ nm, which is in the ultraviolet region of the electromagnetic spectrum.

Problem solving insight
The work function of a metal is the minimum energy needed to eject an electron from the metal. An electron that has received this minimum energy has no kinetic energy once outside the metal.

Concept Simulation 29.1

In this simulation of the photoelectric-effect experiment, light of a particular frequency shines on a metal plate. If the energy of each photon is larger than the work function of the metal, electrons are ejected. By graphing the maximum kinetic energy KE_{max} of the ejected electrons as a function of the frequency f of the light (see Figure 29.5), you can determine the work function of the metal. Three different metals are available—aluminum, silver, and sodium—and each has a different work function. Read the Student Notes accompanying this simulation for instructions on how to run the experiment.

Go to
www.wiley.com/college/cutnell

Another significant feature of the photoelectric effect is that the maximum kinetic energy of the ejected electrons remains the same when the intensity of the light increases, provided the light frequency remains the same. As the light intensity increases, more photons per second strike the metal, and consequently more electrons per second are ejected. However, since the frequency is the same for each photon, the energy of each photon is also the same. Thus, the ejected electrons always have the same maximum kinetic energy.

Whereas the photon model of light explains the photoelectric effect satisfactorily, the electromagnetic wave model of light does not. Certainly, it is possible to imagine that the electric field of an electromagnetic wave would cause electrons in the metal to oscillate and tear free from the surface when the amplitude of oscillation becomes large enough. However, were this the case, higher-intensity light would eject electrons with a greater maximum kinetic energy, a fact that experiment does not confirm. Moreover, in the electromagnetic wave model, a relatively long time would be required with low-intensity light before the electrons would build up a sufficiently large oscillation amplitude to tear free. Instead, experiment shows that even the weakest light intensity causes electrons to be ejected almost instantaneously, provided the frequency of the light is above the minimum value f_0. The failure of the electromagnetic wave model to explain the photoelectric effect does not mean that the wave model should be abandoned. However, we must recognize that the wave model does not account for all the characteristics of light. The photon

(a)

(b)

Figure 29.6 *(a)* Digital cameras like this one use an array of charge-coupled devices instead of film to capture an image. (© Getty Images News and Sport Services) *(b)* Images taken by a digital camera can be easily downloaded to a computer and sent to your friends via the Internet. (© Michael Newman/ PhotoEdit)

The physics of
charge-coupled devices
and digital cameras.

model also makes an important contribution to our understanding of the way light behaves when it interacts with matter.

✔ **Check Your Understanding 1**

In the photoelectric effect, electrons are ejected from the surface of a metal when light shines on it. Which one or more of the following would lead to an increase in the maximum kinetic energy of the ejected electrons? (a) Increasing the frequency of the incident light. (b) Increasing the number of photons per second striking the surface. (c) Using photons whose frequency f_0 is less than W_0/h, where W_0 is the work function of the metal and h is Planck's constant. (d) Selecting a metal that has a greater work function. *(The answer is given at the end of the book.)*

Background: The conservation of energy relates the maximum kinetic energy of the ejected electrons to the energy of the incident photons and the work function of the metal. This relation, Equation 29.3, holds the key to understanding the photoelectric effect.

For similar questions (including calculational counterparts), consult Self-Assessment Test 29.1, which is described at the end of this section.

Because a photon has energy, the photon can eject an electron from a metal surface when it interacts with the electron. However, a photon is different from a normal particle. A normal particle has a mass and can travel at speeds up to, but not equal to, the speed of light. A photon, on the other hand, travels at the speed of light in a vacuum and does not exist as an object at rest. The energy of a photon is entirely kinetic in nature, because it has no rest energy and no mass. To show that a photon has no mass, we rewrite Equation 28.4 for the total energy E as

$$E \sqrt{1 - \frac{v^2}{c^2}} = mc^2$$

The term $\sqrt{1 - (v^2/c^2)}$ is zero because a photon travels at the speed of light, $v = c$. Since the energy E of the photon is finite, the left side of the equation above is zero. Thus, the right side must also be zero, so $m = 0$ kg and the photon has no mass.

One of the most exciting and useful applications of the photoelectric effect is the charge-coupled device (CCD). An array of these devices is used instead of film in digital cameras (see Figure 29.6) to capture images in the form of many small groups of electrons. CCD arrays are also used in digital camcorders and electronic scanners, and they provide the method of choice with which astronomers capture those spectacular images of the planets and the stars. For use with visible light, a CCD array consists of a sandwich of semiconducting silicon, insulating silicon dioxide, and a number of electrodes, as Figure 29.7 shows. The array is divided into many small sections or pixels, sixteen of which are shown in the drawing. Each pixel captures a small part of a picture. Digital cameras for consumers (rather than professionals) have between one and five million pixels, depending on price. The greater the number of pixels, the better is the resolution of the photograph. The blow-up in Figure 29.7 shows a single pixel. Incident photons of visible light strike the silicon and generate electrons via the photoelectric effect. The range of energies of the visible photons is such that one electron is released when a photon interacts with a silicon atom. The electrons are trapped within a pixel because of a positive voltage applied to the electrodes beneath the insulating layer. Thus, the number of electrons that are released and trapped is proportional to the number of photons striking the pixel. In this fashion, each pixel in the CCD array accumulates an accurate representation of the light intensity at that point on the image. Color information is provided using red, green, or blue filters or a system of prisms to separate the colors. Astronomers use CCD arrays not only in the visible region of the electromagnetic spectrum but in other regions as well.

In addition to trapping the photoelectrons, the electrodes beneath the pixels are used to read out the electron representation of the picture. By changing the positive voltages applied to the electrodes, it is possible to cause all of the electrons trapped in one row of pixels to be transferred to the adjacent row. In this fashion, for instance, row 1 in Figure 29.7 is transferred into row 2, row 2 into row 3, and row 3 into the bottom row, which serves a special purpose. The bottom row functions as a horizontal shift register, from which the contents of each pixel can be shifted to the right, one at a time, and read into an analog signal processor. This processor senses the varying number of electrons in each pixel in the shift register as a kind of wave that has a fluctuating amplitude. After another shift in rows, the information in the next row is read out, and so forth. The output of the analog signal processor is sent to an analog-to-digital converter, which produces a digital representation of the image in terms of the zeros and ones that computers recognize.

Another application of the photoelectric effect depends on the fact that the moving photoelectrons in Figure 29.4 constitute a current—a current that changes as the intensity of the light changes. All automatic garage door openers have a safety feature that prevents the door from closing when it encounters an obstruction (person, vehicle, etc.). As Figure 29.8 illustrates, a sending unit transmits an invisible (infrared) beam across the opening of the door. The beam is detected by a receiving unit that contains a photodiode. A photodiode is a type of *p-n* junction diode (see Section 23.5). When infrared photons strike the photodiode, electrons bound to the atoms absorb the photons and become liberated. These liberated, mobile electrons cause the current in the photodiode to increase. When a person walks through the beam, the light is momentarily blocked from reaching the receiving unit, and the current in the photodiode decreases. The change in current is sensed by electronic circuitry that immediately stops the downward motion of the door and then causes it to rise up.

Look back at the spectacular photograph at the beginning of this chapter. It shows the central portion of the Eagle Nebula, a giant star-forming region some 7000 light-years from earth. The photo was taken by the Hubble Space Telescope and reveals towering clouds of molecular gas and dust, in which there is dramatic evidence of the energy carried by photons. These clouds extend more than a light-year from base to tip and are the birthplace of stars. A star begins to form within a cloud when the gravitational force pulls together sufficient gas to create a high-density "ball." When the gaseous ball becomes sufficiently dense, thermonuclear fusion (see Section 32.5) occurs at its core, and the star begins to shine. The newly born stars are buried within the cloud and cannot be seen from earth. However, the process of photoevaporation allows astronomers to see many of the high-density regions where stars are being formed. Photoevaporation is the process in which high-energy, ultraviolet (UV) photons from hot stars outside the cloud heat it up, much like microwave photons heat food in a microwave oven. Figure 29.9 includes a reproduction of the upper left portion of the chapter-opening photo and shows streamers of gas photoevaporating from the cloud as it is illuminated by stars located beyond the photograph's upper edge. As photoevaporation proceeds, globules of gas that are denser than their surroundings are exposed. The globules are known as *evaporating gaseous globules* (EGGs), and they are slightly larger than our solar system. The drawing in part *a* of Figure 29.9 shows that the EGGs shade the gas and dust behind them from the UV photons, creating the many finger-like protrusions seen on the surface of the cloud. Astronomers believe that some of these EGGs contain young stars within them. In some cases, so much gas has boiled off that a newborn star can be seen on the surface of an EGG (see the circled feature in the chapter-opening photograph).

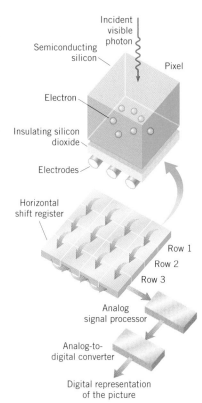

Figure 29.7 A CCD array can be used to capture photographic images using the photoelectric effect.

The physics of a safety feature of garage door openers. (See discussion above.)

The physics of photoevaporation and star formation.

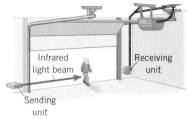

Figure 29.8 When an obstruction prevents the infrared light beam from reaching the receiving unit, the current in the receiving unit drops. This drop in current is detected by an electronic circuit that stops the downward movement of the door and then causes it to rise.

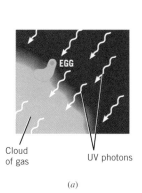

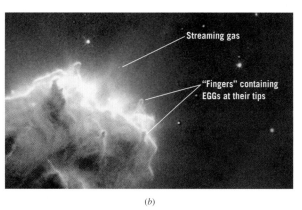

(a) (b)

Figure 29.9 (a) This drawing illustrates the photoevaporation that is occurring in the photograph in part (b). (b) This is the upper left portion of the chapter-opening photograph. Photoevaporation produces finger-like projections on the surface of the gas clouds in the Eagle Nebula. At the fingertips are high-density evaporating gaseous globules (EGGs). (Courtesy NASA)

🖱️ Self-Assessment Test 29.1

Test your understanding of the material in Sections 29.1–29.3:

• The Wave–Particle Duality   • Blackbody Radiation and Planck's Constant
• Photons   • The Photoelectric Effect

Go to **www.wiley.com/college/cutnell**

29.4 *The Momentum of a Photon and the Compton Effect*

Although Einstein presented his photon model for the photoelectric effect in 1905, it was not until 1923 that the photon concept began to achieve widespread acceptance. It was then that the American physicist Arthur H. Compton (1892–1962) used the photon model to explain his research on the scattering of X-rays by the electrons in graphite. X-rays are high-frequency electromagnetic waves and, like light, they are composed of photons.

Figure 29.10 illustrates what happens when an X-ray photon strikes an electron in a piece of graphite. Like two billiard balls colliding on a pool table, the X-ray photon scatters in one direction after the collision, and the electron recoils in another direction. Compton observed that the scattered photon has a frequency f' that is smaller than the frequency f of the incident photon, indicating that the photon loses energy during the collision. In addition, he found that the difference between the two frequencies depends on the angle θ at which the scattered photon leaves the collision. The phenomenon in which an X-ray photon is scattered from an electron, with the scattered photon having a smaller frequency than the incident photon, is called the ***Compton effect.***

In Section 7.3 the collision between two objects is analyzed using the fact that the total kinetic energy and the total linear momentum of the objects are the same before and after the collision. Similar analysis can be applied to the collision between a photon and an electron. The electron is assumed to be initially at rest and essentially free — that is, not bound to the atoms of the material. According to the principle of conservation of energy,

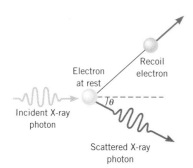

Figure 29.10 In an experiment performed by Arthur H. Compton, an X-ray photon collides with a stationary electron. The scattered photon and the recoil electron depart the collision in different directions.

$$\underbrace{hf}_{\substack{\text{Energy of} \\ \text{incident} \\ \text{photon}}} = \underbrace{hf'}_{\substack{\text{Energy of} \\ \text{scattered} \\ \text{photon}}} + \underbrace{\text{KE}}_{\substack{\text{Kinetic energy} \\ \text{of recoil} \\ \text{electron}}} \tag{29.4}$$

where the relation $E = hf$ has been used for the photon energies. It follows, then, that $hf' = hf - \text{KE}$, which shows that the energy and corresponding frequency f' of the scattered photon are less than the energy and frequency of the incident photon, just as Compton observed. Since $\lambda' = c/f'$ (Equation 16.1), the wavelength of the scattered X-rays is larger than that of the incident X-rays.

For an initially stationary electron, conservation of total linear momentum requires that

$$\begin{array}{c} \text{Momentum of} \\ \text{incident photon} \end{array} = \begin{array}{c} \text{Momentum of} \\ \text{scattered photon} \end{array} + \begin{array}{c} \text{Momentum of} \\ \text{recoil electron} \end{array} \tag{29.5}$$

To find an expression for the magnitude p of the photon's momentum, we use Equations 28.3 and 28.4. According to these equations, the momentum of any particle is $p = mv/\sqrt{1 - (v^2/c^2)}$ and its total energy is $E = mc^2/\sqrt{1 - (v^2/c^2)}$. Dividing these two equations, we find that $p/E = v/c^2$. Since a photon travels at the speed of light, $v = c$ and $p/E = 1/c$. Therefore, the momentum of a photon is $p = E/c$. But the energy of a photon is $E = hf$, while the wavelength is $\lambda = c/f$. Therefore, the magnitude of the momentum is

$$p = \frac{hf}{c} = \frac{h}{\lambda} \tag{29.6}$$

Using Equations 29.4, 29.5, and 29.6, Compton showed that the difference between the wavelength λ' of the scattered photon and the wavelength λ of the incident photon is related to the scattering angle θ by

$$\lambda' - \lambda = \frac{h}{mc}(1 - \cos\theta) \qquad (29.7)$$

In this equation m is the mass of the electron. The quantity $h/(mc)$ is called the **Compton wavelength of the electron** and has the value $h/(mc) = 2.43 \times 10^{-12}$ m. Since $\cos\theta$ varies between $+1$ and -1, the shift $\lambda' - \lambda$ in the wavelength can vary between zero and $2h/(mc)$, depending on the value of θ, a fact observed by Compton.

The photoelectric effect and the Compton effect provide compelling evidence that light can exhibit particle-like characteristics attributable to energy packets called photons. But what about the interference phenomena discussed in Chapter 27, such as Young's double-slit experiment and single-slit diffraction, which demonstrate that light behaves as a wave? Does light have two distinct personalities, in which it behaves like a stream of particles in some experiments and like a wave in others? The answer is yes, for physicists now believe that this wave–particle duality is an inherent property of light. Light is a far more interesting (and complex) phenomenon than just a stream of particles or an electromagnetic wave.

In the Compton effect the electron recoils because it gains some of the photon's momentum. In principle, then, the momentum that photons have can be used to make other objects move. Conceptual Example 3 considers a propulsion system for an interstellar spaceship that is based on the momentum of a photon.

Figure 29.11 A solar sail provides the propulsion for this interstellar spaceship.

Conceptual Example 3
Solar Sails and Interstellar Spaceships

One propulsion method for interstellar spaceships found in science fiction novels uses a large sail. The intent is that sunlight striking the sail creates a force that pushes the ship away from the sun (Figure 29.11), much as the wind propels a sailboat. Does such a design have any hope of working and, if so, should the surface of the sail facing the sun be shiny like a mirror or black, in order to produce the greatest possible force?

Reasoning and Solution There is certainly reason to believe that the design might work. In Conceptual Example 3 in Chapter 7, we found that hailstones striking the roof of a car exert a force on it because they have momentum. Photons also have momentum, so, like the hailstones, they can apply a force to the sail.

As in Chapter 7, we will be guided by the impulse–momentum theorem in assessing the force. This theorem (Equation 7.4) states that when a net force acts on an object, the impulse of the net force is equal to the change in momentum of the object. Greater impulses lead to greater forces for a given time interval. Thus, when a photon collides with the sail, the photon's momentum changes because of the force applied by the sail to the photon. Newton's action–reaction law indicates that the photon simultaneously applies a force of equal magnitude, but opposite direction, to the sail. It is this reaction force that propels the spaceship, and it will be greater when the momentum change experienced by the photon is greater. The surface of the sail facing the sun, then, should be such that it causes the largest possible momentum change for the impinging photons.

As we found in Section 13.3, radiation reflects from a shiny mirror-like surface and is absorbed by a black surface. Now, consider a photon that strikes the sail perpendicularly. In reflecting from a mirror-like surface, its momentum changes from its value in the forward direction to a value of the same magnitude in the reverse direction. This is a greater change than what occurs when the photon is absorbed by a black surface. Then, the momentum changes only from its value in the forward direction to a value of zero. Consequently, the surface of the sail facing the sun should be shiny in order to produce the greatest possible propulsion force. A shiny surface causes the photons to bounce like the hailstones on the roof of the car and, in so doing, apply a greater force to the sail.

Related Homework: *Conceptual Question 10, Problem 20*

✓ **Check Your Understanding 2**

In the Compton effect, an incident X-ray photon of wavelength λ is scattered by an electron, the scattered photon having a wavelength of λ'. Suppose that the incident photon is scattered by a proton instead of an electron. For a given scattering angle θ, is the change $\lambda' - \lambda$ in the wavelength of the photon scattered by the proton greater than, less than, or the same as that scattered by the electron? *(The answer is given at the end of the book.)*

Background: The change in the photon's wavelength is specified by Equation 29.7. How does this change depend on the mass of the particle?

For similar questions (including calculational counterparts), consult Self-Assessment Test 29.2, which is described at the end of Section 29.6.

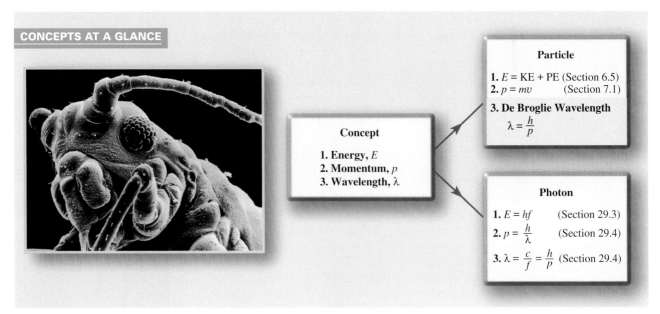

CONCEPTS AT A GLANCE

Concept

1. Energy, E
2. Momentum, p
3. Wavelength, λ

Particle

1. $E = KE + PE$ (Section 6.5)
2. $p = mv$ (Section 7.1)
3. De Broglie Wavelength
$\lambda = \dfrac{h}{p}$

Photon

1. $E = hf$ (Section 29.3)
2. $p = \dfrac{h}{\lambda}$ (Section 29.4)
3. $\lambda = \dfrac{c}{f} = \dfrac{h}{p}$ (Section 29.4)

Figure 29.12 CONCEPTS AT A GLANCE Both a moving particle and a wave possess energy, momentum, and a wavelength. The wavelength of a particle is known as its de Broglie wavelength. This photograph shows a highly magnified view of a spruce aphid, made with a scanning electron microscope. This microscope uses electrons instead of light. The resolution of the fine detail is exceptional because the wavelength of an electron can be made much smaller than that of light. (© S. Lowry/Univ. Ulster/Stone/Getty Images)

29.5 The de Broglie Wavelength and the Wave Nature of Matter

▶ CONCEPTS AT A GLANCE As a graduate student in 1923, Louis de Broglie (1892–1987) made the astounding suggestion that since light waves could exhibit particle-like behavior, particles of matter should exhibit wave-like behavior. De Broglie proposed that all moving matter has a wavelength associated with it, just as a wave does. The Concept-at-a-Glance chart in Figure 29.12, which is a continuation of the chart in Figure 29.3, shows that the notions of energy, momentum, and wavelength are applicable to particles as well as to waves. ◀

De Broglie made the explicit proposal that the wavelength λ of a particle is given by the same relation (Equation 29.6) that applies to a photon:

De Broglie wavelength
$$\lambda = \frac{h}{p} \tag{29.8}$$

where h is Planck's constant and p is the magnitude of the relativistic momentum of the particle. Today, λ is known as the *de Broglie wavelength* of the particle.

Confirmation of de Broglie's suggestion came in 1927 from the experiments of the American physicists Clinton J. Davisson (1881–1958) and Lester H. Germer (1896–1971) and, independently, those of the English physicist George P. Thomson (1892–1975). Davisson and Germer directed a beam of electrons onto a crystal of nickel and observed that the electrons exhibited a diffraction behavior, analogous to that seen when X-rays are diffracted by a crystal (see Section 27.9 for a discussion of X-ray diffraction). The wavelength of the electrons revealed by the diffraction pattern matched that predicted by de Broglie's hypothesis, $\lambda = h/p$. More recently, Young's double-slit experiment has been performed with electrons and reveals the effects of wave interference illustrated in Figure 29.1.

Particles other than electrons can also exhibit wave-like properties. For instance, neutrons are sometimes used in diffraction studies of crystal structure. Figure 29.13 compares the neutron diffraction pattern and the X-ray diffraction pattern caused by a crystal of rock salt (NaCl).

Although all moving particles have a de Broglie wavelength, the effects of this wavelength are observable only for particles whose masses are very small, on the order of the mass of an electron or a neutron, for instance. Example 4 illustrates why.

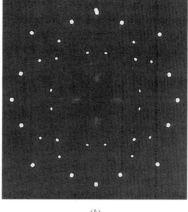

(a)   *(b)*

Figure 29.13 (*a*) The neutron diffraction pattern (Wollan, Shull and Marney, *Phys. Rev.* 73:527, 1948) and (*b*) the X-ray diffraction pattern for a crystal of sodium chloride (NaCl). (Courtesy Edwin Jones, University of South Carolina)

Example 4 The de Broglie Wavelength of an Electron and of a Baseball

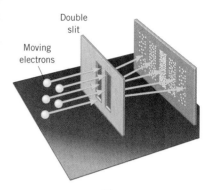

(a)

Determine the de Broglie wavelength for (a) an electron (mass = 9.1×10^{-31} kg) moving at a speed of 6.0×10^{6} m/s and (b) a baseball (mass = 0.15 kg) moving at a speed of 13 m/s.

Reasoning In each case, the de Broglie wavelength is given by Equation 29.8 as Planck's constant divided by the magnitude of the momentum. Since the speeds are small compared to the speed of light, we can ignore relativistic effects and express the magnitude of the momentum as the product of the mass and the speed.

Solution

(a) Since the magnitude p of the momentum is the product of the mass m of the particle and its speed v, we have $p = mv$. Using this expression in Equation 29.8 for the de Broglie wavelength, we obtain

$$\lambda = \frac{h}{p} = \frac{h}{mv} = \frac{6.63 \times 10^{-34} \text{ J} \cdot \text{s}}{(9.1 \times 10^{-31} \text{ kg})(6.0 \times 10^{6} \text{ m/s})} = \boxed{1.2 \times 10^{-10} \text{ m}}$$

A de Broglie wavelength of 1.2×10^{-10} m is about the size of the interatomic spacing in a solid, such as the nickel crystal used by Davisson and Germer, and, therefore, leads to the observed diffraction effects.

(b) A calculation similar to that in part (a) shows that the de Broglie wavelength of the baseball is $\boxed{\lambda = 3.3 \times 10^{-34} \text{ m}}$. This wavelength is incredibly small, even by comparison with the size of an atom (10^{-10} m) or a nucleus (10^{-14} m). Thus, the ratio λ/W of this wavelength to the width W of an ordinary opening, such as a window, is so small that the diffraction of a baseball passing through the window cannot be observed.

(b) After 100 electrons

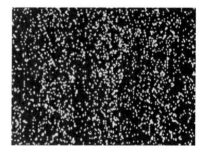

(c) After 3000 electrons

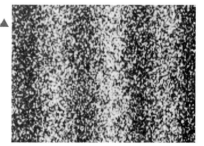

(d) After 70 000 electrons

The de Broglie equation for particle wavelength provides no hint as to what kind of wave is associated with a particle of matter. To gain some insight into the nature of this wave, we turn our attention to Figure 29.14. Part *a* shows the fringe pattern on the screen when electrons are used in a version of Young's double-slit experiment. The bright fringes occur in places where particle waves coming from each slit interfere constructively, while the dark fringes occur in places where the waves interfere destructively.

When an electron passes through the double-slit arrangement and strikes a spot on the screen, the screen glows at that spot, and parts *b*, *c*, and *d* of Figure 29.14 illustrate how the spots accumulate in time. As more and more electrons strike the screen, the spots eventually form the fringe pattern that is evident in part *d*. Bright fringes occur where there is a high probability of electrons striking the screen, and dark fringes occur where there is a low probability. Here lies the key to understanding particle waves. *Particle waves are waves of probability,* waves whose magnitude at a point in space gives an indication of the probability that the particle will be found at that point. At the place where the screen is located, the pattern of probabilities conveyed by the particle waves causes

Figure 29.14 In this electron version of Young's double-slit experiment, the characteristic fringe pattern becomes recognizable only after a sufficient number of electrons have struck the screen. (A. Tonomura, J. Endo, T. Matsuda, and T. Kawasaki, *Am. J. Phys.* 57(2): 117, Feb. 1989)

the fringe pattern to emerge. The fact that no fringe pattern is apparent in part *b* of the figure does not mean that there are no probability waves present; it just means that too few electrons have struck the screen for the pattern to be recognizable.

The pattern of probabilities that leads to the fringes in Figure 29.14 is analogous to the pattern of light intensities that is responsible for the fringes in Young's original experiment with light waves (see Figure 27.4). Section 24.4 discusses the fact that the intensity of the light is proportional to either the square of the electric field strength or the square of the magnetic field strength of the wave. In an analogous fashion in the case of particle waves, the probability is proportional to the square of the magnitude Ψ (Greek letter Psi) of the wave. Ψ is referred to as the *wave function* of the particle.

In 1925 the Austrian physicist Erwin Schrödinger (1887–1961) and the German physicist Werner Heisenberg (1901–1976) independently developed theoretical frameworks for determining the wave function. In so doing, they established a new branch of physics called *quantum mechanics.* The word "quantum" refers to the fact that in the world of the atom, where particle waves must be considered, the particle energy is quantized, so only certain energies are allowed. To understand the structure of the atom and the phenomena related to it, quantum mechanics is essential, and the Schrödinger equation for calculating the wave function is now widely used. In the next chapter, we will explore the structure of the atom based on the ideas of quantum mechanics.

✔ **Check Your Understanding 3**

A beam of electrons passes through a single slit, and a beam of protons passes through a second, but identical, slit. The electrons and the protons have the same speed. Which one of the following correctly describes the beam that experiences the greatest amount of diffraction? (a) The electrons, because they have the smaller momentum and, hence, the smaller de Broglie wavelength. (b) The electrons, because they have the smaller momentum and, hence, the larger de Broglie wavelength. (c) The protons, because they have the smaller momentum and, hence, the smaller de Broglie wavelength. (d) The protons, because they have the larger momentum and, hence, the smaller de Broglie wavelength. (e) Both beams experience the same amount of diffraction, because the electrons and protons have the same de Broglie wavelength. *(The answer is given at the end of the book.)*

Background: The amount of diffraction depends on the ratio of the de Broglie wavelength to the width of the slit (see Section 27.5), and the de Broglie wavelength of a particle depends on its momentum.

For similar questions (including calculational counterparts), consult Self-Assessment Test 29.2, which is described at the end of Section 29.6.

29.6 *The Heisenberg Uncertainty Principle*

As the previous section discusses, the bright fringes in Figure 29.14 indicate the places where there is a high probability of an electron striking the screen. And since there are a number of bright fringes, there is more than one place where each electron has some probability of hitting. As a result, it is not possible to specify in advance exactly where on the screen an individual electron will hit. All we can do is speak of the probability that the electron may end up in a number of different places. No longer is it possible to say, as Newton's laws would suggest, that a single electron, fired through the double slit, will travel directly forward in a straight line and strike the screen. This simple model just does not apply when a particle as small as an electron passes through a pair of closely spaced narrow slits. Because the wave nature of particles is important in such circumstances, we lose the ability to predict with 100% certainty the path that a single particle will follow. Instead, only the average behavior of large numbers of particles is predictable, and the behavior of any individual particle is uncertain.

To see more clearly into the nature of the uncertainty, consider electrons passing through a single slit, as in Figure 29.15. After a sufficient number of electrons strike the screen, a diffraction pattern emerges. The electron diffraction pattern consists of alternating bright and dark fringes and is analogous to that for light waves shown in Figure 27.23. Figure 29.15 shows the slit and locates the first dark fringe on either side of the central bright fringe. The central fringe is bright because electrons strike the screen over

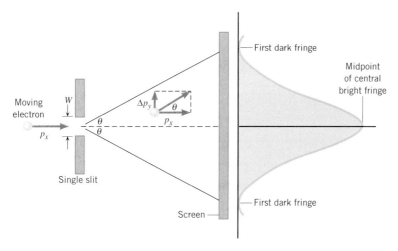

Figure 29.15 When a sufficient number of electrons pass through a single slit and strike the screen, a diffraction pattern of bright and dark fringes emerges. (Only the central bright fringe is shown.) This pattern is due to the wave nature of the electrons and is analogous to that produced by light waves.

the entire region between the dark fringes. If the electrons striking the screen outside the central bright fringe are neglected, the extent to which the electrons are diffracted is given by the angle θ in the drawing. To reach locations within the central bright fringe, some electrons must have acquired momentum in the y direction, despite the fact that they enter the slit traveling along the x direction and have no momentum in the y direction to start with. The figure illustrates that the y component of the momentum may be as large as Δp_y. The notation Δp_y indicates the difference between the maximum value of the y component of the momentum after the electron passes through the slit and its value of zero before the electron passes through the slit. Δp_y represents the *uncertainty* in the y component of the momentum in that the y component may have any value from zero to Δp_y.

It is possible to relate Δp_y to the width W of the slit. To do this, we assume that Equation 27.4, which applies to light waves, also applies to particle waves whose de Broglie wavelength is λ. This equation, $\sin \theta = \lambda/W$, specifies the angle θ that locates the first dark fringe. If θ is small, $\sin \theta \approx \tan \theta$. Moreover, Figure 29.15 indicates that $\tan \theta = \Delta p_y/p_x$, where p_x is the x component of the momentum of the electron. Therefore, $\Delta p_y/p_x \approx \lambda/W$. But $p_x = h/\lambda$ according to de Broglie's equation, so that

$$\frac{\Delta p_y}{p_x} = \frac{\Delta p_y}{h/\lambda} \approx \frac{\lambda}{W}$$

As a result,

$$\Delta p_y \approx \frac{h}{W} \qquad (29.9)$$

which indicates that a smaller slit width leads to a larger uncertainty in the y component of the electron's momentum.

It was Heisenberg who first suggested that the uncertainty Δp_y in the y component of the momentum is related to the uncertainty in the y position of the electron as the electron passes through the slit. Since the electron can pass through anywhere within the width W, the uncertainty in the y position of the electron is $\Delta y = W$. Substituting Δy for W in Equation 29.9 shows that $\Delta p_y \approx h/\Delta y$ or $(\Delta p_y)(\Delta y) \approx h$. The result of Heisenberg's more complete analysis is given below in Equation 29.10 and is known as the ***Heisenberg uncertainty principle***. Note that the Heisenberg principle is a general principle with wide applicability. It does not just apply to the case of single slit diffraction, which we have used here for the sake of convenience.

■ **THE HEISENBERG UNCERTAINTY PRINCIPLE**

Momentum and position

$$(\Delta p_y)(\Delta y) \geq \frac{h}{4\pi} \qquad (29.10)$$

Δy = uncertainty in a particle's position along the y direction
Δp_y = uncertainty in the y component of the linear momentum of the particle

(continues)

Energy and time

$$(\Delta E)(\Delta t) \geq \frac{h}{4\pi} \qquad (29.11)$$

ΔE = uncertainty in the energy of a particle when the particle is in a certain state
Δt = time interval during which the particle is in the state

The Heisenberg uncertainty principle places limits on the accuracy with which the momentum and position of a particle can be specified simultaneously. These limits are not just limits due to faulty measuring techniques. They are fundamental limits imposed by nature, and there is no way to circumvent them. Equation 29.10 indicates that Δp_y and Δy cannot both be arbitrarily small at the same time. If one is small, then the other must be large, so that their product equals or exceeds ($\geq$) Planck's constant divided by 4π. For example, if the position of a particle is known exactly, so that Δy is zero, then Δp_y is an infinitely large number, and the momentum of the particle is completely uncertain. Conversely, if we assume that Δp_y is zero, then Δy is an infinitely large number, and the position of the particle is completely uncertain. In other words, the Heisenberg uncertainty principle states that it is impossible to specify precisely both the momentum and position of a particle at the same time.

There is also an uncertainty principle that deals with energy and time, as expressed by Equation 29.11. The product of the uncertainty ΔE in the energy of a particle and the time interval Δt during which the particle remains in a given energy state is greater than or equal to Planck's constant divided by 4π. Therefore, the shorter the lifetime of a particle in a given state, the greater is the uncertainty in the energy of that state.

Example 5 shows that the uncertainty principle has significant consequences for the motion of tiny particles such as electrons but has little effect on the motion of macroscopic objects, even those with as little mass as a Ping-Pong ball.

Example 5 The Heisenberg Uncertainty Principle

Assume that the position of an object is known so precisely that the uncertainty in the position is only $\Delta y = 1.5 \times 10^{-11}$ m. (a) Determine the minimum uncertainty in the momentum of the object. Find the corresponding minimum uncertainty in the speed of the object, if the object is (b) an electron (mass = 9.1×10^{-31} kg) and (c) a Ping-Pong ball (mass = 2.2×10^{-3} kg).

Reasoning The minimum uncertainty Δp_y in the y component of the momentum is given by the Heisenberg uncertainty principle as $\Delta p_y = h/(4\pi\Delta y)$, where Δy is the uncertainty in the position of the object along the y direction. Both the electron and the Ping-Pong ball have the same uncertainty in their momenta because they have the same uncertainty in their positions. However, these objects have very different masses. As a result, we will find that the uncertainty in the speeds of these objects is very different.

Problem solving insight
The Heisenberg uncertainty principle states that the product of Δp_y and Δy is greater than or equal to $h/4\pi$. The minimum uncertainty occurs when the product is equal to $h/4\pi$.

Solution

(a) The minimum uncertainty in the y component of the momentum is

$$\Delta p_y = \frac{h}{4\pi\Delta y} = \frac{6.63 \times 10^{-34} \text{ J} \cdot \text{s}}{4\pi(1.5 \times 10^{-11} \text{ m})} = \boxed{3.5 \times 10^{-24} \text{ kg} \cdot \text{m/s}} \qquad (29.10)$$

(b) Since $\Delta p_y = m\Delta v_y$, the minimum uncertainty in the speed of the electron is

$$\Delta v_y = \frac{\Delta p_y}{m} = \frac{3.5 \times 10^{-24} \text{ kg} \cdot \text{m/s}}{9.1 \times 10^{-31} \text{ kg}} = \boxed{3.8 \times 10^{6} \text{ m/s}}$$

Thus, the small uncertainty in the y position of the electron gives rise to a large uncertainty in the speed of the electron.

(c) The uncertainty in the speed of the Ping-Pong ball is

$$\Delta v_y = \frac{\Delta p_y}{m} = \frac{3.5 \times 10^{-24} \text{ kg} \cdot \text{m/s}}{2.2 \times 10^{-3} \text{ kg}} = \boxed{1.6 \times 10^{-21} \text{ m/s}}$$

Because the mass of the Ping-Pong ball is relatively large, the small uncertainty in its y position gives rise to an uncertainty in its speed that is much smaller than that for the electron. Thus, in contrast to the electron, we can know simultaneously where the ball is and how fast it is moving, to a very high degree of certainty.

Example 5 emphasizes how the uncertainty principle imposes different uncertainties on the speeds of an electron (small mass) and a Ping-Pong ball (large mass). For objects like the ball, which have relatively large masses, the uncertainties in position and speed are so small that they have no effect on our ability to determine simultaneously where such objects are and how fast they are moving. The uncertainties calculated in Example 5 depend on more than just the mass, however. They also depend on Planck's constant, which is a very small number. It is interesting to speculate about what life would be like if Planck's constant were much larger than 6.63×10^{-34} J·s. Conceptual Example 6 deals with just such speculation.

Conceptual Example 6 What If Planck's Constant Were Large?

A bullet leaving the barrel of a gun is analogous to an electron passing through the single slit in Figure 29.15. With this analogy in mind, what would hunting be like if Planck's constant had a relatively large value instead of its small value of 6.63×10^{-34} J·s?

Reasoning and Solution It takes two to hunt, the hunter and the hunted. Consider the hunter first and what it would mean to take aim if Planck's constant were large. The bullet moving down the barrel of the gun has a de Broglie wavelength given by Equation 29.8 as $\lambda = h/p$. This wavelength is large, since Planck's constant h is now large. (We are assuming that the magnitude p of the bullet's momentum is not abnormally large.) Remember from Section 27.5 that the diffraction of light waves through an opening is greater when the wavelength is greater, other things being equal. We expect, therefore, that with λ being comparable to the opening in the barrel, the hunter will have to contend with appreciable diffraction when the bullet leaves the barrel. This means that, in spite of being aimed directly at the target, the bullet may come nowhere near it. In Figure 29.15, for comparison, diffraction allows the electron to strike the screen on either side of the maximum intensity point of the central bright fringe. Under such circumstances, the hunter might as well not bother aiming. He'll have just as much success by scattering his shots around randomly, hoping to hit something.

Now consider the target and what it means that the hunter "aimed directly at the target." This must mean that the target's location is known, at least approximately. In other words, the uncertainty Δy in the target's position is reasonably small. But the uncertainty principle indicates that the minimum uncertainty in the target's momentum is $\Delta p = h/(4\pi\Delta y)$. Since Planck's constant is now large, Δp is also large. The target, therefore, may have a large momentum and be moving rapidly. Even if the bullet happens to reach the place where the hunter aimed, the target may no longer be there. Hunting would be vastly different indeed, if Planck's constant had a very large value.

Related Homework: *Problem 34*

Self-Assessment Test 29.2

Test your understanding of the material in Sections 29.4–29.6:

- The Momentum of a Photon   • The Compton Effect
- The de Broglie Wavelength and the Wave Nature of Matter
- The Heisenberg Uncertainty Principle

Go to **www.wiley.com/college/cutnell**

29.7 *Concepts & Calculations*

In the photoelectric effect, electrons can be emitted from a metal surface when light shines on it, as we have seen. Einstein explained this phenomenon in terms of photons and the conservation of energy. The same basic explanation applies when the light has a single wavelength or when it consists of a mixture of wavelengths, as sunlight does. Our earlier discussion in Section 29.3 deals with a single wavelength. Example 7, in contrast, considers a mixture of wavelengths and reviews the basic concepts that come into play in this important phenomenon.

Concepts & Calculations Example 7
Sunlight and the Photoelectric Effect

Sunlight, whose visible wavelengths range from 380 to 750 nm, is incident on a sodium surface. The work function is $W_0 = 2.28$ eV. Find the maximum kinetic energy KE_{max} (in joules) of the photoelectrons emitted from the surface and the range of wavelengths that will cause photoelectrons to be emitted.

Concept Questions and Answers Will electrons with the greatest value of KE_{max} be emitted when the incident photons have a relatively greater or a relatively smaller amount of energy?

Answer When a photon ejects an electron from the sodium, the photon energy is used for two purposes. Some of it is used to do the work of ejecting the electron, and what remains is carried away as kinetic energy of the emitted electron. The minimum work to eject an electron is equal to the work function W_0, and in this case the electron carries away a maximum amount KE_{max} of kinetic energy. According to the principle of conservation of energy, the energy of an incident photon is equal to the sum of W_0 and KE_{max}. Thus, for a given value of W_0, greater values of KE_{max} will occur when the incident photon has a greater energy to begin with.

In the range of visible wavelengths, which wavelength corresponds to incident photons that carry the greatest energy?

Answer According to Equation 29.2, the energy of a photon is related to the frequency f and Planck's constant h according to $E = hf$. But the frequency is given by Equation 16.1 as $f = c/\lambda$, where c is the speed of light and λ is the wavelength. Therefore, we have $E = hc/\lambda$, so the smallest value of the wavelength corresponds to the greatest energy.

What is the smallest value of KE_{max} with which an electron can be ejected from the sodium?

Answer The smallest value of KE_{max} with which an electron can be ejected from the sodium is 0 J. This occurs when an incident photon carries an energy that exactly equals the work function W_0, with no energy left over to be carried away by the electron as kinetic energy.

Solution The value given for the work function is in electron volts, so we first convert it into joules by using the fact that 1 eV = 1.60×10^{-19} J:

$$W_0 = (2.28 \text{ eV}) \frac{1.60 \times 10^{-19} \text{ J}}{1 \text{ eV}} = 3.65 \times 10^{-19} \text{ J}$$

The incident photons that have the greatest energy are those with the smallest wavelength or $\lambda = 380$ nm. Then, according to Equation 29.3 and Equation 16.1 ($f = c/\lambda$), the greatest value for KE_{max} is

$$KE_{max} = hf - W_0 = \frac{hc}{\lambda} - W_0$$

$$= \frac{(6.63 \times 10^{-34} \text{ J} \cdot \text{s})(3.00 \times 10^8 \text{ m/s})}{380 \times 10^{-9} \text{ m}} - 3.65 \times 10^{-19} \text{ J} = \boxed{1.58 \times 10^{-19} \text{ J}}$$

As the wavelength increases, the energy of the incident photons decreases until it equals the work function, at which point electrons are ejected with zero kinetic energy. Thus, we have

$$KE_{max} = 0 \text{ J} = \frac{hc}{\lambda} - W_0$$

$$\lambda = \frac{hc}{W_0} = \frac{(6.63 \times 10^{-34} \text{ J} \cdot \text{s})(3.00 \times 10^8 \text{ m/s})}{3.65 \times 10^{-19} \text{ J}} = 5.45 \times 10^{-7} \text{ m} = 545 \text{ nm}$$

The range of wavelengths over which sunlight ejects electrons from the sodium surface is, then, $\boxed{380 \text{ to } 545 \text{ nm}}$.

Kinetic energy and momentum, respectively, are discussed in Chapters 6 and 7. In the present chapter we have seen that moving particles of matter not only have kinetic energy and momentum but are also characterized by a wavelength that is known as the de Broglie wavelength. We conclude with an example that reviews kinetic energy and momentum and focuses on the relation between these fundamental physical quantities and the de Broglie wavelength.

Concepts & Calculations Example 8   The de Broglie Wavelength

▼

An electron and a proton have the same kinetic energy and are moving at speeds much less than the speed of light. Determine the ratio of the de Broglie wavelength of the electron to that of the proton.

Concept Questions and Answers How is the de Broglie wavelength λ related to the magnitude p of the momentum?

Answer   According to Equation 29.8, the de Broglie wavelength is given by $\lambda = h/p$, where h is Planck's constant. The greater the momentum, the smaller the de Broglie wavelength, and vice versa.

How is the magnitude of the momentum related to the kinetic energy of a particle of mass m that is moving at a speed that is much less than the speed of light?

Answer   According to Equation 7.2 the magnitude of the momentum is $p = mv$, where v is the speed. Equation 6.2 gives the kinetic energy as $\text{KE} = \frac{1}{2}mv^2$, which can be solved to show that $v = \sqrt{2(\text{KE})/m}$. Substituting this result into Equation 7.2 gives $p = m\sqrt{2(\text{KE})/m} = \sqrt{2m(\text{KE})}$.

Which has the greater de Broglie wavelength, the electron or the proton?

Answer   According to $\lambda = h/p$, the particle with the greater wavelength is the one with the smaller momentum. But $p = \sqrt{2m(\text{KE})}$ indicates that, for a given kinetic energy, the particle with the smaller momentum is the one with the smaller mass. The masses of the electron and proton are $m_{\text{electron}} = 9.11 \times 10^{-31}$ kg and $m_{\text{proton}} = 1.67 \times 10^{-27}$ kg. Thus, the electron, with its smaller mass, has the greater de Broglie wavelength.

Solution Using Equation 29.8 for the de Broglie wavelength and the fact that the magnitude of the momentum is related to the kinetic energy by $p = \sqrt{2m(\text{KE})}$, we have

$$\lambda = \frac{h}{p} = \frac{h}{\sqrt{2m(\text{KE})}}$$

Applying this result to the electron and the proton gives

$$\frac{\lambda_{\text{electron}}}{\lambda_{\text{proton}}} = \frac{h/\sqrt{2m_{\text{electron}}(\text{KE})}}{h/\sqrt{2m_{\text{proton}}(\text{KE})}} = \sqrt{\frac{m_{\text{proton}}}{m_{\text{electron}}}} = \sqrt{\frac{1.67 \times 10^{-27} \text{ kg}}{9.11 \times 10^{-31} \text{ kg}}} = \boxed{42.8}$$

As expected, the wavelength for the electron is greater than that for the proton.

▲

At the end of the problem set for this chapter, you will find homework problems that contain both conceptual and quantitative parts. These problems are grouped under the heading *Concepts & Calculations, Group Learning Problems*. They are designed for use by students working alone or in small learning groups. The conceptual part of each problem provides a convenient focus for group discussions.

Concept Summary

This summary presents an abridged version of the chapter, including the important equations and all available learning aids. For convenient reference, the learning aids (including the text's examples) are placed next to or immediately after the relevant equation or discussion. The following learning aids may be found on-line at **www.wiley.com/college/cutnell**:

Interactive LearningWare examples are solved according to a five-step interactive format that is designed to help you develop problem-solving skills.	**Concept Simulations** are animated versions of text figures or animations that illustrate important concepts. You can control parameters that affect the display, and we encourage you to experiment.
Interactive Solutions offer specific models for certain types of problems in the chapter homework. The calculations are carried out interactively.	**Self-Assessment Tests** include both qualitative and quantitative questions. Extensive feedback is provided for both incorrect and correct answers, to help you evaluate your understanding of the material.

Topic	*Discussion*	*Learning Aids*
	29.1 The Wave–Particle Duality	
	29.2 Blackbody Radiation and Planck's Constant	
Wave–particle duality	The wave–particle duality refers to the fact that a wave can exhibit particle-like characteristics and a particle can exhibit wave-like characteristics.	
Perfect blackbody	At a constant temperature, a perfect blackbody absorbs and reemits all the electromagnetic radiation that falls on it. Max Planck calculated the emitted radiation intensity per unit wavelength as a function of wavelength. In his theory, Planck assumed that a blackbody consists of atomic oscillators that can have only discrete, or quantized, energies. Planck's quantized energies are given by	
Energies of atomic oscillators	$$E = nhf \qquad n = 0, 1, 2, 3, \ldots \qquad (29.1)$$ where h is Planck's constant (6.63×10^{-34} J·s) and f is the vibration frequency of an oscillator.	
	29.3 Photons and the Photoelectric Effect	
	All electromagnetic radiation consists of photons, which are packets of energy. The energy of a photon is	
Energy of a photon	$$E = hf \qquad (29.2)$$ where h is Planck's constant and f is the frequency of the photon. A photon has no mass and always travels at the speed of light c in a vacuum.	**Example 1** **Interactive LearningWare 29.1**
Photoelectric effect **Work function**	The photoelectric effect is the phenomenon in which light shining on a metal surface causes electrons to be ejected from the surface. The work function W_0 of a metal is the minimum work that must be done to eject an electron from the metal. In accordance with the conservation of energy, the electrons ejected from a metal have a maximum kinetic energy KE_{max} that is related to the energy hf of the incident photon and the work function of the metal by	
Conservation of energy and the photoelectric effect	$$hf = \mathrm{KE}_{max} + W_0 \qquad (29.3)$$	**Examples 2, 7** **Concept Simulation 29.1**

 Use *Self-Assessment Test 29.1* to evaluate your understanding of Sections 29.1–29.3.

	29.4 The Momentum of a Photon and the Compton Effect	
	The magnitude p of a photon's momentum is	
Magnitude of a photon's momentum	$$p = \frac{h}{\lambda} \qquad (29.6)$$ where h is Planck's constant and λ is the wavelength of the photon.	**Interactive Solution 29.17**
Compton effect	The Compton effect is the scattering of a photon by an electron in a material, the scattered photon having a smaller frequency and, hence, a smaller energy than the incident photon. Part of the photon's energy and momentum are transferred to the recoiling electron. The difference between the wavelength λ' of the scattered photon and the wavelength λ of the incident photon is related to the scattering angle θ by	
Wavelength difference in the Compton effect	$$\lambda' - \lambda = \frac{h}{mc}(1 - \cos\theta) \qquad (29.7)$$	**Example 3**

Compton wavelength of the electron

where m is the mass of the electron. The quantity $h/(mc)$ is known as the Compton wavelength of the electron.

29.5 The de Broglie Wavelength and the Wave Nature of Matter

The de Broglie wavelength λ of a particle is

De Broglie wavelength

$$\lambda = \frac{h}{p} \qquad (29.8)$$

Examples 4, 8
Interactive Solution 29.29

where h is Planck's constant and p is the magnitude of the relativistic momentum of the particle. Because of its wavelength, a particle can exhibit wave-like characteristics. The wave associated with a particle is a wave of probability.

29.6 The Heisenberg Uncertainty Principle

The Heisenberg uncertainty principle places limits on our knowledge about the behavior of a particle. The uncertainty principle indicates that

Uncertainty principle— momentum and position

$$(\Delta p_y)(\Delta y) \geq \frac{h}{4\pi} \qquad (29.10)$$

Examples 5, 6

where Δy is the uncertainty in the particle's position along the y direction, and Δp_y is the uncertainty in the y-component of the linear momentum of the particle.

The uncertainty principle also states that

Uncertainty principle— energy and time

$$(\Delta E)(\Delta t) \geq \frac{h}{4\pi} \qquad (29.11)$$

where ΔE is the uncertainty in the energy of a particle when it is in a certain state, and Δt is the time interval during which the particle is in the state.

Use Self-Assessment Test 29.2 to evaluate your understanding of Sections 29.4–29.6.

Conceptual Questions

1. The photons emitted by a source of light do *not* all have the same energy. Is the source monochromatic? Give your reasoning.

2. Which colored light bulb—red, orange, yellow, green, or blue—emits photons with (a) the least energy and (b) the greatest energy? Account for your answers.

3. Does a photon emitted by a higher-wattage red light bulb have more energy than a photon emitted by a lower-wattage red bulb? Justify your answer.

4. When a sufficient number of visible light photons strike a piece of photographic film, the film becomes exposed. An X-ray photon is more energetic than a visible light photon. Yet, most photographic films are not exposed by the X-ray machines used at airport security checkpoints. Explain what these observations imply about the number of photons emitted by the X-ray machines.

5. Radiation of a given wavelength causes electrons to be emitted from the surface of one metal but not from the surface of another metal. Explain why this could be.

6. In the photoelectric effect, suppose that the intensity of the light is increased, while the frequency is kept constant. The frequency is greater than the minimum frequency f_0. State whether each of the following will increase, decrease, or remain constant, and explain your choice: (a) the current in the phototube, (b) the number of electrons emitted per second from the metal surface, (c) the maximum kinetic energy that an electron could have, (d) the maximum momentum that an electron could have, and (e) the minimum de Broglie wavelength that an electron could have.

7. In a Compton scattering experiment, an electron is accelerated straight ahead in the same direction as that of the incident X-ray photon. Which way does the scattered photon move? Explain your reasoning, using the principle of conservation of momentum.

8. A photon can undergo Compton scattering from a molecule such as nitrogen, just as it does from an electron. However, the change in photon wavelength is much less than when the scattering is from an electron. Explain why, using Equation 29.7 for a nitrogen molecule instead of an electron.

9. When the speed of a particle doubles, its momentum doubles, and its kinetic energy becomes four times greater. When the momentum of a photon doubles, does its energy become four times greater? Provide a reason for your answer.

10. Review Conceptual Example 3 as background for this question. The photograph shows a device called a radiometer. The four rectangular panels are black on one side and shiny like a mirror on the other side. In bright light the panel arrangement spins around, in a direction from the black side of a panel toward the

(Courtesy Sargent-Welsch Scientific Co.)

shiny side. Do photon collisions with both sides of the panels (collisions such as those discussed in Conceptual Example 3) cause the observed spinning? Give your reasoning.

11. A stone is dropped from the top of a building. As the stone falls, does its de Broglie wavelength increase, decrease, or remain the same? Provide a reason for your answer.

12. An electron and a neutron have different masses. Is it possible,

according to Equation 29.8, that they can have the same de Broglie wavelength? Account for your answer.

13. In Figure 29.1 replace the electrons with protons that have the same speed. With the aid of Equation 27.1 for the bright fringes in Young's double-slit experiment and Equation 29.8, decide whether the angular separation between the fringes would increase, decrease, or remain the same when compared to that produced by the electrons.

Problems

In working these problems, ignore relativistic effects.

Section 29.3 Photons and the Photoelectric Effect

1. ssm Ultraviolet light is responsible for sun tanning. Find the wavelength (in nm) of an ultraviolet photon whose energy is 6.4×10^{-19} J.

2. The dissociation energy of a molecule is the energy required to break apart the molecule into its separate atoms. The dissociation energy for the cyanogen molecule is 1.22×10^{-18} J. Suppose that this energy is provided by a single photon. Determine the (a) wavelength and (b) frequency of the photon. (c) In what region of the electromagnetic spectrum (see Figure 24.10) does this photon lie?

3. An FM radio station broadcasts at a frequency of 98.1 MHz. The power radiated from the antenna is 5.0×10^{4} W. How many photons per second does the antenna emit?

4. The maximum wavelength that an electromagnetic wave can have and still eject electrons from a metal surface is 485 nm. What is the work function W_0 of this metal? Express your answer in electron volts.

5. ssm Ultraviolet light with a frequency of 3.00×10^{15} Hz strikes a metal surface and ejects electrons that have a maximum kinetic energy of 6.1 eV. What is the work function (in eV) of the metal?

6. Light is shining perpendicularly on the surface of the earth with an intensity of 680 W/m². Assuming all the photons in the light have the same wavelength (in vacuum) of 730 nm, determine the number of photons per second per square meter that reach the earth.

7. A magnesium surface has a work function of 3.68 eV. Electromagnetic waves with a wavelength of 215 nm strike the surface and eject electrons. Find the maximum kinetic energy of the ejected electrons. Express your answer in electron volts.

* **8.** Light is incident on the surface of metallic sodium, whose work function is 2.3 eV. The maximum speed of the photoelectrons emitted by the surface is 1.2×10^{6} m/s. What is the wavelength of the light?

* **9. ssm** Consult **Interactive LearningWare 29.1** at **www.wiley.com/college/cutnell** for background material relating to this problem. An owl has good night vision because its eyes can detect a light intensity as small as 5.0×10^{-13} W/m². What is the minimum number of photons per second that an owl eye can detect if its pupil has a diameter of 8.5 mm and the light has a wavelength of 510 nm?

* **10.** A proton is located at a distance of 0.420 m from a point charge of $+8.30$ μC. The repulsive electric force moves the proton until it is at a distance of 1.58 m from the charge. Suppose that the electric potential energy lost by the system is carried off by a photon that is emitted during the process. What is its wavelength?

** **11. ssm www** (a) How many photons (wavelength = 620 nm) must be absorbed to melt a 2.0-kg block of ice at 0 °C into water at 0 °C? (b) On the average, how many H_2O molecules does one photon convert from the ice phase to the water phase?

** **12.** A laser emits 1.30×10^{18} photons per second in a beam of light that has a diameter of 2.00 mm and a wavelength of 514.5 nm. Determine (a) the average electric field strength and (b) the average magnetic field strength for the electromagnetic wave that constitutes the beam.

Section 29.4 The Momentum of a Photon and the Compton Effect

13. The microwaves used in a microwave oven have a wavelength of about 0.13 m. What is the momentum of a microwave photon?

14. A light source emits a beam of photons, each of which has a momentum of 2.3×10^{-29} kg·m/s. (a) What is the frequency of the photons? (b) To what region of the electromagnetic spectrum do the photons belong? Consult Figure 24.10 if necessary.

15. ssm Incident X-rays have a wavelength of 0.3120 nm and are scattered by the "free" electrons in graphite. The scattering angle in Figure 29.10 is $\theta = 135.0°$. What is the magnitude of the momentum of (a) the incident photon and (b) the scattered photon? (For accuracy, use $h = 6.626 \times 10^{-34}$ J·s and $c = 2.998 \times 10^{8}$ m/s.)

16. In a Compton scattering experiment, the incident X-rays have a wavelength of 0.2685 nm, and the scattered X-rays have a wavelength of 0.2703 nm. Through what angle θ in Figure 29.10 are the X-rays scattered?

17. Refer to **Interactive Solution 29.17** at **www.wiley.com/college/cutnell** for help in solving this problem. An incident X-ray photon of wavelength 0.2750 nm is scattered from an electron that is initially at rest. The photon is scattered at an angle of $\theta = 180.0°$ in Figure 29.10 and has a wavelength of 0.2825 nm. Use the conservation of linear momentum to find the momentum gained by the electron.

* **18.** The X-rays detected at a scattering angle of $\theta = 163°$ in Figure 29.10 have a wavelength of 0.1867 nm. Find (a) the wavelength of an incident photon, (b) the energy of an incident photon, (c) the energy of a scattered photon, and (d) the kinetic energy of the recoil electron. (For accuracy, use $h = 6.626 \times 10^{-34}$ J·s and $c = 2.998 \times 10^{8}$ m/s.)

* **19. ssm www** What is the maximum amount by which the wavelength of an incident photon could change when it undergoes Compton scattering from a nitrogen molecule (N_2)?

20. Review Conceptual Example 3 before attempting this problem. A beam of visible light has a wavelength of 395 nm and shines perpendicularly on a surface. As a result, there are 3.0×10^{18} photons per second striking the surface. By using the impulse–momentum theorem (Section 7.1), obtain the average force that this beam applies to the surface when (a) the surface is a mirror, so the momentum of each photon is reversed after reflection, and (b) the surface is black, so each photon is absorbed and there are no reflected photons.

Section 29.5 The de Broglie Wavelength and the Wave Nature of Matter

21. A honeybee (mass = 1.3×10^{-4} kg) is crawling at a speed of 0.020 m/s. What is the de Broglie wavelength of the bee?

22. The interatomic spacing in a crystal of table salt is 0.282 nm. This crystal is being studied in a neutron diffraction experiment, similar to the one that produced the photograph in Figure 29.13a. How fast must a neutron (mass = 1.67×10^{-27} kg) be moving to have a de Broglie wavelength of 0.282 nm?

23. ssm The de Broglie wavelength of a proton in a particle accelerator is 1.30×10^{-14} m. Determine the kinetic energy (in joules) of the proton.

24. How fast does a proton have to be moving in order to have the same de Broglie wavelength as an electron that is moving at 4.50×10^6 m/s?

25. Recall from Section 14.3 that the average kinetic energy of an atom in a monatomic ideal gas is given by $\overline{KE} = \frac{3}{2}kT$, where $k = 1.38 \times 10^{-23}$ J/K and T is the Kelvin temperature of the gas. Determine the de Broglie wavelength of a helium atom (mass = 6.65×10^{-27} kg) that has the average kinetic energy at room temperature (293 K).

26. Neutrons ($m = 1.67 \times 10^{-27}$ kg) are being emitted by two objects. The neutrons are moving with a speed of 2.80×10^3 m/s and pass through a circular aperture whose diameter is 0.100 mm. This situation is analogous to that shown in Figure 27.29a for light. According to the Rayleigh criterion, what is the minimum angle θ_{min} (in radians) between the two objects, such that they are just resolved using neutrons?

27. ssm The width of the central bright fringe in a diffraction pattern on a screen is identical when either electrons or red light (vacuum wavelength = 661 nm) pass through a single slit. The distance between the screen and the slit is the same in each case and is large compared to the slit width. How fast are the electrons moving?

28. An electron, starting from rest, accelerates through a potential difference of 418 V. What is the final de Broglie wavelength of the electron, assuming that its final speed is much less than the speed of light?

29. Consult **Interactive Solution 29.29** at **www.wiley.com/college/cutnell** to explore a model for solving this problem. In a television picture tube the electrons are accelerated from rest through a potential difference V. Just before an electron strikes the screen, its de Broglie wavelength is 0.900×10^{-11} m. What is the potential difference?

30. The kinetic energy of a particle is equal to the energy of a photon. The particle moves at 5.0% of the speed of light. Find the ratio of the photon wavelength to the de Broglie wavelength of the particle.

Section 29.6 The Heisenberg Uncertainty Principle

31. ssm www In the lungs there are tiny sacs of air, which are called alveoli. The average diameter of one of these sacs is 0.25 mm. Consider an oxygen molecule (mass = 5.3×10^{-26} kg) trapped within a sac. What is the minimum uncertainty in the velocity of this oxygen molecule?

32. An electron is trapped within a sphere whose diameter is 6.0×10^{-15} m (about the size of the nucleus of an oxygen atom). What is the minimum uncertainty in the electron's momentum?

33. Consider a line that is 2.5 m long. A moving object is somewhere along this line, but its position is not known. (a) Find the minimum uncertainty in the momentum of the object. Find the minimum uncertainty in the object's velocity, assuming that the object is (b) a golf ball (mass = 0.045 kg) and (c) an electron.

34. Review Conceptual Example 6 as background for this problem. When electrons pass through a single slit, as in Figure 29.15, they form a diffraction pattern. As Section 29.6 discusses, the central bright fringe extends to either side of the midpoint, according to an angle θ given by $\sin \theta = \lambda/W$, where λ is the de Broglie wavelength of the electron and W is the width of the slit. When λ is the same size as W, $\lambda = 90°$, and the central fringe fills the entire observation screen. In this case, an electron passing through the slit has roughly the same probability of hitting the screen either straight ahead or anywhere off to one side or the other. Now, imagine yourself in a world where Planck's constant is large enough so you exhibit similar effects when you walk through a 0.90-m-wide doorway. Your mass is 82 kg and you walk at a speed of 0.50 m/s. How large would Planck's constant have to be in this hypothetical world?

35. ssm www Particles pass through a single slit of width 0.200 mm (see Figure 29.15). The de Broglie wavelength of each particle is 633 nm. After the particles pass through the slit, they spread out over a range of angles. Use the Heisenberg uncertainty principle to determine the minimum range of angles.

36. Suppose the minimum uncertainty in the position of a particle is equal to its de Broglie wavelength. If the particle has an average speed of 4.5×10^5 m/s, what is the minimum uncertainty in its speed?

Additional Problems

37. A particle has a speed of 1.2×10^6 m/s. Its de Broglie wavelength is 8.4×10^{-14} m. What is the mass of the particle?

38. Two photons have energies of 3.3×10^{-16} J and 1.3×10^{-20} J. Using Figure 24.10, identify the appropriate region in the electromagnetic spectrum for each of these photons.

39. ssm As Section 27.5 discusses, sound waves diffract or bend around the edges of a doorway. Larger wavelengths diffract more than smaller wavelengths. (a) The speed of sound is 343 m/s. With what speed would a 55.0-kg person have to move through a doorway to diffract to the same extent as a 128-Hz bass tone? (b) At the speed calculated in part (a), how long (in years) would it take the person to move a distance of one meter?

40. What is (a) the wavelength of a 5.0-eV photon and (b) the de Broglie wavelength of a 5.0-eV electron?

41. A photon has the same momentum as an electron moving with a speed of 2.0×10^5 m/s. What is the wavelength of the photon?

42. Radiation of a certain wavelength causes electrons with a maximum kinetic energy of 0.68 eV to be ejected from a metal whose work function is 2.75 eV. What will be the maximum kinetic energy (in eV) with which this same radiation ejects electrons from another metal whose work function is 2.17 eV?

* **43. ssm** From a cliff 9.5 m above a lake, a young woman (mass = 41 kg) jumps from rest, straight down into the water. At the instant she strikes the water, what is her de Broglie wavelength?

* **44.** The maximum wavelength for which an electromagnetic wave

can eject electrons from a platinum surface is 196 nm. When radiation with a wavelength of 141 nm shines on the surface, what is the maximum speed of the ejected electrons?

** **45.** An X-ray photon is scattered at an angle of $\theta = 180.0°$ from an electron that is initially at rest. After scattering, the electron has a speed of 4.67×10^6 m/s. Find the wavelength of the incident X-ray photon.

Concepts & Calculations   Group Learning Problems

Note: Each of these problems consists of Concept Questions followed by a related quantitative Problem. They are designed for use by students working alone or in small learning groups. The Concept Questions involve little or no mathematics and are intended to stimulate group discussions. They focus on the concepts with which the problems deal. Recognizing the concepts is the essential initial step in any problem-solving technique.

46. Concept Questions The maximum speed of the electrons emitted from a metal surface is v_A when the wavelength of the light used is λ_A. However, a smaller maximum speed v_B is observed when a wavelength λ_B is used. (a) In which case is there a greater maximum kinetic energy KE_{max}? (b) In which case is the frequency of the light greater? (c) Is wavelength λ_A greater or smaller than λ_B? Explain your answers.

Problem The work function of a metal surface is 4.80×10^{-19} J. The maximum speed of the electrons emitted from the surface is $v_A = 7.30 \times 10^5$ m/s when the wavelength of the light used is λ_A. However, a speed of $v_B = 5.00 \times 10^5$ m/s is observed when a wavelength λ_B is used. Find the wavelengths λ_A and λ_B. Be sure your answers are consistent with your answers to the Concept Questions.

47. Concept Questions Heat is a form of energy. One way to supply the heat needed to warm an object is to irradiate the object with electromagnetic waves. (a) How is the heat Q needed to raise the temperature of an object by ΔT degrees related to its specific heat capacity $c_{specific\ heat}$ and mass m? (b) Is the energy carried by an infrared photon greater than, smaller than, or the same as the energy carried by a visible photon of light? Explain. (c) Assuming that all the photons are absorbed by the object, is the number of infrared photons needed to supply a given amount of heat greater or smaller than the number of visible photons? Explain.

Problem A glass plate has a mass of 0.50 kg and a specific heat capacity of 840 J/(kg·C°). The wavelength of infrared light is 6.0×10^{-5} m, while the wavelength of blue visible light is 4.7×10^{-7} m. Find the number of infrared photons and the number of visible photons needed to raise the temperature of the glass plate by 2.0 C°, assuming that all photons are absorbed by the glass. Verify that your answers are consistent with your answers to the Concept Questions.

48. Concept Questions In the Compton effect, momentum conservation applies, so the total momentum of the photon and the electron is the same before and after the scattering occurs. Suppose in Figure 29.10 that the direction of the incident photon is in the $+x$ direction and that the scattered photon emerges at an angle of $\theta = 90.0°$, which is in the $-y$ direction. (a) Before the scattering, the total momentum has an x component. Does it have a y component? (b) After the scattering, the total momentum has an x component. Does it have a y component? (c) Does the momentum of the scattered electron have an x component? (d) Considering your answer to Question (b), and considering the fact that the scattered photon is moving along

the $-y$ axis, does the momentum of the scattered electron have a y component? In each case justify your answer.

Problem In Figure 29.10 the incident photon has a wavelength of 9.00×10^{-12} m. The scattered photon emerges at $\theta = 90.0°$. Find the x and y components of the momentum of the scattered electron. Check that your answers are consistent with your answers to the Concept Questions.

49. Concept Questions Consider a Young's double-slit experiment that uses electrons. In versions A and B of the experiment the first-order bright fringes are located at angles θ_A and θ_B, where θ_B is greater than θ_A. In these two versions of the experiment, the separation between the slits is the same but the electrons have different momenta. (a) In which case is the wavelength of an electron greater? (b) In which case is the momentum of an electron greater? Explain your answers.

Problem In a Young's double-slit experiment the angle that locates the first-order bright fringes is $\theta_A = 1.6 \times 10^{-4}$ degrees when the magnitude of the electron momentum is $p_A = 1.2 \times 10^{-22}$ kg·m/s. With the same double slit, what momentum magnitude p_B is necessary, so that an angle of $\theta_B = 4.0 \times 10^{-4}$ degrees locates the first-order bright fringes?

* **50. Concept Questions** Particle A is at rest, and particle B collides head-on with it. The collision is completely inelastic, so the two particles stick together and move off after the collision with a common velocity. The masses of the particles are different, and no external forces act on them. (a) What is the total linear momentum of the two-particle system before the collision? Express your answer in terms of the masses and the initial velocities. (b) What is the total linear momentum of the system after the collision? Express your answer in terms of the masses and the final velocity of the particle. (c) Since no external forces act on the two-particle system during the collision, what does this tell you about the total linear momentum of the system? (d) How is the de Broglie wavelength of a particle related to the magnitude of its momentum?

Problem The de Broglie wavelength of particle B before the collision is 2.0×10^{-34} m. What is the de Broglie wavelength of the object that moves off after the collision?

* **51. Concept Questions** A photon of known wavelength strikes a free electron that is initially at rest, and the photon is scattered straight backward. (a) How is the kinetic energy KE of the recoil electron related to the energies E and E' of the incident and scattered photons? (b) If the wavelength λ of the incident photon is known, how can its energy E be determined? (c) How can the wavelength λ' of the scattered photon and, hence, its energy E' be found if we know the wavelength of the incident photon and the fact that the photon is scattered straight backward?

Problem A photon of wavelength 0.45000 nm strikes a free electron that is initially at rest. The photon is scattered straight backward. What is the speed of the recoil electron after the collision?

Chapter 30

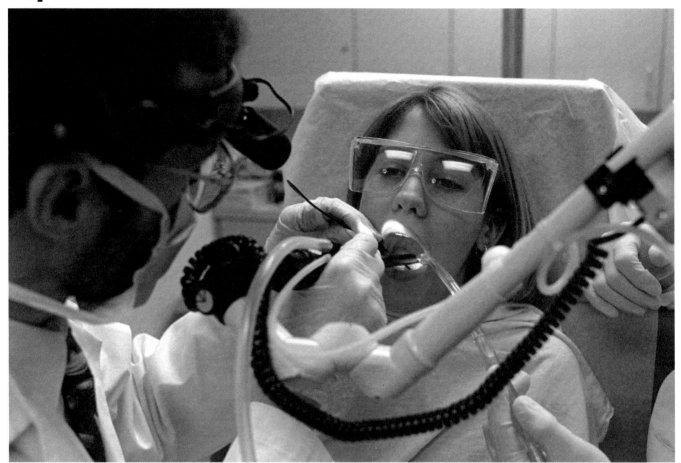

The Nature of the Atom

In their operation, lasers depend on the structure of the atom, as this chapter discusses. Medical diagnostic and surgical procedures use lasers in a wide variety of situations, often with dramatic advantages over conventional techniques. The patient in this photograph is undergoing laser surgery to remove her tonsils and was ready to go out for ice cream ten minutes following the operation. (© AP/Wide World Photos)

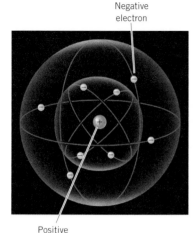

Figure 30.1 In the nuclear atom a small positively charged nucleus is surrounded at relatively large distances by a number of electrons.

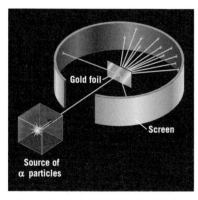

Figure 30.2 A Rutherford scattering experiment in which α particles are scattered by a thin gold foil. The entire apparatus is located within a vacuum chamber (not shown).

30.1 *Rutherford Scattering and the Nuclear Atom*

An atom contains a small, positively charged nucleus (radius $\approx 10^{-15}$ m), which is surrounded at relatively large distances (radius $\approx 10^{-10}$ m) by a number of electrons, as Figure 30.1 illustrates. In the natural state, an atom is electrically neutral because the nucleus contains a number of protons (each with a charge of $+e$) that equals the number of electrons (each with a charge of $-e$). This model of the atom is universally accepted now and is referred to as the ***nuclear atom.***

The nuclear atom is a relatively recent idea. In the early part of the twentieth century a widely accepted model, due to the English physicist Joseph J. Thomson (1856–1940), pictured the atom very differently. In Thomson's view there was no nucleus at the center of an atom. Instead, the positive charge was assumed to be spread throughout the atom, forming a kind of "paste" or "pudding," in which the negative electrons were suspended like "plums."

The "plum-pudding" model was discredited in 1911 when the New Zealand physicist Ernest Rutherford (1871–1937) published experimental results that the model could not explain. As Figure 30.2 indicates, Rutherford and his co-workers directed a beam of alpha particles (α particles) at a thin metal foil made of gold. Alpha particles are positively charged particles (the nuclei of helium atoms, although this was not recognized at the time) emitted by some radioactive materials. If the "plum-pudding" model were correct, the α particles would be expected to pass nearly straight through the foil. After all, there is nothing in this model to deflect the relatively massive α particles, since the electrons have a comparatively small mass and the positive charge is spread out in a diluted "pudding." Using a zinc sulfide screen, which flashed briefly when struck by an α particle, Rutherford and co-workers were able to determine that not all the α particles passed straight through the foil. Instead, some were deflected at large angles, even backward. Rutherford himself said, "It was almost as incredible as if you had fired a fifteen-inch shell at a piece of tissue and it came back and hit you." Rutherford concluded that the positive charge, instead of being distributed thinly and uniformly throughout the atom, was concentrated in a small region called the nucleus.

But how could the electrons in a nuclear atom remain separated from the positively charged nucleus? If the electrons were stationary, they would be pulled inward by the attractive electric force of the nuclear charge. Therefore, the electrons must be moving around the nucleus in some fashion, like planets revolving around the sun. In fact, the nuclear model of the atom is sometimes referred to as the "planetary" model. The dimensions of the atom, however, are such that it contains a larger fraction of empty space than our solar system does, as Conceptual Example 1 discusses.

Conceptual Example 1   Atoms Are Mostly Empty Space

In the planetary model of the atom, the nucleus (radius $\approx 1 \times 10^{-15}$ m) is analogous to the sun (radius $\approx 7 \times 10^8$ m). Electrons orbit (radius $\approx 1 \times 10^{-10}$ m) the nucleus like the earth orbits (radius $\approx 1.5 \times 10^{11}$ m) the sun. If the dimensions of the solar system had the same proportions as those of the atom, would the earth be closer to or farther away from the sun than it actually is?

Reasoning and Solution The radius of an electron orbit is one hundred thousand times larger than the radius of the nucleus: $(1 \times 10^{-10}$ m$)/(1 \times 10^{-15}$ m$) = 10^5$. If the radius of the earth's orbit were one hundred thousand times larger than the radius of the sun, the earth's orbit would have a radius of $10^5 \times (7 \times 10^8$ m$) = 7 \times 10^{13}$ m. This is more than four hundred times greater than the actual orbital radius of 1.5×10^{11} m, so ***the earth would be much farther away from the sun.*** In fact, it would be more than ten times farther from the sun than Pluto is, which is our most distant planet and has an orbital radius of about 6×10^{12} m. An atom, then, contains a much greater fraction of empty space than does our solar system.

Related Homework: *Problem 2*

Although the planetary model of the atom is easy to visualize, it too is fraught with difficulties. For instance, an electron moving on a curved path has a centripetal acceleration, as Section 5.2 discusses. And when an electron is accelerating, it radiates electromagnetic waves, as Section 24.1 discusses. These waves carry away energy. With their energy constantly being depleted, the electrons would spiral inward and eventually collapse into the nucleus. Since matter is stable, such a collapse does not occur. Thus, the planetary model, although providing a more realistic picture of the atom than the "plum-pudding" model, must be telling only part of the story. The full story of atomic structure is fascinating, and the next section describes another aspect of it.

30.2 Line Spectra

We have seen in Sections 13.3 and 29.2 that all objects emit electromagnetic waves, and we will see in Section 30.3 how this radiation arises. For a solid object, such as the hot filament of a light bulb, these waves have a continuous range of wavelengths, some of which are in the visible region of the spectrum. The continuous range of wavelengths is characteristic of the entire collection of atoms that make up the solid. In contrast, individual atoms, free of the strong interactions that are present in a solid, emit only certain specific wavelengths rather than a continuous range. These wavelengths are characteristic of the atom and provide important clues about its structure. To study the behavior of individual atoms, low-pressure gases are used in which the atoms are relatively far apart.

A low-pressure gas in a sealed tube can be made to emit electromagnetic waves by applying a sufficiently large potential difference between two electrodes located within the tube. With a grating spectroscope like that in Figure 27.36, the individual wavelengths emitted by the gas can be separated and identified as a series of bright fringes. The series of fringes is called a *line spectrum* because each bright fringe appears as a thin rectangle (a "line") resulting from the large number of parallel, closely spaced slits in the grating of the spectroscope. Figure 30.3 shows the visible parts of the line spectra for neon and mercury. The specific visible wavelengths emitted by neon and mercury give neon signs and mercury vapor street lamps their characteristic colors.

The simplest line spectrum is that of atomic hydrogen, and much effort has been devoted to understanding the pattern of wavelengths that it contains. Figure 30.4 illustrates in schematic form some of the groups or series of lines in the spectrum of atomic hydro-

The physics of neon signs and mercury vapor street lamps.

Figure 30.3 The line spectra for neon and mercury, along with the continuous spectrum of the sun. The dark lines in the sun's spectrum are called Fraunhofer lines, three of which are marked by arrows. (Courtesy Bausch & Lomb)

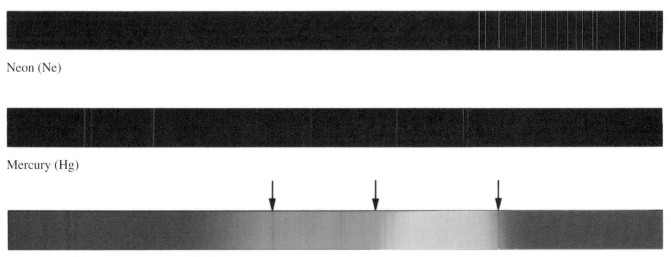

Neon (Ne)

Mercury (Hg)

Solar absorption spectrum (Fraunhofer lines)

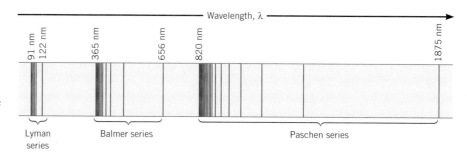

Figure 30.4 Line spectrum of atomic hydrogen. Only the Balmer series lies in the visible region of the electromagnetic spectrum.

gen. The group of lines in the visible region is known as the **Balmer series,** in recognition of Johann J. Balmer (1825–1898), a Swiss schoolteacher who found an empirical equation that gave the values for the observed wavelengths. This equation is given next, along with similar equations that apply to the **Lyman series** at shorter wavelengths and the **Paschen series** at longer wavelengths, which are also shown in the drawing:

Lyman series $\dfrac{1}{\lambda} = R\left(\dfrac{1}{1^2} - \dfrac{1}{n^2}\right)$ $n = 2, 3, 4, \ldots$ (30.1)

Balmer series $\dfrac{1}{\lambda} = R\left(\dfrac{1}{2^2} - \dfrac{1}{n^2}\right)$ $n = 3, 4, 5, \ldots$ (30.2)

Paschen series $\dfrac{1}{\lambda} = R\left(\dfrac{1}{3^2} - \dfrac{1}{n^2}\right)$ $n = 4, 5, 6, \ldots$ (30.3)

In these equations, the constant term R has the value of $R = 1.097 \times 10^7 \text{ m}^{-1}$ and is called the *Rydberg constant.* An essential feature of each group of lines is that there are a long and a short wavelength limit, with the lines being increasingly crowded toward the short wavelength limit. Figure 30.4 also gives the wavelength limits for each series, and Example 2 determines them for the Balmer series.

Example 2   The Balmer Series

Find (a) the longest and (b) the shortest wavelengths of the Balmer series.

Reasoning Each wavelength in the series corresponds to one value for the integer n in Equation 30.2. Longer wavelengths are associated with smaller values of n. The longest wavelength occurs when n has its smallest value of $n = 3$. The shortest wavelength arises when n has a very large value, so that $1/n^2$ is essentially zero.

Solution

(a) With $n = 3$, Equation 30.2 reveals that for the longest wavelength

$$\frac{1}{\lambda} = R\left(\frac{1}{2^2} - \frac{1}{n^2}\right) = (1.097 \times 10^7 \text{ m}^{-1})\left(\frac{1}{2^2} - \frac{1}{3^2}\right) = 1.524 \times 10^6 \text{ m}^{-1}$$

or $\boxed{\lambda = 656 \text{ nm}}$

(b) With $1/n^2 = 0$, Equation 30.2 reveals that for the shortest wavelength

$$\frac{1}{\lambda} = (1.097 \times 10^7 \text{ m}^{-1})\left(\frac{1}{2^2} - 0\right) = 2.743 \times 10^6 \text{ m}^{-1} \quad \text{or} \quad \boxed{\lambda = 365 \text{ nm}}$$

Equations 30.1–30.3 are useful because they reproduce the wavelengths that hydrogen atoms radiate. However, these equations are empirical and provide no insight as to *why* certain wavelengths are radiated and others are not. It was the great Danish physicist, Niels Bohr (1885–1962), who provided the first model of the atom that predicted the discrete wavelengths emitted by atomic hydrogen. Bohr's model started us on the way toward understanding how the structure of the atom restricts the radiated wavelengths to certain values. In 1922 Bohr received the Nobel Prize in physics for his accomplishment.

30.3 *The Bohr Model of the Hydrogen Atom*

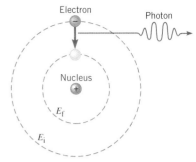

In 1913 Bohr presented a model that led to equations such as Balmer's for the wavelengths that the hydrogen atom radiates. Bohr's theory begins with Rutherford's picture of an atom as a nucleus surrounded by electrons moving in circular orbits. In his theory, Bohr made a number of assumptions and combined the new quantum ideas of Planck and Einstein with the traditional description of a particle in uniform circular motion.

Adopting Planck's idea of quantized energy levels (see Section 29.2), Bohr hypothesized that in a hydrogen atom there can be only certain values of the total energy (electron kinetic energy plus potential energy). These allowed energy levels correspond to different orbits for the electron as it moves around the nucleus, the larger orbits being associated with larger total energies. Figure 30.5 illustrates two of the orbits. In addition, Bohr assumed that an electron in one of these orbits *does not* radiate electromagnetic waves. For this reason, the orbits are called **stationary orbits** or **stationary states.** Bohr recognized that radiationless orbits violated the laws of physics, as they were then known. But the assumption of such orbits was necessary because the traditional laws indicated that an electron radiates electromagnetic waves as it accelerates around a circular path, and the loss of the energy carried by the waves would lead to the collapse of the orbit.

Figure 30.5 In the Bohr model, a photon is emitted when the electron drops from a larger, higher-energy orbit (energy = E_i) to a smaller, lower-energy orbit (energy = E_f).

To incorporate Einstein's photon concept (see Section 29.3), Bohr theorized that a photon is emitted only when the electron *changes* orbits from a larger one with a higher energy to a smaller one with a lower energy, as Figure 30.5 indicates. But how do electrons get into the higher-energy orbits in the first place? They get there by picking up energy when atoms collide, which happens more often when a gas is heated, or by acquiring energy when a high voltage is applied to a gas.

When an electron in an initial orbit with a larger energy E_i changes to a final orbit with a smaller energy E_f, the emitted photon has an energy of $E_i - E_f$, consistent with the law of conservation of energy. But according to Einstein, the energy of a photon is hf, where f is its frequency and h is Planck's constant. As a result, we find that

$$E_i - E_f = hf \qquad (30.4)$$

Since the frequency of an electromagnetic wave is related to the wavelength by $f = c/\lambda$, Bohr could use Equation 30.4 to determine the wavelengths radiated by a hydrogen atom. First, however, he had to derive expressions for the energies E_i and E_f.

THE ENERGIES AND RADII OF THE BOHR ORBITS

▶ CONCEPTS AT A GLANCE To derive an expression for the quantized energy levels, Bohr brought together concepts from the older classical physics and the newer modern physics that emerged at the beginning of the twentieth century. The Concepts-at-a-Glance chart in Figure 30.6 shows the classical and modern physics that he merged to obtain the correct expression for the electron energy levels in atomic hydrogen. ◀

Figure 30.6 CONCEPTS AT A GLANCE By combining ideas from classical and modern physics, Niels Bohr was able to obtain values for the quantized energy levels of the electron in a hydrogen atom. He is shown here with his five sons. (Courtesy A.I.P. Niels Bohr Library, Margrethe Bohr Collection)

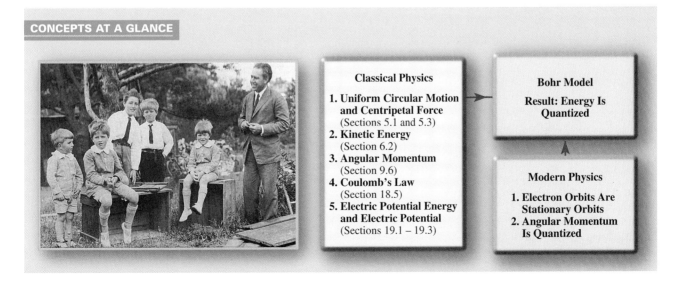

CONCEPTS AT A GLANCE

Classical Physics

1. **Uniform Circular Motion and Centripetal Force** (Sections 5.1 and 5.3)
2. **Kinetic Energy** (Section 6.2)
3. **Angular Momentum** (Section 9.6)
4. **Coulomb's Law** (Section 18.5)
5. **Electric Potential Energy and Electric Potential** (Sections 19.1 – 19.3)

Bohr Model

Result: Energy Is Quantized

Modern Physics

1. **Electron Orbits Are Stationary Orbits**
2. **Angular Momentum Is Quantized**

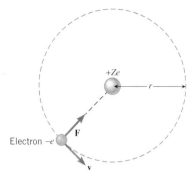

Figure 30.7 In the Bohr model, the electron is in uniform circular motion around the nucleus. The centripetal force **F** is the electrostatic force of attraction that the positive nuclear charge exerts on the electron.

Let's now bring together the concepts outlined in Figure 30.6 to derive Bohr's result. For an electron of mass m and speed v in an orbit of radius r, the total energy is the kinetic energy (KE $= \frac{1}{2}mv^2$) of the electron plus the electric potential energy EPE. The potential energy is the product of the charge $(-e)$ on the electron and the electric potential produced by the positive nuclear charge, in accord with Equation 19.3. We assume that the nucleus contains Z protons,* for a total nuclear charge of $+Ze$. The electric potential at a distance r from a point charge of $+Ze$ is given as $+kZe/r$ by Equation 19.6, where $k = 8.988 \times 10^9 \ \mathrm{N \cdot m^2/C^2}$. The electric potential energy is, then, EPE $= (-e)(+kZe/r)$. Consequently, the total energy E of the atom is

$$E = \mathrm{KE} + \mathrm{EPE}$$

$$= \tfrac{1}{2}mv^2 - \frac{kZe^2}{r} \tag{30.5}$$

But a centripetal force of magnitude mv^2/r (Equation 5.3) acts on a particle in uniform circular motion. As Figure 30.7 indicates, the centripetal force is provided by the electrostatic force of attraction **F** that the protons in the nucleus exert on the electron. According to Coulomb's law (Equation 18.1), the magnitude of the electrostatic force is $F = kZe^2/r^2$. Therefore, $mv^2/r = kZe^2/r^2$, or

$$mv^2 = \frac{kZe^2}{r} \tag{30.6}$$

We can use this relation to eliminate the term mv^2 from Equation 30.5, with the result that

$$E = \frac{1}{2}\left(\frac{kZe^2}{r}\right) - \frac{kZe^2}{r}$$

$$= -\frac{kZe^2}{2r} \tag{30.7}$$

The total energy of the atom is negative because the negative electric potential energy is larger in magnitude than the positive kinetic energy.

A value for the radius r is needed, if Equation 30.7 is to be useful. To determine r, Bohr made an assumption about the orbital angular momentum of the electron. The angular momentum L is given by Equation 9.10 as $L = I\omega$, where $I = mr^2$ is the moment of inertia of the electron moving on its circular path and $\omega = v/r$ is the angular speed of the electron in radians per second. Thus, the angular momentum is $L = (mr^2)(v/r) = mvr$. Bohr conjectured that the angular momentum can assume only certain discrete values; in other words, L is quantized. He postulated that the allowed values are integer multiples of Planck's constant divided by 2π:

$$L_n = mv_n r_n = n\frac{h}{2\pi} \qquad n = 1, 2, 3, \ldots \tag{30.8}$$

Solving this equation for v_n and substituting the result into Equation 30.6 lead to the following expression for the radius r_n of the nth Bohr orbit:

$$r_n = \left(\frac{h^2}{4\pi^2 mke^2}\right)\frac{n^2}{Z} \qquad n = 1, 2, 3, \ldots \tag{30.9}$$

With $h = 6.626 \times 10^{-34} \ \mathrm{J \cdot s}$, $m = 9.109 \times 10^{-31}$ kg, $k = 8.988 \times 10^9 \ \mathrm{N \cdot m^2/C^2}$, and $e = 1.602 \times 10^{-19}$ C, this expression reveals that

Radii for Bohr orbits (in meters)
$$r_n = (5.29 \times 10^{-11} \ \mathrm{m})\frac{n^2}{Z} \qquad n = 1, 2, 3, \ldots \tag{30.10}$$

Therefore, in the hydrogen atom ($Z = 1$) the smallest Bohr orbit ($n = 1$) has a radius of $r_1 = 5.29 \times 10^{-11}$ m. This particular value is called the **Bohr radius.** Figure 30.8 shows the first three Bohr orbits for the hydrogen atom.

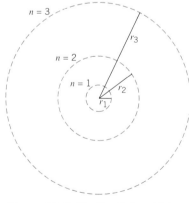

Figure 30.8 The first Bohr orbit in the hydrogen atom has a radius $r_1 = 5.29 \times 10^{-11}$ m. The second and third Bohr orbits have radii $r_2 = 4r_1$ and $r_3 = 9r_1$, respectively.

* For hydrogen, $Z = 1$, but we also wish to consider situations in which Z is greater than 1.

The expression for the radius of a Bohr orbit can be substituted into Equation 30.7 to show that the corresponding total energy for the nth orbit is

$$E_n = -\left(\frac{2\pi^2 m k^2 e^4}{h^2}\right)\frac{Z^2}{n^2} \qquad n = 1, 2, 3, \ldots \qquad (30.11)$$

Substituting values for h, m, k, and e into this expression yields

Bohr energy levels in joules
$$E_n = -(2.18 \times 10^{-18}\ \text{J})\frac{Z^2}{n^2} \qquad n = 1, 2, 3, \ldots \qquad (30.12)$$

Often, atomic energies are expressed in electron volts rather than joules. Since $1.60 \times 10^{-19}\ \text{J} = 1\ \text{eV}$, Equation 30.12 can be rewritten as

Bohr energy levels in electron volts
$$E_n = -(13.6\ \text{eV})\frac{Z^2}{n^2} \qquad n = 1, 2, 3, \ldots \qquad (30.13)$$

ENERGY LEVEL DIAGRAMS

It is useful to represent the energy values given by Equation 30.13 on an *energy level diagram*, as in Figure 30.9. In this diagram, which applies to the hydrogen atom ($Z = 1$), the highest energy level corresponds to $n = \infty$ in Equation 30.13 and has an energy of 0 eV. This is the energy of the atom when the electron is completely removed ($r = \infty$) from the nucleus and is at rest. In contrast, the lowest energy level corresponds to $n = 1$ and has a value of -13.6 eV. The lowest energy level is called the **ground state**, to distinguish it from the higher levels, which are called **excited states.** Observe how the energies of the excited states come closer and closer together as n increases.

The electron in a hydrogen atom at room temperature spends most of its time in the ground state. To raise the electron from the ground state ($n = 1$) to the highest possible excited state ($n = \infty$), 13.6 eV of energy must be supplied. Supplying this amount of energy removes the electron from the atom, producing the positive hydrogen ion H^+. The energy needed to remove the electron is called the **ionization energy.** Thus, the Bohr model predicts that the ionization energy of atomic hydrogen is 13.6 eV, in excellent agreement with the experimental value. In Example 3 the Bohr model is applied to doubly ionized lithium.

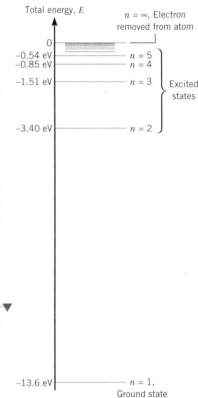

Figure 30.9 Energy level diagram for the hydrogen atom.

Example 3 The Ionization Energy of Li²⁺

The Bohr model does not apply when more than one electron orbits the nucleus because it does not account for the electrostatic force that one electron exerts on another. For instance, an electrically neutral lithium atom (Li) contains three electrons in orbit around a nucleus that includes three protons ($Z = 3$), and Bohr's analysis is not applicable. However, the Bohr model can be used for the doubly charged positive ion of lithium (Li^{2+}) that results when two electrons are removed from the neutral atom, leaving only one electron to orbit the nucleus. Obtain the ionization energy that is needed to remove the remaining electron from Li^{2+}.

Reasoning The lithium ion Li^{2+} contains three times the positive nuclear charge that the hydrogen atom contains. Therefore, the orbiting electron is attracted more strongly to the nucleus in Li^{2+} than to the nucleus in the hydrogen atom. As a result, we expect that more energy is required to ionize Li^{2+} than the 13.6 eV required for atomic hydrogen.

Solution The Bohr energy levels for Li^{2+} are given by Equation 30.13 with $Z = 3$; $E_n = -(13.6\ \text{eV})(3^2/n^2)$. Therefore, the ground state ($n = 1$) energy is

$$E_1 = -(13.6\ \text{eV})\frac{3^2}{1^2} = -122\ \text{eV}$$

In order to remove the electron from Li^{2+}, 122 eV of energy must be supplied: Ionization energy = 122 eV. This value for the ionization energy agrees well with the experimental value of 122.4 eV and, as expected, is greater than the 13.6 eV required for atomic hydrogen.

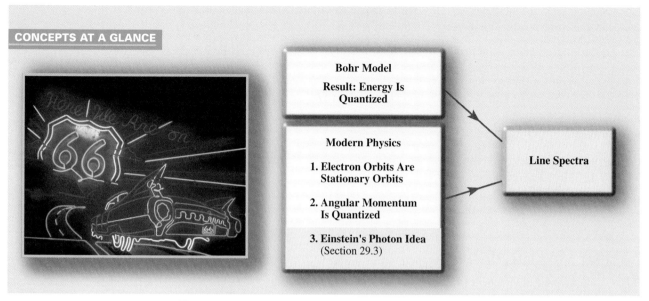

Figure 30.10 CONCEPTS AT A GLANCE This chart extends the chart in Figure 30.6 and shows that Bohr was able to predict the line spectrum of the hydrogen atom by combining his quantized energy level expression with Einstein's photon concept. The visible wavelengths in the line spectrum of atoms are used in producing the light in advertising signs like the one in this photograph. (© Owaki-Kulla/Corbis Images)

THE LINE SPECTRA OF THE HYDROGEN ATOM

▶ CONCEPTS AT A GLANCE   To predict the wavelengths in the line spectrum of the hydrogen atom, Bohr combined his ideas about atoms (electron orbits are stationary orbits and the angular momentum of an electron is quantized) with Einstein's idea of the photon. This synthesis is illustrated in the Concepts-at-a-Glance chart in Figure 30.10, which is an extension of the one in Figure 30.6. ◀

As applied by Bohr, the photon concept is inherent in Equation 30.4, $E_i - E_f = hf$, which states that the frequency f of the photon is proportional to the difference between two energy levels of the hydrogen atom. If we substitute Equation 30.11 for the total energies E_i and E_f into Equation 30.4 and recall from Equation 16.1 that $f = c/\lambda$, we obtain the following result:

$$\frac{1}{\lambda} = \frac{2\pi^2 m k^2 e^4}{h^3 c} (Z^2) \left(\frac{1}{n_f^2} - \frac{1}{n_i^2} \right) \qquad (30.14)$$

$$n_i, n_f = 1, 2, 3, \ldots \quad \text{and} \quad n_i > n_f$$

With the known values for h, m, k, e, and c, it can be seen that $2\pi^2 m k^2 e^4/(h^3 c) = 1.097 \times 10^7 \text{ m}^{-1}$, in agreement with the Rydberg constant R that appears in Equations 30.1–30.3. The agreement between the theoretical and experimental values of the Rydberg constant was a major accomplishment of Bohr's theory.

With $Z = 1$ and $n_f = 1$, Equation 30.14 reproduces Equation 30.1 for the Lyman series. Thus, Bohr's model shows that the Lyman series of lines occurs when electrons make transitions from higher energy levels with $n_i = 2, 3, 4, \ldots$ to the first energy level where $n_f = 1$. Figure 30.11 shows these transitions. Notice that when an electron makes a transition from $n_i = 2$ to $n_f = 1$, the longest wavelength photon in the Lyman series is emitted, since the energy change is the smallest possible. When an electron makes a transition from the highest level where $n_i = \infty$ to the lowest level where $n_f = 1$, the shortest wavelength is emitted, since the energy change is the largest possible. Since the higher energy levels are increasingly close together, the lines in the series become more and more crowded toward the short wavelength limit, as can be seen in Figure 30.4. Figure 30.11 also shows the energy level transitions for the Balmer series, where $n_i = 3, 4, 5, \ldots$, and $n_f = 2$. In the Paschen series (see Figure 30.4) $n_i = 4, 5, 6, \ldots$, and $n_f = 3$. The next example deals further with the line spectrum of the hydrogen atom.

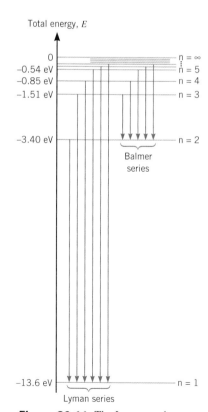

Figure 30.11 The Lyman and Balmer series of lines in the hydrogen atom spectrum correspond to transitions that the electron makes between higher and lower energy levels, as indicated here.

Example 4 The Brackett Series for Atomic Hydrogen

In the line spectrum of atomic hydrogen there is also a group of lines known as the Brackett series. These lines are produced when electrons, excited to high energy levels, make transitions to the $n = 4$ level. Determine (a) the longest wavelength in this series and (b) the wavelength that corresponds to the transition from $n_i = 6$ to $n_f = 4$. (c) Refer to Figure 24.10 and identify the spectral region in which these lines are found.

Reasoning The longest wavelength corresponds to the transition that has the smallest energy change, which is between the $n_i = 5$ and $n_f = 4$ levels in Figure 30.9. The wavelength for this transition, as well as that for the transition from $n_i = 6$ to $n_f = 4$, can be obtained from Equation 30.14.

Solution

(a) Using Equation 30.14 with $Z = 1$, $n_i = 5$, and $n_f = 4$, we find that

$$\frac{1}{\lambda} = (1.097 \times 10^7 \text{ m}^{-1})(1^2)\left(\frac{1}{4^2} - \frac{1}{5^2}\right) = 2.468 \times 10^5 \text{ m}^{-1} \quad \text{or} \quad \boxed{\lambda = 4051 \text{ nm}}$$

(b) The calculation here is similar to that in part (a):

$$\frac{1}{\lambda} = (1.097 \times 10^7 \text{ m}^{-1})(1^2)\left(\frac{1}{4^2} - \frac{1}{6^2}\right) = 3.809 \times 10^5 \text{ m}^{-1} \quad \text{or} \quad \boxed{\lambda = 2625 \text{ nm}}$$

(c) According to Figure 24.10, these lines lie in the $\boxed{\text{infrared region}}$ of the spectrum.

The various lines in the hydrogen atom spectrum are produced when electrons change from higher to lower energy levels and photons are emitted. Consequently, the spectral lines are called **emission lines.** Electrons can also make transitions in the reverse direction, from lower to higher levels, in a process known as absorption. In this case, an atom absorbs a photon that has precisely the energy needed to produce the transition. Thus, if photons with a continuous range of wavelengths pass through a gas and then are analyzed with a grating spectroscope, a series of dark **absorption lines** appear in the continuous spectrum. The dark lines indicate the wavelengths that have been removed by the absorption process. Such absorption lines can be seen in Figure 30.3 in the spectrum of the sun, where they are called Fraunhofer lines, after their discoverer. They are due to atoms, located in the outer and cooler layers of the sun, that absorb radiation coming from the interior. The interior portion of the sun emits a continuous spectrum of wavelengths, since it is too hot for individual atoms to retain their structures.

The physics of absorption lines in the sun's spectrum.

The Bohr model provides a great deal of insight into atomic structure. However, this model is now known to be oversimplified and has been superseded by a more detailed picture provided by quantum mechanics and the Schrödinger equation (see Section 30.5).

Concept Simulation 30.1

The focus of this simulation is on the emission and absorption of photons by a Bohr atom. You can see how the electron changes orbit and energy level due to the emission or absorption of a photon of various energies.

Related Homework: *Problem 8*

Go to **www.wiley.com/college/cutnell**

✔ **Check Your Understanding 1**

An electron in the hydrogen atom is in the $n = 4$ energy level. When this electron makes a transition to a lower energy level, is the wavelength of the photon emitted in (a) the Lyman series, (b) the Balmer series, (c) the Paschen series, or (d) could it be in the Lyman, Balmer, or the Paschen series? *(The answer is given at the end of the book.)*

Background: The Bohr model of the hydrogen atom explains the discrete nature of the wavelengths of the photons that can be emitted by the atom.

For similar questions (including calculational counterparts), consult Self-Assessment Test 30.1, which is described at the end of Section 30.5.

Problem solving insight
In the line spectrum of atomic hydrogen, all lines in a given series (e.g., the Lyman series) are identified by a single value of the quantum number n_f for the *lower energy level into which an electron falls.* Each line in a given series, however, corresponds to a different value of the quantum number n_i for the *higher energy level where an electron originates.*

Need more practice?

Interactive LearningWare 30.1
A stationary hydrogen atom ($m = 1.67 \times 10^{-27}$ kg) emits a photon by making a transition from the $n_i = 3$ to the $n_f = 1$ state. (a) What is the momentum of the emitted photon? (b) Determine the recoil speed of the hydrogen atom.

Related Homework: *Problem 18*

Go to
www.wiley.com/college/cutnell
for an interactive solution.

30.4 De Broglie's Explanation of Bohr's Assumption about Angular Momentum

Of all the assumptions Bohr made in his model of the hydrogen atom, perhaps the most puzzling is the one about the angular momentum of the electron [$L_n = mv_n r_n = nh/(2\pi)$; $n = 1, 2, 3, \ldots$]. Why should the angular momentum have only those values that are integer multiples of Planck's constant divided by 2π? In 1923, ten years after Bohr's work, de Broglie pointed out that his own theory for the wavelength of a moving particle could provide an answer to this question.

In de Broglie's way of thinking, the electron in its circular Bohr orbit must be pictured as a particle wave. And like waves traveling on a string, particle waves can lead to standing waves under resonant conditions. Section 17.5 discusses these conditions for a string. Standing waves form when the total distance traveled by a wave down the string and back is one wavelength, two wavelengths, or any integer number of wavelengths. The total distance around a Bohr orbit of radius r is the circumference of the orbit or $2\pi r$. By the same reasoning, then, the condition for standing particle waves for the electron in a Bohr orbit would be

$$2\pi r = n\lambda \qquad n = 1, 2, 3, \ldots$$

where n is the number of whole wavelengths that fit into the circumference of the circle. But according to Equation 29.8 the de Broglie wavelength of the electron is $\lambda = h/p$, where p is the magnitude of the electron's momentum. If the speed of the electron is much less than the speed of light, the momentum is $p = mv$, and the condition for standing particle waves becomes $2\pi r = nh/(mv)$. A rearrangement of this result gives

$$mvr = n\frac{h}{2\pi} \qquad n = 1, 2, 3, \ldots$$

which is just what Bohr assumed for the angular momentum of the electron. As an example, Figure 30.12 illustrates the standing particle wave on a Bohr orbit for which $2\pi r = 4\lambda$.

De Broglie's explanation of Bohr's assumption about angular momentum emphasizes an important fact—namely, that particle waves play a central role in the structure of the atom. Moreover, the theoretical framework of quantum mechanics provides the basis for determining the wave function Ψ (Greek letter Psi) that represents a particle wave. The next section deals with the picture that quantum mechanics gives for atomic structure, a picture that supersedes the Bohr model. In any case, the Bohr expression for the energy levels (Equation 30.11) can be applied when a single electron orbits the nucleus, whereas the theoretical framework of quantum mechanics can be applied, in principle, to atoms that contain an arbitrary number of electrons.

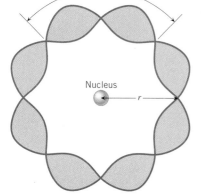

Figure 30.12 De Broglie suggested standing particle waves as an explanation for Bohr's angular momentum assumption. Here, a standing particle wave is illustrated on a Bohr orbit where four de Broglie wavelengths fit into the circumference of the orbit.

30.5 The Quantum Mechanical Picture of the Hydrogen Atom

The picture of the hydrogen atom that quantum mechanics and the Schrödinger equation provide differs in a number of ways from the Bohr model. The Bohr model uses a single integer number n to identify the various electron orbits and the associated energies. Because this number can have only discrete values, rather than a continuous range of values, n is called a **quantum number.** In contrast, quantum mechanics reveals that four different quantum numbers are needed to describe each state of the hydrogen atom. These four are described below.

    1. The principal quantum number _n_. As in the Bohr model, this number determines the total energy of the atom and can have only integer values: $n = 1, 2, 3, \ldots$. In fact, the Schrödinger equation predicts* that the energy of the hydrogen atom is identical to the energy obtained from the Bohr model: $E_n = -(13.6 \text{ eV}) Z^2/n^2$.

---

* This prediction requires that small relativistic effects and small interactions within the atom be ignored, and assumes that the hydrogen atom is not located in an external magnetic field.

2. **The orbital quantum number ℓ.** This number determines the angular momentum of the electron due to its orbital motion. The values that ℓ can have depend on the value of n, and only the following integers are allowed:

$$\ell = 0, 1, 2, \ldots, (n - 1)$$

For instance, if $n = 1$, the orbital quantum number can have only the value $\ell = 0$, but if $n = 4$, the values $\ell = 0, 1, 2$, and 3 are possible. The magnitude L of the angular momentum of the electron is

$$L = \sqrt{\ell(\ell + 1)} \, \frac{h}{2\pi} \qquad (30.15)$$

3. **The magnetic quantum number m_ℓ.** The word "magnetic" is used here because an externally applied magnetic field influences the energy of the atom, and this quantum number is used in describing the effect. Since the effect was discovered by the Dutch physicist Pieter Zeeman (1865–1943), it is known as the *Zeeman effect*. When there is no external magnetic field, m_ℓ plays no role in determining the energy. In either event, the magnetic quantum number determines the component of the angular momentum along a specific direction, which is called the z direction by convention. The values that m_ℓ can have depend on the value of ℓ, with only the following positive and negative integers being permitted:

$$m_\ell = -\ell, \ldots, -2, -1, 0, +1, +2, \ldots, +\ell$$

For example, if the orbital quantum number is $\ell = 2$, then the magnetic quantum number can have the values $m_\ell = -2, -1, 0, -1$, and $+2$. The component L_z of the angular momentum in the z direction is

$$L_z = m_\ell \frac{h}{2\pi} \qquad (30.16)$$

4. **The spin quantum number m_s.** This number is needed because the electron has an intrinsic property called spin angular momentum. Loosely speaking, we can view the electron as spinning while it orbits the nucleus, analogous to the way the earth spins as it orbits the sun. There are two possible values for the spin quantum number of the electron:

$$m_s = +\tfrac{1}{2} \quad \text{or} \quad m_s = -\tfrac{1}{2}$$

Sometimes the phrases "spin up" and "spin down" are used to refer to the directions of the spin angular momentum associated with the values for m_s.

Table 30.1 summarizes the four quantum numbers that are needed to describe each state of the hydrogen atom. One set of values for n, ℓ, m_ℓ, and m_s corresponds to one state. As the principal quantum number n increases, the number of possible combinations of the four quantum numbers rises rapidly, as Example 5 illustrates.

Table 30.1 *Quantum Numbers for the Hydrogen Atom*

Name	Symbol	Allowed Values
Principal quantum number	n	1, 2, 3, ...
Orbital quantum number	ℓ	0, 1, 2, ... , $(n - 1)$
Magnetic quantum number	m_ℓ	$-\ell$, ... , $-2, -1, 0, +1, +2$, ... , $+\ell$
Spin quantum number	m_s	$-\tfrac{1}{2}, +\tfrac{1}{2}$

Example 5 Quantum Mechanical States of the Hydrogen Atom

Determine the number of possible states for the hydrogen atom when the principal quantum number is (a) $n = 1$ and (b) $n = 2$.

Reasoning Each different combination of the four quantum numbers summarized in Table 30.1 corresponds to a different state. We begin with the value for n and find the allowed values for ℓ. Then, for each ℓ value we find the possibilities for m_ℓ. Finally, m_s may be $+\frac{1}{2}$ or $-\frac{1}{2}$ for each group of values for n, ℓ, and m_ℓ.

Solution

(a) The diagram below shows the possibilities for ℓ, m_ℓ, and m_s when $n = 1$:

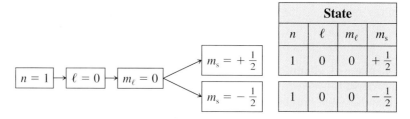

State			
n	ℓ	m_ℓ	m_s
1	0	0	$+\frac{1}{2}$
1	0	0	$-\frac{1}{2}$

Thus, there are two different states for the hydrogen atom. In the absence of an external magnetic field, these two states have the same energy, since they have the same value of n.

(b) When $n = 2$, there are eight possible combinations for the values of n, ℓ, m_ℓ, and m_s, as the diagram below indicates:

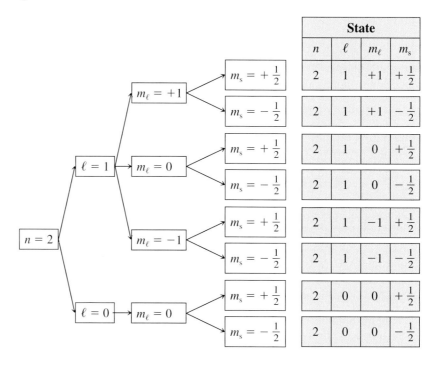

State			
n	ℓ	m_ℓ	m_s
2	1	+1	$+\frac{1}{2}$
2	1	+1	$-\frac{1}{2}$
2	1	0	$+\frac{1}{2}$
2	1	0	$-\frac{1}{2}$
2	1	−1	$+\frac{1}{2}$
2	1	−1	$-\frac{1}{2}$
2	0	0	$+\frac{1}{2}$
2	0	0	$-\frac{1}{2}$

With the same value of $n = 2$, all eight states have the same energy when there is no external magnetic field.

Quantum mechanics provides a more accurate picture of atomic structure than does the Bohr model. It is important to realize that the two pictures differ substantially, as Conceptual Example 6 illustrates.

Conceptual Example 6 The Bohr Model Versus Quantum Mechanics

Consider two hydrogen atoms. There are no external magnetic fields present, and the electron in each atom has the same energy. According to the Bohr model and to quantum mechanics, is it possible for the electrons in these atoms (a) to have zero orbital angular momentum and (b) to have different orbital angular momenta?

Reasoning and Solution

(a) In both the Bohr model and quantum mechanics, the energy is proportional to $1/n^2$, according to Equation 30.13, where n is the principal quantum number. Moreover, the value of n may be $n = 1, 2, 3, \ldots$, and may not be zero. **In the Bohr model, the fact that n may not be zero means that it is not possible for the orbital angular momentum to be zero** because the angular momentum is proportional to n, according to Equation 30.8. In the quantum mechanical picture the magnitude of the orbital angular momentum is proportional to $\sqrt{\ell(\ell + 1)}$, as given by Equation 30.15. Here, ℓ is the orbital quantum number and may take on the values $\ell = 0, 1, 2, \ldots, (n - 1)$. We note that ℓ [and therefore $\sqrt{\ell(\ell + 1)}$] may be zero, no matter what the value for n is. Consequently, **the orbital angular momentum may be zero according to quantum mechanics,** in contrast to the case for the Bohr model.

(b) If the electrons have the same energy, they have the same value for the principal quantum number n. **In the Bohr model, this means that they cannot have different values for the orbital angular momentum** L_n, since $L_n = nh/(2\pi)$, according to Equation 30.8. In quantum mechanics, the energy is also determined by n when external magnetic fields are absent, but the orbital angular momentum is determined by ℓ. Since $\ell = 0, 1, 2, \ldots, (n - 1)$, different values of ℓ are compatible with the same value of n. For instance, if $n = 2$ for both electrons, one of them could have $\ell = 0$, while the other could have $\ell = 1$. **According to quantum mechanics, then, the electrons could have different orbital angular momenta, even though they have the same energy.**

The following table summarizes the discussion from parts (a) and (b):

	Bohr Model	Quantum Mechanics
(a) For a given n, can the angular momentum ever be zero?	No	Yes
(b) For a given n, can the angular momentum have different values?	No	Yes

Related Homework: *Conceptual Question 5, Problem 26*

According to the Bohr model, the nth orbit is a circle of radius r_n, and every time the position of the electron in this orbit is measured, the electron is found exactly at a distance r_n away from the nucleus. This simplistic picture is now known to be incorrect, and the quantum mechanical picture of the atom has replaced it. Suppose the electron is in a quantum mechanical state for which $n = 1$, and we imagine making a number of measurements of the electron's position with respect to the nucleus. We would find that its position is uncertain, in the sense that there is a probability of finding the electron sometimes very near the nucleus, sometimes very far from the nucleus, and sometimes at intermediate locations. The probability is determined by the wave function Ψ, as Section 29.5 discusses. We can make a three-dimensional picture of our findings by marking a dot at each location where the electron is found. More dots occur at places where the probability of finding the electron is higher, and after a sufficient number of measurements, a picture of the quantum mechanical state emerges. Figure 30.13 shows the spatial distribution for an electron in a state for which $n = 1$, $\ell = 0$, and $m_\ell = 0$. This picture is constructed from so many measurements that the individual dots are no longer visible but have merged to form a kind of probability "cloud" whose density changes gradually from place to place. The dense regions indicate places where the probability of finding the electron is higher, and the less dense regions indicate places where the probability is lower. Also indicated in Figure 30.13 is the radius where quantum mechanics predicts the greatest probability per unit radial distance of finding the electron in the $n = 1$ state. This radius matches exactly the radius of 5.29×10^{-11} m found for the first Bohr orbit.

For a principal quantum number of $n = 2$, the probability clouds are different than for $n = 1$. In fact, more than one cloud shape is possible because with $n = 2$ the orbital quantum number can be either $\ell = 0$ or $\ell = 1$. Although the value of ℓ does not affect the energy of the hydrogen atom, the value does have a significant effect on the shape of

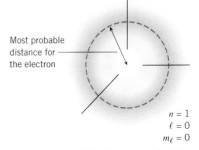

Most probable distance for the electron

$n = 1$
$\ell = 0$
$m_\ell = 0$

Figure 30.13 The electron probability cloud for the ground state ($n = 1$, $\ell = 0$, $m_\ell = 0$) of the hydrogen atom.

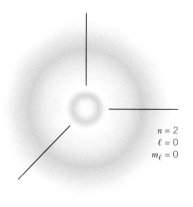

$n = 2$
$\ell = 0$
$m_\ell = 0$

(a)

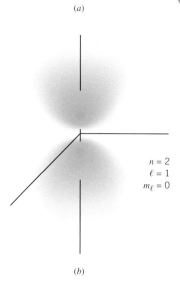

$n = 2$
$\ell = 1$
$m_\ell = 0$

(b)

Figure 30.14 The electron probability clouds for the hydrogen atom when (a) $n = 2$, $\ell = 0$, $m_\ell = 0$ and (b) $n = 2$, $\ell = 1$, $m_\ell = 0$.

the probability clouds. Figure 30.14a shows the cloud for $n = 2$, $\ell = 0$, and $m_\ell = 0$. Part *b* of the drawing shows that when $n = 2$, $\ell = 1$, and $m_\ell = 0$, the cloud has a two-lobe shape with the nucleus at the center between the lobes. For larger values of *n*, the probability clouds become increasingly complex and are spread out over larger volumes of space.

The probability cloud picture of the electron in a hydrogen atom is very different from the well-defined orbit of the Bohr model. The fundamental reason for this difference is to be found in the Heisenberg uncertainty principle, as Conceptual Example 7 discusses.

Conceptual Example 7
The Uncertainty Principle and the Hydrogen Atom

In the Bohr model of the hydrogen atom, the electron in the ground state ($n = 1$) is in an orbit that has a radius of exactly 5.29×10^{-11} m. Furthermore, as calculated in Example 3 in Chapter 18, the speed of the electron in this orbit is exactly 2.18×10^6 m/s. Considering the Heisenberg uncertainty principle, is this a realistic picture of atomic structure?

Reasoning and Solution The Bohr model indicates that the electron is located exactly at a radius of 5.29×10^{-11} m, so the uncertainty Δy in its radial position is zero. We will now show that an uncertainty of $\Delta y = 0$ m is not consistent with the Heisenberg principle, so that the Bohr picture is not realistic. What does the Heisenberg principle [Equation 29.10, $(\Delta p_y)(\Delta y) \geq h/(4\pi)$] say about Δy? According to the principle, the minimum uncertainty in the radial position of the electron is $\Delta y = h/(4\pi \, \Delta p_y)$, where Δp_y is the uncertainty in the momentum. The magnitude of the electron's momentum is the product of its mass *m* and speed *v*, so that $\Delta p_y = \Delta(mv) = m \, \Delta v$. It seems reasonable to assume that the uncertainty Δv in the speed should be less than 2.18×10^6 m/s. This means that the speed is somewhere between zero and twice the value known to apply to the ground state in the Bohr model. A much larger uncertainty would mean that the electron could have so much energy that it would not likely remain in orbit. With $\Delta v = 2.18 \times 10^6$ m/s and $m = 9.11 \times 10^{-31}$ kg for the mass of the electron, we find that the minimum uncertainty in the radial position of the electron is $\Delta y = h/(4\pi m \, \Delta v) = 2.7 \times 10^{-11}$ m. This uncertainty is quite large, because it is about one-half of the Bohr radius. According to the uncertainty principle, then, the radial position of the electron can be anywhere from approximately one-half to one and a half times the Bohr radius. *The single-radius orbit of the Bohr model does not correctly represent this aspect of reality at the atomic level.* Quantum mechanics, however, does correctly represent it in terms of a probability cloud picture of atomic structure.

 Check Your Understanding 2

The magnitude of the orbital angular momentum of the electron in a hydrogen atom is observed to increase. Does this necessarily mean that the energy of the electron also increases? Answer this question from the point of view of (a) the Bohr model and (b) the quantum mechanical picture of the hydrogen atom. *(The answers are given at the end of the book.)*

Background: Both the Bohr model of the hydrogen atom and the quantum mechanical picture of the hydrogen atom predict the same energy levels for the electron in the atom. In other respects, however, these two approaches differ in what they have to say about the hydrogen atom.

For similar questions (including calculational counterparts), consult Self-Assessment Test 30.1, which is described next.

 Self-Assessment Test 30.1

Test your understanding of the material in Sections 30.1–30.5:

• The Nuclear Atom   • Line Spectra   • The Bohr Model of the Hydrogen Atom
• De Broglie's Explanation of Bohr's Assumption about Angular Momentum
• The Quantum Mechanical Picture of the Hydrogen Atom

Go to **www.wiley.com/college/cutnell**

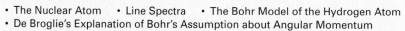

30.6 The Pauli Exclusion Principle and the Periodic Table of the Elements

Except for hydrogen, all electrically neutral atoms contain more than one electron, the number being given by the atomic number Z of the element. In addition to being attracted by the nucleus, the electrons repel each other. This repulsion contributes to the total energy of a multiple-electron atom. As a result, the one-electron energy expression for hydrogen, $E_n = -(13.6 \text{ eV}) Z^2/n^2$, does not apply to other neutral atoms. However, the simplest approach for dealing with a multiple-electron atom still uses the four quantum numbers n, ℓ, m_ℓ, and m_s.

Detailed quantum mechanical calculations reveal that the energy level of each state of a multiple-electron atom depends on both the principal quantum number n and the orbital quantum number ℓ. Figure 30.15 illustrates that the energy generally increases as n increases, but there are exceptions, as the drawing indicates. Furthermore, for a given n, the energy also increases as ℓ increases.

In a multiple-electron atom, all electrons with the same value of n are said to be in the same **shell**. Electrons with $n = 1$ are in a single shell (sometimes called the K shell), electrons with $n = 2$ are in another shell (the L shell), those with $n = 3$ are in a third shell (the M shell), and so on. Those electrons with the same values for both n and ℓ are often referred to as being in the same **subshell**. The $n = 1$ shell consists of a single $\ell = 0$ subshell. The $n = 2$ shell has two subshells, one with $\ell = 0$ and one with $\ell = 1$. Similarly, the $n = 3$ shell has three subshells, one with $\ell = 0$, one with $\ell = 1$, and one with $\ell = 2$.

In the hydrogen atom near room temperature, the electron spends most of its time in the lowest energy level or ground state — namely, in the $n = 1$ shell. Similarly, when an atom contains more than one electron and is near room temperature, the electrons spend most of their time in the lowest energy levels possible. The lowest energy state for an atom is called the **ground state.** However, when a multiple-electron atom is in its ground state, not every electron is in the $n = 1$ shell in general because the electrons obey a principle discovered by the Austrian physicist Wolfgang Pauli (1900–1958).

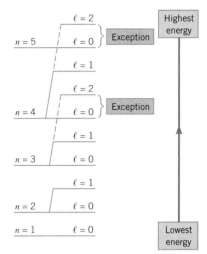

Figure 30.15 When there is more than one electron in an atom, the total energy of a given state depends on the principal quantum number n and the orbital quantum number ℓ. Generally, the energy increases with increasing n and, for a fixed n, with increasing ℓ. There are exceptions to the general rule, however, as indicated here. For clarity, levels for $n = 6$ and higher are not shown.

■ **THE PAULI EXCLUSION PRINCIPLE**

No two electrons in an atom can have the same set of values for the four quantum numbers n, ℓ, m_ℓ, and m_s.

Suppose two electrons in an atom have three quantum numbers that are identical: $n = 3$, $m_\ell = 1$, and $m_s = -\frac{1}{2}$. According to the exclusion principle, it is not possible for each to have $\ell = 2$, for example, since each would then have the same four quantum numbers. Each electron must have a different value for ℓ (for instance, $\ell = 1$ and $\ell = 2$) and, consequently, would be in a different subshell. With the aid of the Pauli exclusion principle, we can determine which energy levels are occupied by the electrons in an atom in its ground state, as the next example demonstrates.

Example 8 Ground States of Atoms

Determine which of the energy levels in Figure 30.15 are occupied by the electrons in the ground state of hydrogen (1 electron), helium (2 electrons), lithium (3 electrons), beryllium (4 electrons), and boron (5 electrons).

Reasoning In the ground state of an atom the electrons are in the lowest available energy levels. Consistent with the Pauli exclusion principle, they fill those levels "from the bottom up" — that is, from the lowest to the highest energy.

Figure 30.16 The electrons (⬤) in the ground state of an atom fill the available energy levels "from the bottom up"—that is, from the lowest to the highest energy, consistent with the Pauli exclusion principle. The ranking of the energy levels in this figure is meant to apply for a given atom only.

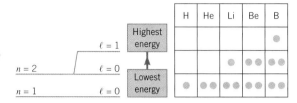

Solution As the colored dot (⬤) in Figure 30.16 indicates, the electron in the hydrogen atom (H) is in the $n = 1$, $\ell = 0$ subshell, which has the lowest possible energy. A second electron is present in the helium atom (He), and both electrons can have the quantum numbers $n = 1$, $\ell = 0$, and $m_\ell = 0$. However, in accord with the Pauli exclusion principle, each electron must have a different spin quantum number, $m_s = +\frac{1}{2}$ for one electron and $m_s = -\frac{1}{2}$ for the other. Thus, the drawing shows both electrons in the lowest energy level.

The third electron that is present in the lithium atom (Li) would violate the exclusion principle if it were also in the $n = 1$, $\ell = 0$ subshell, no matter what the value for m_s is. Thus, the $n = 1$, $\ell = 0$ subshell is filled when occupied by two electrons. With this level filled, the $n = 2$, $\ell = 0$ subshell becomes the next lowest energy level available and is where the third electron of lithium is found (see Figure 30.16). In the beryllium atom (Be), the fourth electron is in the $n = 2$, $\ell = 0$ subshell, along with the third electron. This is possible, since the third and fourth electrons can have different values for m_s.

With the first four electrons in place as just discussed, the fifth electron in the boron atom (B) cannot fit into the $n = 1$, $\ell = 0$ or the $n = 2$, $\ell = 0$ subshell without violating the exclusion principle. Therefore, the fifth electron is found in the $n = 2$, $\ell = 1$ subshell, which is the next available energy level with the lowest energy, as Figure 30.16 indicates. For this electron, m_ℓ can be -1, 0, or $+1$, and m_s can be $+\frac{1}{2}$ or $-\frac{1}{2}$ in each case. However, in the absence of an external magnetic field, all six of these possibilities correspond to the same energy.

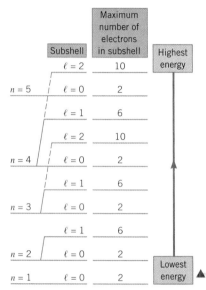

Figure 30.17 The maximum number of electrons that the ℓth subshell can hold is $2(2\ell + 1)$.

Because of the Pauli exclusion principle, there is a maximum number of electrons that can fit into an energy level or subshell. Example 8 shows that the $n = 1$, $\ell = 0$ subshell can hold at most two electrons. The $n = 2$, $\ell = 1$ subshell, however, can hold six electrons because with $\ell = 1$, there are three possibilities for m_ℓ (-1, 0, and $+1$), and for each of these, the value of m_s can be $+\frac{1}{2}$ or $-\frac{1}{2}$. In general, m_ℓ can have the values 0, ± 1, ± 2, . . . , $\pm\ell$, for $2\ell + 1$ possibilities. Since each of these can be combined with two possibilities for m_s, the total number of different combinations for m_ℓ and m_s is $2(2\ell + 1)$. This, then, is the maximum number of electrons the ℓth subshell can hold, as Figure 30.17 summarizes.

For historical reasons, there is a widely used convention in which each subshell of an atom is referred to by a letter rather than by the value of its orbital quantum number ℓ. For instance, an $\ell = 0$ subshell is called an s subshell. An $\ell = 1$ subshell and an $\ell = 2$ subshell are known as p and d subshells, respectively. The higher values of $\ell = 3$, 4, and so on, are referred to as f, g, and so on, in alphabetical sequence, as Table 30.2 indicates.

This convention of letters is used in a shorthand notation that is convenient for indicating simultaneously the principal quantum number n, the orbital quantum number ℓ, and the number of electrons in the n, ℓ subshell. An example of this notation follows:

Table 30.2 The Convention of Letters Used to Refer to the Orbital Quantum Number

Orbital Quantum Number ℓ	Letter
0	s
1	p
2	d
3	f
4	g
5	h

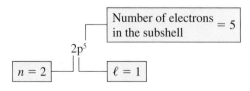

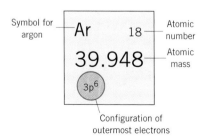

Figure 30.18 The entries in the periodic table of the elements often include the ground-state configuration of the outermost electrons.

With this notation, the arrangement or configuration of the electrons in an atom can be specified efficiently. For instance, in Example 8, we found that the electron configuration for boron has two electrons in the $n = 1$, $\ell = 0$ subshell, two in the $n = 2$, $\ell = 0$ subshell, and one in the $n = 2$, $\ell = 1$ subshell. In shorthand notation this arrangement is expressed as $1s^2 \, 2s^2 \, 2p^1$. Table 30.3 gives the ground-state electron configurations written in

Table 30.3 *Ground-State Electronic Configurations of Atoms*

Element	Number of Electrons	Configuration of the Electrons
Hydrogen (H)	1	$1s^1$
Helium (He)	2	$1s^2$
Lithium (Li)	3	$1s^2\, 2s^1$
Beryllium (Be)	4	$1s^2\, 2s^2$
Boron (B)	5	$1s^2\, 2s^2\, 2p^1$
Carbon (C)	6	$1s^2\, 2s^2\, 2p^2$
Nitrogen (N)	7	$1s^2\, 2s^2\, 2p^3$
Oxygen (O)	8	$1s^2\, 2s^2\, 2p^4$
Fluorine (F)	9	$1s^2\, 2s^2\, 2p^5$
Neon (Ne)	10	$1s^2\, 2s^2\, 2p^6$
Sodium (Na)	11	$1s^2\, 2s^2\, 2p^6\, 3s^1$
Magnesium (Mg)	12	$1s^2\, 2s^2\, 2p^6\, 3s^2$
Aluminum (Al)	13	$1s^2\, 2s^2\, 2p^6\, 3s^2\, 3p^1$

this fashion for elements containing up to thirteen electrons. The first five entries are those worked out in Example 8.

Each entry in the periodic table of the elements often includes the ground-state electronic configuration, as Figure 30.18 illustrates for argon. To save space, only the configuration of the outermost electrons and unfilled subshells is specified, using the shorthand notation just discussed. Originally the periodic table was developed by the Russian chemist Dmitri Mendeleev (1834–1907) on the basis that certain groups of elements exhibit similar chemical properties. There are eight of these groups, plus the transition elements in the middle of the table, including the lanthanide series and the actinide series. The similar chemical properties within a group can be explained on the basis of the configurations of the outer electrons of the elements in the group. Thus, quantum mechanics and the Pauli exclusion principle offer an explanation for the chemical behavior of the atoms. The full periodic table can be found on the inside of the back cover.

 **Check Your Understanding 3**

Using the convention of letters to refer to the orbital quantum number, write down the ground-state configuration of the electrons in krypton (Kr, Z = 36). *(The answer is given at the end of the book.)*

Background: Quantum mechanics and the Pauli exclusion principle determine the electronic configuration of multi-electron atoms.

For similar questions (including calculational counterparts), consult Self-Assessment Test 30.2, which is described at the end of Section 30.9.

30.7 X-Rays

X-rays were discovered by Wilhelm K. Roentgen (1845–1923), a Dutch physicist who performed much of his work in Germany. X-rays can be produced when electrons, accelerated through a large potential difference, collide with a metal target made from molybdenum or platinum, for example. The target is contained within an evacuated glass tube, as Figure 30.19 shows. A plot of X-ray intensity per unit wavelength versus the wavelength looks similar to Figure 30.20 and consists of sharp peaks or lines superimposed on a broad continuous spectrum. The sharp peaks are called characteristic lines or ***characteristic X-rays*** because they are characteristic of the target material. The broad continuous spectrum is referred to as ***Bremsstrahlung*** (German for "braking radiation") and is emitted when the electrons decelerate or "brake" upon hitting the target.

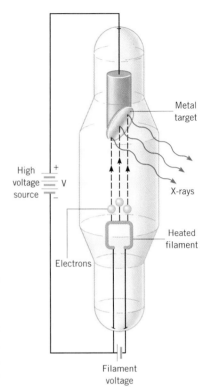

Figure 30.19 In an X-ray tube, electrons are emitted by a heated filament, accelerate through a large potential difference V, and strike a metal target. The X-rays originate when the electrons interact with the target.

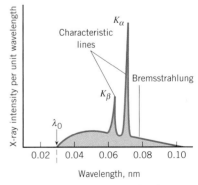

Figure 30.20 When a molybdenum target is bombarded with electrons that have been accelerated from rest through a potential difference of 45 000 V, this X-ray spectrum is produced. The vertical axis is not to scale.

**The physics of
X-rays.**

In Figure 30.20 the characteristic lines are marked K_α and K_β because they involve the $n = 1$ or K shell of a metal atom. If an electron with enough energy strikes the target, one of the K-shell electrons can be knocked out. An electron in one of the outer shells can then fall into the K shell, and an X-ray photon is emitted in the process. Example 9 shows that a large potential difference is needed to operate an X-ray tube, so the electrons impinging on the metal target have sufficient energy to generate the characteristic X-rays.

Example 9   Operating an X-Ray Tube

Strictly speaking, the Bohr model does not apply to multiple-electron atoms, but it can be used to make estimates. Use the Bohr model to estimate the minimum energy that an incoming electron must have to knock a K-shell electron entirely out of an atom in a platinum ($Z = 78$) target in an X-ray tube.

Reasoning According to the Bohr model, the energy of a K-shell electron is given by Equation 30.13 with $n = 1$: $E_n = -(13.6 \text{ eV}) Z^2/n^2$. When striking a platinum target, an incoming electron must have at least enough energy to raise the K-shell electron from this low energy level up to the 0-eV level that corresponds to a very large distance from the nucleus. Only then will the incoming electron knock the K-shell electron entirely out of a target atom.

Problem solving insight
Equation 30.13 for the Bohr energy levels [$E_n = -(13.6 \text{ eV})Z^2/n^2$, $n = 1$] can be used in rough calculations of the energy levels involved in the production of K_α X-rays. In this equation, however, the atomic number Z must be reduced by one, to account approximately for the shielding of one K-shell electron by the other K-shell electron.

Solution The energy of the Bohr $n = 1$ level is

$$E_1 = -(13.6 \text{ eV})\frac{Z^2}{n^2} = -(13.6 \text{ eV})\frac{77^2}{1^2} = -8.1 \times 10^4 \text{ eV}$$

In this calculation we have used 77 rather than 78 for the value of Z. In so doing, we account approximately for the fact that each of the two K-shell electrons applies a repulsive force to the other. This repulsive force tends to balance the attractive force of one nuclear proton. In effect, one electron shields the other from the force of that proton. Therefore, to raise the K-shell electron up to the 0-eV level, the minimum energy for an incoming electron is $\boxed{8.1 \times 10^4 \text{ eV}}$. One electron volt is the kinetic energy acquired when an electron accelerates from rest through a potential difference of one volt. Thus, a potential difference of 81 000 V must be applied to the X-ray tube.

The K_α line in Figure 30.20 arises when an electron in the $n = 2$ level falls into the vacancy that the impinging electron has created in the $n = 1$ level. Similarly, the K_β line arises when an electron in the $n = 3$ level falls to the $n = 1$ level. Example 10 determines an estimate for the K_α wavelength of platinum.

Example 10   The K_α Characteristic X-Ray for Platinum

Need more practice?

Interactive LearningWare 30.2
High-energy electrons impinge on a metal target in an X-ray tube. The target is a mixture of two metals, one of which is cobalt. Both metals emit K_α X-ray photons, and the wavelength that originates from the cobalt is 1.247 times as great as that produced by the unknown metal. What is the unknown metal?

Related Homework: Problem 46

Go to
www.wiley.com/college/cutnell
for an interactive solution.

Use the Bohr model to estimate the wavelength of the K_α line in the X-ray spectrum of platinum ($Z = 78$).

Reasoning This example is very similar to Example 4, which deals with the emission line spectrum of the hydrogen atom. As in that example, we use Equation 30.14, this time with the initial value of n being $n_i = 2$ and the final value being $n_f = 1$. As in Example 9, a value of 77 rather than 78 is used for Z to account approximately for the shielding effect of the single K-shell electron in canceling out the attraction of one nuclear proton.

Solution Using Equation 30.14, we find that

$$\frac{1}{\lambda} = (1.097 \times 10^7 \text{ m}^{-1})(77^2)\left(\frac{1}{1^2} - \frac{1}{2^2}\right) = 4.9 \times 10^{10} \text{ m}^{-1} \quad \text{or}$$

$$\boxed{\lambda = 2.0 \times 10^{-11} \text{ m}}$$

This answer is close to an experimental value of 1.9×10^{-11} m.

Another interesting feature of the X-ray spectrum in Figure 30.20 is the sharp cutoff that occurs at a wavelength of λ_0 on the short-wavelength side of the Bremsstrahlung. This cutoff wavelength is independent of the target material but depends on the energy of the impinging electrons. An impinging electron cannot give up any more than all of its kinetic energy when decelerated by the metal target in an X-ray tube. Thus, at most, an emitted X-ray photon can have an energy equal to the kinetic energy KE of the electron and a frequency given by Equation 29.2 as $f = (KE)/h$, where h is Planck's constant. But the kinetic energy acquired by an electron in accelerating from rest through a potential difference V is eV, according to earlier discussions in Section 19.2; V is the potential difference applied across the X-ray tube. Thus, the maximum photon frequency is $f_0 = (eV)/h$. Since $f_0 = c/\lambda_0$, a maximum frequency corresponds to a minimum wavelength, which is the cutoff wavelength λ_0:

$$\lambda_0 = \frac{hc}{eV} \qquad (30.17)$$

Figure 30.20, for instance, assumes a potential difference of 45 000 V, which corresponds to a cutoff wavelength of

$$\lambda_0 = \frac{(6.63 \times 10^{-34}\ \text{J} \cdot \text{s})(3.00 \times 10^8\ \text{m/s})}{(1.60 \times 10^{-19}\ \text{C})(45\ 000\ \text{V})} = 2.8 \times 10^{-11}\ \text{m}$$

The medical profession began using X-rays for diagnostic purposes almost immediately after their discovery. When a conventional X-ray is obtained, the patient is typically positioned in front of a piece of photographic film, and a single burst of radiation is directed through the patient and onto the film. Since the dense structure of bone absorbs X-rays much more than soft tissue does, a shadow-like picture is recorded on the film. As useful as such pictures are, they have an inherent limitation. The image on the film is a superposition of all the "shadows" that result as the radiation passes through one layer of body material after another. Interpreting which part of a conventional X-ray corresponds to which layer of body material is very difficult.

The technique known as CAT scanning or CT scanning has greatly extended the ability of X-rays to provide images from specific locations within the body. The acronym CAT stands for computerized axial tomography or computer-assisted tomography, and the shorter version CT stands for computerized tomography. In this technique a series of X-ray images are obtained as indicated in Figure 30.21. A number of X-ray beams form a "fanned out" array of radiation and pass simultaneously through the patient. Each of the beams is detected on the other side by a detector, which records the beam intensity. The various intensities are different, depending on the nature of the body material through which the beams have passed. The feature of CAT scanning that leads to dramatic improvements over the conventional technique is that the X-ray source can be rotated to different orientations, so that the "fanned-out" array of beams can be sent through the patient from various directions. Figure 30.21a singles out two directions for illustration. In reality many different orientations are used, and the intensity of each beam in the array is

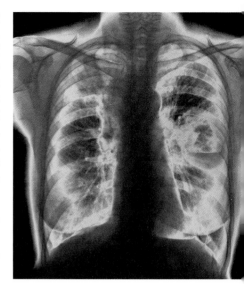

A tumor (red, yellow) of the lung is revealed in this color-enhanced chest X-ray. Microscopic examination of lung tissue is needed to determine whether the tumor is malignant. (© Simon Fraser/Science Photo Library/ Photo Researchers)

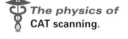

The physics of CAT scanning.

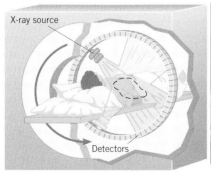

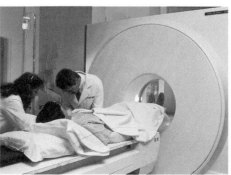

(a)   (b)

Figure 30.21 (a) In CAT scanning, a "fanned-out" array of X-ray beams is sent through the patient from different orientations. (b) A patient being positioned in a CAT scanner. (© Hank Morgan/Photo Researchers)

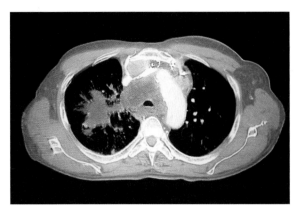

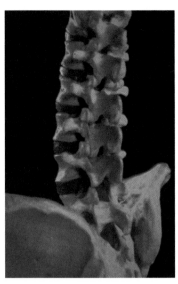

Figure 30.22 (*a*) This color-enhanced two-dimensional CAT scan of the chest shows a view perpendicular to the spine. The heart is in the center of the image, and the lungs are the dark spaces within the yellow region. The irregular pink shape in the lung on the left side of the image is a cancerous growth. (© Scott Camazine/Photo Researchers) (*b*) This color-enhanced three-dimensional CAT scan shows the pelvis and part of the spine, including the intervertebral discs (red). (© Collection CNRI/Phototake)

recorded as a function of orientation. The way in which the intensity of a beam changes from one orientation to another is used as input to a computer. The computer then constructs a highly resolved image of the cross-sectional slice of body material through which the "fan" of radiation has passed. In effect, the CAT scanning technique makes it possible to take an X-ray picture of a cross-sectional "slice" that is perpendicular to the body's long axis. In fact, the word "axial" in the phrase "computerized axial tomography" refers to the body's long axis. Figure 30.22*a* shows a two-dimensional CAT scan of the type discussed here, and part *b* of the figure shows a three-dimensional version.

✔ **Check Your Understanding 4**

Which one of the following statements is true? (a) The K_α wavelength can be smaller than the cutoff wavelength λ_0. (b) The K_α wavelength is produced when an electron undergoes a transition from the $n = 1$ energy level to the $n = 2$ energy level. (c) The K_β wavelength is always smaller than the K_α wavelength for a given metal target. *(The answer is given at the end of the book.)*

Background: X-rays are produced when high-energy electrons collide with a metal target contained within an evacuated glass tube. As a result of the collisions, electrons in the atoms of the target undergo transitions between atomic energy levels.

For similar questions (including calculational counterparts), consult Self-Assessment Test 30.2, which is described at the end of Section 30.9.

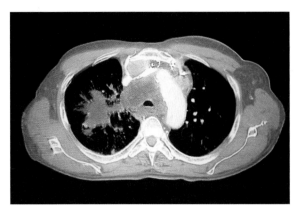

Figure 30.23 (*a*) Spontaneous emission of a photon occurs when the electron (●) makes an unprovoked transition from a higher to a lower energy level, the photon departing in a random direction. (*b*) Stimulated emission of a photon occurs when an incoming photon with the correct energy induces an electron to change energy levels, the emitted photon traveling in the same direction as the incoming photon.

The physics of the laser.

30.8 The Laser

The laser is one of the most useful inventions of the twentieth century. Today, there are many types of lasers, and most of them work in a way that depends directly on the quantum mechanical structure of the atom.

When an electron makes a transition from a higher energy state to a lower energy state, a photon is emitted. The emission process can be one of two types, spontaneous or stimulated. In *spontaneous emission* (see Figure 30.23*a*), the photon is emitted spontaneously, in a random direction, without external provocation. In *stimulated emission* (see Figure 30.23*b*), an incoming photon induces, or stimulates, the electron to change energy levels. To produce stimulated emission, however, the incoming photon must have an energy that exactly matches the difference between the energies of the two levels—namely, $E_i - E_f$. Stimulated emission is similar to a resonant process, in which the incoming photon "jiggles" the electron at just the frequency to which it is particularly sensitive and causes the change between energy levels. This frequency is given by Equation 30.4 as $f = (E_i - E_f)/h$. The operation of lasers depends on stimulated emission.

Stimulated emission has three important features. First, one photon goes in and two photons come out (see Figure 30.23b). In this sense, the process amplifies the number of photons. In fact, this is the origin of the word "laser," which is an acronym for **l**ight **a**mplification by the **s**timulated **e**mission of **r**adiation. Second, the emitted photon travels in the same direction as the incoming photon. Third, the emitted photon is exactly in step with or has the same phase as the incoming photon. In other words, the two electromagnetic waves that these two photons represent are coherent (see Section 17.2) and are locked in step with one another. In contrast, two photons emitted by the filament of an incandescent light bulb are emitted independently. They are not coherent, since one does not stimulate the emission of the other.

Although stimulated emission plays a pivotal role in a laser, other factors are also important. For instance, an external source of energy must be available to excite electrons into higher energy levels. The energy can be provided in a number of ways, including intense flashes of ordinary light and high-voltage discharges. If sufficient energy is delivered to the atoms, more electrons will be excited to a higher energy level than remain in a lower level, a condition known as a ***population inversion.*** Figure 30.24 compares a normal energy level population with a population inversion. The population inversions used in lasers involve a higher energy state that is ***metastable,*** in the sense that electrons remain in it for a much longer period of time than they do in an ordinary excited state (10^{-3} s versus 10^{-8} s, for example). The requirement of a metastable higher energy state is essential, so that there is more time to enhance the population inversion.

Figure 30.25 shows the widely used helium/neon laser. To sustain the necessary population inversion, a high voltage is discharged across a low-pressure mixture of 15% helium and 85% neon contained in a glass tube. The laser process begins when an atom, via spontaneous emission, emits a photon parallel to the axis of the tube. This photon, via stimulated emission, causes another atom to emit two photons parallel to the tube axis. These two photons, in turn, stimulate two more atoms, yielding four photons. Four yield eight, and so on, in a kind of avalanche. To ensure that more and more photons are created by stimulated emission, both ends of the tube are silvered to form mirrors that reflect the photons back and forth through the helium/neon mixture. One end is only partially silvered, however, so that some of the photons can escape from the tube to form the laser beam. When the stimulated emission involves only a single pair of energy levels, the output beam has a single frequency or wavelength and is said to be monochromatic.

A laser beam is also exceptionally narrow. The width is determined by the size of the opening through which the beam exits, and very little spreading-out occurs, except that due to diffraction around the edges of the opening. A laser beam does not spread much because any photons emitted at an angle with respect to the tube axis are quickly reflected out the sides of the tube by the silvered ends (see Figure 30.25). These ends are carefully

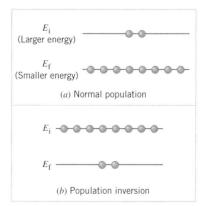

Figure 30.24 (a) In a normal situation at room temperature, most of the electrons in atoms are found in a lower or ground-state energy level. (b) If an external energy source is provided to excite electrons into a higher energy level, a population inversion can be created in which more electrons are in the higher level than in the lower level.

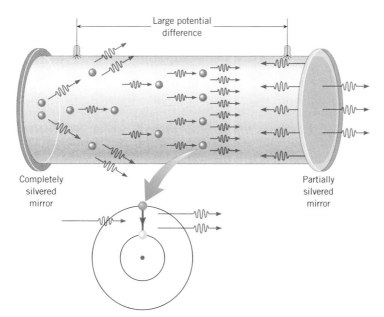

Figure 30.25 A schematic drawing of a helium/neon laser. The blow-up shows the stimulated emission that occurs when an electron in a neon atom is induced to change from a higher- to a lower-energy level.

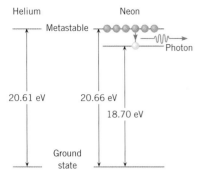

Figure 30.26 These energy levels are involved in the operation of a helium/neon laser.

arranged to be perpendicular to the tube axis. Since all the power in a laser beam can be confined to a narrow region, the intensity, or power per unit area, can be quite large.

Figure 30.26 shows the pertinent energy levels for a helium/neon laser. By coincidence, helium and neon have nearly identical metastable higher energy states, respectively located 20.61 and 20.66 eV above the ground state. The high-voltage discharge across the gaseous mixture excites electrons in helium atoms to the 20.61-eV state. Then, when an excited helium atom collides inelastically with a neon atom, the 20.61 eV of energy is given to an electron in the neon atom, along with 0.05 eV of kinetic energy from the moving atoms. As a result, the electron in the neon atom is raised to the 20.66-eV state. In this fashion, a population inversion is sustained in the neon, relative to an energy level that is 18.70 eV above the ground state. In producing the laser beam, stimulated emission causes electrons in neon to drop from the 20.66-eV level to the 18.70-eV level. The energy change of 1.96 eV corresponds to a wavelength of 633 nm, which is in the red region of the visible spectrum.

The helium/neon laser is not the only kind of laser. There are many different types, including the ruby laser, the argon-ion laser, the carbon dioxide laser, the gallium arsenide solid-state laser, and chemical dye lasers. Depending on the type and whether the laser operates continuously or in pulses, the available beam power ranges from milliwatts to megawatts. Since lasers provide coherent monochromatic electromagnetic radiation that can be confined to an intense narrow beam, they are useful in a wide variety of situations. Today they are used to reproduce music in compact disc players, to weld parts of automobile frames together, to measure distances accurately in surveying, to transmit telephone conversations and other forms of communication over long distances, and to study molecular structure. Many other uses have been found since the laser was invented in 1960, and the next section discusses some of them in the field of medicine.

*30.9 Medical Applications of the Laser

One of the medical areas in which the laser has had a substantial impact is in ophthalmology, which deals with the structure, function, and diseases of the eye. Section 26.10 discusses the human eye and the use of contact lenses and eyeglasses to correct nearsightedness and farsightedness. In these conditions, the eye cannot refract light properly and produces blurred images on the retina.

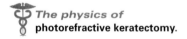

The physics of photorefractive keratectomy.

A laser-based procedure known as photorefractive keratectomy (PRK) offers an alternative treatment for nearsightedness and farsightedness that does not rely on lenses. It involves the use of a laser to remove small amounts of tissue from the cornea of the eye (see Figure 26.34) and thereby change its curvature. As Section 26.10 points out, light enters the eye through the cornea, and it is at the air/cornea boundary that most of the refraction of the light occurs. Therefore, changing the curvature of that boundary can correct deficiencies in the way the eye refracts light, thus causing the image to be focused onto the retina where it belongs. Ideally, the cornea is dome-shaped. If the dome is too steep, however, the rays of light are focused in front of the retina and nearsightedness results. As Figure 30.27a shows, the laser light removes tissue from the center of the cornea, thereby flattening it and increasing the eye's effective focal length. On the other hand, if the shape of the cornea is too flat, light rays would come to a focus behind the retina if they could, and farsightedness occurs. As part b of the drawing illustrates, the center of the cornea is now masked and the laser is used to remove peripheral tissue. This steepens the shape of the cornea, thereby shortening the eye's effective focal length and allowing rays to be focused on the retina.

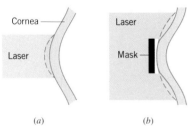

Figure 30.27 (a) To correct for myopia (nearsightedness), the laser vaporizes tissue (dashed line) on the center of the cornea, thereby flattening it. (b) To correct for hyperopia (farsightedness), the laser vaporizes tissue on the peripheral region of the cornea, thereby steepening its contour.

The laser light in the PRK technique is pulsed and comes from an ultraviolet excimer laser that produces a wavelength of 193 nm. The cornea absorbs this wavelength extremely well, so that weak pulses can be used, leading to highly precise and controllable removal of corneal tissue. Typically, 0.1 to 0.5 μm of tissue is removed by each pulse without damaging adjacent layers. During the procedure, eye movement presents a major problem. It can be overcome, however, with the aid of a mechanical device that stabilizes the eye, or, more recently, with the aid of systems capable of tracking eye movements. These movements occur roughly every 15 ms, and the tracking system follows them and retargets the laser accordingly.

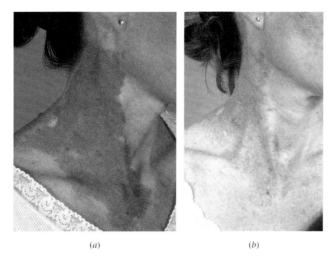

Figure 30.28 A patient with a port-wine stain (*a*) before and (*b*) after treatment using a pulsed dye laser (wavelength = 585 nm). [From Tan, E., and Vinciullo, C., *Pulsed Dye Laser Treatment of Port Wine Stains: A Review of Patients Treated in Western Australia,* The Medical Journal of Australia 164(6):333–336 (1996). Copyright 1996. Reproduced with permission.]

(*a*)   (*b*)

Another medical application of the laser is in the treatment of congenital capillary malformations known as port-wine stains, which affect 0.3% of children at birth. These birthmarks are usually found on the head and neck, as Figure 30.28*a* illustrates. Preferred treatment for port-wine stains now utilizes a pulsed dye laser. Figure 30.28*b* shows an example of an excellent result after irradiation with laser light of wavelength 585 nm, in the form of pulses lasting 0.45 ms and occurring every 3 s. The laser beam was focused to a spot 5 mm in diameter. The light is absorbed by oxyhemoglobin in the malformed capillaries, which are destroyed in the process without damage to adjacent normal tissue. Eventually the destroyed capillaries are replaced by normal blood vessels, which causes the port-wine stain to fade.

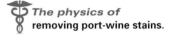 *The physics of* **removing port-wine stains.**

In the treatment of cancer, the laser is being used along with light-activated drugs in photodynamic therapy. The procedure involves administering the drug intravenously, so that the tumor can absorb it from the bloodstream, the advantage being that the drug is then located right near the cancer cells. When the drug is activated by laser light, a chemical reaction ensues that disintegrates the cancer cells and the small blood vessels that feed them. In Figure 30.29 a patient is being treated for cancer of the esophagus. An endoscope that uses optical fibers is inserted down the patient's throat to guide the red laser light to the tumor site and activate the drug. Photodynamic therapy works best with small tumors in their early stages.

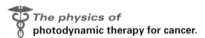 *The physics of* **photodynamic therapy for cancer.**

 Self-Assessment Test 30.2

Test your understanding of the material in Sections 30.6–30.9:

- The Pauli Exclusion Principle and the Periodic Table of the Elements
- X-Rays   • The Laser

Go to **www.wiley.com/college/cutnell**

30.10   Holography

One of the most familiar applications of lasers is in holography, which is a process for producing three-dimensional images. The information used to produce a holographic image is captured on photographic film, which is referred to as a hologram. Figure 30.30 illustrates how a hologram is made. Laser light strikes a half-silvered mirror, or beamsplitter, which reflects part and transmits part of the light. In the drawing, the reflected part is called the *object beam* because it illuminates the object (a chess piece). The transmitted part is called the *reference beam*. The object beam reflects from the chess piece at points such as *A* and *B* and, together with the reference beam, falls on the film. One of the main characteristics of laser light is that it is coherent. Thus, the light from the two beams has a stable phase relationship, like the light from the two slits in Young's double-slit experiment (see Section 27.2). Because of the stable phase relationship and because the two

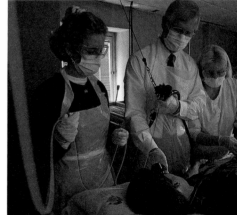

Figure 30.29 Photodynamic therapy to treat cancer of the esophagus is being administered to this patient. Red laser light is routed to the tumor site with an endoscope that incorporates optical fibers. (© Fritz Hoffmann/The Image Works)

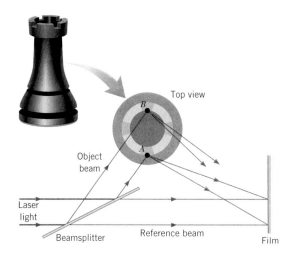

Figure 30.30 An arrangement used to produce a hologram.

The physics of holography.

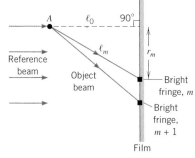

Figure 30.31 This drawing helps to explain how the interference pattern arises on the film when light from point A (see Figure 30.30) and light from the reference beam combine there.

beams travel different distances, an interference pattern is formed on the film. This pattern is the hologram and, although much more complex, is analogous to the pattern of bright and dark fringes formed in the double-slit experiment.

Figure 30.31 shows in greater detail how a holographic interference pattern arises. This drawing considers only the reference beam and the light (wavelength = λ) coming from point A on the chess piece. As we know from Section 27.1, constructive interference between the two light waves leads to a bright fringe; it occurs when the waves, in reaching the film, travel distances that differ by an integer number m of wavelengths. In the drawing, ℓ_m is the distance between point A and the place on the film where the mth-order bright fringe occurs. ℓ_0 is the perpendicular distance that the reference beam would travel from point A to the $m = 0$ bright fringe. In addition, r_m is the distance along the film that locates the bright fringe. In terms of these distances, we know that

$$\ell_m - \ell_0 = m\lambda \qquad \text{(condition for constructive interference)}$$
$$\ell_0^2 + r_m^2 = \ell_m^2 \qquad \text{(Pythagorean theorem)}$$

The first of these equations indicates that $\ell_m = m\lambda + \ell_0$, which can be substituted into the second equation. The result can be rearranged to show that

$$r_m^2 = m\lambda(m\lambda + 2\ell_0)$$

Since ℓ_0 is typically much larger than λ (for instance, $\ell_0 \approx 10^{-1}$ m and $\lambda \approx 10^{-6}$ m), it follows that $r_m \approx \sqrt{m\lambda 2\ell_0}$. In other words, r_m is roughly proportional to $\sqrt{m}$. Therefore, the fringes are farther apart near the top of the film than they are near the bottom. For example, for the $m = 1$ and $m = 2$ fringes, we have $r_2 - r_1 \propto \sqrt{2} - \sqrt{1} = 0.41$, whereas for the $m = 2$ and $m = 3$ fringes, we have $r_3 - r_2 \propto \sqrt{3} - \sqrt{2} = 0.32$.

In addition to the fringe pattern just discussed, the total interference pattern on the hologram includes interference effects that are related to light coming from point B and other locations on the object in Figure 30.30. The total pattern is very complicated. Nevertheless, the fringe pattern for point A alone is sufficient to illustrate the fact that a hologram can be used to produce both a virtual image and a real image of the object, as we will now see.

To produce the holographic images, the laser light is directed through the interference pattern on the film, as in Figure 30.32. The pattern can be thought of as a kind of diffraction grating, with the bright fringes analogous to the spaces between the slits of the grating. Section 27.7 discusses how light passing through a grating is split into higher-order bright fringes that are symmetrically located on either side of a central bright fringe. Figure 30.32 shows the three rays corresponding to the central and first-order bright fringes, as they originate from a spot near the top and a spot near the bottom of the film. The angle θ, which locates the first-order fringes relative to the central fringe, is given by Equation 27.7 as $\sin \theta = \lambda/d$, where d is the separation between the slits of the grating. When the slit separation is greater, as it is near the top of the film, the angle is smaller. When the slit separation is smaller, as it is near the bottom, the angle is larger.

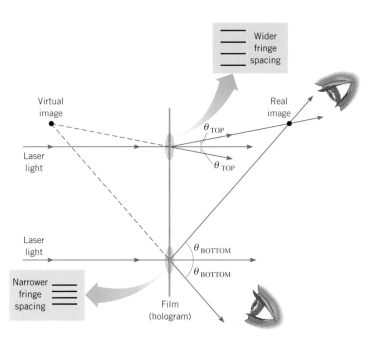

Figure 30.32 When the laser light used to produce a hologram is shone through it, both a real and a virtual image of the object are produced.

Thus, Figure 30.32 has been drawn with θ_{TOP} smaller than θ_{BOTTOM}. Of the three rays emerging from the film at the top and the three at the bottom, we use the uppermost one in each case to locate the real image of point A on the chess piece. The real image is located where these two rays intersect, when extended to the right. To locate the virtual image, we use the lower ray in each of the three-ray bundles at the top and bottom of the film. When projected to the left, they appear to be originating from the spot where the projections intersect—that is, from the virtual image.

A holographic image differs greatly from a photographic image. The most obvious difference is that a hologram provides a three-dimensional image, whereas photographs are two-dimensional. The reason that holographic images are three-dimensional is inherent in the interference pattern formed on the film. In Figure 30.30 part of this pattern arises because the light emitted by point A travels different distances in reaching different spots on the film than does the light in the reference beam. The same can be said about the light emitted from point B and other places as well. As a result, the total interference pattern contains information about how much farther from the film the various points on the object are, and because of this information holographic images are three-dimensional. Furthermore, as Figure 30.33 illustrates, it is possible to "walk around" a holographic image and view it from different angles, as you would the original object.

(a)

(b)

Figure 30.33 A hologram of a galleon spaceship approaching earth. Two views of the same hologram are shown: (a) looking at the hologram perpendicularly and (b) looking at an angle to the right of the perpendicular. (© Holographics North, Inc., and Sprint. Reproduced with permission of Sprint Corporation.)

A vast difference between the methods used to produce holograms and photographs exists. As Section 26.7 discusses, a camera uses a converging lens to produce a photograph. The lens focuses the light rays originating from a point on the object to a corresponding point on the film. To produce a hologram, lenses are not used in this way, and a point on the object *does not correspond* to a single point on the film. In Figure 30.30, light from point A diverges on its way to the film, and there is no lens to make it converge to a single corresponding point. The light falls over the entire exposed region of the film and contributes everywhere to the interference pattern that is formed. Thus, a hologram may be cut into smaller pieces, and each piece will contain some information about the light originating from point A. For this reason, each smaller piece can be used to produce a three-dimensional image of the object. In contrast, it is not possible to reconstruct the entire image in a photograph from only a small piece of the original film.

The holograms discussed here are typically viewed with the aid of the laser light used to produce them. There are also other kinds of holograms. Credit cards, for example, use holograms for identification purposes. This kind of hologram is called a rainbow hologram and is designed to be viewed in white light that is reflected from it. Other applications of holography include head-up displays for instrument panels in high-performance fighter planes, laser scanners at checkout counters, computerized data storage and retrieval systems, and methods for high-precision biomedical measurements.

30.11 Concepts & Calculations

The Bohr model of the hydrogen atom introduces a number of important features that characterize the quantum picture of atomic structure. Among them are the concepts of quantized energy levels and the photon emission that occurs when an electron makes a transition from a higher- to a lower-energy state. Example 11 deals with these ideas.

Concepts & Calculations Example 11
Hydrogen Atom Energy Levels

▼

A hydrogen atom ($Z = 1$) is in the third excited state. It makes a transition to a different state, and a photon is either absorbed or emitted. Determine the quantum number n_f of the final state and the energy of the photon when the photon is (a) emitted with the shortest possible wavelength, (b) emitted with the longest possible wavelength, and (c) absorbed with the longest possible wavelength.

Concept Questions and Answers What is the quantum number of the third excited state?

Answer The lowest-energy state is the ground state, and its quantum number is $n = 1$. The first excited state corresponds to $n = 2$, the second excited state to $n = 3$, and the third excited state to $n = 4$.

When an atom emits a photon, is the final quantum number n_f of the atom greater than or less than the initial quantum number n_i?

Answer Since the photon carries energy away, the final energy of the atom is less than its initial energy. Lower energies correspond to lower quantum numbers (see Figure 30.9). Therefore, the final quantum number is less than the initial quantum number.

When an atom absorbs a photon, is the final quantum number n_f of the atom greater than or less than the initial quantum number n_i?

Answer When an atom absorbs a photon, the atom gains the energy of the photon, so the final energy of the atom is greater than its initial energy. Greater energies correspond to higher quantum numbers, so the final quantum number is greater than the initial quantum number.

How is the wavelength of a photon related to its energy?

Answer The energy E of a photon is given by Equation 29.2 as $E = hf$, where h is Planck's constant and f is the photon's frequency. But the frequency and wavelength λ are related by $f = c/\lambda$, according to Equation 16.1, where c is the speed of light. Combining these relations gives $E = hc/\lambda$, or $\lambda = hc/E$. Thus, we see that the wavelength is inversely proportional to the energy.

Solution

(a) When a photon is emitted with the shortest possible wavelength, it has the largest possible energy, since the wavelength is inversely proportional to the energy. The largest possible energy arises when the electron jumps from the initial state ($n_i = 4$) to the ground state ($n_f = 1$), as shown by transition A in Figure 30.34. Therefore, the quantum number of the final state is $\boxed{n_f = 1}$. The energy E of the photon is the difference between the energies of the two states, so $E = E_4 - E_1$. The energy of the nth state is given by Equation 30.13 as $E_n = -(13.6 \text{ eV}) Z^2/n^2$, so the energy of the photon is

$$E = (-13.6 \text{ eV})(1)^2 \left(\frac{1}{4^2} - \frac{1}{1^2} \right) = \boxed{12.8 \text{ eV}}$$

(b) When a photon is emitted with the longest possible wavelength, it has the smallest possible energy. The smallest possible energy arises when the electron jumps from the initial state ($n_i = 4$) to the next lower state ($n_f = 3$), as shown by transition B in Figure 30.34. Therefore, the quantum number of the final state is $\boxed{n_f = 3}$. The energy E of the photon is the difference between the energies of the two states, so $E = E_4 - E_3$:

$$E = (-13.6 \text{ eV})(1)^2 \left(\frac{1}{4^2} - \frac{1}{3^2} \right) = \boxed{0.661 \text{ eV}}$$

(c) When a photon is absorbed by the hydrogen atom, the electron jumps to a higher energy state. The photon has the longest possible wavelength when its energy is the smallest. The smallest possible energy change in the hydrogen atom arises when the electron jumps from the initial state ($n_i = 4$) to the next higher state ($n_f = 5$), as shown by transition C in the drawing. Therefore, the quantum number of the final state is $\boxed{n_f = 5}$. The energy E of the photon is the difference between the energies of the two states, so $E = E_5 - E_4$:

$$E = (-13.6 \text{ eV})(1)^2 \left(\frac{1}{5^2} - \frac{1}{4^2} \right) = \boxed{0.306 \text{ eV}}$$

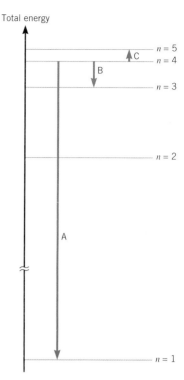

Figure 30.34 A photon is emitted when an electron in the $n = 4$ state jumps to either the $n = 1$ or the $n = 3$ state. A photon is absorbed when an electron makes a transition from the $n = 4$ state to the $n = 5$ state.

The next example reviews the physics of how a K_α X-ray is produced, how its energy is related to the ionization energies of the target atom, and how to determine the minimum voltage needed to produce it in an X-ray tube.

Concepts & Calculations Example 12
The Production of K_α X-Rays

The K-shell and L-shell ionization energies of a metal are 8979 eV and 951 eV, respectively. (a) Assuming that there is a vacancy in the L shell, what must be the minimum voltage across an X-ray tube with a target made from this metal to produce K_α X-ray photons? (b) Determine the wavelength of a K_α photon.

Concept Questions and Answers How is the K_α photon produced and how much energy does it have?

Answer As transition A in Figure 30.35 shows, a K_α photon is produced when an electron in a metal atom jumps from the higher-energy $n = 2$ state to the lower-energy $n = 1$ state. The energy E of the photon is the difference between the energies of these two states: $E = E_2 - E_1$.

What must be the minimum voltage across the X-ray tube to produce a K_α photon?

Answer It takes energy to produce a K_α photon, and this energy comes from the electrons striking the target in the X-ray tube. According to Equation 19.3, the energy possessed by each electron when it arrives at the target is eV, where e is the magnitude of the electron's charge and V is the potential difference across the tube. To cause the target to emit a K_α X-ray photon, the impinging electron must have enough energy to create a vacancy in the K shell. The vacancy is created because the impinging electron provides the energy to move a K-shell electron into a higher energy level or to remove it entirely from the atom. In the present case, with a vacancy in the L shell, the impinging electron can create the K-shell vacancy if it has the energy needed to elevate a K-shell electron to the L shell, the energy needed being $E_2 - E_1$ (see Figure 30.35). This, then, is the minimum energy that the impinging electron must have. Correspondingly, the minimum voltage across the tube must be such that $eV_{min} = E_2 - E_1$, or $V_{min} = (E_2 - E_1)/e$.

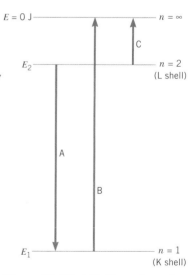

Figure 30.35 A K_α X-ray photon is emitted when an electron jumps from the $n = 2$ state to the $n = 1$ state (transition A). The K-shell ionization energy is the energy required to raise an electron from the $n = 1$ to the $n = \infty$ state, where it has zero total energy (transition B). Similarly, the L-shell ionization energy is the energy needed for the transition from $n = 2$ to $n = \infty$ (transition C). For clarity, the energy levels are not drawn to scale.

What is meant by the phrases "K-shell ionization energy" and "L-shell ionization energy?"

Answer   The K-shell ionization energy is the energy needed to remove an electron in the K shell ($n_i = 1$) completely from the atom (see transition B in Figure 30.35). The removed electron is assumed to have no kinetic energy and is infinitely far away ($n_f = \infty$) so that it has no electric potential energy. Thus, the total energy of the removed electron is zero. Similarly, the L-shell ionization energy is the energy needed to remove an electron in the L shell ($n_i = 2$) completely from the atom (see transition C in Figure 30.35).

What does the difference between the K-shell and L-shell ionization energies represent?

Answer   As transitions B and C in Figure 30.35 suggest, the difference between the two ionization energies is equal to the difference between the total energy (not the ionization energy) E_2 of the L shell and the total energy E_1 of the K shell.

Solution

(a) The minimum voltage across the X-ray tube is $V_{min} = (E_2 - E_1)/e$. But the difference $E_2 - E_1$ in energies between the $n = 2$ and $n = 1$ states is equal to the difference in energies between the K-shell and L-shell ionization energies, or 8979 eV $-$ 951 eV $=$ 8028 eV. Thus, the minimum voltage is

$$V_{min} = \frac{E_2 - E_1}{e} = \frac{(8028 \text{ eV})\left(\dfrac{1.602 \times 10^{-19} \text{ J}}{1 \text{ eV}}\right)}{1.602 \times 10^{-19} \text{ C}} = \boxed{8028 \text{ V}}$$

(b) The wavelength of the K_α photon is $\lambda = c/f$, where f is its frequency. The frequency is related to the energy $E_2 - E_1$ of the photon by Equation 29.2, $f = (E_2 - E_1)/h$, where h is Planck's constant. Combining these relations gives $\lambda = hc/(E_2 - E_1)$, so the wavelength of the emitted photon is

$$\lambda = \frac{hc}{E_2 - E_1} = \frac{(6.626 \times 10^{-34} \text{ J} \cdot \text{s})(2.998 \times 10^8 \text{ m/s})}{(8028 \text{ eV})\left(\dfrac{1.602 \times 10^{-19} \text{ J}}{1 \text{ eV}}\right)} = \boxed{1.545 \times 10^{-10} \text{ m}}$$

At the end of the problem set for this chapter, you will find homework problems that contain both conceptual and quantitative parts. These problems are grouped under the heading *Concepts & Calculations, Group Learning Problems*. They are designed for use by students working alone or in small learning groups. The conceptual part of each problem provides a convenient focus for group discussions.

Concept Summary

This summary presents an abridged version of the chapter, including the important equations and all available learning aids. For convenient reference, the learning aids (including the text's examples) are placed next to or immediately after the relevant equation or discussion. The following learning aids may be found on-line at **www.wiley.com/college/cutnell**:

Interactive LearningWare examples are solved according to a five-step interactive format that is designed to help you develop problem-solving skills.	**Concept Simulations** are animated versions of text figures or animations that illustrate important concepts. You can control parameters that affect the display, and we encourage you to experiment.
Interactive Solutions offer specific models for certain types of problems in the chapter homework. The calculations are carried out interactively.	**Self-Assessment Tests** include both qualitative and quantitative questions. Extensive feedback is provided for both incorrect and correct answers, to help you evaluate your understanding of the material.

Topic	*Discussion*	*Learning Aids*

30.1 Rutherford Scattering and the Nuclear Atom

The idea of a nuclear atom originated in 1911 as a result of experiments by Ernest Rutherford in which α particles were scattered by a thin metal foil. The

Nuclear atom   phrase "nuclear atom" refers to the fact that an atom consists of a small, positively charged nucleus surrounded at relatively large distances by a number of   **Example 1**

Topic	Discussion	Learning Aids
	electrons, whose total negative charge equals the positive nuclear charge when the atom is electrically neutral.	

30.2 Line Spectra

Line spectrum	A line spectrum is a series of discrete electromagnetic wavelengths emitted by the atoms of a low-pressure gas that is subjected to a sufficiently high potential difference. Certain groups of discrete wavelengths are referred to as "series." The following equations can be used to determine the wavelengths in three of the series that are found in the line spectrum of atomic hydrogen:	
Lyman series	*Lyman series* $\dfrac{1}{\lambda} = R\left(\dfrac{1}{1^2} - \dfrac{1}{n^2}\right)$ $n = 2, 3, 4, \ldots$ (30.1)	
Balmer series	*Balmer series* $\dfrac{1}{\lambda} = R\left(\dfrac{1}{2^2} - \dfrac{1}{n^2}\right)$ $n = 3, 4, 5, \ldots$ (30.2)	**Example 2**
Paschen series	*Paschen series* $\dfrac{1}{\lambda} = R\left(\dfrac{1}{3^2} - \dfrac{1}{n^2}\right)$ $n = 4, 5, 6, \ldots$ (30.3)	
	The constant term R is called the Rydberg constant and has the value $R = 1.097 \times 10^7 \text{ m}^{-1}$.	

30.3 The Bohr Model of the Hydrogen Atom

Stationary orbit	The Bohr model applies to atoms or ions that have only a single electron orbiting a nucleus containing Z protons. This model assumes that the electron exists in circular orbits that are called stationary orbits because the electron does not radiate electromagnetic waves while in them. According to this model, a photon is emitted only when an electron changes from an orbit with a higher energy E_i to an orbit with a lower energy E_f. The orbital energies and the photon frequency f are related according to	
Photon frequency	$$E_i - E_f = hf \qquad (30.4)$$	
	where h is Planck's constant. The model also assumes that the orbital angular momentum L_n of the electron can only have the following discrete values:	
Angular momentum	$$L_n = n\frac{h}{2\pi} \qquad n = 1, 2, 3, \ldots \qquad (30.8)$$	
	With these assumptions, it can be shown that the nth Bohr orbit has a radius r_n and is associated with a total energy E_n of	
Orbital radius	$$r_n = (5.29 \times 10^{-11} \text{ m})\frac{n^2}{Z} \qquad n = 1, 2, 3, \ldots \qquad (30.10)$$	
Total orbital energy	$$E_n = -(13.6 \text{ eV})\frac{Z^2}{n^2} \qquad n = 1, 2, 3, \ldots \qquad (30.13)$$	**Example 11**
Ionization energy	The ionization energy is the energy needed to remove an electron completely from an atom. The Bohr model predicts that the wavelengths comprising the line spectrum emitted by a hydrogen atom can be calculated from	**Example 3**
Line spectrum wavelengths	$$\frac{1}{\lambda} = RZ^2\left(\frac{1}{n_f^2} - \frac{1}{n_i^2}\right) \qquad (30.14)$$ $$n_i, n_f = 1, 2, 3, \ldots \quad \text{and} \quad n_i > n_f$$	**Interactive LearningWare 30.1** **Example 4** Concept Simulation 30.1

30.4 De Broglie's Explanation of Bohr's Assumption about Angular Momentum

Louis de Broglie proposed that the electron in a circular Bohr orbit should be considered as a particle wave and that standing particle waves around the orbit offer an explanation of the angular momentum assumption in the Bohr model.

30.5 The Quantum Mechanical Picture of the Hydrogen Atom

Quantum mechanics describes the hydrogen atom in terms of the following four quantum numbers:

Principal quantum number	(1) the principal quantum number n, which can have the integer values $n = 1, 2, 3, \ldots$	**Examples 5, 6**

Topic	Discussion	Learning Aids
Orbital quantum number	(2) the orbital quantum number ℓ, which can have the integer values $\ell = 0$, 1, 2, . . . , $(n-1)$	
Magnetic quantum number	(3) the magnetic quantum number m_ℓ, which can have the positive and negative values $m_\ell = -\ell, . . . , -2, -1, 0, +1, +2, . . . , +\ell$	**Interactive Solution 30.51**
Spin quantum number	(4) the spin quantum number m_s, which, for an electron, can be either $m_s = +\frac{1}{2}$ or $m_s = -\frac{1}{2}$	
	According to quantum mechanics, an electron does not reside in a circular orbit but, rather, has some probability of being found at various distances from the nucleus.	**Example 7**

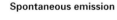 **Use *Self-Assessment Test 30.1* to evaluate your understanding of Sections 30.1–30.5.**

30.6 The Pauli Exclusion Principle and the Periodic Table of the Elements

Pauli exclusion principle	The Pauli exclusion principle states that no two electrons in an atom can have the same set of values for the four quantum numbers n, ℓ, m_ℓ, and m_s. This principle determines the way in which the electrons in multiple-electron atoms are distributed into shells (defined by the value of n) and subshells (defined by the value of ℓ).	**Example 8**
Convention of letters	The following notation is used to refer to the orbital quantum numbers: s denotes $\ell = 0$, p denotes $\ell = 1$, d denotes $\ell = 2$, f denotes $\ell = 3$, g denotes $\ell = 4$, h denotes $\ell = 5$, and so on.	
	The arrangement of the periodic table of the elements is related to the Pauli exclusion principle.	

30.7 X-Rays

X-rays	X-rays are electromagnetic waves emitted when high-energy electrons strike a metal target contained within an evacuated glass tube. The emitted X-ray spectrum of wavelengths consists of sharp "peaks" or "lines", called characteristic X-rays, superimposed on a broad continuous range of wavelengths called Bremsstrahlung. The K_α characteristic X-ray is emitted when an electron in the $n = 2$ level (L shell) drops into a vacancy in the $n = 1$ level (K shell). Similarly, the K_β characteristic X-ray is emitted when an electron in the $n = 3$ level (M shell) drops into a vacancy in the $n = 1$ level (K shell). The minimum wavelength, or cutoff wavelength λ_0, of the Bremsstrahlung is determined by the kinetic energy of the electrons striking the target in the X-ray tube, according to	**Example 9**
Characteristic X-rays		
Bremsstrahlung		**Examples 10, 12**
		Interactive LearningWare 30.2
Cutoff wavelength		**Interactive Solution 30.35**

$$\lambda_0 = \frac{hc}{eV} \qquad (30.17)$$

where h is Planck's constant, c is the speed of light in a vacuum, e is the magnitude of the charge on an electron, and V is the potential difference applied across the X-ray tube.

30.8 The Laser
30.9 Medical Applications of the Laser

Stimulated emission	A laser is a device that generates electromagnetic waves via a process known as stimulated emission. In this process, one photon stimulates the production of another photon by causing an electron in an atom to fall from a higher-energy level to a lower-energy level. The emitted photon travels in the same direction as the photon causing the stimulation. Because of this mechanism of photon production, the electromagnetic waves generated by a laser are coherent and may be confined to a very narrow beam. Stimulated emission contrasts with the process known as spontaneous emission, in which an electron in an atom also falls from a higher- to a lower-energy level, but does so spontaneously, in a random direction, without any external provocation.	
Spontaneous emission		

 Use *Self-Assessment Test 30.2* to evaluate your understanding of Sections 30.6–30.9.

Conceptual Questions

1. At room temperature, most of the atoms of atomic hydrogen contain electrons that are in the ground state or $n = 1$ energy level. A tube is filled with atomic hydrogen. Electromagnetic radiation with a continuous spectrum of wavelengths, including those in the Lyman, Balmer, and Paschen series, enters one end of this tube and leaves the other end. The exiting radiation is found to contain absorption lines. To which one (or more) of the series do the wavelengths of these absorption lines correspond? Assume that once an electron absorbs a photon and jumps to a higher energy level, it does not absorb yet another photon and jump to an even higher energy level. Explain your answer.

2. Refer to Question 1. Suppose the electrons in the atoms are mostly in excited states. Do you expect there to be a greater or lesser number of absorption lines in the exiting radiation, compared to the situation when the electrons are in the ground state? Account for your answer.

3. When the outermost electron in an atom is in an excited state, the atom is more easily ionized than when the outermost electron is in the ground state. Why?

4. In the Bohr model for the hydrogen atom, the closer the electron is to the nucleus, the smaller is the total energy of the atom. Is this also true in the quantum mechanical picture of the hydrogen atom? Justify your answer.

5. Conceptual Example 6 provides background for this question. Consider two different hydrogen atoms. The electron in each atom is in an excited state. Is it possible for the electrons to have different energies but the same orbital angular momentum L, according to (a) the Bohr model and (b) quantum mechanics? In each case, give a reason for your answer.

6. In the quantum mechanical picture of the hydrogen atom, the orbital angular momentum of the electron *may* be zero in any of the possible energy states. For which energy state *must* the orbital angular momentum be zero? Give your reasoning.

7. Can a 5g subshell contain (a) 22 electrons and (b) 17 electrons? Why?

8. An electronic configuration for manganese ($Z = 25$) is written as $1s^2\ 2s^2\ 2p^6\ 3s^2\ 3p^6\ 4s^2\ 3d^4\ 4p^1$. Does this configuration of the electrons represent the ground state or an excited state? Consult Figure 30.17 for the order in which the subshells fill, and account for your answer.

9. The drawing shows the X-ray spectra produced by an X-ray tube when the tube is operated at two different potential differences. Explain why the characteristic lines occur at the same wavelengths in the two spectra, whereas the cutoff wavelength λ_0 shifts to the right when a lower voltage is used to operate the tube.

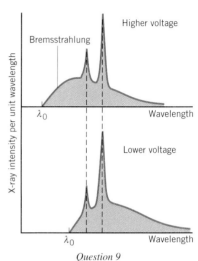

Question 9

10. In the production of X-rays, it is possible to create Bremsstrahlung X-rays without producing the characteristic X-rays. Explain how this can be accomplished by adjusting the electric potential difference used to operate the X-ray tube.

11. The short wavelength side of X-ray spectra ends abruptly at a cutoff wavelength λ_0 (see Figure 30.20). Does this cutoff wavelength depend on the target material used in the X-ray tube? Give your reasoning.

12. Explain why a laser beam focused to a small spot can cut through a piece of metal.

13. Suppose that a laser produces green light instead of the red light produced by a helium/neon laser. Are the photons emitted by this laser more energetic, less energetic, or of equal energy as compared to those emitted by the helium/neon laser? Justify your answer.

Problems

In working these problems, ignore relativistic effects.

ssm Solution is in the Student Solutions Manual.   **www** Solution is available on the World Wide Web at www.wiley.com/college/cutnell

⚕ This icon represents a biomedical application.

Section 30.1 Rutherford Scattering and the Nuclear Atom

1. ssm The nucleus of the hydrogen atom has a radius of about 1×10^{-15} m. The electron is normally at a distance of about 5.3×10^{-11} m from the nucleus. Assuming the hydrogen atom is a sphere with a radius of 5.3×10^{-11} m, find (a) the volume of the atom, (b) the volume of the nucleus, and (c) the percentage of the volume of the atom that is occupied by the nucleus.

2. Review Conceptual Example 1 and use the information therein as an aid in working this problem. Suppose you're building a scale model of the hydrogen atom, and the nucleus is represented by a ball of radius 3.2 cm (somewhat smaller than a baseball). How many miles away (1 mi = 1.61×10^5 cm) should the electron be placed?

3. The nucleus of a hydrogen atom is a single proton, which has a radius of about 1.0×10^{-15} m. The single electron in a hydrogen atom normally orbits the nucleus at a distance of 5.3×10^{-11} m. What is the ratio of the density of the hydrogen nucleus to the density of the complete hydrogen atom?

4. In a Rutherford scattering experiment a target nucleus has a diameter of 1.4×10^{-14} m. The incoming α particle has a mass of 6.64×10^{-27} kg. What is the kinetic energy of an α particle that has a de Broglie wavelength equal to the diameter of the target nucleus? Ignore relativistic effects.

* **5. ssm** The nucleus of a copper atom contains 29 protons and has a radius of 4.8×10^{-15} m. How much work (in electron volts) is done

by the electric force as a proton is brought from infinity, where it is at rest, to the "surface" of a copper nucleus?

* **6.** There are Z protons in the nucleus of an atom, where Z is the atomic number of the element. An α particle carries a charge of $+2e$. In a scattering experiment, an α particle, heading directly toward a nucleus in a metal foil, will come to a halt when all the particle's kinetic energy is converted to electric potential energy. In such a situation, how close will an α particle with a kinetic energy of 5.0×10^{-13} J come to a gold nucleus ($Z = 79$)?

Section 30.2 Line Spectra
Section 30.3 The Bohr Model of the Hydrogen Atom

7. ssm It is possible to use electromagnetic radiation to ionize atoms. To do so, the atoms must absorb the radiation, the photons of which must have enough energy to remove an electron from an atom. What is the longest radiation wavelength that can be used to ionize the ground-state hydrogen atom?

8. Concept Simulation 30.1 at **www.wiley.com/college/cutnell** reviews the concepts on which the solution to this problem depends. The electron in a hydrogen atom is in the first excited state, when the electron acquires an additional 2.86 eV of energy. What is the quantum number n of the state into which the electron moves?

9. In the line spectrum of atomic hydrogen there is also a group of lines known as the Pfund series. These lines are produced when electrons, excited to high energy levels, make transitions to the $n = 5$ level. Determine (a) the longest wavelength and (b) the shortest wavelength in this series. (c) Refer to Figure 24.10 and identify the region of the electromagnetic spectrum in which these lines are found.

10. A singly ionized helium atom (He^+) has only one electron in orbit about the nucleus. What is the radius of the ion when it is in the second excited state?

11. ssm For a doubly ionized lithium atom Li^{2+} ($Z = 3$), what is the principal quantum number of the state in which the electron has the same total energy as a ground-state electron has in the hydrogen atom?

12. In the hydrogen atom the radius of orbit B is sixteen times greater than the radius of orbit A. The total energy of the electron in orbit A is -3.40 eV. What is the total energy of the electron in orbit B?

13. Consider the Bohr energy expression (Equation 30.13) as it applies to singly ionized helium He^+ ($Z = 2$) and doubly ionized lithium Li^{2+} ($Z = 3$). This expression predicts equal electron energies for these two species for certain values of the quantum number n (the quantum number is different for each species). For quantum numbers less than or equal to 9, what are the lowest three energies (in electron volts) for which the helium energy level is equal to the lithium energy level?

14. A hydrogen atom is in the ground state. It absorbs energy and makes a transition to the $n = 3$ excited state. The atom returns to the ground state by emitting two photons. What are their wavelengths?

* **15. ssm** A wavelength of 410.2 nm is emitted by the hydrogen atoms in a high-voltage discharge tube. What are the initial and final values of the quantum number n for the energy level transition that produces this wavelength?

* **16.** Doubly ionized lithium Li^{2+} ($Z = 3$) and triply ionized beryllium Be^{3+} ($Z = 4$) each emit a line spectrum. For a certain series of lines in the lithium spectrum, the shortest wavelength is 40.5 nm. For the same series of lines in the beryllium spectrum, what is the shortest wavelength?

* **17. ssm** The Bohr model can be applied to singly ionized helium He^+ ($Z = 2$). Using this model, consider the series of lines that is produced when the electron makes a transition from higher energy levels into the $n_f = 4$ level. Some of the lines in this series lie in the visible region of the spectrum (380–750 nm). What are the values of n_i for the energy levels from which the electron makes the transitions corresponding to these lines?

* **18. Interactive LearningWare 30.1** at **www.wiley.com/college/cutnell** reviews the concepts that play roles in this problem. A hydrogen atom emits a photon that has momentum with a magnitude of 5.452×10^{-27} kg·m/s. This photon is emitted because the electron in the atom falls from a higher energy level into the $n = 1$ level. What is the quantum number of the level from which the electron falls? Use a value of 6.626×10^{-34} J·s for Planck's constant.

** **19.** A diffraction grating is used in the first order to separate the wavelengths in the Balmer series of atomic hydrogen. (Section 27.7 discusses diffraction gratings.) The grating and an observation screen (see Figure 27.32) are separated by a distance of 81.0 cm. You may assume that θ is small, so $\sin \theta \approx \theta$ when radian measure is used for θ. How many lines per centimeter should the grating have, so that the longest and the next-to-the-longest wavelengths in the series will be separated by 3.00 cm on the screen?

** **20.** (a) Derive an expression for the speed of the electron in the nth Bohr orbit, in terms of Z, n, and the constants k, e, and h. For the hydrogen atom, determine the speed in (b) the $n = 1$ orbit and (c) the $n = 2$ orbit. (d) Generally, when speeds are less than one-tenth the speed of light, the effects of special relativity can be ignored. Are the speeds found in (b) and (c) consistent with ignoring relativistic effects in the Bohr model?

Section 30.5 The Quantum Mechanical Picture of the Hydrogen Atom

21. ssm www The principal quantum number for an electron in an atom is $n = 6$, and the magnetic quantum number is $m_\ell = 2$. What possible values for the orbital quantum number ℓ could this electron have?

22. A hydrogen atom is in its second excited state. Determine, according to quantum mechanics, (a) the total energy (in eV) of the atom, (b) the magnitude of the maximum angular momentum the electron can have in this state, and (c) the maximum value that the z component L_z of the angular momentum can have.

23. The orbital quantum number for the electron in a hydrogen atom is $\ell = 5$. What is the smallest possible value (algebraically) for the total energy of this electron? Give your answer in electron volts.

24. It is known that the possible values for the magnetic quantum number m_ℓ are -4, -3, -2, -1, 0, $+1$, $+2$, $+3$, and $+4$. Determine the orbital quantum number and the smallest possible value of the principal quantum number.

* **25.** The total orbital angular momentum of the electron in a hydrogen atom has a magnitude of $L = 3.66 \times 10^{-34}$ J·s. In the quantum mechanical picture of the atom, what values can the angular momentum component L_z have?

* **26.** Review Conceptual Example 6 as background for this problem. For the hydrogen atom, the Bohr model and quantum mechanics both give the same value for the energy of the nth state. However, they do not give the same value for the orbital angular momentum L. (a) For $n = 1$, determine the values of L [in units of $h/(2\pi)$] predicted by the Bohr model and quantum mechanics. (b) Repeat part (a) for $n = 3$, noting that quantum mechanics permits more than one value of ℓ when the electron is in the $n = 3$ state.

** **27.** ssm www An electron is in the $n = 5$ state. What is the smallest possible value for the angle between the z component of the orbital angular momentum and the orbital angular momentum?

Section 30.6 The Pauli Exclusion Principle and the Periodic Table of the Elements

28. Two of the three electrons in a lithium atom have quantum numbers of $n = 1$, $\ell = 0$, $m_\ell = 0$, $m_s = +\frac{1}{2}$ and $n = 1$, $\ell = 0$, $m_\ell = 0$, $m_s = -\frac{1}{2}$. What quantum numbers can the third electron have if the atom is in (a) its ground state and (b) its first excited state?

29. In the style indicated in Table 30.3, write down the ground-state electronic configuration of manganese Mn ($Z = 25$). Refer to Figure 30.17 for the order in which the subshells fill.

30. Referring to Figure 30.17 for the order in which the subshells fill and following the style used in Table 30.3, determine the ground-state electronic configuration for cadmium Cd ($Z = 48$).

31. ssm When an electron makes a transition between energy levels of an atom, there are no restrictions on the initial and final values of the principal quantum number n. According to quantum mechanics, however, there is a rule that restricts the initial and final values of the orbital quantum number ℓ. This rule is called a *selection rule* and states that $\Delta\ell = \pm 1$. In other words, when an electron makes a transition between energy levels, the value of ℓ can only increase or decrease by one. The value of ℓ may not remain the same or increase or decrease by more than one. According to this rule, which of the following energy level transitions are allowed: (a) 2s → 1s, (b) 2p → 1s, (c) 4p → 2p, (d) 4s → 2p, and (e) 3d → 3s?

* **32.** What is the atom with the smallest atomic number that contains the same number of electrons in its s subshells as it does in its d subshell? Refer to Figure 30.17 for the order in which the subshells fill.

Section 30.7 X-Rays

33. ssm Molybdenum has an atomic number of $Z = 42$. Using the Bohr model, estimate the wavelength of the K_α X-ray.

34. The atomic number of lead is $Z = 82$. According to the Bohr model, what is the energy (in joules) of a K_α X-ray photon?

35. Interactive Solution 30.35 at **www.wiley.com/college/cutnell** provides one model for solving problems such as this one. An X-ray tube is being operated at a potential difference of 52.0 kV. What is the Bremsstrahlung wavelength that corresponds to 35.0% of the kinetic energy with which an electron collides with the metal target in the tube?

36. The K_β characteristic X-ray line for tungsten has a wavelength of 1.84×10^{-11} m. What is the difference in energy between the two energy levels that give rise to this line? Express the answer in (a) joules and (b) electron volts.

* **37.** ssm An X-ray tube contains a silver ($Z = 47$) target. The high voltage in this tube is increased from zero. Using the Bohr model, find the value of the voltage at which the K_α X-ray just appears in the X-ray spectrum.

* **38.** The highest-energy X-rays produced by an X-ray tube have a wavelength of 1.20×10^{-10} m. What is the speed of the electrons in Figure 30.19 just before they strike the metal target?

Section 30.8 The Laser

39. A pulsed laser emits light in a series of short pulses, each having a duration of 25.0 ms. The average power of each pulse is 5.00 mW, and the wavelength of the light is 633 nm. Find (a) the energy of each pulse and (b) the number of photons in each pulse.

40. A laser peripheral iridotomy is a procedure for treating an eye condition known as narrow-angle glaucoma, in which pressure buildup in the eye can lead to loss of vision. A neodymium YAG laser (wavelength = 1064 nm) is used in the procedure to punch a tiny hole in the peripheral iris, thereby relieving the pressure buildup. In one application the laser delivers 4.1×10^{-3} J of energy to the iris in creating the hole. How many photons does the laser deliver?

41. ssm www A laser is used in eye surgery to weld a detached retina back into place. The wavelength of the laser beam is 514 nm, and the power is 1.5 W. During surgery, the laser beam is turned on for 0.050 s. During this time, how many photons are emitted by the laser?

42. The dye laser used in the treatment of the port-wine stain in Figure 30.28 (see Section 30.9) has a wavelength of 585 nm. A carbon dioxide laser produces a wavelength of 1.06×10^{-5} m. What is the minimum number of photons that the carbon dioxide laser must produce to deliver at least as much or more energy to a target as does a single photon from the dye laser?

* **43.** Fusion is the process by which the sun produces energy. One experimental technique for creating controlled fusion utilizes a solid-state laser that emits a wavelength of 1060 nm and can produce a power of 1.0×10^{14} W for a pulse duration of 1.1×10^{-11} s. In contrast, the helium/neon laser used at the checkout counter in a bar-code scanner emits a wavelength of 633 nm and produces a power of about 1.0×10^{-3} W. How long (in days) would the helium/neon laser have to operate to produce the same number of photons that the solid-state laser produces in 1.1×10^{-11} s?

Additional Problems

44. In the style shown in Table 30.3, write down the ground-state electronic configuration for arsenic As ($Z = 33$). Refer to Figure 30.17 for the order in which the subshells fill.

45. ssm www What is the radius for the $n = 5$ Bohr orbit in a doubly ionized lithium atom Li^{2+} ($Z = 3$)?

46. Interactive LearningWare 30.2 at **www.wiley.com/college/cutnell** reviews the concepts that are pertinent to this problem. By using the Bohr model, decide which element is likely to emit a K_α X-ray with a wavelength of 4.5×10^{-9} m.

47. Write down the fourteen sets of the four quantum numbers that correspond to the electrons in a completely filled 4f subshell.

48. What is the minimum potential difference that must be applied to an X-ray tube to knock a K-shell electron completely out of an atom in a copper ($Z = 29$) target? Use the Bohr model as needed.

49. ssm Find the energy (in joules) of the photon that is emitted when the electron in a hydrogen atom undergoes a transition from the $n = 7$ energy level to produce a line in the Paschen series.

* **50.** The energy of the $n = 2$ Bohr orbit is -30.6 eV for an unidentified ionized atom in which only one electron moves about the nucleus. What is the radius of the $n = 5$ orbit for this species?

* **51.** Interactive Solution 30.51 at **www.wiley.com/college/cutnell** offers one approach to problems of this type. For an electron in a hy-

drogen atom, the z component of the angular momentum has a *maximum* value of $L_z = 4.22 \times 10^{-34}$ J·s. Find the three smallest possible values (algebraically) for the total energy (in electron volts) that this atom could have.

* **52.** Consider a particle of mass m that can exist only between $x = 0$ m and $x = +L$ on the x axis. We could say that this particle is confined to a "box" of length L. In this situation, imagine the standing de Broglie waves that can fit into the box. For example, the drawing shows the first three possibilities. Note in this picture that there are either one,

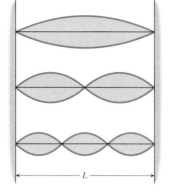

two, or three half-wavelengths that fit into the distance L. Use Equation 29.8 for the de Broglie wavelength of a particle and derive an expression for the allowed energies (only kinetic energy) that the particle can have. This expression involves m, L, Planck's constant, and a quantum number n that can have only the values 1, 2, 3, ...

* **53. ssm** For atomic hydrogen, the Paschen series of lines occurs when $n_f = 3$, whereas the Brackett series occurs when $n_f = 4$ in Equation 30.14. Using this equation, show that the ranges of wavelengths in these two series overlap.

** **54.** A certain species of ionized atoms produces an emission line spectrum according to the Bohr model, but the number of protons Z in the nucleus is unknown. A group of lines in the spectrum forms a series in which the shortest wavelength is 22.79 nm and the longest wavelength is 41.02 nm. Find the next-to-the-longest wavelength in the series of lines.

Concepts & Calculations Group Learning Problems

Note: Each of these problems consists of Concept Questions followed by a related quantitative Problem. They are designed for use by students working alone or in small learning groups. The Concept Questions involve little or no mathematics and are intended to stimulate group discussions. They focus on the concepts with which the problems deal. Recognizing the concepts is the essential initial step in any problem-solving technique.

55. Concept Questions A hydrogen atom is in an excited state. (a) What is meant by the phrase "ionization energy of the excited state"? (b) Is the ionization energy of an excited state greater than, less than, or the same as the ionization energy for the ground state? Why?

Problem (a) What is the ionization energy of a hydrogen atom that is in the $n = 4$ state? (b) Determine the ratio of the ionization energy for $n = 4$ to that of the ground state. Check to make sure that your answer is consistent with your answers to the Concept Questions.

56. Concept Questions Answer the following questions using the quantum mechanical model of the atom. (a) Can the orbital quantum number ℓ be greater than the principal quantum number n? (b) Can the orbital quantum number have negative values? (c) Is it possible for the magnetic quantum number m_ℓ to be greater than the orbital quantum number? (d) Can the magnetic quantum number have negative values?

Problem The table lists the quantum numbers for five states of the hydrogen atom. Which (if any) of them are not possible?

	n	ℓ	m_ℓ
(a)	3	3	0
(b)	2	1	−1
(c)	4	2	3
(d)	5	−3	2
(e)	4	0	0

57. Concept Questions Answer the following questions using the quantum mechanical model of the atom. (a) For a given principal quantum number n, what values of the angular momentum quantum number ℓ are possible? (b) For a given angular momentum, what values of the magnetic quantum number m_ℓ are possible?

Problem Which of the following subshell configurations are not allowed? For those that are not allowed, give the reason why. (a) $3s^1$, (b) $2d^2$, (c) $3s^4$, (d) $4p^8$, (e) $5f^{12}$.

58. Concept Questions Suppose that the molybdenum ($Z = 42$) target in an X-ray tube is replaced by a silver ($Z = 47$) target. Do (a) the cutoff wavelength λ_0 and (b) the wavelength of the K_α X-ray photon increase, decrease, or remain the same? Assume that the voltage across the tube is constant and is sufficient to produce characteristic X-rays from both targets. Provide a reason for each answer.

Problem The voltage across the X-ray tube is 35.0 kV. Determine (a) the cutoff wavelength λ_0 and (b) the wavelengths of the K_α X-ray photons emitted by the molybdenum and silver targets. Verify that your answers are consistent with your answers to the Concept Questions.

59. Concept Questions The drawing shows three energy levels of a laser that are involved in the lasing action. These levels are analogous to the levels in the Ne atoms of a He-Ne laser. The E_2 level is a metastable level, and the E_0 level is the ground state. (a) What two levels are involved when an external source of energy initially causes the electrons to jump to a higher energy state? (b) Between what two levels does the population inversion occur? (c) Between what two levels does the lasing action occur? Provide a reason for each answer.

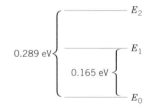

Problem The difference between the energy levels of the laser is shown in the drawing. (a) What energy (in eV per electron) must an external source provide to start the lasing action? (b) What is the wavelength of the laser light? (c) In what region of the electromagnetic spectrum does the laser light lie (see Figure 24.10)? Check your answers for consistency with respect to your answers to the Concept Questions.

* **60. Concept Questions** A sodium atom ($Z = 11$) contains 11 protons in its nucleus. Strictly speaking, the Bohr model does not apply, because the neutral atom contains 11 electrons instead of a single electron. However, we can apply the model to the outermost electron as an approximation, provided that we use an effective value $Z_{effective}$ rather than 11 for the number of protons in

the nucleus. Since electrons are sometimes between the nucleus and the outermost electron, $Z_{effective}$ is less than 11. The repulsion from these "inner" electrons offsets some of the attractive force that the positive nuclear charge exerts on the outermost electron; this is called "shielding." If the 10 "inner" electrons were permanently within the nucleus, we would have $Z_{effective} = 1$. The "inner" electrons are in orbit, however. (a) Since the orbiting "inner" electrons spend part of the time on the side of the nucleus opposite to the outermost electron, they are not always between the nucleus and the outermost electron. Considering this fact, is $Z_{effective}$ less than or greater than 1? (b) Do you expect the radius of the outermost Bohr orbit to be greater with or without the effect of shielding to reduce Z to a value smaller than 11? Explain your answers.

Problem (a) The ionization energy for the outermost electron in a sodium atom is 5.1 eV. Use the Bohr model with $Z = Z_{effective}$ to calculate a value for $Z_{effective}$. (b) Using $Z = 11$ and $Z = Z_{effective}$, determine the corresponding two values for the radius of the outermost

Bohr orbit. Verify that your answers are consistent with your answers to the Concept Questions.

* **61. Concept Questions** The Bohr model, although not strictly applicable, can be used to estimate the minimum energy E_{min} that an incoming electron must have in an X-ray tube in order to knock a K-shell electron entirely out of an atom in the metal target. (a) Explain how the Bohr model is used for this purpose. (b) Explain how the Bohr model can be used to estimate the wavelength of the K_α line in the X-ray spectrum of a metal. (c) If the K_α wavelength of metal A (atomic number = Z_A) is greater than that of metal B (atomic number = Z_B), is Z_A greater than, less than, or equal to Z_B? (d) If the K_α wavelength of metal A (atomic number = Z_A) is greater than that of metal B (atomic number = Z_B), is E_{min} for metal A greater than, less than, or equal to E_{min} for metal B?

Problem The K_α X-ray wavelength of metal A is twice that of metal B. What is the ratio of E_{min} for metal A to E_{min} for metal B? Check to see that your answer is consistent with your answer to Concept Question (d).

Chapter 31

Nuclear Physics and Radioactivity

This is Sue, the largest and most complete *Tyrannosaurus rex* ever found. It was discovered by Susan Hendrickson in 1990 and is currently on display at The Field Museum in Chicago. Paleontologists use the disintegration of radioactive nuclei, one of the subjects of this chapter, to date the fossil remains of extinct species. (© Reuters NewMedia, Inc./ Corbis Images)

31.1 Nuclear Structure

Atoms consist of electrons in orbit about a central nucleus. As we have seen in Chapter 30, the electron orbits are quantum mechanical in nature and have interesting characteristics. Little has been said about the nucleus, however. Since the nucleus is interesting in its own right, we now consider it in greater detail.

The nucleus of an atom consists of neutrons and protons, collectively referred to as **nucleons.** The **neutron,** discovered in 1932 by the English physicist James Chadwick (1891–1974), carries no electrical charge and has a mass slightly larger than that of a proton (see Table 31.1).

The number of protons in the nucleus is different in different elements and is given by the **atomic number Z.** In an electrically neutral atom, the number of nuclear protons equals the number of electrons in orbit around the nucleus. The number of neutrons in the nucleus is N. The total number of protons and neutrons is referred to as the **atomic mass number A** because the total nuclear mass is *approximately* equal to A times the mass of a single nucleon:

$$\underbrace{A}_{\substack{\text{Number of protons} \\ \text{and neutrons} \\ \text{(atomic mass number} \\ \text{or nucleon number)}}} = \underbrace{Z}_{\substack{\text{Number of} \\ \text{protons} \\ \text{(atomic number)}}} + \underbrace{N}_{\substack{\text{Number of} \\ \text{neutrons}}} \tag{31.1}$$

Sometimes A is also called the **nucleon number.** A shorthand notation is often used to specify Z and A along with the chemical symbol for the element. For instance, the nuclei of all naturally occurring aluminum atoms have $A = 27$, and the atomic number for aluminum is $Z = 13$. In shorthand notation, then, the aluminum nucleus is specified as $^{27}_{13}\text{Al}$. The number of neutrons in an aluminum nucleus is $N = A - Z = 14$. In general, for an element whose chemical symbol is X, the symbol for the nucleus is

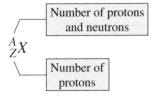

For a proton the symbol is $^{1}_{1}\text{H}$, since the proton is the nucleus of a hydrogen atom. A neutron is denoted by $^{1}_{0}\text{n}$. In the case of an electron we use $^{0}_{-1}\text{e}$, where $A = 0$ because an electron is not composed of protons or neutrons and $Z = -1$ because the electron has a negative charge.

Nuclei that contain the same number of protons, but a different number of neutrons, are known as **isotopes.** Carbon, for example, occurs in nature in two stable forms. In most carbon atoms (98.90%), the nucleus is the $^{12}_{6}\text{C}$ isotope and consists of six protons and six neutrons. A small fraction (1.10%), however, contains nuclei that have six protons and seven neutrons, namely, the $^{13}_{6}\text{C}$ isotope. The percentages given here are the natural abundances of the isotopes. The atomic masses in the periodic table are average atomic masses, taking into account the abundances of the various isotopes.

Table 31.1 *Properties of Select Particles*

Particle	Electric Charge (C)	Mass Kilograms (kg)	Mass Atomic Mass Units (u)
Electron	-1.60×10^{-19}	$9.109\,382 \times 10^{-31}$	$5.485\,799 \times 10^{-4}$
Proton	$+1.60 \times 10^{-19}$	$1.672\,622 \times 10^{-27}$	$1.007\,276$
Neutron	0	$1.674\,927 \times 10^{-27}$	$1.008\,665$
Hydrogen atom	0	$1.673\,534 \times 10^{-27}$	$1.007\,825$

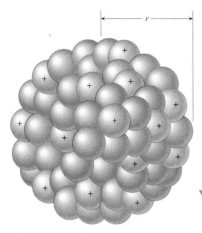

Figure 31.1 The nucleus is approximately spherical (radius = r) and contains protons (⊕) clustered closely together with neutrons (●).

The protons and neutrons in the nucleus are clustered together to form an approximately spherical region, as Figure 31.1 illustrates. Experiment shows that the radius r of the nucleus depends on the atomic mass number A and is given approximately in meters by

$$r \approx (1.2 \times 10^{-15} \text{ m})A^{1/3} \tag{31.2}$$

The radius of the aluminum nucleus ($A = 27$), for example, is $r \approx (1.2 \times 10^{-15} \text{ m})27^{1/3} = 3.6 \times 10^{-15}$ m. Equation 31.2 leads to an important conclusion concerning the nuclear density of different atoms, as Conceptual Example 1 discusses.

Conceptual Example 1 Nuclear Density

It is well known that lead and oxygen contain different atoms and that the density of solid lead is much greater than that of gaseous oxygen. Using Equation 31.2, decide whether the density of the *nucleus* in a lead atom is greater than, approximately equal to, or less than that in an oxygen atom.

Reasoning and Solution We know from Equation 11.1 that density is mass divided by volume. The total mass M of a nucleus is approximately equal to the number of nucleons A times the mass m of a single nucleon, since the masses of a proton and a neutron are nearly the same: $M \approx Am$. The values of A are different for lead and oxygen atoms, however. The volume V of the nucleus is approximately spherical and has a radius r, so that $V = \frac{4}{3}\pi r^3$. But r^3 is proportional to the number of nucleons A, as we can see by cubing each side of Equation 31.2. Therefore, the volume V is also proportional to A. Thus, when the total mass M is divided by the volume V, the factor of A appears in both the numerator and denominator and is eliminated algebraically from the result, no matter what the value of A is. We find, then, that **the density of the nucleus in a lead atom is approximately the same as it is in an oxygen atom.** In general, Equation 31.2 implies that the nuclear density has nearly the same value in all atoms. The difference in densities between solid lead and gaseous oxygen arises mainly because of the difference in how closely the atoms themselves are packed together in the solid and gaseous phases.

Related Homework: *Problems 8, 9*

✔ **Check Your Understanding 1**

Two nuclei have different numbers of protons and different numbers of neutrons. Which one or more of the following statements are true? (a) They are different isotopes of the same element. (b) They have the same electric charge. (c) They could have the same radii. (d) They have approximately the same nuclear density. *(The answer is given at the end of the book.)*

Background: Whether or not a nucleus belongs to a certain element depends on the atomic number of the nucleus. The radius and mass of a nucleus depend on its atomic mass number.

For similar questions (including calculational counterparts), consult Self-Assessment Test 31.1, which is described at the end of Section 31.5.

31.2 The Strong Nuclear Force and the Stability of the Nucleus

Two positive charges that are as close together as they are in a nucleus repel one another with a very strong electrostatic force. What, then, keeps the nucleus from flying apart? Clearly, some kind of attractive force must hold the nucleus together, since many kinds of naturally occurring atoms contain stable nuclei. The gravitational force of attraction between nucleons is too weak to counteract the repulsive electric force, so a different type of force must hold the nucleus together. This force is the *strong nuclear force* and is one of only three fundamental forces that have been discovered, fundamental in the sense that all forces in nature can be explained in terms of these three. The gravitational force is also one of these forces, as is the electroweak force (see Section 31.5).

Many features of the strong nuclear force are well known. For example, it is almost independent of electric charge. At a given separation distance, nearly the same nuclear force of attraction exists between two protons, between two neutrons, or between a proton and a neutron. The range of action of the strong nuclear force is extremely short, with the force of attraction being very strong when two nucleons are as close as 10^{-15} m and essentially zero at larger distances. In contrast, the electric force between two protons decreases to zero only gradually as the separation distance increases to large values and, therefore, has a relatively long range of action.

The limited range of action of the strong nuclear force plays an important role in the stability of the nucleus. For a nucleus to be stable, the electrostatic repulsion between the protons must be balanced by the attraction between the nucleons due to the strong nuclear force. But one proton repels all other protons within the nucleus, since the electrostatic force has such a long range of action. In contrast, a proton or a neutron attracts only its nearest neighbors via the strong nuclear force. As the number Z of protons in the nucleus increases under these conditions, the number N of neutrons has to increase even more, if stability is to be maintained. Figure 31.2 shows a plot of N versus Z for naturally occurring elements that have stable nuclei. For reference, the plot also includes the straight line that represents the condition $N = Z$. With few exceptions, the points representing stable nuclei fall above this reference line, reflecting the fact that the number of neutrons becomes greater than the number of protons as the atomic number Z increases.

As more and more protons occur in a nucleus, there comes a point when a balance of repulsive and attractive forces cannot be achieved by an increased number of neutrons. Eventually, the limited range of action of the strong nuclear force prevents extra neutrons from balancing the long-range electric repulsion of extra protons. The stable nucleus with the largest number of protons ($Z = 83$) is that of bismuth, $^{209}_{83}$Bi, which contains 126 neutrons. All nuclei with more than 83 protons (e.g., uranium, $Z = 92$) are unstable and spontaneously break apart or rearrange their internal structures as time passes. This spontaneous disintegration or rearrangement of internal structure is called *radioactivity*, first discovered in 1896 by the French physicist Henri Becquerel (1852–1908). Section 31.4 discusses radioactivity in greater detail.

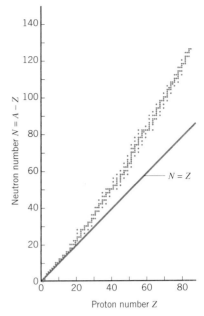

Figure 31.2 With few exceptions, the naturally occurring stable nuclei have a number N of neutrons that equals or exceeds the number Z of protons. Each dot in this plot represents a stable nucleus.

31.3 The Mass Defect of the Nucleus and Nuclear Binding Energy

Because of the strong nuclear force, the nucleons in a stable nucleus are held tightly together. Therefore, energy is required to separate a stable nucleus into its constituent protons and neutrons, as Figure 31.3 illustrates. The more stable the nucleus is, the greater is the amount of energy needed to break it apart. The required energy is called the *binding energy* of the nucleus.

▶ CONCEPTS AT A GLANCE As the Concepts-at-a-Glance chart in Figure 31.4 illustrates, two ideas that we have studied previously come into play as we discuss the binding energy of a nucleus. These are mass (Section 4.2) and the rest energy of an object (Section 28.6). In Einstein's theory of special relativity, mass and energy are equivalent. A change Δm in the mass of a system is equivalent to a change ΔE_0 in the rest energy of the system by an amount $\Delta E_0 = (\Delta m)c^2$, where c is the speed of light in a vacuum. Thus, in Figure 31.3, the binding energy used to disassemble the nucleus appears as extra mass of the separated nucleons. In other words, the sum of the individual masses of the separated protons and neutrons is greater by an amount Δm than the mass of the stable nucleus. The difference in mass Δm is known as the *mass defect* of the nucleus. ◀

As Example 2 shows, the binding energy of a nucleus can be determined from the mass defect according to Equation 31.3:

$$\text{Binding energy} = (\text{Mass defect})c^2 = (\Delta m)c^2 \qquad (31.3)$$

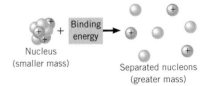

Figure 31.3 Energy, called the binding energy, must be supplied to break the nucleus apart into its constituent protons and neutrons. Each of the separated nucleons is at rest and out of the range of the forces of the other nucleons.

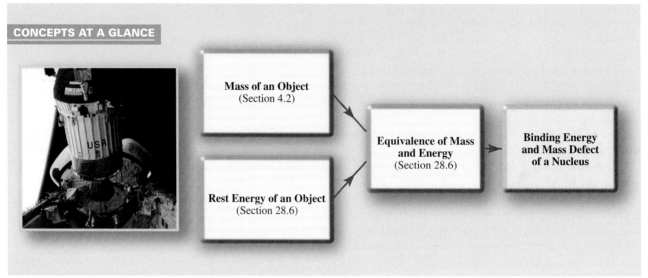

CONCEPTS AT A GLANCE

Mass of an Object
(Section 4.2)

Rest Energy of an Object
(Section 28.6)

Equivalence of Mass
and Energy
(Section 28.6)

Binding Energy
and Mass Defect
of a Nucleus

Figure 31.4 CONCEPTS AT A GLANCE The mass and rest energy of an object are equivalent, in the sense that if one increases (or decreases), the other does too. The binding energy of a nucleus is the mass of the separated nucleons minus the mass of the intact nucleus, expressed in units of energy. This photograph shows NASA's *Galileo* spacecraft, which uses a process called nuclear fission to generate its energy. This process depends on the fact that different nuclei have different binding energies and is discussed in Chapter 32. (Courtesy NASA)

Example 2   The Binding Energy of the Helium Nucleus

The most abundant isotope of helium has a $^4_2$He nucleus whose mass is 6.6447×10^{-27} kg. For this nucleus, find (a) the mass defect and (b) the binding energy.

Reasoning The symbol $^4_2$He indicates that the helium nucleus contains $Z = 2$ protons and $N = 4 - 2 = 2$ neutrons. To obtain the mass defect Δm, we first determine the sum of the individual masses of the separated protons and neutrons. Then we subtract from this sum the mass of the $^4_2$He nucleus. Finally, we use Equation 31.3 to calculate the binding energy from the value for Δm.

Solution

(a) Using data from Table 31.1, we find that the sum of the individual masses of the nucleons is

$$\underbrace{2(1.6726 \times 10^{-27} \text{ kg})}_{\text{Two protons}} + \underbrace{2(1.6749 \times 10^{-27} \text{ kg})}_{\text{Two neutrons}} = 6.6950 \times 10^{-27} \text{ kg}$$

This value is greater than the mass of the intact $^4_2$He nucleus, and the mass defect is

$$\Delta m = 6.6950 \times 10^{-27} \text{ kg} - 6.6447 \times 10^{-27} \text{ kg} = \boxed{0.0503 \times 10^{-27} \text{ kg}}$$

(b) According to Equation 31.3, the binding energy is

$$\frac{\text{Binding}}{\text{energy}} = (\Delta m)c^2 = (0.0503 \times 10^{-27} \text{ kg})(3.00 \times 10^8 \text{ m/s})^2 = 4.53 \times 10^{-12} \text{ J}$$

Usually, binding energies are expressed in energy units of electron volts instead of joules ($1 \text{ eV} = 1.60 \times 10^{-19}$ J):

$$\frac{\text{Binding}}{\text{energy}} = (4.53 \times 10^{-12} \text{ J})\left(\frac{1 \text{ eV}}{1.60 \times 10^{-19} \text{ J}}\right) = 2.83 \times 10^7 \text{ eV} = \boxed{28.3 \text{ MeV}}$$

In this result, one million electron volts is denoted by the unit MeV. The value of 28.3 MeV is more than two million times greater than the energy required to remove an orbital electron from an atom.

In calculations such as that in Example 2, it is customary to use the *atomic mass unit* (u) instead of the kilogram. As introduced in Section 14.1, the atomic mass unit is one-twelfth of the mass of a $^{12}_6$C atom of carbon. In terms of this unit, the mass of a $^{12}_6$C atom is exactly 12 u. Table 31.1 also gives the masses of the electron, the proton, and the neutron in atomic mass units. For future calculations, the energy equivalent of one atomic mass unit can be determined by observing that the mass of a proton is 1.6726×10^{-27} kg or 1.0073 u, so that

$$1 \text{ u} = (1 \text{ u}) \left(\frac{1.6726 \times 10^{-27} \text{ kg}}{1.0073 \text{ u}} \right) = 1.6605 \times 10^{-27} \text{ kg}$$

and

$$\Delta E_0 = (\Delta m)c^2 = (1.6605 \times 10^{-27} \text{ kg})(2.9979 \times 10^8 \text{ m/s})^2 = 1.4924 \times 10^{-10} \text{ J}$$

In electron volts, therefore, one atomic mass unit is equivalent to

$$1 \text{ u} = (1.4924 \times 10^{-10} \text{ J}) \left(\frac{1 \text{ eV}}{1.6022 \times 10^{-19} \text{ J}} \right) = 9.315 \times 10^8 \text{ eV} = 931.5 \text{ MeV}$$

Data tables for isotopes give masses in atomic mass units. Typically, however, the given masses are not nuclear masses. They are *atomic masses*—that is, the masses of neutral atoms, including the mass of the orbital electrons. Example 3 deals again with the ^4_2He nucleus and shows how to take into account the effect of the orbital electrons when using such data to determine binding energies.

Example 3 The Binding Energy of the Helium Nucleus, Revisited

The atomic mass of ^4_2He is 4.0026 u, and the atomic mass of ^1_1H is 1.0078 u. Using atomic mass units instead of kilograms, obtain the binding energy of the ^4_2He nucleus.

Reasoning To determine the binding energy, we calculate the mass defect in atomic mass units and then use the fact that one atomic mass unit is equivalent to 931.5 MeV of energy. The mass of 4.0026 u for ^4_2He *includes the mass of the two electrons in the neutral helium atom*. To calculate the mass defect, we must subtract 4.0026 u from the sum of the individual masses of the nucleons, including the mass of the electrons. As Figure 31.5 illustrates, the electron mass will be included if the masses of two hydrogen atoms are used in the calculation instead of the masses of two protons. The mass of a ^1_1H hydrogen atom is given in Table 31.1 as 1.0078 u, and the mass of a neutron as 1.0087 u.

Solution The sum of the individual masses is

$$\underbrace{2(1.0078 \text{ u})}_{\substack{\text{Two hydrogen} \\ \text{atoms}}} + \underbrace{2(1.0087 \text{ u})}_{\text{Two neutrons}} = 4.0330 \text{ u}$$

The mass defect is $\Delta m = 4.0330 \text{ u} - 4.0026 \text{ u} = 0.0304 \text{ u}$. Since 1 u is equivalent to 931.5 MeV, the binding energy is $\boxed{\text{Binding energy} = 28.3 \text{ MeV}}$, which matches that obtained in Example 2.

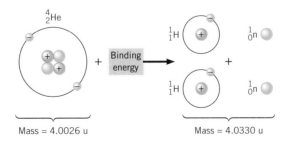

Figure 31.5 Data tables usually give the mass of the neutral atom (including the orbital electrons) rather than the mass of the nucleus. When data from such tables are used to determine the mass defect of a nucleus, the mass of the orbital electrons must be taken into account, as this drawing illustrates for the ^4_2He isotope of helium. See Example 3.

To see how the nuclear binding energy varies from nucleus to nucleus, it is necessary to compare the binding energy for each nucleus on a per-nucleon basis. The graph shown

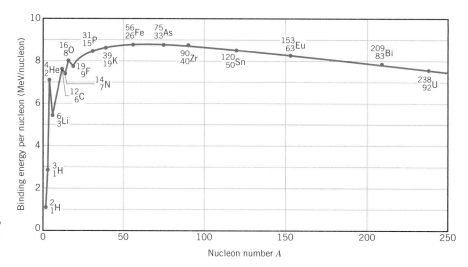

Figure 31.6 A plot of binding energy per nucleon versus the nucleon number A.

in Figure 31.6 shows a plot in which the binding energy divided by the nucleon number A is plotted against the nucleon number itself. In the graph, the peak for the $^4_2$He isotope of helium indicates that the $^4_2$He nucleus is particularly stable. The binding energy per nucleon increases rapidly for nuclei with small masses and reaches a maximum of approximately 8.7 MeV/nucleon for a nucleon number of about $A = 60$. For greater nucleon numbers, the binding energy per nucleon decreases gradually. Eventually, the binding energy per nucleon decreases enough so there is insufficient binding energy to hold the nucleus together. Nuclei more massive than the $^{209}_{83}$Bi nucleus of bismuth are unstable and hence radioactive.

 **Check Your Understanding 2**

The following table gives values for the mass defect Δm for four hypothetical nuclei: A, B, C, and D. Which statement is true regarding the stability of these nuclei? (a) Nucleus D is the most stable, and A is the least stable. (b) Nucleus C is stable, whereas A, B, and D are not stable. (c) Nucleus A is the most stable, and D is not stable. (d) Nuclei A and B are stable, but B is more stable than A. *(The answer is given at the end of the book.)*

	A	B	C	D
Mass defect, Δm	$+6.0 \times 10^{-29}$ kg	$+2.0 \times 10^{-29}$ kg	0 kg	-6.0×10^{-29} kg

Background: The mass defect of a nucleus is the sum of the individual masses of the separated nucleons minus the mass of the intact nucleus. The binding energy of a nucleus is the energy required to separate it into its constituent protons and neutrons. The greater the binding energy, the more stable is the nucleus. The binding energy and the mass defect are related.

For similar questions (including calculational counterparts), consult Self-Assessment Test 31.1, which is described at the end of Section 31.5.

31.4 *Radioactivity*

When an unstable or radioactive nucleus disintegrates spontaneously, certain kinds of particles and/or high-energy photons are released. These particles and photons are collectively called "rays." Three kinds of rays are produced by naturally occurring radioactivity: **α rays, β rays,** and **γ rays.** They are named according to the first three letters of the Greek alphabet, alpha (α), beta (β), and gamma (γ), to indicate the extent of their ability to penetrate matter. α rays are the least penetrating, being blocked by a thin (≈ 0.01 mm) sheet of lead, whereas β rays penetrate lead to a much greater distance (≈ 0.1 mm). γ rays are the most penetrating and can pass through an appreciable thickness (≈ 100 mm) of lead.

CONCEPTS AT A GLANCE

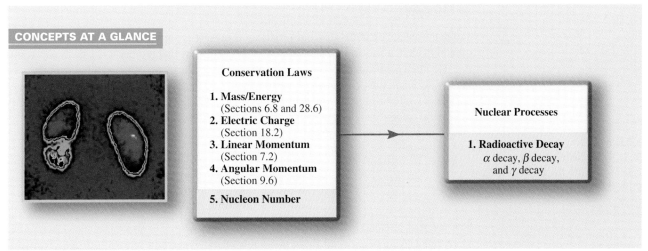

Conservation Laws

1. **Mass/Energy**
   (Sections 6.8 and 28.6)
2. **Electric Charge**
   (Section 18.2)
3. **Linear Momentum**
   (Section 7.2)
4. **Angular Momentum**
   (Section 9.6)
5. **Nucleon Number**

Nuclear Processes

1. **Radioactive Decay**
   α decay, β decay,
   and γ decay

Figure 31.7 CONCEPTS AT A GLANCE The conservation laws listed at the left side of this chart are obeyed when a nucleus undergoes radioactive decay. The three types of naturally occurring decay are α decay, β decay, and γ decay. Nuclear medicine uses radioactive decay to produce scans of organs. This photograph shows a nuclear scan of two kidneys, the one on the left displaying an invasive cancer. (© ISM/Phototake)

▶ CONCEPTS AT A GLANCE The nuclear disintegration process that produces α, β, and γ rays must obey the laws of physics that we have studied in previous chapters. As the Concepts-at-a-Glance chart in Figure 31.7 reminds us, these laws are called conservation laws because each of them deals with a property (such as mass/energy, electric charge, linear momentum, and angular momentum) that is conserved or does not change during a process. To the first four conservation laws in Figure 31.7, we now add a fifth, the conservation of nucleon number. In all radioactive decay processes it has been observed that the number of nucleons (protons plus neutrons) present before the decay is equal to the number of nucleons after the decay. Therefore, the number of nucleons is conserved during a nuclear disintegration. As applied to the disintegration of a nucleus, the conservation laws require that the energy, electric charge, linear momentum, angular momentum, and nucleon number that a nucleus possesses must remain unchanged when it disintegrates into nuclear fragments and accompanying α, β, or γ rays. ◀

The three types of radioactivity that occur naturally can be observed in a relatively simple experiment. A piece of radioactive material is placed at the bottom of a narrow hole in a lead cylinder. The cylinder is located within an evacuated chamber, as Figure 31.8 illustrates. A magnetic field is directed perpendicular to the plane of the paper, and a photographic plate is positioned to the right of the hole. Three spots appear on the developed plate, which are associated with the radioactivity of the nuclei in the material. Since moving particles are deflected by a magnetic field only when they are electrically charged, this experiment reveals that two types of radioactivity (α and β rays, as it turns out) consist of charged particles, whereas the third type (γ rays) does not.

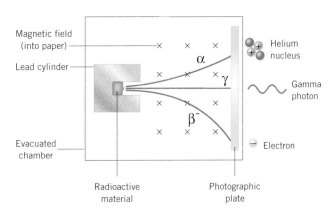

Figure 31.8 α and β rays are deflected by a magnetic field and, therefore, consist of moving charged particles. γ rays are not deflected by a magnetic field and, consequently, must be uncharged.

α DECAY

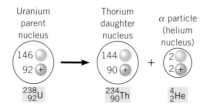

Figure 31.9 α decay occurs when an unstable parent nucleus emits an α particle and in the process is converted into a different, or daughter, nucleus.

When a nucleus disintegrates and produces α rays, it is said to undergo **α decay.** Experimental evidence shows that α rays consist of positively charged particles, each one being the $^4_2$He nucleus of helium. Thus, an α particle has a charge of $+2e$ and a nucleon number of $A = 4$. Since the grouping of 2 protons and 2 neutrons in a $^4_2$He nucleus is particularly stable, as we have seen in connection with Figure 31.6, it is not surprising that an α particle can be ejected as a unit from a more massive unstable nucleus.

Figure 31.9 shows the disintegration process for one example of α decay:

$$^{238}_{92}\text{U} \longrightarrow ^{234}_{90}\text{Th} + ^4_2\text{He}$$

$$\begin{array}{ccc} \text{Parent} & \text{Daughter} & \alpha \text{ particle} \\ \text{nucleus} & \text{nucleus} & \text{(helium} \\ \text{(uranium)} & \text{(thorium)} & \text{nucleus)} \end{array}$$

The original nucleus is referred to as the *parent nucleus* (P), and the nucleus remaining after disintegration is called the *daughter nucleus* (D). Upon emission of an α particle, the uranium $^{238}_{92}$U parent is converted into the $^{234}_{90}$Th daughter, which is an isotope of thorium. The parent and daughter nuclei are different, so α decay converts one element into another, a process known as **transmutation.**

Electric charge is conserved during α decay. In Figure 31.9, for instance, 90 of the 92 protons in the uranium nucleus end up in the thorium nucleus, and the remaining 2 protons are carried off by the α particle. The total number of 92, however, is the same before and after disintegration. α decay also conserves the number of nucleons, because the number is the same before (238) and after (234 + 4) disintegration. Consistent with the conservation of electric charge and nucleon number, the general form for α decay is

α Decay $$^A_Z\text{P} \longrightarrow ^{A-4}_{Z-2}\text{D} + ^4_2\text{He}$$

$$\begin{array}{ccc} \text{Parent} & \text{Daughter} & \alpha \text{ particle} \\ \text{nucleus} & \text{nucleus} & \text{(helium nucleus)} \end{array}$$

When a nucleus releases an α particle, the nucleus also releases energy. In fact, the energy released by radioactive decay is responsible, in part, for keeping the interior of the earth hot and, in some places, even molten. The following example shows how the conservation of mass/energy can be used to determine the amount of energy released in α decay.

Example 4 α Decay and the Release of Energy

The atomic mass of uranium $^{238}_{92}$U is 238.0508 u, that of thorium $^{234}_{90}$Th is 234.0436 u, and that of an α particle $^4_2$He is 4.0026 u. Determine the energy released when α decay converts $^{238}_{92}$U into $^{234}_{90}$Th.

Reasoning Since energy is released during the decay, the combined mass of the $^{234}_{90}$Th daughter nucleus and the α particle is less than the mass of the $^{238}_{92}$U parent nucleus. The difference in mass is equivalent to the energy released. We will determine the difference in mass in atomic mass units and then use the fact that 1 u is equivalent to 931.5 MeV.

Solution The decay and the masses are shown below:

$$^{238}_{92}\text{U} \longrightarrow ^{234}_{90}\text{Th} + ^4_2\text{He}$$

$$\begin{array}{ccc} 238.0508 \text{ u} & \underbrace{234.0436 \text{ u} \qquad 4.0026 \text{ u}} \\ & 238.0462 \text{ u} \end{array}$$

The decrease in mass is 238.0508 u − 238.0462 u = 0.0046 u. As usual, the masses are atomic masses and include the mass of the orbital electrons. But this causes no error here because the same total number of electrons is included for $^{238}_{92}$U, on the one hand, and for $^{234}_{90}$Th plus $^4_2$He, on the other. Since 1 u is equivalent to 931.5 MeV, the released energy is $\boxed{4.3 \text{ MeV}}$.

When α decay occurs as in Example 4, the energy released appears as kinetic energy of the recoiling $^{234}_{90}$Th nucleus and the α particle, except for a small portion carried away as a γ ray. Conceptual Example 5 discusses how the $^{234}_{90}$Th nucleus and the α particle share in the released energy.

Conceptual Example 5
How Energy Is Shared During the α Decay of $^{238}_{92}$U

In Example 4, the energy released by the α decay of $^{238}_{92}$U is found to be 4.3 MeV. Since this energy is carried away as kinetic energy of the recoiling $^{234}_{90}$Th nucleus and the α particle, it follows that $KE_{Th} + KE_\alpha = 4.3$ MeV. However, KE_{Th} and KE_α are not equal. Which particle carries away more kinetic energy, the $^{234}_{90}$Th nucleus or the α particle?

Reasoning and Solution Kinetic energy depends on the mass m and speed v of a particle, since $KE = \frac{1}{2}mv^2$. The $^{234}_{90}$Th nucleus has a much greater mass than the α particle, and since the kinetic energy is proportional to the mass, it is tempting to conclude that the $^{234}_{90}$Th nucleus has the greater kinetic energy. This conclusion is not correct, however, since it does not take into account the fact that the $^{234}_{90}$Th nucleus and the α particle have different speeds after the decay. In fact, we expect the thorium nucleus to recoil with the smaller speed precisely *because* it has the greater mass. The decaying $^{238}_{92}$U is like a father and his young daughter on ice skates, pushing off against one another. The more massive father recoils with much less speed than the daughter. We can use the principle of conservation of linear momentum to verify our expectation.

As Section 7.2 discusses, the conservation principle states that the total linear momentum of an isolated system remains constant. An isolated system is one for which the vector sum of the external forces acting on the system is zero, and the decaying $^{238}_{92}$U nucleus fits this description. It is stationary initially, and since momentum is mass times velocity, its initial momentum is zero. In its final form, the system consists of the $^{234}_{90}$Th nucleus and the α particle and has a final total momentum of $m_{Th}v_{Th} + m_\alpha v_\alpha$. According to momentum conservation, the initial and final values of the total momentum of the system must be the same, so that $m_{Th}v_{Th} + m_\alpha v_\alpha = 0$. Solving this equation for the velocity of the thorium nucleus, we find that $v_{Th} = -m_\alpha v_\alpha/m_{Th}$. Since m_{Th} is much greater than m_α, we can see that the speed of the thorium nucleus is less than the speed of the α particle. Moreover, the kinetic energy depends on the square of the speed and only the first power of the mass. As a result of its much greater speed, ***the α particle has the greater kinetic energy.***

Related Homework: *Problem 24*

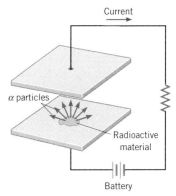

Figure 31.10 A smoke detector.

One widely used application of α decay is in smoke detectors. Figure 31.10 illustrates how a smoke detector operates. Two small and parallel metal plates are separated by a distance of about one centimeter. A tiny amount of radioactive material at the center of one of the plates emits α particles, which collide with air molecules. During the collisions, the air molecules are ionized to form positive and negative ions. The voltage from a battery causes one plate to be positive and the other negative, so that each plate attracts ions of opposite charge. As a result there is a current in the circuit attached to the plates. The presence of smoke particles between the plates reduces the current, since the ions that collide with a smoke particle are usually neutralized. The drop in current that smoke particles cause is used to trigger an alarm.

The physics of
radioactivity and smoke detectors.

β DECAY

The β rays in Figure 31.8 are deflected by the magnetic field in a direction opposite to that of the positively charged α rays. Consequently, these β rays, which are the most common kind, consist of negatively charged particles or β^- particles. Experiment shows that β^- particles are electrons. As an illustration of β^- decay, consider the thorium $^{234}_{90}$Th nucleus, which decays by emitting a β^- particle, as in Figure 31.11:

$$^{234}_{90}\text{Th} \longrightarrow ^{234}_{91}\text{Pa} + ^{\ 0}_{-1}\text{e}$$

| Parent nucleus (thorium) | Daughter nucleus (protactinium) | β^- particle (electron) |

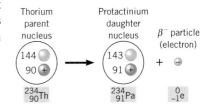

Figure 31.11 β decay occurs when a neutron in an unstable parent nucleus decays into a proton and an electron, the electron being emitted as the β^- particle. In the process, the parent nucleus is transformed into the daughter nucleus.

β^- decay, like α decay, causes a transmutation of one element into another. In this case, thorium $^{234}_{90}$Th is converted into protactinium $^{234}_{91}$Pa. The law of conservation of charge is obeyed, since the net number of positive charges is the same before (90) and after (91 − 1) the β^- emission. The law of conservation of nucleon number is obeyed, since the nucleon number remains at $A = 234$. The general form for β^- decay is

β^- *Decay*
$$_Z^A\text{P} \longrightarrow _{Z+1}^{A}\text{D} + _{-1}^{0}\text{e}$$

Parent   Daughter β^- particle
nucleus   nucleus   (electron)

The electron emitted in β^- decay does *not* actually exist within the parent nucleus and is *not* one of the orbital electrons. Instead, the electron is created when a neutron decays into a proton and an electron; when this occurs, the proton number of the parent nucleus increases from Z to $Z + 1$ and the nucleon number remains unchanged. The electron is usually fast-moving and escapes from the atom, leaving behind a positively charged atom.

Example 6 illustrates that energy is released during β^- decay, just as it is during α decay, and that the conservation of mass/energy applies.

Example 6 β^- Decay and the Release of Energy

The atomic mass of thorium $_{90}^{234}\text{Th}$ is 234.043 59 u, and the atomic mass of protactinium $_{91}^{234}\text{Pa}$ is 234.043 30 u. Find the energy released when β^- decay changes $_{90}^{234}\text{Th}$ into $_{91}^{234}\text{Pa}$.

Reasoning and Solution To find the energy released, we follow the usual procedure of determining how much the mass has decreased because of the decay and then calculating the equivalent energy. The decay and the masses are shown below:

$$\underbrace{_{90}^{234}\text{Th}}_{234.043\ 59\ \text{u}} \longrightarrow \underbrace{_{91}^{234}\text{Pa}}_{234.043\ 30\ \text{u}} + _{-1}^{0}\text{e}$$

Problem solving insight
In β^- decay, be careful not to include the mass of the electron ($_{-1}^{0}$e) twice. As discussed here for the daughter atom ($_{91}^{234}$Pa), the atomic mass already includes the mass of the emitted electron.

When the $_{90}^{234}\text{Th}$ nucleus of a thorium atom is converted into a $_{91}^{234}\text{Pa}$ nucleus, the number of orbital electrons remains the same, so the resulting protactinium atom is missing one orbital electron. However, the given mass includes all 91 electrons of a neutral protactinium atom. In effect, then, the value of 234.043 30 u for $_{91}^{234}\text{Pa}$ already includes the mass of the β^- particle. The mass decrease that accompanies the β^- decay is 234.043 59 u − 234.043 30 u = 0.000 29 u. The equivalent energy (1 u = 931.5 MeV) is $\boxed{0.27\ \text{MeV}}$. This is the maximum kinetic energy that the emitted electron can have.

A second kind of β decay sometimes occurs.* In this process the particle emitted by the nucleus is a *positron* rather than an electron. A positron, also called a β^+ particle, has the same mass as an electron but carries a charge of $+e$ instead of $-e$. The disintegration process for β^+ decay is

β^+ *Decay*
$$_Z^A\text{P} \longrightarrow _{Z-1}^{A}\text{D} + _{1}^{0}\text{e}$$

Parent   Daughter β^+ particle
nucleus   nucleus   (positron)

The emitted positron does *not* exist within the nucleus but, rather, is created when a nuclear proton is transformed into a neutron. In the process, the proton number of the parent nucleus decreases from Z to $Z - 1$, and the nucleon number remains the same. As with β^- decay, the laws of conservation of charge and nucleon number are obeyed, and there is a transmutation of one element into another.

γ DECAY

The nucleus, like the orbital electrons, exists only in discrete energy states or levels. When a nucleus changes from an excited energy state (denoted by an asterisk *) to a lower energy state, a photon is emitted. The process is similar to the one discussed in Section 30.3 for the photon emission that leads to the hydrogen atom line spectrum. With nuclear energy levels, however, the photon has a much greater energy and is called a γ ray.

---

* A third kind of β decay also occurs in which a nucleus pulls in or captures one of the orbital electrons from outside the nucleus. The process is called *electron capture*, or **K capture,** since the electron normally comes from the innermost or K shell.

The γ decay process is written as follows:

γ *Decay*

$$_{Z}^{A}\text{P*} \longrightarrow \, _{Z}^{A}\text{P} \, + \, \gamma$$

Excited   Lower γ ray
energy state   energy state

γ decay does *not* cause a transmutation of one element into another. In the next example the wavelength of one particular γ ray photon is determined.

Example 7 The Wavelength of a Photon Emitted During γ Decay

What is the wavelength of the 0.186 MeV γ-ray photon emitted by radium $_{88}^{226}$Ra?

Reasoning The photon energy is the difference between two nuclear energy levels. Equation 30.4 gives the relation between the energy level separation ΔE and the frequency f of the photon as $\Delta E = hf$. Since $f\lambda = c$, the wavelength of the photon is $\lambda = hc/\Delta E$.

Solution First we must convert the photon energy into joules:

$$\Delta E = (0.186 \times 10^6 \text{ eV}) \left(\frac{1.60 \times 10^{-19} \text{ J}}{1 \text{ eV}} \right) = 2.98 \times 10^{-14} \text{ J}$$

The wavelength of the photon is

$$\lambda = \frac{hc}{\Delta E} = \frac{(6.63 \times 10^{-34} \text{ J} \cdot \text{s})(3.00 \times 10^8 \text{ m/s})}{2.98 \times 10^{-14} \text{ J}} = \boxed{6.67 \times 10^{-12} \text{ m}}$$

Problem solving insight
The energy ΔE of a γ-ray photon, like that of photons in other regions of the electromagnetic spectrum (visible, infrared, microwave, etc.), is equal to the product of Planck's constant h and the frequency f of the photon: $\Delta E = hf$.

Gamma Knife radiosurgery is becoming a very promising medical procedure for treating certain problems of the brain, including benign and cancerous tumors, as well as blood vessel malformations. The procedure, which involves no knife at all, uses powerful, highly focused beams of γ rays aimed at the tumor or malformation. The γ rays are emitted by a radioactive cobalt-60 source. As Figure 31.12*a* illustrates, the patient wears a protective metal helmet that is perforated with 201 small holes. Part *b* of the figure shows that the holes focus the γ rays to a single tiny target within the brain. The target tissue thus receives a very intense dose of radiation and is destroyed, while the surrounding healthy tissue is undamaged. Gamma Knife surgery is a noninvasive, painless, and bloodless procedure that is often performed under local anesthesia. Hospital stays are 70 to 90 percent shorter than with conventional surgery, and patients often return to work within a few days.

An exercise thallium heart scan is a test that uses radioactive thallium to produce images of the heart muscle. When combined with an exercise test, such as walking on a treadmill, the thallium scan helps identify regions of the heart that are not receiving enough blood. The scan is especially useful in diagnosing the presence of blockages in the coronary arteries, the vessels that supply oxygen-rich blood to the heart muscle. During the test, a small amount of thallium is injected into a vein while the patient walks on a

 The physics of
Gamma Knife radiosurgery.

Need more practice?

Interactive LearningWare 31.1
Sodium $_{11}^{24}$Na (atomic mass = 23.99 u) emits a γ ray that has an energy of 0.423 MeV. Assuming that the $_{11}^{24}$Na nucleus is initially at rest, find the speed with which the nucleus recoils. Ignore relativistic effects.

Related Homework: Problem 28

Go to
www.wiley.com/college/cutnell
for an interactive solution.

 The physics of
an exercise thallium heart scan.

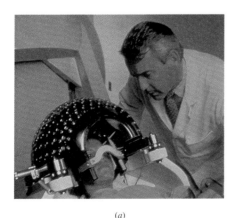

(a)

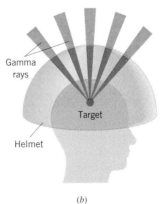

(b)

Figure 31.12 (*a*) In Gamma Knife radiosurgery, a protective metal helmet containing 201 small holes is placed over the patient's head. (Courtesy Elekta Instruments, Inc.) (*b*) The holes focus the beams of γ rays to a tiny target within the brain.

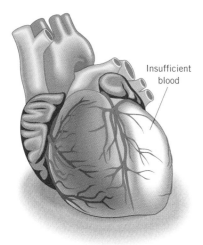

Figure 31.13 An exercise thallium heart scan indicates regions of the heart that receive insufficient blood during exercise.

treadmill. The thallium attaches itself to the red blood cells and is carried throughout the body. The thallium enters the heart muscle by way of the coronary arteries and collects in the cells of the heart muscle that come into contact with the blood. The thallium isotope used, $^{201}_{81}$Tl, emits γ rays, which a special camera records. Since the thallium reaches those regions of the heart that have an adequate blood supply, lesser amounts show up in areas where the blood flow has been reduced due to partially blocked arteries (see Figure 31.13). A second set of images is taken several hours later, while the patient is resting. These images help differentiate between regions of the heart that temporarily do not receive enough blood (the blood flow returns to normal after the exercise) and regions that are permanently damaged due to, for example, a previous heart attack (the blood flow does not return to normal).

✔**Check Your Understanding 3**

Polonium $^{216}_{84}$Po undergoes α decay to produce a daughter nucleus that itself undergoes β^- decay. Which one of the following nuclei is the one that ultimately results: (a) $^{211}_{82}$Pb, (b) $^{211}_{81}$Tl, (c) $^{212}_{81}$Tl, (d) $^{212}_{83}$Bi, (e) $^{213}_{82}$Pb? *(The answer is given at the end of the book.)*

Background: During a nuclear disintegration, the electric charge and the nucleon number are conserved, meaning that these quantities remain unchanged when a nucleus disintegrates into nuclear fragments and the accompanying α and β particles.

For similar questions (including calculational counterparts), consult Self-Assessment Test 31.1, which is described at the end of Section 31.5.

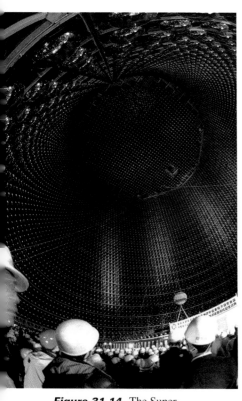

Figure 31.14 The Super-Kamiokande neutrino detector in Japan consists of a steel cylindrical tank containing 12.5 million gallons of ultrapure water. Its inner wall is lined with 11 000 photomultiplier tubes. (© Kyodo News International)

31.5 The Neutrino

When a β particle is emitted by a radioactive nucleus, energy is simultaneously released, as Example 6 illustrates. Experimentally, however, it is found that most β particles do not have enough kinetic energy to account for all the energy released. If a β particle carries away only part of the energy, where does the remainder go? The question puzzled physicists until 1930, when Wolfgang Pauli proposed that part of the energy is carried away by another particle that is emitted along with the β particle. This additional particle is called the ***neutrino,*** and its existence was verified experimentally in 1956. The Greek letter nu (ν) is used to symbolize the neutrino. For instance, the β^- decay of thorium $^{234}_{90}$Th (see Section 31.4) is more correctly written as

$$^{234}_{90}\text{Th} \longrightarrow \,^{234}_{91}\text{Pa} + \,^{0}_{-1}\text{e} + \bar{\nu}$$

The bar above the ν is included because the neutrino emitted in this particular decay process is an antimatter neutrino or antineutrino. A normal neutrino (ν without the bar) is emitted when β^+ decay occurs.

The neutrino has zero electric charge and is extremely difficult to detect because it interacts very weakly with matter. For example, the average neutrino can penetrate one light-year of lead (about 9.5×10^{15} m) without interacting with it. Thus, even though trillions of neutrinos pass through our bodies every second, they have no effect. Although difficult, it is possible to detect neutrinos. Figure 31.14 shows the Super-Kamiokande neutrino detector in Japan. It is located 915 m underground and consists of a steel cylindrical tank, ten stories tall, whose inner wall is lined with 11 000 photomultiplier tubes. The tank is filled with 12.5 million gallons of ultrapure water. Neutrinos colliding with the water molecules produce light patterns that the photomultiplier tubes detect.

One of the major scientific questions of our time is whether neutrinos have mass. The question is important because neutrinos are so plentiful in the universe. Even a very small mass could account for a significant portion of the mass in the universe and, possibly, have an effect on the formation of galaxies. In 1998 the Super-Kamiokande detector yielded the first strong, but indirect, evidence that neutrinos do indeed have mass. (The mass of the electron neutrino appears to be a tiny fraction of the mass of an electron.) This finding implies that neutrinos travel at less than the speed of light. If the neutrino's mass were zero, like that of a photon, it would travel at the speed of light.

The emission of neutrinos and β particles involves a force called *the weak nuclear force* because it is much weaker than the strong nuclear force. It is now known that the weak nuclear force and the electromagnetic force are two different manifestations of a single, more fundamental force, the **electroweak force.** The theory for the electroweak force was developed by Sheldon Glashow (1932–), Abdus Salam (1926–1996), and Steven Weinberg (1933–), who shared a Nobel Prize for their achievement in 1979. The electroweak force, the gravitational force, and the strong nuclear force are the three fundamental forces in nature.

 Self-Assessment Test 31.1

Test your understanding of the material in Sections 31.1–31.5:

• Nuclear Structure   • The Strong Nuclear Force   • The Mass Defect and the Binding Energy of the Nucleus   • Radioactivity   • The Neutrino

Go to **www.wiley.com/college/cutnell**

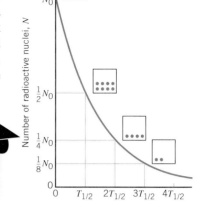

Figure 31.15 The half-life $T_{1/2}$ of a radioactive decay is the time in which one-half of the radioactive nuclei disintegrate.

31.6 *Radioactive Decay and Activity*

The question of which radioactive nucleus in a group of nuclei disintegrates at a given instant is decided like the drawing of numbers in a state lottery; individual disintegrations occur randomly. As time passes, the number N of parent nuclei decreases, as Figure 31.15 shows. This graph of N versus time indicates that the decrease occurs in a smooth fashion, with N approaching zero after enough time has passed. To help describe the graph, it is useful to define the **half-life** $T_{1/2}$ of a radioactive isotope as the time required for one-half of the nuclei present to disintegrate. For example, radium $^{226}_{88}$Ra has a half-life of 1600 years, because it takes this amount of time for one-half of a given quantity of this isotope to disintegrate. In another 1600 years, one-half of the remaining radium atoms will have disintegrated, leaving only one-fourth of the original number intact. In Figure 31.15, the number of nuclei present at time $t = 0$ s is $N = N_0$, and the number present at $t = T_{1/2}$ is $N = \frac{1}{2}N_0$. The number present at $t = 2T_{1/2}$ is $N = \frac{1}{4}N_0$, and so on. The value of the half-life depends on the nature of the radioactive nucleus. Values ranging from a fraction of a second to billions of years have been found. See Table 31.2.

Radon $^{222}_{86}$Rn is a naturally occurring radioactive gas produced when radium $^{226}_{88}$Ra undergoes α decay. There is a nationwide concern about radon as a health hazard because radon in the soil is gaseous and can enter the basement of homes through cracks in the foundation. Once inside, the concentration of radon can rise markedly, depending on the type of housing construction and the concentration of radon in the surrounding soil. Radon gas decays into daughter nuclei that are also radioactive. The radioactive nuclei can attach to dust and smoke particles that can be inhaled, and they remain in the lungs to release tissue-damaging radiation. Prolonged exposure to high levels of radon can lead to lung cancer. Since radon gas concentrations can be measured with inexpensive monitoring devices, it is recommended that all homes be tested for radon. Example 8 deals with the half-life of radon $^{222}_{86}$Rn.

Table 31.2 *Some Half-lives for Radioactive Decay*

Isotope		Half-life
Polonium	$^{214}_{84}$Po	1.64×10^{-4} s
Krypton	$^{89}_{36}$Kr	3.16 min
Radon	$^{222}_{86}$Rn	3.83 da
Strontium	$^{90}_{38}$Sr	28.8 yr
Radium	$^{226}_{88}$Ra	1.6×10^3 yr
Carbon	$^{14}_{6}$C	5.73×10^3 yr
Uranium	$^{238}_{92}$U	4.47×10^9 yr
Indium	$^{115}_{49}$In	4.41×10^{14} yr

The physics of **radioactive radon gas in houses.**

Example 8 The Radioactive Decay of Radon Gas

Suppose 3.0×10^7 radon atoms are trapped in a basement at the time the basement is sealed against further entry of the gas. The half-life of radon is 3.83 days. How many radon atoms remain after 31 days?

Reasoning During each half-life, the number of radon atoms is reduced by a factor of two. Thus, we determine the number of half-lives there are in a period of 31 days and reduce the number of radon atoms by a factor of two for each one.

Solution In a period of 31 days there are (31 days)/(3.83 days) = 8.1 half-lives. In 8 half-lives the number of radon atoms is reduced by a factor of $2^8 = 256$. Ignoring the difference between 8 and 8.1 half-lives, we find that the number of atoms remaining is $(3.0 \times 10^7)/256 = \boxed{1.2 \times 10^5}$.

The **activity** of a radioactive sample is the number of disintegrations per second that occur. Each time a disintegration occurs, the number N of radioactive nuclei decreases. As a result, the activity can be obtained by dividing ΔN, the change in the number of nuclei, by Δt, the time interval during which the change takes place; the average activity over the time interval Δt is the magnitude of $\Delta N/\Delta t$, or $|\Delta N/\Delta t|$. Since the decay of any individual nucleus is completely random, the number of disintegrations per second that occurs in a sample is proportional to the number of radioactive nuclei present, so that

$$\frac{\Delta N}{\Delta t} = -\lambda N \tag{31.4}$$

where λ is a proportionality constant referred to as the **decay constant.** The minus sign is present in this equation because each disintegration decreases the number N of nuclei originally present.

The SI unit for activity is the *becquerel* (Bq); one becquerel equals one disintegration per second. Activity is also measured in terms of a unit called the *curie* (Ci), in honor of Marie (1867–1934) and Pierre (1859–1906) Curie, the discoverers of radium and polonium. Historically, the curie was chosen as a unit because it is roughly the activity of one gram of pure radium. In terms of becquerels,

$$1 \text{ Ci} = 3.70 \times 10^{10} \text{ Bq}$$

The activity of the radium put into the dial of a watch to make it glow in the dark is about 4×10^4 Bq, and the activity used in radiation therapy for cancer treatment is approximately a billion times greater, or 4×10^{13} Bq.

The mathematical expression for the graph of N versus t shown in Figure 31.15 can be obtained from Equation 31.4 with the aid of calculus. The result for the number N of radioactive nuclei present at time t is

$$N = N_0 e^{-\lambda t} \tag{31.5}$$

assuming that the number at $t = 0$ s is N_0. The exponential e has the value $e = 2.718 \ldots$, and many calculators provide the value of e^x. We can relate the half-life $T_{1/2}$ of a radioactive nucleus to its decay constant λ in the following manner. By substituting $N = \frac{1}{2}N_0$ and $t = T_{1/2}$ into Equation 31.5, we find that $\frac{1}{2} = e^{-\lambda T_{1/2}}$. Taking the natural logarithm of both sides of this equation reveals that $\ln 2 = \lambda T_{1/2}$ or

$$T_{1/2} = \frac{\ln 2}{\lambda} = \frac{0.693}{\lambda} \tag{31.6}$$

The following example illustrates the use of Equations 31.5 and 31.6.

Concept Simulation 31.1

This simulation graphically illustrates the decay of radioactive nuclei. The user can select the initial number of nuclei and the decay constant λ. The simulation then plots the number of nuclei remaining as a function of time. The resulting curve, which is described by Equation 31.5, decays exponentially in time.

Go to
www.wiley.com/college/cutnell

Example 9   The Activity of Radon $^{222}_{86}\text{Rn}$

As in Example 8, suppose there are 3.0×10^7 radon atoms ($T_{1/2} = 3.83$ days or 3.31×10^5 s) trapped in a basement. (a) How many radon atoms remain after 31 days? Find the activity (b) just after the basement is sealed against further entry of radon and (c) 31 days later.

Reasoning The number N of radon atoms remaining after a time t is given by $N = N_0 e^{-\lambda t}$, where $N_0 = 3.0 \times 10^7$ is the original number of atoms when $t = 0$ s and λ is the decay constant. The decay constant is related to the half-life $T_{1/2}$ of the radon atoms by $\lambda = 0.693/T_{1/2}$. The activity can be obtained from Equation 31.4, $\Delta N/\Delta t = -\lambda N$.

Solution

(a) The decay constant is

$$\lambda = \frac{0.693}{T_{1/2}} = \frac{0.693}{3.83 \text{ days}} = 0.181 \text{ days}^{-1} \tag{31.6}$$

and the number N of radon atoms remaining after 31 days is

$$N = N_0 e^{-\lambda t} = (3.0 \times 10^7)e^{-(0.181 \text{ days}^{-1})(31 \text{ days})} = \boxed{1.1 \times 10^5} \tag{31.5}$$

This value is slightly different from that found in Example 8 because there we ignored the difference between 8.0 and 8.1 half-lives.

(b) The activity can be obtained from Equation 31.4, provided the decay constant is expressed in reciprocal seconds:

$$\lambda = \frac{0.693}{T_{1/2}} = \frac{0.693}{3.31 \times 10^5 \text{ s}} = 2.09 \times 10^{-6} \text{ s}^{-1} \tag{31.6}$$

Thus, the number of disintegrations per second is

$$\frac{\Delta N}{\Delta t} = -\lambda N = -(2.09 \times 10^{-6} \text{ s}^{-1})(3.0 \times 10^7) = -63 \text{ disintegrations/s} \tag{31.4}$$

The activity is the magnitude of $\Delta N/\Delta t$, so initially $\boxed{\text{Activity} = 63 \text{ Bq}}$.

(c) From part (a), the number of radioactive nuclei remaining at the end of 31 days is $N = 1.1 \times 10^5$, and reasoning similar to that in part (b) reveals that $\boxed{\text{Activity} = 0.23 \text{ Bq}}$.

▲

✔ Check Your Understanding 4

The thallium $^{208}_{81}$Tl nucleus is radioactive with a half-life of 3.053 min. At a given instant, the activity of a certain sample of thallium is 2400 Bq. Using the concept of half-life, and without doing any written calculations, determine whether the activity 9 minutes later is (a) a little less than $\frac{1}{8}(2400 \text{ Bq}) = 300$ Bq, (b) a little more than $\frac{1}{8}(2400 \text{ Bq}) = 300$ Bq, (c) a little less than $\frac{1}{3}(2400 \text{ Bq}) = 800$ Bq, (d) a little more than $\frac{1}{3}(2400 \text{ Bq}) = 800$ Bq. *(The answer is given at the end of the book.)*

Background: The half-life of a radioactive sample is the time required for one-half of the nuclei present to disintegrate. The activity is the product of the number of nuclei present and the decay constant.

For similar questions (including calculational counterparts), consult Self-Assessment Test 31.2, which is described at the end of Section 31.9.

31.7 *Radioactive Dating*

One important application of radioactivity is the determination of the age of archeological or geological samples. If an object contains radioactive nuclei when it is formed, then the decay of these nuclei marks the passage of time like a clock, half of the nuclei disintegrating during each half-life. If the half-life is known, a measurement of the number of nuclei present today relative to the number present initially can give the age of the sample. According to Equation 31.4, the activity of a sample is proportional to the number of radioactive nuclei, so one way to obtain the age is to compare present activity with initial activity. A more accurate way is to determine the present number of radioactive nuclei with the aid of a mass spectrometer.

The present activity of a sample can be measured, but how is it possible to know what the original activity was, perhaps thousands of years ago? Radioactive dating methods entail certain assumptions that make it possible to estimate the original activity. For instance, the radiocarbon technique utilizes the $^{14}_{6}$C isotope of carbon, which undergoes β^- decay with a half-life of 5730 yr. This isotope is present in the earth's atmosphere at an equilibrium concentration of about one atom for every 8.3×10^{11} atoms of normal carbon $^{12}_{6}$C. It is often assumed* that this value has remained constant over the years because $^{14}_{6}$C is created when cosmic rays interact with the earth's upper atmosphere, a production method that offsets the loss via β^- decay. Moreover, nearly all living organisms ingest the equilibrium concentration of $^{14}_{6}$C. However, once an organism dies, metabolism no longer sustains the input of $^{14}_{6}$C, and β^- decay causes half of the $^{14}_{6}$C nuclei to disintegrate every 5730 years. Example 10 illustrates how to determine the $^{14}_{6}$C activity of one gram of carbon in a living organism.

The physics of radioactive dating.

This mummy, with a sun-bleached skull and surrounded by burial artifacts, was found in the arid highlands of southern Peru. The extreme dryness helped to preserve the remains, thought to be around 2000 years old. Radioactive dating is one of the techniques used to determine the age of such artifacts.
(© David Nunuk/Photo Researchers)

---

* The assumption that the $^{14}_{6}$C concentration has always been at its present equilibrium value has been evaluated by comparing $^{14}_{6}$C ages with ages determined by counting tree rings. More recently, ages determined using the radioactive decay of uranium $^{238}_{92}$U have been used for comparison. These comparisons indicate that the equilibrium value of the $^{14}_{6}$C concentration has indeed remained constant for the past 1000 years. However, from there back about 30 000 years, it appears that the $^{14}_{6}$C concentration in the atmosphere was larger than its present value by up to 40%. As a first approximation we ignore such discrepancies.

Example 10 $^{14}_{6}C$ Activity Per Gram of Carbon in a Living Organism

(a) Determine the number of carbon $^{14}_{6}C$ atoms present for every gram of carbon $^{12}_{6}C$ in a living organism. Find (b) the decay constant and (c) the activity of this sample.

Reasoning The total number of carbon $^{12}_{6}C$ atoms in one gram of carbon $^{12}_{6}C$ is equal to the corresponding number of moles times Avogadro's number (see Section 14.1). Since there is only one $^{14}_{6}C$ atom for every 8.3×10^{11} atoms of $^{12}_{6}C$, the number of $^{14}_{6}C$ atoms is equal to the total number of $^{12}_{6}C$ atoms divided by 8.3×10^{11}. The decay constant λ for $^{14}_{6}C$ is $\lambda = 0.693/T_{1/2}$, where $T_{1/2}$ is the half-life. The activity is equal to the magnitude of $\Delta N/\Delta t$, which is equal to the decay constant times the number of $^{14}_{6}C$ atoms present, according to Equation 31.4.

Solution

(a) One gram of carbon $^{12}_{6}C$ (atomic mass = 12 u) is equivalent to 1.0/12 mol. Since Avogadro's number is 6.02×10^{23} atoms/mol and since there is one $^{14}_{6}C$ atom for every 8.3×10^{11} atoms of $^{12}_{6}C$, the number of $^{14}_{6}C$ atoms is

$$\begin{array}{l}\text{Number of } ^{14}_{6}C \\ \text{atoms for every 1.0} \\ \text{gram of carbon } ^{12}_{6}C\end{array} = \left(\frac{1.0}{12}\text{ mol}\right)\left(6.02 \times 10^{23}\frac{\text{atoms}}{\text{mol}}\right)\left(\frac{1}{8.3 \times 10^{11}}\right)$$

$$= \boxed{6.0 \times 10^{10}\text{ atoms}}$$

(b) Since the half-life of $^{14}_{6}C$ is 5730 yr (1.81×10^{11} s), the decay constant is

$$\lambda = \frac{0.693}{T_{1/2}} = \frac{0.693}{1.81 \times 10^{11}\text{ s}} = \boxed{3.83 \times 10^{-12}\text{ s}^{-1}} \tag{31.6}$$

(c) Equation 31.4 indicates that $\Delta N/\Delta t = -\lambda N$, so the magnitude of $\Delta N/\Delta t$ is λN.

$$\begin{array}{l}\text{Activity of } ^{14}_{6}C \text{ for} \\ \text{every 1.0 gram of} \\ \text{carbon } ^{12}_{6}C \text{ in a} \\ \text{living organism}\end{array} = \lambda N = (3.83 \times 10^{-12}\text{ s}^{-1})(6.0 \times 10^{10}\text{ atoms}) = \boxed{0.23\text{ Bq}}$$

An organism that lived thousands of years ago presumably had an activity of about 0.23 Bq per gram of carbon. When the organism died, the activity began decreasing. From a sample of the remains, the current activity per gram of carbon can be measured and compared to the value of 0.23 Bq to determine the time that has transpired since death. This procedure is illustrated in Example 11.

Example 11   The Iceman

Figure 31.16 These remains of the Iceman were discovered in the ice of a glacier in the Italian Alps in 1991. Radiocarbon dating reveals his age (see Example 11). (© Corbis Sygma)

On 19 September 1991, German tourists on a walking trip in the Italian Alps found a Stone Age traveler, later called the Iceman, whose body had become trapped in a glacier. Figure 31.16 shows the well-preserved remains that were dated using the radiocarbon method. Material found with the body had a $^{14}_{6}C$ activity of about 0.121 Bq per gram of carbon. Find the age of the Iceman's remains.

Reasoning According to Equation 31.5, the number of nuclei remaining at time t is $N = N_0 e^{-\lambda t}$. Multiplying both sides of this expression by the decay constant λ and recognizing that the product of λ and N is the activity A, we find that $A = A_0 e^{-\lambda t}$, where $A_0 = 0.23$ Bq is the activity at time $t = 0$ s for one gram of carbon. The decay constant λ can be determined from the value of 5730 yr for the half-life of $^{14}_{6}C$, using Equation 31.6. With known values for A_0 and λ, the given activity of $A = 0.121$ Bq per gram of carbon can be used to find t, the Iceman's age.

Solution For $^{14}_{6}C$, the decay constant is $\lambda = 0.693/T_{1/2} = 0.693/(5730\text{ yr}) = 1.21 \times 10^{-4}\text{ yr}^{-1}$. Since $A = 0.121$ Bq and $A_0 = 0.23$ Bq, the age can be determined from

$$A = 0.121\text{ Bq} = (0.23\text{ Bq})e^{-(1.21 \times 10^{-4}\text{ yr}^{-1})t}$$

Taking the natural logarithm of both sides of this result gives

$$\ln\left(\frac{0.121\text{ Bq}}{0.23\text{ Bq}}\right) = -(1.21 \times 10^{-4}\text{ yr}^{-1})t$$

which gives an age for the sample of $\boxed{t = 5300\text{ yr}}$.

Radiocarbon dating is not the only radioactive dating method. For example, other methods utilize uranium $^{238}_{92}$U, potassium $^{40}_{19}$K, and lead $^{210}_{82}$Pb. For such methods to be useful, the half-life of the radioactive species must be neither too short nor too long relative to the age of the sample to be dated, as Conceptual Example 12 discusses.

Conceptual Example 12 Dating a Bottle of Wine

A bottle of red wine is thought to have been sealed about 5 years ago. The wine contains a number of different kinds of atoms, including carbon, oxygen, and hydrogen. Each of these has a radioactive isotope. The radioactive isotope of carbon is the familiar $^{14}_{6}$C, with a half-life of 5730 yr. The radioactive isotope of oxygen is $^{15}_{8}$O and has a half-life of 122.2 s. The radioactive isotope of hydrogen is $^{3}_{1}$H and is called tritium; its half-life is 12.33 yr. The activity of each of these isotopes is known at the time the bottle was sealed. However, only one of the isotopes is useful for determining the age of the wine accurately. Which is it?

Reasoning and Solution In a dating method that measures the activity of a radioactive isotope, the age of the sample is related to the change in the activity during the time period in question. Here the expected age is about 5 years. This period is only a small fraction of the 5730-yr half-life of $^{14}_{6}$C. As a result, relatively few of the $^{14}_{6}$C nuclei would decay during the wine's life, and the measured activity would change little from its initial value. To obtain an accurate age from such a small change would require prohibitively precise measurements. The $^{15}_{8}$O isotope is not very useful, either; the difficulty is its relatively short half-life of 122.2 s. During a 5-year period, so many half-lives of 122.2 s would occur that the activity would decrease to a vanishingly small level. It would not be even possible to measure it. The only remaining option is ***the tritium isotope of hydrogen.*** The expected age of 5 yr is long enough relative to the half-life of 12.33 yr that a measurable change in activity will occur, but not so long that the activity will have completely vanished for all practical purposes.

Related Homework: *Conceptual Question 14, Problem 42*

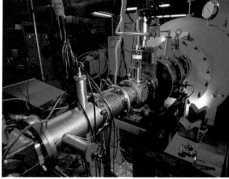

In the radioactive dating technique that uses the carbon $^{14}_{6}$C isotope, the relatively few $^{14}_{6}$C atoms can be detected by measuring their activity, as we have seen. It is also possible to detect these atoms more accurately using an accelerator mass spectrometer. This photograph shows the linear accelerator used in the procedure.
(© James King-Holmes/Photo Researchers)

31.8 *Radioactive Decay Series*

When an unstable parent nucleus decays, the resulting daughter nucleus is sometimes also unstable. If so, the daughter then decays and produces its own daughter, and so on, until a completely stable nucleus is produced. This sequential decay of one nucleus after another is called a ***radioactive decay series.*** Examples 4–6 discuss the first two steps of a series that begins with uranium $^{238}_{92}$U:

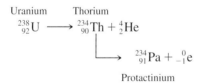

$$\text{Uranium} \quad \text{Thorium}$$
$$^{238}_{92}\text{U} \longrightarrow \,^{234}_{90}\text{Th} + \,^{4}_{2}\text{He}$$
$$\longrightarrow \,^{234}_{91}\text{Pa} + \,^{0}_{-1}\text{e}$$
$$\text{Protactinium}$$

Furthermore, Example 8 deals with radon $^{222}_{86}$Rn, which is formed down the line in the $^{238}_{92}$U radioactive decay series. Figure 31.17 shows the entire series. At several points branches occur because more than one kind of decay is possible for an intermediate species. Ultimately, however, the series ends with lead $^{206}_{82}$Pb, which is stable.

The $^{238}_{92}$U series and other such series are the only sources of some of the radioactive elements found in nature. Radium $^{226}_{88}$Ra, for instance, has a half-life of 1600 yr, which is short enough that all the $^{226}_{88}$Ra created when the earth was formed billions of years ago has now disappeared. The $^{238}_{92}$U series provides a continuing supply of $^{226}_{88}$Ra, however.

31.9 *Radiation Detectors*

There are a number of devices that can be used to detect the particles and photons (γ rays) emitted when a radioactive nucleus decays. Such devices detect the ionization that these particles and photons cause as they pass through matter.

The physics of **radiation detectors.**

The most familiar detector is the ***Geiger counter,*** which Figure 31.18 illustrates. The Geiger counter consists of a gas-filled metal cylinder. The α, β, or γ rays enter the cylin-

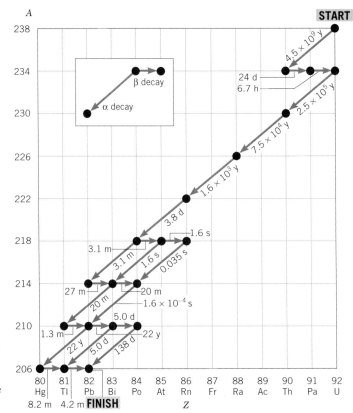

Figure 31.17 The radioactive decay series that begins with uranium $^{238}_{92}$U and ends with lead $^{206}_{82}$Pb. Half-lives are given in seconds (s), minutes (m), hours (h), days (d), or years (y). The inset in the upper left identifies the type of decay that each nucleus undergoes.

der through a thin window at one end. γ rays can also penetrate directly through the metal. A wire electrode runs along the center of the tube and is kept at a high positive voltage (1000–3000 V) relative to the outer cylinder. When a high-energy particle or photon enters the cylinder, it collides with and ionizes a gas molecule. The electron produced from the gas molecule accelerates toward the positive wire, ionizing other molecules in its path. Additional electrons are formed, and an avalanche of electrons rushes toward the wire, leading to a pulse of current through the resistor R. This pulse can be counted or made to produce a "click" in a loudspeaker. The number of counts or clicks is related to the number of disintegrations that produced the particles or photons.

The *scintillation counter* is another important radiation detector. As Figure 31.19 indicates, this device consists of a scintillator mounted on a photomultiplier tube. Often the scintillator is a crystal (e.g., cesium iodide) containing a small amount of impurity (thallium), but plastic, liquid, and gaseous scintillators are also used. In response to ionizing radiation, the scintillator emits a flash of visible light. The photons of the flash then strike the photocathode of the photomultiplier tube. The photocathode is made of a material that emits electrons because of the photoelectric effect. These photoelectrons are then attracted to a special electrode kept at a voltage of about $+100$ V relative to the photocathode. The electrode is coated with a substance that emits several additional electrons for every electron striking it. The additional electrons are attracted to a second similar electrode (voltage $= +200$ V) where they generate even more electrons. Commercial photomultiplier tubes contain as many as 15 of these special electrodes, so photoelectrons resulting from the light flash of the scintillator lead to a cascade of electrons and a pulse of current. As in a Geiger tube, the current pulses can be counted.

Ionizing radiation can also be detected with several types of *semiconductor detectors.* Such devices utilize *n*- and *p*-type materials (see Section 23.5), and their operation depends on the electrons and holes formed in the materials as a result of the radiation. One of the main advantages of semiconductor detectors is their ability to discriminate between two particles with only slightly different energies.

A number of instruments provide a pictorial representation of the path that high-energy particles follow after they are emitted from unstable nuclei. In a *cloud chamber,* a gas is cooled just to the point where it will condense into droplets, provided nucleating

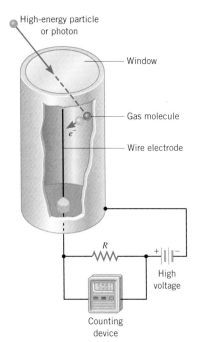

Figure 31.18 A Geiger counter.

agents are available on which the droplets can form. When a high-energy particle, such as an α particle or a β particle, passes through the gas, the ions it leaves behind serve as nucleating agents, and droplets form along the path of the particle. A ***bubble chamber*** works in a similar fashion, except it contains a liquid that is just at the point of boiling. Tiny bubbles form along the trail of a high-energy particle passing through the liquid. The paths revealed in a cloud or bubble chamber can be photographed to provide a permanent record of the event. Figure 31.20 shows a photograph of the tracks in a bubble chamber. A ***photographic emulsion*** also can be used directly to produce a record of the path taken by a particle of ionizing radiation. The ions formed as the particle passes through the emulsion cause silver to be deposited along the track when the emulsion is developed.

Self-Assessment Test 31.2

Test your understanding of the material in Sections 31.6–31.9:

- Radioactive Decay and Activity   • Radioactive Dating
- Radioactive Decay Series   • Radiation Detectors

Go to **www.wiley.com/college/cutnell**

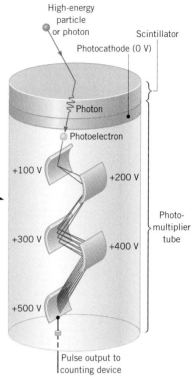

Figure 31.19 A scintillation counter.

31.10 *Concepts & Calculations*

Radioactive decay obeys the conservation laws of physics, and we have studied five of these laws, as the Concepts-at-a-Glance chart in Figure 31.7 illustrates. Two of them are particularly important in understanding the types of radioactivity that occur and the nuclear changes that accompany them. These are the conservation of electric charge and the conservation of nucleon number. Example 13 emphasizes their importance and reviews how they are applied.

Concepts & Calculations Example 13
Electric Charge and Nucleon Number

Thorium $^{228}_{90}$Th produces a daughter nucleus that is radioactive. The daughter, in turn, produces its own radioactive daughter, and so on. This process continues until bismuth $^{212}_{83}$Bi is reached. What are the total number N_α of α particles and the total number N_β of β^- particles that are generated in this series of radioactive decays?

Concept Questions and Answers How many of the 90 protons in the thorium nucleus are carried off by the α particles?

Answer Each α particle is a helium $^4_2$He nucleus and carries off two protons. Therefore, the total number of protons carried off by the α particles is $N_\alpha(2)$.

How many protons are left behind when the β^- particles are emitted?

Answer Each β^- particle is an electron $^0_{-1}$e and is emitted when a neutron in the nucleus decays into a proton and an electron. Therefore, the total number of protons left behind by the β^- particles is N_β.

How many of the 228 nucleons in the thorium nucleus are carried off by the α particles?

Answer Each α particle is a helium $^4_2$He nucleus and carries off four nucleons. Therefore, the total number of nucleons carried off by the α particles is $N_\alpha(4)$.

Does the departure of a β^- particle alter the number of nucleons?

Answer No. Each β^- particle is an electron $^0_{-1}$e and is emitted when a neutron in the nucleus decays into a proton and an electron. In effect, a neutron is replaced by a proton. But since each is a nucleon, the number of nucleons is not changed.

Solution The overall decay process can be written as

$$^{228}_{90}\text{Th} \longrightarrow \,^{212}_{83}\text{Bi} + N_\alpha(^4_2\text{He}) + N_\beta(^0_{-1}\text{e})$$

Since electric charge must be conserved, we know that the charge on the left side of this reaction must be equal to the total charge on the right side, the result being

$$90 = 83 + N_\alpha(2) + N_\beta(-1)$$

Figure 31.20 A photograph showing particle tracks in a bubble chamber. (© LBL/Photo Researchers)

Since the nucleon number must be conserved also, we know that the 228 nucleons on the left must equal the total number of nucleons on the right, so that

$$228 = 212 + N_\alpha(4) + N_\beta(0)$$

Solving this result for the number of α particles gives $\boxed{N_\alpha = 4}$. Substituting $N_\alpha = 4$ into the conservation-of-charge equation, we find that the number of β^- particles is $\boxed{N_\beta = 1}$.

Energy in the form of heat is needed to raise the temperature of an object, as Section 12.7 discusses. One source of the heat can be radioactive decay, as the next example discusses.

Concepts & Calculations Example 14
The Energy from Radioactive Decay

A one-gram sample of thorium $^{228}_{90}$Th contains 2.64×10^{21} atoms and undergoes α decay with a half-life of 1.913 yr (1.677×10^4 h). Each disintegration releases an energy of 5.52 MeV (8.83×10^{-13} J). Assuming that all of the energy is used to heat a 3.8-kg sample of water, find the change in temperature of the water that occurs in one hour.

Concept Questions and Answers How much heat Q is needed to raise the temperature of a mass m of water by ΔT degrees?

Answer According to Equation 12.4, the heat needed is $Q = cm\Delta T$, where the specific heat capacity of water is $c = 4186$ J/(kg·C°).

The energy released by each disintegration is E. What is the total energy E_{Total} released by a number n of disintegrations?

Answer The total energy released is just the number of disintegrations times the energy for each one, or $E_{Total} = nE$.

What is the number n of disintegrations that occur during a time t?

Answer The number of radioactive nuclei remaining after a time t is given by Equations 31.5 and 31.6 as

$$N = N_0 e^{-0.693t/T_{1/2}}$$

where N_0 is the number of radioactive nuclei present at $t = 0$ s and $T_{1/2}$ is the half-life for the decay. Therefore, the number n of disintegrations that occur during the time t is $N_0 - N$ or

$$n = N_0 - N = N_0(1 - e^{-0.693t/T_{1/2}})$$

Solution According to Equation 12.4, the heat needed to change the temperature of the water is $Q = cm\Delta T$. This heat is provided by the total energy released in one hour of radioactive decay, or $E_{Total} = nE$. Setting $Q = E_{Total}$, we obtain $nE = cm\Delta T$, which can be solved for the change in temperature ΔT:

$$\Delta T = \frac{nE}{cm}$$

Using the expression that we obtained in the Concept Questions for the number of disintegrations n that occur in the time t, we obtain

$$\Delta T = \frac{N_0(1 - e^{-0.693t/T_{1/2}})E}{cm}$$

$$= \frac{(2.64 \times 10^{21})[1 - e^{-0.693(1.000 \text{ h})/(1.677 \times 10^4 \text{ h})}](8.83 \times 10^{-13} \text{ J})}{[4186 \text{ J/(kg·C°)}](3.8 \text{ kg})} = \boxed{6.1 \text{ C}°}$$

At the end of the problem set for this chapter, you will find homework problems that contain both conceptual and quantitative parts. These problems are grouped under the heading *Concepts & Calculations, Group Learning Problems*. They are designed for use by students working alone or in small learning groups. The conceptual part of each problem provides a convenient focus for group discussions.

Concept Summary

This summary presents an abridged version of the chapter, including the important equations and all available learning aids. For convenient reference, the learning aids (including the text's examples) are placed next to or immediately after the relevant equation or discussion. The following learning aids may be found on-line at **www.wiley.com/college/cutnell**:

Interactive LearningWare examples are solved according to a five-step interactive format that is designed to help you develop problem-solving skills.	**Concept Simulations** are animated versions of text figures or animations that illustrate important concepts. You can control parameters that affect the display, and we encourage you to experiment.
Interactive Solutions offer specific models for certain types of problems in the chapter homework. The calculations are carried out interactively.	**Self-Assessment Tests** include both qualitative and quantitative questions. Extensive feedback is provided for both incorrect and correct answers, to help you evaluate your understanding of the material.

Topic	Discussion	Learning Aids
	31.1 Nuclear Structure	
Nucleons **Atomic number**	The nucleus of an atom consists of protons and neutrons, which are collectively referred to as nucleons. A neutron is an elecrically neutral particle whose mass is slightly larger than that of the proton. The atomic number Z is the number of protons in the nucleus. The atomic mass number A (or nucleon number) is the total number of protons and neutrons in the nucleus:	
Atomic mass number (or nucleon number)	$$A = Z + N \qquad (31.1)$$ where N is the number of neutrons. For an element whose chemical symbol is X, the symbol for the nucleus is $^{A}_{Z}X$.	
Isotopes	Nuclei that contain the same number of protons, but a different number of neutrons, are called isotopes.	
Radius of a nucleus	The approximate radius (in meters) of a nucleus is given by $$r \approx (1.2 \times 10^{-15} \text{ m}) A^{1/3} \qquad (31.2)$$	**Example 1**
	31.2 The Strong Nuclear Force and the Stability of the Nucleus	
Strong nuclear force	The strong nuclear force is the force of attraction between nucleons (protons and neutrons) and is one of the three fundamental forces of nature. This force balances the electrostatic force of repulsion between protons and holds the nucleus together. The strong nuclear force has a very short range of action and is almost independent of electric charge.	
	31.3 The Mass Defect of the Nucleus and Nuclear Binding Energy	
	The binding energy of a nucleus is the energy required to separate the nucleus into its constituent protons and neutrons. The binding energy is equal to	
Binding energy	$$\text{Binding energy} = (\Delta m)c^2 \qquad (31.3)$$	**Examples 2, 3**
Mass defect	where Δm is the mass defect of the nucleus and c is the speed of light in a vacuum. The mass defect is the amount by which the sum of the individual masses of the protons and neutrons exceeds the mass of the intact nucleus.	
Atomic mass unit	When specifying nuclear masses, it is customary to use the atomic mass unit (u). One atomic mass unit has a mass of 1.6605×10^{-27} kg and is equivalent to an energy of 931.5 MeV.	
	31.4 Radioactivity	
Radioactivity	Unstable nuclei spontaneously decay by breaking apart or rearranging their internal structure in a process called radioactivity. Naturally occurring radioactivity produces α, β, and γ rays. α rays consist of positively charged particles, each particle being the $^{4}_{2}$He nucleus of helium. The general form for α decay is	
α Decay	$$\underbrace{^{A}_{Z}P}_{\substack{\text{Parent} \\ \text{nucleus}}} \longrightarrow \underbrace{^{A-4}_{Z-2}D}_{\substack{\text{Daughter} \\ \text{nucleus}}} + \underbrace{^{4}_{2}He}_{\substack{\alpha \text{ particle} \\ \text{(helium nucleus)}}}$$	**Examples 4, 5** **Interactive Solution 31.25**

Topic	Discussion	Learning Aids
β^- **Decay**	The most common kind of β ray consists of negatively charged particles, or β^- particles, which are electrons. The general form for β^- decay is $$\underbrace{^A_Z P}_{\substack{\text{Parent}\\\text{nucleus}}} \longrightarrow \underbrace{^A_{Z+1} D}_{\substack{\text{Daughter}\\\text{nucleus}}} + \underbrace{^{0}_{-1} e}_{\substack{\beta^-\text{ particle}\\\text{(electron)}}}$$	**Examples 6, 13**
Positron	β^+ decay produces another kind of β ray, which consists of positively charged particles, or β^+ particles. A β^+ particle, also called a positron, has the same mass as an electron, but carries a charge of $+e$ instead of $-e$.	
Transmutation	If a radioactive parent nucleus disintegrates into a daughter nucleus that has a different atomic number, such as that which occurs in α and β decay, one element has been converted into another element, the conversion being referred to as a transmutation.	
γ **Decay**	γ rays are high-energy photons emitted by a radioactive nucleus. The general form for γ decay is $$\underbrace{^A_Z P^*}_{\substack{\text{Excited}\\\text{energy state}}} \longrightarrow \underbrace{^A_Z P}_{\substack{\text{Lower}\\\text{energy state}}} + \underbrace{\gamma}_{\substack{\gamma\text{ ray}}}$$	**Example 7** **Interactive LearningWare 31.1**
	γ decay does not cause a transmutation of one element into another.	

31.5 The Neutrino

Neutrino	The neutrino is an electrically neutral particle that is emitted along with β particles and has a mass that is much, much smaller than the mass of an electron.	

 Use Self-Assessment Test 31.1 to evaluate your understanding of Sections 31.1–31.5.

31.6 Radioactive Decay and Activity

Half-life **Activity**	The half-life of a radioactive isotope is the time required for one-half of the nuclei present to disintegrate or decay. The activity is the number of disintegrations per second that occur. Activity is the magnitude of $\Delta N/\Delta t$, where ΔN is the change in the number N of radioactive nuclei and Δt is the time interval during which the change occurs. In other words, activity is $|\Delta N/\Delta t|$. The SI unit for activity is the becquerel (Bq), one becquerel being one disintegration per second. Activity is also measured in a unit called the curie (Ci); 1 Ci = 3.70×10^{10} Bq.	**Example 8**
	Radioactive decay obeys the following relation $$\frac{\Delta N}{\Delta t} = -\lambda N \qquad (31.4)$$	**Interactive Solution 31.37**
Decay constant	where λ is the decay constant. This equation can be solved by the methods of integral calculus to show that	
Number of nuclei remaining as a function of time	$$N = N_0 e^{-\lambda t} \qquad (31.5)$$ where N_0 is the original number of nuclei. The decay constant λ is related to the half-life $T_{1/2}$ according to	**Concept Simulation 31.1** **Example 14**
Decay constant and half-life	$$\lambda = \frac{0.693}{T_{1/2}} \qquad (31.6)$$	**Example 9**

31.7 Radioactive Dating

	If an object contained radioactive nuclei when it was formed, then the decay of these nuclei can be used to determine the age of the object. One way to obtain the age is to relate the present activity A of an object with its initial activity A_0:	**Example 10**
Activity as a function of time	$$A = A_0 e^{-\lambda t}$$	**Example 11**
	where λ is the decay constant and t is the age of the object. For radiocarbon	

Topic	Discussion	Learning Aids

dating that uses the $^{14}_{6}$C isotope of carbon, the initial activity is often assumed **Example 12**
to be $A_0 = 0.23$ Bq.

31.8 *Radioactive Decay Series*

The sequential decay of one nucleus after another is called a radioactive decay
series. A decay series starts with a radioactive nucleus and ends with a completely stable nucleus. Figure 31.17 illustrates one such series that begins with
uranium $^{238}_{92}$U and ends with lead $^{206}_{82}$Pb.

31.9 *Radiation Detectors*

A number of devices are used to detect α and β particles, as well as γ rays.
These include the Geiger counter, the scintillation counter, semiconductor detectors, cloud and bubble chambers, and photographic emulsions.

  Use *Self-Assessment Test 31.2* **to evaluate your understanding of Sections 31.6–31.9.**

Conceptual Questions

1. A material is known to be an isotope of lead, although the particular isotope is not known. From such limited information,
which of the following quantities can you specify: (a) its atomic
number, (b) its neutron number, (c) its atomic mass number? Explain.

2. Two nuclei have different nucleon numbers A_1 and A_2. Are the
two nuclei necessarily isotopes of the same element? Give your reasoning.

3. Two nuclei have the same radius, even though they contain different numbers of protons and different numbers of neutrons. Explain how this is possible.

4. Using Figure 31.6, rank the following nuclei in ascending order (smallest first) according to the binding energy per
nucleon: phosphorus $^{31}_{15}$P, cobalt $^{59}_{27}$Co, tungsten $^{184}_{74}$W, and thorium $^{232}_{90}$Th.

5. Describe qualitatively how the radius of the daughter nucleus
compares to that of the parent nucleus for (a) α decay and (b) β decay. Justify your answers in terms of Equation 31.2.

6. Why do α and β decay produce new elements, but γ decay does
not?

7. Uranium $^{238}_{92}$U decays into thorium $^{234}_{90}$Th by means of α decay, as
Example 4 in the text discusses. A reasonable question to ask is,
"Why doesn't the $^{238}_{92}$U nucleus just emit a single proton instead of
an α particle?" This hypothetical decay scheme is shown below,
along with the pertinent atomic masses:

$$^{238}_{92}U \longrightarrow \quad ^{237}_{91}Pa \quad + \quad ^{1}_{1}H$$

Uranium	Protactinium	Proton
238.050 78 u	237.051 14 u	1.007 83 u

For a decay to be possible, it must bring the parent nucleus toward a
more stable state by allowing the release of energy. Compare the total mass of the products of this hypothetical decay with the mass of
$^{238}_{92}$U and decide whether the emission of a single proton is possible
for $^{238}_{92}$U. Explain.

8. Explain why unstable nuclei with short half-lives typically have
only a small or zero natural abundance.

9. The half-life of indium $^{115}_{49}$In is 4.41×10^{14} yr. Thus, one-half of
the nuclei in a sample of this isotope will decay in this time, which
is very long. Is it possible for any single nucleus in the sample to decay after only one second? Justify your answer.

10. (a) Is it possible for two different samples that contain different
radioactive elements to have the same activity? (b) Is it possible for
two different samples of the same radioactive element to have different activities? In each case, defend your answer.

11. To which of the following objects, each about 1000 yr old, can
the radiocarbon dating technique *not* be applied: a wooden box, a
gold statue, and some plant seeds? Explain.

12. Two isotopes have half-lives that are short, relative to the known
age of the earth. Today, one of these isotopes is found in nature and
one is not. How is this possible, considering that both should have
long since decayed to zero? Explain.

13. Suppose there were a greater number of carbon $^{14}_{6}$C atoms in a
plant living 5000 yr ago than is currently believed. When the seeds
of this plant are tested today using radiocarbon dating, is the age obtained too small or too large? Give your reasoning.

14. Review Conceptual Example 12 as an aid in answering this
question. Tritium is an isotope of hydrogen and undergoes β^- decay
with a half-life of 12.33 yr. Like carbon $^{14}_{6}$C, tritium is produced in
the atmosphere because of cosmic rays and can be used in a radioactive dating technique. Can tritium dating be used to determine a reli-
able date for a sample that is about 700 yr old? Account for your answer.

15. Because of radioactive decay, one element can be transmuted
into another. Thus, a container of uranium $^{238}_{92}$U ultimately becomes
a container of lead $^{206}_{82}$Pb, as Figure 31.17 indicates. How long does
the transmutation of $^{238}_{92}$U entirely into $^{206}_{82}$Pb require—several
decades, several centuries, thousands of years, millions of years, or
billions of years? State your reasoning.

16. In Figure 31.8 the trajectory of the β^- particles has a greater
curvature than that of the α particles. Assuming that the speeds of
the two types of particles are the same, explain why the curvatures
are different. Consider the effects of both mass and charge.

Problems

The data given for atomic masses in these problems include the mass of the electrons orbiting the nucleus of the electrically neutral atom.

ssm Solution is in the Student Solutions Manual.   www Solution is available on the World Wide Web at www.wiley.com/college/cutnell
☤ This icon represents a biomedical application.

Section 31.1 Nuclear Structure,
Section 31.2 The Strong Nuclear Force and the Stability of the Nucleus

1. ssm In each of the following cases, what element does the symbol X represent and how many neutrons are in the nucleus: (a) $^{195}_{78}X$, (b) $^{32}_{16}X$, (c) $^{63}_{29}X$, (d) $^{11}_{5}X$, and (e) $^{239}_{94}X$? Use the periodic table on the inside of the back cover as needed.

2. In electrically neutral atoms, how many: (a) protons are in the uranium $^{238}_{92}U$ nucleus, (b) neutrons are in the mercury $^{202}_{80}Hg$ nucleus, and (c) electrons are in orbit about the niobium $^{93}_{41}Nb$ nucleus?

3. By what factor does the nucleon number of a nucleus have to increase in order for the nuclear radius to double?

4. What is the radius of a nucleus of titanium $^{48}_{22}Ti$?

5. ssm www The largest stable nucleus has a nucleon number of 209, and the smallest has a nucleon number of 1. If each nucleus is assumed to be a sphere, what is the ratio (largest/smallest) of the surface areas of these spheres?

* **6.** One isotope (X) contains an equal number of protons and neutrons. Another isotope (Y) of the same element has twice the number of neutrons as the first isotope does. Determine the ratio r_Y/r_X of the nuclear radii of the isotopes.

* **7.** An unknown nucleus contains 70 neutrons and has twice the volume of the nickel $^{60}_{28}Ni$ nucleus. Identify the unknown nucleus in the form ^A_ZX. Use the periodic table on the inside of the back cover as needed.

** **8.** Conceptual Example 1 provides some useful background for this problem. (a) Determine an approximate value for the density (in kg/m³) of the nucleus. (b) If a BB (radius = 2.3 mm) from an air rifle had a density equal to the nuclear density, what mass would the BB have? (c) Assuming the mass of a supertanker is about 1.5×10^8 kg, how many "supertankers" of mass would this hypothetical BB have?

** **9.** ssm www Refer to Conceptual Example 1 for a discussion of nuclear densities. A neutron star is composed of neutrons and has a density that is approximately the same as that of a nucleus. What is the radius of a neutron star whose mass is 0.40 times the mass of the sun?

Section 31.3 The Mass Defect of the Nucleus and Nuclear Binding Energy *(Note: The atomic mass for hydrogen ^1_1H is 1.007 825 u, including the mass of one electron.)*

10. Mercury $^{202}_{80}Hg$ has an atomic mass of 201.970 617 u. Obtain the binding energy *per nucleon* (in MeV/nucleon).

11. ssm Use the binding energy per nucleon curve in Figure 31.6 to determine the total binding energy of $^{16}_{8}O$.

12. For lead $^{206}_{82}Pb$ (atomic mass = 205.974 440 u) obtain (a) the mass defect in atomic mass units, (b) the binding energy (in MeV), and (c) the binding energy per nucleon (in MeV).

13. The earth revolves around the sun, and the two represent a bound system that has a binding energy of 2.6×10^{33} J. Suppose the earth and sun were completely separated, so that they were infinitely far apart and at rest. What would be the difference between the mass of the separated system and that of the bound system?

14. Find the binding energy (in MeV) for lithium ^7_3Li (atomic mass = 7.016 003 u).

* **15.** ssm Two isotopes of a certain element have binding energies that differ by 5.03 MeV. The isotope with the larger binding energy contains one more neutron than the other isotope. Find the difference in atomic mass between the two isotopes.

* **16.** (a) Energy is required to separate a nucleus into its constituent nucleons, as Figure 31.3 indicates; this energy is the *total* binding energy of the nucleus. In a similar way one can speak of the energy that binds a single nucleon to the remainder of the nucleus. For example, separating nitrogen $^{14}_7N$ into nitrogen $^{13}_7N$ and a neutron takes energy equal to the binding energy of the neutron, as shown below:

$$^{14}_7N + \text{Energy} \longrightarrow {}^{13}_7N + {}^1_0n$$

Find the energy (in MeV) that binds the neutron to the $^{14}_7N$ nucleus by considering the mass of $^{13}_7N$ (atomic mass = 13.005 738 u) and the mass of ^1_0n (atomic mass = 1.008 665 u), as compared to the mass of $^{14}_7N$ (atomic mass = 14.003 074 u). (b) Similarly, one can speak of the energy that binds a single proton to the $^{14}_7N$ nucleus:

$$^{14}_7N + \text{Energy} \longrightarrow {}^{13}_6C + {}^1_1H$$

Following the procedure outlined in part (a), determine the energy (in MeV) that binds the proton (atomic mass = 1.007 825 u) to the $^{14}_7N$ nucleus. (c) Which nucleon is more tightly bound, the neutron or the proton?

Section 31.4 Radioactivity

17. α decay occurs for each of the following nuclei. Write the decay process for each, including the chemical symbols and values for Z and A for the daughter nuclei: (a) $^{212}_{84}Po$ and (b) $^{232}_{92}U$.

18. Complete the following decay processes by stating what the symbol "X" represents (X = α, β^-, β^+, or γ): (a) $^{211}_{82}Pb \rightarrow {}^{211}_{83}Bi + X$, (b) $^{11}_6C \rightarrow {}^{11}_5B + X$, (c) $^{231}_{90}Th^* \rightarrow {}^{231}_{90}Th + X$, and (d) $^{210}_{84}Po \rightarrow {}^{206}_{82}Pb + X$.

19. For the following nuclei, each undergoing β^- decay, write the decay process, identifying each daughter nucleus with its chemical symbol and values for Z and A: (a) $^{14}_6C$ and (b) $^{212}_{82}Pb$.

20. Osmium $^{191}_{76}Os$ (atomic mass = 190.960 920 u) is converted into Iridium $^{191}_{77}Ir$ (atomic mass = 190.960 584 u) via β^- decay. What is the energy (in MeV) released in this process?

21. ssm Write the β^- decay process for $^{35}_{16}S$, including the chemical symbol and values for Z and A.

22. Find the energy released when lead $^{211}_{82}Pb$ (atomic mass = 210.988 735 u) undergoes β^- decay to become bismuth $^{211}_{83}Bi$ (atomic mass = 210.987 255 u).

23. ssm www Write the β^+ decay process for each of the following nuclei, being careful to include Z and A and the proper chemical symbol for each daughter nucleus: (a) $^{18}_9F$ and (b) $^{15}_8O$.

* **24.** Review Conceptual Example 5 as background for this problem. The α decay of uranium $^{238}_{92}U$ produces thorium $^{234}_{90}Th$ (atomic mass = 234.0436 u). In Example 4, the energy released in this decay is determined to be 4.3 MeV. Determine how much of this energy is carried away by the recoiling $^{234}_{90}Th$ daughter nucleus and how much by the α particle (atomic mass = 4.002 603 u). Assume

that the energy of each particle is kinetic energy, and ignore the small amount of energy carried away by the γ ray that is also emitted. In addition, ignore relativistic effects.

* **25.** Refer to **Interactive Solution 31.25** at **www.wiley.com/college/cutnell** to review a model for solving this type of problem. Polonium $^{210}_{84}$Po (atomic mass = 209.982 848 u) undergoes α decay. Assuming that all the released energy is in the form of kinetic energy of the α particle (atomic mass = 4.002 603 u) and ignoring the recoil of the daughter nucleus (lead $^{206}_{82}$Pb, 205.974 440 u), find the speed of the α particle. Ignore relativistic effects.

* **26.** Determine the symbol ^A_ZX for the parent nucleus whose α decay produces the same daughter as the β^- decay of thallium $^{208}_{81}$Tl.

** **27.** **ssm** Find the energy (in MeV) released when β^+ decay converts sodium $^{22}_{11}$Na (atomic mass = 21.994 434 u) into neon $^{22}_{10}$Ne (atomic mass = 21.991 383 u). Notice that the atomic mass for $^{22}_{11}$Na includes the mass of 11 electrons, whereas the atomic mass for $^{22}_{10}$Ne includes the mass of only 10 electrons.

** **28.** **Interactive LearningWare 31.1** at **www.wiley.com/college/cutnell** reviews the concepts that are involved in this problem. An isotope of beryllium (atomic mass = 7.017 u) emits a γ ray and recoils with a speed of 2.19×10^4 m/s. Assuming that the beryllium nucleus is stationary to begin with, find the wavelength of the γ ray.

Section 31.6 Radioactive Decay and Activity

29. In 9.0 days the number of radioactive nuclei decreases to one-eighth the number present initially. What is the half-life (in days) of the material?

30. The isotope $^{224}_{88}$Ra of radium has a decay constant of 2.19×10^{-6} s^{-1}. What is the half-life (in days) of this isotope?

31. **ssm** How many half-lives are required for the number of radioactive nuclei to decrease to one-millionth of the initial number?

32. Iodine $^{131}_{53}$I is used in diagnostic and therapeutic techniques in the treatment of thyroid disorders. This isotope has a half-life of 8.04 days. What percentage of an initial sample of $^{131}_{53}$I remains after 30.0 days?

33. A device used in radiation therapy for cancer contains 0.50 g of cobalt $^{60}_{27}$Co (59.933 819 u). The half-life of $^{60}_{27}$Co is 5.27 yr. Determine the activity of the radioactive material.

34. Strontium $^{90}_{38}$Sr has a half-life of 28.5 yr. It is chemically similar to calcium, enters the body through the food chain, and collects in the bones. Consequently, $^{90}_{38}$Sr is a particularly serious health hazard. How long (in years) will it take for 99.9900% of the $^{90}_{38}$Sr released in a nuclear reactor accident to disappear?

35. **ssm** If the activity of a radioactive substance is initially 398 disintegrations/min and two days later it is 285 disintegrations/min, what is the activity four days later still, or six days after the start? Give your answer in disintegrations/min.

* **36.** To make the dial of a watch glow in the dark, 1.000×10^{-9} kg of radium $^{226}_{88}$Ra is used. The half-life of this isotope is 1.60×10^3 yr. How many kilograms of radium *disappear* while the watch is in use for fifty years?

* **37.** Refer to **Interactive Solution 31.37** at **www.wiley.com/college/cutnell** for one approach to solving this problem. To see why one curie of activity was chosen to be 3.7×10^{10} Bq, determine the activity (in disintegrations per second) of one gram of radium $^{226}_{88}$Ra ($T_{1/2} = 1.6 \times 10^3$ yr).

* **38.** A sample of ore containing radioactive strontium $^{90}_{38}$Sr has an activity of 6.0×10^5 Bq. The atomic mass of strontium is 89.908 u, and its half-life is 28.5 yr. How many grams of strontium are in the sample?

* **39.** **ssm** Two radioactive nuclei A and B are present in equal numbers to begin with. Three days later, there are three times as many A nuclei as there are B nuclei. The half-life of species B is 1.50 days. Find the half-life of species A.

Section 31.7 Radioactive Dating

40. An archeological specimen containing 9.2 g of carbon has an activity of 1.6 Bq. How old (in years) is the specimen?

41. **ssm** The practical limit to ages that can be determined by radiocarbon dating is about 41 000 yr. In a 41 000-yr-old sample, what percentage of the original $^{14}_{6}$C atoms remains?

42. Review Conceptual Example 12 before starting to solve this problem. The number of unstable nuclei remaining after a time $t = 5.00$ yr is N, and the number present initially is N_0. Find the ratio N/N_0 for (a) $^{14}_{6}$C (half-life = 5730 yr), (b) $^{15}_{8}$O (half-life = 122.2 s; use $t = 1.00$ h, since otherwise the answer is out of the range of your calculator), and (c) $^{3}_{1}$H (half-life = 12.33 yr). Verify that your answers are consistent with the reasoning in Conceptual Example 12.

43. The shroud of Turin is a religious artifact known since the Middle Ages. In 1988 its age was measured using the radiocarbon dating technique, which revealed that the shroud could not have been made before 1200 AD. Of the $^{14}_{6}$C nuclei that were present in the living matter from which the shroud was made, what percentage remained in 1988?

44. The half-life for the α decay of uranium $^{238}_{92}$U is 4.47×10^9 yr. Determine the age (in years) of a rock specimen that contains sixty percent of its original number of $^{238}_{92}$U atoms.

* **45.** **ssm** When any radioactive dating method is used, experimental error in the measurement of the sample's activity leads to error in the estimated age. In an application of the radiocarbon dating technique to certain fossils, an activity of 0.10 Bq per gram of carbon is measured to within an accuracy of $\pm$ten percent. Find the age of the fossils and the maximum error (in years) in the value obtained. Assume that there is no error in the 5730-year half-life of $^{14}_{6}$C nor in the value of 0.23 Bq per gram of carbon in a living organism.

** **46.** (a) A sample is being dated by the radiocarbon technique. If the sample were uncontaminated, its activity would be 0.011 Bq per gram of carbon. Find the true age (in years) of the sample. (b) Suppose the sample is contaminated, so that only 98.0% of its carbon is ancient carbon. The remaining 2.0% is fresh carbon, in the sense that the $^{14}_{6}$C it contains has not had any time to decay. Assuming that the lab technician is unaware of the contamination, what apparent age (in years) would be determined for the sample?

Additional Problems

47. Determine the mass defect of the nucleus for cobalt $^{59}_{27}$Co, which has an atomic mass of 58.933 198 u. Express your answer in (a) atomic mass units and (b) kilograms.

48. For $^{208}_{82}$Pb find (a) the net electrical charge of the nucleus, (b) the number of neutrons, (c) the number of nucleons, (d) the approximate radius of the nucleus, and (e) the nuclear density.

49. ssm The number of radioactive nuclei present at the start of an experiment is 4.60×10^{15}. The number present twenty days later is 8.14×10^{14}. What is the half-life (in days) of the nuclei?

50. Find the energy (in MeV) released when α decay converts radium $^{226}_{88}$Ra (atomic mass = 226.025 40 u) into radon $^{222}_{86}$Rn (atomic mass = 222.017 57 u). The atomic mass of an α particle is 4.002 603 u.

51. In the form $^{A}_{Z}$X, identify the daughter nucleus that results when (a) plutonium $^{242}_{94}$Pu undergoes α decay, (b) sodium $^{24}_{11}$Na undergoes β^- decay, and (c) nitrogen $^{13}_{7}$N undergoes β^+ decay.

52. Bones of the woolly mammoth have been found in North America. The youngest of these bones has a $^{14}_{6}$C activity per gram of carbon that is about 21% of what was present in the live animal. How long ago (in years) did this animal disappear from North America?

* **53. ssm www** The photomultiplier tube in a commercial scintillation counter contains 15 of the special electrodes or dynodes. Each dynode produces 3 electrons for every electron that strikes it. One photoelectron strikes the first dynode. What is the maximum number of electrons that strike the 15th dynode?

* **54.** The isotope $^{198}_{79}$Au (atomic mass = 197.968 u) of gold, which has a half-life of 2.69 days, is used in cancer therapy. What mass (in grams) of this isotope is required to produce an activity of 315 Ci?

* **55.** Radon $^{220}_{86}$Rn produces a daughter nucleus that is radioactive. The daughter, in turn, produces its own radioactive daughter, and so on. This process continues until lead $^{208}_{82}$Pb is reached. What are the total number N_α of α particles and the total number N_β of β^- particles that are generated in this series of radioactive decays?

** **56.** Both gold $^{198}_{79}$Au ($T_{1/2}$ = 2.69 days) and iodine $^{131}_{53}$I ($T_{1/2}$ = 8.04 days) are used in diagnostic medicine related to the liver. At the time laboratory supplies are monitored, the activity of the gold is observed to be five times greater than the activity of the iodine. How many days later will the two activities be equal?

Concepts & Calculations   Group Learning Problems

Note: Each of these problems consists of Concept Questions followed by a related quantitative Problem. They are designed for use by students working alone or in small learning groups. The Concept Questions involve little or no mathematics and are intended to stimulate group discussions. They focus on the concepts with which the problems deal. Recognizing the concepts is the essential initial step in any problem-solving technique.

57. Concept Questions (a) In a nucleus, each proton experiences a repulsive electrostatic force from each of the other protons. Write an expression for the magnitude of the force that one proton (charge = $+e$) applies to another proton that is located a distance r away. (b) The force that acts on either of two particular protons in the nucleus has the smallest possible magnitude. Relative to one another, where in the nucleus must these two protons be located? Explain.

Problem In a nucleus of gold $^{197}_{79}$Au, what is the magnitude of the least possible electrostatic force of repulsion that one proton can exert on another?

58. Concept Questions (a) How is the total released energy related to the decrease in mass that accompanies β^- decay? (b) When a parent nucleus undergoes β^- decay, a daughter nucleus, a β^- particle, and an antineutrino $\overline{\gamma}$ are produced. The released energy is shared among these three particles. Assume that the kinetic energy of the β^- particle is known and that the antineutrino carries away the maximum possible energy. What then can you say about the kinetic energy of the recoiling daughter nucleus?

Problem The β^- decay of phosphorus $^{32}_{15}$P (atomic mass = 31.973 907 u) produces a daughter nucleus that is sulfur $^{32}_{16}$S (atomic mass = 31.972 070 u), a β^- particle, and an antineutrino. The kinetic energy of the β^- particle is 0.90 MeV. Find the maximum possible energy that the antineutrino could carry away.

59. Concept Questions (a) Two radioactive nuclei A and B have half-lives of $T_{1/2, A}$ and $T_{1/2, B}$, where $T_{1/2, A}$ is greater than $T_{1/2, B}$. During the same time period, is the fraction of nuclei A that decay greater than, smaller than, or the same as the fraction of nuclei B that decay? (b) The numbers of these nuclei present initially are $N_{0, A}$ and $N_{0, B}$, the ratio of the two being $N_{0, A}/N_{0, B}$. Is the ratio N_A/N_B of the number of nuclei present at a later time greater than, smaller than, or the same as $N_{0, A}/N_{0, B}$? Justify your answers.

Problem Two waste products from nuclear reactors are strontium $^{90}_{38}$Sr ($T_{1/2}$ = 28.8 yr) and cesium $^{134}_{55}$Cs ($T_{1/2}$ = 2.06 yr). These two species are present initially in a ratio of $N_{0, Sr}/N_{0, Cs} = 7.80 \times 10^{-3}$. What is the ratio N_{Sr}/N_{Cs} fifteen years later? Verify that your answer is consistent with your answers to the Concept Questions.

60. Concept Questions (a) Write an expression for the heat Q needed to melt a mass m of water. (b) In α decay, how is the released energy related to the decrease in mass that accompanies the decay? (c) During α decay, the energy released by each disintegration is E. What is the total energy E_{Total} released by a number n of disintegrations? (d) If the number of radioactive nuclei present initially is N_0, what is the number of disintegrations that occur in one half-life?

Problem A one-gram sample of radium $^{224}_{88}$Ra (atomic mass = 224.020 186 u, $T_{1/2}$ = 3.66 days) contains 2.69×10^{21} nuclei and undergoes α decay to produce radon $^{220}_{86}$Rn (atomic mass = 220.011 368 u). The atomic mass of an α particle is 4.002 603 u. With the energy released in 3.66 days, how many kilograms of ice could be melted at 0 °C?

* **61. Concept Questions** Outside the nucleus, the neutron itself is radioactive and decays into a proton, an electron, and an antineutrino. (a) Suppose that, originally, a number N_0 of neutrons are out-

side the nucleus. What is meant by the statement "The half-life of the neutron is X minutes"? (b) How is the number of neutrons remaining at any time t related to the original number and the half-life of the neutron? (c) Suppose each of the neutrons is moving with the same kinetic energy. How is the speed of each neutron related to its kinetic energy? Ignore relativistic effects. (d) On average, how far would the neutrons go before one-half of them decayed? Express your answer in terms of the neutron's kinetic energy KE, half-life $T_{1/2}$, and mass m.

Problem The half-life of a neutron (mass $= 1.675 \times 10^{-27}$ kg) outside the nucleus is 10.2 min. On average, over what distance (in meters) would a beam of 5.00-eV neutrons travel before the number of neutrons decreases to 75.0% of its initial value? Ignore relativistic effects.

* **62. Concept Questions** (a) Physically, what does the binding energy of a nucleus represent? (b) What is the relationship between the binding energy of a nucleus and its mass defect? (c) How is the mass defect of a nucleus related to the mass of the intact nucleus and the masses of the individual nucleons? (d) A piece of metal, such as a coin, contains a certain number of atoms, and, hence, nuclei. How much energy would be required to break all the nuclei into their constituent protons and neutrons? Give your answer in terms of the number of atoms in the coin and the binding energy of each nucleus. (e) How is the number of atoms in a piece of metal related to its mass? Express your answer in terms of the atomic mass of the atoms and Avogadro's number.

Problem A copper penny has a mass of 3.0 g. Determine the energy (in MeV) that would be required to break all the copper nuclei into their constituent protons and neutrons. Ignore the (small) energy that binds the electrons to the nucleus, and, for simplicity, assume that all the copper nuclei are $^{63}_{29}$Cu (atomic mass = 62.939 598 u).

Chapter 32

Ionizing Radiation, Nuclear Energy, and Elementary Particles

Among other things, this chapter discusses nuclear fusion. One method being tried to produce controlled fusion is called the "Z pinch" and uses a burst of electric current from giant capacitors to create an extremely hot plasma (ionized gas). This photograph shows the electric discharges that illuminate the surface of the water surrounding the fusion process. (© Ray Nelson/Phototake)

32.1 *Biological Effects of Ionizing Radiation*

Ionizing radiation consists of photons and/or moving particles that have sufficient energy to knock an electron out of an atom or molecule, thus forming an ion. The photons usually lie in the ultraviolet, X-ray, or γ-ray regions of the electromagnetic spectrum (see Figure 24.10), whereas the moving particles can be the α and β particles emitted during radioactive decay. An energy of roughly 1 to 35 eV is needed to ionize an atom or molecule, and the particles and γ rays emitted during nuclear disintegration often have energies of several million eV. Therefore, a single α particle, β particle, or γ ray can ionize thousands of molecules.

The physics of the biological effects of ionizing radiation.

Nuclear radiation is potentially harmful to humans because the ionization it produces can alter significantly the structure of molecules within a living cell. The alterations can lead to the death of the cell and even the organism itself. Despite the potential hazards, ionizing radiation is used in medicine for diagnostic and therapeutic purposes, such as locating bone fractures and treating cancer. The hazards can be minimized only if the fundamentals of radiation exposure, including dose units and the biological effects of radiation, are understood.

Exposure is a measure of the ionization produced in air by X-rays or γ rays, and it is defined in the following manner. A beam of X-rays or γ rays is sent through a mass m of dry air at standard temperature and pressure (STP: 0 °C, 1 atm pressure). In passing through the air, the beam produces positive ions whose total charge is q. Exposure is defined as the total charge per unit mass of air: exposure = q/m. The SI unit for exposure is coulombs per kilogram (C/kg). However, the first radiation unit to be defined was the *roentgen* (R), and it is still used today. With q expressed in coulombs (C) and m in kilograms (kg), the exposure in roentgens is given by

$$\text{Exposure (in roentgens)} = \left(\frac{1}{2.58 \times 10^{-4}}\right)\frac{q}{m} \qquad (32.1)$$

Thus, when X-rays or γ rays produce an exposure of one roentgen, $q = 2.58 \times 10^{-4}$ C of positive charge are produced in $m = 1$ kg of dry air:

$$1\ \text{R} = 2.58 \times 10^{-4}\ \text{C/kg} \qquad \text{(dry air, at STP)}$$

Since the concept of exposure is defined in terms of the ionizing abilities of X-rays and γ rays in air, it does not specify the effect of radiation on living tissue. For biological purposes, the *absorbed dose* is a more suitable quantity because it is the energy absorbed from the radiation per unit mass of absorbing material:

$$\text{Absorbed dose} = \frac{\text{Energy absorbed}}{\text{Mass of absorbing material}} \qquad (32.2)$$

The SI unit of absorbed dose is the *gray* (Gy), which is the unit of energy divided by the unit of mass: 1 Gy = 1 J/kg. Equation 32.2 is applicable to all types of radiation and absorbing media. Another unit is often used for absorbed dose—namely, the *rad* (rd). The word "rad" is an acronym for **r**adiation **a**bsorbed **d**ose. The rad and the gray are related by

$$1\ \text{rad} = 0.01\ \text{gray}$$

The amount of biological damage produced by ionizing radiation is different for different kinds of radiation. For instance, a 1-rad dose of neutrons is far more likely to produce eye cataracts than a 1-rad dose of X-rays. To compare the damage caused by different types of radiation, *the relative biological effectiveness* (RBE) is used.* The relative biological effectiveness of a particular type of radiation is the ratio of the dose of 200-keV X-rays needed to produce a certain biological effect to the dose of the radiation needed to produce the same biological effect:

$$\begin{array}{l}\text{Relative biological}\\ \text{effectiveness (RBE)}\end{array} = \frac{\begin{array}{c}\text{The dose of 200-keV X-rays that}\\ \text{produces a certain biological effect}\end{array}}{\begin{array}{c}\text{The dose of radiation that}\\ \text{produces the same biological effect}\end{array}} \qquad (32.3)$$

* The RBE is sometimes called the *quality factor* (QF).

Table 32.1 *Relative Biological Effectiveness (RBE) for Various Types of Radiation*

Type of Radiation	RBE
200-keV X-rays	1
γ rays	1
β^- particles (electrons)	1
Protons	10
α particles	10–20
Neutrons	
Slow	2
Fast	10

The RBE depends on the nature of the ionizing radiation and its energy, as well as the type of tissue being irradiated. Table 32.1 lists some typical RBE values for different kinds of radiation, assuming that an "average" biological tissue is being irradiated. A value of RBE = 1 for γ rays and β^- particles indicates that they produce the same biological damage as do 200-keV X-rays. The larger RBE values for protons, α particles, and fast neutrons indicate that they cause substantially more damage. The RBE is often used in conjunction with the absorbed dose to reflect the damage-producing character of the radiation. The product of the absorbed dose in rads (not in grays) and the RBE is the ***biologically equivalent dose:***

$$\begin{matrix} \text{Biologically equivalent dose} \\ \text{(in rems)} \end{matrix} = \begin{matrix} \text{Absorbed dose} \\ \text{(in rads)} \end{matrix} \times \text{RBE} \qquad (32.4)$$

The unit for the biologically equivalent dose is the *rem*, short for **r**oentgen **e**quivalent, **m**an. Example 1 illustrates the use of the biologically equivalent dose.

Example 1 Comparing Absorbed Doses of γ Rays and Neutrons

A biological tissue is irradiated with γ rays that have an RBE of 0.70. The absorbed dose of γ rays is 850 rd. The tissue is then exposed to neutrons whose RBE is 3.5. The biologically equivalent dose of the neutrons is the same as that of the γ rays. What is the absorbed dose of neutrons?

Reasoning The biologically equivalent doses of the neutrons and the γ rays are the same. Therefore, the tissue damage produced in each case is the same. However, the RBE of the neutrons is larger than the RBE of the γ rays by a factor of 3.5/0.70 = 5.0. Consequently, we will find that the absorbed dose of the neutrons is only one-fifth as great as that of the γ rays.

Solution According to Equation 32.4, the biologically equivalent dose is the product of the absorbed dose (in rads) and the RBE; it is the same for the γ rays and the neutrons. Therefore, we have

$$\begin{matrix} \text{Biologically} \\ \text{equivalent dose} \end{matrix} = (\text{Absorbed dose})_{\gamma \, \text{rays}} \, \text{RBE}_{\gamma \, \text{rays}} = (\text{Absorbed dose})_{\text{neutrons}} \, \text{RBE}_{\text{neutrons}}$$

Solving for the absorbed dose of the neutrons gives

$$(\text{Absorbed dose})_{\text{neutrons}} = (\text{Absorbed dose})_{\gamma \, \text{rays}} \left(\frac{\text{RBE}_{\gamma \, \text{rays}}}{\text{RBE}_{\text{neutrons}}} \right) = (850 \text{ rd}) \left(\frac{0.70}{3.5} \right) = \boxed{170 \text{ rd}}$$

Table 32.2 *Average Biologically Equivalent Doses of Radiation Received by a U. S. Resident*[a]

Source of Radiation	Biologically Equivalent Dose (mrem/yr)[b]
Natural background radiation	
Cosmic rays	28
Radioactive earth and air	28
Internal radioactive nuclei	39
Inhaled radon	$\approx$200
Man-made radiation	
Consumer products	10
Medical/dental diagnostics	39
Nuclear medicine	14
Rounded total:	360

[a] National Council on Radiation Protection and Measurement, Report No. 93, "Ionizing Radiation Exposure of the Population of the United States," 1987.
[b] 1 mrem = 10^{-3} rem.

Everyone is continually exposed to background radiation from natural sources, such as cosmic rays (high-energy particles that come from outside the solar system), radioactive materials in the environment, radioactive nuclei (primarily carbon $^{14}_{6}$C and potassium $^{40}_{19}$K) within our own bodies, and radon. Table 32.2 lists the average biologically equivalent doses received from these sources by a person in the United States. According to this table, radon is a major contributor to the natural background radiation. Radon is an odorless radioactive gas and poses a health hazard because, when inhaled, it can damage the lungs and cause cancer. Radon is found in soil and rocks and enters houses through cracks and crevices in the foundation. The amount of radon in the soil varies greatly throughout the country, with some localities having significant amounts and others having virtually none. Accordingly, the dose that any individual receives can vary widely from the average value of 200 mrem/yr given in Table 32.2 (1 mrem = 10^{-3} rem). In many houses, the entry of radon can be reduced significantly by sealing the foundation against entry of the gas and providing good ventilation so it does not accumulate.

To the natural background of radiation, a significant amount of man-made radiation has been added, mostly from medical/dental diagnostic X-rays. Table 32.2 indicates an average total dose of 360 mrem/yr from all sources.

The effects of radiation on humans can be grouped into two categories, according to the time span between initial exposure and the appearance of physiological symptoms: (1) short-term or acute effects that appear within a matter of minutes, days, or weeks, and (2) long-term or latent effects that appear years, decades, or even generations later.

Radiation sickness is the general term applied to the acute effects of radiation. Depending on the severity of the dose, a person with radiation sickness can exhibit nausea, vomiting, fever, diarrhea, and loss of hair. Ultimately, death can occur. The severity of radiation sickness is related to the dose received, and in the following discussion the biologically equivalent doses quoted are whole-body, single doses. A dose less than 50 rem causes no short-term ill effects. A dose between 50 and 300 rem brings on radiation sickness, the severity increasing with increasing dosage. A whole-body dose in the range of 400–500 rem is classified as an LD_{50} dose, meaning that it is a lethal dose (LD) for about 50% of the people so exposed; death occurs within a few months. Whole-body doses greater than 600 rem result in death for almost all individuals.

Long-term or latent effects of radiation may appear as a result of high-level brief exposure or low-level exposure over a long period of time. Some long-term effects are hair loss, eye cataracts, and various kinds of cancer. In addition, genetic defects caused by mutated genes may be passed on from one generation to the next.

Because of the hazards of radiation, the federal government has established dose limits. The permissible dose for an individual is defined as the dose, accumulated over a long period of time or resulting from a single exposure, that carries negligible probability of a severe health hazard. Federal standards (1991) state that an individual in the general population should not receive more than 500 mrem of man-made radiation each year, *exclusive* of medical sources. A person exposed to radiation in the workplace (e.g., a radiation therapist) should not receive more than 5 rem per year from work-related sources.

✔ **Check Your Understanding 1**

A person faces the possibility of the following absorbed doses of ionizing radiation: 20 rad of γ rays (RBE = 1), 5 rad of neutrons (RBE = 10), and 2 rad of α particles (RBE = 20). Rank the amount of biological damage that these possibilities will cause in decreasing order (greatest damage first). *(The answer is given at the end of the book.)*

Background: The biological damage caused by ionizing radiation is related to the amount of the absorbed dose and the relative biological effectiveness of the radiation.

For similar questions (including calculational counterparts), consult Self-Assessment Test 32.1, which is described at the end of Section 32.7.

32.2 *Induced Nuclear Reactions*

Section 31.4 discusses how a radioactive parent nucleus disintegrates spontaneously into a daughter nucleus. It is also possible to bring about, or induce, the disintegration of a stable nucleus by striking it with another nucleus, an atomic or subatomic particle, or a γ-ray photon. A ***nuclear reaction*** is said to occur whenever the incident nucleus, particle, or photon causes a change to occur in a target nucleus.

In 1919, Ernest Rutherford observed that when an α particle strikes a nitrogen nucleus, an oxygen nucleus and a proton are produced. This nuclear reaction is written as

$$\underbrace{{}^{4}_{2}\text{He}}_{\substack{\text{Incident}\\ \alpha \text{ particle}}} + \underbrace{{}^{14}_{7}\text{N}}_{\substack{\text{Nitrogen}\\ \text{(target)}}} \longrightarrow \underbrace{{}^{17}_{8}\text{O}}_{\text{Oxygen}} + \underbrace{{}^{1}_{1}\text{H}}_{\text{Proton, } p}$$

Because the incident α particle induces the transmutation of nitrogen into oxygen, this reaction is an example of an ***induced nuclear transmutation.***

Nuclear reactions are often written in a shorthand form. For example, the reaction above is designated by ${}^{14}_{7}\text{N}\ (\alpha, p)\ {}^{17}_{8}\text{O}$. The first and last symbols represent the initial and final nuclei, respectively. The symbols within the parentheses denote the incident α particle (on the left) and the small emitted particle or proton p (on the right). Some other induced nuclear transmutations are listed below, together with the equivalent shorthand notations:

Nuclear Reaction	Notation
${}^{1}_{0}\text{n} + {}^{10}_{5}\text{B} \rightarrow {}^{7}_{3}\text{Li} + {}^{4}_{2}\text{He}$	${}^{10}_{5}\text{B}\ (n, \alpha)\ {}^{7}_{3}\text{Li}$
$\gamma + {}^{25}_{12}\text{Mg} \rightarrow {}^{24}_{11}\text{Na} + {}^{1}_{1}\text{H}$	${}^{25}_{12}\text{Mg}\ (\gamma, p)\ {}^{24}_{11}\text{Na}$
${}^{1}_{1}\text{H} + {}^{13}_{6}\text{C} \rightarrow {}^{14}_{7}\text{N} + \gamma$	${}^{13}_{6}\text{C}\ (p, \gamma)\ {}^{14}_{7}\text{N}$

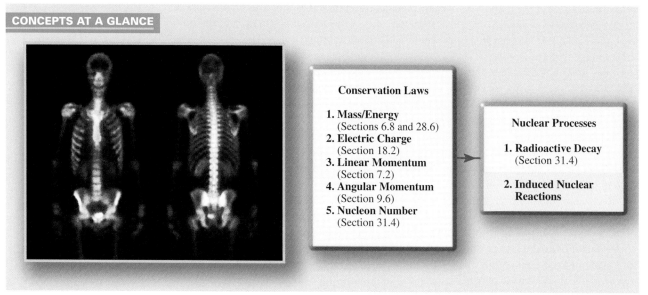

CONCEPTS AT A GLANCE

Figure 32.1 CONCEPTS AT A GLANCE In any nuclear process, such as radioactive decay or an induced nuclear reaction, the conservation laws of physics are obeyed. To obtain images of the human skeleton such as those in the photograph (front view on the left and back view on the right), a radioactive isotope is injected into the body. The isotope accumulates in the bone and decays, emitting gamma rays in the process. The gamma rays are detected to produce the image, which is called a gamma scan. (© RVI Medical Physics, Newcastle/Simon Fraser/Photo Researchers)

▶ CONCEPTS AT A GLANCE Induced nuclear reactions, like the radioactive decay process discussed in Section 31.4, obey the conservation laws of physics. We have discussed these laws in previous chapters, as the Concepts-at-a-Glance chart in Figure 32.1 indicates. This chart is similar to that shown earlier in Figure 31.7 and emphasizes the important role that the conservation laws play in all nuclear processes. In particular, both the total electric charge of the nucleons and the total number of nucleons are conserved during an induced nuclear reaction. The fact that these quantities are conserved makes it possible to identify the nucleus produced in a reaction, as the next example illustrates. ◀

Example 2 An Induced Nuclear Transmutation

An α particle strikes an aluminum $^{27}_{13}\text{Al}$ nucleus. As a result, an unknown nucleus $^{A}_{Z}\text{X}$ and a neutron $^{1}_{0}\text{n}$ are produced:

$$^{4}_{2}\text{He} + ^{27}_{13}\text{Al} \longrightarrow ^{A}_{Z}\text{X} + ^{1}_{0}\text{n}$$

Identify the nucleus produced, including its atomic number Z (the number of protons) and its atomic mass number A (the number of nucleons).

Reasoning The total electric charge of the nucleons is conserved, so that we can set the total number of protons before the reaction equal to the total number after the reaction. The total number of nucleons is also conserved, so that we can set the total number before the reaction equal to the total number after the reaction. These two conserved quantities will allow us to identify the nucleus $^{A}_{Z}\text{X}$.

Solution The conservation of total electric charge and total number of nucleons leads to the equations listed below:

Conserved Quantity	Before Reaction		After Reaction
Total electric charge (number of protons)	2 + 13	=	Z + 0
Total number of nucleons	4 + 27	=	A + 1

Solving these equations for Z and A gives $Z = 15$ and $A = 30$. Since $Z = 15$ identifies the element as phosphorus (P), the nucleus produced is $\boxed{^{30}_{15}\text{P}}$.

Induced nuclear transmutations can be used to produce isotopes that are not found naturally. In 1934, Enrico Fermi suggested a method for producing elements with a higher atomic number than uranium ($Z = 92$). These elements—neptunium ($Z = 93$), plutonium ($Z = 94$), americium ($Z = 95$), and so on—are known as *transuranium elements,* and none occurs naturally. They are created in a nuclear reaction between a suitably chosen lighter element and a small incident particle, usually a neutron or an α particle. For example, Figure 32.2 shows a reaction that produces plutonium from uranium. A neutron is captured by a uranium $^{238}_{92}U$ nucleus, producing $^{239}_{92}U$ and a γ ray. The $^{239}_{92}U$ nucleus is radioactive and decays with a half-life of 23.5 min into neptunium $^{239}_{93}Np$. Neptunium is also radioactive and disintegrates with a half-life of 2.4 days into plutonium $^{239}_{94}Pu$. Plutonium is the final product and has a half-life of 24 100 yr.

The neutrons that participate in nuclear reactions can have kinetic energies that cover a wide range. In particular, those that have a kinetic energy of about 0.04 eV or less are called **thermal neutrons.** The name derives from the fact that such a relatively small kinetic energy is comparable to the average translational kinetic energy of a molecule at room temperature.

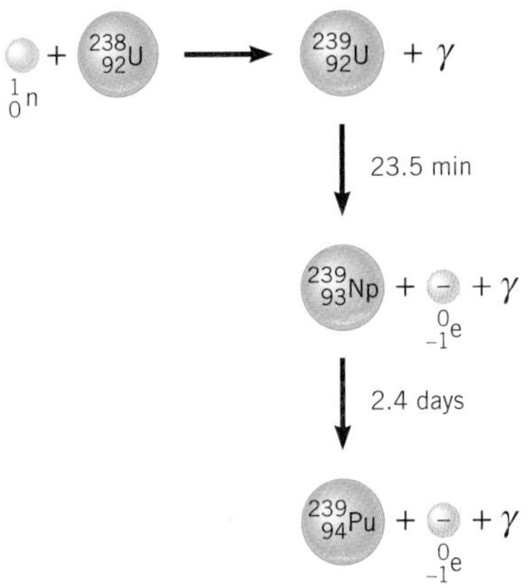

✓ **Check Your Understanding 2**

Which one or more of the following four nuclear reactions could possibly occur?
(a) $^{15}_{7}N\ (\alpha, \gamma)\ ^{18}_{9}F$, (b) $^{14}_{6}C\ (p, n)\ ^{15}_{8}O$, (c) $^{14}_{7}N\ (p, \gamma)\ ^{15}_{8}O$, (d) $^{15}_{8}O\ (n, p)\ ^{13}_{6}C$
[The answer(s) is (are) given at the end of the book.]

Background: Any nuclear reaction that occurs must obey the conservation laws of physics.

For similar questions (including calculational counterparts), consult Self-Assessment Test 32.1, which is described at the end of Section 32.7.

Figure 32.2 An induced nuclear reaction is shown in which $^{238}_{92}U$ is transmuted into the transuranium element plutonium $^{239}_{94}Pu$.

32.3 *Nuclear Fission*

In 1939 four German scientists, Otto Hahn, Lise Meitner, Fritz Strassmann, and Otto Frisch, made an important discovery that ushered in the atomic age. They found that a uranium nucleus, after absorbing a neutron, splits into two fragments, each with a smaller mass than the original nucleus. The splitting of a massive nucleus into two less massive fragments is known as **nuclear fission.**

Figure 32.3 shows a fission reaction in which a uranium $^{235}_{92}U$ nucleus is split into barium $^{141}_{56}Ba$ and krypton $^{92}_{36}Kr$ nuclei. The reaction begins when $^{235}_{92}U$ absorbs a slowly moving neutron, creating a "compound nucleus," $^{236}_{92}U$. The compound nucleus disintegrates quickly into $^{141}_{56}Ba$, $^{92}_{36}Kr$, and three neutrons according to the following reaction:

$$^{1}_{0}n + {}^{235}_{92}U \longrightarrow \underbrace{^{236}_{92}U}_{\substack{\text{Compound} \\ \text{nucleus} \\ \text{(unstable)}}} \longrightarrow \underbrace{^{141}_{56}Ba}_{\text{Barium}} + \underbrace{^{92}_{36}Kr}_{\text{Krypton}} + \underbrace{3\,{}^{1}_{0}n}_{\text{3 neutrons}}$$

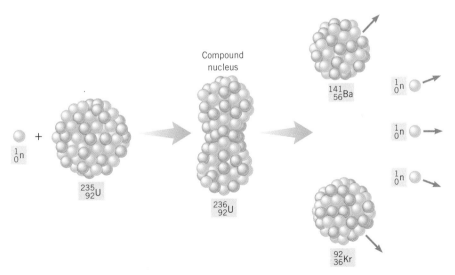

Figure 32.3 A slowly moving neutron causes the uranium nucleus $^{235}_{92}U$ to fission into barium $^{141}_{56}Ba$, krypton $^{92}_{36}Kr$, and three neutrons.

This reaction is only one of the many possible reactions that can occur when uranium fissions. For example, another reaction is

$$\underset{\text{Compound nucleus (unstable)}}{\underbrace{{}^{1}_{0}n + {}^{235}_{92}U \longrightarrow {}^{236}_{92}U}} \longrightarrow \underset{\text{Xenon}}{{}^{140}_{54}Xe} + \underset{\text{Strontium}}{{}^{94}_{38}Sr} + \underset{\text{2 neutrons}}{2{}^{1}_{0}n}$$

Some reactions produce as many as 5 neutrons; however, the average number produced per fission is 2.5.

When a neutron collides with and is absorbed by a uranium nucleus, the uranium nucleus begins to vibrate and becomes distorted. The vibration continues until the distortion becomes so severe that the attractive strong nuclear force can no longer balance the electrostatic repulsion between the nuclear protons. At this point, the nucleus bursts apart into fragments, which carry off energy, primarily in the form of kinetic energy. The energy carried off by the fragments is enormous and was stored in the original nucleus primarily in the form of electric potential energy. An average of roughly 200 MeV of energy is released per fission. This energy is approximately 10^8 times greater than the energy released per molecule in an ordinary chemical reaction, such as the combustion of gasoline or coal. Example 3 demonstrates how to estimate the energy released during the fission of a nucleus.

Example 3   The Energy Released During Nuclear Fission

Estimate the amount of energy released when a massive nucleus ($A = 240$) fissions.

Reasoning Figure 31.6 shows that the binding energy of a nucleus with $A = 240$ is about 7.6 MeV per nucleon. We assume that this nucleus fissions into two fragments, each with $A \approx 120$. According to Figure 31.6, the binding energy of the fragments increases to about 8.5 MeV per nucleon. Consequently, when a massive nucleus fissions, there is a release of about 8.5 MeV − 7.6 MeV = 0.9 MeV of energy per nucleon.

Solution Since there are 240 nucleons involved in the fission process, the total energy released per fission is approximately (0.9 MeV/nucleon)(240 nucleons) $\approx$ $\boxed{200 \text{ MeV}}$.

Need more practice?

Interactive LearningWare 32.1
Consider the following neutron-induced fission reaction:

$$\underset{1.008\ 665\ u}{{}^{1}_{0}n} + \underset{239.052\ 157\ u}{{}^{239}_{94}Pu} \longrightarrow \underset{95.921\ 750\ u}{{}^{96}_{38}Sr} + \underset{139.910\ 581\ u}{{}^{A}_{Z}X} + \underset{4(1.008\ 665\ u)}{4{}^{1}_{0}n}$$

(a) Determine the atomic mass number A and the atomic number Z for the unknown nucleus ${}^{A}_{Z}X$. What element is it? (b) Find the energy (in MeV) released by this reaction.

Related Homework: Problem 20

Go to **www.wiley.com/college/cutnell** for an interactive solution.

Virtually all naturally occurring uranium is composed of two isotopes. These isotopes and their natural abundances are ${}^{238}_{92}U$ (99.275%) and ${}^{235}_{92}U$ (0.720%). Although ${}^{238}_{92}U$ is by far the most abundant isotope, the probability that it will capture a neutron and fission is very small. For this reason, ${}^{238}_{92}U$ is not the isotope of choice for generating nuclear energy. In contrast, the isotope ${}^{235}_{92}U$ readily captures a neutron and fissions, *provided the neutron is a thermal neutron* (kinetic energy ≈ 0.04 eV or less). The probability of a thermal neutron causing ${}^{235}_{92}U$ to fission is about five hundred times greater than a neutron whose energy is relatively high—say, 1 MeV. Thermal neutrons can also be used to fission other nuclei, such as plutonium ${}^{239}_{94}Pu$. Conceptual Example 4 deals with one of the reasons why thermal neutrons are useful for inducing nuclear fission.

Conceptual Example 4
Thermal Neutrons Versus Thermal Protons or Alpha Particles

Why is it possible for a thermal neutron (i.e., one with a relatively small kinetic energy) to penetrate a nucleus, whereas a proton or an α particle would need a much larger amount of energy to penetrate the same nucleus?

Reasoning and Solution To penetrate a nucleus, a particle such as a neutron, a proton, or an α particle must have enough kinetic energy to do the work of overcoming any repulsive force that it encounters along the way. A repulsive force can arise because protons in the nucleus are electrically charged. **Since a neutron is electrically neutral, however, it encounters no electrostatic force of repulsion as it approaches the nuclear protons and, hence, needs relatively little energy to reach the nucleus.** In contrast, a proton and an α particle both carry positive charge, and each encounters an electrostatic force of repulsion as it approaches the nuclear protons. It is true that each also experiences the attractive strong nuclear force from the nuclear protons and neutrons. But the strong nuclear force has an extremely short range of action and, therefore, comes into play only when an impinging particle reaches the nucleus. In comparison, the electrostatic force of repulsion has a long range of action and is encountered by an incoming charged particle throughout the entire journey to the target nucleus. Consequently, **an impinging proton or α particle needs a relatively large amount of kinetic energy to do the work of overcoming the repulsive electrostatic force.**

Related Homework: *Problem 40*

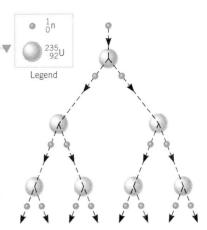

Figure 32.4 A chain reaction. For clarity, it is assumed that each fission generates two neutrons (2.5 neutrons are actually liberated on the average). The fission fragments are not shown.

The fact that the uranium fission reaction releases 2.5 neutrons, on the average, makes it possible for a self-sustaining series of fissions to occur. As Figure 32.4 illustrates, each neutron released can initiate another fission event, resulting in the emission of still more neutrons, followed by more fissions, and so on. A **chain reaction** is a series of nuclear fissions whereby some of the neutrons produced by each fission cause additional fissions. During an uncontrolled chain reaction, it would not be unusual for the number of fissions to increase a thousandfold within a few millionths of a second. With an average energy of about 200 MeV being released per fission, an uncontrolled chain reaction can generate an incredible amount of energy in a very short time, as happens in an atomic bomb (which is actually a *nuclear* bomb).

By limiting the number of neutrons in the environment of the fissile nuclei, it is possible to establish a condition whereby each fission event contributes, on average, only *one neutron* that fissions another nucleus (see Figure 32.5). In this manner, the chain reaction and the rate of energy production are *controlled*. The controlled fission chain reaction is the principle behind nuclear reactors used in the commercial generation of electric power.

Concept Simulation 32.1

This simulation presents animated versions of Figures 32.4 and 32.5 and helps further to illustrate the concepts that underlie controlled and uncontrolled chain reactions.

Go to **www.wiley.com/college/cutnell**

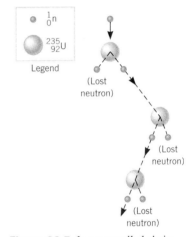

Figure 32.5 In a controlled chain reaction, only one neutron, on average, from each fission event causes another nucleus to fission. As a result, energy is released at a steady or controlled rate.

32.4 Nuclear Reactors

A nuclear reactor is a type of furnace in which energy is generated by a controlled fission chain reaction. The first nuclear reactor was built by Enrico Fermi in 1942, on the floor of a squash court under the west stands of Stagg Field at the University of Chicago. Today, there are a number of kinds and sizes of reactors, and many have the same three basic components: fuel elements, a neutron moderator, and control rods. Figure 32.6 illustrates these components.

The **fuel elements** contain the fissile fuel and, for example, may be thin rods about 1 cm in diameter. In a large power reactor there may be thousands of fuel elements placed close together, the entire region of fuel elements being known as the **reactor core**.

Uranium $^{235}_{92}$U is a common reactor fuel. Since the natural abundance of this isotope is only about 0.7%, there are special uranium-enrichment plants to increase the percent-

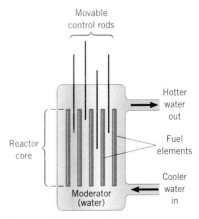

Figure 32.6 A nuclear reactor consists of fuel elements, control rods, and a moderator (in this case, water).

The physics of
nuclear reactors.

age. Most commercial reactors use uranium in which the amount of $^{235}_{92}U$ has been enriched to about 3%.

Whereas neutrons with energies of about 0.04 eV (or less) readily fission $^{235}_{92}U$, the neutrons released during the fission process have significantly greater energies of several MeV or so. Consequently, a nuclear reactor must contain some type of material that will decrease or moderate the speed of such energetic neutrons so they can readily fission additional $^{235}_{92}U$ nuclei. The material that slows down the neutrons is called a *moderator.* One commonly used moderator is water. When an energetic neutron leaves a fuel element, the neutron enters the surrounding water and collides with water molecules. With each collision, the neutron loses an appreciable fraction of its energy and slows down. Once slowed down to thermal energy by the moderator, a process that takes less than 10^{-3} s, the neutron is capable of initiating a fission event upon reentering a fuel element.

If the output power from a reactor is to remain constant, only one neutron from each fission event must trigger a new fission, as Figure 32.5 suggests. When each fission leads to one additional fission—no more or no less—the reactor is said to be *critical.* A reactor normally operates in a critical condition, because then it produces a steady output of energy. The reactor is *subcritical* when, on average, the neutrons from each fission trigger *less than one* subsequent fission. In a subcritical reactor, the chain reaction is not self-sustaining and eventually dies out. When the neutrons from each fission trigger *more than one* additional fission, the reactor is *supercritical.* During a supercritical condition, the energy released by a reactor increases. If left unchecked, the increasing energy can lead to a partial or total meltdown of the reactor core, with the possible release of radioactive material into the environment.

Clearly, a control mechanism is needed to keep the reactor in its normal or critical state. This control is accomplished by a number of **control rods** that can be moved into and out of the reactor core (see Figure 32.6). The control rods contain an element, such as boron or cadmium, that readily absorbs neutrons without fissioning. If the reactor becomes supercritical, the control rods are automatically moved farther into the core to absorb the excess neutrons causing the condition. In response, the reactor returns to its critical state. Conversely, if the reactor becomes subcritical, the control rods are partially withdrawn from the core. Fewer neutrons are absorbed, more neutrons are available for fission, and the reactor again returns to its critical state.

Figure 32.7 illustrates a pressurized water reactor. In such a reactor, the heat generated within the fuel rods is carried away by water that surrounds the rods. To remove as much heat as possible, the water temperature is allowed to rise to a high value (about 300 °C). To prevent boiling, which occurs at 100 °C at 1 atmosphere of pressure, the water is pressurized in excess of 150 atmospheres. The hot water is pumped through a heat exchanger, where heat is transferred to water flowing in a second, closed system. The heat transferred to the second system produces steam that drives a turbine. The turbine is coupled to an electric generator, whose output electric power is delivered to consumers via high-voltage transmission lines. After exiting the turbine, the steam is condensed back into water that is returned to the heat exchanger.

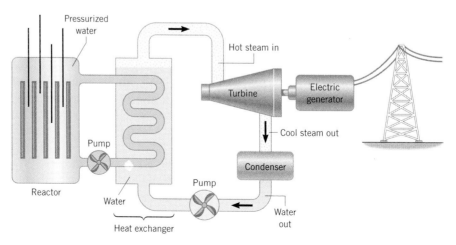

Figure 32.7 Diagram of a nuclear power plant that uses a pressurized water reactor.

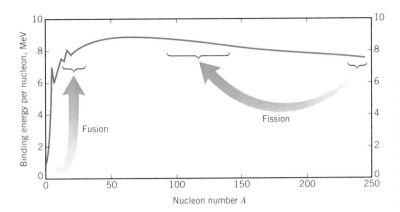

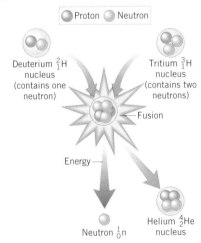

Figure 32.8 When fission occurs, a massive nucleus divides into two fragments whose binding energy per nucleon is greater than that of the original nucleus. When fusion occurs, two low-mass nuclei combine to form a more massive nucleus whose binding energy per nucleon is greater than that of the original nuclei.

32.5 Nuclear Fusion

In Example 3 in Section 32.3, the binding-energy-per-nucleon curve is used to estimate the amount of energy released in the fission process. As summarized in Figure 32.8, the massive nuclei at the right end of the curve have a binding energy of about 7.6 MeV per nucleon. The less massive fission fragments are near the center of the curve and have a binding energy of approximately 8.5 MeV per nucleon. The energy released per nucleon by fission is the difference between these two values, or about 0.9 MeV per nucleon.

A glance at the far left end of the diagram in Figure 32.8 suggests another means of generating energy. Two nuclei with very low mass and relatively small binding energies per nucleon could be combined or "fused" into a single, more massive nucleus that has a greater binding energy per nucleon. This process is called *nuclear fusion.* A substantial amount of energy can be released during a fusion reaction, as Example 5 shows.

Example 5 **The Energy Released During Nuclear Fusion**

Two isotopes of hydrogen, ^2_1H (deuterium, D) and ^3_1H (tritium, T), fuse to form ^4_2He and a neutron according to the following reaction:

$$^2_1\text{H} + ^3_1\text{H} \longrightarrow ^4_2\text{He} + ^1_0\text{n}$$

Determine the energy released by this fusion reaction, which is illustrated in Figure 32.9.

Reasoning Energy is released, so the total mass of the final nuclei is less than the total mass of the initial nuclei. To determine the energy released, we find the amount (in atomic mass units u) by which the total mass has decreased. Then, we use the fact that 1 u is equivalent to 931.5 MeV of energy, as determined in Section 31.3. This approach is the same as that used in Section 31.4 for radioactive decay.

Solution The masses of the initial and final nuclei in this reaction, as well as that of the neutron, are

	Initial Masses		Final Masses
^2_1H	2.0141 u	^4_2He	4.0026 u
^3_1H	3.0161 u	^1_0n	1.0087 u
Total:	5.0302 u	Total:	5.0113 u

The decrease in mass, or the mass defect, is $\Delta m = 5.0302$ u $- 5.0113$ u $= 0.0189$ u. Since 1 u is equivalent to 931.5 MeV, the energy released is $\boxed{17.6 \text{ MeV}}$.

The deuterium nucleus contains 2 nucleons, and the tritium nucleus contains 3. Thus, there are 5 nucleons that participate in the fusion, so the energy released per nucleon is about 3.5 MeV. This energy per nucleon is greater than that released in a fission process ($\approx$0.9 MeV per nucleon). Thus, for a given mass of fuel, a fusion reaction yields more energy than a fission reaction.

Figure 32.9 Deuterium and tritium are fused together to form a helium nucleus (^4_2He). The result is the release of an enormous amount of energy, mainly carried by a single high-energy neutron (^1_0n).

Because fusion reactions release so much energy, there is considerable interest in fusion reactors, although to date no commercial units have been constructed. The difficulties in building a fusion reactor arise mainly because the two low-mass nuclei must be brought sufficiently near each other so that the short-range strong nuclear force can pull them together, leading to fusion. But each nucleus has a positive charge and repels the other electrically. For the nuclei to get sufficiently close in the presence of the repulsive electric force, they must have large kinetic energies, and hence large temperatures, to start with. For example, a temperature of around a hundred million °C is needed to start the deuterium–tritium reaction discussed in Example 5.

Reactions that require such extremely high temperatures are called ***thermonuclear reactions.*** The most important thermonuclear reactions occur in stars, such as our own sun. The energy radiated by the sun comes from deep within its core, where the temperature is high enough to initiate the fusion process. One group of reactions thought to occur in the sun is the *proton–proton* cycle, which is a series of reactions whereby six protons form a helium nucleus, two positrons, two γ rays, two protons, and two neutrinos. The energy released by the proton–proton cycle is about 25 MeV (see problem 33).

Man-made fusion reactions have been carried out in a fusion-type nuclear bomb—commonly called a hydrogen bomb. In a hydrogen bomb, the fusion reaction is ignited by a fission bomb using uranium or plutonium. The temperature produced by the fission bomb is sufficiently high to initiate a thermonuclear reaction where, for example, hydrogen isotopes are fused into helium, releasing even more energy. For fusion to be useful as a commercial energy source, the energy must be released in a steady, controlled manner—unlike that in a bomb. To date, scientists have not succeeded in constructing a fusion device that produces more energy on a continual basis than is expended in operating the device. A fusion device uses a high temperature to start a reaction, and under such a condition, all the atoms are completely ionized to form a *plasma* (a gas composed of charged particles, like $_1^2\text{H}^+$ and e^-). The problem is to confine the hot plasma for a long enough time so that collisions among the ions can lead to fusion.

The physics of nuclear fusion using magnetic confinement.

One ingenious method of confining the plasma is called *magnetic confinement* because it uses a magnetic field to contain and compress the charges in the plasma. Charges moving in the magnetic field are subject to magnetic forces. As the forces increase, the associated pressure builds, and the temperature rises. The gas becomes a superheated plasma, ultimately fusing when the pressure and temperature are high enough.

The physics of nuclear fusion using inertial confinement.

Another type of confinement scheme, known as *inertial confinement,* is also being developed. Tiny, solid pellets of fuel are dropped into a container. As each pellet reaches the center of the container, a number of high-intensity laser beams strike the pellet simultaneously. The heating causes the exterior of the pellet to vaporize almost instantaneously. However, the inertia of the vaporized atoms keeps them from expanding outward as fast as the vapor is being formed. As a result, high pressures, high densities, and high temperatures are achieved at the center of the pellet, thus causing fusion. A variant of inertial confinement fusion, called "Z pinch," is under development at Sandia National Laboratories in New Mexico (see the chapter-opening photo). This device would also implode tiny fuel pellets, but without using lasers. Instead, scientists are using a cylindrical array of fine tungsten wires that are connected to a gigantic capacitor. When the capacitor is discharged, a huge current is sent through the wires. The heated wires vaporize almost instantly, generating a hot gas of ions, or plasma. The plasma is driven inward upon itself by the huge magnetic field produced by the current. The compressed plasma becomes super hot and generates a gigantic X-ray pulse. This pulse, it is hoped, would implode the solid fuel pellets to temperatures and pressures at which fusion would occur.

When compared to fission, fusion has some attractive features as an energy source. As we have seen in Example 5, fusion yields more energy than fission, for a given mass of fuel. Moreover, one type of fuel, $_1^2\text{H}$ (deuterium), is found in the waters of the oceans and is plentiful, cheap, and relatively easy to separate from the common $_1^1\text{H}$ isotope of hydrogen. Fissile materials like naturally occurring uranium $_{92}^{235}\text{U}$ are much less available, and supplies could be depleted within a century or two. However, the commercial use of fusion to provide cheap energy remains in the future.

Which one or more of the following statements concerning fission and fusion are true?
(a) Both fission and fusion reactions are characterized by a mass defect. (b) Both fission and fusion reactions always obey the conservation laws of physics. (c) Both fission and fusion take advantage of the fact that the binding energy per nucleon varies with the nucleon number of the nucleus. (d) All three of the previous statements are true. *(The answer is given at the end of the book.)*

Background: In nuclear fission, a massive nucleus splits into less massive fragments. In nuclear fusion, two nuclei with smaller masses combine to form a single nucleus with a larger mass.

For similar questions (including calculational counterparts), consult Self-Assessment Test 32.1, which is described at the end of Section 32.7.

32.6 *Elementary Particles*

SETTING THE STAGE

By 1932 the electron, the proton, and the neutron had been discovered and were thought to be nature's three ***elementary particles,*** in the sense that they were the basic building blocks from which all matter is constructed. Experimental evidence obtained since then, however, shows that several hundred additional particles exist, and scientists no longer believe that the proton and the neutron are elementary particles.

Most of these new particles have masses greater than the electron's mass, and many are more massive than protons or neutrons. Virtually all the new particles are unstable and decay in times between about 10^{-6} and 10^{-23} s.

Often, new particles are produced by accelerating protons or electrons to high energies and letting them collide with a target nucleus. For example, Figure 32.10 shows a collision between an energetic proton and a stationary proton. If the incoming proton has sufficient energy, the collision produces an entirely new particle, the *neutral pion* (π^0). The π^0 particle lives for only about 0.8×10^{-16} s before it decays into two γ-ray photons. Since the pion did not exist before the collision, it was created from part of the incident proton's energy. Because a new particle such as the neutral pion is often created from energy, it is customary to report the mass of the particle in terms of its equivalent *rest energy* (see Equation 28.5). Often, energy units of MeV are used. For instance, detailed analyses of experiments reveal that the mass of the π^0 particle is equivalent to a rest energy of 135.0 MeV. For comparison, the more massive proton has a rest energy of 938.3 MeV. Analyses of experiments also provide the electric charge and other properties of particles created in high-energy collisions. In the limited space available here, it is not possible to describe all the new particles that have been found. However, we will highlight some of the more significant discoveries.

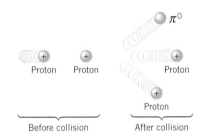

Figure 32.10 When an energetic proton collides with a stationary proton, a neutral pion (π^0) is produced. Part of the energy of the incident proton goes into creating the pion.

NEUTRINOS

In 1930, Wolfgang Pauli suggested that a particle called the ***neutrino*** (now known as the electron neutrino) should accompany the β decay of a radioactive nucleus. As Section 31.5 discusses, the neutrino has no electric charge, has a very small mass (a tiny fraction of the mass of an electron), and travels at speeds approaching (but less than) the speed of light. Neutrinos were finally discovered in 1956. Today, neutrinos are created in abundance in nuclear reactors and particle accelerators and are thought to be plentiful in the universe.

POSITRONS AND ANTIPARTICLES

The year 1932 saw the discovery of the *positron* (a contraction for "positive electron"). The positron has the same mass as the electron but carries an opposite charge of $+e$. A collision between a positron and an electron is likely to annihilate both particles, converting them into electromagnetic energy in the form of γ rays. For this reason, positrons never coexist with ordinary matter for any appreciable length of time. The mutual annihilation of a positron and an electron lies at the heart of an important medical diagnostic technique, as Conceptual Example 6 discusses.

Conceptual Example 6   Positron Emission Tomography

The physics of PET scanning

Positron emission tomography, or PET scanning, as it is known, utilizes positrons in the following way. Certain radioactive isotopes decay by positron emission—for example, oxygen $^{15}_{8}O$. Such isotopes are injected into the body, where they collect at specific sites. The positron ($^{0}_{1}e$) emitted during the decay of the isotope encounters an electron ($_{-1}^{0}e$) in the body tissue almost at once. The resulting mutual annihilation produces two γ-ray photons ($^{0}_{1}e + _{-1}^{0}e \rightarrow \gamma + \gamma$), which are detected by devices mounted on a ring around the patient, as Figure 32.11a shows. The two photons strike oppositely positioned detectors and, in doing so, reveal the line on which the annihilation occurred. Such information leads to a computer-generated image that can be useful in diagnosing abnormalities at the site where the radioactive isotope collected (see Figure 32.12). How does the principle of conservation of linear momentum explain the fact that the photons strike detectors located opposite one another?

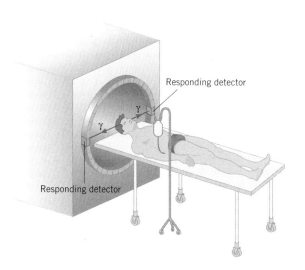

(a)   (b)

Figure 32.11 (a) In positron emission tomography, or PET scanning, a radioactive isotope is injected into the body. The isotope that is used decays by emitting a positron, which annihilates an electron in the body tissue, producing two γ-ray photons. These photons strike detectors mounted on opposite sides of a ring that surrounds the patient. (b) PET scanning is being used to monitor metabolic activity in the brain. (© CC Studio/Photo Researchers)

Reasoning and Solution The momentum conservation principle states that the total linear momentum of an isolated system remains constant (see Section 7.2). Therefore, the total momentum of the γ-ray photons will be equal to the total momentum of the positron and electron before annihilation, provided that the system is isolated. An isolated system is one on which no net external force acts. The positron and the electron do exert electrostatic forces on one another, however, since they carry electric charges. But these are internal, not external, forces and cannot change the total linear momentum of the two-particle system. The total photon momentum, then, must equal the total momentum of the positron and the electron, which is nearly zero, to the extent that the two particles have much less momentum than the photons do. With a total linear momentum of zero, the momentum vector of one photon must point opposite to the momentum vector of the other photon. Thus, the two photons depart from the annihilation site traveling in opposite directions and strike oppositely located detectors.

Related Homework: *Problem 37*

Healthy brain   Diseased brain

Figure 32.12 Positron emission tomography (PET) provides an important medical diagnostic technique. The PET scan of the diseased brain shown here reveals Alzheimer's disease. The colors are computer-generated. (© National Institutes of Health/Photo Researchers)

The positron is an example of an antiparticle, and after its discovery, scientists came to realize that for every type of particle there is a corresponding type of antiparticle. The antiparticle is a form of matter that has the same mass as the particle but carries an opposite electric charge (e.g., the electron–positron pair) or has a magnetic moment that is oriented in an opposite direction relative to the spin (e.g., the neutrino–antineutrino pair). A few electrically neutral particles, like the photon and the neutral pion (π^0), are their own antiparticles.

MUONS AND PIONS

In 1937, the American physicists S. H. Neddermeyer and C. D. Anderson discovered a new charged particle whose mass was about 207 times greater than the mass of the electron. The particle is designated by the Greek letter μ (mu) and is known as a *muon*. There are two muons that have the same mass but opposite charge: the particle μ^- and its antiparticle μ^+. The μ^- muon has the same charge as the electron, whereas the μ^+ muon has the same charge as the positron. Both muons are unstable, with a lifetime of 2.2×10^{-6} s. The μ^- muon decays into an electron (β^-), a muon neutrino (ν_μ), and an electron antineutrino ($\bar{\nu}_e$), according to the following reaction:

$$\mu^- \longrightarrow \beta^- + \nu_\mu + \bar{\nu}_e$$

The μ^+ muon decays into a positron (β^+), a muon antineutrino ($\bar{\nu}_\mu$), and an electron neutrino (ν_e):

$$\mu^+ \longrightarrow \beta^+ + \bar{\nu}_\mu + \nu_e$$

Muons interact with protons and neutrons via the weak nuclear force (see Section 31.5).

The Japanese physicist Hideki Yukawa (1907–1981) predicted in 1935 that *pions* exist, but they were not discovered until 1947. Pions come in three varieties: one that is positively charged, the negatively charged antiparticle with the same mass, and the neutral pion, mentioned earlier, which is its own antiparticle. The symbols for these pions are, respectively, π^+, π^-, and π^0. The charged pions are unstable and have a lifetime of 2.6×10^{-8} s. The decay of a charged pion almost always produces a muon:

$$\pi^- \longrightarrow \mu^- + \bar{\nu}_\mu$$
$$\pi^+ \longrightarrow \mu^+ + \nu_\mu$$

As mentioned earlier, the neutral pion π^0 is also unstable and decays into two γ-ray photons, the lifetime being 0.8×10^{-16} s. The pions are of great interest because, unlike the muons, the pions interact with protons and neutrons via the strong nuclear force.

CLASSIFICATION OF PARTICLES

It is useful to group the known particles into three families—the photons, the leptons, and the hadrons—as Table 32.3 (page 980) summarizes. This grouping is made according to the nature of the force by which a particle interacts with other particles. The *photon family*, for instance, has only one member, the photon. The photon interacts only with charged particles, and the interaction is only via the *electromagnetic force*. No other particle behaves in this manner.

The *lepton family* consists of particles that interact by means of the *weak nuclear force*. Leptons can also exert gravitational and (if the leptons are charged) electromagnetic forces on other particles. The four better-known leptons are the electron, the muon, the electron neutrino ν_e, and the muon neutrino ν_μ. Table 32.3 lists these particles together with their antiparticles. Recently, two other leptons have been discovered, the tau particle (τ) and its neutrino (ν_τ), bringing the number of particles in the lepton family to six.

The *hadron family* contains the particles that interact by means of the *strong nuclear force and the weak nuclear force*. Hadrons can also interact by gravitational and electromagnetic forces, but at short distances ($\leq 10^{-15}$ m) the strong nuclear force dominates. Among the hadrons are the proton, the neutron, and the pions. As Table 32.3 indicates, most hadrons are short-lived. The hadrons are subdivided into two groups, the *mesons* and the *baryons*, for a reason that will be discussed in connection with the idea of quarks.

Table 32.3   *Some Particles and Their Properties*

Family	Particle	Particle Symbol	Antiparticle Symbol	Rest Energy (MeV)	Lifetime (s)
Photon	Photon	γ	Self[a]	0	Stable
Lepton	Electron	e^- or β^-	e^+ or β^+	0.511	Stable
	Muon	μ^-	μ^+	105.7	2.2×10^{-6}
	Tau	τ^-	τ^+	1777	2.9×10^{-13}
	Electron neutrino	ν_e	$\bar{\nu}_e$	≈ 0	Stable
	Muon neutrino	ν_μ	$\bar{\nu}_\mu$	≈ 0	Stable
	Tau neutrino	ν_τ	$\bar{\nu}_\tau$	≈ 0	Stable
Hadron					
Mesons					
	Pion	π^+	π^-	139.6	2.6×10^{-8}
		π^0	Self[a]	135.0	8.4×10^{-17}
	Kaon	K^+	K^-	493.7	1.2×10^{-8}
		K_S^0	$\bar{K}_S^0$	497.7	8.9×10^{-11}
		K_L^0	$\bar{K}_L^0$	497.7	5.2×10^{-8}
	Eta	η^0	Self[a]	547.3	$<10^{-18}$
Baryons					
	Proton	p	$\bar{p}$	938.3	Stable
	Neutron	n	$\bar{n}$	939.6	886
	Lambda	Λ^0	$\bar{\Lambda}^0$	1116	2.6×10^{-10}
	Sigma	Σ^+	$\bar{\Sigma}^-$	1189	8.0×10^{-11}
		Σ^0	$\bar{\Sigma}^0$	1193	7.4×10^{-20}
		Σ^-	$\bar{\Sigma}^+$	1197	1.5×10^{-10}
	Omega	Ω^-	Ω^+	1672	8.2×10^{-11}

[a] The particle is its own antiparticle.

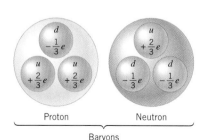

Figure 32.13 According to the original quark model of hadrons, all mesons consist of a quark and an antiparticle, whereas baryons contain three quarks.

QUARKS

As more and more hadrons were discovered, it became clear that they were not all elementary particles. The suggestion was made that the hadrons are made up of smaller, more elementary particles called *quarks.* In 1963, a quark theory was advanced independently by M. Gell–Mann (1929–) and G. Zweig (1937–). The theory proposed that there are three quarks and three corresponding antiquarks, and that hadrons are constructed from combinations of these. Thus, the quarks are elevated to the status of elementary particles for the hadron family. The particles in the photon and lepton families are considered to be elementary, and as such they are not composed of quarks.

The three quarks were named *up* (u), *down* (d), and *strange* (s), and were assumed to have, respectively, fractional charges of $+\frac{2}{3}e$, $-\frac{1}{3}e$, and $-\frac{1}{3}e$. In other words, a quark possesses a charge magnitude smaller than that of an electron, which has a charge of $-e$. Table 32.4 lists the symbols and electric charges of these quarks and the corresponding antiquarks. Experimentally, quarks should be recognizable by their fractional charges, but in spite of an extensive search for them, free quarks have never been found.

According to the original quark theory, the mesons are different from the baryons, because each meson consists of only two quarks—a quark and an antiquark—whereas a baryon contains three quarks. For instance, the π^- pion (a meson) is composed of a d quark and a $\bar{u}$ antiquark, $\pi^- = d + \bar{u}$, as Figure 32.13 shows. These two quarks combine to give the π^- pion a net charge of $-e$. Similarly, the π^+ pion is a combination of the $\bar{d}$ and u quarks, $\pi^+ = \bar{d} + u$. In contrast, protons and neutrons, being baryons, consist of three quarks. A proton contains the combination $d + u + u$, and a neutron contains the combination $d + d + u$ (see Figure 32.13). These groups of three quarks give the correct charges for the proton and neutron.

Table 32.4 *Quarks and Antiquarks*

Name	Quarks		Antiquarks	
	Symbol	Charge	Symbol	Charge
Up	u	$+\frac{2}{3}e$	$\bar{u}$	$-\frac{2}{3}e$
Down	d	$-\frac{1}{3}e$	$\bar{d}$	$+\frac{1}{3}e$
Strange	s	$-\frac{1}{3}e$	$\bar{s}$	$+\frac{1}{3}e$
Charmed	c	$+\frac{2}{3}e$	$\bar{c}$	$-\frac{2}{3}e$
Top	t	$+\frac{2}{3}e$	$\bar{t}$	$-\frac{2}{3}e$
Bottom	b	$-\frac{1}{3}e$	$\bar{b}$	$+\frac{1}{3}e$

The original quark model was extremely successful in predicting not only the correct charges for the hadrons, but other properties as well. However, in 1974 a new particle, the J/ψ meson, was discovered. This meson has a rest energy of 3097 MeV, much larger than the rest energies of other known mesons. The existence of the J/ψ meson could be explained only if a new quark–antiquark pair existed; this new quark was called *charmed* (*c*). With the discovery of more and more particles, it has been necessary to postulate a fifth and a sixth quark; their names are *top* (*t*) and *bottom* (*b*), although some scientists prefer to call these quarks *truth* and *beauty*. Today, there is firm evidence for all six quarks, each with its corresponding antiquark. All of the hundreds of the known hadrons can be accounted for in terms of these six quarks and their antiquarks.

In addition to electric charge, quarks also have other properties. For example, each quark possesses a characteristic called ***color,*** for which there are three possibilities: blue, green, or red. The corresponding possibilities for the antiquarks are antiblue, antigreen, and antired. The use of the term "color" and the specific choices of blue, green, and red are arbitrary, for the visible colors of the electromagnetic spectrum have nothing to do with quark properties. The quark property of color, however, is important, because it brings the quark model into agreement with the Pauli exclusion principle and enables the model to account for experimental observations that are otherwise difficult to explain.

THE STANDARD MODEL

The various elementary particles that have been discovered can interact via one or more of the following four forces: the gravitational force, the strong nuclear force, the weak nuclear force, and the electromagnetic force. In particle physics, the phrase *"the standard model"* refers to the currently accepted explanation for the strong nuclear force, the weak nuclear force, and the electromagnetic force. In this model, the strong nuclear force between quarks is described in terms of the concept of color, the theory being referred to as quantum chromodynamics. According to the standard model, the weak nuclear force and the electromagnetic force are separate manifestations of a single even more fundamental force, referred to as the electroweak force, as we have seen in Section 31.5.

In the standard model, our understanding of the building blocks of matter follows the hierarchical pattern illustrated in Figure 32.14. Molecules, such as water (H_2O) and glucose ($C_6H_{12}O_6$), are composed of atoms. Each atom consists of a nucleus that is surrounded by a cloud of electrons. The nucleus, in turn, is made up of protons and neutrons, which are composed of quarks.

Figure 32.14 The current view of how matter is composed of basic units, starting with a molecule and ending with a quark. The approximate sizes of each unit are also listed.

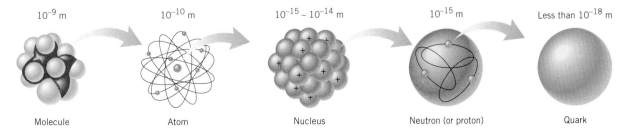

10^{-9} m	10^{-10} m	$10^{-15} - 10^{-14}$ m	10^{-15} m	Less than 10^{-18} m
Molecule	Atom	Nucleus	Neutron (or proton)	Quark

✔ **Check Your Understanding 4**

Of the following particles, which ones are not composed of quarks and antiquarks: (a) a proton, (b) an electron, (c) a neutron, (d) a neutrino? *[The answer(s) is (are) given at the end of the book.]*

Background: Of the hundreds of particles that have been discovered, those in the hadron family are composed of quarks, and those in the lepton family are not.

For similar questions (including calculational counterparts), consult Self-Assessment Test 32.1, which is described at the end of Section 32.7.

32.7 *Cosmology*

Cosmology is the study of the structure and evolution of the universe. In this study, both the very large and the very small aspects of the universe are important. Astronomers, for example, study stars located at enormous distances from the earth, up to billions of light-years away. In contrast, particle physicists focus their efforts on the very small elementary particles (10^{-18} m or smaller) that comprise matter. The synergy between the work of astronomers and that of particle physicists has led to significant advances in our understanding of the universe. Central to that understanding is the belief that the universe is expanding, and we begin by discussing the evidence that justifies this belief.

THE EXPANDING UNIVERSE AND THE BIG BANG

The physics of an expanding universe.

The idea that the universe is expanding originated with the astronomer Edwin P. Hubble (1889–1953). He found that light reaching the earth from distant galaxies is Doppler-shifted toward greater wavelengths—that is, toward the red end of the visible spectrum. As Section 24.5 discusses, this type of Doppler shift results when the observer and the source of the light are moving away from each other. The speed at which a galaxy is receding from the earth can be determined from the measured Doppler shift in wavelength. Hubble found that a galaxy located at a distance d from the earth recedes from the earth at a speed v given by

Hubble's law
$$v = Hd \qquad (32.5)$$

where H is a constant known as the Hubble parameter. In other words, the recession speed is proportional to the distance d, so that more distant galaxies are moving away from the earth at greater speeds. Equation 32.5 is referred to as Hubble's law.

Hubble's picture of an expanding universe does not mean that the earth is at the center of the expansion. In fact, there is no literal center. Imagine a loaf of raisin bread expanding as it bakes. Each raisin moves away from every other raisin, without any single one acting as a center for the expansion. Galaxies in the universe behave in a similar fashion. Observers in other galaxies would see distant galaxies moving away, just as we do.

The physics of "dark energy."

Not only is the universe expanding, it is doing so at an accelerated rate, according to recent astronomical measurements of the brightness of supernovas, or exploding stars. To account for the accelerated rate, astronomers have postulated that a kind of "dark energy" pervades the universe. The normal gravitational force between galaxies slows the rate at which they are moving away from each other. The dark energy, however, gives rise to a force that counteracts gravity and pushes galaxies apart. As yet, little is known about the properties of dark energy.

Experimental measurements by astronomers indicate that an approximate value for the Hubble parameter is

$$H = 0.022 \, \frac{\text{m}}{\text{s} \cdot \text{light-year}}$$

The value for the Hubble parameter is believed to be accurate within 10%. Scientists are very interested in obtaining an accurate value for H because it can be related to an age for the universe, as the next example illustrates.

Example 7 An Age for the Universe

Determine an estimate of the age of the universe using Hubble's law.

Reasoning Consider a galaxy currently located at a distance d from the earth. According to Hubble's law, this galaxy is moving away from us at a speed of $v = Hd$. At an earlier time, therefore, this galaxy must have been closer. We can imagine, in fact, that in the remote past the separation distance was relatively small and that the universe originated at such a time. To estimate the age of the universe, we calculate the time it has taken the galaxy to recede to its present position. For this purpose, time is simply distance divided by speed, or $t = d/v$.

Solution Using Hubble's law and the fact that a distance of 1 light-year is 9.46×10^{15} m, we estimate the age of the universe to be

$$t = \frac{d}{v} = \frac{d}{Hd} = \frac{1}{H}$$

$$t = \frac{1}{0.022 \, \frac{\text{m}}{\text{s} \cdot \text{light-year}}} = \frac{1}{\left(0.022 \, \frac{\text{m}}{\text{s} \cdot \text{light-year}}\right)\left(\frac{1 \text{ light-year}}{9.46 \times 10^{15} \text{ m}}\right)}$$

$$= 4.3 \times 10^{17} \text{ s} \quad \text{or} \quad \boxed{1.4 \times 10^{10} \text{ yr}}$$

The idea presented in Example 7, that our galaxy and other galaxies in the universe were very close together at some earlier instant in time, lies at the heart of the *Big Bang theory.* This theory postulates that the universe had a definite beginning in a cataclysmic event, sometimes called the primeval fireball. Dramatic evidence supporting the theory was discovered in 1965 by Arno A. Penzias (1933–) and Robert W. Wilson (1936–). Using a radio telescope, they discovered that the earth is being bathed in weak electromagnetic waves in the microwave region of the spectrum (wavelength = 7.35 cm, see Figure 24.10). They observed that the intensity of these waves is the same, no matter where in the sky they pointed their telescope, and concluded that the waves originated outside of our galaxy. This microwave background radiation, as it is called, represents radiation left over from the Big Bang and is a kind of cosmic afterglow. Subsequent measurements have confirmed the research of Penzias and Wilson and shown that the microwave radiation is consistent with a perfect blackbody (see Sections 13.3 and 29.2) radiating at a temperature of 2.7 K, in agreement with theoretical analysis of the Big Bang. In 1978, Penzias and Wilson received a Nobel Prize for their discovery.

THE STANDARD MODEL FOR THE EVOLUTION OF THE UNIVERSE

Based on the recent experimental and theoretical research in particle physics, scientists have proposed an evolutionary sequence of events following the Big Bang. This sequence is known as the *standard cosmological model* and is illustrated in Figure 32.15.

Immediately following the Big Bang, the temperature of the universe was incredibly high, about 10^{32} K. During this initial period, the three fundamental forces (the gravitational force, the strong nuclear force, and the electroweak force) all behaved as a single unified force. Very quickly, in about 10^{-43} s, the gravitational force took on a separate identity all its own, as Figure 32.15 indicates. Meanwhile, the strong nuclear force and the electroweak force continued to act as a single force, which is sometimes referred to as the GUT force. GUT stands for the Grand Unified Theory that presumably would explain such a force. Slightly later, at about 10^{-35} s after the Big Bang, the GUT force separated into the strong nuclear force and the electroweak force (see Figure 32.15), the universe expanding and cooling somewhat to a temperature of roughly 10^{28} K. From this point on, the strong nuclear force behaved as we know it today, while the electroweak force maintained its identity. In this scenario, note that the weak nuclear force and the electromagnetic force have not yet manifested themselves as separate entities. The disappearance of the electroweak force and the appearance of the weak nuclear force and the electromag-

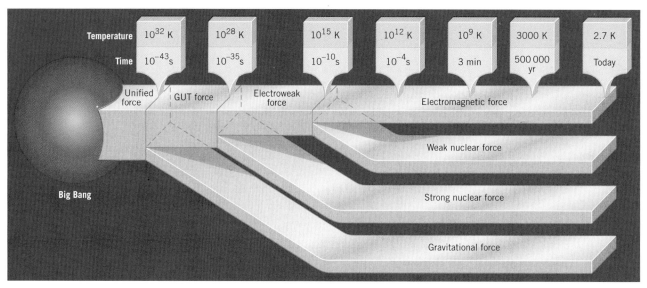

Figure 32.15 According to the standard cosmological model, the universe has evolved as illustrated here. In this model, the universe is presumed to have originated with a cataclysmic event known as the Big Bang. The times shown are those following this event.

netic force eventually occurred at approximately 10^{-10} s after the Big Bang, when the temperature of the expanding universe had cooled to about 10^{15} K.

From the Big Bang up until the strong nuclear force separated from the GUT force at a time of 10^{-35} s, all particles of matter were similar, and there was no distinction between quarks and leptons. After this time, quarks and leptons became distinguishable. Eventually the quarks and antiquarks formed hadrons, such as protons and neutrons and their antiparticles. By a time of 10^{-4} s after the Big Bang, however, the temperature had cooled to approximately 10^{12} K, and the hadrons had mostly disappeared. Protons and neutrons survived only as a very small fraction of the total number of particles, the majority of which were leptons such as electrons, positrons, and neutrinos. Like most of the hadrons before them, most of the electrons and positrons eventually disappeared. However, they did leave behind a relatively small number of electrons to join the small number of protons and neutrons at a time of about 3 min following the Big Bang. At this time the temperature of the expanding universe had decreased to about 10^9 K, and small nuclei such as that of helium began forming. Later, when the universe was about 500 000 years old and the temperature had dropped to near 3000 K, hydrogen and helium atoms began forming. As the temperature decreased further, stars and galaxies formed, and today we find a temperature of 2.7 K characterizing the cosmic background radiation of the universe.

**Self-Assessment Test 32.1**

Test your understanding of the material in Sections 32.1–32.7:

- Biological Effects of Ionizing Radiation   • Induced Nuclear Reactions
- Nuclear Fission   • Nuclear Reactors   • Nuclear Fusion
- Elementary Particles   • Cosmology

Go to **www.wiley.com/college/cutnell**

32.8 Concepts & Calculations

When considering the biological effects of ionizing radiation, the concept of the biologically equivalent dose is especially important. Its importance lies in the fact that the biologically equivalent dose incorporates both the amount of energy per unit mass that is absorbed and the effectiveness of a particular type of radiation in producing a certain biological effect. Example 8 examines this concept, as well as reviewing the notions of power (Section 6.7) and intensity (Section 16.7) of a wave.

Concepts & Calculations Example 8
The Biologically Equivalent Dose

▼

A patient is being given a chest X-ray. The X-ray beam is turned on for 0.20 s, and its intensity is 0.40 W/m². The area of the chest being exposed is 0.072 m², and the radiation is absorbed by 3.6 kg of tissue. The relative biological effectiveness (RBE) of the X-rays for this tissue is 1.1. Determine the biologically equivalent dose received by the patient.

Concept Questions and Answers How is the power of the beam related to the beam intensity?

Answer The power P is equal to the intensity I times the cross-sectional area A of the beam, as indicated by Equation 16.8 ($P = IA$).

How is the energy absorbed by the tissue related to the power of the beam?

Answer The energy E is equal to the product of the power P delivered by the beam and the time t of exposure, according to Equation 6.10b ($E = Pt$).

What is the absorbed dose?

Answer The absorbed dose is defined by Equation 32.2 as the energy E absorbed by the tissue divided by its mass m, or Absorbed dose = E/m. The unit of the absorbed dose is the gray (Gy).

How is the biologically equivalent dose related to the absorbed dose?

Answer The biologically equivalent dose is equal to the product of the absorbed dose and the relative biological effectiveness (RBE), as indicated by Equation 32.4. However, the absorbed dose must be expressed in rad units, not in gray units. The relation between the two units is 1 rad = 0.01 Gy.

Solution The biologically equivalent dose is given by Equation 32.4 as

$$\text{Biologically equivalent dose} = [\text{Absorbed dose (in rad)}](\text{RBE})$$

At this point we need to convert the absorbed dose from Gy units to rad units:

$$\text{Absorbed dose (in rad)} = [\text{Absorbed dose (in Gy)}]\left(\frac{1 \text{ rad}}{0.01 \text{ Gy}}\right)$$

The biologically equivalent dose then becomes

$$\text{Biologically equivalent dose} = [\text{Absorbed dose (in Gy)}]\left(\frac{1 \text{ rad}}{0.01 \text{ Gy}}\right)(\text{RBE})$$

Equation 32.2 indicates that the absorbed dose (in Gy) is equal to the energy E divided by the mass m of the tissue, so that we have

$$\text{Biologically equivalent dose} = \left(\frac{E}{m}\right)\left(\frac{1 \text{ rad}}{0.01 \text{ Gy}}\right)(\text{RBE})$$

From the answers to the first and second Concept Questions, we can conclude that the absorbed energy E is related to the intensity I, the cross-sectional area A, and the time duration t of the beam via $E = Pt = IAt$. Substituting this result into the equation for the biologically equivalent dose gives

$$\begin{aligned}
\text{Biologically equivalent dose} &= \left(\frac{IAt}{m}\right)\left(\frac{1 \text{ rad}}{0.01 \text{ Gy}}\right)(\text{RBE}) \\
&= \frac{(0.40 \text{ W/m}^2)(0.072 \text{ m}^2)(0.20 \text{ s})}{3.6 \text{ kg}}\left(\frac{1 \text{ rad}}{0.01 \text{ Gy}}\right)(1.1) = \boxed{0.18 \text{ rem}}
\end{aligned}$$

▲

The next example illustrates the decay of a particle into two photons, so matter is completely converted into electromagnetic waves. The example discusses how to find the energy, frequency, and wavelength of each photon. It also provides an opportunity to review the principle of conservation of energy and the principle of conservation of linear momentum.

Concepts & Calculations Example 9   The Decay of the π^0 Meson

The π^0 meson is a particle that has a rest energy of 135.0 MeV (see Table 32.3). It lives for a very short time and then decays into two γ-ray photons: $\pi^0 \rightarrow \gamma + \gamma$. Suppose that one of the γ-ray photons travels along the $+x$ axis. If the π^0 is at rest when it decays, find (a) the energy (in MeV), (b) the frequency and wavelength, and (c) the momentum of each γ-ray photon.

Concept Questions and Answers   How is the energy E of each γ-ray photon related to the rest energy E_0 of the π^0 particle?

Answer   According to the principle of conservation of energy, the sum of the energies of the two γ-ray photons must equal the rest energy E_0 of the π^0 particle, since it is at rest. Therefore, the energy E of each photon is $E = \frac{1}{2}E_0$.

How can the frequency and wavelength of a photon be determined from its energy?

Answer   A photon is a packet of electromagnetic energy, and each contains an amount of energy given by Equation 29.2 as $E = hf$, where h is Planck's constant and f is the frequency. Equation 16.1 indicates that the wavelength λ of the photon is inversely proportional to its frequency, $\lambda = c/f$, where c is the speed of light.

How is the total linear momentum of the photons related to the momentum of the π^0 particle, and what is the momentum of each photon?

Answer   The principle of conservation of linear momentum states that the total linear momentum of an isolated system is constant (see Section 7.2). Since the π^0 particle is at rest, the total linear momentum before the decay is zero. Since the π^0 particle is an isolated system, momentum is conserved, and the total linear momentum of the two γ-ray photons must also be zero. The magnitude p of the momentum of a photon is given by Equation 29.6 as $p = h/\lambda$. Thus, if one photon travels along the $+x$ axis with a momentum of $+h/\lambda$, the other must travel along the $-x$ axis with a momentum of $-h/\lambda$, ensuring that the total momentum is zero.

Solution

(a) Since the sum of the energies of the two γ-ray photons must be equal to the rest energy E_0 of the π^0 particle, the energy of each photon is

$$E = \tfrac{1}{2}E_0 = \tfrac{1}{2}(135.0 \text{ MeV}) = \boxed{67.5 \text{ MeV}}$$

(b) The frequency of each photon is given by Equation 29.2 as its energy (in units of joules) divided by Planck's constant. Using the fact that $1 \text{ eV} = 1.60 \times 10^{-19}$ J, we find that the frequency of each photon is

$$f = \frac{E}{h} = \frac{(67.5 \times 10^6 \text{ eV})\left(\dfrac{1.60 \times 10^{-19} \text{ J}}{1 \text{ eV}}\right)}{6.63 \times 10^{-34} \text{ J} \cdot \text{s}} = \boxed{1.63 \times 10^{22} \text{ Hz}}$$

The wavelength of each photon can be found directly from Equation 16.1:

$$\lambda = \frac{c}{f} = \frac{3.00 \times 10^8 \text{ m/s}}{1.63 \times 10^{22} \text{ Hz}} = \boxed{1.84 \times 10^{-14} \text{ m}}$$

(c) The magnitude of each photon's momentum is related to its wavelength by Equation 29.6:

$$p = \frac{h}{\lambda} = \frac{6.63 \times 10^{-34} \text{ J} \cdot \text{s}}{1.84 \times 10^{-14} \text{ m}} = 3.60 \times 10^{-20} \text{ kg} \cdot \text{m/s}$$

Since the photons move along the x axis in opposite directions, their momenta are $\boxed{+3.60 \times 10^{-20} \text{ kg} \cdot \text{m/s}}$ and $\boxed{-3.60 \times 10^{-20} \text{ kg} \cdot \text{m/s}}$.

At the end of the problem set for this chapter, you will find homework problems that contain both conceptual and quantitative parts. These problems are grouped under the heading *Concepts & Calculations, Group Learning Problems.* They are designed for use by students working alone or in small learning groups. The conceptual part of each problem provides a convenient focus for group discussions.

Concept Summary

This summary presents an abridged version of the chapter, including the important equations and all available learning aids. For convenient reference, the learning aids (including the text's examples) are placed next to or immediately after the relevant equation or discussion. The following learning aids may be found on-line at **www.wiley.com/college/cutnell**:

Interactive LearningWare examples are solved according to a five-step interactive format that is designed to help you develop problem-solving skills.	Concept Simulations are animated versions of text figures or animations that illustrate important concepts. You can control parameters that affect the display, and we encourage you to experiment.
Interactive Solutions offer specific models for certain types of problems in the chapter homework. The calculations are carried out interactively.	Self-Assessment Tests include both qualitative and quantitative questions. Extensive feedback is provided for both incorrect and correct answers, to help you evaluate your understanding of the material.

Topic	Discussion	Learning Aids

32.1 Biological Effects of Ionizing Radiation

Ionizing radiation

Ionizing radiation consists of photons and/or moving particles that have enough energy to ionize an atom or molecule. Exposure is a measure of the ionization produced in air by X-rays or γ rays. When a beam of X-rays or γ rays is sent through a mass m of dry air (0 °C, 1 atm pressure) and produces positive ions whose total charge is q, the exposure in coulombs per kilogram (C/kg) is

Exposure (in C/kg)

$$\text{Exposure (in coulombs per kilograms)} = \frac{q}{m}$$

With q in coulombs (C) and m in kilograms (kg), the exposure in roentgens is

Exposure (in roentgens)

$$\text{Exposure (in roentgens)} = \left(\frac{1}{2.58 \times 10^{-4}}\right)\frac{q}{m} \qquad (32.1)$$

The absorbed dose is the amount of energy absorbed from the radiation per unit mass of absorbing material:

Absorbed dose

$$\text{Absorbed dose} = \frac{\text{Energy absorbed}}{\text{Mass of absorbing material}} \qquad (32.2)$$

The SI unit of absorbed dose is the gray (Gy); 1 Gy = 1 J/kg. However, the rad (rd) is another unit that is often used: 1 rd = 0.01 Gy.

The amount of biological damage produced by ionizing radiation is different for different types of radiation. The relative biological effectiveness (RBE) is the absorbed dose of 200-keV X-rays required to produce a certain biological effect divided by the dose of a given type of radiation that produces the same biological effect.

Relative biological effectiveness

$$\text{RBE} = \frac{\begin{array}{c}\text{The dose of 200-keV X-rays that}\\\text{produces a certain biological effect}\end{array}}{\begin{array}{c}\text{The dose of radiation that}\\\text{produces the same biological effect}\end{array}} \qquad (32.3)$$

The biologically equivalent dose (in rems) is the product of the absorbed dose (in rads) and the RBE:

Biologically equivalent dose

$$\begin{array}{c}\text{Biologically equivalent dose}\\\text{(in rems)}\end{array} = \begin{array}{c}\text{Absorbed dose}\\\text{(in rads)}\end{array} \times \text{RBE} \qquad (32.4)$$ **Examples 1, 8**

32.2 Induced Nuclear Reactions

Induced nuclear reaction

An induced nuclear reaction occurs whenever a target nucleus is struck by an incident nucleus, an atomic or subatomic particle, or a γ-ray photon and undergoes a change as a result. An induced nuclear transmutation is a reaction in which the target nucleus is changed into a nucleus of a new element.

Induced nuclear transmutation

Nuclear reactions obey conservation laws

All nuclear reactions (induced or spontaneous) obey the conservation laws of physics as they relate to mass/energy, electric charge, linear momentum, angular momentum, and nucleon number. **Example 2**

Shorthand notation for nuclear reactions

Nuclear reactions are often written in a shorthand form, such as $^{14}_{7}\text{N}(\alpha, p)\,^{17}_{8}\text{O}$. The first and last symbols $^{14}_{7}\text{N}$ and $^{17}_{8}\text{O}$ denote, respectively, the initial and final

Topic	Discussion	Learning Aids
	nuclei. The symbols within the parentheses denote the incident α particle (on the left) and the small emitted particle or proton p (on the right).	
Thermal neutron	A thermal neutron is one that has a kinetic energy of about 0.04 eV.	

32.3 Nuclear Fission

Topic	Discussion	Learning Aids
Nuclear fission	Nuclear fission occurs when a massive nucleus splits into two less massive fragments. Fission can be induced by the absorption of a thermal neutron. When a massive nucleus fissions, energy is released because the binding energy per nucleon is greater for the fragments than for the original nucleus. Neutrons are also released during nuclear fission. These neutrons can, in turn, induce	**Example 3** **Interactive LearningWare 32.1** **Example 4**
Chain reaction **Controlled chain reaction**	other nuclei to fission and lead to a process known as a chain reaction. A chain reaction is said to be controlled if each fission event contributes, on average, only one neutron that fissions another nucleus.	**Concept Simulation 32.1**

32.4 Nuclear Reactors

Topic	Discussion	Learning Aids
Nuclear reactor **Fuel elements** **Neutron moderator** **Control rods** **Critical state** **Subcritical state** **Supercritical state**	A nuclear reactor is a device that generates energy by a controlled chain reaction. Many reactors in use today have the same three basic components: fuel elements, a neutron moderator, and control rods. The fuel elements contain the fissile fuel, and the entire region of fuel elements is known as the reactor core. The neutron moderator is a material (water, for example) that slows down the neutrons released in a fission event to thermal energies so they can initiate additional fission events. Control rods contain material that readily absorbs neutrons without fissioning. They are used to keep the reactor in its normal, or critical, state, in which each fission event leads to one additional fission, no more, no less. The reactor is subcritical when, on average, the neutrons from each fission trigger less than one subsequent fission. The reactor is supercritical when, on average, the neutrons from each fission trigger more than one additional fission.	

32.5 Nuclear Fusion

Topic	Discussion	Learning Aids
Nuclear fusion	In a fusion process, two nuclei with smaller masses combine to form a single nucleus with a larger mass. Energy is released by fusion when the binding energy per nucleon is greater for the larger nucleus than for the smaller nuclei. Fusion reactions are said to be thermonuclear because they require extremely high temperatures to proceed. Current studies of nuclear fusion utilize either	**Example 5** **Interactive Solution 32.29**
Magnetic confinement **Inertial confinement**	magnetic confinement or inertial confinement to contain the fusing nuclei at the high temperatures that are necessary.	

32.6 Elementary Particles

Topic	Discussion	Learning Aids
Photon family **Lepton family** **Hadron family**	Subatomic particles are divided into three families: the photon family (which consists only of the photon), the lepton family (which includes the electron), and the hadron family (which includes the proton and the neutron).	**Examples 6, 9**
Elementary particles	Elementary particles are the basic building blocks of matter. All members of the photon and lepton families are elementary particles.	
Quark theory	The quark theory proposes that the hadrons are not elementary particles but are composed of elementary particles called quarks. Currently, the hundreds of hadrons can be accounted for in terms of six quarks (up, down, strange, charmed, top, and bottom) and their antiquarks.	
The standard model	The standard model consists of two parts: (1) the currently accepted explanation for the strong nuclear force in terms of the quark concept of "color" and (2) the theory of the electroweak interaction.	

32.7 Cosmology

Topic	Discussion	Learning Aids
Cosmology	Cosmology is the study of the structure and evolution of the universe. Our universe is expanding. The speed v at which a distant galaxy recedes from the earth is given by Hubble's law:	
Hubble's law	$$v = Hd \qquad (32.5)$$	**Example 7**
Hubble parameter	where $H = 0.022$ m/(s·light-year) is called the Hubble parameter and d is the distance of the galaxy from the earth.	

Topic	Discussion	Learning Aids
Big Bang theory	The Big Bang theory postulates that the universe had a definite beginning in a cataclysmic event, sometimes called the primeval fireball. The radiation left over from this event is in the microwave region of the electromagnetic spectrum, and it is consistent with a perfect blackbody radiating at a temperature of 2.7 K, in agreement with theoretical analysis of the Big Bang.	
Standard cosmological model	The standard cosmological model for the evolution of the universe is summarized in Figure 32.15.	

Use *Self-Assessment Test 32.1* **to evaluate your understanding of Sections 32.1–32.7.**

Conceptual Questions

1. Two different types of radiation have the same RBE. Does this mean that each type delivers the same amount of energy to the tissue that it irradiates? Justify your answer.

2. When a dentist X-rays your teeth, a lead apron is placed over your chest and lower body. What is the purpose of this apron?

3. State whether the two quantities in each of the following cases are related and, if so, give the relation between them: (a) rads and grays, and (b) rads and roentgens.

4. Explain why the following reactions are *not* allowed: (a) $^{60}_{28}$Ni (α, p) $^{62}_{29}$Cu, (b) $^{27}_{13}$Al (n, n) $^{28}_{13}$Al, (c) $^{39}_{19}$K (p, α) $^{36}_{17}$Cl.

5. Thermal neutrons and thermal protons have nearly the same speed. But thermal neutrons and thermal electrons have very different speeds. Account for this observation, and explain whether the neutrons or the electrons have the greater speed.

6. Would a release of energy accompany the fission of a nucleus of nucleon number 25 into two fragments of about equal mass? Using the curve in Figure 32.8, account for your answer.

7. In the fission of $^{235}_{92}$U there are, on the average, 2.5 neutrons released per fission. Suppose a *different* element is being fissioned and, on the average, only 1.0 neutron is released per fission. If a small fraction of the thermal neutrons absorbed by the nuclei does *not* produce a fission, can a self-sustaining chain reaction be produced using this element? Explain.

8. The mass of coal consumed in a coal-burning electric power plant is about two million times greater than the mass of $^{235}_{92}$U used to fuel a comparable nuclear power plant. Why?

9. Refer to Figure 32.8 and decide whether the fusion of two nuclei, each with a nucleon number of 60, would release energy. Give your reasoning.

10. Explain the difference between fission and fusion and why each process produces energy.

Problems

ssm Solution is in the Student Solutions Manual.   www Solution is available on the World Wide Web at www.wiley.com/college/cutnell
☤ This icon represents a biomedical application.

Section 32.1 Biological Effects of Ionizing Radiation

1. ssm www A 75-kg person is exposed to 45 mrem of α particles (RBE = 12). How much energy has this person absorbed?

2. Neutrons (RBE = 2.0) and α particles have the same biologically equivalent dose. However, the absorbed dose of the neutrons is six times the absorbed dose of the α particles. What is the RBE for the α particles?

3. A film badge worn by a radiologist indicates that she has received an absorbed dose of 2.5×10^{-5} Gy. The mass of the radiologist is 65 kg. How much energy has she absorbed?

4. During an X-ray examination, a person is exposed to radiation at a rate of 3.1×10^{-5} grays per second. The exposure time is 0.10 s, and the mass of the exposed tissue is 1.2 kg. Determine the energy absorbed.

5. ssm A beam of γ rays passes through 4.0×10^{-3} kg of dry air and generates 1.7×10^{12} ions, each with a charge of $+e$. What is the exposure (in roentgens)?

6. A beam of particles is directed at a 0.015-kg tumor. There are 1.6×10^{10} particles per second reaching the tumor, and the energy of each particle is 4.0 MeV. The RBE for the radiation is 14. Find the biologically equivalent dose given to the tumor in 25 s.

*7. A 2.0-kg tumor is being irradiated by a radioactive source. The tumor receives an absorbed dose of 12 Gy in a time of 850 s. Each disintegration of the radioactive source produces a particle that enters the tumor and delivers an energy of 0.40 MeV. What is the activity $\Delta N / \Delta t$ (see Section 31.6) of the radioactive source?

*8. A beam of nuclei is used for cancer therapy. Each nucleus has an energy of 130 MeV, and the relative biological effectiveness (RBE) of this type of radiation is 16. The beam is directed onto a 0.17-kg tumor, which receives a biologically equivalent dose of 180 rem. How many nuclei are in the beam?

*9. ssm What absorbed dose (in rad) of γ rays is required to change a block of ice at 0.0 °C into steam at 100.0 °C?

Section 32.2 Induced Nuclear Reactions

10. Identify the unknown species $^A_Z X$ in the nuclear reaction $^{22}_{11}Na$ (d, α) $^A_Z X$, where d stands for the deuterium isotope $^2_1 H$ of hydrogen.

11. ssm A nitrogen $^{14}_7 N$ nucleus absorbs a deuterium $^2_1 H$ nucleus during a nuclear reaction. What is the name, atomic number, and nucleon number of the compound nucleus?

12. Write the reactions below in the shorthand form discussed in the text.

(a) $^1_0 n + {}^{14}_7 N \longrightarrow {}^{14}_6 C + {}^1_1 H$

(b) $^1_0 n + {}^{238}_{92} U \longrightarrow {}^{239}_{92} U + \gamma$

(c) $^1_0 n + {}^{24}_{12} Mg \longrightarrow {}^{23}_{11} Na + {}^2_1 H$

13. Complete the following nuclear reactions, assuming that the unknown quantity signified by the question mark is a single entity:

(a) $^{43}_{20} Ca$ $(\alpha, ?)$ $^{46}_{21} Sc$ (d) $?$ (α, p) $^{17}_8 O$

(b) $^9_4 Be$ $(?, n)$ $^{12}_6 C$ (e) $^{55}_{25} Mn$ (n, γ) $?$

(c) $^9_4 Be$ (p, α) $?$

14. A neutron causes $^{232}_{90} Th$ to change according to the reaction

$$^1_0 n + {}^{232}_{90} Th \longrightarrow {}^A_Z X + \gamma$$

(a) Identify the unknown nucleus $^A_Z X$, giving its atomic mass number A, its atomic number Z, and the symbol X for the element. (b) The $^A_Z X$ nucleus subsequently undergoes β^- decay, and its daughter does too. Identify the final nucleus, giving its atomic mass number, atomic number, and name.

* **15. ssm** Consider the induced nuclear reaction $^2_1 H + {}^{14}_7 N \rightarrow {}^{12}_6 C + {}^4_2 He$. The atomic masses are $^2_1 H$ (2.014 102 u), $^{14}_7 N$ (14.003 074 u), $^{12}_6 C$ (12.000 000 u), and $^4_2 He$ (4.002 603 u). Determine the energy (in MeV) released when the $^{12}_6 C$ and $^4_2 He$ nuclei are formed in this manner.

* **16.** During a nuclear reaction, an unknown particle is absorbed by a copper $^{63}_{29} Cu$ nucleus, and the reaction products are $^{62}_{29} Cu$, a neutron, and a proton. What is the name, atomic number, and nucleon number of the *compound nucleus?*

Section 32.3 Nuclear Fission, Section 32.4 Nuclear Reactors

17. $^{235}_{92} U$ absorbs a thermal neutron and fissions into rubidium $^{93}_{37} Rb$ and cesium $^{141}_{55} Cs$. What nucleons are produced by the fission, and how many are there?

18. How many neutrons are produced when $^{235}_{92} U$ fissions in the following way: $^1_0 n + {}^{235}_{92} U \rightarrow {}^{132}_{50} Sn + {}^{101}_{42} Mo + \text{neutrons}$?

19. ssm When a $^{235}_{92} U$ (235.043 924 u) nucleus fissions, about 200 MeV of energy is released. What is the ratio of this energy to the rest energy of the uranium nucleus?

20. Interactive LearningWare 32.1 at **www.wiley.com/college/cutnell** reviews the concepts that lie at the heart of this problem. What energy (in MeV) is liberated by the following fission reaction?

$$\underbrace{^1_0 n}_{1.009\ u} + \underbrace{^{235}_{92} U}_{235.044\ u} \longrightarrow \underbrace{^{141}_{56} Ba}_{140.914\ u} + \underbrace{^{92}_{36} Kr}_{91.926\ u} + \underbrace{3^1_0 n}_{3(1.009\ u)}$$

21. Neutrons released by a fission reaction must be slowed by collisions with the moderator nuclei before the neutrons can cause further fissions. Suppose a 1.5-MeV neutron leaves each collision with 65% of its incident energy. How many collisions are required to reduce the neutron's energy to at least 0.040 eV, which is the energy of a thermal neutron?

22. The energy released by each fission within the core of a nuclear reactor is 2.0×10^2 MeV. The number of fissions occurring each

second is 2.0×10^{19}. Determine the power (in watts) that the reactor generates.

* **23. ssm** When 1.0 kg of coal is burned, about 3.0×10^7 J of energy is released. If the energy released per $^{235}_{92} U$ fission is 2.0×10^2 MeV, how many kilograms of coal must be burned to produce the same energy as 1.0 kg of $^{235}_{92} U$?

* **24.** Suppose the $^{239}_{94} Pu$ nucleus fissions into two fragments whose mass ratio is 0.32 : 0.68. With the aid of Figure 32.8, estimate the energy (in MeV) released during this fission.

* **25.** (a) If each fission of a $^{235}_{92} U$ nucleus releases about 2.0×10^2 MeV of energy, determine the energy (in joules) released by the complete fissioning of 1.0 gram of $^{235}_{92} U$. (b) How many grams of $^{235}_{92} U$ are consumed in one year, in order to supply the energy needs of a household that uses 30.0 kWh of energy per day, on the average?

** **26.** A 20.0 kiloton atomic bomb releases as much energy as 20.0 kilotons of TNT (1.0 kiloton of TNT releases about 5.0×10^{12} J of energy). Recall that about 2.0×10^2 MeV of energy is released when each $^{235}_{92} U$ nucleus fissions. (a) How many $^{235}_{92} U$ nuclei are fissioned to produce the bomb's energy? (b) How many grams of uranium are fissioned? (c) What is the equivalent mass (in grams) of the bomb's energy?

** **27. ssm www** A nuclear power plant is 25% efficient, meaning that 25% of the power it generates goes into producing electricity. The remaining 75% is wasted as heat. The plant generates 8.0×10^8 watts of electric power. If each fission releases 2.0×10^2 MeV of energy, how many kilograms of $^{235}_{92} U$ are fissioned per year?

Section 32.5 Nuclear Fusion

28. Consider the fusion of $^1_1 H$ (mass = 1.0078 u) with $^{12}_6 C$ (mass = 12.0000 u) to form $^{13}_7 N$ (mass = 13.0057 u). Determine the energy (in MeV) released in this reaction, part of which is carried away by a γ ray.

29. Interactive Solution 32.29 at **www.wiley.com/college/cutnell** illustrates one way to approach problems of this type. The fusion of two deuterium nuclei ($^2_1 H$, mass = 2.0141 u) can yield a helium nucleus ($^3_2 He$, mass = 3.0160 u) and a neutron ($^1_0 n$, mass = 1.0087 u). What is the energy (in MeV) released in this reaction?

30. Tritium ($^3_1 H$) is a rare isotope of hydrogen that can be produced by the fusion reaction

$$\underbrace{^1_Z X}_{1.0087\ u} + \underbrace{^4_1 Y}_{2.0141\ u} \longrightarrow \underbrace{^3_1 H}_{3.0161\ u} + \gamma$$

(a) Determine the atomic mass number A, the atomic number Z, and the names X and Y of the unknown particles. (b) Using the masses given in the reaction, determine how much energy (in MeV) is released by this reaction.

* **31. ssm** Imagine your car is powered by a fusion engine in which the following reaction occurs: $3^2_1 H \rightarrow {}^4_2 He + {}^1_1 H + {}^1_0 n$. The masses are $^2_1 H$ (2.0141 u), $^4_2 He$ (4.0026 u), $^1_1 H$ (1.0078 u), and $^1_0 n$ (1.0087 u). The engine uses 6.1×10^{-6} kg of deuterium $^2_1 H$ fuel. If one gallon of gasoline produces 2.1×10^9 J of energy, how many gallons of gasoline would have to be burned to equal the energy released by all the deuterium fuel?

* **32.** Deuterium ($^2_1 H$) is an attractive fuel for fusion reactions because it is abundant in the waters of the oceans. In the oceans, about 0.015% of the hydrogen atoms in the water (H_2O) are deuterium atoms. (a) How many deuterium atoms are there in one kilogram of water? (b) If each deuterium nucleus produces about 7.2 MeV in a fusion reaction, how many kilograms of ocean water would be

needed to supply the energy needs of the United States for one year, estimated to be 9.3×10^{19} J?

** **33.** The proton–proton cycle thought to occur in the sun consists of the following sequence of reactions:

$$(1) \ {}_1^1\text{H} + {}_1^1\text{H} \longrightarrow {}_1^2\text{H} + {}_1^0\text{e} + \nu$$
$$(2) \ {}_1^1\text{H} + {}_1^2\text{H} \longrightarrow {}_2^3\text{He} + \gamma$$
$$(3) \ {}_2^3\text{He} + {}_2^3\text{He} \longrightarrow {}_2^4\text{He} + {}_1^1\text{H} + {}_1^1\text{H}$$

In these reactions ${}_1^0\text{e}$ is a positron (mass = 0.000 549 u), ν is a neutrino (mass ≈ 0 u), and γ is a gamma ray photon (mass = 0 u). Note that reaction (3) uses two ${}_2^3\text{He}$ nuclei, which are formed by *two* reactions of type (1) and *two* reactions of type (2). Verify that the proton–proton cycle generates about 25 MeV of energy. The atomic masses are ${}_1^1\text{H}$ (1.007 825 u), ${}_1^2\text{H}$ (2.014 102 u), ${}_2^3\text{He}$ (3.016 030 u), and ${}_2^4\text{He}$ (4.002 603 u). Be sure to account for the fact that there are two electrons in two hydrogen atoms, whereas there is one electron in a single deuterium (${}_1^2\text{H}$) atom. The mass of one electron is 0.000 549 u.

Section 32.6 Elementary Particles

34. The main decay mode for the negative pion is $\pi^- \rightarrow \mu^- + \bar{\nu}_\mu$. Find the energy (in MeV) released in this decay. Consult Table 32.3 as needed.

35. ssm The lambda particle Λ^0 has an electric charge of zero. It is a baryon and, hence, is composed of three quarks. They are all different. One of these quarks is the up quark u, and there are no anti-

quarks present. Make a list of the three possibilities for the quarks contained in Λ^0. (Other information is needed to decide which one of these possibilities is the Λ^0 particle.)

36. A high-energy proton collides with a stationary proton, and the reaction $p + p \rightarrow n + p + \pi^+$ occurs. The rest energy of the π^+ pion is 139.6 MeV. Ignore momentum conservation and find the minimum energy (in MeV) the incident proton must have.

37. Review Conceptual Example 6 as background for this problem. An electron and its antiparticle annihilate each other, producing two γ-ray photons. The kinetic energies of the particles are negligible. For each photon, determine (a) its energy (in MeV), (b) its wavelength, and (c) the magnitude of its momentum.

38. The K^- particle has a charge of $-e$ and contains one quark and one antiquark. (a) Which quarks can the particle *not* contain? (b) Which antiquarks can the particle *not* contain?

* **39. ssm www** Suppose a neutrino is created and has an energy of 35 MeV. (a) Assuming the neutrino, like the photon, has no mass and travels at the speed of light, find the momentum of the neutrino. (b) Determine the de Broglie wavelength of the neutrino.

* **40.** Review Conceptual Example 4 as background for this problem. An energetic proton is fired at a stationary proton. For the reaction to produce new particles, the two protons must approach each other to within a distance of about 8.0×10^{-15} m. The moving proton must have a sufficient speed to overcome the repulsive Coulomb force. What must be the minimum initial kinetic energy (in MeV) of the proton?

Additional Problems

41. A collision between two protons produces three new particles: $p + p \rightarrow p + \pi^+ + \Lambda^0 + K^0$. The rest energies of the new particles are π^+ (139.6 MeV), Λ^0 (1116 MeV), and K^0 (497.7 MeV). Note that one proton disappears during the reaction. How much of the protons' incident energy (in MeV) is transformed into matter during this reaction?

42. A person receives a single whole-body dose of α particles. The absorbed dose is 38 rad, and the RBE of the α particles is 12. (a) Determine the biologically equivalent dose received by this person. (b) With this dose, which one of the following would you expect to happen: no short-term ill effects, the onset of radiation sickness, a 50% chance of dying, or almost certain death?

43. ssm www Write the equation for the reaction ${}_8^{17}\text{O} \ (\gamma, \alpha n) \ {}_6^{12}\text{C}$. The notation "$\alpha n$" means that an α particle and a neutron are produced by the reaction.

44. Uranium ${}_{92}^{235}\text{U}$ fissions into two fragments plus three neutrons: ${}_0^1\text{n} \rightarrow {}_{92}^{235}\text{U} \rightarrow (2 \text{ fragments}) + 3{}_0^1\text{n}$. The mass of a neutron is 1.008 665 u and that of ${}_{92}^{235}\text{U}$ is 235.043 924 u. If 225.0 MeV of energy is released during the fission, what is the combined mass of the two fragments?

* **45. ssm** One proposed fusion reaction combines lithium ${}_3^6\text{Li}$ (6.015 u) with deuterium ${}_1^2\text{H}$ (2.014 u) to give helium ${}_2^4\text{He}$ (4.003 u): ${}_1^2\text{H} + {}_3^6\text{Li} \rightarrow 2{}_2^4\text{He}$. How many kilograms of lithium would be

needed to supply the energy needs of one household for a year, estimated to be 3.8×10^{10} J?

* **46.** A sample of liquid water at 100 °C and 1 atm pressure boils into steam at 100 °C because it is irradiated with a large dose of ionizing radiation. What is the absorbed dose of the radiation in rads?

* **47.** The water that cools a reactor core enters the reactor at 216 °C and leaves at 287 °C. (The water is pressurized, so it does not turn to steam.) The core is generating 5.6×10^9 W of power. Assume that the specific heat capacity of water is 4420 J/(kg·C°) over the temperature range stated above, and find the mass of water that passes through the core each second.

* **48.** The energy consumed in one year in the United States is about 9.3×10^{19} J. When each ${}_{92}^{235}\text{U}$ nucleus fissions, about 2.0×10^2 MeV of energy is released. How many kilograms of ${}_{92}^{235}\text{U}$ would be needed to generate this energy if all the nuclei fissioned?

** **49.** One kilogram of dry air at STP conditions is exposed to 1.0 R of X-rays. One roentgen is defined by Equation 32.1. An equivalent definition can be based on the fact that an exposure of one roentgen deposits 8.3×10^{-3} J of energy per kilogram of dry air. Using the two definitions and assuming that all ions produced are singly charged, determine the average energy (in eV) needed to produce one ion in air.

Concepts & Calculations Group Learning Problems

Note: Each of these problems consists of Concept Questions followed by a related quantitative Problem. They are designed for use by students working alone or in small learning groups. The Concept Questions involve little or no mathematics and are intended to stimulate group discussions. They focus on the concepts with which the problems deal. Recognizing the concepts is the essential initial step in any problem-solving technique.

50. Concept Questions Identical body tissues are being irradiated by two types of radiation (e.g., X-rays and protons). (a) If the absorbed doses are the same, which type of radiation, the one with the larger RBE or the smaller RBE, produces the greater biological effect on the tissue? (b) If the two types of radiation have the same RBE, which type, the one that generates the greater absorbed dose or the smaller absorbed dose, produces the greater biological effect? Provide a reason for each answer.

Problem Someone stands near a radioactive source and receives doses of the following types of radiation: γ rays (20 mrad, RBE = 1), electrons (30 mrad, RBE = 1), protons (5 mrad, RBE = 10), and slow neutrons (5 mrad, RBE = 2). Rank the types of radiation, highest first, as to which produces the largest biologically equivalent dose. Verify that your answers are consistent with your answers to the Concept Questions.

51. Concept Question The definition of the biologically equivalent dose (Equation 32.4) uses the concept of the absorbed dose measured in a unit called the rad. The definition of the absorbed dose measured in a unit called the gray is given by Equation 32.2 as the energy absorbed divided by the mass of the absorbing material. How does one convert from the absorbed dose (in rad) to the absorbed dose (in Gy)?

Problem The biologically equivalent dose for a typical chest X-ray is 2.5×10^{-2} rem. The mass of the exposed tissue is 21 kg, and it absorbs 6.2×10^{-3} J of energy. What is the relative biological effectiveness (RBE) for the radiation on this particular type of tissue?

52. Concept Questions (a) Which of the following are nucleons: protons, electrons, neutrons, γ-ray photons? (b) In a nuclear reaction, what is meant by the statement "The total number of nucleons is conserved"? (c) Which of the following have electric charge: protons, electrons, neutrons, γ-ray photons? (d) In a nuclear reaction,

what is meant by the statement "The total electric charge is conserved"?

Problem For each of the nuclear reactions listed below, determine the unknown particle $^A_Z X$:

(a) $^A_Z X + ^{14}_7 N \rightarrow ^1_1 H + ^{17}_8 O$

(b) $^{15}_7 N + ^A_Z X \rightarrow ^{12}_6 C + ^4_2 He$

(c) $^1_1 H + ^{27}_{13} Al \rightarrow ^A_Z X + ^1_0 n$

(d) $^7_3 Li + ^1_1 H \rightarrow ^4_2 He + ^A_Z X$

53. Concept Questions During a nuclear reaction, whether it is a fission reaction or a fusion reaction, energy is released. (a) Is the total mass of the particles after the reaction greater than or less than the total mass of the particles before the reaction? (b) How is this difference in total mass related to the energy released by the reaction?

Problem In one type of fusion reaction a proton fuses with a neutron to form a deuterium nucleus plus a γ-ray photon: $^1_1 H + ^1_0 n \rightarrow ^2_1 H + \gamma$. The masses are $^1_1 H$ (1.0078 u), $^1_0 n$ (1.0087 u), and $^2_1 H$ (2.0141 u). The γ-ray photon is massless. How much energy (in MeV) is released by this reaction?

*54. Concept Questions** In Example 5 it was determined that 17.6 MeV of energy is released when the following fusion reaction occurs:

$$^2_1 H + ^3_1 H \longrightarrow ^4_2 He + ^1_0 n$$
$$\underbrace{2.0141 \text{ u}} \quad \underbrace{3.0161 \text{ u}} \quad \underbrace{4.0026 \text{ u}} \quad \underbrace{1.0087 \text{ u}}$$

(a) Assuming that the two initial nuclei are at rest when the fusion takes place, what does the conservation of linear momentum tell you about how the momentum of the neutron $^1_0 n$ is related to the momentum of the α particle $^4_2 He$? (b) Which particle has the greater speed, the neutron or the α particle? (c) How is the kinetic energy KE related to the magnitude p of the momentum and the mass m of a particle? (d) Assuming that all of the energy E released in the reaction appears as kinetic energy of the neutron and the α particle, how is KE_n related to KE_α? Account for each of your answers to the Concept Questions.

Problem Determine the kinetic energies of the neutron and the α particle. Verify that your answers are consistent with your answers to the Concept Questions.

Appendix A
Powers of Ten and Scientific Notation

In science, very large and very small decimal numbers are conveniently expressed in terms of powers of ten, some of which are listed below:

$$10^3 = 10 \times 10 \times 10 = 1000 \qquad 10^{-3} = \frac{1}{10 \times 10 \times 10}$$
$$= 0.001$$

$$10^2 = 10 \times 10 = 100 \qquad 10^{-2} = \frac{1}{10 \times 10} = 0.01$$

$$10^1 = 10 \qquad 10^{-1} = \frac{1}{10} = 0.1$$

$$10^0 = 1$$

Using powers of ten, we can write the radius of the earth in the following way, for example:

$$\text{Earth radius} = 6\,380\,000 \text{ m} = 6.38 \times 10^6 \text{ m}$$

The factor of ten raised to the sixth power is ten multiplied by itself six times, or one million, so the earth's radius is 6.38 million meters. Alternatively, the factor of ten raised to the sixth power indicates that the decimal point in the term 6.38 is to be moved six places *to the right* to obtain the radius as a number without powers of ten.

For numbers less than one, negative powers of ten are used. For instance, the Bohr radius of the hydrogen atom is

$$\text{Bohr radius} = 0.000\,000\,000\,0529 \text{ m} = 5.29 \times 10^{-11} \text{ m}$$

The factor of ten raised to the minus eleventh power indicates that the decimal point in the term 5.29 is to be moved eleven places *to the left* to obtain the radius as a number without powers of ten. Numbers expressed with the aid of powers of ten are said to be in *scientific notation.*

Calculations that involve the multiplication and division of powers of ten are carried out as in the following examples:

$$(2.0 \times 10^6)(3.5 \times 10^3) = (2.0 \times 3.5) \times 10^{6+3} = 7.0 \times 10^9$$

$$\frac{9.0 \times 10^7}{2.0 \times 10^4} = \left(\frac{9.0}{2.0}\right) \times 10^7 \times 10^{-4}$$

$$= \left(\frac{9.0}{2.0}\right) \times 10^{7-4} = 4.5 \times 10^3$$

The general rules for such calculations are

$$\frac{1}{10^n} = 10^{-n} \tag{A-1}$$

$$10^n \times 10^m = 10^{n+m} \qquad \text{(Exponents added)} \tag{A-2}$$

$$\frac{10^n}{10^m} = 10^{n-m} \qquad \text{(Exponents subtracted)} \tag{A-3}$$

where n and m are any positive or negative number.

Scientific notation is convenient because of the ease with which it can be used in calculations. Moreover, scientific notation provides a convenient way to express the significant figures in a number, as Appendix B discusses.

Appendix B
Significant Figures

The number of **significant figures** in a number is the number of digits whose values are known with certainty. For instance, a person's height is measured to be 1.78 m, with the measurement error being in the third decimal place. All three digits are known with certainty, so that the number contains three significant figures. If a zero is given as the last digit to the right of the decimal point, the zero is presumed to be significant. Thus, the number 1.780 m contains four significant figures. As another example, consider a distance of 1500 m. This number contains only two significant figures, the one and the five. The zeros immediately to the left of the unexpressed decimal point are not counted as significant figures. However, zeros located between significant figures are significant, so a distance of 1502 m contains four significant figures.

Scientific notation is particularly convenient from the point of view of significant figures. Suppose it is known that a certain distance is fifteen hundred meters, to four significant figures. Writing the number as 1500 m presents a problem because it implies that only two significant figures are known. In contrast, the scientific notation of 1.500×10^3 m has the advantage of indicating that the distance is known to four significant figures.

When two or more numbers are used in a calculation, the number of significant figures in the answer is limited by the number of significant figures in the original data. For instance, a rectangular garden with sides of 9.8 m and 17.1 m has an area of (9.8 m)(17.1 m). A calculator gives 167.58 m² for this product. However, one of the original lengths is known only to two significant figures, so the final answer is limited to only two significant figures and should be rounded off to 170 m². In general, **when numbers are multiplied or divided, the number of significant figures in the final answer equals the smallest number of significant figures in any of the original factors.**

The number of significant figures in the answer to an addition or a subtraction is also limited by the original data. Consider the total distance along a biker's trail that consists of three segments with the distances shown as follows:

$$
\begin{array}{ll}
 & 2.5 \ \ \text{km} \\
 & 11 \ \ \ \ \text{km} \\
 & \underline{5.26 \ \text{km}} \\
\text{Total} & 18.76 \ \text{km}
\end{array}
$$

The distance of 11 km contains no significant figures to the right of the decimal point. Therefore, neither does the sum of the three distances,

and the total distance should not be reported as 18.76 km. Instead, the answer is rounded off to 19 km. In general, **when numbers are added or subtracted, the last significant figure in the answer occurs in the last column (counting from left to right) containing a number that results from a combination of digits that are all significant.** In the answer of 18.76 km, the eight is the sum of $2 + 1 + 5$, each digit being significant. However, the seven is the sum of $5 + 0 + 2$, and the zero is not significant, since it comes from the 11-km distance, which contains no significant figures to the right of the decimal point.

Appendix C
Algebra

C.1 Proportions and Equations

Physics deals with physical variables and the relations between them. Typically, variables are represented by the letters of the English and Greek alphabets. Sometimes, the relation between variables is expressed as a proportion or inverse proportion. Other times, however, it is more convenient or necessary to express the relation by means of an equation, which is governed by the rules of algebra.

If two variables are **directly proportional** and one of them doubles, then the other variable also doubles. Similarly, if one variable is reduced to one-half its original value, then the other is also reduced to one-half its original value. In general, if x is directly proportional to y, then increasing or decreasing one variable by a given factor causes the other to change in the same way by the same factor. This kind of relation is expressed as $x \propto y$, where the symbol $\propto$ means "is proportional to."

Since the proportional variables x and y always increase and decrease by the same factor, the ratio of x to y must have a constant value, or $x/y = k$, where k is a constant, independent of the values for x and y. Consequently, a proportionality such as $x \propto y$ can also be expressed in the form of an equation: $x = ky$. The constant k is referred to as a **proportionality constant.**

If two variables are **inversely proportional** and one of them increases by a given factor, then the other decreases by the same factor. An inverse proportion is written as $x \propto 1/y$. This kind of proportionality is equivalent to the following equation: $xy = k$, where k is a proportionality constant, independent of x and y.

C.2 Solving Equations

Some of the variables in an equation typically have known values, and some do not. It is often necessary to solve the equation so that a variable whose value is unknown is expressed in terms of the known quantities. **In the process of solving an equation, it is permissible to manipulate the equation in any way, as long as a change made on one side of the equals sign is also made on the other side.** For example, consider the equation $v = v_0 + at$. Suppose values for v, v_0, and a are available, and the value of t is required. To solve the equation for t, we begin by subtracting v_0 from *both* sides:

$$
\begin{array}{ccc}
v & = & v_0 + at \\
\underline{-v_0} & = & \underline{-v_0} \\
v - v_0 & = & at
\end{array}
$$

Next, we divide both sides of $v - v_0 = at$ by the quantity a:

$$
\frac{v - v_0}{a} = \frac{at}{a} = (1)t
$$

On the right side, the a in the numerator divided by the a in the denominator equals one, so that

$$
t = \frac{v - v_0}{a}
$$

It is always possible to check the correctness of the algebraic manipulations performed in solving an equation by substituting the answer back into the original equation. In the previous example, we substitute the answer for t into $v = v_0 + at$:

$$
v = v_0 + a\left(\frac{v - v_0}{a}\right) = v_0 + (v - v_0) = v
$$

The result $v = v$ implies that our algebraic manipulations were done correctly.

Algebraic manipulations other than addition, subtraction, multiplication, and division may play a role in solving an equation. The same basic rule applies, however: Whatever is done to the left side of an equation must also be done to the right side. As another example, suppose it is necessary to express v_0 in terms of v, a, and x, where $v^2 = v_0^2 + 2ax$. By subtracting $2ax$ from both sides, we isolate v_0^2 on the right:

$$
\begin{array}{ccc}
v^2 & = & v_0^2 + 2ax \\
\underline{-2ax} & = & \underline{-2ax} \\
v^2 - 2ax & = & v_0^2
\end{array}
$$

To solve for v_0, we take the positive and negative square root of *both* sides of $v^2 - 2ax = v_0^2$:

$$
v_0 = \pm\sqrt{v^2 - 2ax}
$$

C.3 Simultaneous Equations

When more than one variable in a single equation is unknown, additional equations are needed if solutions are to be found for all of the unknown quantities. Thus, the equation $3x + 2y = 7$ cannot be solved by itself to give unique values for both x and y. However, if x and y also (i.e., simultaneously) obey the equation $x - 3y = 6$, then both unknowns can be found.

There are a number of methods by which such simultaneous equations can be solved. One method is to solve one equation for x in terms of y and substitute the result into the other equation to obtain an expression containing only the single unknown variable y. The equation $x - 3y = 6$, for instance, can be solved for x by adding $3y$ to each side, with the result that $x = 6 + 3y$. The substitution of this expression for x into the equation $3x + 2y = 7$ is shown below:

$$3x + 2y = 7$$
$$3(6 + 3y) + 2y = 7$$
$$18 + 9y + 2y = 7$$

We find, then, that $18 + 11y = 7$, a result that can be solved for y:

$$
\begin{array}{r}
18 + 11y = 7 \\
-18 -18 \\
\hline
11y = -11
\end{array}
$$

Dividing both sides of this result by 11 shows that $y = -1$. The value of $y = -1$ can be substituted in either of the original equations to obtain a value for x:

$$
\begin{array}{rcl}
x - 3y & = & 6 \\
x - 3(-1) & = & 6 \\
x + 3 & = & 6 \\
-3 & & -3 \\
\hline
x & = & 3
\end{array}
$$

C.4 The Quadratic Formula

Equations occur in physics that include the square of a variable. Such equations are said to be *quadratic* in that variable, and often can be put into the following form:

$$ax^2 + bx + c = 0 \tag{C-1}$$

where a, b, and c are constants independent of x. This equation can be solved to give the **quadratic formula,** which is

$$x = \frac{-b \pm \sqrt{b^2 - 4ac}}{2a} \tag{C-2}$$

The $\pm$ in the quadratic formula indicates that there are two solutions. For instance, if $2x^2 - 5x + 3 = 0$, then $a = 2$, $b = -5$, and $c = 3$. The quadratic formula gives the two solutions as follows:

Solution 1:
Plus sign

$$x = \frac{-b + \sqrt{b^2 - 4ac}}{2a}$$
$$= \frac{-(-5) + \sqrt{(-5)^2 - 4(2)(3)}}{2(2)}$$
$$= \frac{+5 + \sqrt{1}}{4} = \frac{3}{2}$$

Solution 2:
Minus sign

$$x = \frac{-b - \sqrt{b^2 - 4ac}}{2a}$$
$$= \frac{-(-5) - \sqrt{(-5)^2 - 4(2)(3)}}{2(2)}$$
$$= \frac{+5 - \sqrt{1}}{4} = 1$$

Appendix D
Exponents and Logarithms

Appendix A discusses powers of ten, such as 10^3, which means ten multiplied by itself three times, or $10 \times 10 \times 10$. The three is referred to as an **exponent.** The use of exponents extends beyond powers of ten. In general, the term y^n means the factor y is multiplied by itself n times. For example, y^2, or y squared, is familiar and means $y \times y$. Similarly, y^5 means $y \times y \times y \times y \times y$.

The rules that govern algebraic manipulations of exponents are the same as those given in Appendix A (see Equations A-1, A-2, and A-3) for powers of ten:

$$\frac{1}{y^n} = y^{-n} \tag{D-1}$$

$$y^n y^m = y^{n+m} \quad \text{(Exponents added)} \tag{D-2}$$

$$\frac{y^n}{y^m} = y^{n-m} \quad \text{(Exponents subtracted)} \tag{D-3}$$

To the three rules above we add two more that are useful. One of these is

$$y^n z^n = (yz)^n \tag{D-4}$$

The following example helps to clarify the reasoning behind this rule:

$$3^2 5^2 = (3 \times 3)(5 \times 5) = (3 \times 5)(3 \times 5) = (3 \times 5)^2$$

The other additional rule is

$$(y^n)^m = y^{nm} \quad \text{(Exponents multiplied)} \tag{D-5}$$

To see why this rule applies, consider the following example:

$$(5^2)^3 = (5^2)(5^2)(5^2) = 5^{2+2+2} = 5^{2 \times 3}$$

Roots, such as a square root or a cube root, can be represented with fractional exponents. For instance,

$$\sqrt{y} = y^{1/2} \quad \text{and} \quad \sqrt[3]{y} = y^{1/3}$$

In general, the nth root of y is given by

$$\sqrt[n]{y} = y^{1/n} \tag{D-6}$$

The rationale for Equation D-6 can be explained using the fact that $(y^n)^m = y^{nm}$. For instance, the fifth root of y is the number that, when multiplied by itself five times, gives back y. As shown below, the term $y^{1/5}$ satisfies this definition:

$$(y^{1/5})(y^{1/5})(y^{1/5})(y^{1/5})(y^{1/5}) = (y^{1/5})^5 = y^{(1/5) \times 5} = y$$

Logarithms are closely related to exponents. To see the connection between the two, note that it is possible to express any number y as another number B raised to the exponent x. In other words,

$$y = B^x \tag{D-7}$$

The exponent x is called the *logarithm* of the number y. The number B is called the *base number.* One of two choices for the base number is usually used. If $B = 10$, the logarithm is known as the *common logarithm,* for which the notation "log" applies:

Common logarithm $y = 10^x$ or $x = \log y$ (D-8)

If $B = e = 2.718 \ldots$, the logarithm is referred to as the *natural logarithm,* and the notation "ln" is used:

Natural logarithm $y = e^z$ or $z = \ln y$ (D-9)

The two kinds of logarithms are related by

$$\ln y = 2.3026 \log y \qquad \text{(D-10)}$$

Both kinds of logarithms are often given on calculators.

The logarithm of the product or quotient of two numbers A and C can be obtained from the logarithms of the individual numbers according to the rules below. These rules are illustrated here for natural logarithms, but they are the same for any kind of logarithm.

$$\ln (AC) = \ln A + \ln C \qquad \text{(D-11)}$$

$$\ln \left(\frac{A}{C} \right) = \ln A - \ln C \qquad \text{(D-12)}$$

Thus, the logarithm of the product of two numbers is the sum of the individual logarithms, and the logarithm of the quotient of two numbers is the difference between the individual logarithms. Another useful rule concerns the logarithm of a number A raised to an exponent n:

$$\ln A^n = n \ln A \qquad \text{(D-13)}$$

Rules D-11, D-12, and D-13 can be derived from the definition of the logarithm and the rules governing exponents.

Appendix E
Geometry and Trigonometry

E.1 *Geometry*
ANGLES

Two angles are equal if
1. They are vertical angles (see Figure E1).
2. Their sides are parallel (see Figure E2).

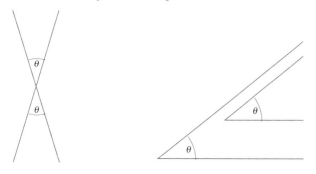

Figure E1   **Figure E2**

3. Their sides are mutually perpendicular (see Figure E3).

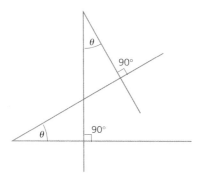

Figure E3

TRIANGLES

1. The *sum of the angles* of any triangle is $180°$ (see Figure E4).

$\alpha + \beta + \gamma = 180°$

Figure E4

2. A *right triangle* has one angle that is $90°$.
3. An *isosceles triangle* has two sides that are equal.
4. An *equilateral triangle* has three sides that are equal. Each angle of an equilateral triangle is $60°$.
5. Two triangles are *similar* if two of their angles are equal (see Figure E5). The corresponding sides of similar triangles are proportional to each other:

$$\frac{a_1}{a_2} = \frac{b_1}{b_2} = \frac{c_1}{c_2}$$

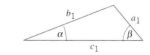

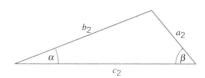

Figure E5

6. Two similar triangles are *congruent* if they can be placed on top of one another to make an exact fit.

CIRCUMFERENCES, AREAS, AND VOLUMES OF SOME COMMON SHAPES

1. Triangle of base b and altitude h (see Figure E6):

$$\text{Area} = \tfrac{1}{2}bh$$

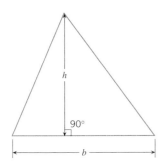

Figure E6

2. Circle of radius r:

$$\text{Circumference} = 2\pi r$$
$$\text{Area} = \pi r^2$$

3. Sphere of radius r:

$$\text{Surface area} = 4\pi r^2$$
$$\text{Volume} = \tfrac{4}{3}\pi r^3$$

4. Right circular cylinder of radius r and height h (see Figure E7):

$$\text{Surface area} = 2\pi r^2 + 2\pi rh$$
$$\text{Volume} = \pi r^2 h$$

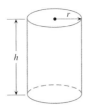

Figure E7

E.2 *Trigonometry*

BASIC TRIGONOMETRIC FUNCTIONS

1. For a right triangle, the sine, cosine, and tangent of an angle θ are defined as follows (see Figure E8):

$$\sin\theta = \frac{\text{Side opposite }\theta}{\text{Hypotenuse}} = \frac{h_\text{o}}{h}$$

$$\cos\theta = \frac{\text{Side adjacent to }\theta}{\text{Hypotenuse}} = \frac{h_\text{a}}{h}$$

$$\tan\theta = \frac{\text{Side opposite }\theta}{\text{Side adjacent to }\theta} = \frac{h_\text{o}}{h_\text{a}}$$

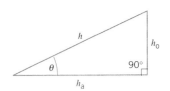

Figure E8

2. The secant ($\sec\theta$), cosecant ($\csc\theta$), and cotangent ($\cot\theta$) of an angle θ are defined as follows:

$$\sec\theta = \frac{1}{\cos\theta} \qquad \csc\theta = \frac{1}{\sin\theta} \qquad \cot\theta = \frac{1}{\tan\theta}$$

TRIANGLES AND TRIGONOMETRY

1. The **Pythagorean theorem** states that the square of the hypotenuse of a right triangle is equal to the sum of the squares of the other two sides (see Figure E8):

$$h^2 = h_\text{o}{}^2 + h_\text{a}{}^2$$

2. The *law of cosines* and the *law of sines* apply to any triangle, not just a right triangle, and they relate the angles and the lengths of the sides (see Figure E9):

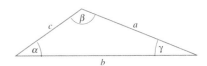

Figure E9

Law of cosines $\qquad c^2 = a^2 + b^2 - 2ab\cos\gamma$

Law of sines $\qquad \dfrac{a}{\sin\alpha} = \dfrac{b}{\sin\beta} = \dfrac{c}{\sin\gamma}$

OTHER TRIGONOMETRIC IDENTITIES

1. $\sin(-\theta) = -\sin\theta$
2. $\cos(-\theta) = \cos\theta$
3. $\tan(-\theta) = -\tan\theta$
4. $(\sin\theta)/(\cos\theta) = \tan\theta$
5. $\sin^2\theta + \cos^2\theta = 1$
6. $\sin(\alpha \pm \beta) = \sin\alpha\cos\beta \pm \cos\alpha\sin\beta$

$$\text{If } \alpha = 90°, \sin(90° \pm \beta) = \cos\beta$$
$$\text{If } \alpha = \beta, \sin 2\beta = 2\sin\beta\cos\beta$$

7. $\cos(\alpha \pm \beta) = \cos\alpha\cos\beta \mp \sin\alpha\sin\beta$

$$\text{If } \alpha = 90°, \cos(90° \pm \beta) = \mp\sin\beta$$
$$\text{If } \alpha = \beta, \cos 2\beta = \cos^2\beta - \sin^2\beta = 1 - 2\sin^2\beta$$

Appendix F
Selected Isotopes[a]

Atomic No. Z	Element	Symbol	Atomic Mass No. A	Atomic Mass u	% Abundance, or Decay Mode If Radioactive	Half-Life (if Radioactive)
0	(Neutron)	*n*	1	1.008 665	β^-	10.37 min
1	Hydrogen	H	1	1.007 825	99.985	
	Deuterium	D	2	2.014 102	0.015	
	Tritium	T	3	3.016 050	β^-	12.33 yr
2	Helium	He	3	3.016 030	0.000 138	
			4	4.002 603	≈ 100	
3	Lithium	Li	6	6.015 121	7.5	
			7	7.016 003	92.5	
4	Beryllium	Be	7	7.016 928	EC, γ	53.29 days
			9	9.012 182	100	
5	Boron	B	10	10.012 937	19.9	
			11	11.009 305	80.1	
6	Carbon	C	11	11.011 432	β^+, EC	20.39 min
			12	12.000 000	98.90	
			13	13.003 355	1.10	
			14	14.003 241	β^-	5730 yr
7	Nitrogen	N	13	13.005 738	β^+, EC	9.965 min
			14	14.003 074	99.634	
			15	15.000 108	0.366	
8	Oxygen	O	15	15.003 065	β^+, EC	122.2 s
			16	15.994 915	99.762	
			18	17.999 160	0.200	
9	Fluorine	F	18	18.000 937	EC, β^+	1.8295 h
			19	18.998 403	100	
10	Neon	Ne	20	19.992 435	90.51	
			22	21.991 383	9.22	
11	Sodium	Na	22	21.994 434	β^+, EC, γ	2.602 yr
			23	22.989 767	100	
			24	23.990 961	β^-, γ	14.659 h
12	Magnesium	Mg	24	23.985 042	78.99	
13	Aluminum	Al	27	26.981 539	100	
14	Silicon	Si	28	27.976 927	92.23	
			31	30.975 362	β^-, γ	2.622 h
15	Phosphorus	P	31	30.973 762	100	
			32	31.973 907	β^-	14.282 days
16	Sulfur	S	32	31.972 070	95.02	
			35	34.969 031	β^-	87.51 days
17	Chlorine	Cl	35	34.968 852	75.77	
			37	36.965 903	24.23	
18	Argon	Ar	40	39.962 384	99.600	
19	Potassium	K	39	38.963 707	93.2581	
			40	39.963 999	β^-, EC, γ	1.277×10^9 yr
20	Calcium	Ca	40	39.962 591	96.941	
21	Scandium	Sc	45	44.955 910	100	
22	Titanium	Ti	48	47.947 947	73.8	

[a] Data for atomic masses are taken from *Handbook of Chemistry and Physics*, 66th ed., CRC Press, Boca Raton, FL. The masses are those for the neutral atom, including the Z electrons. Data for percent abundance, decay mode, and half-life are taken from E. Browne and R. Firestone, *Table of Radioactive Isotopes*, V. Shirley, Ed., Wiley, New York, 1986. α = alpha particle emission, β^- = negative beta emission, β^+ = positron emission, γ = γ-ray emission, EC = electron capture.

APPENDIX F Selected Isotopes *(continued)*

Atomic No. Z	Element	Symbol	Atomic Mass No. A	Atomic Mass u	% Abundance, or Decay Mode If Radioactive	Half-Life (if Radioactive)
23	Vanadium	V	51	50.943 962	99.750	
24	Chromium	Cr	52	51.940 509	83.789	
25	Manganese	Mn	55	54.938 047	100	
26	Iron	Fe	56	55.934 939	91.72	
27	Cobalt	Co	59	58.933 198	100	
			60	59.933 819	β^-, γ	5.271 yr
28	Nickel	Ni	58	57.935 346	68.27	
			60	59.930 788	26.10	
29	Copper	Cu	63	62.939 598	69.17	
			65	64.927 793	30.83	
30	Zinc	Zn	64	63.929 145	48.6	
			66	65.926 034	27.9	
31	Gallium	Ga	69	68.925 580	60.1	
32	Germanium	Ge	72	71.922 079	27.4	
			74	73.921 177	36.5	
33	Arsenic	As	75	74.921 594	100	
34	Selenium	Se	80	79.916 520	49.7	
35	Bromine	Br	79	78.918 336	50.69	
36	Krypton	Kr	84	83.911 507	57.0	
			89	88.917 640	β^-, γ	3.16 min
			92	91.926 270	β^-, γ	1.840 s
37	Rubidium	Rb	85	84.911 794	72.165	
38	Strontium	Sr	86	85.909 267	9.86	
			88	87.905 619	82.58	
			90	89.907 738	β^-	28.5 yr
			94	93.915 367	β^-, γ	1.235 s
39	Yttrium	Y	89	88.905 849	100	
40	Zirconium	Zr	90	89.904 703	51.45	
41	Niobium	Nb	93	92.906 377	100	
42	Molybdenum	Mo	98	97.905 406	24.13	
43	Technecium	Tc	98	97.907 215	β^-, γ	4.2×10^6 yr
44	Ruthenium	Ru	102	101.904 348	31.6	
45	Rhodium	Rh	103	102.905 500	100	
46	Palladium	Pd	106	105.903 478	27.33	
47	Silver	Ag	107	106.905 092	51.839	
			109	108.904 757	48.161	
48	Cadmium	Cd	114	113.903 357	28.73	
49	Indium	In	115	114.903 880	95.7; β^-	4.41×10^{14} yr
50	Tin	Sn	120	119.902 200	32.59	
51	Antimony	Sb	121	120.903 821	57.3	
52	Tellurium	Te	130	129.906 229	38.8; β^-	2.5×10^{21} yr
53	Iodine	I	127	126.904 473	100	
			131	130.906 114	β^-, γ	8.040 days
54	Xenon	Xe	132	131.904 144	26.9	
			136	135.907 214	8.9	
			140	139.921 620	β^-, γ	13.6 s
55	Cesium	Cs	133	132.905 429	100	
			134	133.906 696	β^-, EC, γ	2.062 yr
56	Barium	Ba	137	136.905 812	11.23	
			138	137.905 232	71.70	
			141	140.914 363	β^-, γ	18.27 min
57	Lanthanum	La	139	138.906 346	99.91	

APPENDIX F Selected Isotopes *(continued)*

Atomic No. Z	Element	Symbol	Atomic Mass No. A	Atomic Mass u	% Abundance, or Decay Mode If Radioactive	Half-Life (if Radioactive)
58	Cerium	Ce	140	139.905 433	88.48	
59	Praseodymium	Pr	141	140.907 647	100	
60	Neodymium	Nd	142	141.907 719	27.13	
61	Promethium	Pm	145	144.912 743	EC, α, γ	17.7 yr
62	Samarium	Sm	152	151.919 729	26.7	
63	Europium	Eu	153	152.921 225	52.2	
64	Gadolinium	Gd	158	157.924 099	24.84	
65	Terbium	Tb	159	158.925 342	100	
66	Dysprosium	Dy	164	163.929 171	28.2	
67	Holmium	Ho	165	164.930 319	100	
68	Erbium	Er	166	165.930 290	33.6	
69	Thulium	Tm	169	168.934 212	100	
70	Ytterbium	Yb	174	173.938 859	31.8	
71	Lutetium	Lu	175	174.940 770	97.41	
72	Hafnium	Hf	180	179.946 545	35.100	
73	Tantalum	Ta	181	180.947 992	99.988	
74	Tungsten (wolfram)	W	184	183.950 928	30.67	
75	Rhenium	Re	187	186.955 744	62.60; β^-	4.6×10^{10} yr
76	Osmium	Os	191	190.960 920	β^-, γ	15.4 days
			192	191.961 467	41.0	
77	Iridium	Ir	191	190.960 584	37.3	
			193	192.962 917	62.7	
78	Platinum	Pt	195	194.964 766	33.8	
79	Gold	Au	197	196.966 543	100	
			198	197.968 217	β^-, γ	2.6935 days
80	Mercury	Hg	202	201.970 617	29.80	
81	Thallium	Tl	205	204.974 401	70.476	
			208	207.981 988	β^-, γ	3.053 min
82	Lead	Pb	206	205.974 440	24.1	
			207	206.975 872	22.1	
			208	207.976 627	52.4	
			210	209.984 163	α, β^-, γ	22.3 yr
			211	210.988 735	β^-, γ	36.1 min
			212	211.991 871	β^-, γ	10.64 h
			214	213.999 798	β^-, γ	26.8 min
83	Bismuth	Bi	209	208.980 374	100	
			211	210.987 255	α, β^-, γ	2.14 min
			212	211.991 255	β^-, α, γ	1.0092 h
84	Polonium	Po	210	209.982 848	α, γ	138.376 days
			212	211.988 842	α, γ	45.1 s
			214	213.995 176	α, γ	163.69 μs
			216	216.001 889	α, γ	150 ms
85	Astatine	At	218	218.008 684	α, β^-	1.6 s
86	Radon	Rn	220	220.011 368	α, γ	55.6 s
			222	222.017 570	α, γ	3.825 days
87	Francium	Fr	223	223.019 733	α, β^-, γ	21.8 min
88	Radium	Ra	224	224.020 186	α, γ	3.66 days
			226	226.025 402	α, γ	1.6×10^3 yr
			228	228.031 064	β^-, γ	5.75 yr
89	Actinium	Ac	227	227.027 750	α, β^-, γ	21.77 yr
			228	228.031 015	β^-, γ	6.13 h

APPENDIX F Selected Isotopes *(continued)*

Atomic No. Z	Element	Symbol	Atomic Mass No. A	Atomic Mass u	% Abundance, or Decay Mode If Radioactive	Half-Life (if Radioactive)
90	Thorium	Th	228	228.028 715	α, γ	1.913 yr
			231	231.036 298	β^-, γ	1.0633 days
			232	232.038 054	100; α, γ	1.405×10^{10} yr
			234	234.043 593	β^-, γ	24.10 days
91	Protactinium	Pa	231	231.035 880	α, γ	3.276×10^4 yr
			234	234.043 303	β^-, γ	6.70 h
			237	237.051 140	β^-, γ	8.7 min
92	Uranium	U	232	232.037 130	α, γ	68.9 yr
			233	233.039 628	α, γ	1.592×10^5 yr
			235	235.043 924	0.7200; α, γ	7.037×10^8 yr
			236	236.045 562	α, γ	2.342×10^7 yr
			238	238.050 784	99.2745; α, γ	4.468×10^9 yr
			239	239.054 289	β^-, γ	23.47 min
93	Neptunium	Np	239	239.052 933	β^-, γ	2.355 days
94	Plutonium	Pu	239	239.052 157	α, γ	2.411×10^4 yr
			242	242.058 737	α, γ	3.763×10^5 yr
95	Americium	Am	243	243.061 375	α, γ	7.380×10^3 yr
96	Curium	Cm	245	245.065 483	α, γ	8.5×10^3 yr
97	Berkelium	Bk	247	247.070 300	α, γ	1.38×10^3 yr
98	Californium	Cf	249	249.074 844	α, γ	350.6 yr
99	Einsteinium	Es	254	254.088 019	α, γ, β^-	275.7 days
100	Fermium	Fm	253	253.085 173	EC, α, γ	3.00 days
101	Mendelevium	Md	255	255.091 081	EC, α	27 min
102	Nobelium	No	255	255.093 260	EC, α	3.1 min
103	Lawrencium	Lr	257	257.099 480	α, EC	646 ms
104	Rutherfordium	Rf	261	261.108 690	α	1.08 min
105	Dubnium	Db	262	262.113 760	α	34 s

Answers to Check Your Understanding

Chapter 1
CYU 1: b and d
CYU 2: (a) 11 m
(b) 5 m
CYU 3: (a) A_x is $-$, A_y is $+$
(b) B_x is $+$, B_y is $-$
(c) R_x is $+$, R_y is $+$
CYU 4: a

Chapter 2
CYU 1: His average velocity is 2.7 m/s, due east. His average speed is 8.0 m/s.
CYU 2: c
CYU 3: 1.73
CYU 4: b

Chapter 3
CYU 1: b
CYU 2: c
CYU 3: a and c
CYU 4: 1. $+70$ m/s
2. $+30$ m/s
3. $+40$ m/s
4. -60 m/s

Chapter 4
CYU 1: c
CYU 2: d
CYU 3: 43°
CYU 4: a
CYU 5: a

Chapter 5
CYU 1: (a) The velocity is due south, and the acceleration is due west.
(b) The velocity is due west, and the acceleration is due north.
CYU 2: (a) $4r$
(b) $4r$
CYU 3: less than
CYU 4: (a) less than
(b) equal to

Chapter 6
CYU 1: b
CYU 2: false
CYU 3: b and d
CYU 4: c

Chapter 7
CYU 1: b
CYU 2: a
CYU 3: d
CYU 4: above the halfway point

Chapter 8
CYU 1: B, C, A
CYU 2: 1.0 rev/s
CYU 3: 0.30 m
CYU 4: 8.0 m/s^2

Chapter 9
CYU 1: 0°, 45°, 90°
CYU 2: (a) C
(b) A
(c) B
CYU 3: A, B, C
CYU 4: (a) Both have the same translational speed.
(b) Both have the same translational speed.

Chapter 10
CYU 1: 180 N/m
CYU 2: II
CYU 3: b, c, a
CYU 4: The rod with the square cross section is longer.

Chapter 11
CYU 1: increase, decrease, remain constant
CYU 2: c
CYU 3: d
CYU 4: e

Chapter 12
CYU 1: 178 °X
CYU 2: b and d
CYU 3: c, b, d, a
CYU 4: c, a, b

Chapter 13
CYU 1: c
CYU 2: b

Chapter 14
CYU 1: 66.4%
CYU 2: Xenon has the greatest and argon the smallest temperature.
CYU 3: $\dfrac{v_{\text{rms, new}}}{v_{\text{rms, initial}}} = 0.707$
CYU 4: c, a, b

Chapter 15
CYU 1: d
CYU 2: b
CYU 3: c
CYU 4: d

Chapter 16
CYU 1: a
CYU 2: CO and N_2
CYU 3: (a) $\frac{1}{4}$
(b) 2
CYU 4: (a) The truck driver is driving faster.
(b) minus sign in both places

Chapter 17
CYU 1: (a) 0 cm
(b) -2 cm
(c) $+2$ cm
CYU 2: 387 Hz
CYU 3: (a) 4
(b) 3
(c) node
(d) 110 Hz
CYU 4: (a) antinode
(b) node
(c) $\frac{1}{4}\lambda$
(d) lowered

Chapter 18
CYU 1: b and e
CYU 2: c, a, b
CYU 3: (a) corner C
(b) negative
(c) greater
CYU 4: The fluxes in (a) and (b) are the same.

Chapter 19
CYU 1: (a) Yes.
(b) No.
(c) Yes.
(d) Yes.
CYU 2: b, c, a and d (a tie)
CYU 3: (a) $+2.0$ V
(b) 0 V
(c) $+2.0$ V
CYU 4: (a) decreases
(b) increases
(c) increases
(d) increases

Chapter 20
CYU 1: B, D, and E
CYU 2: c
CYU 3: $3R$, $\frac{3}{2}R$, $\frac{2}{3}R$, and $\frac{1}{3}R$
CYU 4: $I_1 + I_3 = I_2$
$3.0\,\text{V} + 7.0\,\text{V} + I_3R_3 = I_1R_1$
$5.0\,\text{V} = I_3R_3 + 7.0\,\text{V} + I_2R_2$
CYU 5: 0.80 s

Chapter 21
CYU 1: d
CYU 2: 1, 3, 2
CYU 3: B and D (a tie), A, C
CYU 4: A, D, C, B

Chapter 22
CYU 1: d
CYU 2: a
CYU 3: The field points away from you and is decreasing, or the field points toward you and is increasing.
CYU 4: the one-turn coil
CYU 5: b and d

Chapter 23
CYU 1: The ratio increases by a factor of 3.
CYU 2: (a) inductor
(b) resistor
CYU 3: less than
CYU 4: (a) remains the same
(b) decreases

Chapter 24
CYU 1: e
CYU 2: d
CYU 3: B, A, C
CYU 4: unpolarized—c
polarized—b

Chapter 25
CYU 1: 55°
CYU 2: $f_{\text{inside}} = +0.30 \text{ m}$
$f_{\text{outside}} = -0.30 \text{ m}$
CYU 3: The magnitudes of the image distance and the image height both become larger.
CYU 4: A, D, and E

Chapter 26
CYU 1: B
CYU 2: c
CYU 3: B
CYU 4: converging lens, $d_{\text{o}} = \frac{1}{2}f$
CYU 5: 0.042 rad

Chapter 27
CYU 1: (a) constructive
(b) destructive
(c) destructive
CYU 2: (a) d_1 and λ_2
(b) d_2 and λ_1
CYU 3: (a) A and C
(b) B
CYU 4: (a) broaden
(b) contract
CYU 5: (a) the maximum that is closer to the central maximum
(b) away from the central maximum

Chapter 28
CYU 1: d
CYU 2: b
CYU 3: greatest—c
smallest—b
CYU 4: c

Chapter 29
CYU 1: a
CYU 2: less than
CYU 3: b

Chapter 30
CYU 1: d
CYU 2: (a) Yes.
(b) No.
CYU 3: $1s^2 \, 2s^2 \, 2p^6 \, 3s^2 \, 3p^6 \, 4s^2 \, 3d^{10} \, 4p^6$
CYU 4: c

Chapter 31
CYU 1: c, d
CYU 2: c
CYU 3: d
CYU 4: a

Chapter 32
CYU 1: neutrons, α particles, γ rays
CYU 2: c
CYU 3: d
CYU 4: b and d

Answers to Odd-Numbered Problems

Chapter 1

1. (a) 5×10^{-3} g
  (b) 5 mg
  (c) 5×10^3 μg
3. (a) 5700 s
  (b) 86 400 s
5. 1.37 lb
7. (a) correct
  (b) not correct
  (c) not correct
  (d) correct
  (e) correct
9. 3.13×10^8 m^3
11. 5.5 km
13. 80.1 km, 25.9° south of west
15. 3.73 m
17. 35.3°
19. 99° opposite the 190-cm side, 3.0×10^1 degrees opposite the 95-cm side, 51° opposite the 150-cm side
21. 200 N due east or 600 N due west
23. 0.90 km, 56° north of west
25. (a) 5.31 km, south
  (b) 5.31 km, north
27. (a) 5600 newtons
  (b) along the dashed line
29. (a) 8.6 units
  (b) 34.9° north of west
  (c) 8.6 units
  (d) 34.9° south of west
31. (a) 15.8 m/s
  (b) 6.37 m/s
33. (a) 147 km
  (b) 47.9 km
35. (a) 5.70×10^2 newtons
  (b) 33.6° south of west
37. (a) 25.0°
  (b) 34.8 newtons
39. (a) 322 newtons
  (b) 209 newtons
  (c) 279 newtons
41. 7.1 m, 9.9° north of east
43. 3.00 m, 42.8° above the $-x$ axis
45. (a) 2.7 km
  (b) 6.0×10^1 degrees north of east
47. (a) 10.4 units
  (b) 12.0 units
49. (a) 178 units
  (b) 164 units
51. (a) 64 m
  (b) 37° south of east
53. (a) C
  (b) B
55. 30.2 m, 10.2°
57. 222 m, 55.8° below the $-x$ axis
59. 6.88 km, 26.9°

61. 1.2×10^2 m
63. 288 units due west and 156 units due north

Chapter 2

1. 0.80 s
3. (a) 12.4 km
  (b) 8.8 km, due east
5. (a) 464 m/s
  (b) 1040 mi/h
7. 52 m
9. 7.2×10^3 m
11. 2.1 s
13. (a) 4.0 s
  (b) 4.0 s
15. 3.44 m/s, due west
17. +30.0 m/s
19. 4.5 m
21. (a) 11.4 s
  (b) 44.7 m/s^2
23. (a) 1.7×10^2 cm/s^2
  (b) 0.15 s
25. -3.1 m/s^2
27. 0.74 m/s
29. 39.2 m
31. 0.87 m/s^2, in the same direction as the velocity
33. 1.2
35. (a) 13 m/s
  (b) 0.93 m/s^2
37. 91.5 m/s
39. -1.5 m/s^2
41. 1.1 s
43. 6.12 s
45. 1.7 s
47. 8.96 m
49. 0.767 m/s
51. 2.0×10^1 m
53. 10.6 m
55. 6.0 m below the top of the cliff
57. The answer is in graphical form.
59. 1.9 m/s^2 (segment A), 0 m/s^2 (segment B), 3.3 m/s^2 (segment C)
61. -8.3 km/h^2
63. (a) 6.6 s
  (b) 5.3 m/s
65. 44.1 m/s
67. (a) 1.5 m/s^2
  (b) 1.5 m/s^2
  (c) The car travels 76 m farther than the jogger.
69. 5×10^4 y
71. -22 m/s^2
73. (a) 2.67×10^4 m
  (b) 6.74 m/s, due north
75. 0.40 s

77. (a) 11 s
  (b) 3.0×10^1 m
79. 14 s

Chapter 3

1. 8.8×10^2 m
3. 8600 m
5. 242 m/s
7. 5.4 m/s
9. 27.0°
11. (a) 2.99×10^4 m/s
  (b) 2.69×10^4 m/s
13. 2.40 m
15. (a) 1.78 s
  (b) 20.8 m/s
17. 14.1 m/s
19. (a) 6.0×10^1 m
  (b) 290 m
21. 30.0 m
23. (a) 85 m
  (b) 610 m
25. (a) 1.1 s
  (b) 1.3 s
27. 24 buses
29. 1.7 s
31. 48 m
33. 11 m/s
35. 14.7 m/s
37. 14.9 m
39. 21.9 m/s, 40.0°
41. 0.141° and 89.860°
43. $D = 850$ m, $H = 31$ m
45. $v_{0B} = 8.79$ m/s, $\theta_B = 81.5°$
47. (a) 41 m/s, due east
  (b) 41 m/s, due west
49. 31 s
51. (a) 2.0×10^3 s
  (b) 1.8×10^3 m
53. 6.3 m/s, 18° north of east
55. 7.63 m/s, 26.4° north of east
57. 63.9 m/s, 85° west of south
59. 3.05 m/s, 14.8° north of west
61. 4.42 s
63. 14.6 s
65. (a) 239 m/s, 57.1° with respect to the horizontal
  (b) 239 m/s, 57.1° with respect to the horizontal
67. 5.17 s
69. 5.2 m/s, 52° west of south
71. 42°
73. 27.2°

Chapter 4

1. 93 N
3. 3.5×10^4 N

A-12

5. 130 N

7. 37 N

9. (a) 3.6 N

  (b) 0.40 N

11. 1.83 m/s^2, left

13. 30.9 m/s^2, 27.2° above the $+x$ axis

15. 0.78 m, 21° south of east

17. 18.4 N, 68° north of east

19. 1.8×10^{-7} N

21. (a) $W = 1.13 \times 10^3$ N, $m = 115$ kg

  (b) $W = 0$ N, $m = 115$ kg

23. 0.223 m/s^2

25. 1.76×10^{24} kg

27. (a) 3.75 m/s^2

  (b) 2.4×10^2 N

29. 4.7 kg

31. 0.0050

33. 0.414 L

35. 7.3×10^2 N

37. The block will move. $a = 3.72$ m/s^2

39. 0.444

41. (a) 390 N

  (b) 7.7 m/s, direction is toward second base

43. (a) 0.980 m/s^2, direction is opposite to the direction of motion

  (b) 29.5 m

45. 68°

47. (a) 7.40×10^5 N

  (b) 1.67×10^9 N

49. 9.70 N for each force

51. 929 N

53. 1.00×10^2 N, 53.1° south of east

55. 62 N

57. 16.3 N

59. 286 N

61. 406 N

63. 18.0 m/s^2, 56.3° above the $+x$ axis

65. 1730 N, due west

67. (a) 2.99 m/s^2

  (b) 129 N

69. 6.6 m/s

71. (a) 1.3 N

  (b) 6.5 N

73. 29 400 N

75. 0.14 m/s^2

77. (a) 4.25 m/s^2

  (b) 1080 N

79. 0.265 m

81. (a) $\Delta T_A = 0$ N, $\Delta T_B = -4.7$ N, $\Delta T_C = 0$ N

  (b) $\Delta T_A = 0$ N, $\Delta T_B = 0$ N, $\Delta T_C = +4.7$ N

83. 1.2 s

85. (a) 13.7 N

  (b) 1.37 m/s^2

87. (a) 10.5 m/s^2

  (b) 1.07

89. 39 N

91. (a) 447 N

  (b) 241 N

93. (a) 1610 N

  (b) 2640 N

95. 4290 N

97. (a) 1.04×10^3 N

  (b) 1.04×10^3 N

  (c) 2.45 m/s^2

  (d) 1.74×10^{-22} m/s^2

99. (a) 3.56 m/s^2

  (b) 281 N

101. 0.141

103. 8.7 s

105. (a) 5.2 m/s^2

  (b) 11 m/s^2

107. 1.9×10^2 N

109. (a) 0.60 m/s^2

  (b) 104 N (left string), 230 N (right string)

111. 0.665

Chapter 5

1. 160 s

3. 6.9 m/s^2

5. 332 m

7. (a) 5.0×10^1 m/s^2

  (b) zero

  (c) 2.0×10^1 m/s^2

9. 2.2

11. 0.68 m/s

13. (a) 0.189 N

  (b) 4.00

15. 12 m/s

17. 3500 N

19. (a) 3510 N

  (b) 14.9 m/s

21. 2.0×10^1 m/s

23. 184 m

25. 2.12×10^6 N

27. 1.33×10^4 m/s

29. 4.20×10^4 m/s

31. 1/27

33. 2.45×10^4 N

35. (a) 912 m

  (b) 228 m

  (c) 2.50 m/s^2

37. (a) 1.70×10^3 N

  (b) 1.66×10^3 N

39. 4.72×10^3 m

41. 8.48 m/s

43. (a) 1.2×10^4 N

  (b) 1.7×10^4 N

45. 61°

47. 0.71

49. 2.9×10^4 N

51. 2.0×10^2 m/s^2

53. 0.250

55. 23 N at 19.0 m/s and 77 N at 38.0 m/s

Chapter 6

1. -2.6×10^6 J

3. (a) 2980 J

  (b) 3290 J

5. 42.8°

7. (a) 54.9 N

  (b) 1060 J

  (c) -1060 J

  (d) 0 J

9. 45 N

11. 2.07×10^3 N

13. (a) 3.1×10^3 J

  (b) 2.2×10^2 J

15. (a) 38 J

  (b) 3.8×10^3 N

17. 6.4×10^5 J

19. 18%

21. 1.4×10^{11} J

23. 10.9 m/s

25. (a) -3.0×10^4 J

  (b) The resistive force is not a conservative force.

27. 2.39×10^5 J

29. 5.24×10^5 J

31. 2.3×10^4 J

33. (a) 28.3 m/s

  (b) 28.3 m/s

  (c) 28.3 m/s

35. 4.13 m

37. 4.8 m/s

39. 3.29 m/s

41. 6.33 m

43. 40.8 kg

45. -4.51×10^4 J

47. 16.5 m

49. -1.21×10^6 J

51. (a) 2.8 J

  (b) 35 N

53. 4.17 m/s

55. (a) 3.3×10^4 W

  (b) 5.1×10^4 W

57. 3.6×10^6 J

59. 3.0×10^3 W

61. 6.7×10^2 N

63. (a) 93 J

  (b) 0 J

  (c) 2.3 m/s

65. (a) Bow 1 requires more work.

  (b) 25 J

67. (a) 1.50×10^2 J

  (b) 7.07 m/s

69. At $h = 20.0$ m, KE = 0 J, PE = 392 J, and E = 392 J. At $h = 10.0$ m, KE = 196 J, PE = 196 J, and E = 392 J. At $h = 0$ m, KE = 392 J, PE = 0 J, and E = 392 J.

71. 2.2×10^3 J

73. (a) 1.80×10^3 J

  (b) -1.20×10^3 J

75. 2450 N

77. 1.7 m/s

79. 13.5 m

Chapter 7

1. -8.7 kg·m/s

3. (a) $+1.7$ kg·m/s

  (b) $+570$ N

5. 11 N·s, in the same direction as the average force

7. $+69$ N

9. 3.7 N $\cdot$ s

11. 4.28 N $\cdot$ s, upward

13. 960 N

15. -1.5×10^{-4} m/s

17. (a) Bonzo, since he has the recoil velocity with the smaller magnitude
    (b) 1.7

19. (a) 77.9 m/s
    (b) 45.0 m/s

21. $m_1 = 1.00$ kg, $m_2 = 1.00$ kg

23. $+547$ m/s

25. 84 kg

27. 3.00 m

29. $+9.09$ m/s

31. (a) -0.400 m/s (5.00-kg ball), $+1.60$ m/s (7.50-kg ball)
    (b) $+0.800$ m/s

33. (a) $+8.9$ m/s
    (b) -3.6×10^4 N $\cdot$ s
    (c) 5.9 m

35. (a) $73.0°$
    (b) 4.28 m/s

37. (a) $+3.17$ m/s
    (b) 0.0171

39. (a) 5.56 m/s
    (b) -2.83 m/s (1.50-kg ball), $+2.73$ m/s (4.60-kg ball)
    (c) 0.409 m (1.50-kg ball), 0.380 m (4.60-kg ball)

41. 4.67×10^6 m

43. 6.46×10^{-11} m

45. (a) -1.5 m/s
    (b) $+1.1$ m/s

47. $+9.3$ m/s

49. 3.0 m/s

51. $+4500$ m/s, in the same direction as the rocket before the explosion

53. 1.5×10^{-10} m/s

55. 0.34 s

57. 0.097 m

Chapter 8

1. 13 rad/s

3. 63.7 grad

5. 6.4×10^{-3} rad/s^2

7. 492 rad/s

9. 825 m

11. 336 m/s

13. 6.05 m

15. 25 rev

17. 157.3 rad/s

19. (a) 1.2×10^4 rad
    (b) 1.1×10^2 s

21. 28 rad/s

23. 12.5 s

25. 1.95×10^4 rad

27. 7.37 s

29. 157 m/s

31. 22 rev/s

33. (a) 4.66×10^2 m/s
    (b) $70.6°$

35. (a) 1.25 m/s
    (b) 7.98 rev/s

37. 14.8 rad/s

39. (a) 9.00 m/s^2
    (b) radially inward

41. (a) 2.5 m/s^2
    (b) 3.1 m/s^2

43. $1/\sqrt{3}$

45. 1.00 rad

47. 8.71 rad/s^2

49. 693 rad

51. 28.0 rad/s

53. 11.8 rad

55. 20.6 rad

57. 380 m/s^2

59. (a) 4.00×10^1 rad
    (b) 15.0 rad/s

61. 0.62 m

63. 0.300 m/s

65. 7.45 s

67. 2 revolutions

Chapter 9

1. 4.2 N $\cdot$ m

3. 843 N

5. 1.3

7. (a) $\tau = FL$
   (b) $\tau = FL$
   (c) $\tau = FL$

9. 0.667 m

11. 196 N (force on each hand), 96 N (force on each foot)

13. 0.591

15. (a) 27 N, to the left
    (b) 27 N, to the right
    (c) 27 N, to the right
    (d) 143 N, downward and to the left ($79°$ below the horizontal)

17. 1200 N, to the left

19. (a) 1.60×10^5 N
    (b) 4.20×10^5 N

21. $37.6°$

23. (a) 1.21×10^3 N
    (b) 1.01×10^3 N, downward

25. 51.4 N

27. 1.7 m

29. 1.25 kg $\cdot$ m^2

31. (a) -11 N $\cdot$ m
    (b) -9.2 rad/s^2

33. (a) 0.131 kg $\cdot$ m^2
    (b) 3.6×10^{-4} kg $\cdot$ m^2
    (c) 0.149 kg $\cdot$ m^2

35. (a) 5.94 rad/s^2
    (b) 44.0 N

37. 0.78 N

39. (a) 2.67 kg $\cdot$ m^2
    (b) 1.16 m

41. 2.12 s

43. (a) $v_{T1} = 12.0$ m/s, $v_{T2} = 9.00$ m/s, $v_{T3} = 18.0$ m/s
    (b) 1.08×10^3 J
    (c) 60.0 kg $\cdot$ m^2
    (d) 1.08×10^3 J

45. 6.1×10^5 rev/min

47. $2/7$

49. $3/4$

51. 2.0 m

53. 4.4 kg $\cdot$ m^2

55. 0.26 rad/s

57. 8% increase

59. 0.17 m

61. 0.25 m

63. (a) 0.14 rad/s
    (b) A net external torque must be applied in a direction opposite to the angular deceleration.

65. 0.060 kg $\cdot$ m^2

67. (a) 7.40×10^2 N
    (b) 0.851 N

69. 34 m/s

71. 69 N

Chapter 10

1. 237 N

3. (a) 7.44 N
   (b) 7.44 N

5. 0.012 m

7. 2.29×10^{-3} m

9. 0.79

11. 0.240 m

13. (a) 1.00×10^3 N/m
    (b) 0.340

15. 3.5×10^4 N/m

17. (a) 0.080 m
    (b) 1.6 rad/s
    (c) 2.0 N/m
    (d) 0 m/s
    (e) 0.20 m/s^2

19. 696 N/m

21. 4.3 kg

23. 0.806

25.

h (meters)	KE	PE (gravity)	PE (elastic)	E
0	0 J	0 J	8.76 J	8.76 J
0.200	1.00 J	3.92 J	3.84 J	8.76 J
0.400	0 J	7.84 J	0.92 J	8.76 J

27. (a) 58.8 N/m
    (b) 11.4 rad/s

29. 7.18×10^{-2} m

31. 14 m/s

33. (a) 9.0×10^{-2} m
    (b) 2.1 m/s

35. 1.25 m/s (11.2-kg block), 0.645 m/s (21.7-kg block)

37. 2.37×10^3 N/m

39. 0.99 m

41. 6.0 m/s^2

43. 0.816

45. $7R/5$

47. 3.7×10^{-5} m

49. 1.6×10^5 N

51. 260 m

53. 1.4×10^{-6}

55. -2.8×10^{-4}

57. 6.6×10^4 N

59. (a) 6.3×10^{-2} m
   (b) 7.3×10^{-2} m
61. 4.6×10^{-4}
63. 1.0×10^{-3} m
65. -4.4×10^{-5}
67. (a) 140 m/s
   (b) 1.7×10^{15} m/s^2
69. 21 m
71. 7.7×10^{-5} m
73. (a) 4.9×10^6 N/m^2
   (b) 6.0×10^{-6} m
75. (a) 2.66 Hz
   (b) 0.0350 m
77. 24.2 rad/s
79. 33.4 m/s
81. 4.0×10^{-5} m
83. (a) 0.25 s
   (b) 0.75 s

Chapter 11

1. 317 m^2
3. 3400 N
5. 6.6×10^6 kg
7. 1.9 gal
9. 63%
11. 1.1×10^3 N
13. 4.33×10^6 Pa
15. 24 blocks
17. 0.750 m
19. 1.2×10^5 Pa
21. (a) 3.5×10^6 N
   (b) 1.2×10^6 N
23. (a) 2.45×10^5 Pa
   (b) 1.73×10^5 Pa
25. 7.0×10^5 Pa
27. 1.19×10^5 Pa
29. 0.74 m
31. 3.8×10^5 N
33. (a) 93.0 N
   (b) 94.9 N
35. 8.50×10^5 N·m
37. 5.7×10^{-2} m
39. 4.89 m
41. 2.7×10^{-4} m^3
43. 2.04×10^{-3} m^3
45. 0.20 m
47. 7.6×10^{-2} m
49. 5.28×10^{-2} m and 6.20×10^{-2} m
51. 1.91 m/s
53. (a) 0.18 m
   (b) 0.14 m
55. (a) 1.6×10^{-4} m^3/s
   (b) 2.0×10^1 m/s
57. (a) 150 Pa
   (b) The pressure inside the roof is greater than that outside the roof, so there is a net outward force.
59. 1.92×10^5 N
61. (a) 2.48×10^5 Pa
   (b) 1.01×10^5 Pa
   (c) 0.342 m^3/s
63. 33 m/s

65. (a) 14 m/s
   (b) 0.98 m^3/s
67. 9600 N
69. 7.78 m/s
71. (a) 1.01×10^5 Pa
   (b) 1.19×10^5 Pa
73. 1.7 m
75. (a) 2.8×10^{-5} N
   (b) 1.0×10^1 m/s
77. 10.3 m
79. 59 N
81. 7.0×10^{-2} m
83. 4.5×10^{-5} kg/s
85. (a) 1.26×10^5 Pa
   (b) 19.4 m
87. 0.816
89. 6.3×10^{-3} kg
91. 31.3 rad/s
93. 0.19 m
95. 5.75×10^4 Pa

Chapter 12

1. 0.2 C°
3. (a) 10.0 °C and 40.0 °C
   (b) 283.2 K and 313.2 K
5. -459.67 °F
7. -164 °C
9. 0.084 m
11. 1500 m
13. 110 C°
15. -2.82×10^{-4} m
17. 49 °C
19. 2.0027 s
21. (a) tension
   (b) 9.6×10^7 N/m^2
23. 0.6
25. 3.1×10^{-3} m^3
27. 2.5×10^{-7} m^3
29. 18
31. 0.33 gal
33. 9.0 mm
35. 45 atm
37. 6.9
39. 36.2 °C
41. $230
43. 21.03 °C
45. 940 °C
47. 21 barrels
49. 0.016 C°
51. 3.9×10^5 J
53. 9.3 kg
55. 9.49×10^{-3} kg
57. 1.85×10^5 J
59. (a) 3.0×10^{20} J
   (b) 3.2 years
61. 1.9×10^4 J/kg
63. 2.6×10^{-3} kg
65. 3.50×10^2 m/s
67. 0 °C
69. 33%
71. 2400 Pa
73. 28%

75. 39%
77. 4.9×10^{-2} m
79. 3.9×10^{-3} kg
81. 8.0×10^{-4}
83. one penny
85. 10 °C
87. 0.223
89. 650 W
91. 6.7×10^2 W
93. 1.3×10^5 J

Chapter 13

1. 8.0×10^2 J/s
3. 14 h
5. 2.0×10^{-3} m
7. 17
9. (a) 21 °C
   (b) 18 °C
11. (a) 130 °C
   (b) 830 J
   (c) 237 °C
13. 287 °C
15. (a) 2.0
   (b) 0.61
17. 14.5 da
19. 0.3
21. 320 K
23. 0.70
25. 12
27. 12 J
29. 1.2×10^4 s
31. 532 K
33. (a) 101.2 °C
   (b) 110.6 °C
35. 0.74

Chapter 14

1. Aluminum
3. (a) 294.307 u
   (b) 4.887×10^{-25} kg
5. 1.07×10^{-22} kg
7. 2.6×10^{-10} m
9. 1.0×10^3 kg
11. 2.3×10^{-2} mol
13. (a) 201 mol
   (b) 1.21×10^5 Pa
15. 2.5×10^{21}
17. 882 K
19. 0.93 mol/m^3
21. 5.9×10^4 g
23. 0.205
25. 6.19×10^5 Pa
27. 308 K
29. (a) 46.3 m^2/s^2
   (b) 40.1 m^2/s^2
31. 1.2×10^4 m/s
33. 3.9×10^5 J
35. 2820 m
37. 4.0×10^1 Pa
39. 11 s
41. (a) 2.1 s
   (b) 1.6×10^{-5} s

43. (a) 5.00×10^{-13} kg/s
   (b) 5.8×10^{-3} kg/m^3
45. 2.3×10^6 s
47. 925 K
49. 1.34×10^{-7} kg
51. 1.6×10^{-15} kg
53. 1.02×10^{20}
55. 11.7
57. 440 K

Chapter 15

1. (a) -261 J
   (b) Work is done on the system.
3. (a) -87 J
   (b) $+87$ J
5. 32 miles
7. 13 000 J
9. (a) 3100 J
   (b) negative
11. 0.24 m
13. 3.0×10^5 Pa
15. The answer is a proof.
17. 4.99×10^{-6}
19. (a) 0 J
   (b) -6.1×10^3 J
   (c) 310 K
21. 0.66
23. 1.81
25. (a) -8.00×10^4 J
   (b) Heat flows out of the gas.
27. 19.3
29. (a) 327 K
   (b) 0.132 m^3
31. 2400 J
33. (a) 1.1×10^4 J
   (b) 1.8×10^4 J
35. 5/2
37. (a) 60.0%
   (b) 40.0%
39. 2.38×10^4 J
41. 0.631
43. (a) 8600 J
   (b) 3100 J
45. $e = e_1 + e_2 - e_1e_2$
47. 0.21
49. (a) 1260 K
   (b) 1.74×10^4 J
51. lowering the temperature of the cold reservoir
53. (a) 0.360
   (b) 1.3×10^{13} J
55. 5.7 C°
57. 3.24×10^4 J
59. 5.86×10^5 J
61. 9.03
63. 1.4 K
65. (a) 2.0×10^1
   (b) 1.5×10^4 J
67. 1.26×10^3 K
69. 11.6 J/K
71. (a) $+3.68 \times 10^3$ J/K
   (b) $+1.82 \times 10^4$ J/K

(c) The vaporization process creates more disorder.
73. (a) $+8.0 \times 10^2$ J/K
   (b) The entropy of the universe should increase.
75. (a) -2.1×10^2 K
   (b) decrease
77. 0 J, The process is an adiabatic process.
79. (a) reversible
   (b) -125 J/K
81. (a) 1050 J
   (b) 2.99
83. 1.2×10^7 Pa
85. 44.3 s
87. 8.49×10^5 Pa
89. 7.5 s
91. (a) 5/9
   (b) 1/3

Chapter 16

1. 0.083 Hz
3. 0.49 m
5. 0.25 m
7. 78 cm
9. 5.0×10^1 s
11. (a) 1.1 m/s
   (b) 6.55 m
13. 64 N
15. 600 m/s
17. 7.7 m/s^2
19. $m_1 = 28.7$ kg, $m_2 = 14.3$ kg
21. 3.26×10^{-3} s
23. $y = (0.37$ m$)$ sin $(2.6 \, \pi t + 0.22 \, \pi x)$, where x and y are in meters and t is in seconds.
25. (a) $A = 0.45$ m, $f = 4.0$ Hz, $\lambda = 2.0$ m, $v = 8.0$ m/s
   (b) $-x$ direction
27. 2.5 N
29. 1730 m/s
31. (a) 7.87×10^{-3} s
   (b) 4.12
33. 2.06
35. (a) metal wave first, water wave second, air wave third
   (b) Second sound arrives 0.059 s later, and third sound arrives 0.339 s later.
37. 690 rad/s
39. tungsten
41. 650 m
43. 8.0×10^2 m/s
45. 239 m/s
47. 0.404 m
49. 1.98%
51. 2.4×10^{-5} W/m^2
53. 6.5 W
55. 7.6×10^3 W/m^2
57. 2.6 s
59. 0.316 W/m^2
61. 1.3
63. 56.2 W

65. 2.6
67. 0.84 s
69. 2.39 dB
71. 17 m/s
73. 860 Hz
75. 1350 Hz
77. 22 m/s
79. 209 m
81. 25
83. (a) 431 m/s
   (b) 322 m/s
85. 8.68×10^{-3} kg/m
87. 1000
89. 8.19×10^{-2} m
91. 56 m/s
93. -6.0 dB
95. 77.0 dB
97. 153 N
99. (a) 2.20×10^2 m/s
   (b) 9.19 m/s
101. 57% argon, 43% neon

Chapter 17

1. (a) 2 cm
   (b) 1 cm
3. The answer is a series of drawings.
5. 107 Hz
7. 3.89 m
9. 3.90 m, 1.55 m, 6.25 m
11. (a) 44°
   (b) 0.10 m
13. 1.5×10^4 Hz
15. 3.7°
17. 8 Hz
19. (a) 50 kHz
   (b) 90 kHz
21. 8 Hz
23. 1.10×10^2 Hz
25. 0.46 m
27. 171 N
29. (a) 180 m/s
   (b) 1.2 m
   (c) 150 Hz
31. 0.077 m
33. 20.8° and 53.1°
35. (a) $f_2 = 800$ Hz, $f_3 = 1200$ Hz, $f_4 = 1600$ Hz
   (b) $f_2 = 800$ Hz, $f_3 = 1200$ Hz, $f_4 = 1600$ Hz
   (c) $f_3 = 1200$ Hz, $f_5 = 2000$ Hz, $f_7 = 2800$ Hz
37. 0.50 m
39. 602 Hz
41. 6.1 m
43. 1.68×10^5 Pa
45. 2.4 m/s
47. 5.06 m
49. 0.35 m
51. 3.93×10^{-3} kg/m
53. 28 Hz and 42 Hz
55. 12 Hz

Chapter 18
1. 1.5×10^{13}
3. 1.6×10^{13}
5. (a) $+1.5\,q$
   (b) $+4\,q$
   (c) $+4\,q$
7. 120 N
9. 0.14 N
11. (a) both positive or both negative
   (b) 1.7×10^{-16} C
13. (a) 4.56×10^{-8} C
   (b) 3.25×10^{-6} kg
15. 3.8×10^{12}
17. $+2.0\ \mu C$
19. -3.3×10^{-6} C
21. 3.5×10^{-5} C
23. (a) $15.4°$
   (b) 0.813 N
25. 1.8 N due east
27. 6.5×10^3 N/C downward
29. (a) -6.2×10^7 N/C
   (b) $+2.9 \times 10^8$ N/C
31. 1.3 m
33. (a) positive
   (b) 2.53×10^7 protons
35. 3.9×10^6 N/C, in the $+y$ direction
37. 35 N/C
39. $|q_1| = 0.716\,q,\ |q_2| = 0.0895\,q$
41. 0.577
43. 3.25×10^{-8} C
45. $61°$
47. (a) 350 N·m²/C
   (b) 460 N·m²/C
49. (a) 2.3×10^5 N·m²/C
   (b) 2.3×10^5 N·m²/C
   (c) 2.3×10^5 N·m²/C
51. (a) The flux through the face in the x, z plane at $y = 0$ m is -6.0×10^1 N·m²/C. The flux through the face parallel to the x, z plane at $y = 0.20$ m is $+6.0 \times 10^1$ N·m²/C. The flux through each of the remaining four faces is zero.
   (b) 0 N·m²/C
53. The answer is a proof.
55. The answer is a drawing.
57. $-q$ on interior surface and $+3\,q$ on exterior surface
59. 0.45 N due east
61. 0.38 N directed $49°$ below the $-x$ axis
63. 5.53×10^{-2} m
65. 2.2×10^5 N/C along the $-x$ axis
67. 92.0 N/m

Chapter 19
1. 1.1×10^{-20} J
3. (a) $+2.00 \times 10^{-14}$ J
   (b) 2.00×10^{-14} J
5. 9.4×10^7 m/s
7. 67 hp
9. 339 V
11. $+3.6 \times 10^{-9}$ C

13. -4.35×10^{-18} J
15. -4.7×10^{-2} J
17. $-q/\sqrt{2}$
19. -0.746 J
21. 1.53×10^{-14} m
23. 0.0342 m
25. (a) $-2q/3$
   (b) $-2q$
27. 18 000 V
29. 3.5×10^4 V
31. $+9.0 \times 10^3$ V
33. (a) 179 V
   (b) 143 V
   (c) 155 V
35. 0.213 J
37. 1.1×10^3 V
39. 5.3
41. (a) 33 J
   (b) 8500 W
43. 52 V
45. 1.2×10^{-8} J
47. The answer is a proof.
49. 2.4
51. (a) 1.3×10^{-12} C
   (b) 8.1×10^6
53. 8.0×10^{-5} C
55. 4.2×10^3 V/m, directed from A to B
57. 1.3×10^{-4} C
59. 2.77×10^6 m/s

Chapter 20
1. (a) 2.6 C
   (b) 310 J
3. 0.21 A
5. 82 Ω
7. 6.2×10^4 J
9. (a) 4.7×10^{13}
   (b) $17\ C°$
11. 1.64
13. 9.9×10^{-3} m
15. $-34.6\ °C$
17. 189 Ω
19. $L_{tungsten}/L_{carbon} = 70$
21. 6.0×10^2 W
23. 8.9 h
25. (a) 1300 W
   (b) 480 W
   (c) 1.4
27. 250 °C
29. 33 °C
31. 1.77 A
33. (a) 9.0×10^2 W
   (b) 1.8×10^3 W
35. $92
37. 2.0 h
39. 32 Ω
41. (a) 10.0 V
   (b) 5.00 V
43. (a) 145 Ω
   (b) 74 V
45. (a) 15.5 V
   (b) 14.2 W

47. (a) 35 Ω
   (b) $5.0 \times 10^1\ \Omega$
49. 190 Ω
51. 5.3 Ω
53. (a) 4.57 A
   (b) 1450 W
55. (a) 3.6 Ω
   (b) 33 A, the breaker will open
57. 3.58×10^{-8} m²
59. $1.0 \times 10^2\ \Omega$
61. 4.6 Ω
63. 25 Ω
65. 6.00 Ω, 0.545 Ω, 3.67 Ω, 2.75 Ω, 2.20 Ω, 1.50 Ω, 1.33 Ω, 0.833 Ω
67. 0.054 Ω
69. 30
71. 24.0 V
73. (a) 0.38 A
   (b) 2.0×10^1 V
   (c) B
75. 33 A
77. 0.75 V; left end
79. 0.94 V; D
81. $3.43 \times 10^3\ \Omega$
83. 0.450 Ω
85. (a) 30.0 V
   (b) 28.1 V
87. 2.0 μF
89. 1.7 μF
91. The answer is a proof.
93. C_0
95. 1.2×10^{-2} s
97. 4.1×10^{-7} F
99. 0.29 s
101. 4.69×10^{-4} C
103. 22 A
105. 38 °C
107. 5.01 A
109. (a) 0.750 A
   (b) 2.11 A
111. (a) 1.2 Ω
   (b) 110 V
113. 360 °C
115. 11 V

Chapter 21
1. $75.1°$ and $105°$
3. 3.7×10^{-12} N
5. 4.1×10^{-3} m/s
7. $19.7°$
9. (a) due south
   (b) 2.55×10^{14} m/s²
11. (a) The answer is a drawing.
   (b) 2.6×10^{-2} T
   (c) 2.7×10^{16} m/s²
13. 1.5×10^{-8} s
15. (a) 7.2×10^6 m/s
   (b) 3.5×10^{-13} N
17. 1.63×10^{-2} m
19. (a) $\theta = 0°$
   (b) 0.29 m
21. 0.16 T

23. 8.7×10^{-3} s
25. 9.6×10^4 m/s
27. 8.1 N
29. 0.96 N (top and bottom sides), 0 N (left and right sides)
31. 57.6°
33. (a) left to right
    (b) 1.1×10^{-2} m
35. 14 A
37. (a) 24 A $\cdot$ m^2
    (b) 4.8 N $\cdot$ m
39. 0.062 m
41. (a) 170 N $\cdot$ m
    (b) increase
43. 8.3 N
45. 1.2×10^{-5} A $\cdot$ m^2
47. 8.0×10^{-5} T
49. 1.9×10^{-4} N $\cdot$ m
51. 190 A
53. $H = 2.1\,R$
55. 8.6 A, The current in the outer coil has an opposite direction relative to that in the inner coil.
57. 1.04×10^{-2} T
59. I_3 is directed out of the paper; $I_3/I = 2$.
61. (a) 1.1×10^{-5} T
    (b) 4.4×10^{-6} T
63. The answer is a proof.
65. 1.3×10^{-2} T
67. 0.023 N $\cdot$ m
69. 8.1×10^{-5} T
71. 0.0904 m
73. 140 V/m, directed toward the bottom of the page
75. (a) 6.8×10^{-3} N
    (b) 36°
77. 9.3×10^{-24} A $\cdot$ m^2

Chapter 22

1. 150 m/s
3. 0.065 V
5. (rod A) emf = 0 V; (rod B) emf = 1.6 V and end 2 is positive; (rod C) emf = 0 V
7. (a) 3.3 m/s
   (b) 4.6 N
9. (a) 0.23 kg
   (b) -1.8 J
   (c) 1.8 J
11. 70.5°
13. (a) 1.2×10^{-4} Wb
    (b) 3.2×10^{-4} Wb
    (c) 2.0×10^{-4} Wb
15. (two triangular ends) 0 Wb; (bottom surface) 0 Wb, (1.2 m × 0.30 m surface) 0.090 Wb; (1.2 m × 0.50 m surface) 0.090 Wb
17. 1.5 m^2/s
19. 8.6×10^{-5} T
21. 4.8 s
23. 6.6×10^{-2} J
25. (a) 3.6×10^{-3} V
    (b) 2.0×10^{-3} m^2/s; area must be shrunk

27. 2100 rad/s
29. (a) left to right
    (b) right to left
31. (a) clockwise
    (b) clockwise
33. There is no induced current.
35. 12 V
37. 0.30 T
39. (a) 2.4 Hz
    (b) 15 rad/s
    (c) 0.62 T
41. 102 V
43. (a) 0.80 A
    (b) 8.00 A
    (c) 4.40 A
45. 1.5×10^9 J
47. 1.4 V
49. 220
51. 2.80×10^{-4} H
53. $M = \mu_0\,\pi N_1 N_2\,R_2^2/(2R_1)$
55. 1.0×10^1 W
57. step-down transformer, 1/12
59. (a) 7.0×10^5 W
    (b) 7.0×10^1 W
61. The answer is a proof.
63. 3.0×10^5
65. (Figure 22.1b) right to left; (Figure 22.1c) left to right
67. 0.150 m
69. 0.14 V
71. 0.459 T
73. 1.6×10^{-5} A

Chapter 23

1. 126 Hz
3. 2.7×10^{-7} F
5. 5.00×10^{-2} s
7. 3/2
9. 8.0×10^1 Hz
11. 176 mH
13. 0.17 V
15. 83.9 V
17. 38 V
19. (a) 0.925 A
    (b) 31.8°
21. 270 Hz
23. (a) 29.0 V
    (b) -0.263 A
25. 0.651 W
27. (a) 352 Hz
    (b) 15.5 A
29. 3.1 kHz
31. 2.8 kHz
33. (a) 2.94×10^{-3} H
    (b) 4.84 Ω
    (c) 0.163
35. (a) $4/\sqrt{3}$
    (b) $1/\sqrt{3}$
37. (a) $2.00\ \mu$F
    (b) 0.77 A
39. (a) 113 Ω
    (b) $+21°$

41. 0.44 A
43. 8
45. 0.707

Chapter 24

1. 0.015 m
3. 1.5×10^{-4} H
5. The answers are in graphical form.
7. 1.4×10^{17} Hz
9. 3.7×10^4
11. 1.25 m
13. 1.5×10^{10} Hz
15. 1.3×10^6 m
17. 540 rev/s
19. 8.75×10^5
21. 3.8×10^2 W/m^2
23. 0.07 N/C
25. (a) 183 N/C
    (b) 6.10×10^{-7} T
27. 5600 W
29. 3.93×10^{26} W
31. (a) receding
    (b) 3.1×10^6 m/s
33. (a) 6.175×10^{14} Hz
    (b) 6.159×10^{14} Hz
35. 71.6°
37. 14 W/m^2
39. 20
41. 11.118 m
43. 4.1×10^{16} m
45. 5.6×10^9 J
47. (a) 0.82
    (b) 0.18
49. (a) 2.4×10^9 Hz
    (b) 0.063 m
51. 920 W
53. 6.25×10^{-9} J

Chapter 25

1. 55°
3. 14°
5. 10°
7. 11.3°, 45.0°, 71.6°, 135°, 162°
9. 33.7°
11. (a) The image distance is 3.0×10^1 cm behind the mirror.
    (b) The image height is 5.0 cm.
13. (a) The image distance is 16.7 cm behind the mirror.
    (b) The image height is 6.67 cm.
15. 10.9 cm
17. +74 cm
19. (a) 290 cm
    (b) -8.9 cm
    (c) upside down
21. +22 cm
23. (a) convex
    (b) 24.0 cm
25. -3
27. +42.0 cm
29. (a) The answer is a proof.
    (b) The answer is a proof.

31. (a) R
    (b) -1
    (c) inverted
33. (a) $+46$ cm
    (b) 2.4
35. (a) $+62$ cm
    (b) $+0.35$
    (c) upright
    (d) smaller
37. 0.533 m
39. 80.0 cm, toward the mirror
41. 1.73

Chapter 26
1. 2.00×10^8 m/s
3. 1.198
5. 2.0×10^{-11} s
7. 1.82
9. (a) $43°$
    (b) $31°$
11. 3.0 m
13. 1.92×10^8 m/s
15. 1.65
17. 1.19 mm
19. The answer is a derivation.
21. 3.23 cm
23. 1.54
25. 1.51
27. (a) B
    (b) A
29. $42.67°$
31. $25.0°$
33. $34.39°$
35. 1.52
37. The answer is a proof.
39. $0.35°$
41. (red ray) $44.6°$; (violet ray) $45.9°$
43. (red ray) $52.7°$; (violet ray) $56.2°$
45. $d_i = 18$ cm
47. 2.8
49. (a) $d_i = -75$ cm, $m = 2.5$
    (b) $d_i = -75.0$ cm, $m = 2.50$
51. (a) -0.00625 m
    (b) -0.0271 m
53. (a) converging
    (b) $2f$
    (c) $2f$
55. 48 cm
57. $+35$ cm and $+90.5$ cm
59. 0.13 m to the right of the second lens
61. (a) 4.00 cm to the left of the diverging
       lens
    (b) -0.167
    (c) virtual
    (d) inverted
    (e) smaller
63. 11.8 cm
65. (a) 18.1 cm
    (b) real
    (c) inverted
67. (a) 35.2 cm
    (b) 32.9 cm

69. -220 cm
71. -9.2 m
73. (a) 11.8 cm
    (b) 47.8 cm
75. (a) -4.5 m
    (b) 0.50 m
77. 3.7
79. 0.13 m
81. 6.3 cm
83. 0.64 rad
85. 0.81 cm
87. 0.435 cm
89. 0.261 cm
91. 1.1 m
93. (a) 1.3-diopter lens
    (b) 0.86 m
    (c) -8.5
95. -24
97. 61.1 mm
99. (a) 1.50
    (b) 1.27
101. (right eye) -0.20 diopters;
     (left eye) -0.15 diopters
103. (a) -15 cm
     (b) virtual
105. 2.46×10^8 m/s
107. (a) converging
     (b) farsighted
     (c) 96.3 cm
109. (a) 6.74×10^{-7} m^2
     (b) 7.86×10^5 W/m^2
111. (a) 4.52×10^{-4} m
     (b) 6.12×10^{-2} m
113. 26.9 cm
115. -181
117. (a) 22.4 cm
     (b) 28.4 cm

Chapter 27
1. 6.0×10^{-5} m
3. $17.8°$
5. 660 nm
7. 0.0248 m
9. 487 nm
11. 102 nm
13. 207 nm
15. 115 nm
17. 427 nm
19. (a) $0.21°$
    (b) $22°$
21. (a) $24°$
    (b) $39°$
23. 490 nm
25. 0.447
27. 0.013
29. (a) 220 m
    (b) No.
31. 1.0×10^4 m
33. 2.3 m
35. (a) 1.22λ
    (b) shorter

37. (a) $37°$
    (b) $22°$
39. 630 nm
41. 4.0×10^{-6} m
43. 3/4
45. 1.95 m
47. constructive
49. (a) 4.2×10^{-5} degrees
    (b) $59°$
51. 0.0254 m
53. 3.2×10^3 m
55. (a) $7.9°$ (violet), $13°$ (red)
    (b) $16°$ (violet), $26°$ (red)
    (c) $24°$ (violet), $41°$ (red)
    (d) The second and third orders overlap.
57. (a) 2
    (b) $m_B = 4$ and $m_A = 2$;
      $m_B = 6$ and $m_A = 3$

Chapter 28
1. 0.15 rad/s
3. 2.4×10^8 m/s
5. 2.28 s
7. 16
9. 2.60×10^8 m/s
11. 530 m
13. 4.0 light-years
15. 3.0 m $\times$ 1.3 m
17. 2.83×10^8 m/s
19. 1.80×10^8 m/s
21. -2.0 m/s
23. (a) 1.0
    (b) 6.6
25. 1.1 kg
27. 5.0×10^{-13} J
29. 1.40×10^{-15} m
31. $0.31\,c$
33. 42 m
35. (a) 2.82×10^8 m/s
    (b) 1.8×10^{-16} kg·m/s
37. 1.3
39. (a) 5.1×10^{15} kg·m/s
    (b) 9.7×10^{15} kg·m/s
41. $-0.406\,c$
43. (a) 4.3 y
    (b) the twin traveling at $0.500\,c$

Chapter 29
1. 310 nm
3. 7.7×10^{29} photons/s
5. 6.3 eV
7. 2.10 eV
9. 73 photons/s
11. (a) 2.1×10^{24} photons
    (b) 32 molecules/photon
13. 5.1×10^{-33} kg·m/s
15. (a) 2.124×10^{-24} kg·m/s
    (b) 2.096×10^{-24} kg·m/s
17. 4.755×10^{-24} kg·m/s
19. 9.50×10^{-17} m
21. 2.6×10^{-28} m

23. 7.77×10^{-13} J
25. 7.38×10^{-11} m
27. 1.10×10^{3} m/s
29. 1.86×10^{4} V
31. 4.0×10^{-6} m/s
33. (a) 2.1×10^{-35} kg·m/s
     (b) 4.7×10^{-34} m/s
     (c) 2.3×10^{-5} m/s
35. $-0.0144° \leq \theta \leq +0.0144°$
37. 6.6×10^{-27} kg
39. (a) 4.50×10^{-36} m/s
     (b) 7.05×10^{27} years
41. 3.6×10^{-9} m
43. 1.2×10^{-36} m
45. 3.09×10^{-10} m

Chapter 30

1. (a) 6.2×10^{-31} m^3
     (b) 4×10^{-45} m^3
     (c) 7×10^{-13}%
3. 1.5×10^{14}
5. -8.7×10^{6} eV
7. 91.2 nm
9. (a) 7458 nm
     (b) 2279 nm
     (c) infrared
11. 3
13. -13.6 eV, -3.40 eV, -1.51 eV
15. $n_i = 6$ and $n_f = 2$
17. $6 \leq n_i \leq 19$
19. 2180 lines/cm
21. 2, 3, 4, and 5
23. -0.378 eV
25. $\pm 3.16 \times 10^{-34}$ J·s,
     $\pm 2.11 \times 10^{-34}$ J·s,
     $\pm 1.05 \times 10^{-34}$ J·s, 0 J·s
27. 26.6°
29. $1s^2\, 2s^2\, 2p^6\, 3s^2\, 3p^6\, 4s^2\, 3d^5$
31. (a) not permitted
     (b) permitted
     (c) not permitted
     (d) permitted
     (e) not permitted
33. 7.230×10^{-11} m
35. 6.83×10^{-11} m
37. 21 600 V
39. (a) 1.25×10^{-4} J
     (b) 3.98×10^{14} J
41. 1.9×10^{17}

43. 21 days
45. 4.41×10^{-10} m
47.

n	ℓ	m_ℓ	m_s
4	3	3	1/2
4	3	3	$-1/2$
4	3	2	1/2
4	3	2	$-1/2$
4	3	1	1/2
4	3	1	$-1/2$
4	3	0	1/2
4	3	0	$-1/2$
4	3	-1	1/2
4	3	-1	$-1/2$
4	3	-2	1/2
4	3	-2	$-1/2$
4	3	-3	1/2
4	3	-3	$-1/2$

49. 1.98×10^{-19} J
51. -0.544 eV, -0.378 eV, -0.278 eV
53. The answer is a proof.

Chapter 31

1. (a) X = Pt, 117 neutrons
     (b) X = S, 16 neutrons
     (c) X = Cu, 34 neutrons
     (d) X = B, 6 neutrons
     (e) X = Pu, 145 neutrons
3. 8
5. 35.2
7. $^{120}_{50}$Sn
9. 9.4×10^{3} m
11. 128 MeV
13. 2.9×10^{16} kg
15. 1.003 27 u
17. (a) $^{212}_{84}$Po $\rightarrow$ $^{208}_{82}$Pb $+$ $^{4}_{2}$He
     (b) $^{232}_{92}$U $\rightarrow$ $^{228}_{90}$Th $+$ $^{4}_{2}$He
19. (a) $^{14}_{6}$C $\rightarrow$ $^{14}_{7}$N $+$ $^{0}_{-1}$e
     (b) $^{212}_{82}$Pb $\rightarrow$ $^{212}_{83}$Bi $+$ $^{0}_{-1}$e
21. $^{35}_{16}$S $\rightarrow$ $^{35}_{17}$Cl $+$ $^{0}_{-1}$e
23. (a) $^{18}_{9}$F $\rightarrow$ $^{18}_{8}$O $+$ $^{0}_{+1}$e
     (b) $^{15}_{8}$O $\rightarrow$ $^{15}_{7}$N $+$ $^{0}_{+1}$e
25. 1.61×10^{7} m/s
27. 1.82 MeV
29. 3.0 days
31. 19.9
33. 2.1×10^{13} Bq
35. 146 disintegrations/min
37. 3.7×10^{10} Bq

39. 7.23 days
41. 0.70%
43. 90.9%
45. 6900 yr, maximum error is 900 yr
47. (a) 0.555 357 u
     (b) 9.2217×10^{-28} kg
49. 8.00 days
51. (a) $^{238}_{92}$U
     (b) $^{24}_{12}$Mg
     (c) $^{13}_{6}$C
53. 4 782 969 electrons
55. $N_\alpha = 3$, $N_\beta = 2$

Chapter 32

1. 2.8×10^{-3} J
3. 1.6×10^{-3} J
5. 0.26 roentgens
7. 4.4×10^{11} s^{-1}
9. 3.01×10^{8} rd
11. oxygen $^{16}_{8}$O
13. (a) proton, $^{1}_{1}$H
     (b) alpha particle, $^{4}_{2}$He
     (c) lithium, $^{6}_{3}$Li
     (d) nitrogen, $^{14}_{7}$N
     (e) manganese, $^{56}_{25}$Mn
15. 13.6 MeV
17. 2 neutrons
19. 9.0×10^{-4}
21. 41
23. 2.7×10^{6} kg
25. (a) 8.2×10^{10} J
     (b) 0.48 g
27. 1200 kg
29. 3.3 MeV
31. 1.0 gal
33. released energy = 24.7 MeV
35. possibility #1 = u, d, s
     possibility #2 = u, d, b
     possibility #3 = u, s, b
37. (a) 0.513 MeV
     (b) 2.43×10^{-12} m
     (c) 2.73×10^{-22} kg·m/s
39. (a) 1.9×10^{-20} kg·m/s
     (b) 3.5×10^{-14} m
41. 815 MeV
43. $\gamma + {}^{17}_{8}$O $\rightarrow$ $^{12}_{6}$C $+$ $^{4}_{2}$He $+$ $^{1}_{0}$n
45. 1.1×10^{-4} kg
47. 1.8×10^{4} kg/s
49. 32 eV

Index

SI Units

Quantity	Name of Unit	Symbol	Expression in Terms of Other SI Units	Quantity	Name of Unit	Symbol	Expression in Terms of Other SI Units
Length	meter	m	Base unit	Pressure, stress	pascal	Pa	N/m^2
Mass	kilogram	kg	Base unit	Viscosity	—	—	$Pa \cdot s$
Time	second	s	Base unit	Electric charge	coulomb	C	$A \cdot s$
Electric current	ampere	A	Base unit	Electric field	—	—	N/C
Temperature	kelvin	K	Base unit	Electric potential	volt	V	J/C
Amount of substance	mole	mol	Base unit	Resistance	ohm	Ω	V/A
Velocity	—	—	m/s	Capacitance	farad	F	C/V
Acceleration	—	—	m/s^2	Inductance	henry	H	$V \cdot s/A$
Force	newton	N	$kg \cdot m/s^2$	Magnetic field	tesla	T	$N \cdot s/(C \cdot m)$
Work, energy	joule	J	$N \cdot m$	Magnetic flux	weber	Wb	$T \cdot m^2$
Power	watt	W	J/s	Specific heat capacity	—	—	$J/(kg \cdot K)$ or $J/(kg \cdot C°)$
Impulse, momentum	—	—	$kg \cdot m/s$	Thermal conductivity	—	—	$J/(s \cdot m \cdot K)$ or $J/(s \cdot m \cdot C°)$
Plane angle	radian	rad	m/m	Entropy	—	—	J/K
Angular velocity	—	—	rad/s	Radioactive activity	becquerel	Bq	s^{-1}
Angular acceleration	—	—	rad/s^2	Absorbed dose	gray	Gy	J/kg
Torque	—	—	$N \cdot m$	Exposure	—	—	C/kg
Frequency	hertz	Hz	s^{-1}				
Density	—	—	kg/m^3				

The Greek Alphabet

Alpha	A	α	Iota	I	ι	Rho	P	ρ	
Beta	B	β	Kappa	K	κ	Sigma	Σ	σ	
Gamma	Γ	γ	Lambda	Λ	λ	Tau	T	τ	
Delta	Δ	δ	Mu	M	μ	Upsilon	Y	υ	
Epsilon	E	ϵ	Nu	N	ν	Phi	Φ	ϕ	
Zeta	Z	ζ	Xi	Ξ	ξ	Chi	X	χ	
Eta	H	η	Omicron	O	o	Psi	Ψ	ψ	
Theta	Θ	θ	Pi	Π	π	Omega	Ω	ω	